T0138708

Fundamentals of Ceramics

Series in Materials Science and Engineering

The series publishes cutting edge monographs and foundational textbooks for interdisciplinary materials science and engineering. It is aimed at undergraduate and graduate level students, as well as practicing scientists and engineers. Its purpose is to address the connections between properties, structure, synthesis, processing, characterization and performance of materials.

Automotive Engineering
Lightweight, Functional, and Novel Materials
Brian Cantor, P. Grant, C. Johnston

Multiferroic Materials
Properties, Techniques, and Applications
Junling Wang, Ed.

2D Materials for Nanoelectronics
Michel Houssa, Athanasios Dimoulas, Alessandro Molle

Skyrmions
Topological Structures, Properties, and Applications
J. Ping Liu, Zhidong Zhang, Guoping Zhao, Eds.

Computational Modeling of Inorganic Nanomaterials
Stefan T. Bromley, Martijn A. Zwijnenburg, Eds.

Physical Methods for Materials Characterisation, Third Edition
Peter E. J. Flewitt, Robert K. Wild

Conductive Polymers
Electrical Interactions in Cell Biology and Medicine
Ze Zhang, Mahmoud Rouabhia, Simon E. Moulton, Eds.

Silicon Nanomaterials Sourcebook, Two-Volume Set
Klaus D. Sattler, Ed.

Advanced Thermoelectrics
Materials, Contacts, Devices, and Systems
Zhifeng Ren, Yucheng Lan, Qinyong Zhang

Fundamentals of Ceramics, Second Edition
Michel W. Barsoum

Fundamentals of Ceramics

Second Edition

Michel W. Barsoum

CRC Press
Taylor & Francis Group
Boca Raton London New York

CRC Press is an imprint of the
Taylor & Francis Group, an **informa** business

Cover: "The Wave" designed and colorized by Patricia Lyons, Moorstown, NJ. This scanning electron microscope micrograph of a Cr_2AlC single crystal deformed at high temperatures was taken by Prof. Thierry Ouisse, LMGP, INP-Grenoble, France.

CRC Press
Taylor & Francis Group
6000 Broken Sound Parkway NW, Suite 300
Boca Raton, FL 33487-2742

First issued in paperback 2022

© 2020 by Taylor & Francis Group, LLC
CRC Press is an imprint of Taylor & Francis Group, an Informa business

No claim to original U.S. Government works

ISBN-13: 978-1-498-70813-5 (hbk)
ISBN-13: 978-1-03-233730-2 (pbk)
DOI: 10.1201/9781498708166

This book contains information obtained from authentic and highly regarded sources. Reasonable efforts have been made to publish reliable data and information, but the author and publisher cannot assume responsibility for the validity of all materials or the consequences of their use. The authors and publishers have attempted to trace the copyright holders of all material reproduced in this publication and apologize to copyright holders if permission to publish in this form has not been obtained. If any copyright material has not been acknowledged please write and let us know so we may rectify in any future reprint.

Except as permitted under U.S. Copyright Law, no part of this book may be reprinted, reproduced, transmitted, or utilized in any form by any electronic, mechanical, or other means, now known or hereafter invented, including photocopying, microfilming, and recording, or in any information storage or retrieval system, without written permission from the publishers.

For permission to photocopy or use material electronically from this work, please access www.copyright.com (http://www.copyright.com/) or contact the Copyright Clearance Center, Inc. (CCC), 222 Rosewood Drive, Danvers, MA 01923, 978-750-8400. CCC is a not-for-profit organization that provides licenses and registration for a variety of users. For organizations that have been granted a photocopy license by the CCC, a separate system of payment has been arranged.

Trademark Notice: Product or corporate names may be trademarks or registered trademarks, and are used only for identification and explanation without intent to infringe.

Publisher's Note

The publisher has gone to great lengths to ensure the quality of this reprint but points out that some imperfections in the original copies may be apparent.

Visit the Taylor & Francis Web site at
http://www.taylorandfrancis.com

and the CRC Press Web site at
http://www.crcpress.com

With abiding love and deep gratitude to my treasured and unusual family, Patricia, Michael, Kate, Eric and, last but not least, Quinn.

CONTENTS

SERIES PREFACE

The series publishes cutting edge monographs and foundational textbooks for interdisciplinary materials science and engineering.

Its purpose is to address the connections between properties, structure, synthesis, processing, characterization and performance of materials. The subject matter of individual volumes spans fundamental theory, computational modeling and experimental methods used for design, modeling and practical applications. The series encompasses thin films, surfaces and interfaces, and the full spectrum of material types, including biomaterials, energy materials, metals, semiconductors, optoelectronic materials, ceramics, magnetic materials, superconductors, nanomaterials, composites and polymers.

It is aimed at undergraduate and graduate level students, as well as practicing scientists and engineers.

PREFACE TO THE SECOND EDITION

McGraw Hill first printed this book in 1996. The Institute of Physics (acquired by Taylor and Francis) printed the second printing in 2003. The major differences between the two were the many less typos in the latter. Since 2003, I found about 200 more typos—entropy is a powerful force—that have been eliminated in this second edition. This is not to say that the second edition is perfect, but it at least should have a few less typos. The figures have also been spruced and now, most, include red.

The world is changing quite rapidly indeed and we have to change with it. Fortuitously, for me, the fundamentals have not changed. It follows that writing this second edition was not as long and protracted as the first that was written over 11 years. To bring this edition to the 21st century, at the ends of each chapter I added one or more sections that I labeled Case Histories and/or Computational Materials Science.

If a picture is worth a thousand words, then a one-minute video probably contains as many words as are in this book. In today's world videos are ubiquitous. When combined with computational materials science we can now not only imagine what is happening at the atomic scale, but actually simulate it.

The Case History sections were added to highlight some of the new developments and exciting directions the field, broadly defined as ceramics, is embarking on. This includes the advent of nanotechnology in general and the two dimensional materials revolution in particular. In these sections I did not restrict myself to the future, but also looked back to highlight some tremendous achievements in the field, again broadly defined as ceramics.

Two dimensional solids, including MXenes, are briefly described in Chap. 3. In Chap. 7, I discuss solid oxide fuel cells and the kinetics of alumina forming materials that are key to most high temperature applications in air. In Chap. 9, the focus is on glass-ceramics, that are not new, but have been commercially quite successful. A much more recent development that has been phenomenally successful is the manufacturing of very strong and thin glasses that, among other applications, cover all our cell phones. Electric field assisted and microwave sintering are described in Chap. 10. In Chap. 11, I introduce wear resistant ceramics, strong and tough ceramics and how crack deflection enhances the toughness. In Chap. 12, I highlight the incredible strengths and environmental stability of glass fibers—that are routinely manufactured today by the millions of kilometers and without which the internet would slow to a crawl. These fibers are not only incredibly strong, but are also of such high purity as to possess unparalleled transparency. I also overview ceramic matrix composites that are currently used in jet engines. Talking of the latter, a ceramic-based technology that allows jet engines to run significantly hotter and hence more efficiently are thermal barrier coatings discussed in Chap. 13. Also in Chap. 13, I describe the space shuttle tile, not new, but still noteworthy. In Chap. 14, I overview electrochemical impedance spectroscopy. Cobalites, manganites and colossal magnetoresistance are reviewed in Chap. 15. In the last chapter I discuss optical fibers and how point defects can lead to color.

The Computational Materials Science sections reflect the tremendous progress made and continuing to be made in computational materials science. In Chap. 2, density functional theory, DFT, and molecular dynamics are introduced. Chapter 4 deals with elastic tensor properties and surface energies, two of the more challenging physical properties to measure. How DFT can be used to determine the phonon distributions—another physical property that is non-trivial to measure experimentally—in solids is presented. The energies needed to form point defects are another facet of the solid state that is not easy to measure and where DFT calculations have come to the rescue. This is outlined in Chap. 6.

Our world is currently facing unprecedented problems. Amongst the most challenging is how to power an ever more energy hungry world without generating greenhouse gases that are acidifying our oceans, melting our polar caps and resulting in more extreme weather. How do we insure that every human being has access to clean water and enough food.

In some way or other, I am convinced that new materials will allow us to solve, or at least ameliorate, some of these problems. It is my sincere hope that this textbook will in some small way inspire some young and creative minds to this quest. I, for one, would prefer to live in a world that is not self-destructing. One of my favorite sayings is: There is no hope, but I could be wrong.

I would be remiss if I did not acknowledge and thank all the students I have interacted with and learned from over the years. So thank you to my PhD students, in chronological order, D. Brodkin, T. El-Raghy, M. Radovic, P. Finkel, A. Murugaiah, T. Zhen, A. Ganguly, S. Gupta, E. Hoffman, S. Basu, A. Zhou, S. Amini, A. R. Sakulich, T. Scabarozi, A. Moseson, N. Lane, M. Naguib, B. Anasori, J. Griggs, D. Tallman, J. Halim, M. Ghidiu, S. Kota, V. Natu, M. Carey, H. Badr and T. El-Melegy. I would like to thank the visiting scientists and post-docs, L. Verger, M. Sokol, C. Hu, G. Ying, C. Li and D. Zhao. The list would not be complete without thanking my MSc students, I. Albaryak, S. Chakraborty, J. Spencer, I. Salama and A. Procopio.

I would also like to thank V. Natu and M. Sokol for helping with some of the figures. Finally I would like to thank my wife and son for their help with many of the figures and also for putting up with me for all the time I spent on this book.

PREFACE TO FIRST EDITION

It is a mystery to me why, in a field as interesting, rich and important as ceramics, a basic fundamental text does not exist. My decision to write this text was made almost simultaneously with my having to teach my first introductory graduate class in ceramics at Drexel a decade ago. Naturally, I assigned Kingery, Bowen and Uhlmann's *Introduction to Ceramics* as the textbook for the course. A few weeks into the quarter, however, it became apparent that KBU's book was difficult to teach from and more importantly to learn from. Looking at it from the student's point of view it was easy to appreciate why—few equations are derived from first principles. Simply writing down a relationship, in my opinion, does not constitute learning; true understanding only comes when the trail that goes back to first principles is made clear. However, to say that this book was influenced by KBU's book would be an understatement—the better word would be inspired by it, and for good reason—it remains an authoritative, albeit slightly dated, text in the field.

In writing this book I had a few guiding principles. First, nearly all equations are derived, usually from first principles, with the emphasis being on the physics of the problem, sometimes at the expense of mathematical rigor. However, whenever that trade-off is made, which is not often, it is clearly noted in the text. I have kept the math quite simple, nothing more complicated than differentiation and integration. The aim in every case was to cover enough of the fundamentals, up to a level deep enough to allow the reader to continue his or her education by delving, without too much difficulty, into the most recent literature. In today's fast-paced world, it is more important than ever to understand the fundamentals.

Second, I wanted to write a book that more or less "stood alone" in the sense that it did not assume much prior knowledge of the subject from the reader. Basic chemistry, physics, mathematics and an introductory course in materials science or engineering are the only prerequisites. In that respect, I believe this book will appeal to, and could be used as a textbook in, other than material science and engineering departments, such as chemistry or physics.

Pedagogically I have found that students in general understand concepts and ideas best if they are given concrete examples rather than generalized treatments. Thus maybe, at the expense of elegance and brevity, I have opted for that approach. It is hoped that once the concepts are well understood, for at least one system, the reader will be able to follow more advanced and generalized treatments that can be found in many of the references that I have included at the end of every chapter.

Successive drafts of this book have been described by some reviewers as being arid, a criticism that I believe has some validity and that I have tried to address. Unfortunately, it was simply impossible to cover the range of topics, at the depth I wanted to, and be flowery and descriptive at the same time (the book is already over 650 pages long).

Another area where I think this book falls short is in its lack of what I would term a healthy skepticism (à la Feynman lectures, for instance). Nature is too complicated, and ceramics in particular, to be neatly packaged into monosize dispersed spheres and their corresponding models, for example.

I thus sincerely hope that these two gaps will be filled in by the reader and especially the instructor. First, a little bit of "fat" should make the book much more appetizing—examples from the literature or the instructor's own experience would be just what is required. Second, a dose of skepticism concerning some of the models and their limitation is required. Being an experimentalist, I facetiously tell my students that when theory and experiment converge one of them is probably wrong.

This book is aimed at junior, senior and first-year graduate students in any materials science and engineering program. The sequence of chapters makes it easy to select material for a one-semester course. This might include much of the material in Chaps. 1–8, with additional topics from later chapters. The book is also ideally suited to a two-quarter sequence, and I believe there may even be enough material for a two-semester sequence.

The book can be roughly divided into two parts. The first nine chapters deal with bonding, structure and the physical and chemical properties that are influenced mostly by the type of bonding rather than the microstructure, such as defect structure and the atomic and electronic transport in ceramics. The coverage of the second part, Chaps. 11–16, deals with properties that are more microstructure dependent, such as fracture toughness, optical, magnetic and dielectric properties. In between the two parts lies Chap. 10, which deals with the science of sintering and microstructural development. The technological aspects of processing have been deliberately omitted for two reasons. The first is that there are a number of good undergraduate texts that deal with the topic. Second, it is simply not possible to discuss that topic and do it justice in a section of a chapter.

Chapter 8 on phase diagrams was deliberately pushed back until the notions of defects and nonstoichiometry (Chap. 6) and atom mobility (Chap. 7) were introduced. The chapter on glasses (Chap. 9) follows Chap. 8 since once again the notions introduced in Chaps. 6, 7 and 8 had to be developed in order to explain crystallization.

And while this is clearly not a ceramics handbook, I have included many important properties of binary and ternary ceramics collected over 10 years from numerous sources. In most chapters I also include, in addition to a number of well-tested problem sets with their numerical answers, worked examples to help the student through some of the trickier concepts. Whenever a property or phenomenon is introduced, a section clearly labeled experimental details has been included. It has been my experience that many students lacked knowledge of how certain physical properties or phenomena are measured experimentally, which needless to say makes it rather fruitless to even try to attempt to explain them. These sections are not intended, by any stretch of the imagination, to be laboratory guides or procedures.

Finally, it should also be pointed out that Chaps. 2, 5 and 8 are by no means intended to be comprehensive—but are rather included for the sake of completion, and to highlight aspects that are referred to later in the book as well as to refresh the reader's memory. It is simply impossible to cover inorganic chemistry, thermodynamics and phase equilibria in three chapters. It is in these chapters that a certain amount of prior knowledge by the reader is assumed.

I would like to thank Dr. Joachim Maier for hosting me, and the Max-Planck Institute fur Festkorperforchung in Stuttgart for its financial support during my sabbatical year, when considerable progress was made on the text. The critical readings of some of the chapters by C. Schwandt, H. Naefe, N. Nicoloso and G. Schaefer is also gratefully acknowledged. I would especially like to thank Dr. Rowland M. Cannon for helping me sort out, with the right spirit I may add, Chaps. 10 through 12—his insight, as usual, was invaluable.

I would also like to thank my colleagues in the Department of Materials Engineering and Drexel University for their continual support during the many years it took to finish this work. I am especially indebted to Profs. Roger Doherty and Antonious Zavaliangos with whom I had many fruitful and illuminating discussions. Finally I would like to take this opportunity to thank all those who have, over the many years I was a student, first at the American University in Cairo, Egypt, followed by the ones at the University of Missouri-Rolla and, last but not least, MIT, taught and inspired me. One has only to leaf through the book to appreciate the influence Profs. H. Anderson, R. Coble, D. Kingery, N. Kreidl, H. Tuller, D. Uhlmann, B. Wuench and many others had on this book.

Comments, criticisms, suggestions and corrections, from all readers, especially students, for whom this book was written, are most welcome. Please send them to me at the Department of Materials Engineering, Drexel University, Philadelphia, PA 19104, or by e-mail at Barsoumw@drexel.edu.

Finally, I would like to thank my friends and family, who have been a continuous source of encouragement and support.

Michel W. Barsoum

AUTHOR

Prof. Michel W. Barsoum is Distinguished Professor in the Department of Materials Science and Engineering at Drexel University. As the author of two entries on the MAX phases in the *Encyclopedia of Materials Science*, and the book *MAX Phases* published in 2013, he is an internationally recognized leader in the area of MAX phases. In 2011, he and colleagues at Drexel, selectively etched the A-group layers from the MAX phases to produce an entirely new family of 2D solids that they labeled MXenes, that have sparked global interest because of their potential in a multitude of applications. He has authored the book *MAX Phases: Properties of Machinable Carbides and Nitrides*, published by Wiley VCH in 2013. He has published over 450 refereed papers, including ones in top-tier journals such as *Nature* and *Science*. According to Google Scholar his h-index is >100 with over 44,000 citations. He made ISI's most cited researchers list in 2018 and 2019. He is a foreign member of the Royal Swedish Academy of Engineering Sciences, a fellow of the American Ceramic Society and the World Academy of Ceramics. The latter awarded him the quadrennial International Ceramics Prize 2020, one of the highest honors in the field. In 2000, he was awarded a Humboldt-Max Planck Research Award for Senior US Research Scientists and spent a sabbatical year at the Max Planck Institute in Stuttgart, Germany. In 2008, he spent a sabbatical at the Los Alamos National Laboratory as the prestigious Wheatly Scholar. He has been a visiting professor at Linkoping University in Sweden since 2008. In 2017, he received a Chair of Excellence from the Nanoscience Foundation in Grenoble, France. He is co-editor of Materials Research Letters, published by Taylor & Francis.

1

INTRODUCTION

All that is, at all
Lasts ever, past recall,
Earth changes,
But thy soul and God stand sure,
Time's wheel runs back or stops:
Potter and clay endure.

Robert Browning

1.1 INTRODUCTION

The universe is made up of elements, which in turn consist of neutrons, protons and electrons. There are roughly 100 elements, each possessing a unique electronic configuration determined by their atomic number, Z, and the spatial distribution and energies of their electrons. What determines the latter requires some understanding of quantum mechanics and is discussed in greater detail in the next chapter.

One of the major triumphs of quantum theory was a rational explanation of the **periodic table** (see inside front cover) of the elements that had been determined from experimental observation long before the advent of quantum mechanics. The periodic table places the elements in horizontal rows of increasing atomic number and vertical columns or groups, so that all elements in a group display similar chemical properties. For instance, all elements of group 17, known as halides, exist as diatomic gases characterized by very high reactivity. Conversely, the elements of group 18, the noble gases, are monoatomic and are chemically extremely inert.

A large fraction of the elements are solids at room temperature, and because they are shiny, ductile, and good electrical and thermal conductors, they are considered *metals*. A fraction of the elements—most notably, N, O, H, the halides and the noble gases—are gases at room temperature. The remaining elements are predominantly covalently bonded solids that, at room temperature, are either insulators (B, P, S, C[1]) or semiconductors (Si, Ge). These elements, are typically referred to as metalloids.

Few elements are used in their pure form; most often they are alloyed with other elements to create engineering materials. The latter can be broadly classified as metals, polymers, semiconductors or ceramics, with each class having distinctive properties that reflect the differences in the nature of their bonding.

In metals, the bonding is predominantly metallic, where delocalized electrons provide the "glue" that holds the positive ion cores together. This delocalization of the bonding electrons has far-reaching ramifications since it is responsible for properties most associated with metals: ductility, thermal and electrical conductivity, reflectivity and other distinctive properties.

Polymers consist of very long, for the most part, C-based chains, to which other organic atoms (for example, C, H, N, Cl, F) and molecules are attached. The bonding within the chains is strong, directional and covalent, while the bonding between chains is relatively weak. Thus, the properties of polymers, as a class, are dictated by the weaker bonds, and consequently they possess lower melting points, have higher thermal expansion coefficients and are less stiff than most metals or ceramics.

Semiconductors are covalently bonded solids that, in addition to Si and Ge, already mentioned, include GaAs, CdTe and InP, among many others. The usually strong covalent bonds holding semiconductors together render their mechanical properties similar to those of ceramics (i.e., brittle and hard).

Now that these distinctions have been made, it is possible to answer the nontrivial question: What is a ceramic?

1.2 DEFINITION OF CERAMICS

Ceramics can be defined as *solid compounds that are formed by the application of heat, and sometimes heat and pressure, comprising at least two elements provided one of them is a non-metal or a metalloid. The other element(s) may be a metal(s) or another metalloid(s).* A somewhat simpler definition was given by Kingery, who defined ceramics as "the art and science of making and using solid articles, which have, as their essential component, and are composed in large part of inorganic, nonmetallic materials". In other words, what is neither a metal, a semiconductor or a polymer is a ceramic.

To illustrate, consider the following examples: Magnesia,[2] or MgO, is a ceramic since it is a solid compound of a metal, Mg, bonded to the nonmetal, oxygen, O. Silica is also a ceramic since it combines a

[1] In the form of diamond. It is worth noting that although graphite is a good electrical conductor, it is not considered a metal since it is neither shiny nor ductile.

[2] A note on nomenclature: The addition of the letter a to the end of an element name implies one is referring to its oxide. For example, while silicon refers to the element, silica is SiO_2 or the oxide of silicon. Similarly, alumina is the oxide of aluminum or Al_2O_3; magnesium, magnesia; etc.

metalloid, Si, with a nonmetal. Similarly, TiC and ZrB_2 are ceramics since they combine metals (Ti, Zr) and a metalloid (C, B). SiC is a ceramic because it combines two metalloids. Ceramics are not limited to binary compounds: $BaTiO_3$, $YBa_2Cu_3O_3$, and Ti_3SiC_2 are all perfectly respectable class members.

It follows that the oxides, nitrides, borides, carbides, and silicides (not to be confused with silicates) of all metals and metalloids are ceramics; which, needless to say, leads to a large number of compounds. This number becomes even more daunting when it is appreciated that the silicates are also, by definition, ceramics. Because of the abundance of oxygen and silicon in nature, silicates are ubiquitous; rocks, dust, clay, mud, mountains, sand, in short, the vast majority of the earth's crust is composed of silicate-based minerals. When it is also appreciated that cement, bricks, and concrete are essentially silicates, a case could be made that we live in a ceramic world.

In addition to their ubiquitousness, silicates were singled out above for another reason, namely, as the distinguishing chemistry between traditional and modern ceramics. Before that distinction is made, however, it is important to briefly explore how atoms are arranged in three dimensions.

1.2.1 CRYSTALLINE VERSUS AMORPHOUS SOLIDS

The arrangement of atoms in solids, in general, and ceramics, in particular, will exhibit **long-range order**, only **short-range order**, or a combination of both.[3] Solids that exhibit long-range order[4] are referred to as **crystalline**, while those in which that periodicity is lacking are known as **amorphous**, **glassy** or **noncrystalline solids.**

The difference between the two is illustrated schematically in Fig. 1.1. From the figure, it is obvious that a solid possesses long-range order when the atoms repeat with a periodicity that is much greater than the bond lengths or the distance between the atoms. Most metals and ceramics, with the exception of glasses and glass-ceramics (see Chap. 9), are crystalline.

Since, as discussed throughout this book, the details of the lattice patterns can strongly influence the macroscopic properties of ceramics, it is imperative to understand the rudiments of crystallography.

1.3 ELEMENTARY CRYSTALLOGRAPHY

As noted above, long-range order requires that atoms be arrayed in a three-dimensional (3D) pattern that repeats. The simplest way to describe a pattern is to describe a **unit cell** within that pattern. A *unit cell* is defined as the smallest region in space that, when repeated, completely describes the 3D pattern of atoms in a crystal. Geometrically, it can be shown that there are only seven unit cell *shapes*, or **crystal systems**, that can be stacked together to fill three-dimensional space. The seven systems, shown in Fig. 1.2, are cubic, tetragonal, orthorhombic, rhombohedral, hexagonal, monoclinic and triclinic. These systems are

[3] Strictly speaking, only solids in which grain boundaries are absent, i.e., single crystals, can be considered to possess only long-range order. As discussed below, the vast majority of crystalline solids possess grain boundaries that are areas in which the long-range order breaks down, and thus should be considered as a combination of amorphous and crystalline areas. However, given that in most cases the volume fraction of the grain boundary regions is much less than 0.01, it is customary to describe polycrystalline materials as possessing only long-range order.

[4] Any solid that exhibits long-range order must also exhibit short-range order, but not vice versa.

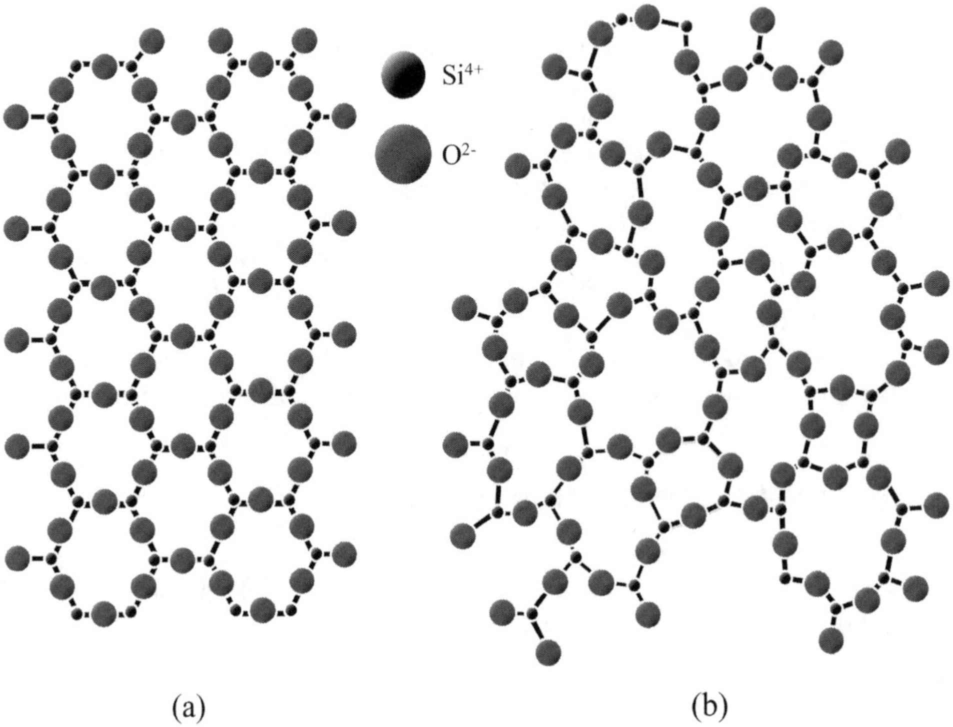

FIGURE 1.1 (*a*) Long-range order; (*b*) short-range order in silica.

distinguished from one another by the lengths of the unit cell edges and the angles between the edges, collectively known as the **lattice parameters** or **lattice constants** (a, b, c, α, β and γ in Fig. 1.2).

It is useful to think of a given crystal system as a "brick" of a certain shape. For example, the bricks can be cubes, hexagons, parallelepipeds, etc. And while the shape of the bricks is an important descriptor of a crystal structure, it is insufficient. In addition to the brick shape, it is important to know the *symmetry* of the lattice pattern within each brick, as well as the actual location of the atoms on these lattice sites. Only then would the description be complete.

It turns out that if one considers only the symmetry within each unit cell, the number of possible permutations is limited to 14. The 14 arrangements, shown in Fig. 1.2, are also known as the **Bravais lattices**. A **lattice** can be defined as an indefinitely extending arrangement of points, each of which is surrounded by an identical grouping of neighboring points. To carry the brick analogy a little further, the Bravais lattice represents the *symmetry* of the *pattern* found on the bricks.

Finally, to describe the atomic arrangement, one must describe the symmetry of the *basis*, defined as the atom or grouping of atoms located at each lattice site. When the basis is added to the lattices, the total number of possibilities increases to 32 *point groups*.[5]

---

[5] For more information, see, for example, A. Kelly and G. W. Groves, *Crystallography and Crystal Defects*, Longmans, London, 1970.

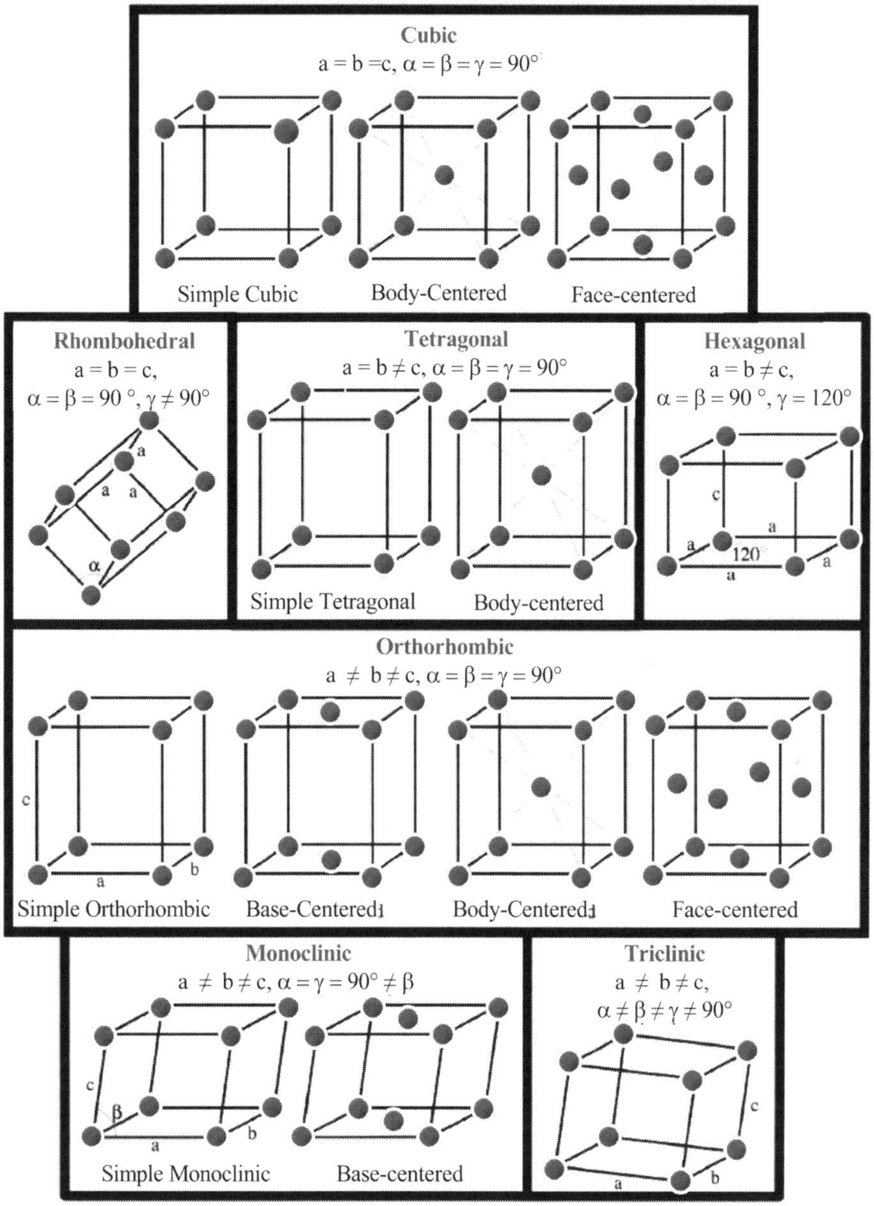

FIGURE 1.2 Geometric characteristics of 7 crystal systems and 14 Bravais lattices.

1.4 CERAMIC MICROSTRUCTURES

Crystalline solids exist as either single crystals or polycrystalline solids. A **single crystal** is a solid in which the periodic and repeated arrangement of atoms is perfect and extends throughout the entirety of the specimen without interruption. A **polycrystalline solid**, Fig. 1.3, is comprised of a collection of many single crystals, termed **grains**, separated from one another by areas of disorder known as **grain boundaries** (see Chap. 6 for more details).

Typically, in ceramics the grains are in the range of 1 to 50 μm and are visible only under a microscope. The shape and size of the grains, together with the presence of porosity, second phases, etc., and their distribution describe what is termed the **microstructure**. As discussed in later chapters, many of the properties of ceramics are microstructure-dependent.

1.5 TRADITIONAL VERSUS ADVANCED CERAMICS

Many people associate the word *ceramics* with pottery, sculpture, sanitary ware, tiles, etc. And whereas this view is not incorrect, it is incomplete because it considers only the traditional, or silicate-based, ceramics. Today the field of ceramic science and engineering encompasses much more than silicates and can be divided into traditional and modern ceramics. Before the distinction is made, however, it is worthwhile to trace the history of ceramics and people's association with them.

It has long been appreciated by our ancestors that some muds, when wet, were easily moldable into shapes that upon heating became rigid. The formation of useful articles from fired mud must constitute one of the oldest and more fascinating of human endeavors. Fired-clay articles have been traced to the dawn of civilization. The usefulness of these new materials, however, was limited by the fact that when fired, they

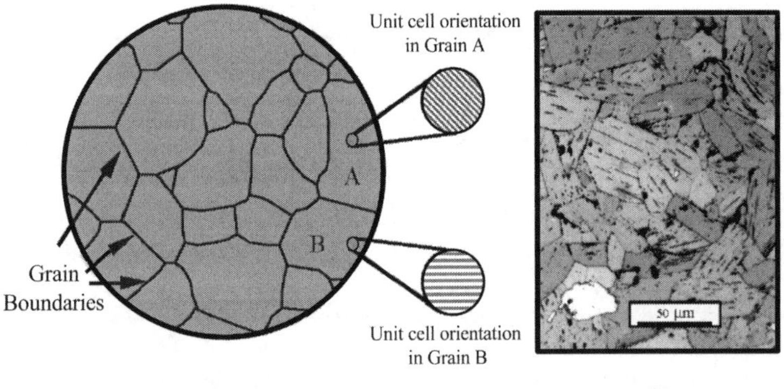

FIGURE 1.3 (*a*) Schematic of a polycrystalline sample. A polycrystal is made up of many grains separated from one another by regions of disorder known as grain boundaries. (*b*) Typical microstructure as seen through an optical microscope.

were porous and thus could not be used to carry or store liquids. Later the serendipitous discovery was made that when heated and slowly cooled, some sands tended to form a transparent, water-impervious solid, known today as glass. From that point on, it was simply a matter of time before glazes were developed that rendered clay objects not only watertight, but also quite beautiful.

With the advent of the industrial revolution, structural clay products such as bricks and heat-resistant refractory materials for the large-scale smelting of metals were developed. And with the discovery of electricity and the need to distribute it, a market was developed for electrically insulating, silicate-based ceramics.

Traditional ceramics are characterized by mostly silicate-based, porous microstructures that are quite coarse, nonuniform and multiphase. They are typically formed by mixing clays and feldspars, followed by forming either by slip casting or on a potter's wheel, firing in a kiln to sinter them and finally glazing.

In a much later stage of development, other ceramics that were not clay- or silicate-based depended on much more sophisticated raw materials, such as binary oxides, carbides, perovskites and even completely synthetic materials for which there are no natural equivalents. The microstructures of these modern ceramics were at least an order of magnitude finer, more homogeneous and much less porous than those of their traditional counterparts. It is the latter—the **modern** or **technical ceramics**—with which this book is mainly concerned.

1.6 GENERAL CHARACTERISTICS OF CERAMICS

As a class, ceramics are hard, wear-resistant, nonmachinable, brittle, prone to thermal shock, refractory, electrically and thermally insulative, intrinsically transparent, nonmagnetic, chemically stable and oxidation-resistant. As with all generalizations, there will be exceptions; some ceramics are electrically and thermally quite conductive, while others are even superconducting. An entire industry is based on the fact that some ceramics are magnetic.

One of the main goals of this book is to answer the question of why ceramics exhibit the properties they do. And while this goal will have to wait until later chapters, at this point it is worthwhile to list some of the applications for which ceramics have been, or are being, developed.

1.7 APPLICATIONS

Traditional ceramics are quite common, from sanitary ware to fine chinas and porcelains to glass products. Currently ceramics are being considered for uses that a few decades ago were inconceivable; applications ranging from ceramic engines to optical communications, from electrooptic applications to laser materials and from substrates in electronic circuits to electrodes in photoelectrochemical devices. Some of the recent applications for which ceramics are used and/or are prime candidates are listed in Table 1.1.

Historically, ceramics were mostly exploited for their electrical insulative properties, for which electrical porcelains and aluminas are prime examples. Today, the so-called electrical and electronic ceramics still play a pivotal role in any modern technological society. For example, their insulative properties together

TABLE 1.1 Properties and applications of advanced ceramics

Property	Applications (Examples)
Thermal	
Insulation	High-temperature furnace linings for insulation (oxide fibers such as SiO_2, Al_2O_3 and ZrO_2)
Refractoriness	High-temperature furnace linings for insulation and containment of molten metals and slags
Thermal conductivity	Heat sinks for electronic packages (AlN)
Electrical and dielectric	
Conductivity	Heating elements for furnaces (SiC, ZrO_2, $MoSi_2$)
Ferroelectricity	Capacitors (Ba-titanate-based materials)
Low-voltage insulators	Ceramic insulation (porcelain, steatite, forsterite)
Insulators in electronic circuits	Substrates for electronic packaging and electrical insulators in general (Al_2O_3, AlN)
Insulators in harsh environments	Spark plugs (Al_2O_3)
Ion-conducting	Sensor and fuel cells (ZrO_2, Al_2O_3, etc.)
Semiconducting	Thermistors and heating elements (oxides of Fe, Co, Mn)
Nonlinear I-V characteristics	Current surge protectors (bi-doped ZnO, SiC)
Gas-sensitive conductors	Gas sensors (SnO_2, ZnO)
Magnetic and superconductive	
Hard magnets	Ferrite magnets [(Ba, Sr)O·6Fe_2O_3]
Soft magnets	Transformer cores [(Zn, M)Fe_2O_3, with M = Mn, Co, Mg]; magnetic tapes (rare-earth garnets)
Superconductivity	Wires and SQUID magnetometers ($YBa_2Cu_3O_7$)
Optical	
Transparency	Windows (soda-lime glasses), cables for optical communication (ultrapure silica)
Translucency and chemical inertness	Heat- and corrosion-resistant materials, usually for Na lamps (Al_2O_3MgO)
Nonlinearity	Switching devices for optical computing ($LiNbO_3$)
IR transparency	Infrared laser windows (CaF_2, SrF_2, NaCl)
Nuclear applications	
Fission	Nuclear fuel (UO_2, UC), fuel cladding (C, SiC) and neutron moderators (C, BeO)
Fusion	Tritium breeder materials (zirconates and silicates of Li, Li_2O); fusion reactor lining (C, SiC, Si_3N_4)
Chemical	
Catalysis	Filters (zeolites); purification of exhaust gases
Anticorrosion	Heat exchangers (SiC), chemical equipment in corrosive environments
Biocompatibility	Artificial joint prostheses (Al_2O_3)
Mechanical	
Hardness	Cutting tools (SiC whisker-reinforced Al_2O_3, Si_3N_4)
High-temperature strength retention	Stators and turbine blades, ceramic engines (Si_3N_4)
Wear resistance	Bearings (Si_3N_4)

with their low loss factors and excellent thermal and environmental stability make them the materials of choice for substrate materials in electronic packages. The development of the perovskite family, with exceedingly large dielectric constants (Chap. 15) holds a significant market share of capacitors produced worldwide. Similarly, the development of magnetic ceramics based on the spinel ferrites is today a mature technology. Other electronic/electrical properties of ceramics that are being commercially exploited include piezoelectric ceramics for sensors and actuators, nonlinear *I-V* characteristics for circuit protection and ionically conducting ceramics for use as solid electrolytes in high-temperature fuel cells, batteries and as chemical sensors.

Mechanical applications of ceramics at room temperature usually exploit their hardness, wear and corrosion resistance. The applications include cutting tools, nozzles, valves and ball bearings in aggressive environments. However, it is the refractoriness of ceramics and their ability to sustain high loads at high temperatures, together with their low densities, that has created the most interest. Applications in that area include ceramic engines for transportation and turbines for energy production.

1.8 THE FUTURE

Paradoxically, because interest in modern ceramics came later than interest in metals and polymers, ceramics are simultaneously mankind's oldest and newest solids. Consequently, working in the field of ceramics can be quite rewarding and exciting. There are a multitude of compounds that have never been synthesized, let alone characterized. Amazing discoveries are always around the corner, as the following examples illustrate.

In 1986, the highest temperature at which any material became superconducting, i.e., the ability to conduct electricity with virtually no loss, was around −250°C, or 23 K. In that year a breakthrough came when Bednorz and Muller[6] shattered the record by demonstrating that a layered lanthanum, strontium copper oxide became superconducting at the relatively balmy temperature of 46 K. This discovery provoked a worldwide frenzy in the subject, and a few months later the record was again almost doubled, to about 90 K. The record today is in excess of 120 K.

Toward the end of 1995, we identified a new class of solids best described as machinable, thermodynamically stable polycrystalline nanolaminates[7,8] (Fig. 1.4*a*). These solids are ternary layered hexagonal early transition metal carbides and nitrides with the general formula $M_{n+1}AX_n$, where $n = 1$ to 3, M is an early transition metal, A is an A-group element (mostly IIIA and IVA) and X is C and/or N. Today this family numbers over 150, with more being still discovered on a routine basis.

Thermally, elastically, chemically and electrically these so-called MAX phases share many of the advantageous attributes of their respective stoichiometric binary transition metal carbides or nitrides: they are electrically and thermally conductive and chemically stable. Mechanically they cannot be more different. When deformed or fractured the basal planes readily kink, bend and delaminate, not unlike how wood

[6] T. G. Bednorz and K. A. Muller, *Z. Phys. B*, 64, 189 (1986).
[7] M. W. Barsoum, *MAX Phases: Properties of Machinable Carbides and Nitrides*, Wiley VCH GmbH & Co., Weinheim, 2013.
[8] M. W. Barsoum and T. El-Raghy, *American Scientist*, 89, 336–345 (2001).

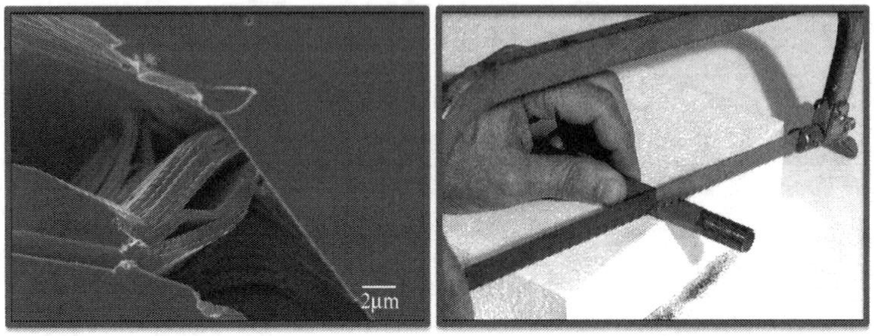

(a) (b)

FIGURE 1.4 (*a*) Example of delaminations possible in Ti_3SiC_2. (*b*) Despite being quite stiff and lightweight, Ti_3SiC_2 is still most readily machinable. (M. W. Barsoum, MAX Phases, VCH-Wiley, 2013.)

would deform (Fig. 1.4*a*). These processes occur at room and elevated temperatures. It is this ability to delaminate, almost at will, that led us to describe them as thermodynamically stable nanolaminates. The MAX phases are also relatively soft and are most readily machinable (Fig. 1.4*b*).

Furthermore, some of these compounds (e.g., Ti_3SiC_2) combine many of the best attributes of metals and ceramics. Like metals, they are excellent electrical and thermal conductors, are *not* susceptible to thermal shock and behave plastically at higher temperatures. Like ceramics, they possess high specific stiffness values (Ti_3SiC_2 is roughly three times as stiff as Ti metal, with the same density) and yet as noted above are machinable with nothing more sophisticated than a manual hack-saw (Fig. 1.4*b*). Some also have good creep and fatigue properties; Ti_2AlC is exceptionally oxidation resistant.

My third example is also related to the MAX phases. In 2011, we wanted to use the MAX phases as anodes for Li ions to replace graphite, the current anode of choice for Li batteries. The one problem we faced was that the Li refused to enter the MAX phase structure. After roughly 9 months of trying several approaches, however, we figured out that all we had to do was immerse an Al-containing MAX phase in a solution containing F anions, such as HF or HCl and LiF. The latter selectively etched the Al layers replacing them with O, OH and F terminations. Once the Al layers are etched away it is not difficult to disperse the two-dimensional (2D) flakes in water to form an aqueous colloidal suspension. Once a colloid suspension is created, the possibilities are endless. I labeled these material MXenes for two reasons; the first is to indicate the removal of the A-layers from MAX and the second is to make the connection to graphene and other 2D materials. The name has stuck and today MXenes are bona fide members of the 2D family of materials that have generated tremendous interest. What renders them quite attractive for myriad applications is their unique combination of good electrical conductivity and hydrophilicity. MXenes can be described in a variety of ways such as 2D metals, conductive clays or hydrophilic graphene.

Traditional ceramics have served humanity well for at least the past ten millennia. However, the nature of modern technology, with its ever-mounting demands on materials, has prompted researchers to take a second look at these stone-age materials, and it now is clear that this oldest class of materials are shaping up to be truly the material of the future. It is my sincerest hope that this book will inspire a new generation of talented and dedicated researchers to embark on a voyage of discovery in this most exciting of fields.

PROBLEMS

1.1. (*a*) According to the definition of a ceramic given in the text, would you consider Si_3N_4 a ceramic? How about CCl_4, $SiCl_4$ or SiF_4? Explain.

(*b*) Would you consider $TiAl_3$ a ceramic? How about Al_3C_4, BN, CN or SiB_6? Explain.

1.2. (*a*) How many crystal systems would you expect in two dimensions? Draw and characterize them by their lattice parameters.

Answer: 4

How many Bravais lattices are there in two dimensions? Draw them.

Answer: 5

ADDITIONAL READING

1. W. D. Kingery, H. K. Bowen, and D. R. Uhlmann, *Introduction to Ceramics*, 2nd ed., Wiley, New York, 1976.
2. A. R. West, *Solid State Chemistry and Its Applications*, Wiley, Chichester, UK, 1984.
3. R. J. Brook, Ed., *Concise Encyclopedia of Advanced Ceramic Materials*, Pergamon, NY, 1991.
4. D. W. Richerson, *Modern Ceramic Engineering*, 3rd ed., Taylor and Francis, NY, 2006.
5. P. A. Cox, *The Electronic Structure and Chemistry of Solids*, Oxford University Press, New York, 1987.
6. J. P. Schaffer, A. Saxena, S. D. Antolovich, T. H. Sanders, and S. B. Warner, *The Science and Design of Engineering Materials*, Irwin, Chicago, 1995.
7. C. Kittel, *Introduction to Solid State Physics*, 8th ed., Wiley, NY, 2004.
8. Y.-M. Chiang, D. P. Birnie, and W. D. Kingery, *Physical Ceramics: Principles for Ceramic Science and Engineering*, John Wiley and Sons, NY, 1997.
9. N. N. Greenwood, *Ionic Crystals, Lattice Defects and Non-Stoichiometry*, Butterworth, London, 1968.
10. L. Pauling, *The Nature of the Chemical Bond*, Cornell University Press, Ithaca, NY, 1960.
11. L. Azaroff, *Introduction to Solids*, McGraw-Hill, New York, 1960.
12. T. Ohji, M. Singh, Eds., *Engineered Ceramics: Current Status and Future Products*, Wiley, Hoboken, NJ, 2016.

2

BONDING IN CERAMICS

All things are Atoms: Earth and Water, Air And Fire, all,
Democritus foretold. Swiss Paracelsus, in's alchemic lair,
Saw Sulfur, Salt, and Mercury unfold Amid Millennial
hopes of faking Gold. Lavoisier dethroned
Phlogiston; then Molecular Analysis made bold
Forays into the gases: Hydrogen
Stood naked in the dazzled sight of Learned Men.

John Updike; *The Dance of the Solids**

2.1 INTRODUCTION

The properties of a solid and the way its atoms are arranged in 3D are determined primarily by the nature and directionality of the bonds holding the atoms together. Consequently, to understand variations in properties, it is imperative to appreciate how and why a solid is "glued" together.

This glue can be strong, which gives rise to *primary bonds,* which can be ionic, covalent or metallic. Usually van der Waals and hydrogen bonds are referred to as secondary bonds and are weaker. *In all cases, however, it is the attractive electrostatic interaction between the positive charges of the nuclei and the negative charges of the electrons that is responsible for the cohesion of solids.*

Very broadly speaking, ceramics can be classified as being either ionically or covalently bonded and, for the sake of simplicity, this notion is maintained throughout this chapter. However, that this simple view

* J. Updike, *Midpoint and Other Poems*, A. Knopf, Inc., New York, 1969. Reprinted with permission.

needs some modification will become apparent in Chap. 4; bonding in ceramics is neither purely covalent nor purely ionic, but a mixture of both. Interestingly, in some ceramics, such as the MAX phases and transition metal carbides and nitrides, the bonding is a combination of metallic and covalent.

Before the intricacies of bonding are described, a brief review of the shape of atomic orbitals is presented in Sec. 2.2. The concept of electronegativity and how it determines the nature of bonding in a ceramic is introduced in Sec. 2.3. In Secs. 2.4 and 2.5, respectively, the ionic bond is treated by a simple electrostatic model, and how such bonds lead to the formation of ionic solids is discussed.

The more complex covalent bond, which occurs by the overlap of electronic wave functions, is discussed in Secs. 2.6 and 2.7. In Sec. 2.8, how the interaction of wave functions of more than one atom results in the formation of energy bands in crystalline solids is elucidated.

It is important to point out, at the outset, that much of this chapter is only intended as review of what the reader is assumed to be familiar with from basic chemistry. Most of the material in this chapter is covered in college-level chemistry textbooks.

2.2 STRUCTURE OF ATOMS

Before bonding between atoms is discussed, it is essential to appreciate the energetics and shapes of single atoms. Furthermore, since bonding involves electrons that obey the laws of quantum mechanics, it is crucial to review the following major conclusions of quantum theory as they apply to bonding.

1. The confinement of a particle results in the quantization of its energy levels. Said otherwise, whenever a particle is attracted to, or confined in space to a certain region, its energy levels are necessarily quantized. As discussed shortly, this follows directly from *Schrödinger's wave equation.*
2. A given quantum level cannot accept more than two electrons, which is *Pauli's exclusion principle.*
3. It is impossible to simultaneously know with certainty both the momentum and the position of a moving particle, which is the *Heisenberg uncertainty principle.*

The first conclusion explains the shape of orbitals and their energies; the second why higher-energy orbitals are stable and populated; and the third elucidates, among other things, why an electron does not spiral continually and fall into the nucleus.

In principle, the procedure for determining the shape of an atomic or molecular orbital is quite simple and involves solving Schrödinger's equation—with the appropriate boundary conditions—from which one obtains the all-important electronic wave function. The latter in turn, allows us to calculate the probability of finding an electron in a given volume. To illustrate, consider the simplest possible case, that of the hydrogen atom, which consists of a proton and an electron.

2.2.1 THE HYDROGEN ATOM

Schrödinger's time-independent equation in one dimension is given by:

$$\frac{\partial^2\psi}{\partial x^2}+\frac{8\pi^2 m_e}{h^2}(E_{tot}-E_{pot})\psi=0 \tag{2.1}$$

where m_e is the mass of the electron, 9.11×10^{-31} kg; h is Planck's constant, 6.625×10^{-34} J·s; and E_{tot} is the total (kinetic + potential) energy of the electron. The potential energy of the electron E_{pot} is nothing but the Coulombic attraction between an electron and a proton,[9] given by:

$$E_{pot} = \frac{z_1 z_2 e^2}{4\pi\varepsilon_0 r} = -\frac{e^2}{4\pi\varepsilon_0 r} \tag{2.2}$$

where z_1 and z_2 are the charges on the electron and nucleus, −1 and +1, respectively; e is the elementary electronic charge, 1.6×10^{-19} C; ε_0 is the permittivity of free space, 8.85×10^{-12} $C^2/(J \cdot m)$ or F/m; r is the distance between the electron and the nucleus.

Now ψ is the wave function of the electron and by itself has no physical meaning, but $|\psi(x,y,z;t)|^2$ gives the probability of finding an electron in a volume element dxdydz. The higher ψ^2 is in some volume in space, the more likely the electron is to be found there.

For the simplest possible case of the hydrogen atom, the orbital is spherically symmetric; and so it is easier to work in spherical coordinates. Thus instead of Eq. (2.1), the differential equation to solve is

$$\frac{h^2}{8\pi^2 m_e}\left(\frac{\partial^2\psi}{\partial r^2} + \frac{2}{r}\frac{\partial\psi}{\partial r}\right) + \left(E_{tot} + \frac{e^2}{4\pi\varepsilon_0 r}\right)\psi = 0 \tag{2.3}$$

where E_{pot} was replaced by the value given in Eq. (2.2). The solution of this equation yields the functional dependence of ψ on r, and it can be shown that (see Prob. 2.1)

$$\psi = \exp(-c_0 r) \tag{2.4}$$

satisfies Eq. (2.3), but *only* provided the energy of the electron is given by

$$E_{tot} = -\frac{m_e e^4}{8\varepsilon_0^2 h^2} \tag{2.5}$$

and

$$c_0 = \frac{\pi m_e e^2}{\varepsilon_0 h^2} \tag{2.6}$$

As noted above, ψ by itself has no physical significance, but ψ^2 is the probability of finding an electron in a given volume element. It follows that the probability distribution function, W, of finding the electron in a thin spherical shell between r and r + dr is obtained by multiplying $|\psi|^2$ by the volume of that shell (see hatched area in Fig. 2.1*a*), or

$$W = 4\pi r^2 |\psi|^2 dr \tag{2.7}$$

[9] For the hydrogen atom z_1 and z_2 are both unity. In general, however, the attraction between an electron and a nucleus has to reflect the total nuclear charge, viz. an element's atomic number.

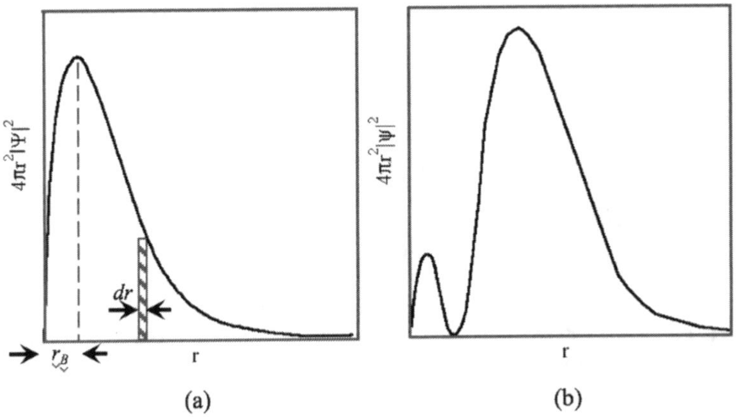

FIGURE 2.1 (*a*) Radial distribution function of 1s state electron. The crosshatched strip has a volume $4\pi r^2 dr$, which, when multiplied by $|\psi|^2$, gives the probability of finding the electron between r and r + dr. The probability of finding the electron very near or very far from the nucleus approaches zero. The most probable position for the electron is at a distance $r_B = 1/c_0$. (*b*) Radial distribution function of a 2s electron, whose energy is one-fourth that of the 1s state.

In other words, the y-axis is simply a measure of the probability of finding the electron at any distance r.

Figure 2.1*a* shows that the probability of finding an electron at the nucleus, or very far from the nucleus, is negligible, but that somewhere in between that probability is at a maximum. This distance is known as the **Bohr radius**, r_B (see Fig. 2.1*a*). The importance of this result lies in appreciating that (1) while the electron spends most of its time at a distance r_B, its spatial extent is clearly not limited to that value and (2) the best one can hope for when discussing the location of an electron is to talk about the probability of finding it in some volume. It is worth noting here that by combining Eqs. (2.4) to (2.7) and finding the location of the maximum, it can be shown that $r_B = 1/c_0$.

Worked Example 2.1

Calculate the ground-state energy level of the electron in the hydrogen atom, and compare the result with the experimentally derived value of −13.6 eV.

ANSWER[10]

Using Eq. (2.5) gives

$$E_{tot} = -\frac{me^4}{8\varepsilon_0^2 h^2} = -\frac{(9.1\times10^{-31})(1.6\times10^{-19})^4}{8(8.85\times10^{-12})^2(6.63\times10^{-34})^2} \tag{2.8}$$

$$= -2.165\times10^{-18}\,J = -13.6\ eV$$

[10] In all problems and throughout this book, SI units are used almost exclusively.

This value is the lowest energy level of a hydrogen electron, a fact that was experimentally known well before the advent of quantum mechanics. This result was one of the first and greatest successes of quantum theory. It is important to note that since this energy is negative, it follows that the electron energy, in the vicinity of the proton, is lower than at an infinite distance away (which corresponds to zero energy).[11]

Equation (2.4) is but one of many possible solutions. For example, it can also be shown that

$$\psi(r) = A(1 + c_1 r)\exp\left(-\frac{rc_0}{2}\right) \tag{2.9}$$

is another perfectly legitimate solution to Eq. (2.3), provided that Eq. (2.5) is divided by 4. The corresponding radial distribution function is plotted in Fig. 2.1*b*. It follows that the energy of this electron is −13.6/4 and it will spend most of its time at a distance given by the second maximum in Fig. 2.1*b*.

To generalize, for a spherically symmetric wave function, the solution (given here without proof) is

$$\psi_n(r) = e^{-c_n r} L_n(r)$$

where L_n is a polynomial. The corresponding energies are given by

$$E_{tot} = \frac{-me^4}{8n^2\varepsilon_0^2h^2} = -\frac{13.6eV}{n^2} \tag{2.10}$$

where n is known as the *principal quantum number*. As n increases, the energy of the electron increases (i.e., becomes less negative) and its spatial extent increases.

2.2.2 ORBITAL SHAPE AND QUANTUM NUMBERS

Equations (2.4) and (2.9) were restricted to spherical symmetry. An even more generalized solution is

$$\psi_{n,I,m} = R_{nl}(r)Y_l^m(\theta,\pi)$$

where Y_l depends on θ and π. Consequently, the size and shape of an orbital will depend on the specific solution considered. It can be shown that each orbital will have associated with it three characteristic

[11] A thorny question that had troubled physicists as they were developing the theories of quantum mechanics was: What prevented an electron from continually losing energy, spiraling into the nucleus and releasing an infinite amount of energy? Originally, the classical explanation was that the angular momentum of the electron gives rise to the apparent repulsion—this explanation is invalid here, because s electrons have *no* angular momentum (see Chap. 15). The actual reason is related to the Heisenberg uncertainty principle and goes something like this: As an electron is confined to a smaller and smaller volume, the uncertainty in its position Δx decreases. But since $\Delta x \Delta p = h$ is a constant, it follows that its momentum p, or, equivalently, its kinetic energy, will have to increase as Δx decreases. Given that the kinetic energy scales with $1/r^2$, but the potential energy scales only as $1/r$, an energy minimum has to be established at a given equilibrium distance.

interrelated quantum numbers, labeled n, l and m_l, known as the *principal*, *angular* and *magnetic quantum numbers*, respectively.

The **principal quantum number, n**, determines the *spatial extent* and *energy* of the orbital.

The **angular momentum quantum number,**[12] **l**, determines the *shape* of the orbital for any given value of n and can only assume the values 0,1,2,3, . . . n − 1. For example, for n = 3, the possible values of l are 0, 1 and 2.

The **magnetic quantum number, m_l**, is related to the *orientation* of the orbital in space. For a given value of l, m_l can take on values from −l to +l. For example, for l = 2, m_l can be +2, +1, 0, −1 or −2. Thus for any value of l there are 2l + 1 values of m_l.

All orbitals with l = 0 are called **s *orbitals*** and are spherically symmetric (Fig. 2.1). When l = 1, the orbital is called a **p *orbital***, and there are three of these (Fig. 2.2*a*), each corresponding to a different value of m_l associated with l = 1, that is, m_l = −1, 0, +1. These three orbitals are labeled p_x, p_y and p_z because their lobes of maximum probability lie along the x, y and z axes, respectively. Note the electron spins in the two lobes are opposite to each other. This is why the orbitals are colored red and gray. It is worth noting that although each of the p orbitals is nonspherically symmetric, their sum gives a spherically symmetric distribution of ψ^2.

When l = 2, there are five possible m_l values corresponding to the **d *orbitals***, shown schematically in Fig. 2.2*b*. These orbitals are labeled, d_{xy}, d_{xz}, d_{yz}, d_{z^2} and $d_{x^2-y^2}$. The latter two are oriented along the main axes, while the former three plot lobes between the axes. Here again the different colors denote different spins. Table 2.1 summarizes orbital notation up to n = 3. The physical significance of l and m_l and their relationships to an atom's angular momenta are discussed in greater detail in Chap. 15.

One final note: The conclusions arrived at so far tend to indicate that all sublevels with the same n have exactly the same energy, when in reality they have slightly different energies. Also, a fourth quantum number, the **spin quantum number, m_s**, which denotes the direction of electron spin, was not mentioned. Both of these omissions are a direct result of ignoring relativistic effects which, when taken into account, are fully accounted for.

2.2.3 POLYELECTRONIC ATOMS AND THE PERIODIC TABLE

Up to now the discussion has been limited to the simplest possible case, namely, that of the hydrogen atom—the only case for which an exact solution to Schrödinger's equation exists. The solution for a polyelectronic atom is similar to that of the hydrogen atom except that the former is inexact and is much more difficult to obtain. Fortunately, the basic shapes of the orbitals do *not* change, the concept of quantum numbers remains useful, and, with some modifications, the hydrogen-like orbitals can account for the electronic structure of atoms having many electrons.

The major modification involves the energy of the electrons. As the nuclear charge or atomic number, Z, increases, the potential energy of the electron has to decrease accordingly, since a large positive nuclear charge now attracts the electron more strongly. This can be accounted for, as a first and quite crude

[12] Sometimes l is referred to as the *orbital-shape quantum number*.

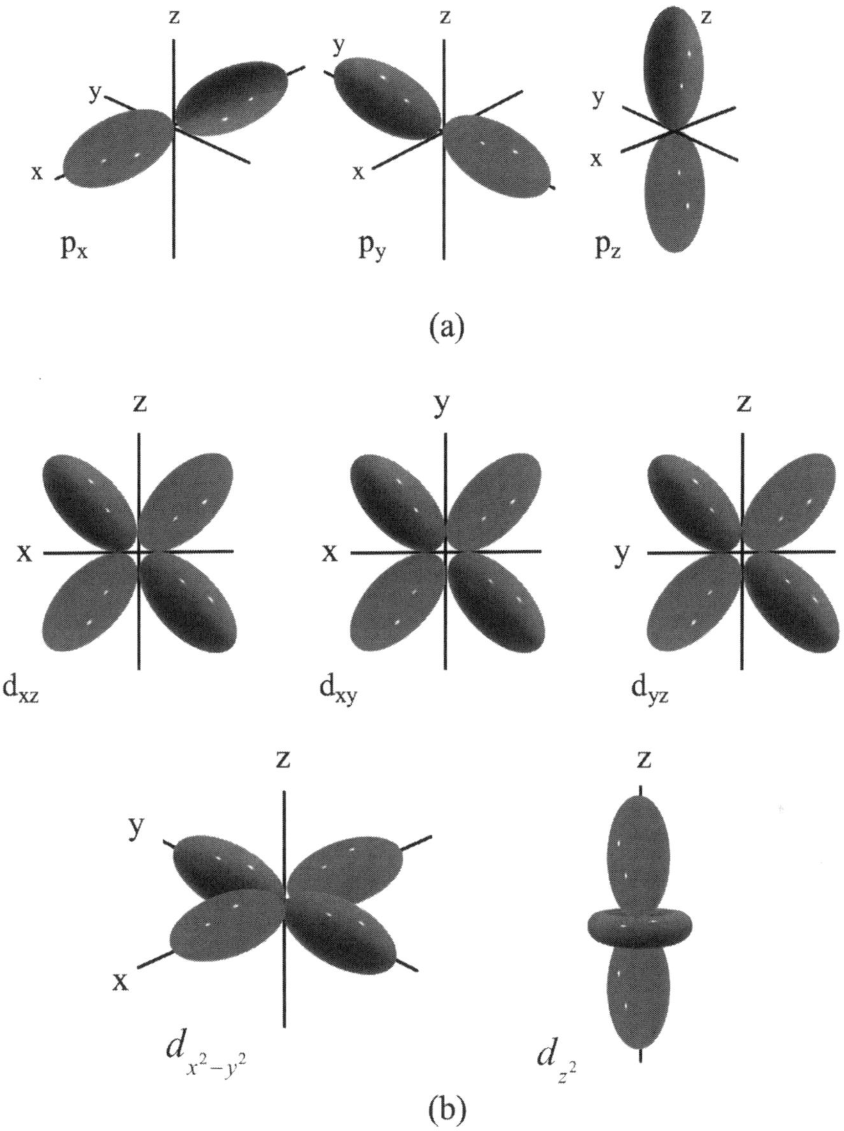

FIGURE 2.2 (*a*) Shape of p orbitals (top three) and (*b*) d orbitals (lower five). Note that the spin of the electrons in the red orbitals is opposite of that in the gray ones.

TABLE 2.1 Summary of orbitals and their notation

n	l	Orbital name	No. of m_l orbitals	Full designation of orbitals
1	0	1s	1	1s
2	0	2s	1	2s
	1	2p	3	$2p_x$, $2p_y$, $2p_z$
3	0	3s	1	3s
	1	3p	3	$3p_x$, $3p_y$, $3p_z$
	2	3d	5	$3d_{z^2}$, $3d_{x^2-y^2}$, $3d_{xy}$, $3d_{xz}$, $3d_{yz}$

approximation, by assuming that the electrons are noninteracting, in which case it can be shown that the energy of an electron is given by

$$E_n = -13.6\frac{Z^2}{n^2}\text{ eV}$$

The actual situation is more complicated, however, due to electron–electron repulsions and electron screening—with both effects contributing to an increase in E_n. Conceptually this is taken into account by introducing the **effective nuclear charge, Z_{eff},** which takes into account the notion that the actual nuclear charge experienced by an electron is always less than, or equal to, the actual charge on the nucleus. This can be easily grasped by comparing the experimental first ionization energy, IE, of helium, viz. −24.59 eV (see Table 2.2), for which $Z=2$ and $n=1$, to what one would expect had there been no electron–electron interaction, or $-13.6\times(2^2)/1^2$, or −54.4 eV. This simple example illustrates the dramatic effect of electron–electron interactions on the ionization energy of He and the importance of the concept of effective charge. Note that the measured second ionization energy for He listed in Table 2.2 is exactly 54.4 eV!

As the number of electrons increases, they are forced by virtue of Pauli's exclusion principle to occupy higher and higher-energy levels, i.e., higher n values. This in turn leads to the **aufbau principle**, the periodic table (see inside front cover) and a unique electronic configuration for each element. The electronic structures of the first 83 elements are summarized in Table 2.2.

Worked Example 2.2

(a) What are the electronic configurations of He, Li and F? (b) Identify the first transition metal series. What feature do these elements have in common?

ANSWER

(a) Helium ($Z=2$) has two electrons, which can be accommodated in the 1s state as long as their spins are opposite. Hence the configuration is $1s^2$. Since this is a closed shell configuration, He is a very inert gas. Lithium ($Z=3$) has three electrons; two are accommodated in the 1s shell and the third has to occupy a higher-energy state, namely, $n=2$ and $l=0$. The electronic configuration is thus: $1s^22s^1$. Similarly, the nine electrons of fluorine are distributed as follows: $1s^22s^22p^5$.

TABLE 2.2 Electronic configuration and first and second ionization energies of the elements

Z	Atom	Orbital electronic configuration	First IE, eV	Second IE, eV
1	H	$1s^1$	13.598	—
2	He	$1s^2$	24.587	54.416
3	Li	$(He)2s^1$	5.392	75.638
4	Be	$(He)\ 2s^2$	9.322	18.211
5	B	$(He)2s^22p^1$	8.298	25.154
6	C	$(He)2s^22p^2$	11.260	24.383
7	N	$(He)2s^22p^3$	14.534	29.601
8	O	$(He)2s^22p^4$	13.618	35.116
9	F	$(He)2s^22p^5$	17.422	34.970
10	Ne	$(He)2s^22p^6$	21.564	40.962
11	Na	$(Ne)3s^1$	5.139	47.286
12	Mg	$(Ne)3s^2$	7.646	15.035
13	Al	$(Ne)3s^23p^1$	5.986	18.828
14	Si	$(Ne)3s^23p^2$	8.151	16.345
15	P	$(Ne)3s^23p^3$	10.486	19.725
16	S	$(Ne)3s^23p^4$	10.360	23.330
17	Cl	$(Ne)3s^23p^5$	12.967	23.810
18	Ar	$(Ne)3s^23p^6$	15.759	27.630
19	K	$(Ar)4s^1$	4.340	31.625
20	Ca	$(Ar)4s^2$	6.113	11.871
21	Sc	$(Ar)4s^23d^1$	6.540	12.800
22	Ti	$(Ar)4s^23d^2$	6.820	13.580
23	V	$(Ar)4s^23d^3$	6.740	14.650
24	Cr	$(Ar)4s^13d^5$	6.766	16.500
25	Mn	$(Ar)4s^23d^5$	7.435	15.640
26	Fe	$(Ar)4s^23d^6$	7.870	16.180
27	Co	$(Ar)4s^23d^7$	7.860	17.060
28	Ni	$(Ar)4s^23d^8$	7.635	18.168
29	Cu	$(Ar)4s^13d^{10}$	7.726	20.292
30	Zn	$(Ar)4s^23d^{10}$	9.394	17.964
31	Ga	$(Ar)4s^23d^{10}4p^1$	5.999	20.510
32	Ge	$(Ar)4s^23d^{10}4p^2$	7.899	15.934
33	As	$(Ar)4s^23d^{10}4p^3$	9.810	18.633
34	Se	$(Ar)4s^23d^{10}4p^4$	9.752	21.190
35	Br	$(Ar)4s^23d^{10}4p^5$	11.814	21.800
36	Kr	$(Ar)4s^23d^{10}4p^6$	13.999	24.359
37	Rb	$(Kr)\ 5s^1$	4.177	27.280
38	Sr	$(Kr)\ 5s^2$	5.695	11.030
39	Y	$(Kr)\ 5s^24d^1$	6.380	12.240
40	Zr	$(Kr)\ 5s^24d^2$	6.840	13.130
41	Nb	$(Kr)\ 5s^14d^4$	6.880	14.320

(Continued)

TABLE 2.2 (CONTINUED) Electronic configuration and first and second ionization energies of the elements

Z	Atom	Orbital electronic configuration	First IE, eV	Second IE, eV
42	Mo	(Kr) $5s^1 4d^5$	7.099	16.150
43	Tc	(Kr)$5s^2 4d^5$	7.280	15.260
44	Ru	(Kr)$5s^1 4d^7$	7.370	16.760
45	Rh	(Kr)$5s^1 4d^8$	7.460	18.080
46	Pd	(Kr)$4d^{10}$	8.340	19.430
47	Ag	(Kr)$5s^1 4d^{10}$	7.576	21.490
48	Cd	(Kr)$5s^2 4d^{10}$	8.993	16.908
49	In	(Kr)$5s^2 4d^{10} 5p^1$	5.786	18.869
50	Sn	(Kr)$5s^2 4d^{10} 5p^2$	7.344	14.632
51	Sb	(Kr)$5s^2 4d^{10} 5p^3$	8.641	16.530
52	Te	(Kr)$5s^2 4d^{10} 5p^4$	9.009	18.600
53	I	(Kr)$5s^2 4d^{10} 5p^5$	10.451	19.131
54	Xe	(Kr)$5s^2 4d^{10} 5p^6$	12.130	21.210
55	Cs	(Xe)$6s^1$	3.894	25.100
56	Ba	(Xe)$6s^2$	5.212	10.004
57	La	(Xe)$6s^2 5d^1$	5.577	11.060
58	Ce	(Xe)$6s^2 4f^1 5d^1$	5.470	10.850
59	Pr	(Xe)$6s^2 4f^3$	5.420	10.560
60	Nd	(Xe)$6s^2 4f^4$	5.490	10.720
61	Pm	(Xe)$6s^2 4f^5$	5.550	10.900
62	Sm	(Xe)$6s^2 4f^6$	5.630	11.070
63	Eu	(Xe)$6s^2 4f^7$	5.670	11.250
64	Gd	(Xe)$6s^2 4f^7 5d^1$	5.426	13.900
65	Tb	(Xe)$6s^2 4f^9$	5.850	11.520
66	Dy	(Xe)$6s^2 4f^{10}$	5.930	11.670
67	Ho	(Xe)$6s^2 4f^{11}$	6.020	11.800
68	Er	(Xe)$6s^2 4f^{12}$	6.100	11.930
69	Tm	(Xe)$6s^2 4f^{13}$	6.180	12.050
70	Yb	(Xe)$6s^2 4f^{14}$	6.254	12.170
71	Lu	(Xe)$6s^2 4f^{14} 5d^1$	5.426	13.900
72	Hf	(Xe)$6s^2 4f^{14} 5d^2$	7.000	14.900
73	Ta	(Xe)$6s^2 4f^{14} 5d^3$	7.890	—
74	W	(Xe)$6s^2 4f^{14} 5d^4$	7.980	—
75	Re	(Xe)$6s^2 4f^{14} 5d^5$	7.880	—
76	Os	(Xe)$6s^2 4f^{14} 5d^6$	8.700	—
77	Ir	(Xe)$6s^2 4f^{14} 5d^7$	9.100	—
78	Pt	(Xe)$6s^1 4f^{14} 5d^9$	9.000	—
79	Au	(Xe)$6s^1 4f^{14} 5d^{10}$	9.225	—
80	Hg	(Xe)$6s^2 4f^{14} 5d^{10}$	10.437	18.756
81	Tl	(Xe)$6s^2 4f^{14} 5d^{10} 6p^1$	6.108	20.428
82	Pb	(Xe)$6s^2 4f^{14} 5d^{10} 6p^2$	7.416	15.032
83	Bi	(Xe)$6s^2 4f^{14} 5d^{10} 6p^3$	7.289	16.600

(b) The first series transition metals are Sc, Ti, V, Cr, Mn, Fe, Co and Ni. They all have partially filled d orbitals. Note that Cu and Zn, which have completely filled d orbitals, are sometimes also considered to be transition metals, although strictly speaking, they would not be since their d orbitals are totally filled (see Table 2.2).

2.3 IONIC VERSUS COVALENT BONDING

In the introduction to this chapter, it was stated that ceramics, very broadly speaking, can be considered to be either ionically or covalently bonded. The next logical question is: What determines the nature of a bond?

Ionic compounds generally form between quite active metallic elements and active nonmetals. For reasons that will become clear shortly, the requirements for an AB ionic bond to form are that A be able to lose electrons readily (i.e., with as little a penalty as possible) and B be able to accept electrons without too much energy input. This restricts ionic bonding to mostly metals from groups 1, 2 and 3, as well as some of the transition metals

For covalent bonding to occur, ionic bonding must be unfavorable. This is tantamount to saying that the energies of the bonding electrons of A and B must be comparable because if the electron energy on one of the atoms were much lower than that on the other, then electron transfer from one to the other would occur and ionic bonds would tend to form instead.

These qualitative requirements, while shedding some light on the problem, do not have much predictive capability as to the nature of the bond that will form. In an attempt to semi-quantify the answer, Pauling[13] established a scale of relative **electronegativity** or "electron greed" of atoms and defined electronegativity to be *the power of an atom to attract electrons to itself.* Pauling's electronegativity scale—listed in Table 2.3—was obtained by arbitrarily fixing the value of H at 2.2. With this scale, it becomes relatively simple to predict a bond's nature. If two elements forming a bond have similar electronegativities, they will tend to share the electrons between them and will form covalent bonds. However, if the electronegativity difference, ΔX, between them is large (indicating that one element is much greedier than the other), the electron will be attracted to the more electronegative element, forming ions which, in turn, attract each other. Needless to say, the transition between ionic and covalent bonding is far from sharp and, except for homopolar bonds that are purely covalent, all bonds will have both an ionic and a covalent character (see Prob. 2.16). However, as a quite rough guide, a bond is considered predominantly ionic when $\Delta X > 1.7$ and predominantly covalent if $\Delta X < 1.7$.

Each type of bond and how it leads to the formation of a solid is discussed separately below, starting with the simpler of the two, namely, the ionic bond.

2.4 IONIC BONDING

Ionically bonded solids are made up of charged particles—positively charged ions, called **cations**, and negatively charged ions, called **anions.** Their mutual attraction holds the solid together. Ionic bonds

[13] L. Pauling, *The Nature of the Chemical Bond*, 3rd ed., Cornell University Press, Ithaca, NY, 1960.

TABLE 2.3 Relative electronegativity scale of the elements

Element	Electronegativity	Element	Electronegativity
1. H	2.20	42. Mo(II)	2.16
2. He		Mo(III)	2.19
3. Li	0.98	43. Tc	1.90
4. Be	1.57	44. Ru	2.20
5. B	2.04	45. Rh	2.28
6. C	2.55	46. Pd	2.20
7. N	3.04	47. Ag	1.93
8. O	3.44	48. Cd	1.69
9. F	3.98	49. In	1.78
10. Ne		50. Sn(II)	1.80
11. Na	0.93	Sn(IV)	1.96
12. Mg	1.31	51. Sb	2.05
13. Al	1.61	52. Te	2.10
14. Si	1.90	53. I	2.66
15. P	2.19	54. Xe	2.60
16. S	2.58	55. Cs	0.79
17. Cl	3.16	56. Ba	0.89
18. Ar		57. La	1.10
19. K	0.82	58. Ce	1.12
20. Ca	1.00	59. Pr	1.13
21. Sc	1.36	60. Nd	1.14
22. Ti(II)	1.54	62. Sm	1.17
23. V(II)	1.63	64. Gd	1.20
24. Cr(II)	1.66	66. Dy	1.22
25. Mn(II)	1.55	67. Ho	1.23
26. Fe(II)	1.83	68. Er	1.24
Fe(III)	1.96	69. Tm	1.25
27. Co(II)	1.88	71. Lu	1.27
28. Ni(II)	1.91	72. Hf	1.30
29. Cu(I)	1.90	73. Ta	1.50
Cu(II)	2.00	74. W	2.36
30. Zn(II)	1.65	75. Re	1.90
31. Ga(III)	1.81	76. Os	2.20
32. Ge(IV)	2.01	77. Ir	2.20
33. As(III)	2.18	78. Pt	2.28
34. Se	2.55	79. Au	2.54
35. Br	2.96	80. Hg	2.00

(Continued)

TABLE 2.3 (CONTINUED) Relative electronegativity scale of the elements

Element	Electronegativity	Element	Electronegativity
36. Kr	2.90	81. Tl(I)	1.62
37. Rb	0.82	82. Pb(II)	1.87
38. Sr	0.95	83. Bi	2.02
39. Y	1.22	90. Th	1.30
40. Zr(II)	1.33	92. U	1.70
41. Nb	1.60		

Source: https://en.wikipedia.org/wiki/Electronegativity.

are omnidirectional. Ionic compounds are typically hard and brittle and are poor electrical and thermal conductors.

To illustrate the energetics of ionic bonding consider the bond formed between Na and Cl. The electronic configuration of Cl (atomic number Z = 17) is $[1s^2 2s^2 2p^6]3s^2 3p^5$, while that of Na (Z = 11) is $[1s^2 2s^2 2p^6]3s^1$. For reasons that will become evident in a moment, when Na and Cl atoms are brought into close proximity, a bond will form by the transfer of an electron from the Na atom to the Cl atom, as shown schematically in Fig. 2.3. The Na atom configuration becomes $[1s^2 2s^2 2p^6]$ and is now +1 positively charged; the Cl atom gains an electron, acquires a −1 negative charge, and its electronic structure becomes $[1s^2 2s^2 2p^6]3s^2 3p^6$. Note that after the charge transfer, the configuration of the Na and Cl ions correspond to those of the noble gases, Ne and Ar, respectively.

The work done to bring the ions from infinity to a distance r apart is once again given by Coulomb's law [Eq. (2.2)]:

$$E_{pot} = \frac{z_1 z_2 e^2}{4\pi\varepsilon_0 r} \tag{2.11}$$

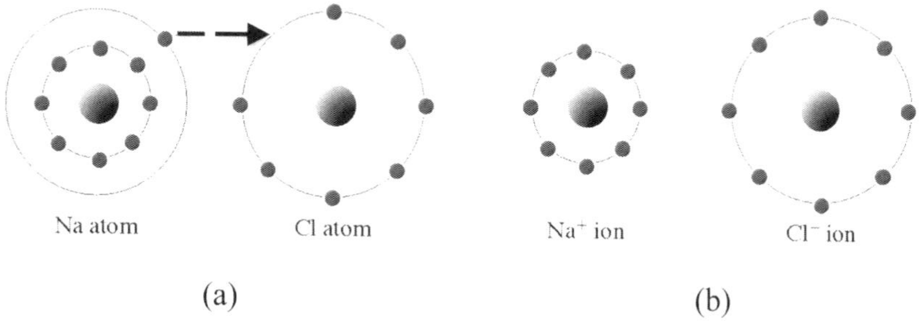

FIGURE 2.3 (*a*) Electron transfer from a Na atom to Cl atom results in the formation of, (*b*) a cation and an anion. Note that a cation is smaller than its atom and vice versa for the anion.

In this case, z_1 and z_2 are the *net* charges on the ions (+1 and −1 for NaCl, −2 and +3 for Al_2O_3, etc.). When z_1 and z_2 are of opposite signs, E_{pot} is negative, which is consistent with the fact that energy is released as the ions are brought together from infinity. A plot of Eq. (2.11) is shown in Fig. 2.4*a* (lower curve), from which it is clear that when the ions are infinitely separated, the interaction energy vanishes, as one would expect. Equation (2.11) also predicts, however, that as the distance between the ions goes to zero, the ions should fuse together and release an infinite amount of energy! That this does not happen is obvious; NaCl does, and incidentally we also, exist.

It follows that for a stable lattice to result, a repulsive force must come into play at short distances. As discussed above, the attraction occurs from the *net* charges on the ions. These ions, however, are themselves made up of positive and negative entities, namely, the nuclei of each ion, but more importantly, the electron cloud surrounding each nucleus. As the ions approach each other, these like charges repel and prevent the ions from coming any closer.

The repulsive energy term is positive by definition and is usually given by the empirical expression:

$$E_{rep} = \frac{B}{r^n} \tag{2.12}$$

where B and n are empirical constants that depend on the material in question. Sometimes referred to as the **Born exponent, n**, usually lies between 6 and 12. Equation (2.12) is also plotted in Fig. 2.4*a* (top curve), from which it is clear that the repulsive component dominates at small r, but decreases quite rapidly as r increases. The Born exponent should not be confused with the principle quantum number, n; they are not related in any way.

The net energy E_{net} of the system is the sum of the attractive and repulsive terms, or

$$E_{net} = \frac{z_1 z_2 e^2}{4\pi\varepsilon_0 r} + \frac{B}{r^n} \tag{2.13}$$

When E_{net} is plotted as a function of r (middle red curve in Fig. 2.4*a*), it goes through a minimum, at a distance denoted by r_0. The minimum in the curve corresponding to the equilibrium situation can be found readily from

$$\left.\frac{dE_{net}}{dr}\right|_{r=r_0} = 0 = -\frac{z_1 z_2 e^2}{4\pi\varepsilon_0 r_0^2} - \frac{nB}{r_0^{n+1}} \tag{2.14}$$

By evaluating the constant B and removing it from Eq. (2.13), it can be shown that the depth of the energy well E_{bond} is given by

$$\boxed{E_{bond} = \frac{z_1 z_2 e^2}{4\pi\varepsilon_0 r_0}\left(1 - \frac{1}{n}\right)} \tag{2.15}$$

Here, r_0 is the equilibrium separation between the ions. The occurrence of this minimum is of paramount importance since it defines a bond; i.e., when two ions are brought closer together from infinity, they will

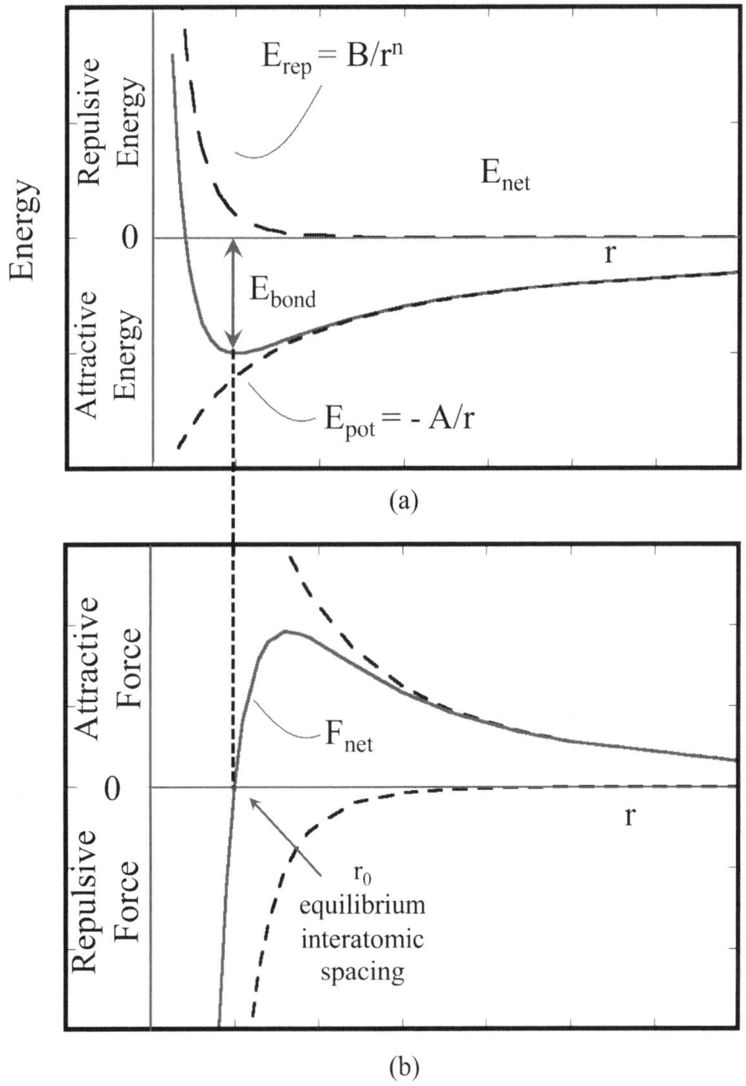

FIGURE 2.4 (*a*) Energy versus distance curves for an ionic bond. The net energy is the sum of attractive and repulsive energies, which gives rise to an energy well. (*b*) Corresponding force versus distance curve. This curve is the derivative of the net energy curve shown in (*a*). Note that when the energy is at a minimum, the net force is zero.

attract each up to an **equilibrium distance, r_0**, and liberate an amount of energy given by Eq. (2.15). Conversely, E_{bond} can be thought of as the energy required to pull the ions apart from a distance r_0 to infinity.

It is important to note that Eq. (2.14) is also an expression for the *net* force between the ions, since by definition

$$F_{net} = \frac{dE_{net}}{dr} = -\frac{z_1 z_2}{4\pi\varepsilon_0 r^2} - \frac{nB}{r^{n+1}} \tag{2.16}$$

F_{net} is plotted in Fig. 2.4*b*. For distances $> r_0$, F_{net} on the ions is attractive; for distances $< r_0$, the net force is repulsive. At r_0 the net force on the ions is zero [Eq. (2.14)], which is why r_0 is the equilibrium interatomic spacing. Figure 2.4*a* and *b* illustrate a fundamental law of nature, namely, that at equilibrium the energy is minimized and the net force on a system is zero.

2.5 IONICALLY BONDED SOLIDS

The next logical question is, how do such bonds lead to the formation of a solid? After all, a solid is made up of roughly 10^{23} bonds. The other related important question has to do with the energy of the solid lattice. The latter is related to the stability of a given structure and directly, or indirectly, determines such properties as melting temperatures, thermal expansion, stiffness and others, discussed in Chap. 4. This section addresses how the lattice energy is calculated and experimentally verified, starting with the simple electrostatic model that led to Eq. (2.15).

2.5.1 LATTICE ENERGY CALCULATIONS

The **lattice energy, E_{latt}**, is defined as the energy released when x moles of cations, A, and y moles of anions, B, react to form the solid A_xB_y. To calculate E_{latt} a structure or packing arrangement of the ions has to be assumed,[14] and all the interactions between the ions have to be taken into account. To illustrate consider NaCl, which has one of the simplest ionic structures known (Fig. 2.5*a*), wherein each Na ion is surrounded by six Cl ions and vice versa. Referring to Fig. 2.5*b*, the central cation—depicted in light gray—is attracted to 6 Cl^- anions at a distance r_0, repelled by 12 Na^+ cations at distance $\sqrt{2}r_0$ (Fig. 2.5*c*), attracted to 8 Cl^- anions at $\sqrt{3}r_0$ (Fig. 2.5*d*), etc. Summing up the electrostatic interactions,[15] one obtains

[14] This topic is discussed in greater detail in the next chapter and depends on the size of the ions involved, the nature of the bonding, etc.

[15] Strictly speaking, this is not exact, since in Eq. (2.17) the repulsive component of the ions that were not nearest neighbors was neglected. If that interaction is taken into account, an exact expression for E_{sum} is given by

$$E_{sum} = \frac{-z_1 z_2 e^2 \alpha}{4\pi\varepsilon_0 r} + \frac{B\beta}{r^n}$$

where β is another infinite series. It is important to note that such a refinement does not in any way alter the result, namely, Eq. (2.18).

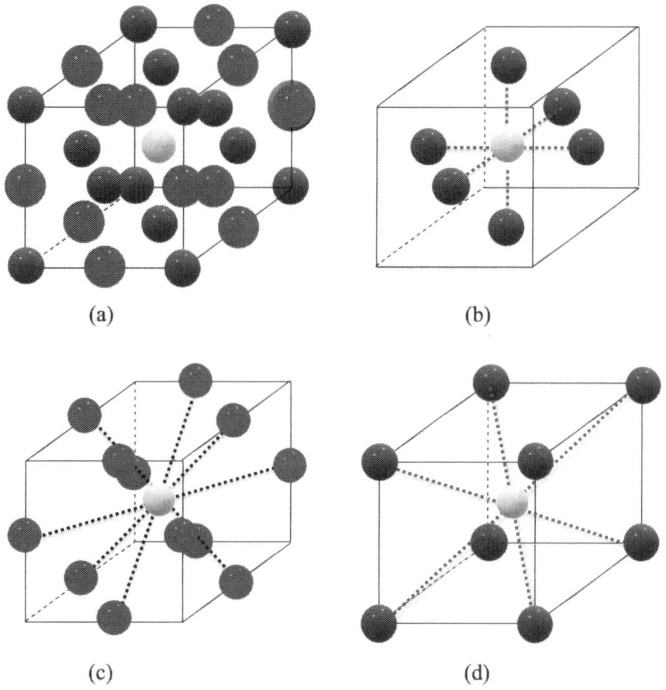

FIGURE 2.5 (*a*) Schematic of the NaCl structure. (*b*) The first six nearest neighbors are attracted to the central cation, (*c*) the second 12 nearest neighbors at a distance $\sqrt{2}r_0$ are repelled, (*d*) third eight nearest neighbors are attracted, etc.

$$\begin{aligned} E_{sum} &= \frac{z_1 z_2 e^2}{4\pi\varepsilon_0 r_0}\left(1-\frac{1}{n}\right)\left(\frac{6}{1}-\frac{12}{\sqrt{2}}+\frac{8}{\sqrt{3}}-\frac{6}{\sqrt{4}}+\frac{24}{\sqrt{5}}-\cdots\right) \\ &= \frac{z_1 z_2 e^2}{4\pi\varepsilon_0 r_0}\left(1-\frac{1}{n}\right)\alpha \end{aligned} \quad (2.17)$$

The second term in parentheses is an alternating series that converges to some value α, known as the **Madelung constant.** Evaluation of this constant, though straightforward, is tedious because the series converges quite slowly. The Madelung constants for a number of crystal structures are listed in Table 2.4.

The total electrostatic attraction or lattice energy to form 1 mol of NaCl in which there are twice Avogadro's number N_{Av} of ions but only N_{Av} bonds is

$$E_{latt} = \frac{N_{Av} z_1 z_2 e^2 \alpha}{4\pi\varepsilon_0 r_0}\left(1-\frac{1}{n}\right) \quad (2.18)$$

According to this equation, sometimes referred to as the *Born–Lande equation*, the information required to calculate E_{latt} is the crystal structure, which determines α, the equilibrium interionic spacing, both easily

TABLE 2.4 Madelung constants for some common ceramic crystal structures (see Chap. 3)

Structure	Coordination number	α^a	$\alpha_{conv}{}^b$
NaCl	6: 6	1.7475	1.7475
CsCl	8: 8	1.7626	1.7626
Zinc blende	4: 4	1.6381	1.6381
Wurtzite	4: 4	1.6410	1.6410
Fluorite	8: 4	2.5190	5.0387
Rutile	6: 3	2.4080[c]	4.1860[c]
Corundum	6: 4	4.1719[c]	25.0312[c]

[a] Assumes structure is made of isocharged ions that factor out.

[b] The problem of structures with more than one charge, such as Al_2O_3, can be addressed by making use of the relationship

$$E_{sum} = \alpha_{conv} \frac{(Z\pm)^2 e^2}{4\pi\varepsilon_0 r_0}\left(1-\frac{1}{n}\right)$$

where $Z\pm$ is the *highest common factor* of z_1 and z_2, i.e., 1 for NaCl, CaF_2 and Al_2O_3, 2 for MgO, TiO_2, ReO_3, etc.

[c] Exact value depends on c/a ratio.

obtainable from X-ray diffraction and n, which is obtainable from compressibility data. Note that E_{latt} is not greatly affected by small errors in n.

In deriving Eq. (2.18), a few terms were ignored. A more exact expression for E_{latt} is

$$E_{latt} = \frac{dE_{net}}{dr} = -\frac{A}{r_0} + \frac{B}{r^n} + \left(\frac{C}{r_0^6} + \frac{D}{r_0^8}\right) + \frac{9}{4} h\nu_{max} \tag{2.19}$$

The first two terms, discussed in detail up to this point, dominate. Note the term $z_1 z_2 e^2/4\pi\varepsilon_0$ in Eq. (2.13) is replaced by the constant A in Eq. (2.19). The term in parentheses represents dipole–dipole and dipole–quadrapole interactions between the ions. The last term represents the zero-point correction, with ν_{max} being the highest frequency of the lattice vibration modes. Lastly, in this section it is worth noting that this ionic model is a poor approximation for crystals containing large anions and small cations, where the covalent contribution to the bonding becomes significant (see Chap. 3).

Worked Example 2.3

(a) Calculate the lattice energy of NaCl given that n = 8.
(b) Repeat part (a) assuming the structure is CsCl instead and comment on the difference.
(c) Repeat part (a) for MgO.

ANSWER

(a) To calculate E_{latt}, r_o, n and the structure of NaCl all are needed. As noted above, the structure of NaCl is the rock salt structure (Fig. 2.5) and hence its Madelung constant is 1.748 (Table 2.4). The equilibrium interionic distance, r_o, is simply the sum of the radii of the Na^+ and Cl^- ions. The values are listed at the end of Chap. 3 in Appendix 3A. Looking up the values, the equilibrium interionic distance $r_o = 181 + 102 = 283$ pm. Applying Eq. 2.18, it follows that

$$E_{latt} = \frac{(-1)(+1)(6.02\times10^{23})(1.6\times10^{-19})^2(1.748)}{4\pi(8.85\times10^{-12})(283\times10^{-12})}\left(1-\frac{1}{8}\right) \approx -749 \text{ kJ/mol}$$

(b) If one assumes the structure is CsCl, then the only terms that change in the above equation are the Madelung constant and the radii of the ions. In this case the Na^+ ion is in 8 fold coordination and its radius is 118 pm. The radius on Cl^- in 8 fold coordination is not given in Appendix 3A, so we assume it to remain unchanged. It follows that $r_o = 181 + 118 = 299$ pm and

$$E_{latt} = \frac{(-1)(+1)(6.02\times10^{23})(1.6\times10^{-19})^2(1.7626)}{4\pi(8.85\times10^{-12})(299\times10^{-12})}\left(1-\frac{1}{8}\right) \approx -714.8 \text{ kJ/mol}$$

These calculations make the following points amply clear: (i) E_{latt} of NaCl in the rock salt structure is lower in energy—more stable—than had it crystallized in the CsCl structure, which is why it crystallizes in the former. (ii) The differences in energy is less than 5%. It follows that it is crucial that our models and the parameters used in them be quite accurate. In many cases, the differences in energy between different polymorphs are even smaller. (iii) Knowing the exact values of the ionic radii is crucial when considering which polymorph is more stable.

(c) For MgO, $r_o = (72+140) = 212$ pm, and assuming $n = 8$, then

$$E_{latt} = \frac{(-2)(+2)(6.02\times10^{23})(1.6\times10^{-19})^2(1.748)}{4\pi(8.85\times10^{-12})(212\times10^{-12})}\left(1-\frac{1}{8}\right) \approx -4000 \text{ kJ/mol}$$

2.5.2 BORN–HABER CYCLE

So far, a rather simple model has been introduced in which it was assumed that an ionic solid is made up of ions attracted to each other by Coulombic attractions. How can such a model be tested? The simplest thing to do would be to compare E_{latt} to experimental results. This is easier said than done, however, given that E_{latt} is the energy released when 1 mol of gaseous *cations and anions* condense into a solid—an experiment that, needless to say, is not easy, if not impossible, to perform.

An alternate method is to make use of the first law of thermodynamics, namely, that energy can be neither created nor destroyed. If a cycle can be devised where all the energies are experimentally known

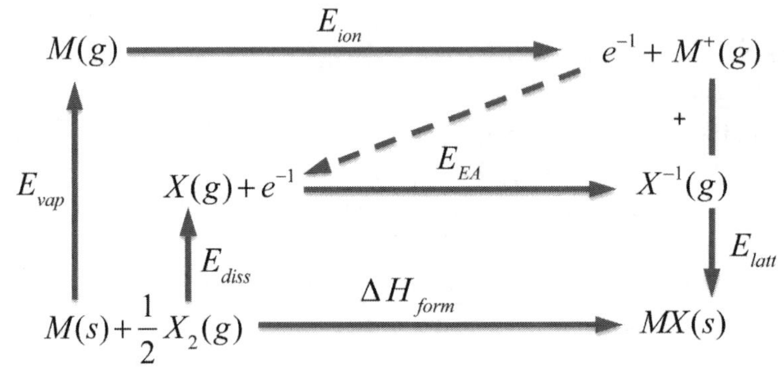

FIGURE 2.6 The Born–Haber cycle.

except E_{latt}, then it can be calculated. For such a cycle, known as the *Born–Haber cycle*, shown in Fig. 2.6, it is necessary that

$$\Delta H_{form}(\text{exo}) = E_{latt}(\text{exo}) + E_{ion}(\text{endo}) + E_{EA}(\text{endo or exo})$$
$$+ E_{diss}(\text{endo}) + E_{vap}(\text{endo})$$

Each of these terms is discussed in greater detail below with respect to NaCl.

ENTHALPY OF FORMATION OR REACTION

The enthalpy of formation, ΔH_{form} of a reaction is the experimental energy absorbed or released when that reaction occurs. In the case of NaCl, the reaction is

$$Na(s) + \tfrac{1}{2}Cl_2(g) \rightarrow NaCl(s)$$

Since energy is liberated, this reaction is exothermic and thus by convention ΔH_{form} is negative. For NaCl at 298 K, the experimentally determined $\Delta H_{form} = -411$ kJ/mol. The enthalpies of formation of most compounds are exothermic.

DISSOCIATION ENERGY

The dissociation energy, E_{diss}, of a molecule is the energy needed to break it up into its constituent atoms. For NaCl, it is energy change for the reaction

$$\tfrac{1}{2}Cl_2 \rightarrow Cl\,(g)$$

This energy is always endothermic and thus positive. For this reaction as written, $E_{diss} = +121$ kJ/mol.

HEAT OF VAPORIZATION

The **latent heat of vaporization, E_{vap}**, is the energy required for the reaction

$$M\,(s) \rightarrow M\,(g)$$

which is always endothermic. In this case of Na, this value is 107.3 kJ/mol.

Values of ΔH_{form}, E_{diss} and E_{vap} can be found in various sources[16] and are well documented for most elements and compounds.

IONIZATION ENERGY

The **ionization energy**, $\mathbf{E_{ion}}$, is the energy required to completely remove an electron from an isolated atom in the gas phase. Ionization energies are endothermic since in all cases work has to be done to remove an electron from its nucleus. Table 2.2 lists the first and second ionization potentials for select elements. For Na, that value is 5.14 eV or 495.8 kJ/mol.

ELECTRON AFFINITY

The electron affinity, $\mathbf{E_{EA}}$, is a measure of the energy change that occurs when an electron is added to the valence shell of an atom. Some selected values of E_{EA} for nonmetals are listed in Table 2.5. The addition of the first electron is usually exothermic (e.g., oxygen, sulfur); further additions, when they occur, are by necessity endothermic since the second electron is now approaching a negatively charged entity. The electron affinity of Cl is −348.7 kJ/mol.

The lattice energy of NaCl was calculated in Worked Example 2.3 to be −750 kJ/mol. If we put all the pieces together, the Born–Haber summation for NaCl yields

$$
\begin{aligned}
\Delta H_{form}(\text{exo}) &= E_{latt}(\text{exo}) + E_{ion}(\text{endo}) + E_{EA}(\text{exo}) \\
&\quad + E_{diss}(\text{endo}) + E_{vap}(\text{endo}) \\
&= -750 + 495.8 - 348.7 + 121 + 107.3 = -374.6\ \text{kJ/mol}
\end{aligned}
$$

TABLE 2.5 Electron affinities[a] of selected nonmetals at 0 K

Element	EA (kJ/mol)	Element	EA (kJ/mol)
$O \rightarrow O^-$	141 (exo)	$Se \rightarrow Se^-$	195 (exo)
$O^- \rightarrow O^{2-}$	780 (endo)	$Se^- \rightarrow Se^{2-}$	420 (endo)
$F \rightarrow F^-$	322 (exo)	$Br \rightarrow Br^-$	324.5 (exo)
$S \rightarrow S^-$	200 (exo)	$I \rightarrow I^-$	295 (exo)
$S^- \rightarrow S^{2-}$	590 (endo)	$Te \rightarrow Te^-$	190.1 (exo)
$Cl \rightarrow Cl^-$	348.7 (exo)		

[a] Electron affinity is usually defined as the energy *released* when an electron is added to the valence shell of an atom. This can be quite confusing. To avoid confusion, the values listed in this table clearly indicate whether the addition of an electron is endo- or exothermic. Data taken from https://en.wikipedia.org/wiki/Electron_affinity_(data_page).

[16] A reliable source for thermodynamic data is JANAF *Thermochemical Tables*, 4th ed., which lists the thermodynamic data of over 1800 substances. http://kinetics.nist.gov/janaf/.

which compares favorably with the experimentally determined value of −411 kJ/mol. If Eq. (2.19) is used, even better agreement is obtained.

This is an important result for two reasons. First, it confirms that our simple model for the interaction between ions in a solid is, for the most part, correct. Second, it supports the notion that NaCl can be considered an ionically bonded solid.

2.6 COVALENT BOND FORMATION

The second important type of primary bond is the covalent bond. Whereas ionic bonds involve electron transfer to produce oppositely charged species, covalent bonds arise as a result of electron sharing. In principle, the energetics of the covalent bond can be understood if it is recognized that electrons spend more time in the area *between* the nuclei than anywhere else. The mutual attraction between the nuclei and the electrons between them lowers the potential energy of the system, resulting in a bond. Several theories and models have been proposed to explain the formation of covalent bonds. Of these, **molecular orbital theory** has been particularly successful and is the one discussed in some detail below. As the name implies, molecular orbital (MO) theory treats a molecule as a single entity and assigns orbitals to the *molecule as a whole*. In principle, the idea is similar to that used to determine the energy levels of isolated atoms, except that now the wave functions have to satisfy Schrödinger's equation with the appropriate expression for the potential energy, which has to include *all* the charges making up the molecule. The solutions, in turn, give rise to various MOs, with the number of filled orbitals determined by the number of electrons needed to balance the nuclear charge of the molecule as a whole, subject to Pauli's exclusion principle.

To illustrate, consider the simplest possible molecule, namely, the H_2^+ molecule, which has one electron but two nuclei. This molecule is chosen in order to avoid the complications arising from electron–electron repulsions alluded to earlier.

2.6.1 HYDROGEN ION MOLECULE

The procedure is similar to that used to solve for the electronic wave function of the H atom [i.e., the wave functions have to satisfy Eq. (2.1)] except that the potential energy term has to account for the presence of *two* positively charged nuclei rather than one. The Schrödinger equation for the H_2^+ molecule thus reads

$$\frac{\partial^2\psi}{\partial x^2}+\frac{8\pi^2 m_e}{h^2}\left(E_{tot}+\frac{e^2}{4\pi\varepsilon_0 r_a}+\frac{e^2}{4\pi\varepsilon_0 r_b}-\frac{e^2}{4\pi\varepsilon_0 R}\right)\psi=0 \tag{2.20}$$

where the distances, r_a, r_b and R are defined in Fig. 2.7*a*. If the distance R between the two nuclei is fixed, then an exact solution exists, which is quite similar to that of the H atom, except that now *two solutions* or *wave functions* emerge. One solution results in an increase in the electron density between the nuclei (Fig. 2.7*c*) whereas the second solution decreases it (Fig. 2.7*d*). In the first case, both nuclei are attracted to the electron between them, which results in the lowering of the energy of the system relative to the isolated

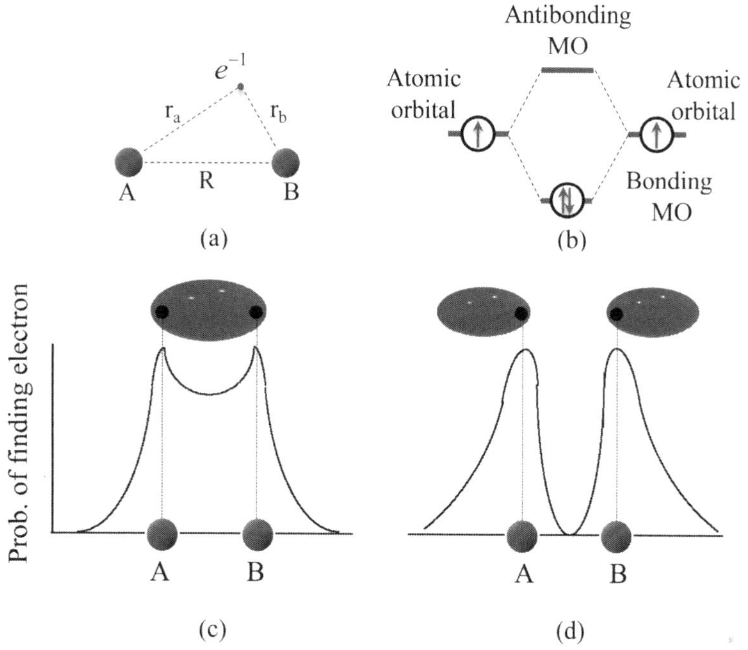

FIGURE 2.7 (*a*) Coordinates for the H_2^+ molecule used in Eq. (2.20). (*b*) Interaction of the two atomic orbitals results in bonding and antibonding orbitals. (*c*) Probability function for the bonding case in which the electron density between the nuclei is enhanced. (*d*) Probability function for the antibonding case, where the probability of finding the electron in the volume between the nuclei is decreased resulting in a higher-energy orbital.

atom case and is thus known as a **bonding orbital** (Fig. 2.7*b*). The second results in an increase in energy relative to the isolated atoms because now the unsheathed, or partially bared, nuclei repel one another. This is known as the **antibonding orbital**, also shown in Fig. 2.7*b*.

It is important to note that in order to obtain an energy-distance curve typical of a bond (see Fig. 2.4*a*), Eq. (2.20) would have to be solved for many values of R, the distance between the protons. It is only by doing so, that an energy distance curve is obtained from which the equilibrium interatomic distance can be predicted and ultimately compared with experimental values.

The solution for the H_2 molecule is quite similar, except that now an extra potential energy term for the repulsion between the two electrons has to be included in Schrödinger's equation. This is nontrivial, but fortunately the end result is similar to that of the H_2^+ case; the individual energy levels split into a bonding and an antibonding orbital. The atomic orbital overlap results in an increased probability of finding the electron between the nuclei. Note that in the case of the H_2 molecule, the two electrons are accommodated in the bonding orbital. A third electron, i.e., H_2^-, would have to go into the antibonding orbital because of Pauli's exclusion principle.

Before proceeding, it is illustrative to consider another slightly more complicated example: the HF molecule.

2.6.2 HF MOLECULE

In the preceding section, the electronegativities of the two atoms and the shapes (both spherical) of the interacting orbitals making up the bond were identical. The situation becomes more complicated when one considers bonding between dissimilar atoms. The HF molecule provides a good example. The electron configuration of H is $1s^1$, and that of F is (He)$2s^2 2p^5$. The valence orbitals of the F atom are shown in Fig. 2.8*a* (the inner core electrons are ignored since they are not involved in bonding). The atoms are held at the distance that separates them, which can either be calculated or obtained experimentally, and the molecular orbitals of HF are calculated. The calculations are nontrivial and beyond the scope of this book; the result, however, is shown schematically in Fig. 2.8*b*. The total number of electrons that have to be accommodated in the MOs is eight (seven from F and one from H). Placing two in each orbital fills the first four orbitals and results in an energy for the molecule that is *lower* (more negative) than that of the sum of the two noninteracting atoms, which in turn renders the HF molecule more stable relative to the isolated atoms.

Figure 2.8 can also be interpreted as follows: the F 2s electrons, by virtue of being at a much lower energy than hydrogen (because of the higher charge on the F nucleus), remain unperturbed by the hydrogen atom.[17] The 1s electron wave function of the H atom and one of the 2p orbitals on the F atom overlap to form a primary σ bond (Fig. 2.8*d*). The remaining electrons on the F atom (the so-called lone pairs) remain unperturbed in energy and in space.

As mentioned above, the calculation for Fig. 2.8 was made for a given interatomic distance. The same calculation can be repeated for various interatomic separations. At infinite separation, the atoms do not

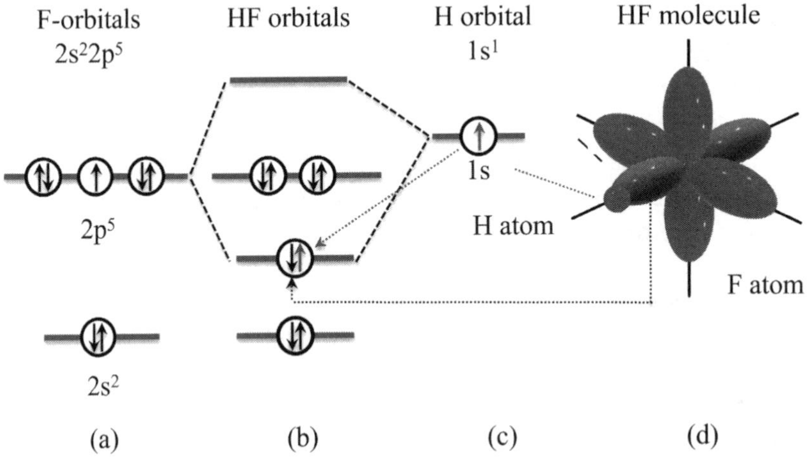

FIGURE 2.8 (*a*) F atomic orbitals. (*b*) HF molecular orbitals. (*c*) H atomic orbital. (*d*) Interaction of H 1s orbital with one of the F p orbitals. The overlap of these two orbitals results in a lowering of the system's energy. Dotted lines joining (*b*) to (*d*) emphasize that it is only the F p orbital that overlaps with the H orbital that lowers its energy. The two pairs of unpaired electrons (red lobes) have the same energy in the molecule that they did on the F atom, since these so-called lone pairs are unperturbed by the presence of the H atom.

[17] For orbitals to overlap, they must be relatively close to each other in energy.

interact, and the system's energy is just the sum of the energies of the electrons on the separate atoms. As the atoms are brought closer together, the attractive potential energy due to the mutual attraction between the electrons and the nuclei decreases the energy of the system up to a point beyond which a repulsive component comes into play and the energy starts increasing again. In other words, at some interatomic distance, a minimum in the energy occurs, and a plot of energy versus interatomic distance results in an energy well that is not unlike the one shown in Fig. 2.4*a*.

2.7 COVALENTLY BONDED SOLIDS

Up to this point the discussion was focused on the energetics of a single covalent bond between two atoms. Such a bond, however, will not lead to the formation of a strong solid, i.e., one in which all the bonds are primary. To form such a solid, each atom has to be simultaneously bonded to at least two other atoms. For example, at room temperature, HF does not form a solid because once a HF bond is formed, both atoms attain their most stable configuration—He for H and Ne for F—which in turn implies that there are no electrons available to form covalent bonds with other atoms. This is why HF is a gas at room temperature, despite the fact that the HF bond is quite strong.[18]

As discussed in greater detail in the next chapter, many predominantly covalently bonded ceramics, especially Si-based ones such as silicon carbide, silicon nitride, and the silicates, are composed of Si atoms simultaneously bonded to four other atoms in a tetrahedral arrangement. Examining the ground-state configuration of Si, that is, (Ne) $3s^2$ $3p^2$ (Fig. 2.9*a*), one would expect only two primary bonds to form, when in fact four bonds are known to form. This apparent contradiction has been explained by postulating that **hybridization** between the s and p wave functions occurs. Hybridization consists of a mixing of, or linear combinations of, s and p orbitals in an atom in such a way as to form new hybrid orbitals. This hybridization can occur between one s orbital and one *p* orbital (forming a sp orbital), or one s and two p orbitals (forming a sp^2 trigonal orbital). In the case of Si, the s orbital hybridizes with all three p orbitals to form what is known as **sp^3 hybrid orbitals.** The latter possesses both s and p character and directionally reaches out in space as lobes in a tetrahedral arrangement, with a bond angle of 109°, as shown in Fig. 2.9*c*. If each of these orbitals is populated by one electron (Fig. 2.9*b*), then each Si atom can now bond to *four* other Si atoms, or any other *four* atoms for that matter, which in turn can lead to three-dimensional structures. Note that the promotion of the electron from the s to the sp^3 hybrid orbital requires some energy, which is more than compensated for by the formation of four primary bonds.

2.8 BAND THEORY OF SOLIDS

One of the more successful theories developed to explain a wide variety of electrical and optical properties in solids is the **band theory of solids.** In this model, the electrons are consigned to bands. Bands that are incompletely filled (Fig. 2.10*a*) are termed **conduction bands**, while those that are full are called

[18] If sufficiently cooled, HF will form a solid as a result of secondary bonds.

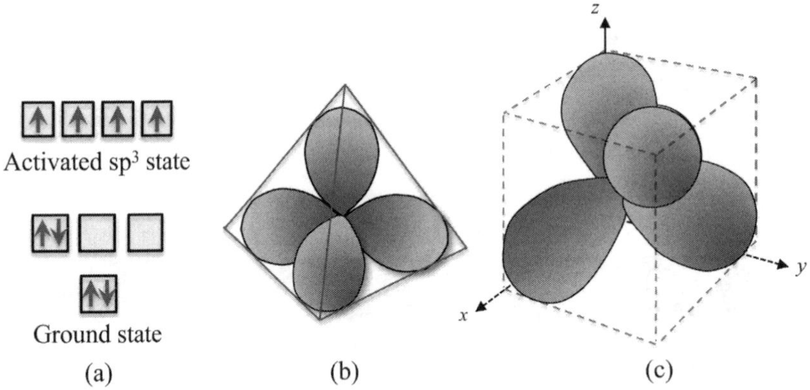

FIGURE 2.9 (*a*) Electronic configuration of ground state of the Si atom (bottom) and after hybridization (top). (*b*) Directionality of *sp*³ bonds relative to a tetrahedral. (*c*) *Same as* (b) *but now embedded in a cube, where each lobe points to a different corner*. By so doing the electrons are kept as far away from each other as possible. Note that each bond lobe contains one electron, and thus the atom can now form *four* covalent bonds with other atoms.

valence bands. The electrons occupying the highest energy in a conduction band can rapidly adjust to an applied electric or magnetic field and give rise to the properties characteristic of a metal, such as high electrical and thermal conductivities, ductility and reflectivity. Solids where the valence bands are completely filled (Fig. 2.10*b*), on the other hand, are poor conductors of electricity and, at 0 K, are perfect insulators. It follows that understanding this model of the solid state is of paramount importance if the electrical and optical properties of solids in general, and ceramics in particular, are to be understood.

The next three subsections address the not-so-transparent concept of how, and why, bands form in solids. Three approaches are discussed. The first is a simple qualitative model. The second is slightly more quantitative and sheds some light on the relationship between the properties of the atoms making up a solid and

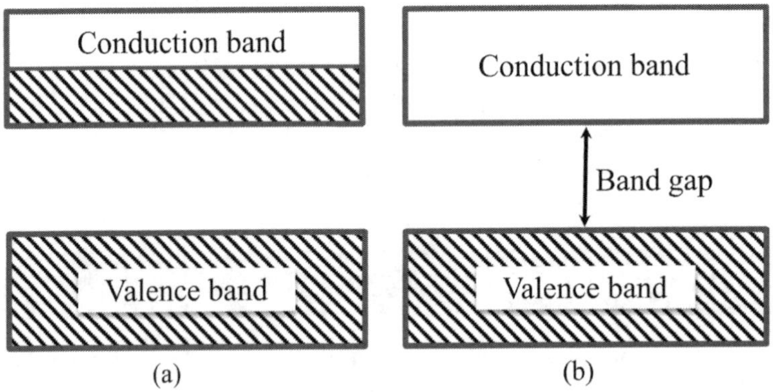

FIGURE 2.10 Band structure of, (*a*) a metal with an incompletely filled conduction band and (*b*) an insulator or semiconductor. At 0 K, such a solid is an insulator because the valence band is totally filled and the conduction band is totally empty. As the temperature is raised, some electrons are promoted into the conduction band and the material starts to conduct.

its band gap. The last model is included because it is physically the most tangible and because it relates the formation of bands to the total internal reflection of electrons by the periodically arranged atoms.

2.8.1 INTRODUCTORY BAND THEORY

In the same way that the interaction between two hydrogen atoms gave rise to two orbitals—one bonding and the other antibonding—the interaction or overlap of the wave functions of $\approx 10^{23}$ atoms in a solid gives rise to energy bands. To illustrate, consider 10^{23} atoms of Si in their ground state (Fig. 2.11*a*). The band model is constructed as follows:

1. Assign four localized tetrahedral sp^3 hybrid orbitals to each Si atom, for a total of 4×10^{23} hybrid orbitals (Fig. 2.11*b*).
2. The overlap of each of two neighboring sp^3 lobes forms one bonding and one antibonding orbital, as shown in Fig. 2.11*d*.
3. The two electrons associated with these two lobes are accommodated in the bonding orbitals (Fig. 2.11*d*).
4. As the crystal grows, every new atom added brings *one orbital to the bonding and one to the antibonding orbital set*. As the orbitals or electron wave functions overlap, they must *broaden* as shown in Fig. 2.11*c*, because of the Pauli exclusion principle.

   Thus, in the solid a spread of orbital energies develops within each orbital set, and the separation between the **highest occupied molecular orbital** (HOMO) and the **lowest unoccupied molecular orbital** (LUMO) in the molecule becomes the **energy band gap, E_g** (Fig. 2.11*c*). Note that the new orbitals are created near the original diatomic bonding σ and antibonding σ^* energies (Fig. 2.11*d*) and move toward the band edges as the size of the crystal increases.
5. In the case of Si, each atom starts with four valence electrons, and the total number of electrons that has to be accommodated in the valence band is 4×10^{23}. But since there are 2×10^{23} levels in that band

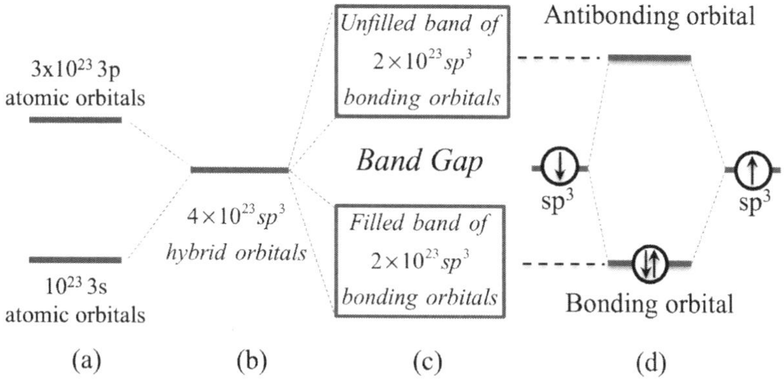

FIGURE 2.11 (*a*) Ground state of Si atoms. (*b*) sp^3 hybrid orbitals. (*c*) Interaction of sp^3 orbitals to form energy bands. (*d*) Localized orbital energy levels between two Si atoms to form a Si_2 molecule. Note that the energy bands are centered on the energy of the diatomic bonds.

and each level can accommodate two electrons, *it follows that at 0 K the valence band is completely filled and the conduction band is empty.*[19]

This last statement has far-reaching implications. If the band gap, usually denoted by E_g, lies somewhere between ≈0.02 and 3 eV, the material is considered to be a semiconductor. For higher values of E_g, the solid is considered an insulator. Said otherwise, if all the electrons are used in bonding, none are left to move freely and conduct electricity. Table 2.6, in which the band gaps of a number of binary and ternary ceramics are listed, clearly indicates that most ceramics are insulators.

Note that the degree of interaction between the orbitals depends on the interatomic distance or the spatial delocalization of the interacting electrons (the two are not unrelated). For example, the band gaps of C (diamond), Si and Ge are, respectively, 5.33, 1.12 and 0.74 eV. In C, the interaction is between the n = 2 electrons, whereas for Si and Ge one is dealing with the n = 3 and n = 4 electrons, respectively. As the interacting atoms become larger, the interaction of their orbitals increases, widening the bands that in turn reduces the band gap.[20]

Orbital overlap, while important, is not the only determinant of band gap width. Another important factor is how tightly the lattice binds the electrons. This is dealt with in the following model.

2.8.2 TIGHT BINDING APPROXIMATION[21]

In this approach, not unlike the one used to explain the formation of a covalent bond, Schrödinger's equation

$$\frac{\partial^2 \psi}{\partial x^2} + \frac{8\pi^2 m_e}{h^2}[E_{tot} - E_{pot}(x)]\psi = 0 \tag{2.21}$$

is solved by assuming that the electrons are subject to a periodic potential E_{pot}, which has the same periodicity as the lattice. By simplifying the problem—to one dimension, with interatomic spacing a and assuming that $E_{pot}(x) = 0$ for regions near the nuclei and $E_{pot} = E_0$ for regions in between, and further assuming that the width of the barrier to be w (see Fig. 2.12*a*)—Eq. (2.21) can be solved. Despite these simplifications, the details of the solution are still beyond the scope of this discussion, and only the final results are presented.[22] It turns out that solutions are possible only if the following *restricting* conditions are satisfied:

---

[19] As discussed subsequently, this is only true at 0 K. As the temperature is raised, the thermal energy will promote some of the electrons into the conduction band.

[20] Interestingly enough, a semiconducting crystal can be made conductive by subjecting it to enormous pressures, which increase the level of interaction of the orbitals to such a degree that the bands widen and eventually overlap.

[21] Also known as the Kronig–Penney model.

[22] The method of solving this problem lies in finding the solution for the case when E = 0, that is,
$\psi_0 = A\exp(i\phi x) + B\exp(i\phi x)$
with $\phi = \sqrt{2\pi m E_{tot}}/h$. And the solution for the case where $E = E_0$, that is,
$\psi_v = C\exp\beta x + D\exp(-\beta x)$
where $\beta = 2\pi\sqrt{2\pi m(E_0 - E_{tot})/h}$. By using the appropriate boundary conditions, namely, continuity of the wave function at the boundaries, and ensuring that the solution is periodic, A, B, C and D can be solved for. If it is further assumed that the barrier area, i.e., the product of wE_0, is a constant, Eqs. (2.22) and (2.23) follow. See R. Bube, *Electrons in Solids*, 2nd ed., Academic Press, New York, 1988, for more details.

TABLE 2.6 Band gaps, E_g, for various ceramics

Material	Band gap, eV	Material	Band gap, eV
Halides			
AgBr	2.8	MgF_2	11.0
BaF_2	8.8	MnF_2	15.5
CaF_2	12.0	NaCl	7.3
KBr	7.4	NaF	6.7
KCl	7.0	SrF_2	9.5
LiF	12.0	TlBr	2.5
Binary oxides, carbides and nitrides			
AlN	6.2	Ga_2O_3	4.6
Al_2O_3 parallel	8.8	MgO (periclase)	7.7
Al_2O_3 perpendicular	8.85	SiC (α)	2.6–3.2
BN	4.8	SiO_2 (fused silica)	8.3
C (diamond)	5.3	UO_2	5.2
CdO	2.1		
Transition metal oxides			
Binaries		Ternaries	
CoO	4.0	$BaTiO_3$	2.8–3.2
CrO_3	2.0	$KNbO_3$	3.3
Cr_2O_3	3.3	$LiNbO_3$	3.8
CuO	1.4	$LiTaO_3$	3.8
Cu_2O	2.1	$MgTiO_3$	3.7
FeO	2.4	$NaTaO_3$	3.8
Fe_2O_3	3.1	$SrTiO_3$	3.4
MnO	3.6	$SrZrO_3$	5.4
MoO_3	3.0	$Y_3Fe_5O_{12}$	3.0
Nb_2O_5	3.9		
NiO	4.2		
Ta_2O_5	4.2		
TiO_2 (rutile)	3.0–3.4		
V_2O_5	2.2		
WO_3	2.6		
Y_2O_3	5.5		
ZnO	3.2		

$$\cos ka = P\frac{\sin\phi a}{\phi a} + \cos\phi a \tag{2.22}$$

where

$$P = \frac{4\pi^2 ma}{h^2} E_o w \tag{2.23}$$

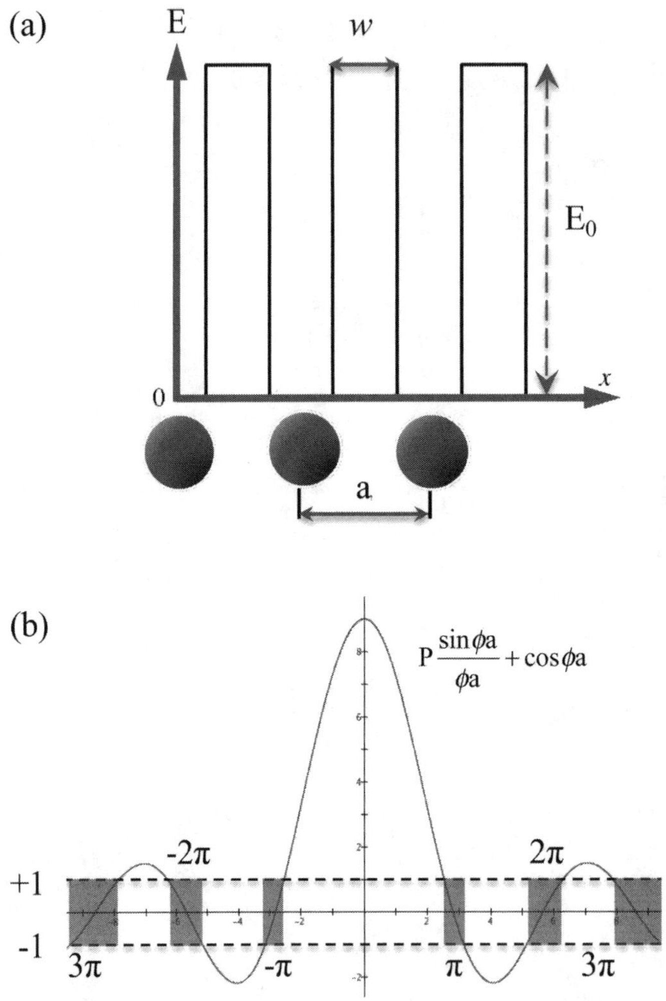

FIGURE 2.12 (*a*) Approximation of periodic potential that an electron is subjected to in a one-dimensional crystal of periodicity a. Here w is the width of the barrier and E_0 is the depth of the energy well. (*b*) Plot of right-hand side of Eq. (2.22) versus ϕa. The x-axis is proportional to the energy of the electron, and the shaded areas denote energies that are permissible, whereas the energies between the these areas are not permissible.

and

$$\phi = \frac{2\pi}{h}\sqrt{2mE_{tot}} \tag{2.24}$$

k is the wave number, defined as:

$$k = \frac{2\pi}{\lambda} \tag{2.25}$$

where λ is the electron's wavelength. Recall the wave number of an electron is a direct measure of its momentum, p, since, according to DeBroglie, $p = 2\pi k/h$ [see Eq. (2A.3)].

Since the left-hand side of Eq. (2.22) can take only values between +1 and −1, the easiest way to find possible solutions to this equation is to do it graphically by plotting the right-hand side of Eq. (2.22) as a function of ϕ a, as shown in Fig. 2.12*b*. Whenever that function lies between +1 and −1 (shaded areas in Fig. 2.12*b*), that represents a solution. Given that ϕ is proportional to the energy of the electron [Eq. (2.24)], what is immediately apparent from Fig. 2.12*b* is that there are regions of energy that are permissible (shaded areas in Fig. 2.12*b*) and regions that are forbidden. This implies that *an electron moving in a periodic potential can only move in so-called allowed energy bands that are separated from each other by forbidden energy zones*. Furthermore, the solution clearly indicates that the E_{tot} *of the electron is a periodic function of the wave function*, k.

The advantage of this model over others is that a semiquantitative relationship between the bonding of an electron to its lattice and the size of the band gap can be construed. This is reflected in the term P—for atoms that are highly electronegative, E_0, and consequently P, is large. As P increases, the right-hand side of Eq. (2.22) becomes steeper, the *bands narrow and the regions of forbidden energy widen*. It follows that if this model is correct, an empirical relationship between the electronegativities of the atoms, or ions, making up a solid and its band gap should exist. That such a relationship, namely,

$$E_g\,(\text{eV}) \approx -15 + 3.75\left(\sqrt{10X_A - 17.5} + \sqrt{10X_B - 17.5}\right)$$

does exist is illustrated nicely in Fig. 2.13. Here X_A and X_B represent the electronegativities of the atoms making up the solid.

Before moving on, it is instructive to look at two limits of the solution arrived at above:

1. The interaction between the electrons and the lattice vanishes, In that case, E_0 or P approaches 0. If $P = 0$ in Eq. (2.22), then $\cos ka = \cos k\phi$, and $k = \phi$, which when substituted in Eq. (2.24) and upon rearranging yields

$$E_{tot} = \frac{h^2k^2}{8\pi^2 m} \tag{2.26}$$

which is nothing but the well-known relationship for the energy of a free electron (see App. 2A).

2. At the boundary of an allowed band, i.e., when $\cos ka = \pm 1$ or

$$k = \frac{n\pi}{a} \quad \text{where } n = 1,2,3 \tag{2.27}$$

This implies that discontinuities in energy occur whenever this condition is fulfilled. When this result is combined with Eq. (2.26) and the energy is plotted versus k, Fig. 2.14 results. The essence of this figure lies in appreciating that at the bottom of the bands the electron dependence on k is parabolic; in other words, the electrons are behaving as if they were free. However, as their k increase, periodically, Eq. (2.27) will be satisfied and a band gap develops. The reason for the formation of such a gap is discussed in the next section.

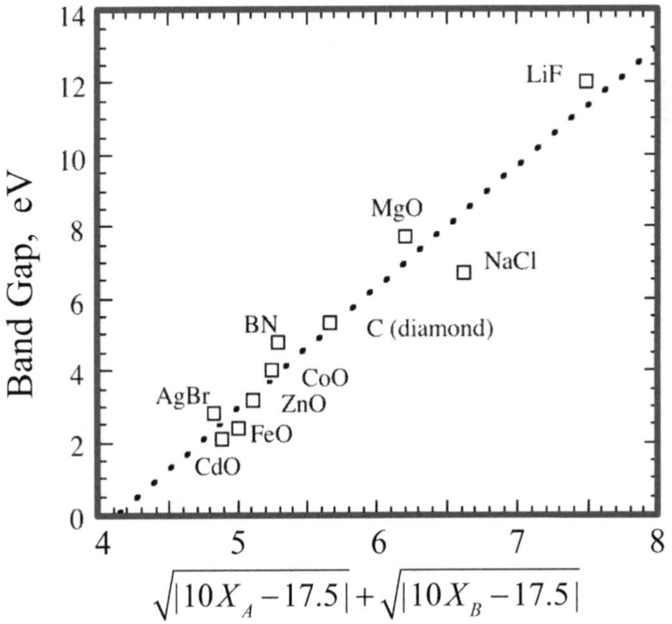

FIGURE 2.13 Empirical correlation between the electronegativities of the atoms making up a solid and its band gap; X_A and X_B are the electronegativities of the constituent atoms or ions.

2.8.3 NEARLY FREE ELECTRON APPROXIMATION

The physical origin of the band gap predicted in the previous model can be understood as follows: as a totally empty band is filled with electrons, they have to populate levels of higher energies or wave numbers, k. As they do, their momentum, p, increases and their wavelength decreases (see Eq. 2A.2). Consequently, at some point the condition $k = n\pi/a$ will be fulfilled, which is another way of saying that a pattern of *standing waves* is set up, and the electrons no longer propagate freely through the crystal because as the waves propagate to the right, they are reflected to the left, and vice versa.[23]

It can be shown further that[24] these standing waves occur with amplitude maxima either at the positions of the lattice points, that is $\psi^2 = (\text{const})\cos^2(n\pi x/a)$ (bottom curve in Fig. 2.15), or in between the lattice points, that is $\psi^2 = (\text{const})\sin^2(n\pi x/a)$ (top curve in Fig. 2.15). In the former case, the attraction of the electrons to the cores reduces the energy of the system—an energy that corresponds to the top of the valence band. In the latter case, the energy is higher and corresponds to that at the bottom of the conduction band. The difference in energy between the two constitutes E_g.

[23] The condition $k = n\pi/a$ is nothing but the well-known Bragg reflection condition, $n\lambda = 2a \cos\theta$, for $\theta = 0$. See Chap. 3 for more details.

[24] See, e.g., L. Solymar and D. Walsh, *Lectures on the Electrical Properties of Materials*, 4th ed., Oxford University Press, New York, 1988, p. 130.

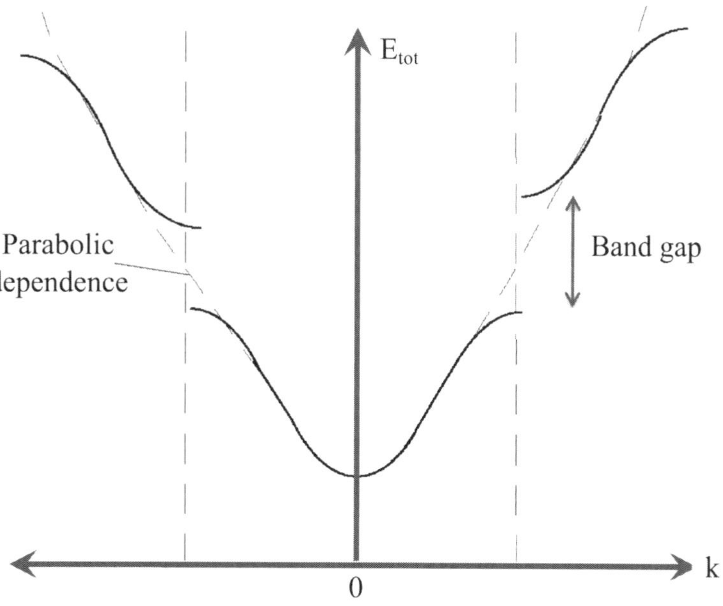

FIGURE 2.14 Functional dependence of E_{tot} on k. Discontinuities occur at $k = n\pi/a$, $n = 1, 2, 3, \ldots$.

It is important to emphasize that the band model of solids, while extremely successful, is simply one approach among several that can be used to describe the properties of solids. It is an approach that is elegant, powerful and amenable to quantification. However, the same conclusions can be deduced by starting from other assumptions. For instance, the band gap can be viewed simply as the energy required to break a covalent bond in a covalently bonded solid, or to ionize the anions in an ionic solid. At absolute zero, the electrons are trapped, in the bonds and the solid is an insulator. At finite temperatures, the lattice atoms will vibrate randomly, and occasionally the amplitude of vibration can be such as to break a bond and release an electron. The higher the temperature, the greater the probability of breaking the bond and the more likely the electron is to be delocalized and thus capable of conducting current.

COMPUTATIONAL MATERIALS SCIENCE 2.1: DENSITY FUNCTIONAL THEORY

Over the past three or so decades, there has been a veritable revolution in our understanding of materials science, solid-state physics and chemistry, as a direct result of the tremendous progress that has been made in what can be generally described as computational materials science. At this time, the latter can be generally divided into at least two general subfields. In **molecular dynamics** (MD, see next section) Newton's law ($F = ma$) is solved. The second—referred to as **density functional theory** (**DFT**)—seeks to solve Schrödinger's equation for systems containing multiple electrons. In 1998, Walter Kohn was awarded with the Nobel Prize in Chemistry for his development of DFT.

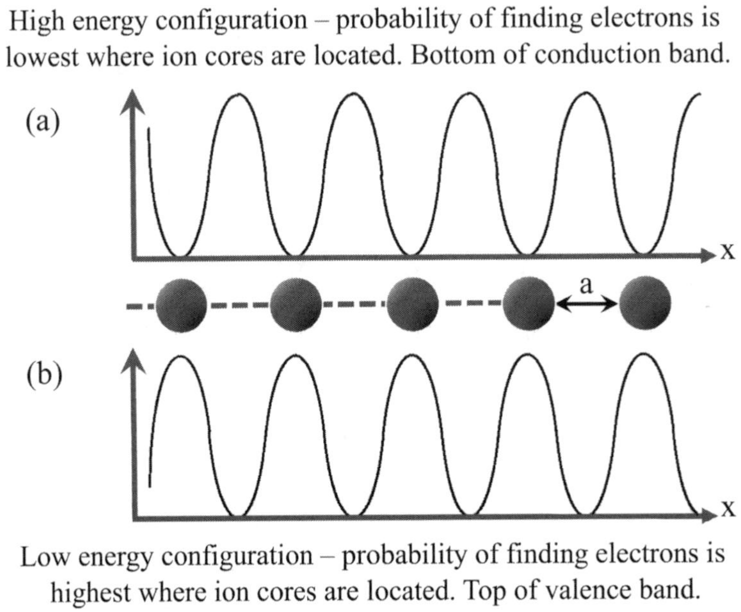
High energy configuration – probability of finding electrons is lowest where ion cores are located. Bottom of conduction band.
(a)
X
a
(b)
X
Low energy configuration – probability of finding electrons is highest where ion cores are located. Top of valence band.

FIGURE 2.15 Probability of finding electrons relative to location of the cores for the two standing waves that form when $k = n\pi/a$. In the bottom case, the standing waves distribute the charge over the ion cores, and the attraction between the negative electrons and the positive cores reduces the system's energy relative to the top situation, where the electrons spend most of their time between the ion cores. Note that in such a representation the atomic cores are ignored.

In DFT theory, the intractable many-body problem of interacting electrons in a static external potential is reduced to a tractable problem of noninteracting electrons moving in an effective potential. In solids, the latter includes the potentials due to the nuclei and the effects of interactions between electrons. DFT is based on two theories by Kohn and Sham. The first states that if the electron density, n(r), is found that minimizes the energy of the system, E_{min}, then all properties are *uniquely* determined. The second states that the n(r) that *minimizes* the energy of the system is the true electron density corresponding to the solution to Schrödinger's equation. Said otherwise, finding n(r) that minimizes a solid's energy can in turn be used to make many predictions concerning its properties, which is the ultimate goal.

A good place to start explaining DFT is Eq. (2.20). However, we will recast it in the form preferred by those working in the field. For simplicity, we also assume the presence of only two electrons. Under those conditions Eq. (2.20) can be recast to read

$$\left[-\frac{\hbar^2}{2m_e}\sum_{i=1}^{2}\nabla_i^2+\sum_{i=1}^{2}V_{e-n}(r_i)+E_{n-n}+V_{e-e}(r_1,r_2)\right]\psi=E_{tot}\psi \tag{2.28}$$

The first term in brackets is the kinetic energy of the two electrons, the second is the potential energy between the electrons and their nuclei, and the third term, E_{n-n}, takes care of repulsions between nuclei.

The last term represents electron–electron interactions, a term that was not included in Eq. (2.20) since it was set up to deal with only *one* electron (Fig. 2.7*a*). Note that in this case, the summation is for two electrons. In a solid, or cluster with N electrons, that number is many orders of magnitude higher. The crux of DFT calculations is to come up with approximations for the last term, which is the most difficult to quantify.

Before doing so, we recast Eq. (2.28) into a less transparent formalism that is mathematically closer to what one needs to solve, viz

$$\left[-\frac{\hbar^2}{2m_e}\sum_{i=1}^{N}\nabla_i^2+\sum_{i=1}^{N}V(r_i)+E_{nn}+\sum_{i=1}^{N}\sum_{j<i}U(r_i,r_j)\right]\psi=E\psi \tag{2.29}$$

Here, the first term is, again, the kinetic energy of the N electrons; the second is that of the interaction between *each* of these electrons and a collection of atomic nuclei; the fourth term is the interaction between the different electrons. The term in brackets is called the Hamiltonian and in many solid-state physics books it is shortened to read:

$$H\psi=E\psi$$

The solution of Eq. (2.29) yields a wave function, ψ, which is a function of the spatial coordinates of *all* N electrons, and E is the ground, or lowest, state energy of all the electrons, E_{min}. The ultimate goal of DFT is to find a set of wave functions that minimizes E. The different solution methods for Eq. (2.29) differ in how the electron–electron interactions are accounted for. For example, in one approach, Eq. (2.29) is solved for each electron, while smearing out, or averaging, the repulsion effects of all other electrons. Others use more sophisticated approximations. The connection is

$$n(r)=2\sum_i \psi_i^* * (r)\psi_i(r) \tag{2.30}$$

where the asterisk denotes the complex conjugate of the wave function. Here the summation is for all wave functions occupied by electrons. The factor of 2 comes about because each wave function can be occupied by 2 electrons of opposite spins.

Equation (2.30) can be reversed such that for a ground-state density, it is possible to calculate the ground-state wave functions. The solution is iterative in nature; first, one assumes a n(r) and calculates resulting the potentials, V(r), which is in turn are used to calculate the orbital wave functions, that reproduce, n(r). The process is repeated until the E is minimized.

DFT calculations are now ubiquitous in materials science. Figure 2.16 summarizes some of the physical and chemical properties DFT can predict. They include defect energies (Chap. 6), phase stabilities and prediction of new phases, diffusion coefficients (Chap. 7), phonon modes (Chap. 5), elastic properties (Chap. 4), absorption and surface energies (Chap. 4), among many others.

Another caveat here is the fact that some have touted DFT calculations in predicting new materials with given properties that have subsequently been discovered. This claim to fame has to be tempered, however, with the fact that thousands of predictions have been made, most of which have *not* come true. Needless to add, solid DFT calculations by knowledgeable teams can be invaluable (Chap. 15)

These comments notwithstanding, it is important to appreciate that a significant number of DFT papers in the literature are really quite poor and are in general more harmful than useful in that they pollute the literature with bad papers. Unfortunately, this is even truer today, given the ease by which these calculations can be carried out. Luckily there is a solution to the pollution: do your homework concerning the area of interest. Good teams/papers are cited much more often than bad ones. Let that be one guide.

COMPUTATIONAL MATERIALS SCIENCE 2.2: MOLECULAR DYNAMICS

Another advance in computational modeling has been molecular dynamics (MD). In this approach, N atoms are assumed to be connected to each other by miniature springs and Newton's law of motion (F = ma = m dv/dt, where v is the velocity) in the form:

$$F(X) = -\nabla E_{pot}(X) = M\frac{V(t)}{dt} \tag{2.31}$$

$$V(t) = \frac{dX}{dt} \tag{2.32}$$

where X represents the coordinates (x,y,z) of each particle and V their velocities in three dimensions, viz. v_x, v_y and v_z. $E_{pot}(X)$ is the potential energy of the *pairwise interaction*, which is a function of X. The central idea of MD is rather simple and quite intuitive. One way to describe how MD works is to imagine what happens when the rack is broken on a billiards or pool table. Here the rack is hit hard with the cue ball. Now imagine this process in very slow motion or in tiny time steps—in solids the timescale in typically in picoseconds. In our case, we do not need to take such tiny steps. Nevertheless, since the billiard balls obey Newton's law, then we can map the trajectory of all 16 balls on the table as a function of time. In this particular example, after a time the balls will have stopped because of friction and our MD simulation would give us their final locations. (To carry out this particular simulation, a friction term needs to be added to Eq. (2.31).) On a pool table where there are no pockets and no friction, the coordinates of the balls and their velocities will continually evolve in time forever. MD maps this evolution.

Physicists refer to E_{pot} in Eq. (2.31) as a potential, chemists call it a force field. Here we will follow the physicists lead and refer to it—as we have been doing all along in this chapter—as E_{pot}. For example, for an ionic bond, E_{pot} is given—in radial coordinates—by Eq. (2.13). In all cases, *bar none*, the potential energy looks like the curve labeled net energy in Fig. 2.4*a*. In other words, it is in the shape of an energy well. In 3D that energy well would look like an ice cream cone without the ice cream.

Another example is the so-called Lennard–Jones potential used to describe weak van der Waals bonds,

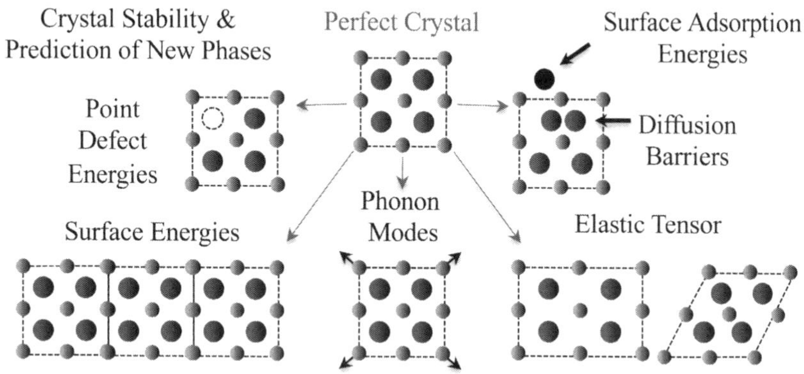

FIGURE 2.16 Partial set of physical and chemical properties DFT can predict in crystalline solids.

$$E_{net} = 4\varepsilon\left[\left(\frac{\sigma}{r}\right)^{12} - \left(\frac{\sigma}{r}\right)^{6}\right] \tag{2.33}$$

where ε and σ are constants.

There are several ways the MD calculations can be carried out, depending on what is being modeled. In one, referred to as microcanonical ensemble (NVE), the system is isolated from changes in moles (N), volume (V) and energy (E). This system is thus an adiabatic one in which no heat exchange is allowed. In the canonical ensemble (NVT), N, V and the temperature (T) are conserved. This is sometimes referred to as constant temperature MD. If endothermic or exothermic reactions take place the energy is exchanged with a thermostat.

Listed below are a few links to some interesting MD simulations.

2.9 SUMMARY

1. The attraction, or confinement, of an electron to a nucleus results in the quantization of its energy. The probability of finding the location of an electron in space then becomes a function of its energy and the orbital it has to populate. The shapes of the orbitals differ and depend on the quantum number of the electron. The s orbitals are spherically symmetric, while the p orbitals are lobed and orthogonal to each other.
2. Ionic bonds are formed by the transfer of electrons from electropositive to electronegative atoms. The long-range Coulombic attraction of these charged species for each other, together with a short-range repulsive energy component, results in the formation of an ionic bond at an equilibrium interatomic distance.
3. Covalent bonds form by the overlap of atomic wave functions. For two wave functions to overlap, they must be close in energy and be able to overlap in space as well.
4. The sp^3 hybridization results in the formation of four energetically degenerate bonds, arranged tetrahedrally with each containing one electron. This allows an atom to bond to four other atoms simultaneously.

5. The interactions and overlap of the wave functions of many atoms or ions in a solid give rise to energy bands. If the outermost band is not filled, the electrons are said to be delocalized, and the solid is considered to be a metal. If the bands are separated from each other by a band gap, E_g, the solid is considered a semiconductor or insulator depending on the size of that gap.

APPENDIX 2A: KINETIC ENERGY OF FREE ELECTRONS

The total energy of a free electron, i.e., one for which $E_{pot} = 0$, is simply its kinetic energy or

$$E_{tot} = \frac{1}{2}mv^2 = \frac{p^2}{2m} \tag{2A.1}$$

where p is its momentum and v its velocity. The momentum, p, in turn, is related to the de Broglie wavelength λ of the electron by

$$p = \frac{h}{\lambda} \tag{2A.2}$$

Combining this equation with Eq. (2.25), the wave number of an electron, k, results in

$$k = \frac{2\pi p}{h} \tag{2A.3}$$

Said otherwise, k is directly proportional to its an electron's momentum. Combining these three equations, it follows that for a free electron

$$E_{tot} = \frac{h^2k^2}{8\pi^2 m} \tag{2A.4}$$

Note that in the presence of a periodic field, the electron's energy can have nonzero values despite the fact that its velocity could be zero.

PROBLEMS

2.1. (*a*) Show that Eq. (2.4) is indeed a solution to Eq. (2.3), provided E_{tot} is given by Eq. (2.5) and c_0 is given by Eq. (2.6).
(*b*) Calculate the radius of the first Bohr orbit.
Answer: 0.0528 nm.
(*c*) Consider two hydrogen atoms. The electron in the first is in the $n = 1$ state, whereas in the second the electron is in the $n = 3$ state. (i) Which atom is in the ground-state configuration? Why? (ii) Which orbit has the larger radius? (iii) Which electron is moving faster? (iv) Which electron has the lower potential energy? (v) Which atom has the higher ionization energy?

2.2. (*a*) Show that

$$\psi(r) = A(1 + c_1 r)\exp\left(\frac{-rc_0}{2}\right)$$

is also a solution to the Schrödinger equation [i.e., Eq. (2.1)], and find an expression for c_1.

(*b*) Show that the energy of this level is equal to −3.4 eV.

(*c*) Determine the value of A. *Hint*: The total probability of finding an electron everywhere must be unity.

2.3. Starting with Eq. (2.13), derive Eq. (2.15).

2.4. Calculate the third ionization of Li. Explain why this calculation can be carried out exactly, with no approximations required.

Answer: −122.4 eV.

2.5. Assuming the Born exponent = ∞ and given 1 mol of Na^+ and 1 mol of Cl^- ions, calculate the energy released when these ions condense as:

(*a*) Noninteracting ion pairs; i.e., consider only one pairwise interactions.

Answer: −490 kJ/mol.

(*b*) Noninteracting ion squares; i.e., every four ions, 2Na and 2Cl interact with each other, but not with others.

Answer: −633 kJ/mol.

(*c*) 1/8 unit cell of NaCl; i.e., eight atoms interact.

Answer: −713 kJ/mol.

(*d*) Compare your answer in (*c*) with the NaCl lattice energy, calculated in Worked Example 2.3, viz. −755 kJ/mol, and comment on how fast or slow you think this cluster method converges to the correct answer as compared to say the Madelung constant approach.

Hint: For (*b*) and (*c*) you need to include *all* pairwise attractions and repulsions and also count the final number of squares/cubes you have.

2.6. Assuming NeCl crystallizes in the NaCl structure, using the Born–Haber cycle, show why NeCl does not exist. Make any necessary assumptions.

2.7. (*a*) Plot the attractive, repulsive and net energy between Mg^{2+} and O^{2-} from 0.18 to 0.24 nm in increments of 0.01 nm. The following information may be useful: n = 9, $B = 0.4 \times 10^{-105}$ J m^9.

(*b*) Assuming that Mg^+O^- and $Mg^{2+}O^{2-}$ both crystallize in the rock salt structure and that the ionic radii are not a strong function of ionization and taking n = ∞, calculate the difference in the enthalpies of formation ΔH_{form} of $Mg^{2+}O^{2-}$ and Mg^+O^-. Which is more stable?

Answer: $Mg^{2+}O^{2-}$ is more stable by 1200 kJ/mol.

(*c*) Why is MgO not written as $Mg^{3+}O^{3-}$?

2.8. (*a*) Calculate the Madelung constant for an infinite chain of alternating positive and negative ions + − + − + − and so on.

Answer: 2 ln 2.

(*b*) Write down the first four terms of the Madelung constant of a two-dimensional NaCl crystal.

2.9. Write the first three terms of the Madelung constant for the NaCl and the CsCl structures. How does the sum of these terms compare to the numbers listed in Table 2.4? What are the implications, if any, if the Madelung constant comes out negative?

2.10. (*a*) Explain in terms of molecular orbital theory why He_2 is unstable. What does this statement imply about the energies of the bonding and antibonding orbitals relative to those of the isolated atoms?

(*b*) Explain in terms of molecular orbital theory why He_2^+ is stable and has a bond energy (relative to the isolated atoms) that is $\approx H_2^-$.

2.11. (*a*) Boron reacts with oxygen to form B_2O_3. (i) How many oxygen atoms are bonded to each B, and vice versa? (ii) Given the electronic ground states of B and O, propose a hybridization scheme that would explain the resulting bonding arrangement.
(*b*) Repeat part (*a*) for BN.

2.12. The total energy (electronic) of an atom or molecule can be taken to be the sum of the energies of the individual electrons. Convince yourself that the sum of the energies of the electrons in the HF molecule shown in Fig. 2.8*a* is indeed lower (more negative) than the sum of the energies of the two isolated atoms.

2.13. (*a*) Which has the higher ionization energy—Li or Cs; Li or F; F or I? Explain.
(*b*) Which has the higher electron affinity—Cl or Br; O or S; S or Se? Explain.

2.14. The symbol n has been used in this chapter to represent two completely distinct quantities. Name them and clearly differentiate between them by discussing each.

2.15. (*a*) To what inert gases do the ions Ca^{2+} and O^{2-} correspond?
(*b*) Estimate the equilibrium interionic spacing of the $Ca^{2+}O^{2-}$ bond.
(*c*) Calculate the force of attraction between a Ca^{2+} ion and an O^{2-} ion if the ion centers are separated by 1 nm. State all assumptions.

2.16. The fraction ionic character of a bond between elements A and B can be approximated by

$$\text{Fraction ionic character} = 1 - e^{-(X_A - X_B)^2/4}$$

where X_A and X_B are the electronegativities of the respective elements.
(*a*) Using this expression, compute the fractional ionic character for the following compounds: NaCl, MgO, FeO, SiO_2 and LiF.
(*b*) Explain what is meant by saying that the bonding in a solid is 50% ionic and 50% covalent.

ADDITIONAL READING

W. D. Kingery, H. K. Bowen, and D. R. Uhlmann, *Introduction to Ceramics*, 2nd ed., Wiley, New York, 1976.
R. West, *Solid State Chemistry and Its Applications*, Wiley, Chichester, UK, 1984.
J. Maier, *Physical Chemistry of Ionic Materials*, Wiley, Chichester, UK, 2004.
N. N. Greenwood, *Ionic Crystals, Lattice Defects and Non-Stoichiometry*, Butterworth, London, 1968.
P. W. Atkins, *Physical Chemistry*, 4th ed., Oxford University Press, New York, 1990.
M. Gerloch, *Orbitals, Terms and States*, Wiley, Chichester, UK, 1986.
D. S. Sholl and J. A. Steckel, *Density Functional Theory: A Practical Introduction*, Wiley, Hoboken, NJ, 2009.
P. A. Cox, *The Electronic Structure and Chemistry of Solids*, Oxford University Press, Oxford, UK, 1987.
C. Kittel, *Introduction to Solid State Physics*, 6th ed., Wiley, New York, 1986.
L. Solymar and D. Walsh, *Lectures on the Electrical Properties of Materials*, 4th ed., Oxford University Press, New York, 1988.
J. Huheey, *Inorganic Chemistry*, 2nd ed., Harper & Row, New York, 1978.
R. J. Borg and G. D. Dienes, *The Physical Chemistry of Solids*, Academic Press, New York, 1992.
F. Wells, *Structural Inorganic Chemistry*, 4th ed., Clarendon Press, Oxford, UK, 1975.
J. C. Slater, *Introduction to Chemical Physics*, McGraw-Hill, New York, 1939.
L. Pauling, *The Nature of the Chemical Bond*, Cornell University Press, Ithaca, NY, 1960.

L. Pauling and E. B. Wilson, *Introduction to Quantum Mechanics with Applications to Chemistry*, McGraw-Hill, New York, 1935.
C. A. Coulson, *Valence*, Clarendon Press, Oxford, UK, 1952.
L. Azaroff, *Introduction to Solids*, McGraw-Hill, New York, 1960.
G. S. Rohrer, *Structure and Bonding in Crystalline Materials*, Cambridge University Press, Cambridge, UK, 2001.
C. B. Carter and M. G. Norton, *Ceramic Materials*, 2nd ed., Springer, New York, 2013.

OTHER REFERENCES

1. Introduction to molecular dynamics: https://www.youtube.com/watch?v=lLFEqKl3sm4.
2. MD simulations of six particles in a box where the potential between the particles is given by the Lennard–Jones potential [Eq. (2.33)] https://www.youtube.com/watch?v=YyHjYkeUex4.
3. To visualize the various atomic and molecular orbitals go to http://www.chemtube3d.com/Organic%20Structures%20and%20Bonding.html.
4. This website is recommended for anybody interested in materials science and chemistry. It is excellently well laid out, intuitive and includes a large number of atomic structures: http://www.chemtube3d.com/index.html.
5. A compilation of MD animations: https://www.youtube.com/watch?v=x8Fo2slT2WA.
6. Database: https://materialsproject.org/.

   This free database is quite useful and easy to use. To start the following tutorials are useful:

   https://www.youtube.com/playlist?list=PLTjFYVNE7LTjHJwV994iQHS-bIkh3UXKJ.

   It includes the structures, lattice parameters, electronic and band structures, elastic constants, XRD diffraction patterns and much more of a large number of compounds.

3

STRUCTURE OF CERAMICS

The Solid State, however, kept its grains
of microstructure coarsely veiled until
X-ray diffraction pierced the Crystal Planes
That roofed the giddy Dance, the taut Quadrille
Where Silicon and Carbon Atoms will
Link Valencies, four-figured, hand in hand
With common Ions and Rare Earths to fill
The lattices of Matter, Glass or Sand
With tiny Excitations, quantitatively grand.

John Updike; *The Dance of the Solids**

3.1 INTRODUCTION

The previous chapter dealt with how atoms form bonds with one another. This chapter is devoted to the next level of structure, namely, the arrangement of ions and atoms in crystalline ceramics. This topic is of vital importance because many properties, including thermal, electrical, dielectric, optical, and magnetic, can be quite sensitive to crystal structure.

Ceramics, by definition, are composed of at least two elements, and consequently their structures are, in general, more complicated than those of metals. While most metals are face-centered cubic (FCC), body-centered cubic (BCC) or hexagonal close-packed (HCP), ceramics exhibit a much wider variety of structures. Furthermore, and in contrast to metals where the structure is descriptive of the atomic arrangement, ceramic structures are named after the mineral for which the structure was first decoded. For example, compounds where the anions and cations are arranged as they are in the rock salt structure, such as NiO

* J. Updike, *Midpoint and Other Poems*, A. Knopf, Inc., New York, 1969. Reprinted with permission.

and FeO, are described to have the rock salt structure. Similarly, any compound that crystallizes in the arrangement shown by corundum—the mineral name for Al_2O_3—has the corundum structure, and so forth.

Figure 3.1 illustrates a number of common ceramic crystal structures with varying anion-to-cation radius ratios. These can be further categorized into the following:

∞ AX-type structures, which include the rock salt, cesium chloride, zinc blende and wurtzite structures. The rock salt structure (Fig. 3.1*a*), named after NaCl, is the most common of the binary structures, with over one-half of the 400 compounds so far investigated having this structure. In this structure, the **coordination number**—defined as the number of *nearest neighbors*—for both cations and anions is 6. In the CsCl structure (Fig. 3.1*b*), the coordination number for both ions is 8. ZnS exists

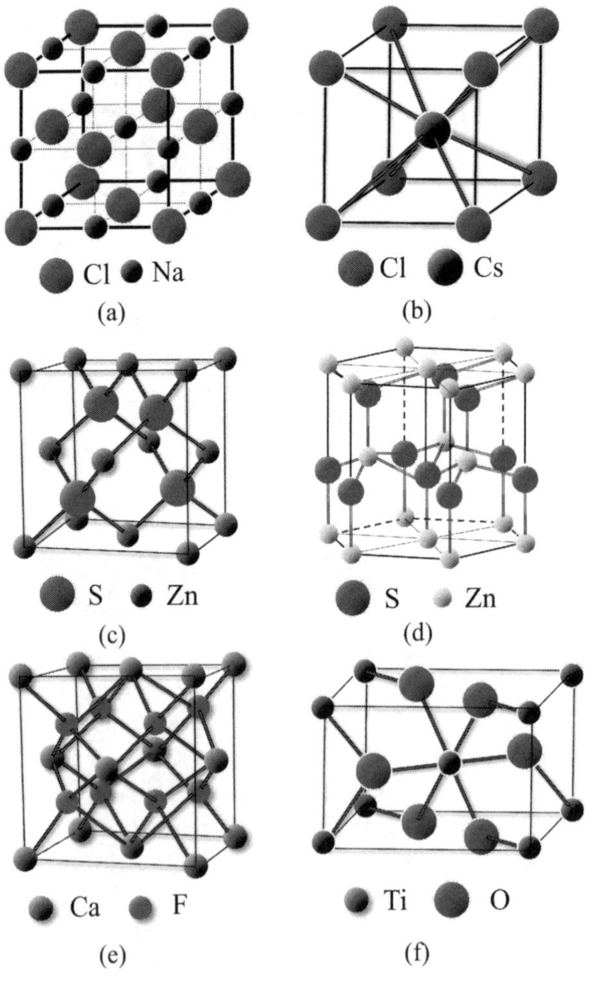

FIGURE 3.1 Some common ceramic structures: (*a*) rock salt, (*b*) cesium chloride, (*c*) zinc blende, (*d*) wurtzite, (*e*) calcium fluorite, (*f*) rutile.

in two polymorphs, the zinc blende and wurtzite structures, shown in Fig. 3.1*c* and *d*, respectively. In these structures, the coordination number is 4; that is, all ions are tetrahedrally coordinated.

- AX_2-type structures. Calcium fluorite (CaF_2) and rutile (TiO_2) shown, respectively, in Fig. 3.1*e* and *f*, are two examples of this type of structure.
- $A_mB_nX_p$ structures in which more than one cation, A and B (or the same cation with differing valences), are incorporated in an anion sublattice. Spinels (Fig. 3.10) and perovskites (Fig. 3.9) are two of the more ubiquitous ones.

At the end of this brief introduction, it is important to note that the structures shown in Fig. 3.1 represent but a few of a much larger number of possible ones. Since a comprehensive survey of ceramic structures would be impossible within the scope of this book, instead some of the underlying principles that govern the way atoms and ions arrange themselves in crystals, which in turn can aid in understanding the multitude of structures that exist, are outlined. This chapter is structured as follows: the next section outlines some of the more important and obvious factors that determine the *local* atomic structure (i.e., the coordination number of the cations and anions) and how these factors can be used to predict the type of structure a certain compound can assume. In Sec. 3.3, the binary ionic structures are dealt with from the perspective of ion packing. In Sec. 3.4, more complex ternary structures are briefly described. Sections 3.5 and 3.6 deal with Si-based covalently bonded ceramics such as SiC and Si_3N_4 and the silicates. The structure of glasses will be dealt with separately in Chap. 9. The last section deals with lattice parameters and density.

3.2 CERAMIC STRUCTURES

3.2.1 FACTORS AFFECTING STRUCTURE

Three factors are critical in determining the structure of ceramic compounds: crystal stoichiometry, radius ratios, and the propensity for covalency and tetrahedral coordination.

CRYSTAL STOICHIOMETRY

Any crystal has to be electrically neutral, i.e., the sum of its positive charges must be balanced by an equal number of negative charges, a fact that is reflected in its chemical formula. For example, in alumina, every two Al^{3+} cations have to be balanced by three O^{2-} anions, hence the chemical formula Al_2O_3. This requirement places severe limitations on the type of structure the ions can assume. For instance, an AX_2 compound cannot crystallize in the rock salt structure because the stoichiometry of the latter is AX, and vice versa.

RADIUS RATIO[25]

When cations and anions combine in a structure, they do so by maximizing attractions and minimizing repulsions. Attractions are maximized when each cation surrounds itself with as many anions as possible, with the proviso that neither the cations nor the anions "touch." To illustrate, consider the four anions—of

[25] This radius ratio scheme was first proposed by L. Pauling. See, e.g., *The Nature of the Chemical Bond*, 3rd ed., Cornell University Press, Ithaca, NY, 1960.

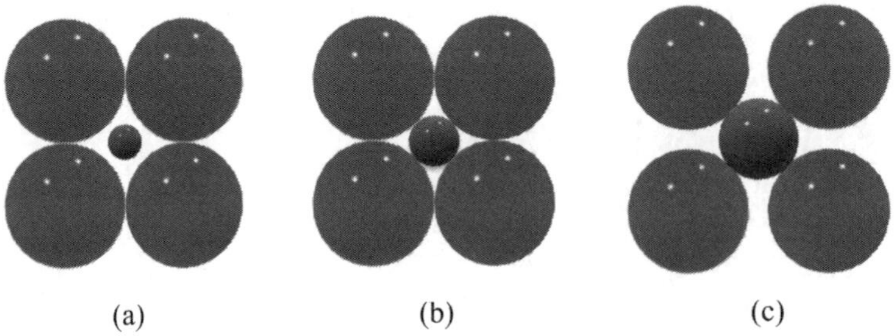

FIGURE 3.2 Stability criteria used to determine critical radius ratios.

a fixed size—surrounding cations of increasing radii as shown in Fig. 3.2. The atomic arrangement in Fig. 3.2*a* is not stable because of the obvious anion–anion repulsions. Figure 3.2*c*, however, is stabilized by the mutual attraction of the cation and the anions. When the anions are just touching (Fig. 3.2*b*), the configuration is termed *critically stable* and is used to calculate the critical radii at which one structure becomes unstable with respect to another (see Worked Example 3.1).

Since cations are usually smaller than anions, the crystal structure is usually determined by the maximum number of anions that can be packed around the cations, which, for a given anion size, will increase as the size of the cation increases. Geometrically, this can be expressed in terms of the radius ratio r_c/r_a, where r_c and r_a are the cation and anion radii, respectively. The critical radius ratios for various coordination numbers are shown in Fig. 3.3. Even the smallest cation can be surrounded by two anions and results in a linear arrangement (not shown in Fig. 3.3). As the size of the cation increases, i.e., as r_c/r_a increases, the number of anions that can be accommodated around a given cation increases to three and a triangular arrangement (top of Fig. 3.3) becomes stable. For $r_c/r_a \geq 0.225$, the tetrahedral arrangement becomes stable, and so forth.

PROPENSITY FOR COVALENCY AND TETRAHEDRAL COORDINATION

In many compounds, tetrahedral coordination is observed despite the fact that the radius ratio would predict otherwise. For example, many compounds with radius ratios greater than 0.414 still crystallize with tetrahedral arrangements such as zinc blende and wurtzite. This situation typically arises when the covalent character of the bond is enhanced, such as when

- Cations with high polarizing power (for example, Cu^{2+}, Al^{3+}, Zn^{2+}, Hg^{2+}) are bonded to anions that are readily polarizable[26] (I^-, S^{2-}, Se^{2-}). As discussed in greater detail in Chap. 4, this combination tends to increase the covalent character of the bond and favor tetrahedral coordination.
- Atoms that favor sp^3 hybridization, such as Si, C and Ge, tend to stabilize the tetrahedral coordination for obvious reasons.

Worked Example 3.1

Derive the critical radius ratio for the tetrahedral arrangement (second from top in Fig. 3.3).

[26] Polarizing power and polarizability are discussed in Chap. 4.

Coordination number	Arrangement of ions around central ion	Range of cation/anion ratios	Structure
3	corners of a triangle	≥ 0.155	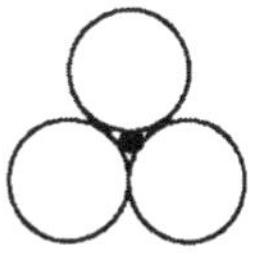
4	corners of a tetrahedron	≥ 0.225	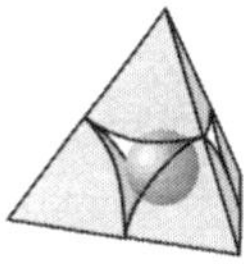
6	corners of a octahedron	≥ 0.414	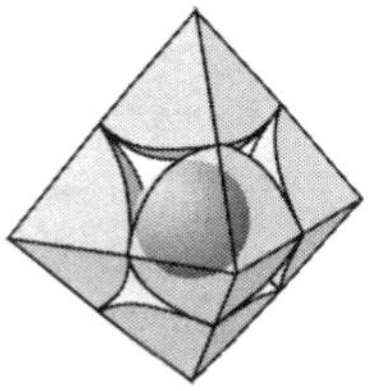
8	corners of a cube	≥ 0.732	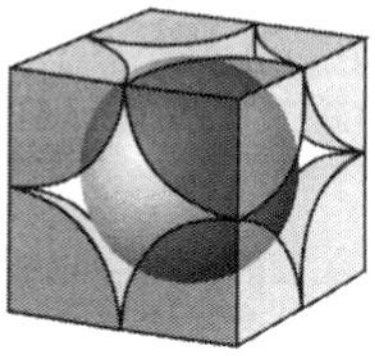
12	corners of a cuboctahedron	≈ 1.000	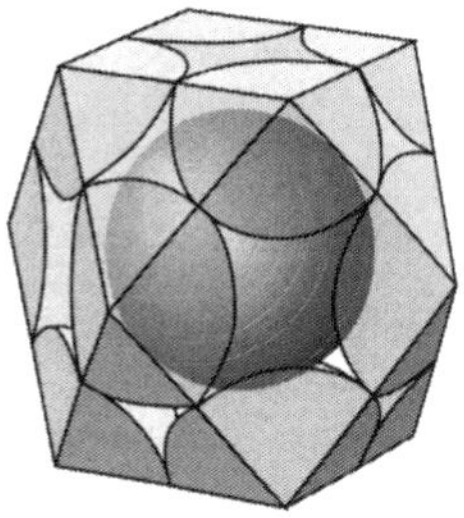

FIGURE 3.3 Critical radius ratios for various coordination numbers. The most stable structure is usually the one with the maximum coordination allowed by the radius ratio.

ANSWER

The easiest way to derive this ratio is to appreciate that when the radius ratio is critical, the cations just touch the anions, while the latter, in turn, are just touching one another (i.e., the anions are closely packed). Since the coordinates of the tetrahedral position in a close-packed arrangement (Fig. 3.4*b*) are ¼, ¼ and ¼, it follows that the distance between anion and cation centers is

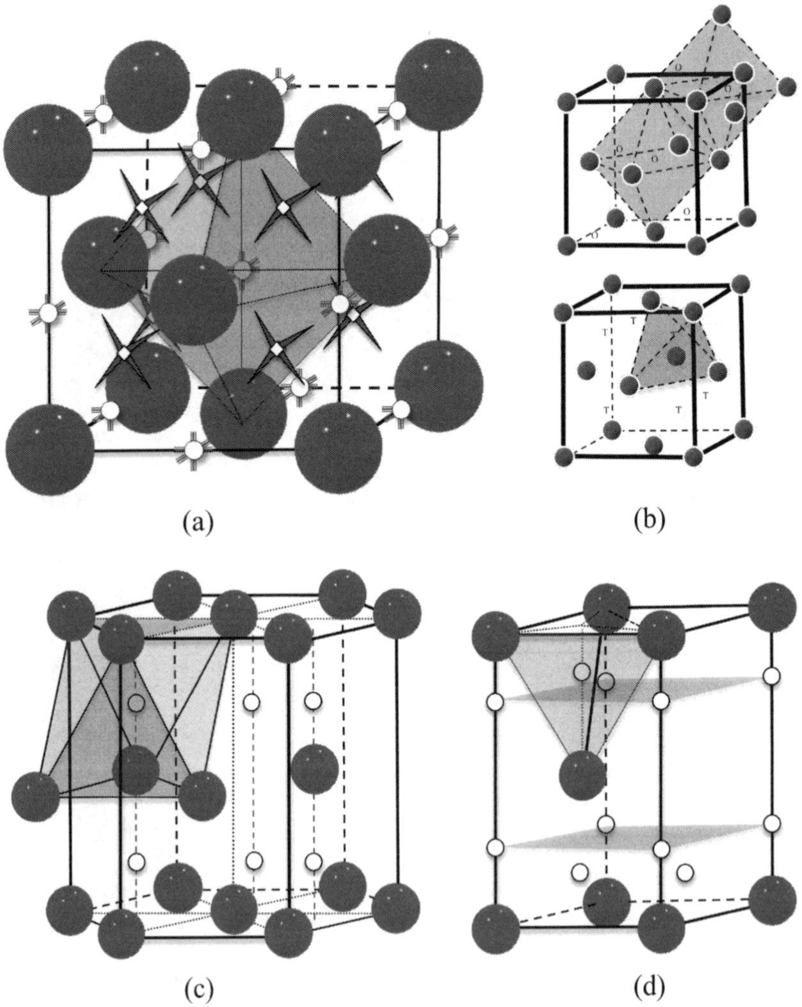

FIGURE 3.4 (*a*) Location of tetrahedral and octahedral interstitial sites within the cubic close-packed arrangement. The number of octahedral sites is always equal to the number of atoms, while the number of tetrahedral sites is always double the number of atoms. Note octahedron around ion in center of unit cell. (*b*) When two close-packed planes of spheres are stacked, one on top of the other, they define octahedral (O) and tetrahedral (T) sites between them. (*c*) Location of octahedral sites in the hexagonal close-packed arrangement. (*d*) Location of tetrahedral sites.

$$r_{cation} + r_{anion} = \sqrt{\left(\frac{a}{4}\right)^2 + \left(\frac{a}{4}\right)^2 + \left(\frac{a}{4}\right)^2} = \sqrt{3}\frac{a}{4} \tag{3.1}$$

where a is the lattice parameter. Referring to Fig. 3.4*b*, the critical condition implies that the anions are just touching along the face diagonal, thus $4r_{anion} = \sqrt{2}a$. Combining these two equations yields $r_c/r_a = 0.225$.

3.2.2 PREDICTING STRUCTURES

It follows from the foregoing discussion that, at least in principle, it should be possible to predict the *local* arrangement of ions in a crystal if the r_c/r_a ratio is known. To illustrate the general validity of this statement, consider the oxides of group 14 elements. The results are summarized in Table 3.1, and in all cases, the observed structures are what one would predict based on the radius ratios.

This is not to say that the radius ratio should be taken absolutely; there are notable exceptions. For instance, according to the radius ratios, the Cs in CsCl should be octahedrally coordinated, when in fact it is not.

Clearly one of the more important parameters needed for understanding crystal structures and carrying out lattice energy calculations, etc., is the ionic radius. From X-ray diffraction, the distance between ions—i.e., $r_c + r_a$ can be measured with great precision. However, knowing where one ion ends and where the other begins is a more difficult matter. When careful X-ray diffraction measurements have been used to map out the electron density between ions, the point at which the electron density is a minimum is taken as the operational definition of the limits of the ions involved.

Over the years there have been a number of compilations of ionic radii, probably the most notable among them being the one by Pauling.[27] Shannon[28] compiled a comprehensive set of radii—listed in App. 3A—that are slightly different than those of Pauling (see Table 3.2). When the latter are compared with the Shannon radii, the match is quite good, especially for larger ions as shown in Table 3.2. Currently, the Shannon radii are considered to be closer to representing the real size of ions in crystals than those of other compilations

TABLE 3.1 Comparison of predicted and observed structures based on radius ratio r_c/r_a

Compound	Radius ratio[a]	Prediction	Observed structure
CO_2	0.11	Linear coordination	CO_2 linear molecule
SiO_2	0.19	Tetrahedral coordination	Quartz—tetrahedral
GeO_2	0.28	Tetrahedral coordination	Quartz—tetrahedral
SnO_2	0.49	Octahedral coordination	Rutile—octahedral
PbO_2	0.75	Cubic arrangement	Rutile—octahedral
ThO_2	0.85	Cubic arrangement	Fluorite—cubic

[a] The radii used are the ones listed in App. 3A (Table 3A.1)

[27] L. Pauling, *The Nature of the Chemical Bond*, 3rd ed., Cornell University Press, Ithaca, New York, 1960, pp. 537–540.
[28] R. D. Shannon, *Acta Crstallogr.*, A32, 751 (1976).

TABLE 3.2 Comparison of ionic radii, all in pm, (crystal radii) with those measured from X-ray diffraction, XRD

Crystal	r_{M-X}	Minimum electron density distance from XRD	Pauling radii	Shannon radii
LiF	201	$r_{Li}=92$	$r_{Li}=60$	$r_{Li}=59$
		$r_F=109$	$r_F=136$	$r_F=131$
NaCl	281	$r_{Na}=117$	$r_{Na}=95$	$r_{Na}=102$
		$r_{Cl}=164$	$r_{Cl}=181$	$r_{Cl}=181$
KCl	314	$r_K=144$	$r_K=133$	$r_K=138$
		$r_{Cl}=170$	$r_{Cl}=181$	$r_{Cl}=181$
KBr	330	$r_K=157$	$r_K=133$	$r_K=138$
		$r_{Br}=173$	$r_{Br}=195$	$r_{Br}=196$

and are the ones in this book. As noted above, Shannon's comprehensive set of radii are listed at the end of this chapter in Table 3A.1 in App. 3A.

3.3 BINARY IONIC COMPOUNDS

The close packing of spheres occurs in one of two stacking sequences: ABABAB or ABCABC. The former results in a hexagonal close-packed (HCP) arrangement, while the latter results in the cubic close-packed or face-centered cubic (FCC) arrangement. Geometrically, and regardless of the stacking sequence, both arrangements create **octahedral and tetrahedral sites**, with coordination numbers 6 and 4, respectively (Fig. 3.4*a*[29]). The locations of these interstitial sites relative to the position of the atoms are shown in Fig. 3.4*b* for the FCC and in Fig. 3.4*c* and *d* for the HCP arrangements.

The importance of this aspect of packing lies in the fact that many ceramic structures can be succinctly described by characterizing the *anion packing together with the fractional occupancy of each of the interstitial sites that are defined by that anion packing.* Table 3.3 summarizes the structure of the most prevalent ceramic materials according to that scheme. When grouped in this manner, it becomes immediately obvious that for most structures the *anions are in a close-packed arrangement* (second column), with the cations (fourth column) occupying varying fractions of the interstitial sites defined by the anion packing. How this results in the various ceramic structures is described below. Before we tackle that subject, however, it is useful to examine one of the simplest ionic structures: CsCl.

3.3.1 CSCL STRUCTURE

In this structure, shown in Fig. 3.1*b*, the anions are in a simple cubic arrangement, and the cations occupy the centers of each unit cell.[30] Note that this is not a BCC structure because two different kinds of ions are involved.

[29] Sites are named for the number of faces of the shapes that form around the interstitial site.

[30] Unit cells and lattice parameters are discussed in Chap. 1 and Sec. 3.8.

TABLE 3.3 Ionic structures grouped according to anion packing

Structure name	Anion packing	Coordination # of M and X	Sites occupied by cations	Examples
Binary compounds				
Rock salt	Cubic close-packed	6:6 MX	All oct.	NaCl, KCl, LiF, KBr, MgO, CaO, SrO, BaO, CdO, VO, MnO, FeO, CoO, NiO, EuO
Rutile	Distorted cubic close-packed	6:3 MX_2	1/2 oct.	TiO_2, GeO_2, SnO_2, PbO_2, VO_2, NbO_2, TeO_2, MnO_2, RuO_2, OsO_2, IrO_2
Zinc blende	Cubic close-packed	4:4 MX	1/2 tet.	ZnS, BeO, SiC
Antifluorite	Cubic close- packed	4: 8 M_2X	All tet.	Li_2O, Na_2O, K_2O, Rb_2O, sulfides
Wurtzite	Hexagonal close-packed	4:4 MX	1/2 tet.	ZnS, ZnO, SiC, ZnTe
Nickel arsenide	Hexagonal close-packed	6:6 MX	All oct.	NiAs, FeS, FeSe, CoSe
Cadmium iodide	Hexagonal close-packed	6:3 MX_2	1/2 oct.	CdI_2, TiS_2, ZrS_2, MgI_2, VBr_2
Corundum	Hexagonal close-packed	6:4 M_2X_3	2/3 oct.	Al_2O_3, Fe_2O_3, Cr_2O_3, Ti_2O_3, V_2O_3, Ga_2O_3, Rh_2O_3
CsCl	Simple cubic	8:8 MX	All cubic	CsCl, CsBr, CsI
Fluorite	Simple cubic	8:4 MX_2	1/2 cubic	ThO_2, CeO_2, UO_2, ZrO_2, HfO_2, NpO_2, PuO_2, AmO_2, PrO_2, CaF_2
Silica types	Connected tetrahedra	4:2 MO_2	—	SiO_2, GeO_2
Complex structures				
Perovskite	Cubic close-packed	12:6:6 ABO_3	1/4 oct. (B)	$CaTiO_3$, $SrTiO_3$, $SrSnO_3$, $SrZrO_3$, $SrHfO_3$, $BaTiO_3$
Spinel (normal)	Cubic close-packed	4:6:4 AB_2O_4	1/8 tet. (A) 1/2 oct. (B)	$FeAl_2O_4$, $ZnAl_2O_4$, $MgAl_2O_4$
Spinel (inverse)	Cubic close-packed	4:6:4 $B(AB)O_4$	1/8 tet. (B) 1/2 oct. (A,B)	$FeMgFeO_4$, $MgTiMgO_4$
Illmenite	Hexagonal close-packed	6:6:4 ABO_3	2/3 oct. (A,B)	$FeTiO_3$, $NiTiO_3$, $CoTiO_3$
Olivine	Hexagonal close-packed	6:4:4 AB_2O_4	1/2 oct. (A) 1/8 tet. (B)	Mg_2SiO_4, Fe_2SiO_4

Source: Adapted from W. D. Kingery, H. K. Bowen, and D. R. Uhlmann, ***Introduction to Ceramics***, 2d ed., Wiley, New York, 1976.

3.3.2 BINARY STRUCTURES BASED ON CLOSE PACKING OF ANIONS

CUBIC CLOSE-PACKED

The structures in which the anions are in an FCC arrangement are many and include rock salt, rutile, zinc blende, antifluorite (Fig. 3.6), perovskite (Fig. 3.9) and spinel (Fig. 3.10). To see how this scheme works, consider the rock salt structure in which, according to Table 3.3, the anions are in FCC arrangement. When cations are placed on each of the octahedral sites in Fig. 3.4*b*, the resulting structure is the rock salt structure shown in Fig. 3.1*a*. Similarly, the zinc blende (Fig. 3.1*c*) structure is one in which half the tetrahedral sites are filled.

HEXAGONAL CLOSE-PACKED

Up to this point we only considered cubic unit cells. However, as Table 3.3 shows there are many hexagonal-based binary ceramics. Arguably the most important technologically is corundum, which is a form

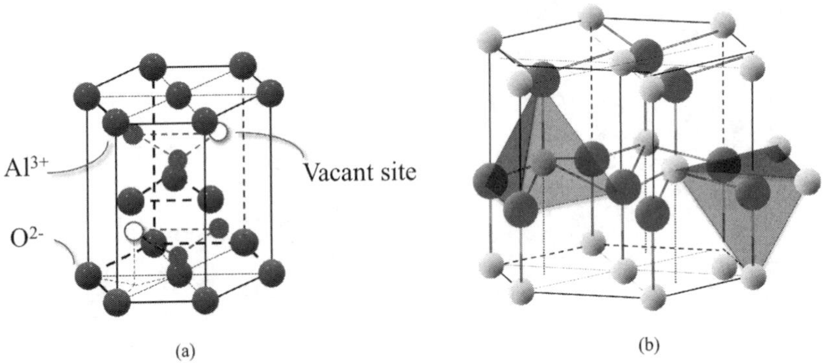

FIGURE 3.5 Hexagonal-based binary ceramics. (*a*) Unit cell of α-Al_2O_3. (*b*) Wurtzite (ZnS) structure. The structure is drawn in this way to emphasize the tetrahedral sites.

of Al_2O_3. It occurs as a mineral in schist and gneiss and other metamorphic rocks among others. When in single crystal form it is referred to as sapphire or ruby depending on its color. Rubies are red due to the presence of Cr ions; sapphires exhibit a range of colors that depend on the nature of the transition metal impurity. Interestingly corundum is derived from the Tamil word "Kurundam," which, in turn, derives from the ancient Sanskrit word "Kuruvindam."

The structure of corundum, also referred to as α-Al_2O_3, is shown in Fig. 3.5*a*. Here the oxygen anions are in an HCP arrangement and two-thirds of the Al sites are occupied by Al. The a and c lattice constants are, respectively, 0.478 and 1.299 nm.

Another common hexagonal-based structure is wurtzite (Fig. 3.5*b*), the chemical prototype of which is ZnS. According to Table 3.3, in this structure the anions are in HCP arrangement and the cations occupy half of the tetrahedral sites. To make this abundantly clear, it is left as an exercise to the reader to show that

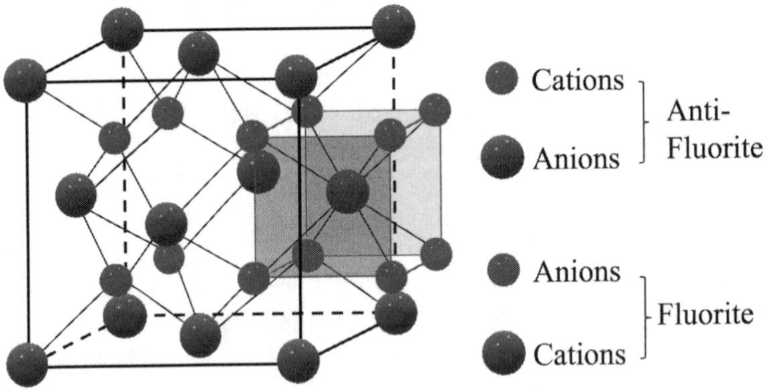

FIGURE 3.6 Relationship between fluorite and antifluorite structures. Note that in the fluorite structure the coordination number of the cations is 8, and the anions are in a simple cubic arrangement. Both structures can be viewed as two interlaced structures: an FCC and a simple cubic.

the Zn:S ratio in this structure is indeed 1:1 and that the CN of both the cations and anions is 4. Note that ZnS comes in a cubic polymorph viz. zinc blende shown in Fig. 3.1*c*.

In addition to corundum and wurtzite, nickel arsenide, cadmium iodide, illmenite and olivine are all structures in which the anion arrangement is HCP.

FLUORITE AND ANTIFLUORITE STRUCTURES

The antifluorite structure is best visualized by placing the anions in an FCC arrangement and filling all the tetrahedral sites with cations, as shown in Fig. 3.6. The resulting stoichiometry is M_2X. The oxides and chalcogenides of the alkali metals, such as Li_2O, Na_2O, Li_2S and Li_2Se, crystallize in this structure.

In the fluorite structure, also shown in Fig. 3.6, the situation is reversed with the anions filling all the tetrahedral interstices of the close-packed *cation* sublattice. The resulting compound is MX_2. The oxides of large quadrivalent cations such as Zr, Hf, Th and the fluorides of large divalent cations, such as Ca, Sr, Ba, Cd, Hg and Pb, crystallize in this structure.

Another way to view this structure is to focus on the anions, that are in a *simple cubic* arrangement (see Fig. 3.6) with alternate cubic body centers occupied by cations. If viewed from this perspective, the eightfold coordination of the cations becomes obvious, which is not surprising since r_c/r_a now approaches 1, that—according to Table 3.3—renders the cubic arrangement stable.

RUTILE AND OTHER STRUCTURES

The idealized rutile, TiO_2 unit cell, shown in Fig. 3.1*e*, is neither cubic nor close-packed. In this structure each Ti^{4+} is octahedrally surrounded by six O^{2-} ions and each O^{2-} ion is surrounded by three Ti^{4+} ions (denoted by dashed lines in Fig. 3.7*a*) distributed in a plane triangular fashion. The structure also can be viewed as rectilinear ribbons of edge-shared TiO_6 octahedra joined together by similar ribbons, with the orientations of the adjacent ribbons differing by 90° (Fig. 3.7*b*). The relationship between the unit cell (shown by dashed lines) and the stacking of the octahedra is shown in Fig. 3.7*a*. It should be noted that the actual structure comprises distorted octahedra rather than the regular ones shown here.

Table 3.3 does not include all binary oxides. However, in many cases, those not listed are derivatives of the ones that are. To illustrate, consider the structure of yttria, Y_2O_3, shown in Fig. 3.7*c*. Here each cation is surrounded by six anions located at six of the eight corners of a cube. In half the cubes, the missing oxygens lie at the end of a face diagonal, and for the remaining half the missing oxygen falls on a body diagonal. The unit cell contains 48 O^{2-} ions and 32 Y^{3+} ions; i.e., the full unit cell contains four layers of these minicubes, of which only the first row is shown here for clarity's sake. The unit cell chemistry is $Y_{32}O_{48}$, which is simplified to Y_2O_3. All the sesquioxides of the rare earths such as Gd_2O_3, Pr_2O_3, Yb_2O_3, etc. belong to this system, as do Ga_2O_3, In_2O_3, Tl_2O_3 and the mineral bixbyite, $(Fe,Mn)_2O_3$.

Worked Example 3.2

Consider two hypothetical compounds MX and MX_2 with r_c/r_a values of 0.3 and 0.5, respectively. What possible structures can either of them adopt?

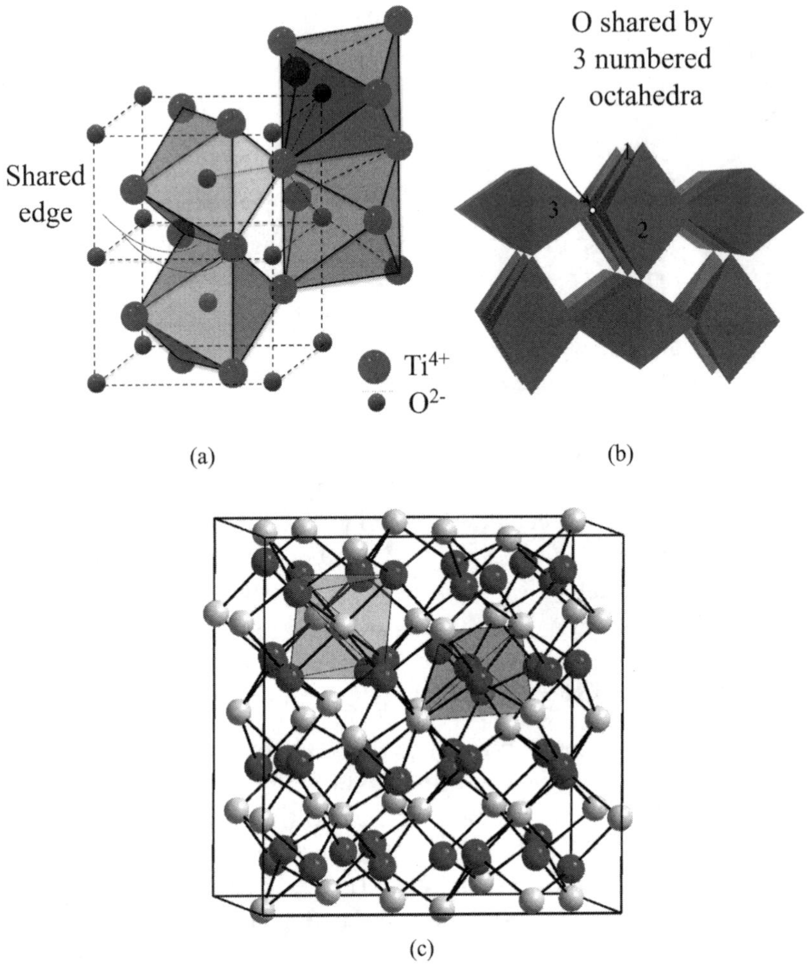

FIGURE 3.7 (*a*) Idealized stacking of TiO_6 octahedra in rutile. (*b*) Stacking of TiO_6 octahedra. Two unit cells are shown in (a) by dotted lines. (*c*) Idealized unit cell of Y_2O_3. Note two types of octahedral cation sites in alternating layers. The regular octahedron is tinted gray; the irregular one is tinted red. In this unit cell the Y ions are red and the oxygen ions are gray. Note that O ions are in FCC arrangement.

ANSWER

For the MX compound, r_c/r_a predicts that the tetrahedral arrangement is the most stable. From Table 3.3, the only structures that would simultaneously satisfy the radius ratio requirements and the chemistry are zinc blende and wurtzite; all others would be eliminated. Which of these two structures is more stable is a more difficult question to answer and is a topic of ongoing research and depends subtly on interactions between ions.

Using similar arguments, the case can be made that the only possible structures for the MX_2 compound are rutile and cadmium iodide.

3.4 COMPOSITE CRYSTAL STRUCTURES

In the preceding section, the structures of binary ceramics were discussed. As the number of elements in a compound increases, however, the structures naturally become more complex since the size and charge requirements of each ion differ. And while it is possible to describe the structures of ternary compounds by the scheme shown in Table 3.1, an alternative approach, which is sometimes more illustrative of the coordination number of the cations, is to imagine the structure to be made of the various building blocks shown in Fig. 3.3. In other words, the structure can be viewed as a three-dimensional jigsaw puzzle. Examples of such composite crystal structures are shown in Fig. 3.8. Two of the more important complex structures are spinels and perovskites, described below.

3.4.1 PEROVSKITE STRUCTURE

Perovskite—a naturally occurring mineral named after a nineteenth-century Russian mineralogist, Count Perovski—has the composition $CaTiO_3$. Hence the general perovskite formula is ABX_3. Its idealized cubic

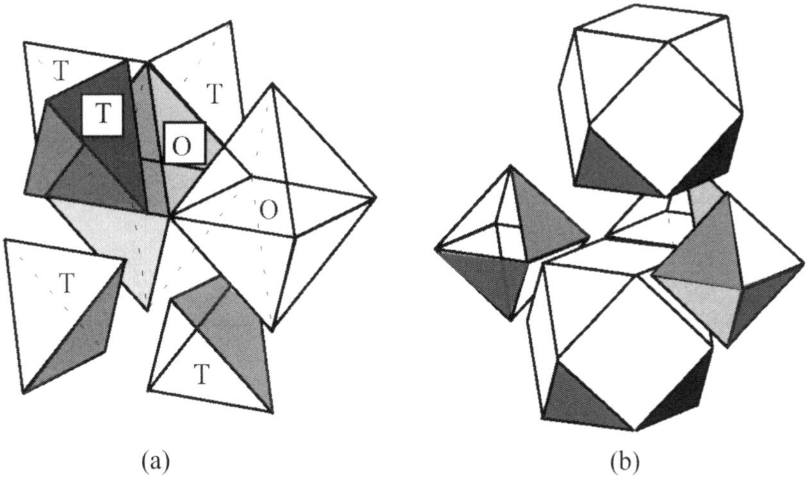

FIGURE 3.8 Examples of composite crystal structures. (a) Antifluorite structure, provided the octahedra are not occupied. (*b*) Perovskite structure ($CaTiO_3$). At the center of each cuboctahedron is a Ca ion. Each Ca cuboctahedron is surrounded by eight titania octahedra. Also see Fig. 3.9.

structure is shown in Figs. 3.8*b* and 3.9*a*, where the larger A cations, Ca^{2+} in this case, are surrounded by 12 oxygens, and the smaller B, or Ti^{4+}, ions are coordinated by 6 oxygens.

Perovskites, like the spinels discussed in the next subsection, are able to accommodate a large number of cationic combinations as long as the overall crystal is neutral. For instance, $NaWO_3$, $CaSnO_3$ and $YAlO_3$ all crystallize in that structure or modified versions of it. The modified versions usually occur when the larger cation is small, which tends to tilt the axes of the B octahedra with respect to their neighbors. This results in puckered networks of linked B octahedra that are the basis for one of the unusual electrical properties of perovskites, namely, piezoelectricity, discussed in greater detail in Chap. 15.

Several AB_3 structures can be easily derived from the perovskite structure by simply removing the atoms in the cuboctahedron positions. Several oxides and fluorides, such as ReO_3, WO_3, NbO_3, NbF_3 and TaF_3, and other oxyfluorides such as $TiOF_2$ and $MoOF_2$ crystallize in that structure shown in Fig. 3.9*b*. In this structure a large vacant cuboctahedron (inset in Fig. 3.9*b*) exists in the center of the unit cell. This site allows for the intercalation of Li^+ and protons and other cations.

3.4.2 SPINEL STRUCTURE

This structure—named after the naturally occurring mineral $MgAl_2O_4$—has a general formula of AB_2O_4, where the A and B cations are in the +2 and +3 oxidation states, respectively. There are many ways to envision this structure. We will consider a few here. One way to look at the structure is to focus on the FCC stacking[31] of the oxygen ions; the cations, on the other hand, occupy one-eighth of the tetrahedral and one-half of the octahedral sites (see Table 3.3). The same structure can also be viewed from a unit cell perspective. The spinel unit cell is comprised of two subunits (Fig. 3.10*b* and *c*). In one subunit (Fig. 3.10*b*), 2 of 8 tetrahedral and 1.5 octahedral sites are filled. In the second subunit (Fig. 3.10*c*), none of the tetrahedral and 2.5 of the octahedral sites are filled. The way the two subunits fit together in the unit cell is shown in the inset of Fig. 3.10*a*. The full unit cell is comprised of 32 O^{2-} anions, 8 A^{2+} cations in the tetrahedral sites and 16 A^{3+} cations in the octahedral sites.

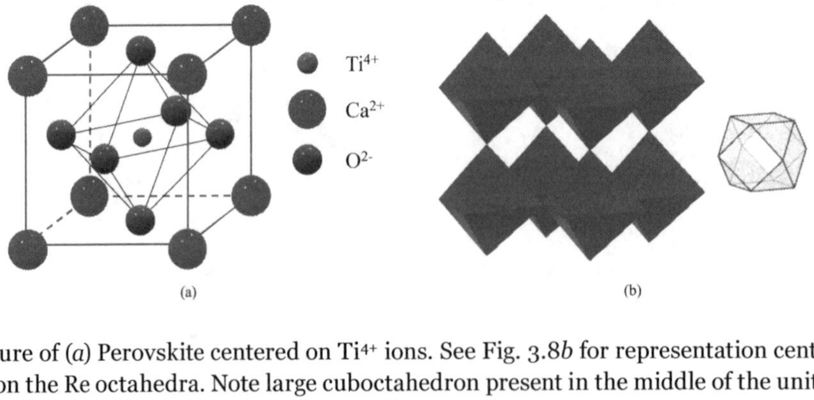

FIGURE 3.9 Structure of (*a*) Perovskite centered on Ti^{4+} ions. See Fig. 3.8*b* for representation centered on a Ca^{2+} ion. (*b*) ReO_3 centered on the Re octahedra. Note large cuboctahedron present in the middle of the unit cell.

[31] Figure 3.10*b* illustrates nicely the ABCABC or FCC stacking sequence of the anions and how that stacking defines two types of interstitial sites.

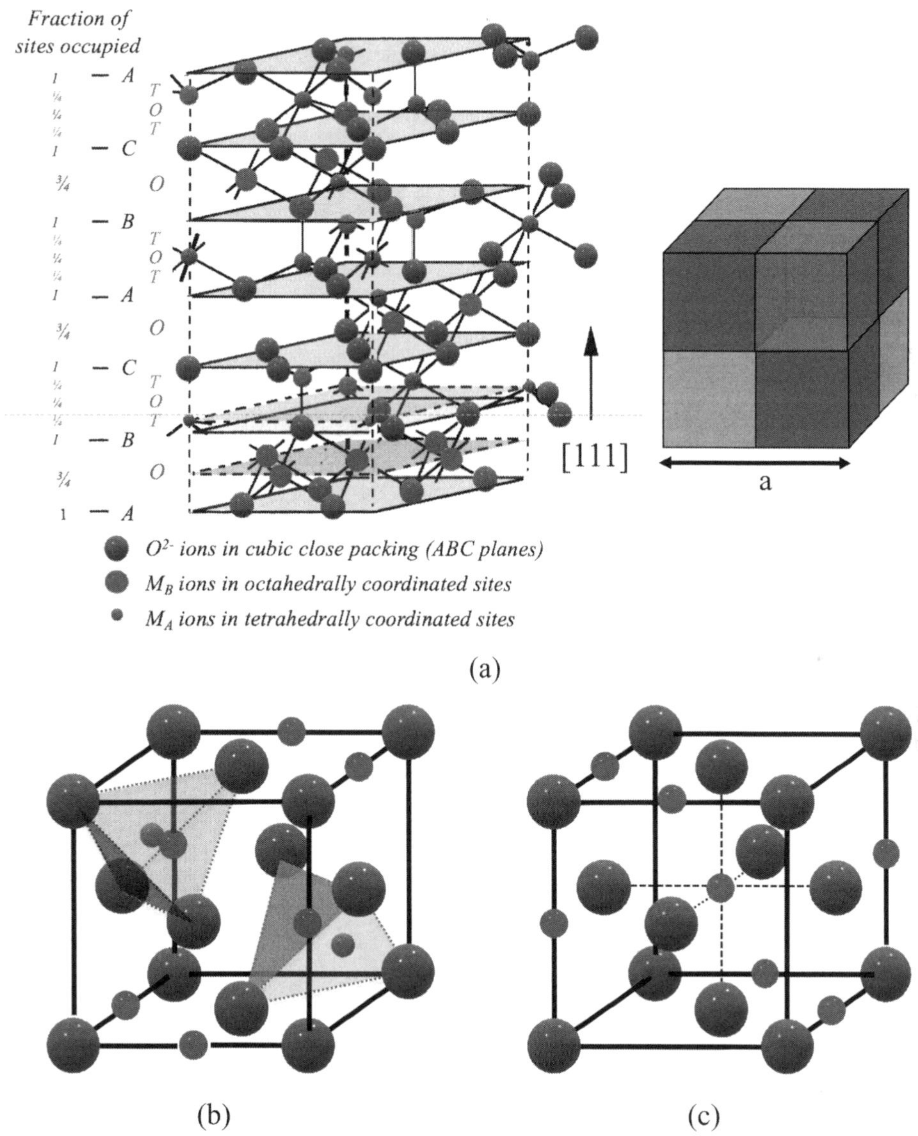

FIGURE 3.10 Spinel structure, (*a*) viewed by stacking of O in FCC packing. (*b*) and (*c*) Two octants or subunits of the spinel structure. Inset in (*a*) shows how the two subunits fit together to form a unit cell with lattice parameter a.

When the A^{2+} ions exclusively occupy the tetrahedral sites and the B^{3+} ions occupy the octahedral sites, the spinel is labeled a **normal spinel.** Usually the larger cations tend to populate the larger octahedral sites, and vice versa. In the **inverse spinel**, the A^{2+} ions and one-half of the B^{3+} ions occupy the octahedral sites, while the other half of the B^{3+} ions occupy the tetrahedral sites.

As discussed in greater detail in Chap. 6, the oxidation states of spinel cations need not be restricted to +2 and +3, but may be any combination as long as the crystal remains neutral. This important class of ceramics is revisited in Chap. 15, when magnetic ceramics are dealt with.

3.5 STRUCTURE OF COVALENT CERAMICS

The building block of silicon-based covalent ceramics, which include among others, SiC and Si_3N_4 and the silicates (dealt with separately in the next section), is in all cases the Si tetrahedron: SiO_4 in the case of silicates, SiC_4 for SiC and SiN_4 for Si_3N_4. The reason Si bonds tetrahedrally was discussed in the last chapter.

Silicon nitride, Si_3N_4 exists in two polymorphs α and β. The structure of the β polymorph is shown in Fig. 3.11, where a fraction of the nitrogen atoms is linked to two Si atoms and another fraction is linked to three Si atoms. Because of its high hardness, excellent wear chemical and shock resistances, Si_3N_4 has many uses prime among them are ball bearings. Figure 11.25 is a picture of various bearings made of Si_3N_4.

The structure of SiC also exists in many polymorphs, the simplest of which is cubic SiC, which has the zinc blende structure and is shown in Figs. 3.12 and 3.1c.

3.6 STRUCTURE OF LAYERED CERAMICS

Many ceramics are layered. At some level, and depending on perspective, one can claim that most ceramics are indeed layered. For example, looking at the NaCl structure along the [111] direction, it is not too difficult to describe the structure as being comprised of alternating near-close-packed, of pure Cl anion

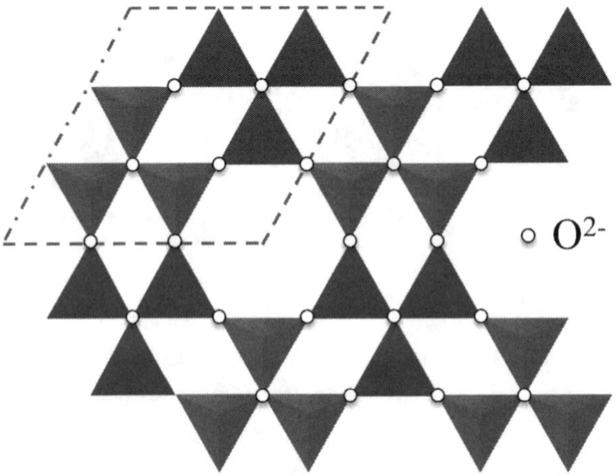

FIGURE 3.11 Structure of β-Si_3N_4 is hexagonal and made up of puckered six-member rings linked together at corners. The red tetrahedra, at whose center the Si ions reside, stick out of the plane of the paper while the gray ones are pointed into the plane of the paper. The unit cell is dashed.

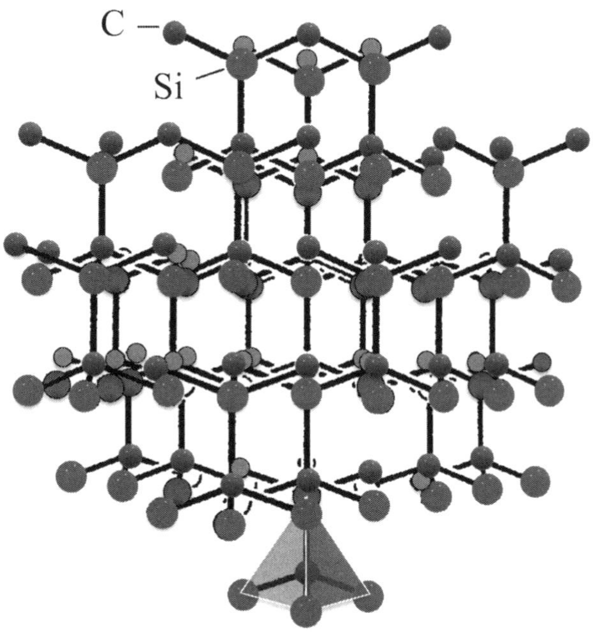

FIGURE 3.12 Structure of hexagonal SiC, which crystallizes in the wurtzite structure. The structure is comprised of tetrahedra.

layers interspersed with Na cation layers that fit in the octahedral sites defined by the Cl^- packing. Another example, described above, are spinels (Fig. 3.10*a*).

Gibbsite, $Al(OH)_3$, is comprised of AlO_6 octahedra layers, in which every O anion is bonded to a proton. Brucite, $Mg(OH)_2$, is another layered material, except in this case the octahedra are Mg-based. The layers are bonded together by weak hydrogen bonds.

Layered ceramics have garnered much interest recently since many of them are central to the ongoing nanomaterials revolution. These are discussed below in Case Study 3.1. Another example, discussed in the next section, are layered silicates, bedrocks of geology.

3.7 STRUCTURE OF SILICATES

The earth's crust is about 48 wt.% oxygen, 26% silicon, 8% aluminum, 5% iron and 11% calcium, sodium, potassium and magnesium combined. Thus, it is not surprising that the earth's crust and mantle consist mainly of silicate minerals. The chemistry and structure of silicates can be quite complex indeed and cannot possibly be covered in detail here. Instead, a few guidelines to understanding their structure are given below.

Before proceeding much further, it is important to distinguish between two types of oxygens that exist in silicate structures, namely, **bridging** and **nonbridging oxygens**. An oxygen atom that is bonded to two

TABLE 3.4 Relationship between silicate structure and the O/Si ratio

Structure	O/Si ratio	No. of O per Si: Bridg.	No. of O per Si: NBO	Structure and e.gs.
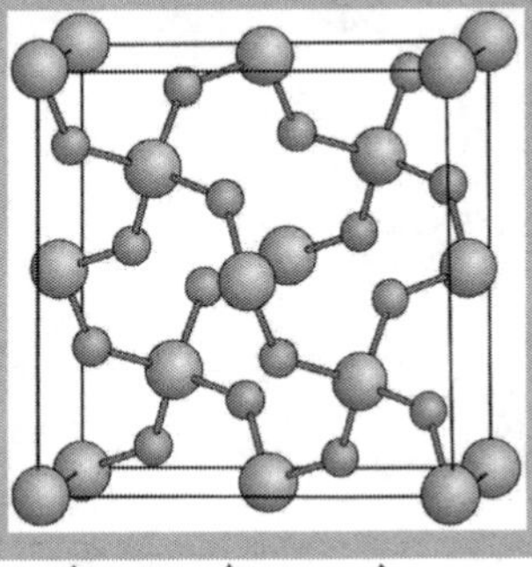	2.00	4.0	0.0	Three-dimensional network quartz, tridymite, cristobalite are all polymorphs of silica
 Repeat unit $(Si_4O_{10})^{4-}$	2.50	3.0	1.0	Infinite sheets $Na_2Si_2O_5$ clays (kaolinite), mica, talc
 Repeat unit $(Si_4O_{11})^{6-}$	2.75	2.5	1.5	Double chains, example, asbestos
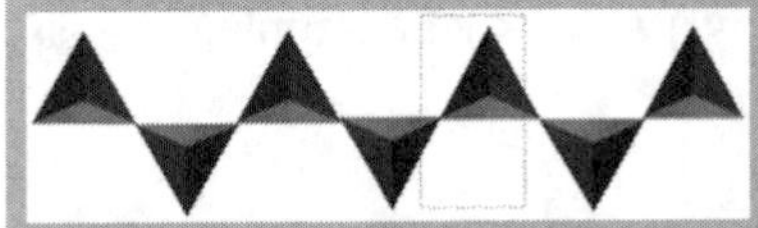 Repeat unit $(SiO_3)^{2-}$	3.00	2.0	2.0	Chains $(SiO_3)_n^{2n-}$ Na_2SiO_3, $MgSiO_3$
 Repeat unit $(SiO_4)^{4-}$	4.00	0.0	4.0	Isolated SiO_4^{4-}, tetrahedra, Mg_2SiO_4 olivine, Li_4SiO_4

[a] The simplest way to determine the number of nonbridging oxygens per Si is to divide the charge on the repeat unit by the number of Si atoms in the repeat unit.

Si atoms is a bridging oxygen, whereas one that is bonded to only one Si atom is nonbridging. Nonbridging oxygens (NBOs) are formed by the addition of, for the most part, either alkali or alkali–earth metal oxides to silica according to the following chemical equation:

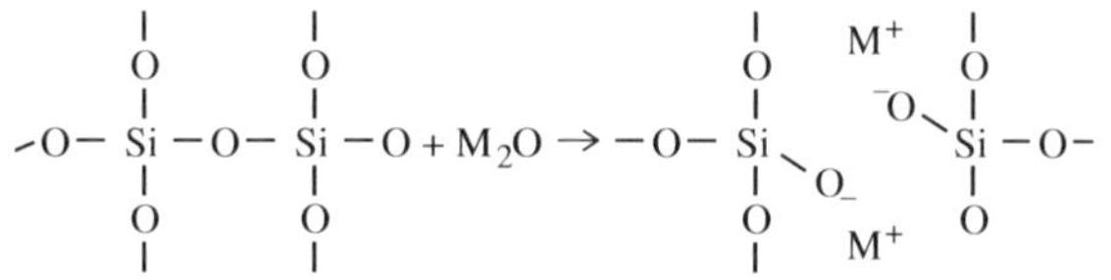

(3.2)

where O^- denotes a NBO. It is worth noting here that NBOs are negatively charged and that local charge neutrality is maintained by having the cations end up adjacent to the NBOs. Furthermore, based on this equation, the following points are noteworthy:

1. The number of NBOs is proportional to the number of moles of alkali or alkali–earth metal oxide added (see Worked Example 3.3).
2. The addition of alkali or alkali–earth metal oxides to silica must increase the overall *O/Si ratio* of the silicate.
3. Increasing the number of NBOs results in the progressive breakdown of the silicate structure into smaller units.

It thus follows that *a critical parameter that determines the silicate structures is the number of NBOs per tetrahedron, which in turn is determined by the O/Si ratio.* How this ratio determines structure is discussed below. Before addressing this point, it is important to appreciate that in general the following principles also apply:

1. The basic building block is the SiO_4 tetrahedron. The Si–O bond is partly covalent and the tetrahedron satisfies both the bonding requirements of covalent directionality and the relative size ratio.
2. Because of the high charge on the Si^{4+} ion, the tetrahedral units are rarely joined edge-to-edge and never face-to-face, but almost always share corners, with no more than two tetrahedra sharing one corner. The reason behind this rule, first stated by Pauling, is demonstrated in Fig. 3.13, where it is obvious that the cation separation distance decreases in going from corner to edge to face sharing. This in turn results in cation–cation repulsions and a decrease in the stability of the structure.

The relationship between the O/Si ratio, which can only vary between 2 and 4, and silicate structures is illustrated in Table 3.4.[32] Depending on the shape of the repeat units, these structures have been classified as three-dimensional networks, infinite sheets, chains and isolated tetrahedra. Each of these structures is discussed in some detail below.

[32] Implicit in Table 3.4 is that each O atom **not** shared between two Si tetrahedra, i.e., the nonbridging oxygens, is **negatively** charged.

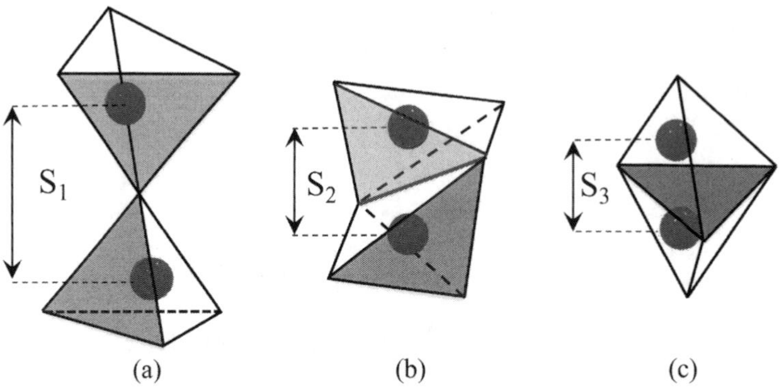

FIGURE 3.13 Effect of corner, edge and face sharing of tetrahedra on cation–cation separation. The distances S_1:S_2:S_3 are in the ratio 1:0.58:0.33. It follows that cation–cation repulsion increases on going from left to right, which tends to destabilize the structure.

3.7.1 SILICA

For a ratio of 2, that is, SiO_2, each oxygen atom is linked to two Si atoms and each Si is linked to four oxygens, resulting in a three-dimensional network, as shown at the top of Table 3.4. The resulting structures are all allotropes of silicas, which, depending on the exact arrangement of the tetrahedra, include among others quartz, tridymite and cristobalite. If, however, long-range order is lacking, the resulting solid is labeled *amorphous silica* or *fused quartz* (see Chap. 9 for more details concerning the structure of fused silica).

3.7.2 SHEET SILICATES

When three out of four oxygens are shared, i.e., for an O/Si ratio of 2.5, a sheet structure results (Table 3.4). Such minerals are collectively labeled phyllosilicates—phyllo means leaf in Greek. The structure of many minerals have this O/Si ratio and the differences between them is in simply how the sheets are connected. Clays, talcs and micas are three such examples. Let us explore each of them.

Clays: Clays come in two flavors: 1:1 and 2:1. The former consists of one Al-based octahedral sheet and one Si-based tetrahedral sheet (Fig. 3.14a). The latter consists of one Al-based octahedral sheet sandwiched between two Si-based tetrahedral sheets (Fig. 3.14*b*). Kaolinite clay, $Al_2(OH)_4{\cdot}Si_2O_5$, is an example of a 1:1 clay. Its structure helps explain why clays absorb water so readily; the polar water molecule is easily absorbed between the top of the positive sheets and the bottom of the silicate sheets. If a single layer of water is present between the layers the mineral is called halloysite. Montmorillonite, $Al_2(OH)_2{\cdot}Si_4O_{10}$, is an example of a 2:1 clay (Fig. 3.14*b*). The individual layers are relatively weakly bound, which allows water to readily penetrate between the layers causing the clay to swell. The water content of montmorillonite is variable and the mineral increases greatly in volume when it absorbs water. Its cation exchange capacity is due to isomorphous substitution of Mg for Al in the central alumina plane. When the Mg substitutes for the Al cations, a deficiency of positive charge occurs. This deficiency is balanced by the presence of Na^+ and Ca^{++} cations between the sheets, for a typical chemistry of $(Na,Ca)_{0.33}(Al,Mg)_2(Si_4O_{10})(OH)_2{\cdot}nH_2O$.

Mica: In mica—typical chemistry $KAl_2(OH)_2(AlSi_3)O_{10}$—shown in Fig. 3.14*c*, Al^{3+} ions substitute for one-fourth of the Si atoms in the sheets, requiring an alkali ion, such as K^+, in order for the structure to remain electrically neutral. The alkali ions fit in the hexagonal hole of the tetrahedral sheets, bonding the sheets together with an ionic bond that is somewhat stronger than that in clays. Thus, whereas micas does not absorb water as readily as clays do, little effort is required to cleave off very thin layers of the material.

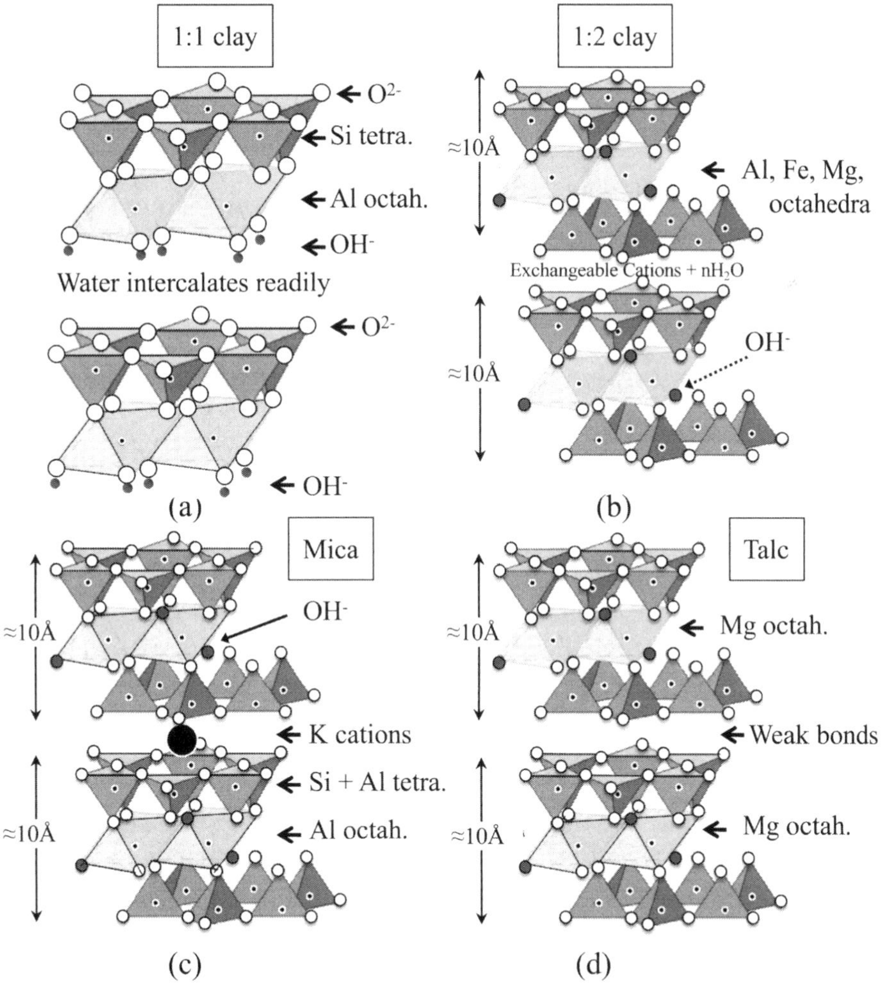

FIGURE 3.14 Structure of (*a*) 1:1 kaolinite clay, (*b*) 1:2 kaolinite clay, (*c*) mica and (*d*) talc.

Talc: Talc—typical chemistry, $Mg_3(OH)_2{\cdot}Si_4O_{10}$—has a structure related to the one described above with two silicate sheets connected by Mg^{+2} octahedra (Fig. 3.14*d*). Here three of these bridging Mg^{2+} cations are present for every two $[Si_2O_5]^{2-}$. The additional hydroxide groups (that are shared between Mg cations and are colored red in Fig. 3.14) are necessary for charge balance. It follows that in contradistinction to mica, the sandwich units are neutral. They are thus connected by quite weak van der Waals bonds. This gives talc (and talcum powder that is made from it) a slippery feel.

3.7.3 CHAIN SILICATES

For O/Si ratios of 3.0, infinite chains or ring structures result. The most notorious of this class is asbestos, in which the silicate chains are held together by weak electrostatic forces that are easier to pull apart than the bonds holding the chains together. This results in the stringy, fibrous structures that embed themselves in the human lung with devastating long-term consequences.

3.7.4 ISLAND SILICATES

When the O/Si ratio is 4, the structural units are isolated $(SiO_4)^{4-}$ tetrahedra, which cannot join to each other but are bonded together by the positive ions in the crystal structure. The resulting structure is termed an *island silicate*, for which garnets $(Mg,Fe^{2+}, Mn, Ca)_3(Cr,Al,Fe^{3+})_2(SiO_4)_3$ and olivines $(Mg,Fe^{2+})_2SiO_4$ are examples.[33] Here the $(SiO_4)^{4-}$ tetrahedron behaves as an anion and the resulting pseudobinary structure is ionically bonded.

3.7.5 ALUMINOSILICATES

Aluminum can play two roles in silicates. The first is that the Al^{3+} ions substitute for the Si^{4+} ions in the network, the missing charge has to be compensated by additional cations (e.g., mica). In the second role it can simply occupy octahedral and/or tetrahedral holes between the silicate network, as in the case for clays. In other words, it acts as any other cation.

When Al substitutes for Si in the network, the appropriate ratio for determining the silicate structure is the O/(Al + Si) ratio. So, for example, in albite ($NaAlSi_3O_8$), anorthite ($CaAl_2Si_2O_8$), eucryptite ($LiAlSiO_4$), orthoclase ($KAlSi_3O_8$) and spodune ($LiAlSi_2O_6$), the O/(Al + Si) ratio is 2. In all cases a three-dimensional structure is expected and indeed observed. As a result of this three-dimensionality, the melting points of some of these silicates are among the highest known.

It should be obvious from the preceding discussion that, with the notable exception of silica and some of the aforementioned aluminosilicates, most silicates exhibit mixed bonding, with the bonding within the silicate network, i.e., the Si–O–Si bonds, being quite different from those bonds holding the units together, which can be either ionic or weak secondary bonds, depending on the material.

33 Separating elements by a comma denotes that these elements can be found in various proportions without changing the basic structure. For example, the end members $Mg_2(SiO_4)$ and $Fe_2(SiO_4)$ and any combination in between denoted as $(Mg,Fe)_2SiO_4$ would all exhibit the same structure.

Worked Example 3.3

(*a*) Derive a generalized expression relating the number of nonbridging oxygens per Si atom present in a silicate structure to the mole fraction of metal oxide added.
(*b*) Calculate the number of bridging and NBOs per Si atom for $Na_2O{\cdot}2SiO_2$. What is the most likely structure for this compound?

ANSWER

(*a*) The simplest way to obtain the appropriate expression is to realize that in order to maintain charge neutrality, the number of NBOs has to equal the total cationic charge. Hence starting with a basis of y moles of SiO_2, the addition of η moles of $M_\zeta O$ results in the formation of $z(\zeta\eta)$ NBOs, where z is the charge on the modifying cation. Thus the number of nonbridging oxygens per Si atom is simply:

$$\text{NBO} = \frac{z(\zeta\eta)}{y}$$

The corresponding O/Si ratio, denoted by R, is

$$R = \frac{2y + \eta}{y}$$

(*b*) For $Na_2O{\cdot}2SiO_2$, $\eta = 1$, $\zeta = 2$ and $y = 2$. Consequently, $\text{NBO} = (2 \times 1 \times 1)/2 = 1$, and the number of bridging oxygens per Si atom is thus $4 - 1 = 3$. Furthermore, since $R = 2.5$, it follows that the most likely structure of this silicate is a sheet structure (Table 3.4).

3.8 LATTICE PARAMETERS AND DENSITY

3.8.1 LATTICE PARAMETERS

As noted in Chap. 1, every unit cell can be characterized by six **lattice parameters**—three edge lengths (a, b and c) and three interaxial angles (α, β and γ). On this basis, there are seven possible combinations of a, b, and c and α, β, and γ that correspond to seven crystal systems (see Fig. 1.2). In order of decreasing symmetry, they are cubic, hexagonal, tetragonal, rhombohedral, orthorhombic, monoclinic and triclinic. In the remainder of this section, for the sake of simplicity, the discussion is restricted to the cubic system for which $a = b = c$ and $\alpha = \beta = \gamma = 90°$. Consequently, this system is characterized by only one parameter, usually denoted by **a**.

The *lattice parameter* is the length of the unit cell, which is defined to be the smallest repeat unit that satisfies the **symmetry** of the crystal. For example, the rock salt unit cell shown in Fig. 3.1*a* contains four cations and four anions, because this is the smallest repeat unit that also satisfies the requirements that the crystal possess a fourfold symmetry (in addition to the threefold symmetry along the body diagonal). It is not difficult to appreciate that if only one quadrant of the unit cell shown in Fig. 3.1*a* were chosen as the unit cell, such a unit would *not* possess the required symmetry. Similar arguments can be made as to why the unit cell of Y_2O_3 is the one depicted in Fig. 3.7c, or that of spinel is the one shown in Fig. 3.10, etc.

3.8.2 DENSITY

One of the major attributes of ceramics is that as a class of materials, they are less dense than metals and hence are attractive when specific (i.e., per unit mass) properties are important. The main factors that determine density are, first, the masses of the atoms that make up the solid. Clearly, the heavier the atomic mass, the denser the solid, which is why, for example, NiO is denser than NaCl. The second factor relates to the nature of the bonding and its directionality. Covalently bonded ceramics are more "open" structures and tend to be less dense, whereas the near-close-packed ionic structures, such as NaCl, tend to be denser. For example, MgO and SiC have very similar molecular weights (≈40 g) but the density of SiC is less than that of MgO (see Worked Example 3.4 and Table 4.3).

Worked Example 3.4

Starting with the radii of the ions or atoms, calculate the theoretical densities of MgO and SiC.

ANSWER

The density of any solid can be determined from a knowledge of the unit cell. The density can be calculated from the following relationship:

$$\rho = \frac{\text{weight of ions within unit cell}}{\text{volume of unit cell}} = \frac{n'(\Sigma M_C + \Sigma M_A)}{V_C N_{Av}}$$

where:

n' = number of formula units within the unit cell
$\sum M_C$ = sum of atomic weights of all cations within unit cell
$\sum M_A$ = sum of atomic weights of all anions within unit cell
V_C = unit cell volume
N_{Av} = Avogadro's number

MgO has the rock salt structure, which implies that the ions touch along the side of the unit cell as shown in Fig. 3.1*a*. The lattice parameter, a, is thus

$$2r_{Mg} + 2r_O = 2(72 + 140) = 242 \text{ pm}.$$

The atomic weight of Mg is 24.31 g/mol, whereas the atomic weight of O is 16 g/mol. Since there are four Mg and four O ions within the unit cell, it follows that

$$\rho = \frac{4(16 + 24.31)}{(6.022\times10^{23})(424\times10^{-10})^3} = 3.51 \text{ g/cm}^3$$

To calculate the lattice parameter a for SiC (Fig. 3.1*c*) is a little trickier since the atoms touch along the body diagonal with length $\sqrt{3}a$. The Si–C distance is thus equal to one-fourth the length of the body diagonal. The atomic radius of Si is 118 pm, while that of C is 71 pm. It follows that

$$\frac{\sqrt{3}}{4}a = 118 + 71$$

$$a = 4365 \text{ pm}$$

Given that each unit cell contains four C and four Si atoms, with molecular weights of 12 and 28.09, respectively, the density is

$$\rho = \frac{4(12+28.09)}{(6.022\times10^{23})(436.5\times10^{-10})^3} = 3.2 \text{ g/cm}^3.$$

Note that while the weights of the atoms in the unit cell are quite comparable, the lower density of SiC is a direct consequence of the larger lattice parameter that reflects the more "open" structure of covalently bonded solids. It is also important to point out that the use of ionic radii listed in App. 3A is inappropriate in this case because the bonding is almost purely covalent.

EXPERIMENTAL DETAILS: DETERMINING CRYSTAL STRUCTURES, LATTICE PARAMETERS AND DENSITY

Crystal Structures and Lattice Parameters

By far the most powerful technique to determine crystal structure employs X-ray or neutron diffraction. The essentials of the technique are shown in Fig. 3.15 where a collimated X-ray beam strikes a crystal. The electrons of the crystal scatter the beam through a wide angle, and for the most part the scattered rays will interfere with each other destructively and will cancel. At various directions, however, the scattered X-rays will interfere constructively and will give rise to a strong reflection.

The condition for constructive interference corresponds to that when the scattered waves are in phase. In Fig. 3.15*a*, the wave front labeled 1 would have to travel a distance AB + BC farther than the wave front labeled 2. Thus if and when AB + BC is a multiple of the wavelength of the incident X-ray λ, that is,

$$AB+BC=n\lambda$$

coherent reflection will result. It is a trivial exercise in trigonometry to show that

$$AB+BC = 2d\sin\theta$$

where θ is the angle of incidence of the X-ray on the crystal surface defined in Fig. 3.15*a*. Combining the two equations results in the diffraction condition, also known as **Bragg's law**:

$$2d_{hkl}\sin\theta = n\lambda, \quad n = 1,2,... \tag{3.3}$$

where d_{hkl} is the distance between adjacent planes in a crystal.

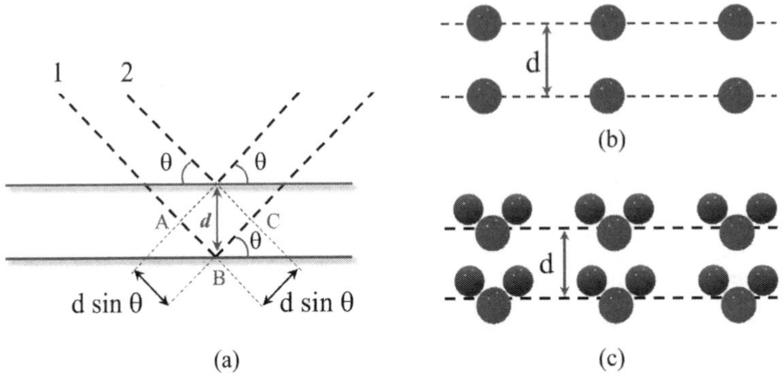

FIGURE 3.15 (a) Scattering of waves by crystal planes. While the angle of the scattered waves will depend only on d, the intensity will depend on the nature of the scatterers, as is clear when one compares (*b*) and (*c*). They both have the same lattice spacing, d, but quite different crystal structures, which in turn is reflected in the intensity of the scattered waves.

There are numerous X-ray diffraction techniques. The idea behind them all, however, is similar: Either the beam is moved relative to the diffracting crystals and the intensity of the diffracted beam is measured as a function of angle θ; or the beam is fixed, and the crystal is rotated, and the angles at which the diffraction occurs are recorded.

Note that the angle at which diffraction occurs is only part of the information that is needed and used to determine crystal structures—the intensity of the diffracted beam is also an indispensable clue. This can be easily grasped by comparing the two lattices in Fig. 3.15*b* and *c*. If the only information available were the diffraction angles, then these two quite different structures would be indistinguishable. However, constructive or destructive interference between the atoms *within* the molecules, in Fig. 3.15*c*, would clearly result in X-ray peaks with intensities that would be different from the ones shown in Fig. 3.15*b*, for example. In short, while the angle at which scattering occurs depends on the lattice type, the intensity depends on the nature of the scatterers.

DENSITY

Measuring the density of a fully dense ceramic is relatively straightforward. If the sample is uniform in shape, then the volume is calculated from the dimensions, and the weight is accurately measured using a sensitive balance. The ratio of mass to volume is the density.

A more accurate method for measuring the volume of a sample is to make use of Archimedes' principle, where the difference between the sample weight in air w_{air} and its weight in a fluid w_{fluid}, divided by the density of the fluid ρ_{fluid}, gives the volume of the liquid displaced, which is identical to the volume of the sample. The density of the sample is then simply

$$\rho = \frac{w_{air}}{(w_{air} - w_{fluid})/\rho_{fluid}} \tag{3.4}$$

Ceramics are not always fully dense, however, and open porosities can create problems in measuring their densities. Immersion of a porous body in a fluid can result in the fluid penetrating the pores, reducing the volume of fluid displaced, which consequently results in densities that appear higher than actual. Several techniques can be used to overcome this problem. One is to coat the sample with a very thin layer of molten paraffin wax, to seal the pores prior to immersion in the fluid. Another is to carry out the measurement as described previously, remove the sample from the fluid, wipe any excess liquid with a cloth saturated with the fluid and then measure the weight of the fluid-saturated sample. The difference in weight $w_{sat} - w_{air}$ is a measure of the weight of the liquid trapped in the pores, which when divided by ρ_{fluid} yields the volume of the pores. For greater detail it is best to refer to the ASTM test methods.

COMPUTATIONAL MATERIALS SCIENCE 3.1: TO BE OR NOT TO BE

When discovering new materials, DFT calculations can be quite useful indeed. Consider as an example, Mo_2GaC and Mo_2Ga_2C. The former is a MAX phase discovered in 1967; the latter was discovered in 2015.[34] After synthesizing this phase, XRD was used to determine that the order of the layers was Mo-C-Mo-Ga-Ga-Mo-C-Mo. However, high-resolution transmission electron microscope (TEM) images (Fig. 3.16) suggested a very peculiar arrangement for the Ga atoms: instead of being in a close-packed arrangement as one

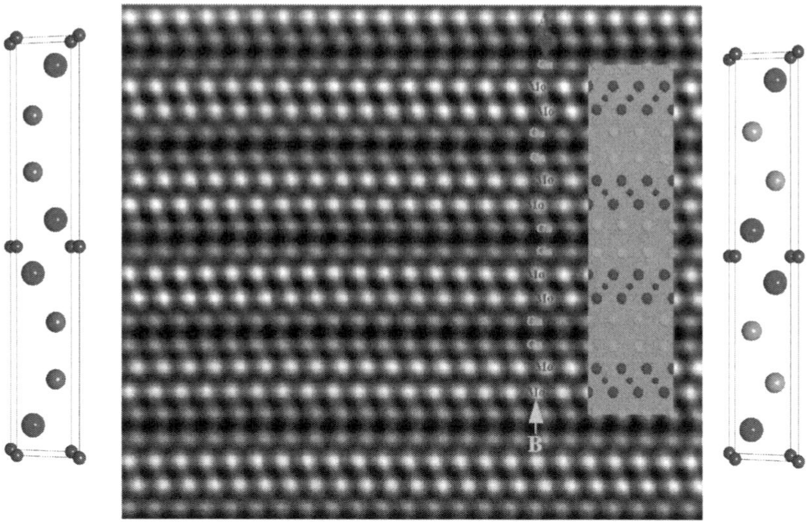

FIGURE 3.16 HRTEM of Mo_2Ga_2C showing the arrangement of the atoms. Unit cells shown on left and right of the main panel were plausible competing structures. DFT calculations confirmed that indeed the one on the left is the more stable one, consistent with the TEM images. (Taken from C.-C. Lai *et al.*, Acta Mater. **99**, 157, 2015.)

[34] Hu, C. et al. Mo_2Ga_2C: A new ternary nanolaminated carbide, *Chem. Commun.*, 51, 6560 (2015).

would expect (right unit cell in Fig. 3.16), they appeared to be in a simple cubic arrangement, i.e., one on top of the other (left unit cell in Fig. 3.16). To determine whether this unusual arrangement was real—and not say an artifact of producing thin TEM foils—the energies of two crystals that were identical in every way but for the arrangement of the Ga atoms were calculated by DFT and it was concluded that the simple cubic arrangement was indeed the more stable one! The difference in energy was −130 meV/atom more negative for the structure where the Ga atoms were in a simple cubic arrangement.

It is important to note here that before the advent of DFT theory, and concomitant advances in accuracy, the question of why a collection of atoms crystallized one way and not another was simply an unanswerable question. This is especially true given that typically the energy differences between polymorphs can be quite small indeed (see Problem 2.3). Lastly, it is important to note that in general the agreement between experimental and DFT-derived lattice parameters is for the most part excellent.

CASE STUDY 3.1: TWO-DIMENSIONAL SOLIDS

In 1959 Richard Feynman foretold the nanorevolution we are currently living through when he gave a lecture titled: "There's plenty of room at the Bottom." In that wide-ranging lecture he said: "What could we do with layered structures with just the right layers? What would the properties of materials be if we could really arrange the atoms the way we want them? They would be very interesting to investigate theoretically. I can't see exactly what would happen, but I can hardly doubt that when we have some *control* of the arrangement of things on a small scale we will get an enormously greater range of possible properties that substances can have, and of different things that we can do." As usual he was very prescient.

Interest in two-dimensional (2D) solids went ballistic, pun intended, when in 2004 physicists—who were eventually awarded a Noble Prize in 2010—were able to isolate a single graphite layer, also known as graphene, and characterized its properties. Since then, interest in 2D solids has exploded. Table 3.5 lists the classes of 2D materials known. In this scheme, there are basically four groups:

TABLE 3.5 **Summary of known 2D materials. Bold font denotes materials that are unstable in air. TMD stands for transition metal chalcogenides. RE stands for rare earth**

Graphene family	Graphene, C_2N	h-BN	BCN	**Phosphorene**	Graphene oxide
TMDs	*Semiconducting*: MoS_2, $MoTe_2$, ZrS_2, WS_2, WSe_2		***Metallic***: **$NbSe_2$, TaS_2, WTe_2**		***Semiconducting***: **GaSe, Bi_2Se_3, InSe**
Oxides & hydroxides	*Oxides*: Micas/clays MoO_3, WO_3, $YBa_2Cu_3O_7$ TiO_2, MnO_2, RuO_2, TaO_3		*Perovskite-type oxides*: $LaNb_2O_7$, $Ca_2Ta_2TiO_{10}$, $(Sr,Ca)_2Nb_3O_{10}$, $Bi_4Ti_3O_{12}$,		*Hydroxides*: $M^{2+}_{1-x}M^{3+}_x(OH)^{x+}_2$ $RE(OH)_{2.5}xH_2O^{0.5+}$
MXenes	*Mostly metallic*: Ti_2CT_x, $Ti_3C_2T_x$, Mo_2CT_x, Nb_2CT_x, $Nb_4C_3T_x$, $Mo_{1.33}CT_x$, etc.				

(a) Graphene family (top row): This family is one atomic layer thick. In graphene and hexagonal BN, the layer is flat (left in Fig. 3.17*a*); in other cases, like in phosphorene (right in Fig. 3.17*a*) it is puckered. The graphene and BN structures are identical except h-BN is comprised of two elements.

(b) Transition metal chalcogenides or TMDs (second row in Table 3.2): The chalcogens are a group 16 elements and include sulfur (S), selenium (Se), and tellurium (Te). Many of them, such as MoS_2, have a simple structure wherein the transition metal occupies the octahedral sites between chalcogens (left sketch in Fig. 3.17*b*). What renders this family intriguing is that some members are semiconducting while others are conducting. Furthermore, by intercalating cations between the sheets, i.e., by doping, it is possible to change the nature of the conductivity from semiconducting to metallic and back. Some like $NbSe_2$ are superconducting. Others like Bi_2Se_3 (right sketch in Fig. 3.17*b*) are topological insulators.

(c) Oxides and hydroxides (third row in Table 3.2): The oldest known—clays, talcs and micas—were discussed previously. However, there are many oxides that are naturally layered and with the right intercalant can be separated from each other to yield 2D flakes. Examples are shown in Fig. 3.17*c*. Here the sheets—like clays—are negatively charged and thus cations are attracted to the interlayer space. The same is true of hydroxides (Fig. 3.17*d*). Some of them are layered and, with the right chemistry/solvent, the layers can be induced to peel off forming colloidal suspensions. Interestingly, in this case, the sheets are positively charged, which is why the interlayer space is amenable to anions.

(d) MXenes: In 2011, we discovered that by simply immersing Al-containing MAX phases (see Chap. 1) in HF results in the selective etching of the Al layers and their replacement with surface terminations. In other words, by simply immersing the MAX phase in an acid a 3D to 2D transformation was induced. The Al layers are selectively etched and replaced by surface terminations, T, that are a combination of –OH, –F and –O. I labeled the resulting 2D materials MXenes to denote the loss of the A-layers from the MAX phases and I added the –ene suffix to make the connection to graphene, and other 2D materials. Today MXenes are well-recognized family members of the 2D material world.

Since 2011, when the first member, $Ti_3C_2T_x$ was discovered, the number of MXenes has grown by leaps and bounds. The reason for this state of affairs is directly related to the large number of configurations in which the MAX phases exist. This is best demonstrated in the schematics shown in Fig. 3.17*e*. First, Al-containing monoatomic MAX phases are quite numerous and come in three flavors: $n=1$, $n=2$ and $n=3$ (shown from left to right in Fig. 3.17*e*). Next, they readily form random solid solutions (second row in Fig. 3.17*e*). Lastly, they also form ordered solid solutions. For $n=2$ and 3, there is *out-of-plane* order. I refer to those as being derived from o-MAX phases. For the $n=1$, very recently a group in Sweden discovered that some MAX phases, with $n=1$, are *in-plane* ordered. These are referred to as *i-MAX*. Today, roughly 8 years after discovery, 30 separate MXene compounds, such as Ti_3C_2, Mo_2C, $TiNbC_2$, etc., have been synthesized and dozens more predicted. If not already, in due time, MXenes will become the largest family of 2D materials known.

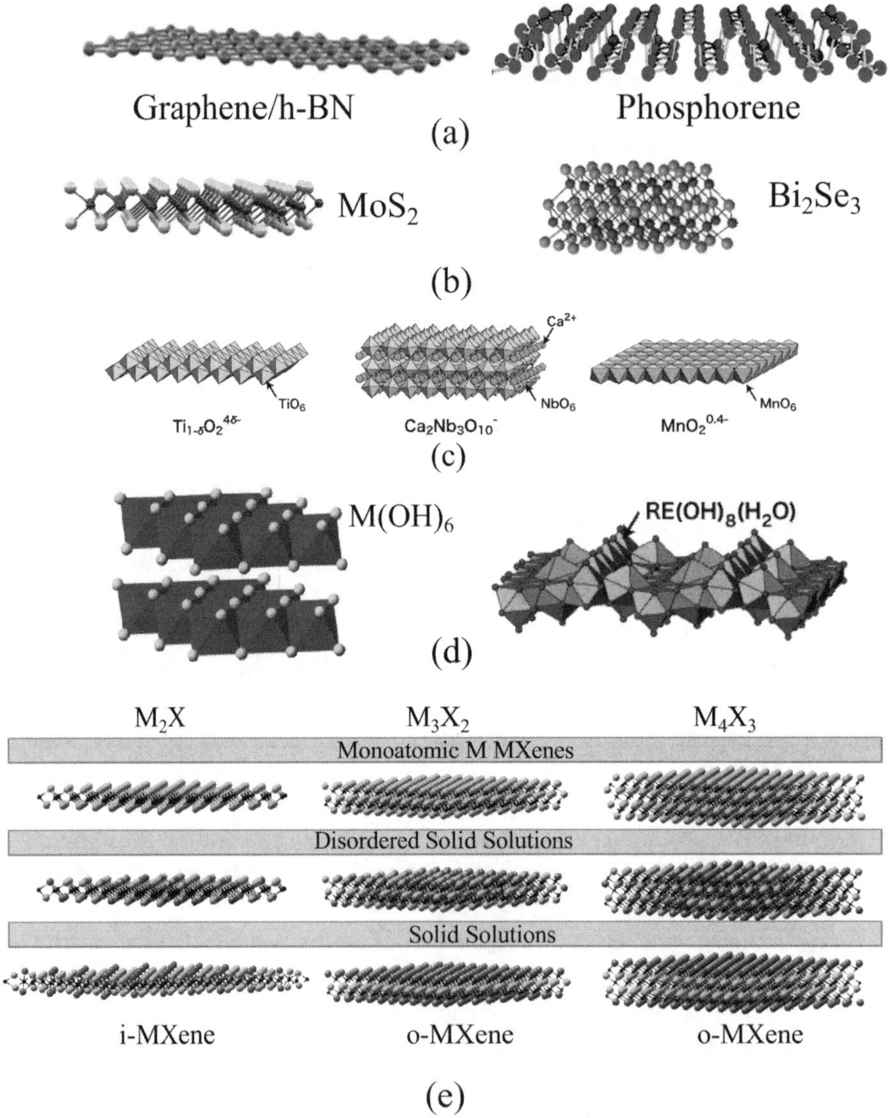

FIGURE 3.17 Schematics of some 2D materials: (*a*) graphene/BN (left) phosphorene (right); (*b*) MoS_2 (left) and Bi_2Se_3 (right); (*c*) $[Ti_{1-\delta}O_2]^{4\delta-}$ (left), $[Ca_2Nb_3O_{10}]^{-1}$ (center), MnO_2 (right); (*d*) $M^{2+}_{1-x}M^{3+}_{x}(OH)^{x+}_{2}$ (right) and $RE(OH)_{2.5}xH_2O^{0.5+}$ (left) where RE stands for rare earth. (*e*) MXene structures. MXene structures with increasing n values (left to right). Top row depicts MXenes with single M elements; middle row depicts disordered quaternary MXenes with 2 M elements. The bottom row depicts *i-MXene* (left) and two *o-MXene* structures (right). For the i-MXene case, two structures can be obtained: with mild etching, the structure shown here results; with aggressive etching, ordered divacancies are created. Note, surface terminations are not shown here.

MXenes stand out in several ways because they are (i) conductive, with a high density of states at the Fermi level and metal-like carrier densities, (ii) chemically diverse and tailorable, allowing for systematic variation of both their intrinsic composition and their post-synthetically modified surface chemistries, (iii) hydrophilic, allowing for co-assembly with polar species and enabling sustainable, green processability.

There are several descriptors of MXene that can be used, all of which are reasonable. One is to think of them as 2D metals, the other as conductive clays and a third as hydrophilic graphene. The potential applications for MXenes appear to be quite broad; they range from electrodes in energy storage devices, such as batteries and supercapacitors, to conductive transparent electrodes, to cationic absorption, electromagnetic shielding, among many others.

3.9 SUMMARY

Ceramic structures can be quite complicated and diverse, and for the most part depend on the type of bonding present. For ionically bonded ceramics, the stoichiometry and the radius ratio of the cations to the anions are critical determinants of structure. The former narrows the possible structures, and the latter determines the local arrangement of the anions around the cations. The structures can be best visualized by focusing first on the anion arrangement, which, for the vast majority of ceramics, is FCC, HCP or simple cubic. Once the anion sublattice is established, the structures that arise will depend on the fractional cationic occupancy of the various interstitial sites—octahedral vs. tetrahedral—defined by the anion sublattice.

The structures of covalent ceramics that are Si-based are based on the SiX_4 tetrahedron. These tetrahedra are usually corner linked to each other. For silicates, the building block is the SiO_4 tetrahedron. The most important parameter in determining the structure of silicates is the O/Si ratio. The minimum ratio is 2 and results in a three-dimensional network. The addition of modifier oxides to silica increases that ratio and results in both the formation of nonbridging oxygens and the progressive breakdown of the structure. As the O/Si ratio increases, the structure changes to sheets, chains and finally—when the ratio is 4—to island silicates.

APPENDIX 3A

TABLE 3A.1 Effective ionic radii of the elements

Ion	Coord. #	Effective radii, pm
Ac^{3+}	6	112.0
Ag^{+}	2	67.0
	4	100.0
	4 SQ	102.0
	5	109.0
	6	115.0
	7	122.0
	8	128.0
Ag^{2+}	4 SQ	79.0
	6	94.0
Ag^{3+}	4 SQ	67.0
	6	75.0
Al^{3+}	4	39.0
	5	48.0
	6	53.5
Am^{2+}	7	121.0
	8	126.0
	9	131.0
Am^{3+}	6	97.5
	8	109.0
Am^{4+}	6	85.0
	8	95.0
As^{3+}	6	58.0
As^{5+}	4	33.5
	6	46.0
At^{7+}	6	62.0
Au^{+}	6	137.0
Au^{3+}	4 SQ	68.0
	6	85.0
Au^{5+}	6	57.0
B^{3+}	3	1.0
	4	11.0
	6	27.0
Ba^{2+}	6	135.0
	7	138.0
	8	142.0
	9	147.0
	10	152.0
	11	157.0
	12	161.0
Be^{2+}	3	16.0
	4	27.0
	6	45.0
Bi^{3+}	5	96.0
	6	103.0
	8	117.0
Bi^{5+}	6	76.0
Bk^{3+}	6	96.0
Bk^{4+}	6	83.0
	8	93.0
Br^{-}	6	196.0
Br^{3+}	4 SQ	59.0
Br^{5+}	3 PY	31.0
Br^{7+}	4	25.0
	6	39.0
C^{4+}	3	−8.0
	4	15.0
	6	16.0
Ca^{2+}	6	100.0
	7	106.0
	8	112.0
	9	118.0
	10	123.0
	12	134.0
Cd^{2+}	4	78.0
	5	87.0
	6	95.0
	7	103.0
	8	110.0
	12	131.0
Ce^{4+}	6	87.0
	8	97.0
	10	107.0
	12	114.0
Cf^{3+}	6	95.0
Cf^{4+}	6	82.1
	8	92.0
Cl^{-}	6	181.0
Cl^{5+}	3 PY	12.0
Cl^{7+}	4	8.0
	6	27.0
Cm^{3+}	6	97.0
Cm^{4+}	6	85.0
	8	95.0
Cd^{2+}	4 HS	58.0
	5	67.0
	6 LS	65.0
	6 HS	74.5
	8	90.0
Cd^{3+}	6 LS	54.5
	6 HS	61.0
Cd^{4+}	4	40.0
	6 HS	53.0
Cr^{2+}	6 LS	73.0
	6 HS	80.0
Cr^{3+}	6	61.5
Cr^{4+}	4	41.0
	6	55.0
Cr^{5+}	4	34.5
	6	49.0
	8	57.0
Cr^{6+}	4	26.0
	6	44.0
Cs^{+}	6	167.0
	8	174.0
	9	178.0
	10	181.0
	11	185.0
	13	188.0
Cu^{+}	2	46.0
	4	60.0

(Continued)

TABLE 3A.1 (CONTINUED) Effective ionic radii of the elements

Ion	Coord. #	Effective radii, pm	Ion	Coord. #	Effective radii, pm	Ion	Coord. #	Effective radii, pm
	6	77.0	Fr^{+}	6	180.0		12	164.0
Cu^{2+}	4	57.0	Ga^{3+}	4	47.0	La^{3+}	6	103.2
	4 SQ	57.0		5	55.0		7	110.0
	5	65.0		6	62.0		8	116.0
	6	73.0	Gd^{3+}	6	93.8		9	121.6
Cu^{3+}	6 LS	54.0		7	100.0		10	127.0
D^{+}	2	−10.0		8	105.3		12	136.0
Dy^{2+}	6	107.0		9	110.7	Li^{+}	4	59.0
	7	113.0	Ge^{2+}	6	73.0		6	76.0
	8	119.0	Ge^{4+}	4	39.0		8	92.0
Dy^{3+}	6	91.2		6	53.0	Lu^{3+}	6	86.1
	7	97.0	H^{+}	1	−38.0		8	97.7
	8	102.7		2	−18.0		9	103.2
	9	108.3	Hf^{4+}	4	58.0	Mg^{2+}	4	57.0
Er^{3+}	6	89.0		6	71.0		5	66.0
	7	94.5		7	76.0		6	72.0
	8	100.4		8	83.0		8	89.0
	9	106.2	Hg^{+}	3	97.0	Mn^{2+}	4	66.0
Eu^{2+}	6	117.0		6	119.0		5	75.0
	7	120.0	Hg^{2+}	2	69.0		6	67.0
	8	125.0		4	96.0		6	83.0
	9	130.0		6	102.0		7	90.0
	10	135.0		8	114.0		8	96.0
Eu^{3+}	6	94.7	Ho^{3+}	6	90.1	Mn^{3+}	5	58.0
	7	101.0		8	101.5		6	58.0
	8	106.6		9	107.2		6	64.5
	9	112.0		10	112.0	Mn^{4+}	4	39.0
F^{-}	2	128.5	I^{-}	6	220.0		6	53.0
	3	130.0	I^{5+}	3 PY	44.0	Mn^{5+}	4	33.0
	4	131.0		6	95.0	Mn^{6+}	4	25.5
	6	133.0	I^{7+}	4	42.0	Mn^{7+}	4	25.0
F^{7+}	6	8.0		6	53.0		6	46.0
Fe^{2+}	4 HS	63.0	In^{3+}	4	62.0	Mo^{3+}	6	69.0
	4 SQ HS	64.0		6	80.0	Mo^{4+}	6	65.0
	6 LS	61.0		8	92.0	Mo^{5+}	4	46.0
	6 HS	78.0	Ir^{3+}	6	68.0		6	61.0
	8 HS	92.0	Ir^{4+}	6	62.5	Mo^{6+}	4	41.0
Fe^{3+}	4 HS	49.0	Ir^{5+}	6	57.0		5	50.0
	5	58.0	K^{+}	4	137.0		6	59.0
	6 LS	55.0		6	138.0		7	73.0
	6 HS	64.5		7	146.0	N^{3-}	4	146.0
	8 HS	78.0		8	151.0	N^{3+}	6	16.0
Fe^{4+}	6	58.5		9	155.0	N^{5+}	3	−104.0
Fe^{6+}	4	25.0		10	159.0		6	13.0

(Continued)

TABLE 3A.1 (CONTINUED) Effective ionic radii of the elements

Ion	Coord. #	Effective radii, pm	Ion	Coord. #	Effective radii, pm	Ion	Coord. #	Effective radii, pm
Na^{+}	4	99.0	Os^{4+}	6	63.0	Pt^{2+}	4 SQ	60.0
	5	100.0	Os^{5+}	6	57.5		6	80.0
	6	102.0	Os^{6+}	5	49.0	Pt^{4+}	6	62.5
	7	112.0		6	54.5	Pt^{5+}	6	57.0
	8	118.0	Os^{7+}	6	52.5	Pu^{3+}	6	100.0
	9	124.0	Os^{8+}	4	39.0	Pu^{4+}	6	86.0
	12	139.0	P^{3+}	6	44.0		8	96.0
Nb^{3+}	6	72.0	P^{5+}	4	17.0	Pu^{5+}	6	74.0
Nb^{4+}	6	68.0		5	29.0	Pu^{6+}	6	71.0
	8	79.0		6	38.0	*Ra^{2+}*	8	148.0
Nb^{5+}	4	48.0	Pa^{3+}	6	104.0		12	170.0
	6	64.0	Pa^{4+}	6	90.0	Rb^{+}	6	152.0
	7	69.0		8	101.0		7	156.0
	8	74.0	Pa^{5+}	6	78.0		8	161.0
Nd^{2+}	8	129.0		8	91.0		9	163.0
	9	135.0		9	95.0		10	166.0
Nd^{3+}	6	98.3	Pb^{2+}	4 PY	98.0		11	169.0
	8	110.9		6	119.0		12	172.0
	9	116.3		7	123.0		14	183.0
	12	127.0		8	129.0	Re^{4+}	6	63.0
Ni^{2+}	4	55.0		9	135.0	Re^{5+}	6	58.0
	4 SQ	49.0		10	140.0	Re^{6+}	6	55.0
	5	63.0		11	145.0	Re^{7+}	4	38.0
	6	69.0		12	149.0		6	53.0
Ni^{3+}	6 LS	56.0	Pb^{4+}	4	65.0	Rh^{3+}	6	66.5
	6 HS	60.0		5	73.0	Rh^{4+}	6	60.0
Ni^{4+}	6 LS	48.0		6	77.5	Rh^{5+}	6	55.0
No^{2+}	6	110.0		8	94.0	Ru^{3+}	6	68.0
Np^{2+}	6	110.0	Pd^{+1}	2	59.0	Ru^{4+}	6	62.0
Np^{3+}	6	101.0	Pd^{2+}	4 SQ	64.0	Ru^{5+}	6	56.5
Np^{4+}	6	87.0		6	86.0	Ru^{7+}	4	38.0
	8	98.0	Pd^{3+}	6	76.0	Ru^{8+}	4	36.0
Np^{5+}	6	75.0	Pd^{4+}	6	61.5	S^{2-}	6	184.0
Np^{6+}	6	72.0	Pm^{3+}	6	97.0	S^{4+}	6	37.0
Np^{7+}	6	71.0		8	109.3	S^{6+}	4	12.0
O^{2-}	2	135.0		9	114.4		6	29.0
	3	136.0	Po^{4+}	6	94.0	Sb^{3+}	4 PY	76.0
	4	138.0		8	108.0		5	80.0
	6	140.0	Po^{6+}	6	67.0		6	76.0
	8	142.0	Pr^{3+}	6	99.0	Sb^{5+}	6	60.0
OH^{-}	2	132.0		8	112.6	Sc^{3+}	6	74.5
	3	134.0		9	117.9		8	87.0
	4	135.0	Pr^{4+}	6	85	Se^{2-}	6	198.0
	6	137.0		8	96	Se^{4+}	6	50

(Continued)

TABLE 3A.1 (CONTINUED) Effective ionic radii of the elements								
Ion	Coord. #	Effective radii, pm	Ion	Coord. #	Effective radii, pm	Ion	Coord. #	Effective radii, pm
Se^{6+}	4	28.0	Te^{4+}	3	52.0		7	81.0
	6	42.0		4	66.0		8	86.0
Si^{4+}	4	26.0		6	97.0	V^{2+}	6	79.0
	6	40.0	Te^{6+}	4	43.0	V^{3+}	6	64.0
Sm^{2+}	7	122.0		6	56.0	V^{4+}	5	53.0
	8	127.0	Th^{4+}	6	94.0		6	58.0
	9	132.0		8	105.0		8	72.0
Sm^{3+}	6	95.8		9	109.0	V^{5+}	4	35.5
	7	102.0		10	113.0		5	46.0
	8	107.9		11	118.0		6	54.0
	9	113.2		12	121.0	W^{4+}	6	66.0
	12	124.0	Ti^{2+}	6	86.0	W^{5+}	6	62.0
Sn^{4+}	4	55.0	Ti^{3+}	6	67.0	W^{6+}	4	42.0
	5	62.0	Ti^{4+}	4	42.0		5	51.0
	6	69.0		5	51.0		6	60.0
	7	75.0		6	60.5	Xe^{8+}	4	40.0
	8	81.0		8	74.0		6	48.0
Sr^{2+}	6	118.0	Tl^{+}	6	150.0	Y^{3+}	6	90.0
	7	121.0		8	159.0		7	96.0
	8	126.0		12	170.0		8	101.9
	9	131.0	Tl^{3+}	4	75.0		9	107.5
	10	136.0		6	88.5	Yb^{2+}	6	102.0
	12	144.0		8	98.0		7	108.0
Ta^{3+}	6	72.0	Tm^{2+}	6	103.0		8	114.0
Ta^{4+}	6	68.0		7	109.0	Yb^{3+}	6	86.8
Ta^{5+}	6	64.0	Tm^{3+}	6	88.0		7	92.5
	7	69.0		8	99.4		8	98.5
	8	74.0		9	105.2		9	104.2
Tb^{3+}	6	92.3	U^{3+}	6	102.5	Zn^{2+}	4	60.0
	7	98.0	U^{4+}	6	89.0		5	68.0
	8	104.0		7	95.0		6	74.0
	9	109.5		8	100.0		8	90.0
Tb^{4+}	6	76.0		9	105.0	Zr^{4+}	4	59.0
	8	88.0		12	117.0		5	66.0
Tc^{4+}	6	64.5	U^{5+}	6	76.0		6	72.0
Tc^{5+}	6	60.0		7	84.0		7	78.0
Tc^{7+}	4	37.0	U^{6+}	2	45.0		8	84.0
	6	56.0		4	52.0		9	89.0
Te^{2-}	6	221.0		6	73.0			

HS = high spin, LS = low spin; SQ = square, PY = pyramid

Source: R. D. Shannon, ***Acta. Crystallogr.***, A32, 751 (1976).

PROBLEMS

3.1. (*a*) Show that the minimum cation/anion radius ratio for a coordination number of 6 is 0.414.
(*b*) Repeat part (*a*) for coordination number 3.
(*c*) Which interstitial site is larger: tetrahedral or octahedral? Calculate the ratio of the size of the tetrahedral to the octahedral interstitial sites.
(*d*) When oxygen ions are in a hexagonal close-packed arrangement, what is the ratio of the octahedral sites to oxygen ions? What is the ratio of the tetrahedral sites to oxygen ions?

3.2. Starting with the cubic close packing of oxygen ions:
(*a*) How many tetrahedral and how many octahedral sites are there per unit cell?
(*b*) What is the ratio of octahedral sites to oxygen ions? What is the ratio of tetrahedral sites to oxygen ions?
(*c*) What oxide would you get if one-half of the octahedral sites are filled? Two-thirds? All?
(*d*) Locate all the tetrahedral sites, and fill them up with cations. What structure do you obtain? If the anions are oxygen, what must be the charge on the cation for charge neutrality to be maintained?
(*e*) Locate all the octahedral sites, fill them with cations and repeat part (*d*). What structure results?

3.3. Using information in Table 3.3, draw the zinc blende structure. What, if anything, does this structure have in common with the diamond cubic structure? Explain.

3.4. The structure of lithia, Li_2O, has anions in cubic close packing, with Li ions occupying all tetrahedral positions.
(*a*) Draw the structure and calculate the density of Li_2O. *Hint*: Oxygen ions do not touch, but O–Li–O ions do.
Anwers: $\rho = 1.99$ g cm^{-3}.
(*b*) What is the maximum radius of a cation which can be accommodated in the vacant interstice of the anion array in Li_2O?
Answer: $r_c = 1.04$ Å.

3.5. Look up the radii of Ti^{4+}, Ba^{2+} and O^{2-} listed in App. 3A, and making use of Pauling's size criteria, choose the most suitable cage for each cation. Based on your results, choose the appropriate composite crystal structure and draw the unit cell of $BaTiO_3$. How many atoms of each element are there in each unit cell?

3.6. Garnets are semiprecious gems with the chemical composition $Ca_3Al_2Si_3O_{12}$. The crystal structure is cubic and is made up of three building blocks: tetrahedra, octahedra and dodecahedra (distorted cubes).
(*a*) Which ions do you think occupy which building block?
(*b*) In a given unit cell, what must the ratio of the number of blocks be?

3.7. Using information in Table 3.3, draw the NiAs structure. Find the structure in the literature and compare it to the one you drew.

3.8. Beryllium oxide (BeO) may form a structure in which the oxygen ions are in an FCC arrangement. Look up the ionic radius of Be^{2+} and determine which type of interstitial site it will occupy. What fraction of the available interstitial sites will be occupied? Does your result agree with that shown in Table 3.3? If not, explain possible reasons for the discrepancy.

3.9. Cadmium sulfide has a density of 4.82 g cm^3. Using the radii of the ions show that: (*a*) a cubic unit cell is not possible.

(*b*) Propose a likely structure(s) for CdS? How many Cd^{2+} and S^{2-} ions are there per unit cell?

3.10. The compound MX has a density of 2.1 g/cm^3 and a cubic unit cell with a lattice parameter of 0.57 nm. The atomic weights of M and X are, respectively, 28.5 and 30 g/mol. Based on this information, which of the following structures is (are) possible: NaCl, CsCl or zinc blende? Justify your choices.

3.11. What complex anions (i.e., sheets, chain, island, etc.) are expected in the following compounds?

(*a*) Tremolite or $Ca_2Mg_5(OH)_2Si_8O_{22}$.

(*b*) Mica or $CaAl_2(OH)_2(Si_2Al_2)O_{10}$.

(*c*) Kaolinite $Al_2(OH)_4Si_2O_5$.

3.12. Determine the expected crystal structure including the ion positions of the hypothetical salt AB_2, where the radius of A is 154 pm and that of B is 49 pm. Assume that A has a charge of +2.

3.13. (*a*) The electronic structure of N is $1s^22s^22p^3$. The structure of Si_3N_4 is based on the SiN_4 tetrahedron. Propose a way by which these tetrahedra can be joined together in three dimensions to form a solid, maintaining the 3:4 ratio of Si to N, other than the one shown in Fig. 3.11.

(*b*) Repeat part (*a*) for SiC. How many carbons are attached to each Si, and vice versa? What relationship, if any, do you think this structure has to the diamond cubic structure?

3.14. (*a*) Write an equation for the formation of a nonbridging oxygen (NBOs). Explain what is meant by a NBO. How does one change their number? What do you expect would happen to the properties of a glass as the number of NBOs increases?

(*b*) What happens to silicates as the O/Si ratio increases.

3.15. What would be the formulas (complete with negative charge) of the silicate units shown in Fig. 3.18?

3.16. (*a*) Derive an expression relating the mole fractions of alkali–earth oxides to the number of nonbridging oxygens per Si atom present in a silicate structure.

(*b*) Repeat Worked Example 3.3*b* for the composition Na_2O $0.5CaO{\cdot}2SiO_2$.

Answer: 1.5.

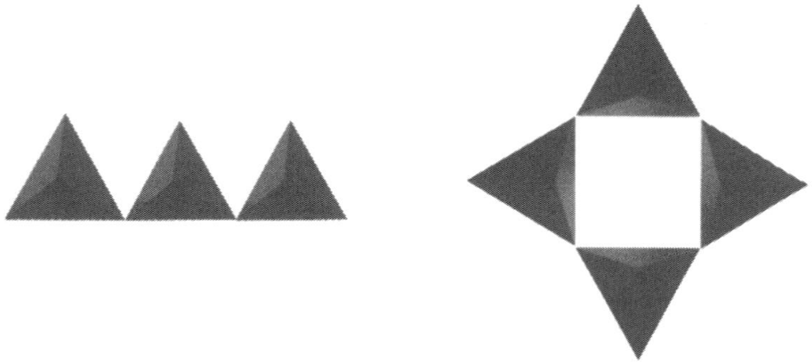

FIGURE 3.18 Silicate units. In this representation any corner that is *not* shared is a nonbridging oxygen that is negatively charged.

(*c*) Show that chains of infinite length would occur at a mole fraction of Na_2O of 0.5, balance SiO_2. What do you think the structure would be for a composition in between 0.33 and 0.5?

(*d*) Show that for any silicate structure the number of nonbridging oxygens per Si is given by NBO = $2R - 4$ and the number of bridging oxygens is $8 - 2R$, where R is the O/Si ratio.

3.17. (*a*) Talc, $Mg_3(OH)_2(Si_2O_5)_2$, is slippery. Based on its structure why is that?

(*b*) Consider the following two micas: $CaAl_2(OH)_2(Si_2Al_2)O_{10}$ and $KAl_2(OH)_2(AlSi_3)O_{10}$. The Mohs hardness of one of them is double the other. From your knowledge of their chemistries and structures, which one do you conclude would be the harder mineral. Explain your rationale.

(*c*) Along the same lines where do you think the hardness of the mineral pyrophyillite, $Al_2(OH)_2(Si_2O_5)_2$, falls in comparison to the micas in part (*b*)? Explain your rationale.

ADDITIONAL READING

1. R. W. G. Wyckoff, *Crystal Structures*, vols. 1 to 6, Wiley, New York, 1971.
2. F. D. Bloss, *Crystallography and Crystal Chemistry, An Introduction*, Holt, Rinehart and Winston, New York, 1971.
3. W. D. Kingery, H. K. Bowen, and D. R. Uhlmann, *Introduction to Ceramics*, 2nd ed., Wiley, New York, 1976.
4. L. Van Vlack, *Elements of Materials Science and Engineering*, 5th ed., Addison-Wesley, Reading, MA, 1985.
5. O. Muller and R. Roy, *The Major Ternary Structural Families*, Springer-Verlag, Berlin, 1974.
6. N. N. Greenwood, *Ionic Crystals, Lattice Defects and Non-Stoichiometry*, Butterworth, London, 1968.
7. R. J. Borg and G. D. Dienes, *The Physical Chemistry of Solids*, Academic Press, New York, 1992.
8. A. F. Wells, *Structural Inorganic Chemistry*, 4th ed., Clarendon Press, Oxford, UK, 1975.
9. N. B. Hannay, ed., *Treatise on Solid Chemistry*, vols. 1 to 6, Plenum, New York, 1973–1976.
10. C. B. Carter and M. G. Norton, *Ceramic Materials*, 2nd ed., Springer, New York, 2013.

OTHER REFERENCES

1. Data Base: https://materialsproject.org/.

   This free database is quite useful and easy to use. To start the following tutorials are useful:

   https://www.youtube.com/playlist?list=PLTjFYVNE7LTjHJwV994iQHS-bIkh3UXKJ.

   In addition to structures, lattice parameters, etc. one can find the electronic and band structures, elastic properties, XRD diffraction patterns, phase diagrams and much more.
2. This website is good for visualizing roughly 60 atomic structures, many of them discussed in this chapter and throughout the book. To see the structures, click on gray column on top left titled Inorganic Chemistry. The site is interactive and allows the viewer to rotate the structure and different visualizations are possible (ball and stick, space filling, etc.) Compare the table found when clicking the URL (http://www.chemtube3d.com/solidstate/_table.htm) to Table 3.2.
3. This website (https://en.wikipedia.org/wiki/Lattice_constant) lists a number of lattice parameters of select materials.Figure 3.6 Hexagonal-based binary ceramics. (*a*) Unit cell of α-Al_2O_3. (*b*) Wurtzite (ZnS) structure. The structure is drawn in this way to emphasize the tetrahedral sites.

4

EFFECT OF CHEMICAL FORCES ON PHYSICAL PROPERTIES

Now how curiously our ideas expand by watching these conditions of the attraction of cohesion! — how many new phenomena it gives us beyond those of the attraction of gravitation! See how it gives us great strength.

Michael Faraday, *On the Various Forces of Nature*

4.1 INTRODUCTION

The forces of attraction between the various ions or atoms in solids determine many of their properties. Intuitively, it is not difficult to appreciate that a strongly bonded material would have a high melting point and be stiff. In addition, it can be shown, as is done below, that its theoretical strength and surface energy will also increase, with a concomitant decrease in thermal expansion. In this chapter, semiquantitative relationships between these properties and the depth and shape of the energy well, described in Chap. 2, are developed.

In Sec. 4.2, the importance of bond strengths on the melting points of ceramics is elucidated. In Sec. 4.3, how strong bonds result in solids with low coefficients of thermal expansion is discussed. In Sec. 4.4, the relationships between bond strengths, stiffness and theoretical strengths is developed. Sec. 4.5 relates bond strengths to surface energies.

4.2 MELTING POINTS

Fusion, evaporation and sublimation result when sufficient thermal energy is supplied to a crystal to overcome the potential energy holding its atoms together. Experience has shown that, at constant pressure, a pure substance will melt at a fixed temperature, with the absorption of heat. The amount of heat absorbed is known as the **heat of fusion, ΔH_f**, and it is the heat required for the transformation

$$\text{Solid} \rightarrow \text{Liquid}$$

ΔH_f is a measure of the enthalpy difference between the solid and liquid states at the melting point. The entropy difference, ΔS_f, between the liquid and solid is defined by

$$\Delta S_f = \frac{\Delta H_f}{T_m} \tag{4.1}$$

where T_m is the melting point in degrees Kelvin. The entropy difference ΔS_f is a direct measure of the degree of disorder that arises in the system during the melting process and is by necessity positive, since the liquid state is always more disordered than the solid. The melting points and ΔS_f values for a number of ceramics are listed in Table 4.1. Inspection of Table 4.1 reveals that there is quite a bit of variability in the melting points or ceramics.[35] To understand this variability, one needs to understand the various factors that influence the melting point.

4.2.1 FACTORS AFFECTING THE MELTING POINTS OF PREDOMINANTLY IONICALLY BONDED CERAMICS

4.2.1.1 IONIC CHARGE

The most important factor determining T_m of an ionic ceramic is the bond strength holding the ions in place. In Eq. (2.15), the strength of an ionic bond, E_{bond}, was found to be proportional to the product of the ionic charges z_1 and z_2 making up the solid. It follows that the greater the ionic charges, the stronger the attraction between ions, and consequently the higher the melting point. For example, both MgO and NaCl crystallize in the rock salt structure, but their melting points are, respectively, 2852 and 800°C—a difference directly attributable to the fact that in MgO the ions are doubly ionized, whereas in NaCl they are singly ionized. Said otherwise, everything else being equal, the energy well of MgO is roughly four times deeper than that of NaCl. It is therefore not surprising that it requires more thermal energy to melt MgO than it does to melt NaCl.

4.2.1.2 COVALENT CHARACTER OF THE IONIC BOND

Based on Eq. (4.1), T_m is proportional to ΔH_f, and consequently whatever reduces one reduces the other. It turns out, as discussed subsequently, that increasing the covalent character of an ionic bond tends to

[35] Interestingly enough, for most solids including metals, the entropy of fusion per ion lies in the narrow range between 10 and 12 J/(mol·deg). This is quite remarkable, given the large variations in melting points, and strongly suggests that the structural changes on the atomic scale due to melting are similar for most substances. This observation is even more remarkable when the data for noble gas solids such as Ar are included. The melting point of Ar is 83 K and ΔS_f = 14 J/mol·K.

TABLE 4.1 Melting points and entropies of fusion for selected inorganic compounds

Compound	Melting point, °C	ΔS_f J/mol·K	Compound	Melting point, °C	ΔS_f J/mol·K
Oxides					
Al_2O_3	2054 ± 6	47.7	Mullite	1850	
BaO	2013	25.8	Na_2O (α)	1132	33.9
BeO	2780 ± 100	30.5	Nb_2O_5	1512 ± 30	58.4
Bi_2O_3	825		Sc_2O_3	2375 ± 25	
CaO	2927 ± 50	24.8	SrO	2665 ± 20	25.6
Cr_2O_3	2330 ± 15	49.8	Ta_2O_5	1875 ± 25	
EU_2O_3	2175 ± 25		ThO_2	3275 ± 25	
Fe_2O_3	Decomposes at 1735 K to Fe_3O_4 and oxygen		TiO_2 (rutile)	1857 ± 20	31.5
			UO_2	2825 ± 25	
Fe_3O_4	1597 ± 2	73.8	V_2O_5	2067 ± 20	
Li_2O	1570	32.0	Y_2O_3	2403	≈38.7
Li_2ZrO_3	1610		ZnO	1975 ± 25	
Ln_2O_3	2325 ± 25		ZrO_2	2677	29.5
MgO	2852	25.8			
Halides					
AgBr	434		LiBr	550	
AgCl	455		LiCl	610	22.6
CaF_2	1423		LiF	848	
CsCl	645	22.2	LiI	449	
KBr	730		NaCl	800	25.9
KCl	776	25.2	NaF	997	
KF	880		RbCl	722	23.8
Silicates and other glass-forming oxides					
B_2O_3	450 ± 2	33.2	$Na_2Si_2O_5$	874	31.0
$CaSiO_3$	1544	31.0	Na_2SiO_3	1088	38.5
GeO_2	1116		P_2O_5	569	
$MgSiO_3$	1577	40.7	SiO2 (high quartz)	1423 ± 50	4.6
Mg_2SiO_4	1898	32.8			
Carbides, nitrides, borides and silicides					
B_4C	2470 ± 20	38.0	ThN	2820	
HfB_2	2900		TiB_2	2897	
HfC	3900		TiC	3070	
HfN	3390		TiN	2947	
HfSi	2100		$TiSi_2$	1540	

(*Continued*)

TABLE 4.1 (Continued) Melting points and entropies of fusion for selected inorganic compounds

Compound	Melting point, °C	ΔS_f J/mol·K	Compound	Melting point, °C	ΔS_f J/mol·K
$MoSi_2$	2030		UC	2525	
NbC	3615		UN	2830	
NbN	2204		VB_2	2450	
SiC	2837		VC	2650	
Si_3N_4	At 2151 K P_{N2} over Si_3N_4 = 1 atm		VN	2177	
			WC	2775	
			ZrB_2	3038	
TaB_2	3150		ZrC	3420	
TaC	3985		ZrN	2980 ± 50	
$TaSi_2$	2400		$ZrSi_2$	1700	
ThC	2625				

reduce ΔH_f by stabilizing discrete units in the melt, which in turn reduces the number of bonds that have to be broken during melting, which is ultimately reflected in lower melting points.

It is important to note that covalency per se does not necessarily favor either higher or lower melting points. The important consideration depends on the melt structure; if the strong covalent bonds have to be broken in order for melting to occur, extremely high melting temperatures can result. Conversely, if the strong bonds do not have to be broken for melting to occur, the situation can be quite different.[36]

The effect of covalency on the structures of three MX_2 compounds is best shown graphically in Fig. 4.1. In the figure, the covalent character of the bonds increases in going from left to right, which results in changes in the structure from three-dimensional in TiO_2, to a layered structure for CdI_2, to a molecular lattice in the case of CO_2. Also shown in Fig. 4.1 are the corresponding melting points; the effect of the structural changes on the latter is obvious.

It follows from this brief introduction that in order to understand the subtleties in melting point trends, one needs to somewhat quantify the extent of covalency present in an ionic bond. In Chap. 2, the bonds between ions were assumed to be either predominantly covalent or ionic. As noted then, and reiterated here, the reality of the situation is more complex—ionic bonds possess covalent character and vice versa. Historically, this complication has been addressed by means of one of two approaches. The first was to assume that the bond is purely covalent, and then consider the effect of shifting the electron cloud toward the more electronegative atom. The second approach, discussed subsequently, was to assume the bond is purely ionic and then impart a covalent character to it.

[36] An extreme example of this phenomenon occurs in polymers, where the bonding is quite strong within the chains and yet the melting points are quite low, because these bonds are *not* broken during melting.

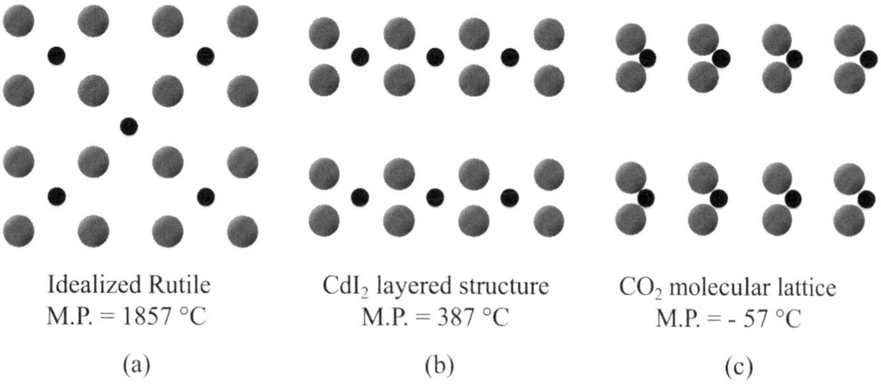

FIGURE 4.1 Effect of polarization on crystal structure and melting temperature.

The latter approach was championed by Fajans[37] and is embodied in Fajans' rules, whose basic premise is summarized in Fig. 4.2. In Fig. 4.2*a* an idealized ion pair is shown for which the covalent character is nonexistent (i.e., the ions are assumed to be hard spheres). In Fig. 4.2*b* some covalent character is imparted to the bond by shifting the electron cloud of the more polarizable anion toward the polarizing cation. In the extreme case, that the cation is totally embedded in the electron cloud of the anion (Fig. 4.2*c*), a strong covalent bond is formed. The extent to which the electron cloud is distorted and shared between the two ions is thus a measure of the covalent character of that bond. The latter thus defined depends on the three following factors:

Polarizing power of cation. High charge and small sizes increase the polarizing power of cations. Over the years many functions have been proposed to quantify the effect, and one of the simplest is to define the **ionic potential** of a cation as:

$$\phi = \frac{z^+}{r}$$

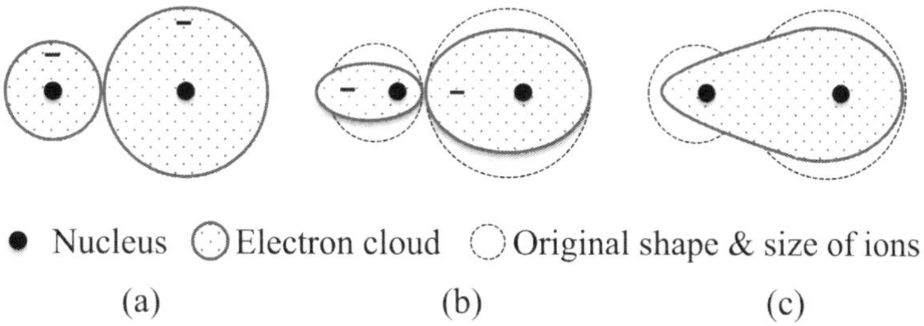

FIGURE 4.2 Polarization effects: (a) idealized ion pair with no polarization; (b) polarized ion pair; (c) polarization sufficient to form covalent bond.

37 K. Fajans, *Struct. Bonding*, 2, 88 (1967).

TABLE 4.2 Ionic potentials of selected cations in nm^{-1}

Li^+	17.0	Be^{2+}	64.0	B^{3+}	150.0
Na^+	10.5	Mg^{2+}	31.0	Al^{3+}	60.0
K^+	7.0	Ca^{2+}	20.0	Si^{4+}	100.0

where z^+ is the charge on the cation and r its radius. The ionic powers of a few selected cations are listed in Table 4.2, where it is clear that high charge and small size greatly enhance ϕ and consequently a bond's covalent character.

To illustrate compare MgO and Al_2O_3. On the basis of ionic charge alone, one would expect the melting point of Al_2O_3 (+3, −2) to be higher than that of MgO (−2, +2), and yet the reverse is observed. However, based on the relative polarizing power of Al^{3+} and Mg^{2+}, it is reasonable to conclude that the covalent character of the Al-O bond is greater than that of the Mg-O bond. This greater covalency *appears to stabilize discrete units in the liquid state and lowers the melting point.* Further evidence that the Al_2O_3 melt is more "structured" than MgO is reflected in the fact that ΔS_f *per ion* for Al_2O_3 [9.54 J/(mol·K)] is smaller than that of MgO [12.9 J/(mol·K)].

Polarizability of anions. The *polarizability* of an ion is a measure of the ease with which its electron cloud can be pulled away from its nucleus, which, as discussed in greater detail in Chap. 14, scales with its volume. Increasing anionic polarizability increases the covalent character of the bond, which once again results in lower melting points. For example, the melting points of LiCl, LiBr and LiI are, respectively, 613, 547 and 446°C.[38]

Electron configuration of cations. The d electrons are less effective in shielding the nuclear charge than the s or p electrons and are thus more polarizing. Thus, ions with d electrons tend to form more covalent bonds. For example, Ca^{2+} and Hg^{2+} have quite similar radii (100 and 102 pm, respectively), and yet $HgCl_2$ melts at 276°C, whereas $CaCl_2$ melts at 782°C.

4.2.2 COVALENT CERAMICS

The discussion so far has focused on understanding the relationship between the interatomic forces holding atoms together and the melting points of mostly ionic ceramics. The melting points and general thermal stability of covalent ceramics are quite high as a result of the very strong primary bonds that typically form between Si and C, N or O. Covalent ceramics are interesting materials in that some do not melt, but rather decompose at higher temperatures. For example, Si_3N_4 decomposes at temperatures in excess of 2000°C, when the nitrogen partial pressure reaches 1 atm.

4.2.3 GLASS-FORMING LIQUIDS

Glass-forming oxides such as SiO_2, many of the silicates, B_2O_3, GeO_2 and P_2O_5 among others, possess anomalously low entropies of fusion. For example, for SiO_2, ΔS_f is 4.6 J/(mol·K). This signifies that at T_m, the solid

38 Another contributing factor to the lowering of melting points that cannot be ignored is the increase in the radii of the anions that decreases E_{bond} by increasing r_0. This is a second-order effect, however.

and liquid structures are quite similar. Given that glasses can be considered supercooled liquids, it is not surprising that these oxides, called *network formers*, are the basis of many inorganic glasses (see Chap. 9).

4.3 THERMAL EXPANSION

It is well known that solids expand upon heating. The extent of the expansion is characterized by a **coefficient of linear expansion, *α***, defined as the fractional change in length, with change in temperature at constant pressure, or

$$\alpha = \frac{1}{l_0}\left(\frac{\partial l}{\partial T}\right)_p \tag{4.2}$$

where l_0 is the original length.

The origin of thermal expansion can be traced to the **anharmonicity,** or asymmetry of the energy–distance curves discussed in Chap. 2 and reproduced in Fig. 4.3. This asymmetry reflects the fact that it is easier to pull two atoms apart than to push them together. This in turn is due to the fact that to form a bond it is imperative that repulsive forces are short range as opposed to the longer ranged attractive forces (see Chap. 2)

To understand why the asymmetry results in thermal expansion one needs to follow the average locations of the atoms at various temperatures starting from 0 K. At 0 K, the total energy is potential in nature, and the atoms are sitting at the bottom of the well (point a in Fig. 4.3). As the temperature is raised to, say T_1, the average energy of the system increases correspondingly by a value of kT_1, where k is Boltzmann's constant. At T_1 the atoms vibrate between positions x_1 and x_2, and their energy fluctuates between purely

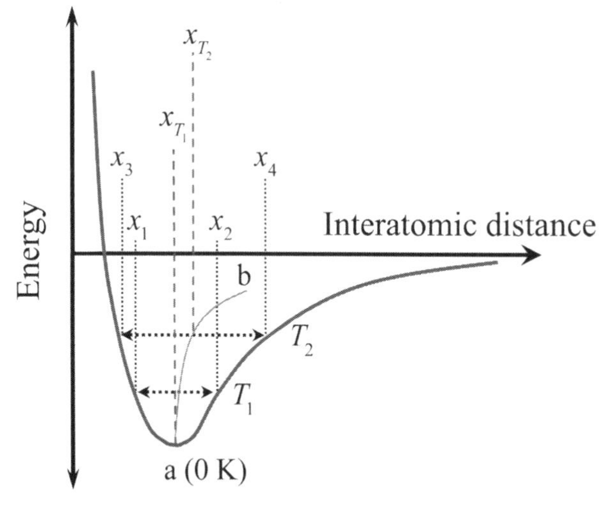

FIGURE 4.3 Effect of heat on interatomic distance between atoms. Note that asymmetry of well is responsible for thermal expansion. The average position of the atoms in a perfectly symmetric well would not change with temperature.

potential at x_1 and x_2 (i.e., zero kinetic energy) and speed up somewhere in between. In other words, the atoms behave as if they were attached to each other by springs. The average location of the atoms at T_1 will thus be midway between x_1 and x_2, that is, at x_{T_1}. If the temperature is raised to, say, T_2, the average position of the atoms will move to x_{T_2}, etc. It follows that, with increasing temperatures, the atoms' average positions will move along line ab, shown in Fig. 4.3, which, *per force*, expands the crystal dimensions.

In general, the asymmetry of the energy well increases with decreasing bond strength, and consequently a solid's thermal expansion scales inversely with its bond strength or melting point. For example, α, of solid Ar is on the order of 1×10^{-3} K^{-1}, whereas for most metals and ceramics (see below) it is closer to 1×10^{-5} K^{-1}.

Perusal of Table 4.3, in which the mean α of a number of ceramics are listed, makes it clear that α for most ceramics lies between 3 and 10×10^{-6} K^{-1}. The temperature dependence of the fractional increase in length for a number of ceramics and metals is shown in Fig. 4.4. Given that the slope of these lines is α, the following generalizations can be made:

1. Ceramics have lower α values than metals.
2. α increases with increasing temperature. This reflects the fact that as one moves up the energy well—i.e., with increasing temperature—it becomes more asymmetric. Thus, it is important to specify the temperature range reported, since as the temperature range expands, the mean α also increases.
3. Covalently bonded ceramics, such as SiC and Si_3N_4, have lower α values than more close-packed ceramic structures such as NaCl and MgO. This is a reflection of the influence of atomic packing. In contradistinction to close-packed structures, where all vibrations increase in the dimensions of the crystal, the more open structures of covalent ceramics allow for other modes of vibration that do not necessarily contribute to thermal expansion. In other words, the added thermal energy can result in changes in bond angles without significant changes in bond lengths. (Think of the atoms as vibrating into the "open spaces" rather than against each other.)

   One of the most striking examples of the importance of atomic packing on α is vitreous silica, which has an extremely low α (Fig. 4.5). Quartz and cristobalite, on the other hand, have much higher α values, as shown in Fig. 4.5.
4. Although not explicitly stated, the discussion so far was only strictly true for isotropic, e.g., cubic, polycrystalline materials. Crystals that are noncubic—and consequently are anisotropic in their thermal expansion—can behave quite differently. In some cases, a crystal can actually shrink in one direction as it expands in another. When a polycrystal is made up of such crystals, the average thermal expansion can be quite low indeed. Cordierite and lithium-aluminosilicate (LAS) (see Fig. 4.4) are good examples of this class of materials. As discussed in greater detail in Chap. 13, this anisotropy in thermal expansion, which has been exploited to fabricate very low-α materials, also results in the buildup of large thermal residual stresses that can be quite detrimental to the strength and integrity of ceramic parts.

4.4 YOUNG'S MODULUS AND THE STRENGTH OF PERFECT SOLIDS

In addition to understanding the response of ceramics to heating, it is important to understand their behavior when they are loaded or stressed. The objective of this section is to relate the shape of the energy

TABLE 4.3 Mean thermal expansion coefficients and theoretical densities, TDs, of various ceramics

Ceramic	TD (g/cm^3)	α (°C^{-1})×10^6	Ceramic	TD (g/cm^3)	α (°C^{-1})×10^6
Binary Oxides					
α-Al_2O_3	3.98	7.2–8.8	Nb_2O_5	4.47	
BaO	5.72	17.8	SiO_2 (low cristobalite)	2.32	
$Bi_2O_3(\alpha)$	8.90	14.0			
		(RT–730°C)	ThO_2	9.86	9.2
Bi_2O_3 (δ)	8.90	24.0	TiO_2	4.25	8.5
		(650–825°C)	UO_2	10.96	10.0
CeO_2	7.20		WO_2	7.16	
Cr_2O_3	5.22		Y_2O_3	5.03	9.3[a]
Gd_2O_3	7.41	10.5	ZnO	5.61	8.0 (c axis)
Fe_3O_4	5.24				4.0 (a axis)
Fe_2O_3	5.18		ZrO_2 (monoclinic)	5.83	7.0
HfO_2	9.70	9.4–12.5			
MgO	3.60	13.5	ZrO_2 (tetragonal)	6.10	12.0
Na_2O	2.27				
Mixed Oxides					
Al_2O_3·TiO_2		9.7 (average)	Cordierite	2.51	2.1
Al_2O_3·MgO	3.58	7.6	MgO·SiO_2		10.8[a]
		(25–1400)	2MgO·SiO_2		11.0[a]
BaO·ZrO_2		8.5[a]	MgO·TiO_2		7.9
BeO·Al_2O_3	3.69	3.2–6.7	MgO·ZrO_2		12.0[a]
		(25–1000)	2SiO_2·3Al_2O_3	3.20	5.1[a]
CaO·SiO_2(β)		5.9	(mullite)		
		(25–700)	$SiZrO_4$	4.20	4.5[a]
CaO·SiO_2(α)		11.2	(zircon)		
		(25–700)	SrO·TiO_2		9.4[a]
CaO·ZrO_2		10.5	SrO·ZrO_2		9.6
2CaO·SiO_2(β)		14.4	TiO_2·ZrO_2??		7.9[a]
Borides, Nitrides, Carbides and Silicides					
AlN	3.26	5.6[a]	TaC	14.48	6.3
B_4C	2.52	5.5	TiC	4.95	7.7–9.5
BN	2.27	4.4	TiN	5.40	9.4
Cr_3C_2	6.68	10.3	$TiSi_2$	4.40	10.5
$CrSi_2$	4.40		Ti_3SiC_2	4.51	9.1
HfB_2	11.20	5.0	WC	15.70	4.3
HfC	12.60	6.6	ZrB_2	6.11	5.7–7.0
$HfSi_2$	7.98		ZrC	6.70	6.9[a]
β-Mo_2C	9.20	7.8	$ZrSi_2$	4.90	7.6[a]

(Continued)

TABLE 4.3 (Continued) Mean thermal expansion coefficients and theoretical densities, TDs, of various ceramics

Ceramic	TD (g/cm³)	α (°C⁻¹)×10⁶	Ceramic	TD (g/cm³)	α (°C⁻¹)×10⁶
Si_3N_4	3.20	3.1–3.7	ZrN	7.32	7.2
SiC	3.20	4.3–4.8			
Halides					
CaF_2	3.20	24.0	LiCl	2.07	12.2
LiF	2.63	9.2	LiI	4.08	16.7
LiBr	3.46	14.0	MgF_2		16.0
KI	3.13		NaCl	2.16	11.0
Glasses					
Soda-lime glass	2.52	9.0	Fused silica	2.20	0.55
Pyrex	≈2.23	3.2			

[a] Denotes range of 25 to 1000°C.

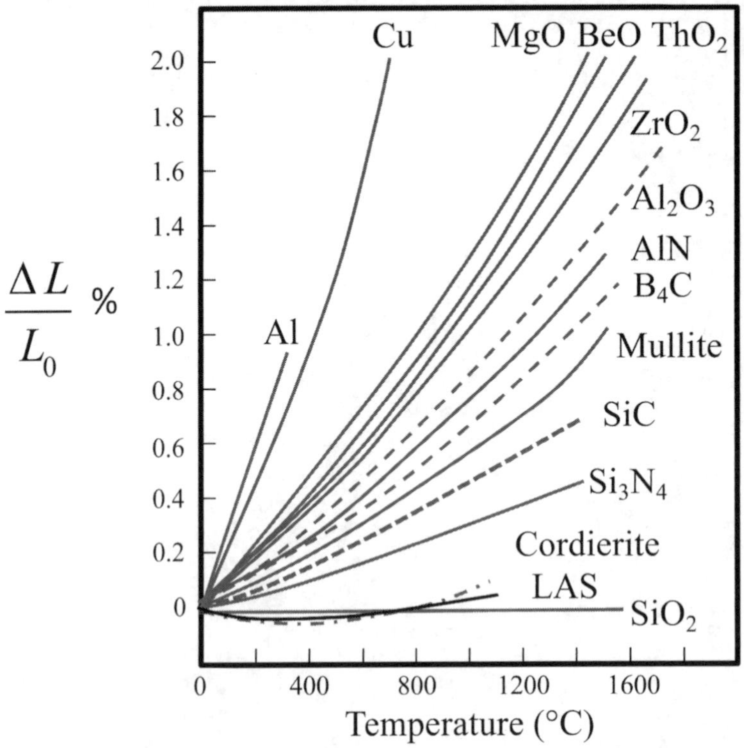

FIGURE 4.4 Temperature dependence of $\Delta L/L_0$ (%) for select materials. The slopes of these lines at any temperature yields α. For most ceramics, α is more or less constant with temperature. For anisotropic solids, the c-axis expansion is reported. (Adapted from J. Chermant, *Les Ceramiques Thermomechaniques*, CNRS Presse, France, 1989.)

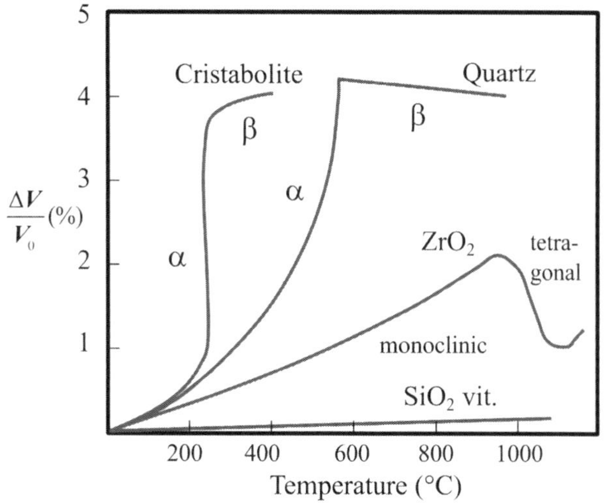

FIGURE 4.5 Temperature dependence of $\Delta V/V_0$ (%) for cristobalite, quartz, zirconia, and vitreous or amorphous SiO_2. The abrupt changes in behavior with temperature are a result of phase transformations (see Chap. 8). (Adapted from W. D. Kingery, H. K. Bowen, and D. R. Uhlmann, *Introduction to Ceramics*, 2d ed., Wiley, New York, 1976.)

versus distance curve E(r), discussed in Chap. 2, to the elastic modulus, which is a measure of the stiffness and the theoretical strength of that material. To accomplish this goal, one needs to examine the forces F(r) that develop between atoms as a result of externally applied stresses. As noted in Sec. 2.4, F(r) is defined as

$$F(r) = \frac{dE(r)}{dr} \tag{4.3}$$

From the general shape of the E(r) curve, one can easily sketch the shape of a typical force, F(r), versus distance curve, as shown in Fig. 4.6. The following features are noteworthy:

- The net force between the atoms or ions is zero at equilibrium, i.e., at $r = r_0$.
- Pulling the atoms apart results in the development of an *attractive restoring force* between them that tends to pull them back together. The opposite is true if one tries to push the atoms together.
- In the region around $r = r_0$ the response can be considered, to a very good approximation, linear (inset in Fig. 4.6). In other words, the atoms act as if they are connected by miniature springs. It is in this region that Hooke's law (see below) applies.
- The force pulling the atoms apart cannot be increased indefinitely. Beyond some separation r_{fail}, the bond will fail. The force at which this occurs represents the maximum force F_{max} that a bond can withstand before failing.

In the remainder of this section, the relationships between stiffness and theoretical strength, on one hand, and E(r) and F(r), on the other hand, are developed.

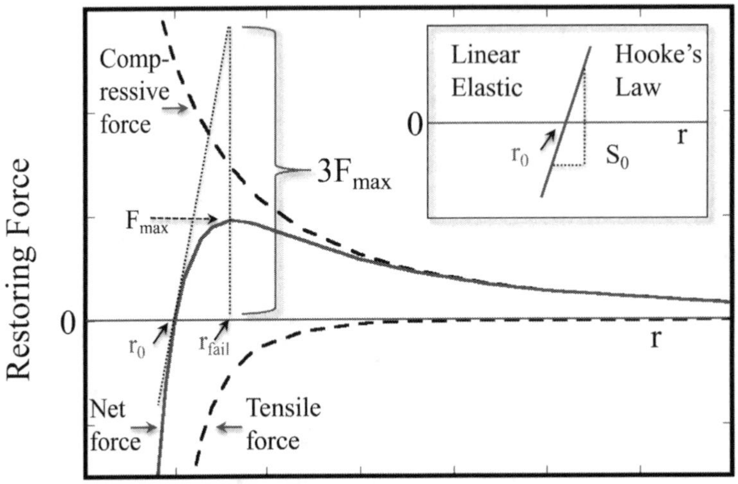

FIGURE 4.6 Typical force–distance curve. Slope of the line going through r_0 is the stiffness of the bond, S_0. In this construction, it is assumed that the maximum force is related to the stiffness as shown. This is quite approximate but serves to illustrate the relationship between stiffness and theoretical strength, i.e., Eq. (4.12). Inset emphasizes that around equilibrium the response is linear elastic.

4.4.1 ATOMIC VIEW OF YOUNG'S MODULUS

When a solid is subjected to a small **stress**, σ, it will respond by deforming in proportion to the stress applied, a phenomenon described by **Hooke's law:**

$$\sigma = Y\varepsilon \tag{4.4}$$

where Y is Young's modulus and ε is the **strain** experienced by the material, defined as

$$\varepsilon = \frac{L - L_0}{L_0} \tag{4.5}$$

Here L is the length under the applied stress and L_0 is the original length.

Refer once more to inset shown in Fig. 4.6. In the vicinity of r_0, the following excellent approximation can be made:

$$F = S_0(r - r_0) \tag{4.6}$$

where S_0 is the **stiffness** of the bond, defined as

$$S_0 = \left(\frac{dF}{dr}\right)_{r=r_0} \tag{4.7}$$

Note that Eq. (4.6) describes the behavior of a linear spring. Dividing both sides of Eq. (4.6) by r_0^2 and noting that F/r_0^2 is approximately the stress on the bond, while $(r - r_0)/r_0$ is the strain on the same bond, and comparing the resulting expression with Eq. (4.4), one can see immediately that

$$\boxed{Y \approx \frac{S_0}{r_0}} \tag{4.8}$$

Combining this result with Eqs. (4.3) and (4.7), one can further show that

$$Y = \frac{1}{r_0}\left(\frac{dF}{dr}\right)_{r=r_0} = \frac{1}{r_0}\left(\frac{d^2E}{dr^2}\right)_{r=r_0} \tag{4.9}$$

This is an important result because it relates the stiffness of a solid, as measured by Y, to the curvature of its energy/distance curve. (In math, the second derivative of a function is its curvature.) Furthermore, it implies that strong bonds will be stiffer than weak bonds, a result that is not in the least surprising. It also explains why, in general, given their high melting temperatures, ceramics are quite stiff solids.

4.4.2 THEORETICAL STRENGTHS OF SOLIDS

The next task is to estimate a solid's theoretical strength or the stress that would be required to *simultaneously* break all the bonds across a fracture plane. It can be shown (see Prob. 4.2) that most primary bonds will fail when they are stretched by about 25%, i.e., when $r_{fail} \approx 1.25 r_0$. It follows from the geometric construction shown in Fig. 4.6 that

$$S_0 \approx \frac{3F_{max}}{r_{fail} - r_0} \approx \frac{3F_{max}}{1.25r_o - r_o} \tag{4.10}$$

Dividing both sides of this equation by r_0 and noting that

$$\frac{F_{max}}{r_0^2} \approx \sigma_{max} \tag{4.11}$$

i.e., the force divided by the area over which it operates, one obtains

$$\sigma_{max} \approx \frac{Y}{12} \tag{4.12}$$

For a more exact calculation, one can start with the energy/interatomic distance function in its most general form, i.e.,

$$E_{bond} = \frac{C}{r^n} - \frac{D}{r^m} \tag{4.13}$$

where C and D are constants and $n > m$. Assuming $\sigma_{max} \approx F_{max}/r_0^2$, one can show (see Prob. 4.2) that σ_{max} is better approximated by

$$\boxed{\sigma_{max} = \frac{Y}{[(n+1)/(m+1)]^{(m+1)/(n-m)}} \frac{1}{n+1}} \tag{4.14}$$

Substituting typical values for m and n, say, $m = 1$ and $n = 9$, for an ionic bond yields $\sigma_{max} \approx Y/15$. This value is not too far off our simple derivation that resulted in Eq. (4.12).

Based on these results, one may conclude that a solid's theoretical strength should be roughly one-tenth of its Young's modulus. Experience has shown, however, that the actual strengths of ceramics are significantly lower; they are closer to Y/100 to Y/1000. The reason for this state of affairs is discussed in greater detail in Chap. 11, and reflects the fact that real solids are not perfect, as assumed here, but contain many flaws and defects that tend to locally concentrate the applied stress, which in turn significantly weakens the material.

4.5 SURFACE ENERGY

The surface energy, γ, of a solid is the energy needed to create a unit area of surface. The process is shown in Fig. 4.7*a*, where two new surfaces are created by breaking a solid in two. Given this picture, γ is simply the product of the number of bonds broken per unit area, N_s, of a crystal surface and the energy per bond, E_{bond}, or

$$\gamma = -N_s E_{bond} \tag{4.15}$$

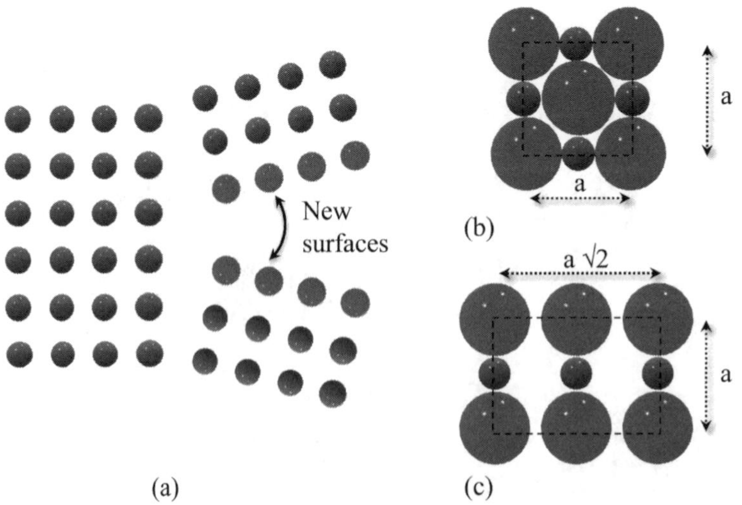

FIGURE 4.7 (*a*) Creation of new surface entails the breaking of bonds across that surface. (*b*) Structure of (100) and (c) (110) planes in the rock salt structure.

Since E_{bond} is a negative quantity, a negative sign is introduced to insure that γ is positive, i.e., endothermic. In what follows for simplicity's sake, only first-neighbor interactions will be considered, which implies that E_{bond} is given by Eq. (2.15).

Since N_s is a function of crystallography, it follows that γ is also a function of crystallography. To calculate N_s the following expression is useful:

$$N_s = \left(\frac{CN - CN_p}{2}\right)\left(\frac{\text{\# of atoms/plane}}{2A}\right) \tag{4.16}$$

where CN is the coordination number, i.e., number of nearest neighbors and CN_p is the coordination number of *ions in the plane* of the surface formed. Note that Eq. (4.16) is only valid when the cation and anion coordination numbers are equal. When they are not this relationship needs to be modified (see Prob. 4.11).

Calculations of γ based on Eq. (4.16) invariably yield values that are substantially greater than the measured ones (see Table 4.4). The reason for this discrepancy comes about because in our simple model surface relaxation and atomic rearrangement upon the formation of the new surface were not accounted for. When the surface is allowed to relax, some of the energy needed to form it is recovered, and the theoretical predictions do indeed approach the experimentally measured values (see below).

How to calculate the surface energies of various crystallographic planes is best explained by the following worked example.

Worked Example 4.1

Calculate γ of the (a) (100) and (b) (110) planes[39] in the rock salt structure and compare your results with those listed in Table 4.4.

TABLE 4.4 Measured free surface energies of select solids

Substance	Surface	Environment	Temp., K	Surface energy, J/m²
Mica	(0001)	Air	298	0.4
		Vacuum	298	5.0
MgO	(100)	Air	298	1.1
KCl	(100)	Air	298	0.1
Si	(111)	Liquid N_2	77	1.2
NaCl	(100)	Liquid N_2	77	0.3
CaF_2	(111)	Liquid N_2	77	0.4
LiF	(100)	Liquid N_2	77	0.3
$CaCO_3$	$(10\bar{1}0)$	Liquid N_2	77	0.2

39 It is assumed here that the reader is familiar with Miller indices, a topic covered in almost all introductory materials science textbooks.

ANSWER

(*a*) The (100) plane in NaCl is shown in Fig. 4.7*b*. In an area of $(2r_0)^2$, where r_0 is the equilibrium interionic distance, two cations and two anions are present. Note, however, that the total surface area created is twice that, or $2\times(2r_0)^2$ since upon cleavage, two surfaces are created. The CN in rock salt is 6 and that in the (100) plane is 4. The number of atoms in the plane is 4; two cations and two anions. It follows that

$$N_s\ (\text{bonds/m}^2) = \left(\frac{6-4}{2}\right)\left(\frac{4}{2(2r_0)^2}\right) = \frac{1}{2r_0^2}$$

Combining this result with Eqs. (2.15) and (4.15) yields

$$\gamma_{100} \approx -E_{bond}\left[\frac{1}{2r_0}\right] \approx -\frac{z_1 z_2 e^2}{8\pi\varepsilon_0 r_0^3}\left(1-\frac{1}{n}\right) \qquad (4.17)$$

For NaCl, $r_0 = 283$ pm, which when substituted in Eq. (4.16), assuming $n = 8$, yields a value of γ of 4.44 J/m². Comparing this value with the experimentally measured value listed in Table 4.4, it is immediately obvious that it is off by more than an order of magnitude, for reasons alluded to below.

$$\gamma_{100} \approx -\frac{(1.6\times10^{-19})^2}{8\pi 8.85\times10^{-12}(283\times10^{-12})^3}\left(1-\frac{1}{8}\right) = 4.4\ \text{J/m}^2$$

(*b*) The (110) plane (Fig. 4.7*c*) has an area of $\sqrt{2}(2r_0)(2r_0)$, but still contains two Na and two Cl ions. However, the CN of each of the atoms in the plane is now 2 instead of 4, which implies that each ion is coordinated to *two* other ions above and two ions below the plane (here, once again for simplicity, all but first-neighbor interactions are considered). In other words, to create the plane, one needs to break two bonds per ion. It follows that

$$N_s\ (\text{bonds/m}^2) = \left(\frac{6-2}{2}\right)\left(\frac{4}{\sqrt{2}(2r_0)^2}\right) = \frac{\sqrt{2}}{r_0^2}$$

$$\gamma_{110} \approx -\frac{(1.6\times10^{-19})^2}{4\pi 8.85\times10^{-12}(283\times10^{-12})^3}\left(1-\frac{1}{8}\right)\sqrt{2} = 12.6\ \text{J/m}^2$$

4.6 FREQUENCIES OF ATOMIC VIBRATIONS

If an atomic bond can be reasonably and accurately be modeled by a spring, then like a mass attached to a spring, it should possess a natural frequency of vibration, ν_0. It follows that ν_0 is related to a bond's spring constant, S_0, discussed above by

$$\omega_0 = 2\pi\nu_0 \approx \sqrt{\frac{S_0}{M_{red}}} \qquad (4.18)$$

where ω_0 is the angular frequency in rad s^{-1} and M_{red} is the reduced mass of the oscillator system. For a monoatomic solid, M_{red} is simply the mass of the vibrating atoms. For a ceramic with two different atoms of masses m_1 and m_2, $M_{red} = m_1 m_2/(m_1 + m_2)$. For a more detailed calculation see Worked Example 5.3.

At this stage, it is useful to estimate the order of magnitude of ν_0. The simplest way to do so is to equate the average thermal energy of the atoms with their vibrational energy. In other words, assume: $3/2\ kT \approx \frac{1}{2}\ h\nu_0$. It follows that at 300 K, $\nu_0 \approx 9.4 \times 10^{12}\ s^{-1}$ or $\approx 10^{13}\ s^{-1}$. This is a remarkable result since it implies that, on average, the atoms in our bodies and all those surrounding us are vibrating back and forth roughly 10^{13} times every second!

EXPERIMENTAL DETAILS: MELTING POINTS

Several methods can be used to measure the melting point of a solid. One of the simplest is to use a **differential thermal analyzer**, or DTA for short. The basic arrangement of a DTA is shown schematically in Fig. 4.8*a*. The sample and an inert reference (usually alumina powder) are placed side by side in a furnace, and identical thermocouples are placed below each. The temperature of the furnace is then slowly ramped, and the difference in temperature $\Delta T = T_{sample} - T_{ref}$ is measured as a function of the temperature of the furnace, which is measured by a third thermocouple (thermocouple 3 in Fig. 4.8*a*). Typical results are shown in Fig. 4.8*b* and are interpreted as follows. As long as both the sample and the reference are inert, they should have the same temperature and $\Delta T = 0$. However, if for any reason the sample absorbs (endothermic process) or gives off (exothermic process) heat, its temperature vis-à-vis the reference thermocouple will change accordingly. For example, melting, being an endothermic process, will appear as a trough upon heating. The melting point is thus the temperature at which the trough appears. In contrast, upon cooling, freezing, being an exothermic process, will appear as a peak.

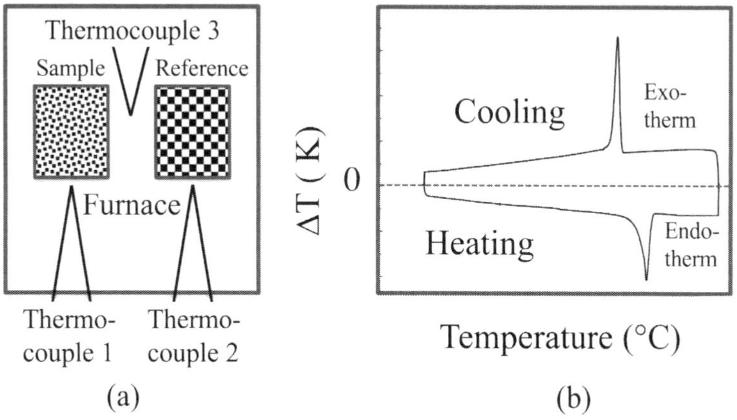

FIGURE 4.8 (*a*) Schematic of a DTA setup. (*b*) Typical DTA traces upon heating (bottom curve) and cooling (top curve).

ELASTIC CONSTANTS

The simplest and also the best way to measure a solid's elastic properties is to measure the velocity of sound in that solid. This is typically carried out using piezoelectric transducers that can generate either a normal or a shear wave in the solid. After the wave is generated, the time it takes to travel from one end of the sample to the other is measured by say another transducer. That information is then used to calculate the longitudinal, v_L, and shear, v_S, velocities of sound in the solid. The Young's, Y, and shear, G, moduli are then calculated assuming:

$$Y = \rho v_L^2 \left(\frac{3v_L^2 - 4v_S^2}{v_L^2 - v_S^2} \right) \tag{4.19}$$

and

$$G = \rho v_S^2 \tag{4.20}$$

where ρ is the solid's density.

THERMAL EXPANSION COEFFICIENTS

Thermal expansion coefficients are measured with a dilatometer, which is essentially a high-temperature furnace from which a rod sticks out (Fig. 4.9). One side of the rod is pushed against the sample for which the thermal expansion is to be measured; the other side is attached to a device that can measure the displacement of the rod quite accurately, such as a linear variable differential transformer or LVDT. In a typical experiment, the sample is placed inside the furnace and is heated at a constant rate. Simultaneously the push rod's displacement is measured. Typical curves for a number of ceramics and metals are shown in Fig. 4.4.

SURFACE ENERGIES

A variety of methods can be used to measure the surface energy of ceramics. One technique, of limited applicability (see below), is to measure the force needed to cleave a crystal by starting with an atomically sharp notch of length c. If all the mechanical energy is used to create new surfaces then the following relationship between the surface energy, Young's modulus and the applied stress at fracture σ_{app} can be derived:

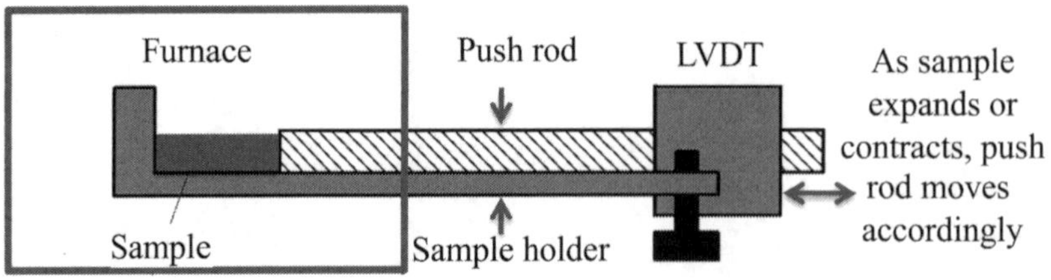

FIGURE 4.9 Schematic of a dilatometer.

$$\gamma = \frac{A'c\sigma_{app}^2}{2Y}$$

where A′ is a geometric factor that depends on the loading conditions and specimen geometry. Once σ_{app} is measured for a given c, γ can be calculated from the equation above if the modulus is known. In deriving this equation, it is implicit that all the mechanical energy supplied by the testing rig goes into creating new surfaces. Also implicit is that there are no energy-consuming mechanisms occurring at the crack tip, such as dislocation movements, i.e., the failure was a pure brittle failure. It is important to note that this condition is only satisfied for a small number of ionic and covalent ceramics, some of which are listed in Table 4.4. Fortuitously, as discussed below DFT calculations are well suited for calculating γ.

COMPUTATIONAL MATERIALS SCIENCE 4.1: ELASTIC CONSTANTS

In deriving Eq. (4.8) some crude approximations were made. A single value of Young's modulus was assumed. In reality, the elastic properties are a tensor that depend on the symmetry of the crystal; the higher the symmetry the less constants are needed. In general, it is nontrivial to measure a solid's elastic constants because, for the most part, good quality, large, single crystals are required. Fortunately, today DFT calculations can be used to calculate the elastic constants, c_{ij}, quite accurately. This is typically carried out by first modeling the solid and then displacing the atoms in various directions and generating energy/distance curves, not unlike the one shown in Fig. 4.6. The bond stiffness in that direction is then simply obtained from Eq. (4.7).

To convert c_{ij}s, to moduli, the so-called Voigt shear modulus, G_v, is calculated assuming:

$$G_v = \frac{1}{15}(2c_{11} + c_{33} - c_{12} - 2c_{13}) + \frac{1}{5}\left(2c_{44} + \frac{1}{2}(c_{11} - c_{12})\right) \quad (4.21)$$

The corresponding Voigt bulk modulus, B_v, is given by:

$$B_v = \frac{2}{9}\left(c_{11} + c_{12} + 2c_{13} + \frac{c_{33}}{2}\right) \quad (4.22)$$

The corresponding Voigt modulus, E_v, is then calculated assuming:

$$E_v = \frac{9G_vB_v}{3B_v + G_v} \quad (4.23)$$

Table 4.5 summarizes the average c_{ij} values—calculated using DFT by various research groups—for Ti_2SC, a hexagonal MAX phase. The last three columns compare the values of E_v and G_v calculated using Eqs. (4.21) to (4.23), to those determined experimentally by measuring v_L and v_S in polycrystalline Ti_2SC samples (bottom row). From these results, it is clear that the agreement is within 10%, which is quite typical.

TABLE 4.5 Elastic constants, c_{ij} (in GPa) for Ti_2SC calculated from DFT. Also listed in the last column are the values of G_v, E_v and B_v calculated from the c_{ij}s using Eqs. (4.21) to (4.23). The last row lists the experimentally determined values using Eqs. (4.19) and (4.20)
***Source*: M. Shamma et al. *Scr. Mater.*, 65, 573–576 (2011)**

c_{11}	c_{12}	c_{13}	c_{33}	c_{44}	c_{66}	E_v	G_v	B_v
335±4	98±8	99±2	362±19	161±3	119±5	328	137	180
Experimental Values →						292	128	140

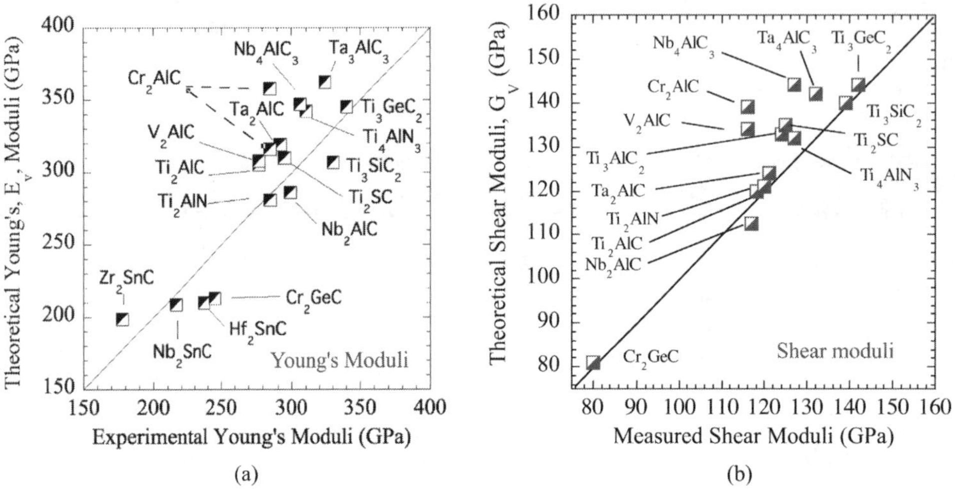

FIGURE 4.10 Comparison between theoretical and experimental values for select MAX phases. (Taken from M. W. Barsoum, *MAX Phases*, VCH-Wiley, 2013.)

More often than not, the experimental values are lower than the theoretical ones. Reasons for the discrepancies include defects in the solid such as pores and vacancies that tend to lower the measured moduli. The fact that the calculations are typically carried out at 0 K, but the experiments are carried out at room temperature is also a contributing factor in the discrepancy.

Figures 4.10 compare the measured and DFT calculated Y and G values for a number of MAX phases; with a few exceptions, the agreement is quite good.

COMPUTATIONAL MATERIALS SCIENCE 4.2: SURFACE ENERGY CALCULATIONS

As noted previously, measuring γ is nontrivial. Today, DFT calculations are routinely used to estimate γ. A typical procedure is to first calculate the energy of the atoms in the bulk per unit cell, E_{bulk}. Then the energy of a slab, E_{slab}, comprised of a relatively small number of layers (see Fig. 4.7*a*), is calculated. If the slab is far enough removed from the one adjacent to it—so the atoms in the two slabs do not interact—then the surface energy is simply given as

$$\gamma = \frac{1}{A}\left[E_{slab} - nE_{bulk}\right] \tag{4.24}$$

where A is the total surface area—top and bottom denoted by red atoms in Fig. 4.7*a*—created and n is the number of unit cells per slab. Typically, γ values obtained from DFT calculations are in eV/Å^2; the measured values in J/m^2. The conversion factor is: 1 J/m^2 = 16 eV/Å^2.

Two types of DFT calculations of surface energies are typically carried out. The faster ones simply calculate γ from Eq. (4.24), without allowing the surface atoms to reconstruct or rearrange. The second approach—which always leads to lower γs—allows the surface atoms to relax.

In monoatomic solids, both surfaces created when a crystal is cleaved are identical. In ceramics, where the cut is made can make a stark difference. To illustrate, refer back to Fig. 4.7*b* or *c*. Cleaving the NaCl crystal to create the (100) or (110) planes results in neutral or nonpolar surfaces. However, if one were to cleave and create a (111) plane, one surface would be comprised of Cl^- anions solely, the other Na^+ cations. For obvious reasons, these surfaces are called polar. Whether a surface is polar or not can affect many of its properties, least of which is the surface adsorption of molecules and reactivity during catalysis.

For ternary, and higher, compounds there is an extra wrinkle especially if they are layered. For example, in Ti_2AlC, where one makes the cut determines its surface energy. If the cut is made between the Ti and C layers, the unrelaxed γ is 6.2 J/m^2, which, when relaxed, drops to 5.3 J/m^2. If the cut is made between the Ti and Al layers, on the other hand, the relaxed and unrelaxed γs are 2 J/m^2 (Music et al. 2007).[40]

4.7 SUMMARY

1. The bond strengths between atoms or ions in solids, by and large, determine many of their properties, such as melting and boiling points, stiffnesses, thermal expansions, surface energies and theoretical strengths.
2. The stronger the bond, the higher the melting point. However, partial covalency to an ionic bond will tend to stabilize discrete units in the melt and lower the melting point.
3. Thermal expansion originates from the anharmonic vibrations of atoms in a solid. The asymmetry of the energy well is a measure of the thermal expansion coefficient α, with stronger bonds resulting in more symmetric energy wells and consequently lower α values. In addition, the atomic arrangement can play an important role in determining α.
4. As a first approximation, the curvature of the energy/distance well is a good measure of the stiffness or Young's modulus of a solid. In general, the stronger the bond, the stiffer the solid. Other factors such as atomic arrangement are also important, however.
5. The theoretical strength of a bond is on the order of Y/10. The actual strengths of ceramics, however, are much lower for reasons to be discussed in Chap. 11.

[40] Music D. et al., *Surface Science* 601 (2007) 896–899.

6. The surface energy of a solid not only scales with the bond energy but also depends on crystallographic orientation.
7. Bonds have natural frequencies of vibration that are of the order of 10^{13} Hz.

PROBLEMS

4.1. (*a*) The equilibrium interatomic spacings of the Na halides and their melting points are listed below. Explain the trend observed.

	NaF	NaCl	NaBr	NaI
Spacing, nm	0.23	0.28	0.29	0.32
Melting point, °C	988	801	740	660

(*b*) Explain the melting point trends observed for the alkali metal chlorides as one goes from HCl (–115.8°C) to CsCl.

(*c*) Which of these pairs of compounds would you expect to have the higher melting points: CaF_2 versus ZrO_2; UO_2 versus CeO_2; CaF_2 versus CaI_2? Explain.

4.2. Starting with Eq. (4.13) in text, do the following:

(*a*) Derive the following relationship:

$$S_0 = \frac{mD}{r_0^{m+2}}(n-m)$$

Using this equation, calculate S_0 for NaCl. Assume $n = 9$.

Hint: Show that for an ionic bond $D = \frac{z_1 z_2 e^2}{4\pi\varepsilon_0}$

Ans: 81 N/m.

(*b*) Derive the following expression for Young's modulus. State all assumptions.

$$Y \approx \frac{mD}{r_0^{m+3}(n-m)}$$

(*c*) Show that the distance at which the bond will break r_{fail} is given by

$$r_{fail} = \left(\frac{n+1}{m+1}\right)^{\frac{1}{n-m}} r_0$$

For ionic bonds, $m = 1$ and $n \approx 9$; for van der Waals bonds, $m = 6$ and $n = 12$. Calculate the strain at failure for each bond. Which is higher? Is that the expected result?

(*d*) Derive Eq. (4.14) in the text and show that for an ionic bond $\sigma_{fail} \approx Y/15$.

4.3. Estimate the order of magnitude of the maximum displacement of Na and Cl ions in NaCl from their equilibrium position at 300 and at 900 K.

4.4. Prove that the linear expansion coefficient, α, with very little loss in accuracy, can be assumed to be one-third that of the volume coefficient for thermal expansion α_v. You can assume that $l = l_0(1+\alpha)$ and $v = v_0(1+\alpha_v)$ and $v_0 = l_0^3$.

4.5. (*a*) "A solid for which the energy distance curve is perfectly symmetric would have a large thermal expansion coefficient." Do you agree with this statement? Explain.

(*b*) The potential energy U(x) of a pair of atoms that are displaced by x from their equilibrium position can be written as $U(x) = \alpha x^2 - \beta x^3 - \gamma x^4$, where the last two terms represent the anharmonic part of the well. At any given temperature, T, the probability of displacement occurring relative to that it will not occur is given by the Boltzmann factor $e^{-U/(kT)}$, from which it follows that the average displacement at this temperature is

$$\bar{x} = \frac{\int_{-\infty}^{\infty} x e^{-U/(kT)}}{\int_{\infty}^{\infty} e^{-U/(kT)}}$$

Show that at small displacements, the average displacement is

$$\bar{x} = \frac{3\beta kT}{4\alpha^2}$$

What does this final result imply about the effect of bond strength on thermal expansion?

4.6. (*a*) Show that for the rock salt structure $\gamma_{(111)}/\gamma_{(100)} = \sqrt{3}$.

(*b*) Calculate from first principles the surface energies of the (100) and (111) planes of MgO. How do your values compare with those shown in Table 4.4? Discuss all assumptions.

(*c*) It has been observed that NaCl crystals cleave more easily along the (100) planes than along the (110) planes. Show, using calculations, why you think that is so.

4.7. Calculate the number of broken bonds per square centimeter for Ge (which has a diamond cubic structure identical to the one shown in Fig. 3.1c except that all the atoms are identical) for the (100) and (111) surfaces. Which surface do you think has the lower surface energy? Why? The lattice constant of Ge is 0.565 nm and its density is 5.32 g/cm³.

Ans: For (100), 1.25×10^{15} bonds/cm² for (111), 0.72×10^{15} bonds/cm².

4.8. Take the C–C bond energy to be 376 kJ/mol. Calculate the surface energy of the (111) plane in diamond. Repeat for the (100) plane. Which plane do you think would cleave more easily? Information you may find useful: density of diamond is 3.51 g/cm³ and its lattice parameter is 0.356 nm.

Ans: $\gamma_{(111)} = 9.82$ J/m².

4.9. Would you expect the surface energies of noble gas solids to be greater than, about the same as, or smaller than those of ionic crystals? Explain.

4.10. Estimate the thermal expansion coefficient of alumina from Fig. 4.4 and compare with value listed in Table 4.3. Does your answer depend on the temperature range over which you carry out the calculation? Explain.

4.11. (*a*) Eq. (4.16) assumes that the CN of the cations and anions are identical. Derive a more generalized expression that can account for the fact that the CN of the cations and anions may be different.
(*b*) Show that the expression derived in a is identical to Eq. (4.16) when the CNs are equal.
(*c*) Use your expression to calculate the surface energy of the (100) plane in rutile.

ADDITIONAL READING

1. L. Van Vlack, *Elements of Materials Science and Engineering*, 5th ed., Addison-Wesley, Reading, MA, 1985.
2. N. N. Greenwood, *Ionic Crystals, Lattice Defects and Non-Stoichiometry*, Butterworth, London, 1968.
3. L. Azaroff, *Introduction to Solids*, McGraw-Hill, New York, 1960.
4. J. Huheey, *Inorganic Chemistry*, 2nd ed., Harper & Row, New York, 1978.
5. L. Solymar and D. Walsh, *Lectures on the Electrical Properties of Materials*, 4th ed., Oxford University Press, New York, 1988.
6. C. Kittel, *Introduction to Solid State Physics*, 6th ed., Wiley, New York, 1986.
7. B. H. Flowers and E. Mendoza, *Properties of Matter*, Wiley, New York, 1970.
8. A. H. Cottrell, *The Mechanical Properties of Matter*, Wiley, New York, 1964.
9. M. F. Ashby and R. H. Jones, *Engineering Materials*, 4th ed., Butterworth-Heinemann, Oxford, UK, 2011.
10. C. B. Carter and M. G. Norton, *Ceramic Materials*, 2nd ed., Springer, New York, 2013.

MULTIMEDIA REFERENCES AND DATABASES

1. Database: https://materialsproject.org/
   This free database is quite useful and easy to use. To start the following tutorials are useful:
   https://www.youtube.com/playlist?list=PLTjFYVNE7LTjHJwV994iQHS-bIkh3UXKJ.
   In addition to structures, lattice parameters, etc. one can find the electronic and band structures, elastic properties, XRD diffraction patterns phase diagrams and much more.
2. MD of melting of ice: https://www.youtube.com/watch?v=6sob_keOiOU.
3. MD of water freezing to form ice Part I: https://www.youtube.com/watch?v=gmjLXrMaFTg.
4. MD of water freezing to form ice Part II: https://www.youtube.com/watch?v=RIW65QLWsjE.
5. MD of water vaporization: https://www.youtube.com/watch?v=B3cXuisH8PI.
6. Y. Fei, *Mineral Physics and Crystallography: A Handbook of Physical Constants*, AGU Reference Shelf 2, 1995. This reference lists the thermal expansion coefficients of over 200 compounds.

5

THERMODYNAMIC AND KINETIC CONSIDERATIONS

$$S = k \ln \Omega$$

Boltzmann

5.1 INTRODUCTION

Most of the changes that occur in solids in general and ceramics in particular, especially as a result of heating or cooling, come about because they lead to a reduction in the free energy of the system. For any given temperature and pressure, kinetics permitting, every system strives to attain its lowest possible free energy. The beauty of thermodynamics lies in the fact that while it will not predict what can happen, it most certainly will predict what cannot happen. In other words, if calculations show that a certain process would increase the free energy of a system, then one can with utmost confidence dismiss that process as impossible.

Unfortunately thermodynamics, for the most part, is made confusing and very abstract. In reality, thermodynamics, while not being the easiest of subjects, is not as difficult as generally perceived. As somebody once noted, some people use thermodynamics as a drunk uses a lamppost—more for support than illumination. The purpose of this chapter is to dispel some of the mystery surrounding thermodynamics and hopefully illuminate and expose some of its beauty. It should be emphasized, however, that one chapter cannot, by any stretch of the imagination, cover a subject as complex and subtle as thermodynamics. This chapter is included more for the sake of completion. It should be more a reminder of what the reader should already be familiar with than an attempt to cover the subject in any but a cursory manner.

This chapter is structured as follows. In the next three subsections, enthalpy, entropy and free energy are defined and explained. Section 5.3 deals with the conditions of equilibrium and the corresponding mass action expressions. The chemical stability of binary compounds is discussed in Sec. 5.4. In Sec. 5.5 the concept of electrochemical potentials is presented, which is followed by the closely related notion of charged interfaces and Debye lengths. In Sec. 5.7 the Gibbs–Duhem relation for binary oxides is introduced. In the final section a few remarks are made concerning the kinetics and driving forces of various processes that occur in solids.

5.2 FREE ENERGY

If the condition for equilibrium were simply that the energy content, or enthalpy of a system, be minimized, one would be hard pressed to explain many commonly occurring phenomena, especially endothermic processes. For example, during melting, the energy content of the melt is greater than that of the solid it is replacing, and yet experience has shown that when heated to sufficiently high temperatures, most solids will melt. Gibbs was the first to appreciate that it was another function that had to be minimized before equilibrium could be achieved. This function, called the **Gibbs free-energy function**, is dealt with in this section and comprises two terms, namely, enthalpy H and entropy S.

5.2.1 ENTHALPY

When a substance absorbs a quantity of heat dq, its temperature will rise accordingly by an amount dT. The ratio of the two is the **heat capacity**, defined as

$$c = \frac{dq}{dT} \tag{5.1}$$

Since dq is not a state function, c will depend on the path. The problem can be simplified by introducing the **enthalpy function**, defined as

$$H = E + PV \tag{5.2}$$

where E, P and V are, respectively, the internal energy, pressure and volume of the system. By differentiating Eq. (5.2) and noting that, from the first law of thermodynamics, $dE = dq + dw$, where dw is the work done on the system, it follows that

$$dH = d(E + PV) = dq + dw + PdV + VdP \tag{5.3}$$

If the heat capacity measurement is carried out at constant pressure, $dP = 0$, and since by definition $dw = -P\,dV$, it follows from Eq. (5.3) that $dH = dq|_p$. In other words, the heat absorbed or released by any substance at constant pressure is a measure of its enthalpy.

From this result it follows from Eq. (5.1) that

$$c_p = \left(\frac{dq}{dT}\right)_p = \left(\frac{dH}{dT}\right)_p \tag{5.4}$$

where c_p is the heat capacity measured at constant pressure. Integrating Eq. (5.4) shows that the enthalpy content of a crystal is given by

$$H^T - H^{298}_{elem} = \int_{298}^{T} c_{p,elem}\,dT \tag{5.5}$$

Given that there is no absolute scale for energy, the best that can be done is to arbitrarily define a standard state and relate all other changes to that state—thermodynamics deals only with relative changes. Consequently, and by convention, the formation enthalpy of the *elements* in their standard state at 298 K is assumed to be zero; i.e., for any element, $H^{298} = 0$.

The heat liberated or consumed during the formation of a *compound from its elements* can be determined experimentally using calorimetry, for example. It follows that the enthalpies of formation of many compounds at 298 K, denoted by ΔH^{298}_{form}, are known and tabulated. At any temperature other than 298 K, the heat content of a compound ΔH^T is given by

$$\Delta H^T = \Delta H^{298}_{form} + \int_{298}^{T} c_{p,comp}\,dT \tag{5.6}$$

Finally, it is worth noting that the heat capacity data are often expressed in the empirical form

$$c_p = A + BT + \frac{C}{T^2}$$

and the heat content of a solid at any temperature is thus simply determined by substituting this expression in (5.5) or (5.6) and integrating.

Worked Example 5.1

The c_p of Al is given by: $c_p = 20.7 + 0.0124T$ in the 298 to 932 K temperature range and that of Al_2O_3 is given by: $c_p = 106.6 + 0.0178T - 2{,}850{,}000T^{-2}$ in the 298 to 1800 K temperature range. At 298 K the enthalpy of formation of Al_2O_3 from its elements is −1675.7 kJ/mol, calculate the enthalpy content of Al and Al_2O_3 at 298 and 900 K.

ANSWER

Enthalpy content of Al at 298 is zero by definition. Heat content of Al at 900 K is thus

$$H_{Al}^{900} - H_{Al}^{298} = H_{Al}^{900} = \int_{298}^{900} (20.7 + 0.0124T)\,dT = 16.93 \text{ kJ/mol}$$

At 298 K, the heat content of Al_2O_3 is simply its enthalpy of formation from its elements, or −1675.7 kJ/mol. At 900 K,

$$H_{Al_2O_3}^{900} - H_{Al_2O_3}^{298} = \Delta H_{form}^{298} + \int_{298}^{900} (106.6 + 0.0178\,T - 2{,}850{,}000\,T^{-2})dT$$

$$H_{Al_2O_3}^{900} = -1675.7 + 64.2 = -1611.1 \text{ kJ/mol}$$

5.2.2 ENTROPY

Disorder constitutes entropy, macroscopically defined as

$$dS = \frac{dq_{rev}}{T} \tag{5.7}$$

where q_{rev} is the heat absorbed in a reversible process. Boltzmann, in one of his most brilliant insights, related entropy to the microscopic domain by the following expression:[41]

$$\boxed{S = k \ln \Omega_\beta} \tag{5.8}$$

where k is Boltzmann's constant and Ω_β is the total number of different configurations in which the system can be arranged at constant energy. There are several forms of entropy, which include:

- ∞ Configurational, where the entropy is related to the number of configurations in which the various atoms and/or defects can be arranged on a given number of lattice sites.
- ∞ Thermal, where Ω_β is the number of possible different configurations in which the particles (e.g., atoms or ions) can be arranged over existing energy levels.
- ∞ Electronic.
- ∞ Other forms of entropy, such as that arising from the randomization of magnetic or dielectric moments.

Each will be discussed in some detail in the following sections.

[41] This expression is inscribed on Boltzmann's tomb in Vienna.

5.2.2.1 CONFIGURATIONAL ENTROPY

This contribution refers to the entropy associated with atomic disorder. To illustrate, let's consider the entropy associated with the formation of n point defects or vacancies (see Chap. 6 for more details). Combinatorially, it can be shown that the number of ways of distributing n vacant sites and N atoms on n + N sites is given by:[42]

$$\Omega_\beta = \frac{(n+N)!}{n!N!} \tag{5.9}$$

When Eq. (5.9) is substituted in Eq. (5.8) and Stirling's approximation[43] is applied, which for large x reduces to

$$\ln x! \approx x \ln x - x$$

the following expression for the configuration entropy

$$S_{config} = -k\left(N \ln \frac{N}{N+n} + n \ln \frac{n}{n+N}\right) \tag{5.10}$$

is obtained (see Prob. 5.1).

It is worth noting here that a similar expression, namely,

$$S_{config} = -R(x_A \ln x_A + x_B \ln x_B) \tag{5.11}$$

results for the mixing of two solids, A and B, that form an ideal solution. Here x_A and x_B are the mole fractions of A and B, respectively. R is the universal gas constant, where $R = kN_{Av}$.

Worked Example 5.2

(*a*) Calculate the total number of possible configurations for eight atoms and one vacancy. Draw the various configurations. (*b*) Calculate the entropy change when introducing 1×10^{18} vacancies in a mole of a perfect crystal. Does the entropy increase or decrease?

ANSWER

(*a*) Substituting N = 8 and n = 1 in Eq. (5.9) yields $\Omega_\beta = 9$. The possible configurations are depicted in Fig. 5.1.

(*b*) Applying Eq. (5.10) gives

[42] See, e.g., C. Newey and G. Weaver, eds., *Materials Principles and Practice*, Butterworth, London, 1990, p. 212.

[43] $\ln x! \cong x \ln x - x + 1/2 \ln 2\pi x$.

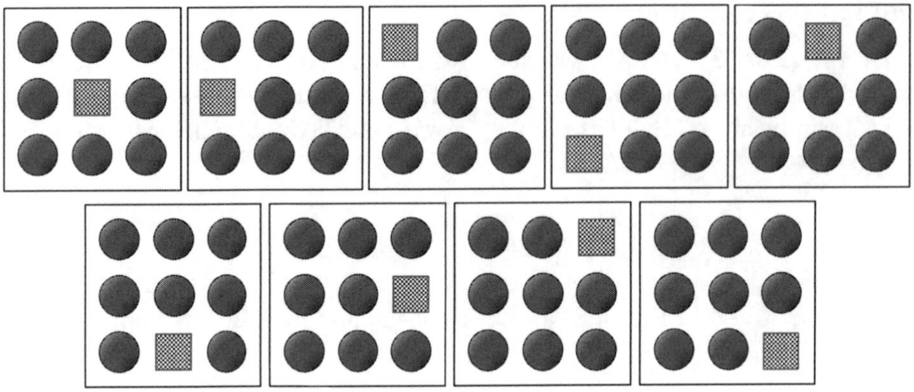

FIGURE 5.1 Various configurations for arranging eight atoms (circles) and one vacancy (square). Note that an identical picture would have emerged had the circles been A atoms and the squares B atoms.

$$\Delta S_{config} = (-1.38 \times 10^{-23}) \times \left(6.02 \times 10^{23} \ln \frac{6.02 \times 10^{23}}{6.02 \times 10^{23} + 10^{18}} + 10^{18} \ln \frac{10^{18}}{6.02 \times 10^{23} + 10^{18}} \right) = 0.0002 \, J/K$$

Since ΔS_{config} is positive, it follows that the entropy of the defective crystal that is higher than that of the perfect one. In other words, introducing vacancies increases S_{config} of the crystal.

5.2.2.2 THERMAL ENTROPY

As the atoms or ions vibrate in a solid, the uncertainty in the exact value of their energy constitutes thermal entropy, S_T. Combining (5.4) and (5.7) it follows that:

$$dS_T \equiv \frac{dq_{rev}}{T} = \frac{c_p}{T} dT$$

from which it directly follows that for any substance[44]

$$\Delta S_T = \int_0^T \frac{c_p}{T} dT \tag{5.12}$$

[44] In contrast to energy, one can assign an absolute value for entropy if it is postulated that the entropy of a perfect (i.e., defect-free) solid goes to zero at absolute zero (third law of thermodynamics). One of the implications of the third law is that every substance has a certain amount of "S" associated with it at any given temperature above absolute 0 K.

Microscopically, to understand the concept of thermal entropy, or heat capacity for that matter, one needs to appreciate that the vibrational energy levels of atoms in a crystal are quantized. If the atoms are assumed to behave as **simple harmonic oscillators**, i.e., miniature springs (see Chap. 4), it can be shown that their energy will be quantized, with a spacing between energy levels given by

$$\varepsilon = \left(n + \frac{1}{2}\right)h\nu \quad \text{where } n = 0,1,2,\ldots \tag{5.13}$$

where h, n and ν are, respectively, Planck's constant, an integer and the bond's characteristic vibration frequency, ν_0. Combining Eqs. (4.18) and (5.13), it is straightforward to show that

$$\varepsilon = \left(n + \frac{1}{2}\right)\frac{h}{2\pi}\sqrt{\frac{S_0}{M_{red}}} \tag{5.14}$$

where S_0[45] is the bond stiffness given by Eq. (4.7). From Eq. (5.14) it is obvious that the spacing between energy levels, $\Delta\varepsilon$, for solids, (i) with strong bonds—i.e., high S_0—will be greater than those for weakly bonded ones, and (ii) comprised of larger heavier atoms will be small as compared to those with lighter elements. Since it is well established that the natural frequency of vibration of a mass on a spring is higher the stiffer the spring and the lighter the mass (Eq. 4.18), and vice versa, none of this should be in the least bit surprising.

At absolute zero, the atoms populate the lowest energy levels available, and only one configuration exists. Upon heating, however, the probability of exciting atoms to higher energy levels increases, which in turn increases the number of possible configurations of the system—which is another way of saying that S_T has increased.

To proceed with the discussion, the main result of one of the simpler models, namely, the **Einstein solid**, are given below without proof.[46] As discussed in Computational Materials Science 5.1, the actual situation is more complicated.

By assuming the solid to consist of Avogadro's number N_{Av} of independent harmonic oscillators, all oscillating with the same frequency v_e at all temperatures, Einstein showed that the thermal entropy, S_T, per mole is given by

$$S_T = 3N_{Av}k\left[\frac{h\nu_e}{kT(e^{h\nu_e/kT} - 1)} - \ln(1 - e^{-h\nu_e/kT})\right] \tag{5.15}$$

For temperatures $kT >> h\nu_e (e^x \approx 1 + x)$, Eq. (5.15) simplifies to

$$\boxed{S_T = 3R\left(\ln\frac{kT}{h\nu_e} + 1\right)} \tag{5.16}$$

45 S_0 is not to be confused with entropy, S.

46 For more details see, e.g., K. Denbigh, The *Principles of Chemical Equilibrium*, 4th ed., Cambridge University Press, New York, 1981, Chap. 13.

Based on this important result,[47] the following conclusions concerning S_T can be reached:

1. S_T is a monotonically increasing function of T. Said otherwise, S_T increases as T increases. This comes about because as T is raised, atoms can populate higher and higher energy levels. The uncertainty of distributing these atoms—among the larger number of *accessible* energy levels—constitutes *thermal* entropy, S_T.
2. S_T decreases with increasing ν_e. Given that v_e scales with S_0 [Eq. (4.18)], which in turn scales with the bond strength, it follows that *for a given temperature, a solid with weaker bonds will have a higher* S_T. The reason is simple: If the bonds are strong, that is, if S_0 is large, then the spacing between energy levels [Eq. (4.3)] will also be large, and for a given ΔT increase, only a few levels are accessible and S_T will be low. In a weakly bound solid, on the other hand, for the same ΔT, many more levels are accessible and the uncertainty increases. As discussed in greater detail later, this conclusion is important when one is dealing with temperature-induced polymorphic transformations since they tend to occur in the direction of increased S_T. In other words, polymorphic transformations will tend to occur from phases of higher cohesive energy (e.g., close-packed structures) to those of lower cohesive energies (more open structures).

Another implication of Eq. (5.16) is that if the vibrational frequency of the atoms changes from, say, a frequency ν to ν', as for example a result of a phase transformation or the formation of defects (see Chap. 6), then the associated entropy change is

$$\Delta S_T^{trans} = 3R\ln\left(\frac{\nu}{\nu'}\right) \quad (5.17)$$

Note that if $\nu > \nu'$, ΔS_T^{trans} will be positive and vice versa.

Worked Example 5.3

(*a*) Sketch the various possible configurations for three particles distributed over three energy levels subject to the constraint that the system's total energy is constant at 3 units.

(*b*) By defining the Einstein characteristic temperature to be $\theta_e = h\nu_e/k$, it can be shown that

$$c_v = 3N_{Av}k\left(\frac{\theta_e}{T}\right)^2\left[\frac{e^{\theta_e/T}}{(e^{\theta_e/T}-1)^2}\right]$$

where c_v is the molar heat capacity at constant volume. Usually θ_e is determined by choosing the value that best fits the experimental c_p vs. T data. For KCl, $\theta_e \approx 230$ K. Estimate ν_e for KCl from the c_p data and compare your result with that calculated based on Eq. (4.7). Assume the Born exponent n = 9 for KCl. Atomic weight of Cl is 35.5 g/mol and that of K is 39.1 g/mol.

[47] The more accurate Debye model that assumes a distribution of frequencies rather than a single frequency yields virtually the same result at higher temperatures. See Computational Materials Science 5.1 for more details.

ANSWER

(*a*) The various configurations—totaling 10—are shown in Fig. 5.2.

(*b*) The interatomic distance for KCl is 319 pm. It follows that for KCl (see Prob. 4.2)

$$S_0 = \frac{z_1 z_2 e^2}{4\pi\varepsilon_0 r_0^3}(n-m) = 56.7\,\frac{N}{m}$$

$$M_{red} = \frac{m_1 m_2}{(m_1 + m_2)} = \frac{18.6}{1000 N_{Av}} = 3.1\times 10^{-26}\ \text{kg}$$

Applying Eq. (4.7):

$$\nu = \frac{1}{2\pi}\sqrt{\frac{S_0}{M_{red}}} = \frac{1}{2\pi}\sqrt{\frac{56.7}{3.1\times 10^{-26}}} = 6.8\times 10^{12}\ s^{-1}$$

Since $\theta_e = h\nu_e/k = 230$ K, then $\nu_e = 4.8\times 10^{12}\ s^{-1}$.

Considering the many simplifying assumptions made to arrive at Eq. (4.7), the agreement has to be considered quite good. Note these results confirm that ions in solids vibrate at frequencies on the order of 10^{12} to $10^{13}\ s^{-1}$.

5.2.2.3 ELECTRONIC ENTROPY

In the same manner as the randomization of atoms over available energy levels constitutes entropy, the same can be said about the distribution of electrons over their energy levels. At 0 K, electrons and holes in semiconductors and insulators are in their lowest energy state, and only one configuration exists. As the temperature is raised, however, they are excited to higher energy levels, and the uncertainty of finding them in any number of excited energy levels constitutes a form of entropy. This point will be dealt with in greater detail in Chaps. 6 and 7.

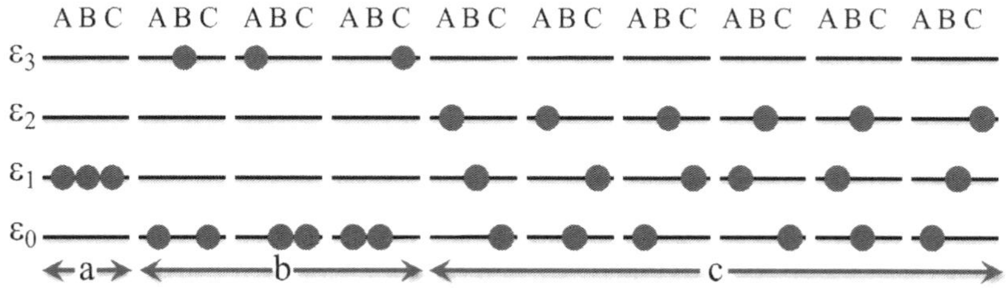

FIGURE 5.2 Possible configurations of arranging three particles in a system with a total energy of 3 units. Here it is assumed that $\varepsilon_0 = 0$, $\varepsilon_1 = 1$ unit, $\varepsilon_2 = 2$ units, etc. (See D. Gaskell, *Introduction to Metallurgical Thermodynamics*, 2nd ed., Hemisphere, New York, 1981, Chap. 4, for more details.)

5.2.2.4 OTHER FORMS OF ENTROPY

Some elements and compounds have magnetic or dielectric moments. These moments can be randomly oriented, or they may be ordered. For example, when they are ordered, the magnetic entropy is zero since there is only one configuration. As the temperature is increased, however, the entropy increases as the number of possible configurations increases. The same argument can be made for dielectric moments (see Chap. 14).

5.2.2.5 TOTAL ENTROPY

Since entropies are additive, it follows that the total entropy of a system is given by

$$S_{tot} = S_{config} + S_T + S_{elec} + S_{other} \tag{5.18}$$

5.2.3 FREE ENERGY, CHEMICAL POTENTIALS AND EQUILIBRIUM

As noted at the outset of this section, the function that defines equilibrium, or lack thereof, is neither enthalpy nor entropy, but rather the free-energy function G, defined by Gibbs as

$$G = H - TS \tag{5.19}$$

It follows that the free energy changes occurring during any reaction or transformation are given by

$$\Delta G = \Delta H - T\Delta S \tag{5.20}$$

where ΔS now includes all forms of entropy changes.

Furthermore, it can be shown that at equilibrium $\Delta G = 0$. To illustrate, consider changes occurring in a system as a function of a given reaction variable ξ that affects its free energy as shown schematically in Fig. 5.3. The variable ξ can be the number of vacancies in a solid, the number of atoms in the gas phase, the extent of a reaction, the number of nuclei in a supercooled liquid, etc. As long as $\xi \neq \xi_0$ (Fig. 5.3), then $\Delta G \neq 0$ and the reaction will proceed. When $\xi = \xi_0$, ΔG is at a minimum and the system is said to be in equilibrium, since the driving force for change vanishes. The equilibrium condition can thus be simply stated as

$$\boxed{\Delta G\big|_{P,T,n_i} = \frac{\Delta G}{\Delta \xi} = 0} \tag{5.21}$$

Equation (5.21), despite its apparent simplicity, is an extremely powerful relationship because once the free energy of a system is formulated as a function of ξ, the state of equilibrium can simply be determined by differentiation of that function[48] (i.e., locating the minimum). In Chap. 6 how this simple but powerful method is used to determine the equilibrium number of vacancies in a solid is developed. It is important

[48] Needless to say, the real difficulty does not lie in determining the location of the minimum—that is the easy part. The hard part is determining the relationship between G and the reaction variable—therein lies the challenge.

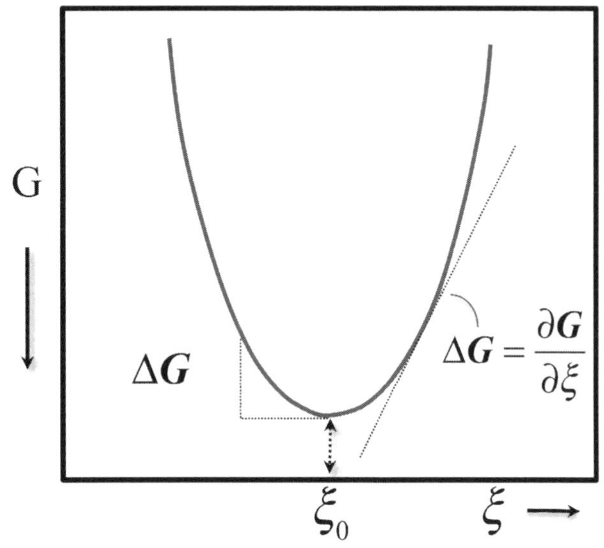

FIGURE 5.3 Schematic of free-energy versus reaction coordinate curve. At $\xi = \xi_0$, that is, when $\Delta G = 0$, the system is said to be in equilibrium.

to emphasize that this equilibrium condition is valid only when the changes are occurring at constant temperature and pressure.

Free-energy change, ΔG, is an extensive property, i.e., it depends on the size of the system. If normalized, however, to a per-mole or per-atom basis, it becomes known as the **chemical potential**. The formal definition of the chemical potential of species, i, is

$$\mu_i = \left.\frac{\partial G}{\partial n_i}\right|_{P,T,j} \tag{5.22}$$

The *chemical potential thus defined is the work that would be required to remove an atom from the bulk of an uncharged solid to infinity at constant pressure and temperature, while keeping all other chemical components, j, in the system fixed.*

Once again, as in the case of enthalpy, since one is dealing with energy, there are no absolute values. To circumvent this problem, the **standard chemical potential** *of a pure element or compound* μ_i° is defined, and all changes that occur in a system are then referred to that standard state.[49]

To take into account the fact that an element or compound is not in its standard state, the concept of activity has to been introduced. Mathematically μ_i can be described by

$$\mu_i = \mu_i^\circ + RT \ln a_i \tag{5.23}$$

[49] This value is unknown. This is not a problem, however, because what is of interest is the difference: $\mu_i - \mu_o$.

where a_i is the activity of that species, which can be further defined by

$$a_i = \gamma_i X_i \tag{5.24}$$

where X_i and γ_i are the mole fraction and activity coefficient, respectively. It follows directly from the definition of the standard state that a_i of a pure element in its standard state is 1, in which case not surprisingly $\mu_i = \mu_i^\circ$.

The activity coefficient is generally a function of composition. However, if a solution is dilute enough such that the solute atoms do not interact with each other, the activity coefficient can be assumed to be constant and

$$a_i = \gamma_i^\circ X_i \tag{5.25}$$

where γ_i° is known as the *Henrian activity coefficient*, which is not a function of composition. It also follows that by definition for an ideal solution, $\gamma_i = 1$ and the activity is simply equal to the mole fraction. This simplifying assumption is made throughout this book.

EXPERIMENTAL DETAILS: MEASURING ACTIVITIES

Whereas it is possible to define activities mathematically by using Eq. (5.23), it is only when it is appreciated how a_i is measured that a better understanding of that concept emerges. There are several ways a_i can be measured; the most tangible entails measuring the partial pressure P_i of the species for which the activity is to be determined and comparing that value to the partial pressure of the same species when it is in its pure standard state. The activity is then related to the partial pressures by[50]

$$a_i = \frac{P_i}{P^\circ} \tag{5.26}$$

where P° is the partial pressure of the same species i in its standard state, i.e., pure. Note that for gases P° is taken to be 1 atm or 0.1 MPa.

To further illustrate, consider the following thought experiment. Take an element M, place it in an evacuated and sealed container, and heat the system to a given temperature until equilibrium is attained, then measure the pressure of the gas atoms in the container. By definition, the measured quantity represents P_M° of pure M. This value, which is solely a function of temperature, is well documented and can be looked up.

By proceeding further with the thought experiment and alloying M with a second element N, such that the molar ratio is, say, 50:50 and repeating the aforementioned experiment, one of the following three outcomes will occur:

[50] This can be easily seen by noting that the work done in transferring one mole of atoms from a region where the pressure is P_i to one where the pressure is P° is simply $\Delta\mu = RT \ln(P_i/P^\circ)$. This work has to be identical to the energy change for the reaction $M_{pure} \Rightarrow M_{alloy}$ for which $\Delta\mu = RT \ln(a_i/1)$. This is essentially how Eq. (5.23) is obtained.

1. The fraction of M atoms in the gas phase is equal to their fraction in the alloy, or 0.5, in which case the solution is termed *ideal* and $a_i = P_i/P° = 0.5 = X_i$ and $\gamma_i = 1$.
2. The fraction of M atoms in the gas phase is less than 0.5. So $a_i = P_i/P° < 0.5$, hence $\gamma_i < 1$. This is termed *negative deviation* from ideality and implies that the M atoms prefer being in the solid, or melt, to being in the gas phase relative to the ideal mixture.
3. The fraction of M atoms in the gas phase is greater than 0.5. So $a_i = P_i/P° > 0.5$ and $\gamma_i > 1$. This is termed *positive deviation* from ideality and implies that the M atoms prefer being in the gas phase relative to the ideal mixture.

Thus, by measuring the partial pressure of an element, or a compound, in its pure state and by repeating the measurement with the element or compound combined with some other material, its activity can be calculated.

5.3 CHEMICAL EQUILIBRIUM AND THE MASS ACTION EXPRESSION

Consider the reaction

$$M(s) + \tfrac{1}{2}X_2(g) \Rightarrow MX(s) \quad \Delta G_{rxn} \tag{I}$$

where ΔG_{rxn} represents the free-energy change associated with this reaction. Clearly ΔG_{rxn} will depend on the state of the reactants. For instance, one would expect ΔG_{rxn} to be greater if the partial pressure of X_2 were 1 atm than if it were lower, and vice versa.

Mathematically, this is taken into account by appreciating that the driving force for any reaction is composed of two terms: The first is how likely one expects the reaction to occur under *standard conditions* and the second factor takes into account the fact that the reactants may or may not be in their standard states. In other words, it can be shown (App. 5A) that the driving force ΔG_{rxn} for any reaction is given by

$$\Delta G_{rxn} = \Delta G^{\circ}_{rxn} + RT \ln K \tag{5.27}$$

where ΔG°_{rxn} is the free-energy change associated with the reaction when the reactants are in their standard state. K is known as the **equilibrium constant** of the reaction. For reaction (I),

$$\boxed{K = \frac{a_{MX}}{a_M (P_{X_2})^{1/2}}} \tag{5.28}$$

where a_{MX}, a_M and P_{X_2} are, respectively, the activities of MX and M, and the partial pressure of X_2 at any time during the reaction. Equation (5.28) is also known as the **mass action expression** for reaction (I).

At equilibrium, $\Delta G_{rxn} = 0$, and Eq. (5.27) simplifies to the well-known result

$$\boxed{\Delta G^{\circ}_{rxn} = -RT \ln K_{eq}} \tag{5.29}$$

At equilibrium, $K = K_{eq} = \exp-[\Delta G^{\circ}_{rxn}/RT)]$.

Before one proceeds further, it is instructive to dwell briefly on the ramifications of Eq. (5.27). First, this equation says that if the reactants and products are in their standard state,[51] that is, $P_{X_2} = a_M = a_{MX} = 1$, then $K = 1$ and $\Delta G_{rxn} = \Delta G^\circ_{rxn}$, which is how ΔG°_{rxn} was defined in the first place! The other extreme occurs when the driving force for the reaction is zero, that is, $\Delta G_{rxn} = 0$, which by definition is the equilibrium state, in which case Eq. (5.29) applies.

For the generalized reaction

$$aA + bB \Rightarrow cC + dD$$

the equilibrium constant is given by

$$\boxed{K = \frac{a_C^c a_D^d}{a_A^a a_B^b}} \tag{5.30}$$

where the a_i values represent the activities of the various species raised to their respective stoichiometric coefficients a, b, c, etc.

Armed with these important relationships, it is now possible to tackle the next important topic, namely, the delineation of the chemical stability domains of ceramic compounds.

5.4 CHEMICAL STABILITY DOMAINS

A compound's chemical stability domain represents the range of activity or gaseous partial pressure over which that compound is stable. For example, under sufficiently reducing conditions, all oxides are unstable and are reducible to their parent metals. Conversely, all metals, with the notable exception of the noble ones, are unstable in air—their oxides are more stable. From a practical point of view, it is important to be able to predict the stability, or lack thereof, of a ceramic in a given environment. A related question, whose answer is critical for the successful reduction of ores, is: At what oxygen partial pressure, P_{O_2}, will an oxide no longer be stable?

To illustrate, consider an oxide MO_z, for which a higher oxide MO_y also exists (that is, $y > z$) and let us calculate its stability domain. The equilibrium partial pressure of the oxide that is in equilibrium with the parent metal is determined by applying Eq. (5.29) to the following reaction:

$$\frac{z}{2} O_2 + M \leftrightarrow MO_z \quad \Delta G_f^I \tag{II}$$

or

$$\ln P_{O_2} = +\frac{2\Delta G_f^I}{zRT} \tag{5.31}$$

[51] As noted above, the standard state of a gas is chosen to be the state of 1 mol of pure gas at 1 atm (0.1 MPa) pressure and the temperature of interest. One should thus realize that whenever a partial pressure P_i appears in an expression such as Eq. (5.28), it is implicit that one is dealing with the *dimensionless ratio*, P_i/1 atm. Said otherwise, if the mass action expression is dimensionless, then it is implicit that the partials pressures are all given in atmospheres.

Further oxidation of MO_z to MO_y occurs by the following reaction:

$$O_2 + \frac{2}{y-z}MO_z \leftrightarrow \frac{2}{y-z}MO_y \quad \Delta G_f^{II} \tag{III}$$

and the corresponding equilibrium oxygen partial pressure is given by

$$\ln P_{O_2} = \frac{\Delta G_f^{II}}{RT} \tag{5.32}$$

where

$$\Delta G_f^{II} = \frac{2}{y-z}\left[\Delta G_{f,MO_y} - G_f^{I}\right]$$

It follows that the P_{O_2} regime over which MO_z is stable is bounded by the values obtained from Eqs. (5.31) and (5.32). The following worked example should clarify the concept. Needless to say, carrying out the type of calculations just described would be impossible without a knowledge of the temperature dependence of the standard free energies of formation of the oxides involved. Figure 5.4 plots such data for a number of binary oxides.

Worked Example 5.4

Calculate the chemical stability domains for the phases in the Fe–O system at 1000 K, given the following standard free energies of formation:[52]

$$\Delta G^{\circ}_{FeO} @1000\,K = -206.95\,kJ\,mol$$

$$\Delta G^{\circ}_{Fe_3O_4} = -792.6\,kJ\,mol$$

$$\Delta G^{\circ}_{Fe_2O_3} = -561.8\,kJ\,mol$$

ANSWER

At equilibrium between Fe and FeO, the pertinent reaction is:

$$Fe + \tfrac{1}{2}O_2 \Rightarrow FeO \quad \Delta G_{FeO}$$

Applying Eq. (5.31) and solving for the equilibrium P_{O_2} at 1000 K yields 2.4×10^{-22} atm.

As the P_{O_2} is further increased, Fe_3O_4 becomes the stable phase according to[53]

$$3FeO + \tfrac{1}{2}O_2 \Rightarrow Fe_3O_4 \quad \Delta G_{r,1} = \Delta G_{Fe_3O_4} - 3\Delta G_{FeO}$$

---

52 One of the more comprehensive and reliable sources of thermodynamic data is the JANAF thermochemical tables.

53 The stoichiometry of the phases of interest can be read easily from the pertinent phase diagram.

Once again solving for P_{O_2} gives 1.14×10^{-18} atm.

Similarly, Fe_3O_4 is stable up to a P_{O_2} given by its equilibrium with Fe_2O_3, or

$$\tfrac{2}{3}Fe_3O_4 + \tfrac{1}{6}O_2 \Rightarrow Fe_2O_3 \quad \Delta G_{r,2} = \Delta G_{Fe_2O_3} - \tfrac{2}{3}\Delta G_{Fe_3O_4}$$

Solving for P_{O_2} yields 3.4×10^{-11} atm.

To summarize at 1000 K: Below a P_{O_2} of 2.4×10^{-22} atm, Fe is the stable phase, between 2.4×10^{-22} and 1.14×10^{-18}, FeO is stable; Fe_3O_4 is stable between 1.14×10^{-18} and 3.4×10^{-11}. At P_{O_2}s greater than 3.4×10^{-11} atm, Fe_2O_3 is the stable phase up to 1 atm (see Fig. 6.8*c* for a graphical representation of these results as a function of temperature).

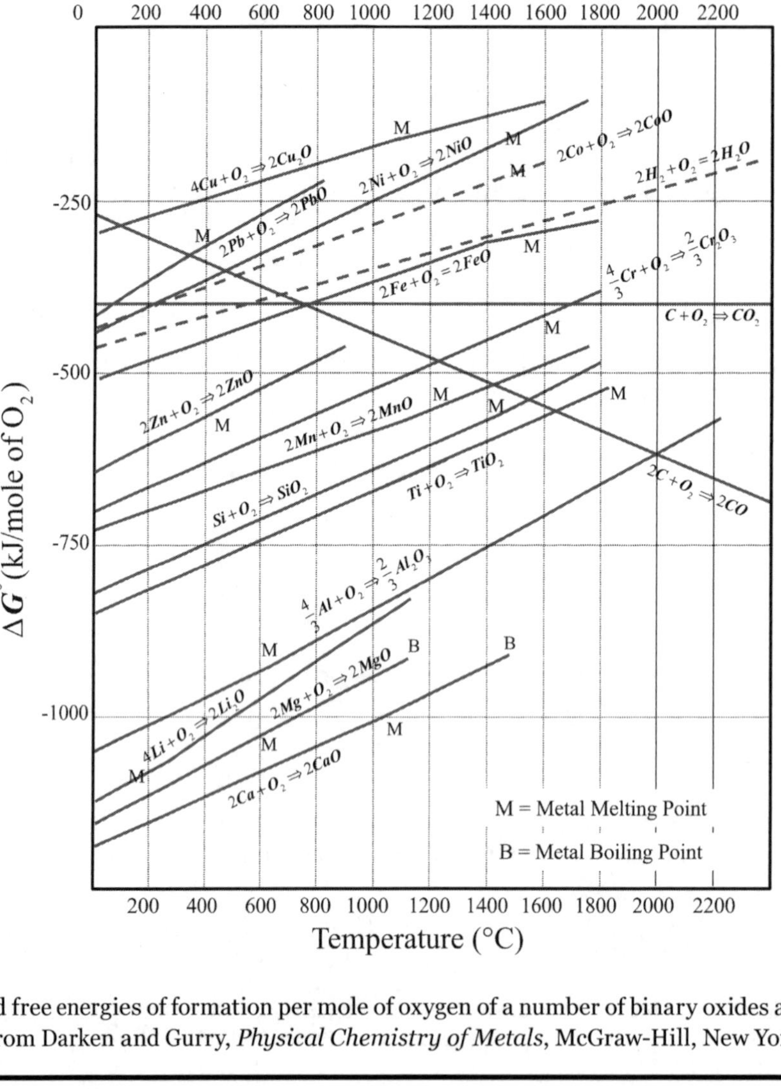

FIGURE 5.4 Standard free energies of formation per mole of oxygen of a number of binary oxides as a function of temperature. (Adapted from Darken and Gurry, *Physical Chemistry of Metals*, McGraw-Hill, New York, 1953.)

5.5 ELECTROCHEMICAL POTENTIALS

In the previous section, the chemical potential of species i in a given phase was defined as the work needed to bring a mole of that species from infinity into the bulk of that phase. This concept is of limited validity for ceramics however, since it only applies to *neutral* specie or uncharged media, where, in either case, the electric work is zero. Clearly, the charged nature of ionic ceramics and ions in solutions renders that definition invalid. Instead the pertinent function that is applicable in this case is the **electrochemical potential** η_i defined for a particle of net charge z_i by:

$$\boxed{\eta_i = \frac{\mu_i}{N_{Av}} + z_i e\phi} \tag{5.33}$$

where μ_i is the chemical potential per mole and ϕ is the electric potential. On a molar basis, this expression reads

$$\boxed{\eta_i^{molar} = \mu_i + z_i F\phi} \tag{5.34}$$

where F is Faraday's constant ($F = N_{Av}e = 96{,}500$ C/equivalent). In other words, Eq. (5.33) states that η_i is the sum of the *chemical and electrical* work needed to bring a particle of charge z_ie from infinity to that phase. If z_i were zero, the electrochemical and chemical potentials would be identical, or $\eta_i = \mu_i$, which is the case for metals and other electronically conducting materials. This conclusion is also valid when one is dealing with the insertion or the removal of a *neutral* species from charged media, such as ionic ceramics or liquid electrolytes.[54] The fundamental problem in dealing with ionic ceramics arises, however, if the problem involves *charged* species. In that case, one has to grapple with η_i.

It can be shown (see Chap. 7) that the driving force on a charged species is the gradient in its electrochemical potential. It follows directly that the condition for equilibrium for any given species i is that all gradient vanishes, i.e., when

$$\boxed{\frac{d\eta_i}{dx} = 0} \tag{5.35}$$

[54] An interesting ramification of this statement is that it is impossible to measure the activities or chemical potentials of individual *ions* in a compound, for the simple reason that it is impossible to indefinitely add or remove only one type of ion, without having a charge buildup. For example, if one starts removing cations from an MX compound, it will very quickly acquire a net negative charge that will render removing further ions more and more difficult. In other words, because it is impossible to measure the "partial pressure"' of, say, Na *ions* above an NaCl crystal, it follows that it is impossible to measure their activity. Interestingly enough, it is, in principle, possible to measure the partial pressure of Na metal, Cl_2 gas or NaCl vapor over an NaCl crystal. Said otherwise, it is only possible to measure the activity of **neutral** entities. This problem is by no means restricted to ionic solids. The problem was historically first looked at in liquid electrolytic solutions. For an excellent exposition of this problem, see J. Bockris and A. K. N. Reddy, *Modern Electrochemistry*, vol. 2, Plenum, New York, 1970, Chap. 7.

In other words, *at equilibrium the electrochemical potential gradient of every species everywhere must vanish.* It follows that for charged species the condition for equilibrium occurs *not* when $d\mu = 0$, but rather when $d\eta = 0$.

The astute reader may argue at this point that since the bulk of any material has to be neutral, it follows that ϕ is constant across that material and therefore the electric work is also a constant that could be included in μ^o, for instance. The fundamental problem with this approach, however, is that in order to insert a charged particle into a given phase, an interface has to be crossed. It follows that if that interface is charged with respect to the bulk, the electric work cannot be neglected.

5.6 CHARGED INTERFACES, DOUBLE LAYERS AND DEBYE LENGTHS

The next pertinent question is, are interfaces charged, and if so, why? The answer to the first part is simple: almost all interfaces and surfaces are indeed charged. The answer to the second part is more complicated; it depends on the type of interface, class of material, etc., and is clearly beyond the scope of this book. However, to illustrate the concept the following idealized and simplified thought experiment is useful. Consider the bulk of an MO oxide depicted in Fig. 5.5*a*. Focus on the central ion. It is obvious that this ion is being tugged at equally in all directions. Imagine, now, that the crystal is sliced in two such that an interface is created in the near vicinity of the aforementioned ion. This cutting process bares two surfaces and causes an imbalance of the forces acting on ions that are near the surface, depicted in Fig. 5.5*b*.

This force asymmetry in turn induces the ion to migrate one way or another. If it is further assumed, again for the sake of simplicity, that in this case the O ions are immobile and that the driving force is such as to induce the M ions to migrate to the surface, then it follows that the interface or surface will now be positively charged with respect to the bulk—a charge that has to be balanced by a negative one in the bulk. For a pure MO compound,[55] this is accomplished automatically, because as the ions migrate to the surface, the vacancies that are left behind are negatively charged (see Chap. 6). The formation of a surface sheet of charge that is balanced by a concentration of oppositely charged bulk entities constitutes a **double layer** (Fig. 5.5*c*).

For reasons that will become apparent in Chap. 7 (namely, diffusion), it can be shown that the compensating charges to the interfacial ones i.e., the cation vacancies in this case, are not concentrated in a plane but rather are diffusely distributed in the bulk of the solid, as shown in Fig. 5.5*c* and *d*. One measure of the thickness of this so-called double layer, also known as the **Debye length**, λ, is given by[56]

$$\lambda = \left(\frac{e^2 n_i z_i^2}{k' \varepsilon_0 kT} \right)^{-1/2} \tag{5.36}$$

where z_i and n_i are the charge and number density (particles per cubic meter) of the defects in the *bulk* of the material and ε_0 and k' are, respectively, the permittivity of free space and the relative dielectric constant of the solvent (see Chap. 14). All other symbols have their usual meaning. The distance at which the

55 Most oxides contain impurities that, in an attempt to reduce the system's strain energy, tend to migrate to the interfaces, grain boundaries and surfaces. It is usually the segregation of these impurities that is responsible for the surface charges. This charge is usually compensated, however, with bulk ionic defects (see Chap. 6).

56 See, for example, J. Bockris and A. K. N. Reddy, *Modern Electrochemistry,* Plenum, New York, 1970.

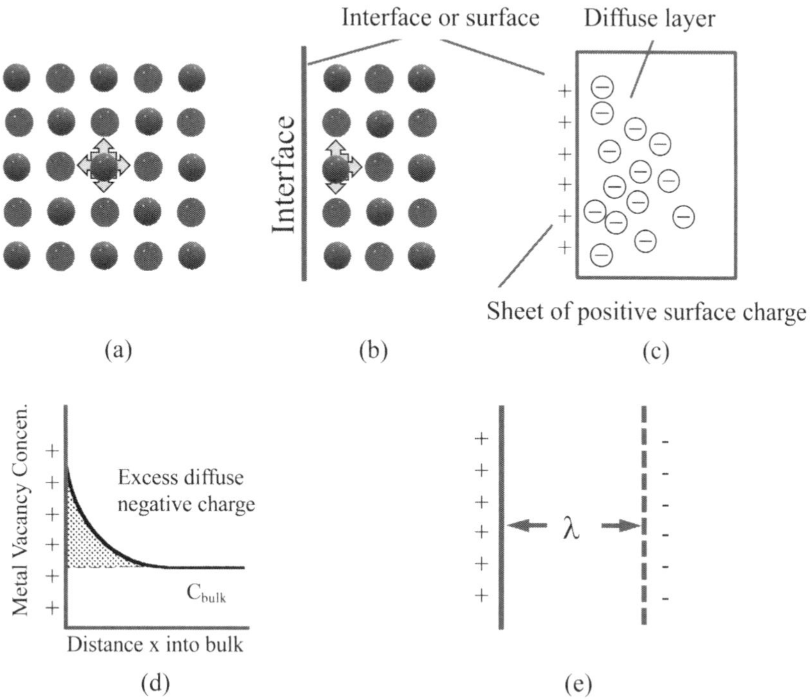

FIGURE 5.5 (*a*) An ion in bulk is subjected to a symmetric force. (*b*) Near an interface the forces are no longer symmetric, and the ions migrate one way or another. (*c*) Schematic of diffuse layer extending into the bulk from the interface. In this case, for simplicity it is assumed the cations move toward the surface and are responsible for the sheet of positive charge depicted. The net positive charge is compensated by a distribution of negatively charged cation *vacancies* in the bulk (see Chap. 6). (*d*) Charge distribution of cation vacancies. (*e*) Sheet of charge equal in magnitude to surface charge at a distance λ from interface.

diffuse charge can be replaced by an equivalent sheet of charge (Fig. 5.5*e*) that would result in the same capacitance as that of the diffuse charge defines λ. Note that Eq. (5.36) is only applicable to dilute solutions and breaks down at higher concentrations.

Charged interfaces are created not only at free surfaces, but whenever two dissimilar phases come into contact. Electrified interfaces are at the heart of much of today's technology, from drug manufacturing to integrated circuits. Life itself would be impossible without them. More specifically in ceramics, the electric double layer is responsible for such diverse phenomena as varistor behavior, chemical sensing, catalysis, etc. to name but a few.

5.7 GIBBS–DUHEM RELATION FOR BINARY OXIDES

The chemical potentials of the various components in a multicomponent system are interrelated by the **Gibbs–Duhem equation**, expounded on in this section. Its applicability and usefulness, however, will only become apparent in Chap. 7.

In terms of the building blocks of a binary MO_ξ compound, one can write

$$MO_\xi \Leftrightarrow M^{\xi(2+)} + \xi O^{2-} \tag{5.37}$$

from which it follows that

$$\eta_{MO_\xi} = \eta_{M^{\xi(2+1)}} + \xi\eta_{O^{2-}}$$

At equilibrium, by definition, $d\eta_{MO_\xi} = 0$, and consequently, $d\eta_{M^{\xi(2+)}} = -\xi d\eta_{O^{2-}}$ or

$$d\mu_{M^{\xi(2+1)}} + 2\xi e d\phi = -\xi(d\mu_{O^{2-}} - 2e d\phi)$$

Furthermore, since locally the anions and cations are subjected to the same potential ϕ, it follows that for a binary oxide

$$d\mu_{M^{\xi(2+)}} = -\xi d\mu_{O^{2-}} \tag{5.38}$$

This expression is known as the **Gibbs–Duhem relationship** and it expresses the fact that the changes in the chemical potentials of the building blocks of a binary crystal (i.e., anions and cations) are interrelated.[57]

Eq. (5.37) could have also been written as

$$MO_\xi \Leftrightarrow M + \frac{\xi}{2} O_2 \tag{5.39}$$

from which it follows that at equilibrium

$$\boxed{d\mu_M = -\frac{\xi}{2} d\mu_{O_2}} \tag{5.40}$$

As noted above, the importance and applicability of these relationships will become apparent in Chap. 7.

WORKED EXAMPLE 5.5

(*a*) Assume a crystal of MgO is placed between Mg metal on one side and pure oxygen at 1 atm on the other side as shown in Fig. 5.6. Calculate the chemical potential of each species at each interface at 1000 K, given that at that temperature $\Delta G^\circ_{MgO} = -492.95$ kJ/mol.

(*b*) Show that the Gibbs–Duhem relationship holds for MgO crystal described in part (*a*).

57 In solution thermodynamics textbooks, the Gibbs–Duhem expression is usually written as $x_A\, d\mu_A + x_B\, d\mu_B = 0$, where x_A and x_B denote the mole fractions of A and B, respectively.

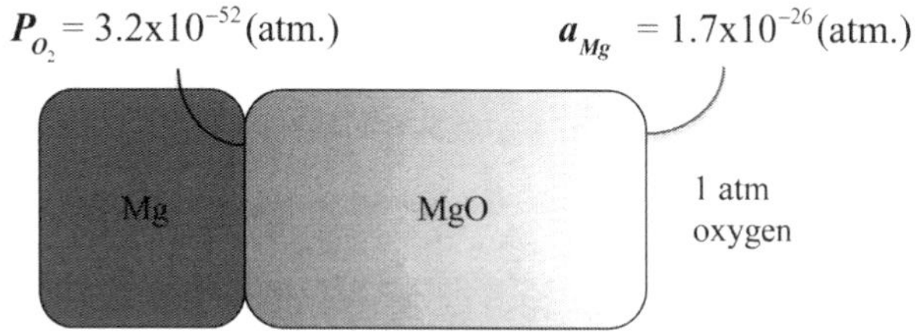

FIGURE 5.6 Equilibrium conditions for an MgO crystal simultaneously in contact with Mg metal on one side and pure oxygen at 1 atm on the other.

ANSWER

(*a*) The pertinent reaction and its corresponding equilibrium mass action expression are, respectively,

$$Mg + \tfrac{1}{2}O_2 \Rightarrow MgO(s)$$

$$\Delta G^\circ = -RT\ln K = -RT\,\text{In}\,\frac{a_{MgO}}{a_{Mg}P_{O/2}^{1/2}}$$

Since the MgO that forms in this case is pure, i.e., in its standard state (e.g., not in solid solution), it follows that by definition $a_{MgO} = 1$ on either side. On the metal side, $a_{Mg} = 1.0$, and solving for P_{O_2} at 1000 K yields 3.2×10^{-52} atm, or 3.2×10^{-53} MPa.

Conversely, on the oxygen side, $P_{O_2} = 1$ atm, and a_{Mg} is calculated to be

$$a_{Mg} = 1.8\times10^{-26}$$

The results are summarized in Fig. 5.6.

(*b*) The Gibbs–Duhem expression simply expresses the fact that the chemical potentials of the constituents of a binary compound are interrelated. Referring to Fig. 5.6, the following applies:

$$\mu_{MgO}\big|_{oxygen} = \mu_{MgO}\big|_{metal}$$

or

$$\mu^{o}_{O_2} + \frac{RT}{2}\ln P_{O_2} + \mu^{o}_{Mg} + RT\ln a_{Mg}\Big|_{oxygen}$$

$$= \mu^{o}_{O_2} + \frac{RT}{2}\ln P_{O_2}\Big|_{metal} + \mu^{o}_{Mg} + RT\ln a_{Mg}\Big|_{metal}$$

which simplifies to

$$\ln a_{Mg}\Big|_{oxygen} = \tfrac{1}{2}\ln P_{O_2}\Big|_{metal}$$

Insertion of the appropriate values for the activity of Mg and P_{O_2} at each interface shows that this identity is indeed fulfilled.

5.8 KINETIC CONSIDERATIONS

In the preceding sections, the fundamental concept of equilibrium was discussed. Given sufficient time, all systems tend to their lowest energy state. Experience has shown, however, that many systems do not exist in their most stable configuration, but rather in a metastable form. Most materials are generally neither produced nor used in their equilibrium states. For example, glasses are metastable with respect to their crystalline counterparts, yet are of great utility because at the temperatures at which they are typically used, the kinetics of the glass-crystal transformation are negligible.

In general, as a first approximation the kinetics or the rate of any transformation is assumed to be proportional to a driving force F

$$\text{Rate} = \beta F \tag{5.41}$$

where the proportionality constant β is a system property that depends on the process involved. For instance, β can be a diffusion coefficient, a reaction rate constant or a conductance of any sort.

The driving force is a measure of how far a system is from equilibrium. Referring to Fig. 5.3, the driving force is nothing but $\partial G/\partial \xi$, or ΔG. Thus the importance of thermodynamics lies not only in defining the state of equilibrium, but also in quantifying the driving force—it is only by knowing the final equilibrium state that the rate at which a system will approach that state can be estimated.

All changes and transformations require a driving force, the nature and magnitude of which can vary over many orders of magnitude depending on the process involved (see Table 5.1). For example, the driving forces for chemical reactions, such as oxidation, are usually quite large, in the range of a few hundred kilojoules per mole. On the other hand, the driving forces for boundary migration, coarsening and densification are much smaller, on the order of 100 J/mol or less. This, in turn, partially explains why it is much easier to oxidize a fine metal powder than it is to sinter it.

The four most important driving forces operative in materials science are those due to

1. Reduction in free energies of formation as a result of chemical reactions and phase transformations, e.g., oxidation or crystallization
2. Reduction of energy due to applied stresses, e.g., creep
3. Reduction of surface or interfacial energy, e.g., sintering and grain growth
4. Reduction of strain energy, e.g., fracture, segregation

TABLE 5.1 Typical orders of magnitude of driving forces governing various phenomena discussed in this book

Process	Driving force	Typical values, J/mol[a]	Comments
Fracture (Chap. 11)	$V_m\sigma^2/(2Y)$	0.5	σ is stress at failure and Y is Young's modulus
Grain growth (Chap. 10)	$2V_m\gamma_{gb}/r$	20	γ_{gb} is grain boundary energy and r is radius of a particle
Sintering or coarsening (Chap. 10)	$2V_m\gamma/r$	20	γ is surface energy term (Chap. 4)
Creep (Chap. 12)	σV_m	1000	σ is applied stress and V_m molar volume
Crystallization (Chap. 9)	$\Delta H_m\,\Delta T/T_m$	2400	ΔH is enthalpy of transformation, ΔT is undercooling and T_m is melting point
Interdiffusion (Chap. 7)	$RT(x_a \ln x_a + x_b \ln x_b)$	5000	Assuming ideal solution [see Eq. (5.11)]
Oxidation (Chap. 7)	ΔG^o_{form}	50,000–500,000	ΔG^o_{form} free energy of formation of oxide-normalized to a per-mole-of-oxygen basis

[a] Assumptions: T = 1000 K, molar volume, $V_m = 10^{-5}$ m³/mol (10 cm³/mol); r = 1 μm, $\gamma = \gamma_{gb} = 1$ J/m²; σ = 100 MPa, Y = 100 GPa; ΔS_m = 12 J/mol K; ΔT = 200 K.

At this point the expressions and order-of-magnitude values of these driving forces are simply listed (Table 5.1). However, each will be revisited and discussed in detail in subsequent chapters. Fracture is dealt with in Chap. 11, grain growth and sintering in Chap. 10, crystallization in Chap. 9, creep in Chap. 12, and oxidation and interdiffusion in Chap. 7.

The second important parameter that determines the rate at which a given process will occur is the factor β. And since, with the notable exception of fracture, all the processes listed in Table 5.1 require the movement of atoms, *β is usually equated to the rate at which an atom or ion will make a jump.* This concept is discussed in greater detail in Chap. 7, where diffusion is discussed.

COMPUTATIONAL MATERIALS SCIENCE 5.1: PHONON MODES

When solids are heated their atoms vibrate. At room temperature the frequencies of vibration are of the order of 10^{13} Hz. It can be shown that these vibrations can be described by a combination of waves—referred to as **phonons**—with various frequencies. At any T, various atoms of different masses and bond strengths will vibrate differently and therefore a distribution of these vibrations ensues. To understand the thermal properties of a solid—heat capacities, thermal conductivities, thermal stabilities, etc.—that all important frequency distribution must be known. As discussed in Sec. 5.2.2, Einstein was the first to tackle the problem; he assumed all atoms vibrated at the same frequency, we labeled ν_e in Eq. (5.15). Later, Debye made a more realistic assumption: he assumed the distribution is a parabola, depicted by the dashed black line in Fig. 5.7*b*.

The actual situation in solids is more complicated, especially in multi-elemental compounds like ceramics, because every atomic species is bonded slightly differently and has a different mass and will thus vibrate differently. In other words, it is important to obtain the **distribution of the phonon density of**

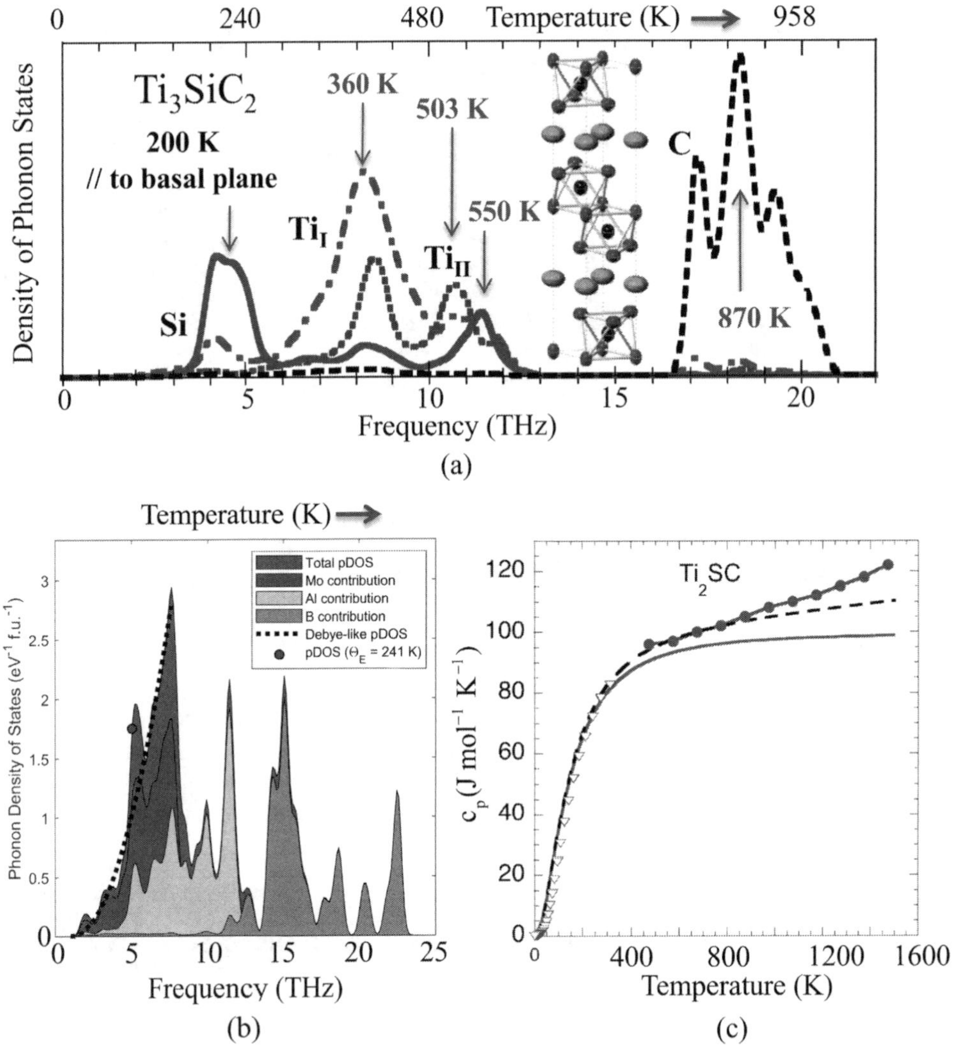

FIGURE 5.7 Phonon density of states for (*a*) Ti_3SiC_2. Inset shows relative vibration amplitudes of atoms at 900°C, and (*b*) MoAlB. (*c*) Heat capacity of Ti_2SC. Blue triangles and red solid circles are experimental results; solid green line is what DFT predicts; dashed black line is sum of DFT predictions and experimentally determined electronic heat capacities.

states D(ω), for the solid as a whole. Figure 5.7*a* is an example of D(ω), determined by DFT, for Ti_3SiC_2 a layered MAX phase. In Ti_3SiC_2, the Si atoms are relatively weakly bound to the structure; the C atoms much more so. The Ti_3SiC_2 unit cell is shown in inset of Fig. 5.7*a*. The corresponding results for MoAlB, a layered ternary transition metal boride, are shown in Fig. 5.7*b*.

To understand how to interpret D(ω) results such as those shown in Fig. 5.7*a* and *b*, it is useful to start at 0 K. At that temperature, the only vibrations are the zero-point energy. (The latter are needed to ensure that $\Delta x \Delta p > h$.) Referring to Fig. 5.7*a*, as Ti_3SiC_2 is heated, the first atoms to vibrate—at around 200 K—are the most loosely bound, viz. Si. However, since these atoms are less bonded within the basal planes, than normal to them, they start vibrating parallel to the basal planes first! At that temperature, the Ti and C atoms are not vibrating very much. However, between ≈360 and ≈500 K, the Ti atoms start vibrating. At 550 K, the Si atoms start vibrating normal to the basal planes and their overall vibration is now more isotropic than at lower temperatures. Because the C atoms are the strongest bonded and the lightest, they only start vibrating appreciably only when the temperatures approach 800 K (Fig. 5.7*a*). Beyond ≈1000 K, all the atoms are vibrating fully and each is contributing 3k to the heat capacity. Said otherwise, beyond 1000 K the Dulong and Petit limit of 3k per atom, or 3R per mole, is applicable (see below). Using appropriate software, it is now possible to convert the D(ω) results to schematics in which the relative atomic displacement parameters are compared. Inset in Fig. 5.7*a* shows the results of such an exercise at 900°C for Ti_3SiC_2. Note that at that T the amplitudes of vibration of the Si (green) atoms are significantly higher than those of the Ti (red) or C (black) atoms.

The same arguments can be made about MoAlB. Here the least bound atoms are the Mo (red region in Fig. 5.7*b*). The next are the Al atoms (gray region Fig. 5.7*b*). The best bound are the B atoms and they only start vibrating at temperatures above 700 K. Note that the Debye model—shown for this solid as a dashed black line—does a decent job describing the phonon DOS, but only up to about 300 K.

Once D(ω) is known, the total number of modes is given by:

$$N = \int_0^\infty D(\omega)\,d\omega \tag{5.42}$$

In turn, the internal energy, U, is given as:

$$U = \int_0^\infty D(\omega)\langle E_\omega \rangle d\omega = \int_0^\infty D(\omega)\frac{\hbar\omega}{\exp(\hbar\omega/k_BT)-1}d\omega \tag{5.43}$$

where E_ω is the energy carried by a wave with frequency ω. Once U is calculated, then c_v is given by:

$$c_v = \left[\frac{\partial U}{\partial T}\right]_V \tag{5.44}$$

It follows that to calculate the temperature dependence of all important c_v of a solid all that is required is D(ω). The fact that DFT calculations can quite accurately determine D(ω) has been a great boon since experimentally measuring D(ω) is nontrivial. One method involves nonelastic neutron scattering, which will not be discussed here.

Experimentally it is much easier to measure the heat capacity at constant pressure, c_p—viz. at 1 atm—and compare the results with those predicted by DFT calculations. For solids the difference between the calculated, c_v, and measured c_p is negligible and can be safely ignored. Figure 5.7*c* shows an example where c_v and c_p are compared for Ti_2SC, another MAX phase. Here three sets of lines are shown: The experimental points are represented by symbols (open triangle for low and solid circles for high temperatures). The lower solid line is that predicted by DFT for phonons only. Since Ti_2SC is a decent metallic conductor, its conduction electrons also absorb heat. This electronic heat capacity can be readily obtained from low temperature c_p results. The dashed black line is the sum of the heat capacities of the conduction electrons and the phonons, which is ultimately compared to the experimental results. The agreement between theory and experiment is excellent up to ≈900 K. Note that the phonon calculations asymptotically approach ≈100 J/mol K, viz. the product of 3R per atom and four atoms per formula unit.

To conclude, like elastic properties, the heat capacities of most ceramics can be well predicted by DFT and when experimental results are unknown can be used with a good degree of confidence. As always with DFT calculations, large discrepancies between theory and experiment raise red flags. If the scientists carrying out the DFT calculations are reputable and know what they are doing, it implies that the samples used for the measurements were far from the perfect solid that DFT calculations typically start with.

5.9 SUMMARY

The free energy is a function made up of two terms: an enthalpy term and an entropy term. Entropy can be of various kinds, but fundamentally it is a measure of the disorder in a system.

At constant pressure and temperature, the state of equilibrium of a system is defined as that state for which the free energy is at a minimum.

For a chemical reaction equilibrium dictates that $\Delta G_{rxn} = 0$ and consequently

$$\Delta G^o = -RT \ln K$$

where K is the equilibrium constant for that reaction.

In ionic ceramics, it is not the chemical but the electrochemical potential that defines equilibrium.

The thermodynamic parameters of any solid can be readily calculated if the density of state of the phonon is either calculated or measured.

APPENDIX 5A: DERIVATION OF EQ. (5.27)

Reaction (I) in text reads

$$M(s) + \tfrac{1}{2}X_2(g) \Rightarrow MX(s) \quad \Delta G_{rxn} \tag{5A.I}$$

Applying Eq. (5.23) to reactants and products, one obtains

$$\mu_{MX} = \mu_{MX}^{o} + RT \ln a_{MX} \tag{5A.1}$$

$$\mu_{M} = \mu_{M}^{o} + RT \ln a_{M} \tag{5A.2}$$

$$\mu_{X_2} = \tfrac{1}{2}\mu^{\circ}_{X_2} + RT\ln P^{1/2}_{X_2} \tag{5A.3}$$

It follows that the free-energy change associated with this reaction is

$$\Delta G_{rxn} = \mu_{MX} - \mu_{M} - \tfrac{1}{2}\mu_{X_2} \tag{5A.4}$$

Combining Eqs. (5A.1) to (5A.4) gives

$$\Delta G_{form} = (\mu^{o}_{MX} - \mu^{o}_{M} - \tfrac{1}{2}\mu^{o}_{X_2}) + RT\ln\frac{a_{MX}}{a_M P^{1/2}_{X_2}} \tag{5A.5}$$

If one defines ΔG° as

$$\Delta G^{o} = \mu^{o}_{MX} - \mu^{o}_{M} - \tfrac{1}{2}\mu^{o}_{X_2} \tag{5A.6}$$

and K by Eq. (5.28), Eqs. (5A.5) and (5.27) are identical. Furthermore, since at equilibrium $\Delta G_{form} = 0$, Eq. (5A.5) now reads

$$\Delta G^{o} = -RT\ln K_{eq} = -RT\ln\frac{a_{MX}}{a_M P^{1/2}_{X_2}} \tag{5A.7}$$

PROBLEMS

5.1. Starting with Eq. (5.9), and making use of Stirling's approximation, derive Eq. (5.10).

5.2. Pure stoichiometric ZnO is heated to 1400 K in an evacuated chamber of a vapor deposition furnace. What is the partial pressure of Zn and O_2 generated by the thermal decomposition of ZnO? Information you may find useful: ΔG^{o}_{ZnO} at 1400 K = −183 kJ/mol. Hint: the partial pressures are related.
Ans: $\log P_{O_2} = -4.75$, $\log P_{Zn} = -4.45$.

5.3. Calculate the driving force for the oxidation of pure Mg subjected to an oxygen partial pressure of 10^{-12} atm at 1000 K. Compare that value to the driving force if the oxygen partial pressure was 1 atm.
Ans: −378.1 kJ/mol.

5.4. *(a)* If ΔG^{o} at 1000 K for the $Si + ½\,O_2 = SiO_2$ is −729.1 kJ/mol, calculate the equilibrium partial pressure of oxygen in the Si-silica system at 1000 K.
(b) If the oxidation is occurring in the presence of water vapor, calculate the equilibrium constant and the H_2/H_2O ratio in equilibrium with Si and silica at 1000 K.
(c) Compute the equilibrium partial pressure of oxygen for a gas mixture with an H_2/H_2O ratio calculated in part *(b)*. Compare your result with the oxygen partial pressure calculated in part *(a)*.

5.5. Given at 1623 K:

$$\Delta G^{o}_{SiO_2} = -623\,kJ/mol$$

$$\Delta G^{o}_{Si_2N_2O} = -446\,kJ/mol$$

$$\Delta G^{o}_{Si_3N_4} = -209\,kJ/mol$$

Confirm that the stability diagram in the Si-N-O system at 1623 K is the one shown in Fig. 5.8.

5.6. (*a*) Can Al reduce Fe_2O_3 at 1200°C? Explain.

Ans: Yes.

(*b*) Is it possible to oxidize Ni in a CO/CO_2 atmosphere with a ratio of 0.1?

Ans:

(*c*) Will silica oxidize zinc at 700°C. Explain.

Ans: No.

5.7. Calculate the chemical stability domains of NiO and CoO at 1000 K. Compare your results with those listed in Table 6.1. Hint: The information needed to solve this problem is deliberately omitted; you need to find it. Hint: Look up JANAF Tables.

Ans:

5.8. A crucible of BN is heated in a gas stream containing N_2, H_2 and H_2O at 1200 K. The partial pressure of nitrogen is kept fixed at 0.5 atm. What must the ratio P_{H_2}/P_{H_2O} have to be or exceed in order for B_2O_3 not to form? Information you may find useful: at 1200 K

$$\Delta G^{o}_{BN} = -743\ kJ/mol;\ \Delta G^{o}_{B_2O_3} = -957.47\ kJ/mol;\ \Delta G^{o}_{H_2O} = -181.425\ kJ/mol.$$

Ans: 1.

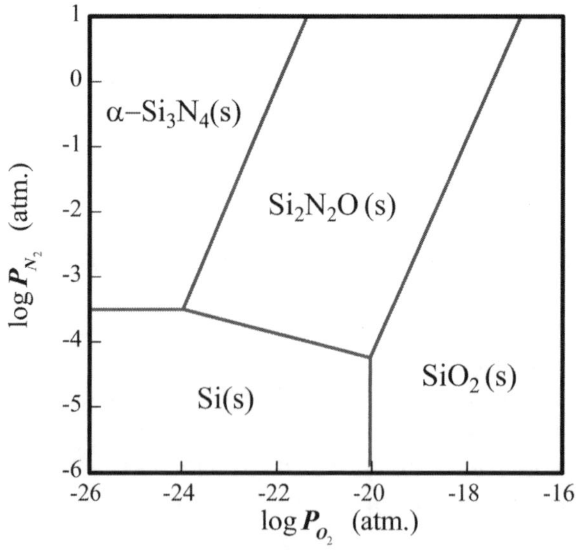

FIGURE 5.8 Si-N-O stability diagram at 1623 K.

ADDITIONAL READING

1. R. A. Swalin, *Thermodynamics of Solids*, 2nd ed., Wiley, New York, 1972.
2. D. R. Gaskell, *Introduction to Metallurgical Thermodynamics*, 2nd ed., Hemisphere Publishing, New York, 1981.
3. K. Denbigh, *The Principles of Chemical Equilibrium*, 4th ed., Cambridge University Press, New York, 1981.
4. C. H. P. Lupis, *Chemical Thermodynamics of Materials*, North-Holland, Amsterdam, 1983.
5. J. W. Gibbs, *The Scientific Papers of J. W. Gibbs, vol. 1, Thermodynamics*, Dover, New York, 1961.
6. J. Bockris and A. K. N. Reddy, *Modern Electrochemistry*, Plenum, New York, 1970.
7. P. Shewmon, *Transformation in Metals*, McGraw-Hill, New York, 1969.
8. R. DeHoff, *Thermodynamics in Materials Science*, McGraw-Hill, New York, 1993.
9. L.S. Darken and R. W. Gurry, *Physical Chemistry of Metals*, McGraw-Hill, New York, 1951.

THERMODYNAMIC DATA

10. M. W. Chase, C. A. Davies, J. R. Downey, D. J. Frurip, R. A. McDonald and A. N. Syverud, JANAF thermodynamic tables, 3rd ed., *J. Phys. Chem. Ref. Data*, **14**, Supp. 1 (1985).
11. J. D. Cox, D. D. Wagman, and V. A. Medvedev, *CODATA Key Values of Thermodynamics*, Hemisphere Publishing, New York, 1989.
12. Online access to the SOLGASMIX program is available through F*A*C*T (Facility for the Analysis of Chemical Thermodynamics), Ecole Polytechnique, CRCT, Montreal, Quebec, Canada.
13. I. Barin, O. Knacke, and O. Kubaschewski, *Thermodynamic Properties of Inorganic Substances Supplement*, Springer-Verlag, New York, 1977.

6

DEFECTS IN CERAMICS

Textbooks and Heaven only are Ideal;
Solidity is an imperfect state.
Within the cracked and dislocated Real
Nonstoichiometric crystals dominate.
Stray Atoms sully and precipitate;
Strange holes, excitons, wander loose, because
Of Dangling Bonds, a chemical Substrate
Corrodes and catalyzes—surface Flaws
Help Epitaxial Growth to fix adsorptive claws.

John Updike, *The Dance of the Solids**

6.1 INTRODUCTION

Alas, as John Updike so eloquently points out, only textbooks (present company excepted) and heaven are ideal. Real crystals, however, are not perfect but contain imperfections that are classified according to their geometry and shape into point, line and planar defects. A *point defect* can be defined as any lattice point that is not occupied by the proper ion, or atom, needed to preserve the structure's long-range periodicity. Dislocations are defects that cause lattice distortions centered on a line and are thus classified as *linear defects*. *Planar defects* are surface imperfections in polycrystalline solids that separate grains or domains of different orientations and include grain and twin boundaries. In addition, there are three-dimensional bulk defects such as pores, cracks and inclusions; these are not treated in this chapter, however, but are

* J. Updike, *Midpoint and Other Poems*, A. Knopf, Inc., New York, New York, 1969. Reprinted with permission.

considered in Chap. 11, where it is shown that these defects are critical in determining the strengths of ceramics.

The importance of defects in general, and point defects in particular, cannot be overemphasized. As will become apparent in subsequent chapters, many of the properties considered are strongly affected by the presence or absence of these defects. For instance, in Chap. 7, the one-to-one correlation between the concentration of point defects and atom movement or diffusion is elucidated. In metals, but less so in ceramics except at higher temperatures, it is the presence and movement of dislocations that is responsible for ductility and creep. In Chap. 11, the correlation between grain size and mechanical strength is made. As discussed in Chap. 16, the scattering of light by pores is responsible for their opacity.

Generally speaking, in ceramic systems more is known about point defects than about the structure of dislocations, grain boundaries or free surfaces—a fact that is reflected in the coverage of this chapter in which the lion's share is devoted to point defects.

6.2 POINT DEFECTS

In contrast to pure metals and elemental crystals for which point defects are rather straightforward to describe (because only one type of atom is involved and charge neutrality is not an issue), the situation in ceramics is more complex. One overriding constraint operative during the formation of ceramic defects is the preservation of electroneutrality at all times. Consequently, the defects occur in neutral "bunches" and fall in one of three categories:

6.2.1 STOICHIOMETRIC DEFECTS

These are defined as ones in which the crystal chemistry, i.e., the ratio of the cations to anions, does *not* change, and they include, among others, Schottky and Frenkel defects (Fig. 6.3).

6.2.2 NONSTOICHIOMETRIC DEFECTS

These defects are formed by the selective addition, or loss, of one (or more) of the constituents of the crystal and consequently lead to a *change* in crystal chemistry and the notion of nonstoichiometry discussed subsequently. The basic notion that a compound's composition is a constant with simple ratios between the numbers of constituent atoms is one that is reiterated in most chemistry courses. For instance, in MgO the cation/anion ratio is unity, that for Al_2O_3 is 2/3, etc. In reality, however, it can be rigorously shown using thermodynamic arguments that the composition of *every* compound *must* vary within its existence regime.[58]

[58] The existence regime defines the range of chemical potentials of the constituents of a compound over which it is thermodynamically stable. For example, it was shown in Worked Example 5.5 that MgO was stable between the P_{O_2} of 1 atm and 3.2×10^{-52} atm—below 3.2×10^{-52}, MgO decomposed to Mg metal and oxygen.

A material accommodates those compositional changes by selectively losing one of its constituents to its environment that in turn creates or eliminates defects (see e.g. Fig. 6.4). In so doing, a compound will adjust its composition to reflect the externally imposed thermodynamic parameters. This leads to the idea of **nonstoichiometry** where the simple ratio between the numbers of the constituent atoms of a compound breaks down. For example, if an MO oxide is annealed in a high oxygen partial pressure, P_{O_2}, it would be fair to assume that the number of O anions would be relatively greater than the number of M cations. Conversely, if the P_{O_2} is very low, one would expect the cation concentration to be higher.

The importance of nonstoichiometry lies in the fact that many physical properties such as color, diffusivity, electrical conductivity, superconductivity, photoconductivity and magnetic susceptibility can vary markedly with small changes in composition.

6.2.3 EXTRINSIC DEFECTS

These are defects created as a result of the presence of *impurities* in the host crystal.

The remainder of section 6.2 attempts to answer, among others, the following questions: Why do point defects form? What are the different types of defects that can form? And how is their concentration influenced by temperature and externally imposed thermodynamic parameters, such as P_{O_2}? Before we proceed, however, it is imperative to describe in greater detail the various defects that can form and to formulate a scheme by which they can be notated.

6.2.4 POINT DEFECTS AND THEIR NOTATION

In a *pure* binary compound, the following lattice defects, shown schematically in Fig. 6.1, can exist:

1. *Vacancies*: sites where an atom is missing; these can occur on either sublattice.
2. *Interstitial atoms*: atoms found in sites that are normally unoccupied.

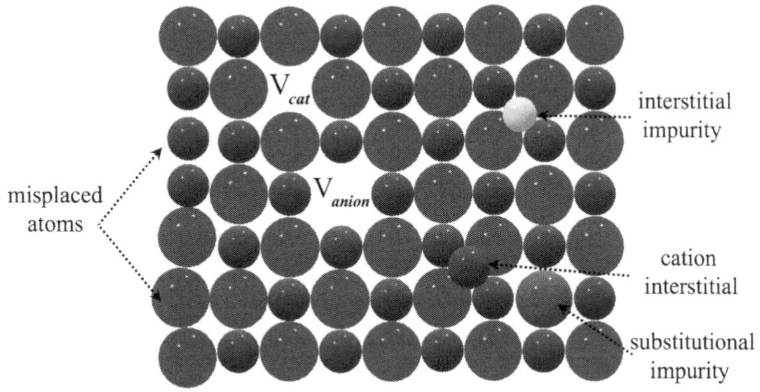

FIGURE 6.1 Various types of defects typically found in a binary ceramics. Misplaced atoms can only occur in covalent ceramics due to charge considerations.

3. *Misplaced atoms*: types of atoms found at a site normally occupied by other types. This defect is only possible in covalent ceramics, however, where the atoms are not charged.
   The following electronic defects also exist:
4. *Free electrons*: electrons that are in the conduction band of the crystal.
5. *Electron holes*: positive mobile electronic carriers that are present in the valence band of the crystal (see Chap. 7).
   In addition to the aforementioned, an *impure* crystal will contain
6. *Interstitial and substitutional impurities*: As depicted in Fig. 6.1, these can occur on either sublattice.

Over the years, several schemes have been proposed to describe defects in ceramics. The one that is now used almost universally is the **Kroger–Vink notation** and is thus the one adopted here. In this notation, the defect is represented by a main symbol followed by a superscript and a subscript.

Main symbol. The main symbol is either the species involved, i.e., chemical symbol of an element, or the letter V for vacancy.
Subscript. The subscript is either the crystallographic position occupied by the species involved or the letter i for interstitial.
Superscript. The superscript denotes the **effective electric charge** on the defect, defined as the difference between the real charge of the defect species and that of the species that would have occupied that site in a perfect crystal.[59] The superscript is a prime for each negative charge, a dot for every positive charge or an x for a zero effective charge.

The best way to explain how the notation works is through a series of examples.

EXAMPLE 6.1

Consider the possible defects that can occur in a *pure* NaCl crystal.

(*a*) Vacancy on the Na^+ sublattice: $V'_{Na\Rightarrow \text{site on which vacancy resides}}$. The symbol V is always used for a vacancy. The superscript here is a prime - representing a single negative charge - because the effective charge on the vacancy is $0-(+1)=-1$.
(*b*) Vacancy on Cl^- sublattice: $V^{\bullet}_{Cl}$. In this case the superscript is a small dot (which denotes a positive charge) because the effective charge on the vacancy is $0-(-1)=+1$.
(*c*) Interstitial on Na sublattice: $Na^{\bullet}_{i}$. The main symbol here is the misplaced Na ion; the subscript i denotes the interstitial position; the effective charge is: $+1-0=+1$.
(*d*) Interstitial position on Cl sublattice: Cl'_{i}. The main symbol here is the misplaced Cl anion; the subscript i denotes the interstitial position; the effective charge is: $-1+0=-1$.

[59] The charge is so called because it denotes the effective – not actual – charge on the defect *relative* to the perfect crystal. It is this effective charge that determines the direction in which the defect will move in response to an electric field (see Chap. 7). It also denotes the type of interaction between defects, for instance, whether two defects would attract or repel each other.

EXAMPLE 6.2

Consider the addition of $CaCl_2$ to NaCl. The Ca cation can substitute for a Na ion or go interstitial (needless to say, because of charge considerations, only cations will substitute for cations and only anions for anions). In the first case, the defect notation is ⊗, and the effective charge $[+2-(+1)=1]$ is +1. Conversely, an interstitial Ca ion is denoted as $Ca_i^{\bullet\bullet}$.

EXAMPLE 6.3

Instead of adding $CaCl_2$, consider KCl. If the K ion, which has the same charge as Na, substitutes for a Na ion, the notation is K_{Na}^{x}, since the effective charge in this case is 0 (denoted by an x). If the K anion goes interstitial, the notation is $K_i^{\bullet}$.

EXAMPLE 6.4

Dope the NaCl crystal with Na_2S. Again only anions can substitute for anions, or they can go interstitial. Two possibilities are S'_{Cl} and S''_i.

EXAMPLE 6.5

One would expect to find the following defects in pure Al_2O_3: $Al_i^{\bullet\bullet\bullet}$, O''_i, V'''_{Al}, and $V_O^{\bullet\bullet}$.

After this brief introduction to defects and their notation, it is pertinent to ask why point defects form in the first place. Before the more complicated case of defects in binary ionic ceramics is tackled in Sec. 6.2.6, the simpler situation involving vacancy formation in elemental crystals such as Si, Ge or pure metals is treated.

6.2.5 THERMODYNAMICS OF POINT DEFECT FORMATION IN ELEMENTAL CRYSTALS

There are several ways by which vacancy formation can be envisioned. A particularly useful and instructive one is to remove an atom from the crystal bulk and place it on its surface. The enthalpy change, Δh, associated with such a process has to be endothermic because more bonds are broken than are reformed. This brings up the legitimate question: If it costs energy to form defects, why do they form? The answer lies in the fact that at equilibrium, as discussed in Chap. 5, it is the free energy rather than the enthalpy that is minimized. In other words, it is only when the entropy changes associated with the formation of the defects are taken into account that it becomes clear why vacancies are thermodynamically stable and their equilibrium concentration can be calculated. It follows that if it can be shown that, at any given temperature, the Gibbs free energy associated with a perfect crystal G_{perf} is *higher* than that of a crystal containing n_v defects, i.e., that $G_{def} - G_{perf} < 0$, where G_{def} is the free energy of the defective crystal, then the defective crystal has to be more stable. The procedure is as follows:

6.2.5.1 FREE ENERGY OF A PERFECT CRYSTAL[60]

For a perfect crystal,

$$G_{perf} = H_{perf} - TS_{perf}$$

where H is the enthalpy; S, the entropy; and T, the absolute temperature of the crystal.

As noted in Chap. 5, the total entropy of a collection of atoms is the sum of a configuration term and a vibration entropy term, or

$$S = S_{config} + S_T$$

For a perfect crystal, $S_{config} = 0$ since there is only one way of arranging N atoms on N lattice sites. The vibration component, however, can be approximated by Eq. (5.16), or

$$S_T = Nk\left(\ln\frac{kT}{h\nu} + 1\right)$$

where N is the number of atoms involved, k is Boltzmann's constant and ν is the vibration frequency of atoms in the perfect crystal. Adding the various terms, one obtains

$$G_{perf} = H_{perf} - TS_{perf} = H_{perf} - NkT\left(\ln\frac{kT}{h\nu} + 1\right) \tag{6.1}$$

6.2.5.2 FREE ENERGY OF A DEFECTIVE CRYSTAL

If one assumes that it costs h_d to create *one* defect, it follows that the enthalpy of the crystal upon formation of n_v vacancies increases (i.e., becomes less negative) by $n_v h_d$. Hence the enthalpy of the defective crystal is

$$H_{def} = H_{perf} + n_v h_d \tag{6.2}$$

Furthermore, the configurational entropy is no longer zero because the n_v vacancies and N atoms can now be distributed on $N + n_v$ total atomic sites. The corresponding configuration entropy [see Eq. (5.10)] is given by

$$S_{config} = -k\left(N\ln\frac{N}{N+n_v} + n_v\ln\frac{n_v}{N+n_v}\right) \tag{6.3}$$

As a first, and last approximation, here we assume that (i) only atoms in the near vicinity of each vacancy will vibrate at a different frequency ν' than the rest, (ii) the remaining atoms will be unaffected and will continue to vibrate with the same frequency as given in Eq. 6.1.

60 This approach is not strictly orthodox because G_{perf} cannot be calculated on an absolute scale. However, the approach here is still valid because before the final result is reached, that energy will be subtracted from G_{def}.

In other words, we only assume nearest-neighbor interactions. In that case, for a coordination number ζ of the vacancies, the total number of atoms affected is simply ζn_v. It follows that the vibration entropy term is given by

$$S = k(N - \zeta n_v)\left(\ln\frac{kT}{h\nu} + 1\right) + n_v \zeta k\left(\ln\frac{kT}{h\nu'} + 1\right) \tag{6.4}$$

where the first term represents atoms whose vibration frequencies have been unaffected by the vacancies and the second term represents those that have, and are now vibrating with, a new frequency, ν'.

Combining Eqs. (6.2) to (6.4) yields

$$\begin{aligned} G_{def} &= H_{perf} + n_v h_d \\ &- kT\left[(N - n_v\zeta)\left(\ln\frac{kT}{hv} + 1\right) + n_v\zeta\left(\ln\frac{kT}{hv'} + 1\right)\right. \\ &\left. - N\ln\frac{N}{n_v + N} - n_v \ln\frac{n_v}{n_v + N}\right] \end{aligned} \tag{6.5}$$

Subtracting Eq. (6.1) from Eq. (6.5) yields the sought-after result

$$\begin{aligned} \Delta G &= G_{def} - G_{perf} \\ &= n_v h_d + kTn_v\zeta \ln\frac{\nu'}{\nu} \\ &+ kT\left(N\ln\frac{N}{n_v + N} + n_v \ln\frac{n_v}{n_v + N}\right) \end{aligned} \tag{6.6}$$

This is an important result because it shows that the free-energy change upon the introduction of n_v defects, in an otherwise perfect crystal, is a function of both n_v and T.

If T is kept constant and ΔG is plotted versus n_v, as shown in Fig. 6.2*a*, it is immediately obvious that this function goes through a minimum.[61] In other words, the addition of vacancies to a perfect crystal will initially lower its free energy up to a point beyond which further increases in the number of vacancies is no longer energetically favorable, and the free energy increases once again.[62] The number of vacancies at which the minimum in ΔG occurs - i.e., when $\partial G/\partial n_v = 0$ - is the equilibrium number of vacancies n_{eq} at that temperature and is given by (see Prob. 6.1)

$$\boxed{\frac{n_{eq}}{n_{eq} + N} \approx \frac{n_{eq}}{N} \approx \exp\left(-\frac{h_d - \Delta s_{vib}}{kT}\right) = \exp\left(-\frac{\Delta g_d}{kT}\right)} \tag{6.7}$$

where $\Delta g_d = h_d - T\Delta s_{vib}$ and $\Delta s_{vib} = \zeta\, k \ln(\nu/\nu')$. Note that the final expression does not contain any configuration entropy terms, but depends solely on the free energy associated with the formation of a single defect Δg_d.

[61] For the sake of simplicity, the second term in Eq. (6.6) was omitted from Fig. 6.2.

[62] Note here n_v is the reaction variable discussed in Chap. 5 (see Fig. 5.3).

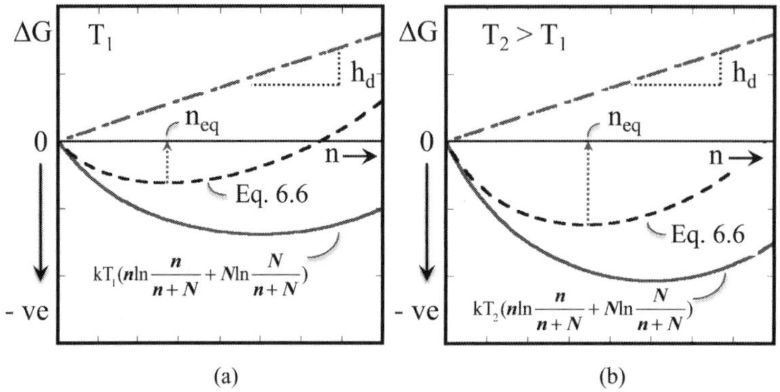

FIGURE 6.2 (*a*) Free-energy change as a function of number of defects n_v. The top line represents energy needed to create the defects. The lowest curve is the free energy gained as a result of the configuration entropy multiplied by temperature. The dashed black centerline represents the sum of the two components [i.e., a plot of Eq. (6.6)], which clearly goes through a minimum. (*b*) Same plot using the same scale as in (*a*) except at a higher temperature. Note that the equilibrium number of defects increases with increasing temperature.

Equation (6.7) predicts that n_{eq} increases exponentially with temperature. To understand why, it is instructive to compare Fig. 6.2*a* and *b*, where Eq. (6.6) is plotted, on the same scale, for two different temperatures. At higher temperatures (Fig. 6.2*b*), the configurational entropy term becomes more important relative to the enthalpy term, which in turn shifts n_{eq} to higher values.

At this point, the slightly more complicated problem of defects in ceramics is dealt with. The complications arise, as noted above, because the charges on the defects preclude their forming separately—they always form in "bunches" so as to maintain charge neutrality. In the following section, defect formation in binary ionic compounds is dealt with by writing down balanced-defect reactions. Expressions for the equilibrium concentration of these defects are then calculated using two approaches. The first uses the statistical approach used to derive Eq. (6.7). The second (Sec. 6.2.8) makes use of the mass action expression of the pertinent defect reactions. Needless to say, the two approaches should, and do, yield identical results.

6.2.6 DEFECT REACTIONS

The formation of various point defects is best described by chemical reactions for which the following rules have to be followed:

- ∞ *Mass balance*: mass cannot be created or destroyed. Vacancies have zero mass.
- ∞ *Electroneutrality* or charge balance: charges cannot be created or destroyed.
- ∞ *Preservation of regular site ratio*: the ratio between the numbers of regular cation and anion sites must remain constant and equal to the ratio of the parent lattice.[63] Thus if a normal lattice site of one constituent is created or destroyed, a corresponding number of normal sites of the other constituent

[63] Interstitial sites are not considered to be regular sites.

must be simultaneously created or destroyed so as to preserve a compound's site ratio. This requirement recognizes that one cannot create one type of lattice site without the other and indefinitely extend the crystal. For instance, for a MX compound, if a number of cation lattice sites are created or destroyed, then an equal number of anion lattice sites have to be created or destroyed. Conversely, for a M_2X compound, the ratio must be maintained at 2:1, etc.

To generalize, for a M_aX_b compound, the following relationship has to be maintained at all times:

$$a(X_X + V_X) = b(M_M + V_M)$$

that is, the ratio of the sum of the number of atoms and vacancies on each sublattice has to be maintained at the stoichiometric ratio, or

$$\frac{M_M + V_M}{X_X + V_X} = \frac{a}{b}$$

Note this does not imply the number of atoms or ions have to maintain that ratio, only that the number of sites do. In the following subsections these rules are applied to the various types of defects present in ceramics.

6.2.6.1 STOICHIOMETRIC DEFECT REACTIONS

A stoichiometric defect reaction by definition is one where the chemistry of the crystal does not change as a result of the reaction. Said otherwise, a stoichiometric reaction is one in which no mass is transferred across the crystal boundaries. The three most common stoichiometric defects are Schottky defects, Frenkel defects and antistructure disorder or misplaced atoms.

Schottky defect reactions. In the Schottky defect reaction, electric-charge-equivalent numbers of vacancies are formed on each sublattice. In NaCl, for example, a Schottky defect entails the formation of Na and Cl vacancy pairs (Fig. 6.3*a*). In general, for an MO oxide, the reaction reads[64]

$$\text{Null (or perfect crystal)} \Rightarrow V_M'' + V_O^{\bullet\bullet} \quad \Delta g_S \qquad (6.8)$$

where Δg_S is the free-energy change associated with the formation of the Schottky defect.

Similarly, for an M_2O_3 oxide,

[64] To see how that occurs, consider the formation of a defect pair in an MO oxide by the migration of a cation and an anion to the surface. In that case, one can write

$O_O^x + M_M^x \Rightarrow O_{O,s}^x + M_{M,S}^x + V_O^{\bullet\bullet} + V_M''$

where the subscript s refers to the surface sites. But since the ions that migrated to the surface covered ions previously located at the surface, this equation is abbreviated to Eq. (6.8).

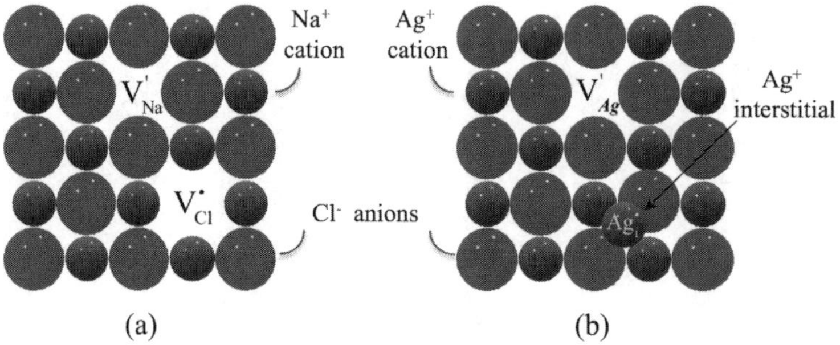

FIGURE 6.3 (*a*) Schottky defect in NaCl; (*b*) Frenkel defect in AgCl.

$$\text{Null (or perfect crystal)} \Rightarrow 2V_M''' + 3V_O^{\bullet\bullet}$$

In general, for an M_aO_b oxide,

$$\text{Null (or perfect crystal)} \Rightarrow aV_M^{b-} + bV_O^{a+}$$

It is left as an exercise to the reader to ascertain that as written, these reactions satisfy the aforementioned rules.

Equation (6.7) was derived with the implicit assumption that only one type of vacancy forms. The thermodynamics of Schottky defect formation is slightly more complicated, however, because disorder can now occur on both sublattices. This is taken into account as follows: Assuming the number of ways of distributing the cation vacancies V_{cat} on $N_{cat} + V_{cat}$ sites is Ω_1, and the number of ways of distributing V_{an} anion vacancies on $N_{an} + V_{an}$ sites is Ω_2, it can be shown that the configuration entropy change upon the introduction of both defects is given by

$$\Delta S = k \ln \Omega = k \ln \Omega_1 \Omega_2$$

where

$$\Omega = \frac{(N_{cat} + V_{cat})!(N_{an} + V_{an})!}{(N_{cat})!(V_{cat})!(N_{an})!(V_{an})!}$$

where N_{cat} and N_{an} are the total numbers of cations and anions in the crystal, respectively. Following a derivation similar to the one shown for the elemental crystal and taking an MO oxide as an example, i.e., subject to the constraint that $(N_{cat} + n_{cat})/(N_{an} + n_{an}) = 1$, one can show that (see Prob. 6.1*c*) at *equilibrium*

$$\boxed{\frac{V_{an}^{eq} V_{cat}^{eq}}{(N_{an} + V_{an}^{eq})(V_{cat}^{eq} + N_{cat})} \approx \frac{V_{an}^{eq} V_{cat}^{eq}}{N_{an} N_{cat}} = \exp\left(-\frac{\Delta h_s - T\Delta s_S}{kT}\right)} \quad (6.9)$$

where V_{cat}^{eq} and V_{cat}^{eq} are, respectively, the *equilibrium* numbers of cation and anion vacancies. And Δs_S and Δh_S are, respectively, the entropy and enthalpy associated with the formation of a Schottky pair, or $\Delta g_S = \Delta h_S - T\,\Delta s_S$.

This result predicts that the product of the cation and anion vacancy concentrations is a *constant* that depends only on temperature and holds true as long as equilibrium can be assumed.[65] In certain cases, discussed in greater detail subsequently, when Schottky defects dominate, that is, $V_{cat}^{eq} = V_{cat}^{eq} >>$ the sum of all other defects, Eq. (6.9) simplifies to

$$[V_a] = [V_c] = \exp\frac{\Delta s_S}{2k}\exp\left(-\frac{\Delta h_S}{2kT}\right) \tag{6.10}$$

where:

$$[V_c] = \frac{V_{cat}}{V_{cat} + N_{cat}} \quad \text{and} \quad [V_a] = \frac{V_{an}}{V_{an} + N_{an}} \tag{6.11}$$

Note that from hereon, in equations in which defects are involved, square brackets will be used exclusively to represent the mole or site fraction of defects.

Frenkel defect reactions. The Frenkel defect (Fig. 6.3*b*) is one in which a vacancy is created by having an ion in a regular lattice site migrate into an interstitial site. This defect can occur on either sublattice. For instance, the Frenkel reaction for a trivalent cation is

$$M_M^x \Rightarrow V_M''' + M_i^{\bullet\bullet\bullet} \tag{6.12}$$

while that on the oxygen sublattice is

$$O_O^x \Rightarrow O_i'' + V_O^{\bullet\bullet} \tag{6.13}$$

Note that since interstitial sites do *not constitute regular lattice sites*, the Frenkel reaction as written does not violate rule 3 above. The oxides FeO, NiO, CoO and Cu_2O are examples of oxides that exhibit Frenkel defects.

Similar to the Schottky formulation, the number of ways of distributing n_i interstitials on N^* interstitial *sites* is

$$\Omega_1 = \frac{N^*!}{(N^* - n_i)!\,n_i!}$$

Similarly, the number of configurations of distributing V_{cat} vacancies on N_T total sites is

[65] A good analogy comes from chemistry, where it is known that for water at room temperature the product of the concentrations of H^+ and OH^- ions is a constant equal to 10^{14}, a result that is always valid. Increasing the proton concentration decreases the OH^- concentration, and vice versa.

$$\Omega_2 = \frac{N_T!}{(N_T - V_{cat})!V_{cat}!}$$

The configurational entropy is once again $\Delta S = k \ln \Omega_1\Omega_2$. At equilibrium,

$$\boxed{\frac{V_{cat}^{eq} n_i^{eq}}{N_T N^*} \approx \exp\left(-\frac{\Delta g_F}{kT}\right)} \tag{6.14}$$

where Δg_F is the free-energy change associated with the formation of a Frenkel defect.

Note that in general, N^* will depend on crystal structure. For instance, for 1 mol of NaCl, if the ions migrate to tetrahedral sites, $N^* \approx 2N_{Av}$; if they migrate to octahedral sites, then $N^* \approx N_{Av}$.

Worked Example 6.1

Estimate the number of Frenkel defects in AgBr (NaCl structure) at 500°C. The enthalpy of formation of the defect is 110 kJ/mol, and the entropy of formation is 6.6 R. The density and molecular weights are 6.5 g/cm³ and 187.8 g/mol, respectively. State all necessary assumptions.

ANSWER

By taking a basis of 1 mol, assuming that the Frenkel disorder occurs on the cation sublattice, and further assuming that the silver ions go into the tetrahedral sites (i.e., number of interstitial sites = double the number of lattice sites $\approx 2N_{Av}$), it follows that

$$\frac{V_{cat}^{eq} n_i^{eq}}{2(6.02\times10^{23})^2} = \exp\frac{6.6R}{R}\exp\left\{-\frac{110\times10^3}{8.314(500+273)}\right\} = 2.7\times10^{-5}$$

or

$$V_{cat}^{eq} n_i^{eq} \approx 1.96\times10^{43}\,\text{defects/mol}^2$$

As long as the crystal is in equilibrium, this expression is *always* valid; i.e., the left-hand side of the equation will always be equal to 2.7×10^{-5}. Under certain conditions, discussed subsequently, the Frenkel defects can dominate, in which case $V_{cat}^{eq} = n_i^{eq}$ and

$$V_{cat}^{eq} = n_i^{eq} = 4.43\times10^{21}\,\text{defects/mol}$$

and the corresponding number of defects per volume is $4.43\times10^{21}\times6.5/187.7 = 1.5\times10^{20}$ cm^{-3}.

Antistructure disorder or misplaced atoms. These are sites where one type of atom is found at a site normally occupied by another. This defect does not occur in ionic ceramics, but it has been postulated to occur in covalent ceramics like SiC. The notation for such a defect would be Si_C or C_{Si}, and the corresponding defect reaction is

$$C_c + Si_{Si} \Rightarrow Si_C + C_{Si}$$

where the effective charge is assumed to be zero throughout.

Finally, note that for a stoichiometric reaction, all that is happening is the rearrangement of the ions, comprising the crystal on a larger number of lattice sites, which consequently increases the crystal's configurational entropy. In a stoichiometric reaction, the ratio of the atoms comprising the crystal does *not* change.

6.2.6.2 NONSTOICHIOMETRIC DEFECTS

In nonstoichiometric defect reactions, the composition of the crystal changes. Said otherwise, a nonstoichiometric reaction is one in which mass is transferred across the boundaries of the crystal. The possible number of nonstoichiometric defect reactions is quite large, and covering even a fraction of them is not feasible here. The best that can be done is to touch on some of their more salient points.

One of the more common nonstoichiometric reactions that occur at low oxygen partial pressure, is shown in Fig. 6.4, where one of the components (oxygen in this case) leaves the crystal. The corresponding defect reaction is

$$O_O^x \Rightarrow \tfrac{1}{2}O_2(g) + V_O^x \tag{6.15}$$

As the oxygen atom escapes, an oxygen vacancy is created. Given that the oxygen has to leave as a neutral species,[66] it has to leave two electrons (the ones that belonged to the cations in the first place!) behind (Fig.

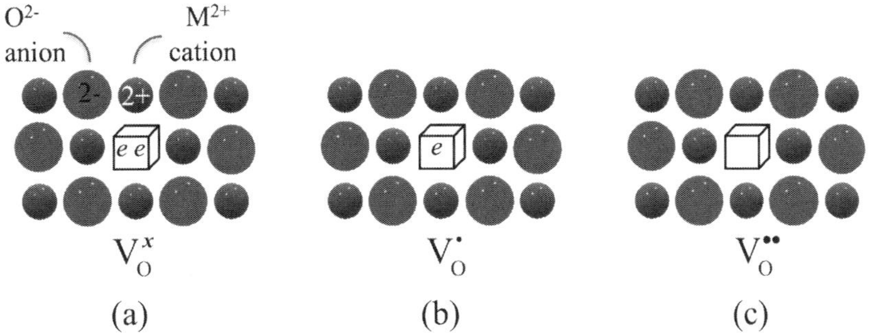

FIGURE 6.4 (*a*) Formation of an oxygen vacancy by the loss of an oxygen atom to the gas phase. This is a nonstoichiometric reaction because the crystal chemistry changes as a result. Note that as drawn, the electrons are *localized* at the vacancy site, rendering its effective charge zero. (*b*) $V_O^{\bullet}$ site is formed when one of these electrons is excited into the conduction band. (*c*) Escape of the second electron into the conduction band creates a $V_O^{\bullet\bullet}$ site.

[66] The reason for this is quite simple: if charged entities were to escape, a charge would build up near the surface that would very rapidly prevent any further escape of ions. See the section on electrochemical potentials in Chap. 5.

6.4*a*). As long as these electrons remain localized at the vacant site, it is effectively neutral {−2 − (−2) = 0}. However, the electrons in this configuration are usually weakly bound to the defect site and are easily excited into the conduction band; i.e., V_O^x acts as a donor—see Chap. 7. The ionization reaction can be envisioned to occur in two stages:

$$V_O^x \Rightarrow V_O^{\bullet} + e'$$

$$V_O^{\bullet} \Rightarrow V_O^{\bullet\bullet} + e'$$

and the net reaction is

$$O_O^x \Rightarrow 1/2 O_2\,(g) + V_O^{\bullet\bullet} + 2e' \tag{6.16}$$

At that stage the oxygen vacancy is doubly ionized (Fig. 6.4*c*) and carries an effective charge of +2 as indicated.

Another possible nonstoichiometric defect reaction is one in which oxygen is incorporated into the crystal interstitially, i.e.,

$$\tfrac{1}{2} O_2(g) \Rightarrow O_i^x \tag{6.17}$$

Ionization can also occur in this case, creating holes in the valence band (i.e., the defect acts as an acceptor) such that

$$O_i^x \Rightarrow O_i' + h^{\bullet}$$

$$O_i' \Rightarrow O_i'' + h^{\bullet}$$

for a net reaction of:

$$\tfrac{1}{2} O_2\,(g) \Rightarrow O_i'' + 2h^{\bullet} \tag{6.18}$$

Nonstoichiometric defect reactions, with the selective addition, or removal, of one of the constituents, naturally lead to the formation of nonstoichiometric compounds. The type of defect reaction that occurs will determine whether an oxide is oxygen- or metal-deficient. For example, reaction (6.16) will result in an oxygen-deficient oxide,[67] whereas reaction (6.18) will result in an oxygen-rich oxide.

REDOX REACTIONS

When one assumes that the electrons or holes generated as a result of **redox reactions**, such as Eqs. (6.16) or (6.17), end up delocalized (i.e., in the conduction or valence bands, see Chap. 7), the implicit assumption is that the cations are only stable in one oxidation state (e.g., +3 for Al or +2 for Mg). For oxides in which

---

[67] Note that oxygen deficiency is also equivalent to the presence of excess metal. One possible such reaction is $M_M + O_O \Rightarrow M_i^x + \tfrac{1}{2} O_2\,(g)$.

the cations can exist in more than one oxidation state, such as transition metal ions, an alternate possibility exists.

As long as the energy associated with changing the oxidation state of the cations is not too large, the electronic defects can—instead of being promoted to the conduction band—change the oxidation state of the cations. To illustrate, consider magnetite, Fe_3O_4, which has a spinel structure with two-thirds of the Fe ions in the +3 state and one-third in the +2 state. One can express the oxidation of Fe_3O_4 in two steps as follows:

$$\tfrac{1}{2}O_2(g) \Leftrightarrow O_O^x + V_{Fe}'' + 2h^{\bullet}$$

$$2Fe^{2+} + 2h^{\bullet} \Rightarrow 2Fe^{3+}$$

The net reaction is

$$\tfrac{1}{2}O_2(g) + 2Fe^{2+} \Rightarrow 2Fe^{3+} + O_O^x + V_{Fe}''$$

In other words, the holes created as a result of the oxidation change the valence state of the cations from +2 to +3.[68] How this idea is used to understand conductivity in La-manganates is discussed in Chaps. 7 and 15.

EXTRINSIC DEFECTS

The discussion so far only applied to pure crystals. Most crystals, however, are far from pure, and their properties, especially electrical and optical, are often dominated by the presence of trace amounts of impurities (see Worked Example 6.3). These impurities cannot be avoided; and even if the starting raw materials are exceptionally pure, it is difficult to maintain the purity levels during subsequent high-temperature processing. The next task is thus to consider impurity incorporation reactions— a task that very rapidly gets out of hand, what with literally thousands of compounds and reactions. What is attempted here instead is to present some simple guidelines for addressing the issue.

First and foremost, impurities usually substitute for the host ion of electronegativity nearest their own, even if the sizes of the ions differ. In other words, cations substitute for cations and anions for anions, irrespective of size differences.[69] For example, in NaCl, Ca and O would be expected to occupy the cation and anion sites, respectively. In more covalent compounds where the electronegativities may be similar, size may play a more important role. Whether an impurity will occupy an interstitial site is more difficult to predict. Most interstitial atoms are small, but even large ions are sometimes found in interstitial sites.

In writing an extrinsic defect incorporation reaction, the following simple bookkeeping operation is of help:

1. Sketch a unit or multiple units of the host (solvent) crystal, as shown in Fig. 6.5*a*.
2. Place a unit, or multiple units, of the dopant (solute) crystal on top of the sketch drawn in step 1, such that the cations are placed on top of the cations and the anions on top of the anions. *It is important to note that the locations of the ions so placed are not where they end up in the crystal. This is simply a bookkeeping operation.*

68 Magnetite can be considered a solid solution of FeO and Fe_2O_3. Thus, upon oxidation, it makes sense that the average oxidation state should move toward Fe_2O_3, that is, more Fe^{3+} ions should be present.

69 This topic is addressed again in Chap. 8, where solid solutions and phase diagrams are considered.

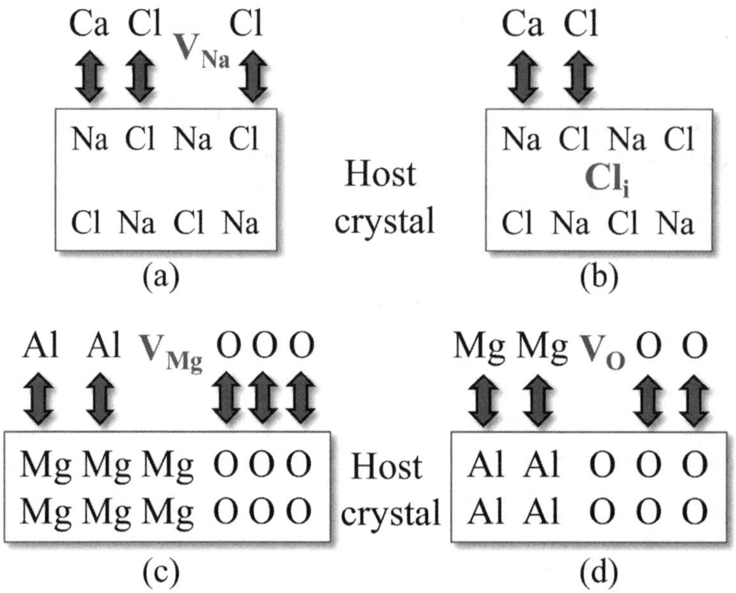

FIGURE 6.5 Bookkeeping technique for impurity incorporation reactions. (*a*) $CaCl_2$ in NaCl leaves a vacancy on cation sublattice. (*b*) An alternate reaction is for the extra Cl ion to go interstitial, which is unlikely, however, given the large size of the Cl ion. (*c*) Al_2O_3 in MgO creates a vacancy on the cation sublattice. (*d*) MgO in Al_2O_3 creates a vacancy on the anion sublattice. Note the effective charges on the defects are not shown.

3. Whatever is left over is the defect that arises, with the caveat that one should try to *minimize* the total number of defects formed.

To illustrate, consider the following examples.

EXAMPLE 6.6

Incorporate $CaCl_2$ into NaCl. From Fig. 6.5*a*, it is obvious that one possible incorporation reaction is

$$CaCl_2 \underset{2NaCl}{\Rightarrow} Ca^{\bullet}_{Na} + V'_{Na} + 2Cl^{x}_{Cl}$$

A second perfectly legitimate incorporation reaction is shown in Fig. 6.5*b*, for which the corresponding defect reaction is

$$CaCl_2 \underset{NaCl}{\Rightarrow} Ca^{\bullet}_{Na} + Cl'_i + Cl^{x}_{Cl}$$

Note that in both cases the overriding concern was the preservation of the regular site ratios of the host crystal. In the first case, two Cl lattice sites were created by the introduction of the dopant, and hence the same number of lattice sites had to be created on the cationic sublattice. But since only one Ca cation was available, a vacancy on the Na sublattice had to be created. In the second case (Fig. 6.5*b*), there is no need to create vacancies because the number of lattice sites created does not change the *regular site ratios* of the host crystal (recall interstitial sites are not considered regular sites).

EXAMPLE 6.7

Doping MgO with Al_2O_3 (Fig. 6.5*c*):

$$Al_2O_3 \underset{3MgO}{\Rightarrow} 2Al_{Mg}^{\bullet} + V_{Mg}'' + 3O_O^x$$

EXAMPLE 6.8

When doping Al_2O_3 with MgO (Fig. 6.5*d*), one possible incorporation reaction is

$$2MgO \underset{Al_2O_3}{\Rightarrow} 2Mg_{Al}' + V_O^{\bullet\bullet} + 2O_O^x$$

It should be emphasized at this point that it is difficult to predict a priori the actual incorporation reactions. In the past, the latter were determined from experiments such as density measurements (see Probs. 6.8 and 6.9) and XRD diffraction. More recently, theoretical calculations have been used to shed light on the problem.

In some oxides, the structure is such as to be able to simultaneously accommodate various types of foreign cations. These multiple substitutions are allowed as long as charge neutrality is maintained. The preservation of site ratios is no longer an issue because the distinction blurs between a regular lattice site and a regular lattice site that is vacant. Good examples are clays, spinels (Fig. 3.10) and the β-alumina structure (Fig. 7.9).

Consider the clay structure shown in Fig. 3.14*b*. The substitution of divalent cations for the trivalent Al ions between the sheets occurs readily as long as for every Al^{3+} substituted, the additional incorporation of a singly charged cation, usually an alkali-metal ion from the surrounding, occurs to maintain charge neutrality such that at any time the reaction

$$Al_2(OH)_4(Si_2O_5) \Rightarrow (Al_{2-x}Na_xMg_x)(OH)_4(Si_2O_5)$$

holds.

The chemistry of spinels is also similar in that multiple substitutions are possible, as long as the crystal remains neutral. For instance, the unit cell of normal spinel, $Mg_8Al_{16}O_{32}$, can be converted to an inverse spinel by substituting the eight Mg ions by four Li and four Al ions to give $Li_4Al_{20}O_{32}$, where the Li ions now reside on the octahedral sites and the Al ions are distributed on the remaining octahedral and tetrahedral sites. It is worth noting here that the vast number of possible structural and chemical combinations in spinels and the corresponding changes in their magnetic, electric and dielectric properties have rendered them indispensable to the electronics industry. In essence, spinels can be considered to be cationic "garbage cans," and, within reasonable size constraints, any combination of cations is possible as long as, at the end, the crystal remains neutral. In that respect, spinels can be compared to another "universal" solvent, namely, glass (see Chap. 9).

6.2.7 ELECTRONIC DEFECTS

In a perfect semiconductor or insulating crystal at 0 K, all the electrons are localized and are firmly in the grasp of the nuclei, and free electrons and holes do not exist. At finite temperatures, however, some of these electrons are knocked loose as a result of lattice vibrations and end up in the conduction band. As elaborated in Chap. 7, for an *intrinsic* semiconductor the liberation of an electron also results in the formation of an electron hole such that the intrinsic electronic defect reaction can be written as

$$\text{Null} \Leftrightarrow e' + h^{\bullet} \tag{6.19}$$

Given that the energy required to excite an electron from the valence to the conduction band is the band gap energy E_g (see Chap. 2), by a derivation similar to the one used to arrive at Eq. (6.14), it can be shown that

$$\frac{np}{N_v N_c} = \exp\left(-\frac{E_g}{kT}\right) = K_i \tag{6.20}$$

where n and p are, respectively, the numbers of free electrons and holes per unit volume; N_c and N_v are the density of states per unit volume in the conduction and valence bands, respectively. It can be shown (App. 7B) that for an intrinsic semiconductor, N_c and N_v are given by

$$N_c = 2\left(\frac{2\pi m_e^* kT}{h^2}\right)^{3/2} \quad \text{and} \quad N_v = 2\left(\frac{2\pi m_h^* kT}{h^2}\right)^{3/2} \tag{6.21}$$

where m_e^* and m_h^* are the effective masses of the electrons and holes, respectively, h is Planck's constant and all other terms have their usual meanings.

Note that the mathematical treatment for the formation of a Frenkel defect pair is almost identical to that of an electron–hole pair. A Frenkel defect forms when an ion migrates to an interstitial site, leaving a hole or a vacancy behind. Similarly, an electron–hole pair forms when the electron escapes into the conduction band, leaving an electron hole, or vacancy, in the valence band. Conceptually, N_c and N_v (in complete analogy to N^* and N_T) can be considered to be the number of energy levels or "sites" over which the electrons and holes can be distributed. The multiplicity of configurations over which the electronic defects can populate these levels is the source of the configurational entropy necessary to lower the free energy of the system.

6.2.8 DEFECT EQUILIBRIA AND KROGER–VINK DIAGRAMS

One of the main aims of this chapter is to relate the concentration of defects to temperature and other externally imposed thermodynamic parameters such as P_{O_2}, a goal that is now almost at hand. This is accomplished by considering defects to be structural elements that possess a chemical potential[70] and hence activity and expressing their equilibrium concentrations by a mass action expression similar to Eq. (5.30)

$$\frac{x_C^c x_D^d}{x_A^a x_B^b} = \exp\left(-\frac{\Delta G^\circ}{kT}\right) = K^{eq}$$

This expression is almost identical to Eq. (5.30), except that here ideality has been assumed and the activities have been replaced by the mole fractions x_i.

To illustrate, consider an MO oxide subjected to the following P_{O_2} regimes:

6.2.8.1 LOW OXYGEN PARTIAL PRESSURE

At very low P_{O_2}, it is plausible to assume that oxygen vacancies will form according to reaction (6.16), or

$$O_O^x \Rightarrow V_O^{\bullet\bullet} + 2e' + \tfrac{1}{2}O_2(g) \quad \Delta g_{red} \tag{I}$$

The corresponding mass action expression is:[71]

$$\frac{[V_O^{\bullet\bullet}][n]^2 P_{O_2}^{1/2}}{[O_O^x]} = K_{red} \tag{6.23}$$

where $K_{red} = \exp(-\Delta g_{red}/kT)$. Note that as long as $V_{an} << N_{an}$ then $[O_O^x] = N_{an}/(N_{an} + V_{an}) \approx 1$.

INTERMEDIATE OXYGEN PARTIAL PRESSURE

Here it is assumed that Schottky equilibrium dominates, i.e.,

$$M_M^x + O_O^x \Leftrightarrow V_M'' + V_O^{\bullet\bullet} \quad \Delta g_s \tag{II}$$

Applying the mass action law yields:

$$\frac{[V_M''][V_O^{\bullet\bullet}]}{[M_M^x][O_O^x]} = K_s = \exp\left(-\frac{\Delta g_s}{kT}\right) \tag{6.24}$$

which, not surprisingly, is identical to Eq. (6.9), since $[O_O^x] \approx [M_M^x] \approx 1$.

---

[70] A distinction has to be made here between chemical and virtual potentials. As discussed at some length in Chap. 5, since the activity or chemical potential of an individual ion or charged defect cannot be defined, it follows that its chemical potential is also undefined. The distinction, however, is purely academic, because defect reactions are always written so as to preserve site ratios and electroneutrality, in which case it is legitimate to discuss their chemical potentials.

[71] In keeping with the notation scheme outlined above, the following applies for electronic defects:

$$[n] = \frac{n}{N_c} \text{ and } [p] = \frac{p}{N_v}$$

It is worth emphasizing once more that [n] and [p] are *dimensionless*, whereas p and n have the units of m^{-3}. The advantage of using site fractions, instead of actual concentrations, in the mass action expression is that the left-hand side of the mass action expression [for example, Eq. (6.23)] would be dimensionless and thus equal to $\exp\{-\Delta g/kT\}$. If other units are used for concentration, the K values have to change accordingly (see Worked Examples 6.2 and 6.3).

HIGH OXYGEN PARTIAL PRESSURE

In this region, a possible defect reaction is:[72]

$$\tfrac{1}{2}O_2(g) \Leftrightarrow O_O^x + 2h^{\bullet} + V_M'' \quad \Delta g_{oxid} \tag{III}$$

for which

$$\frac{[O_O^x][V_M''][p^2]}{P_{O_2}^{1/2}} = K_{oxid} \tag{6.25}$$

where $K_{oxid} = \exp\{-\Delta g_{oxid}/(kT)\}$. Here, increasing P_{O_2} increases the number of cation vacancies.[73]

In addition to Eqs. (6.23) to (6.25), the following reaction

$$\text{Null} \Leftrightarrow e' + h^{\bullet} \tag{IV}$$

is relevant, and at equilibrium

$$[n][p] = K_i = \exp\left(-\frac{E_g}{kT}\right) \tag{6.26}$$

which, not surprisingly, is identical to Eq. (6.20).

At equilibrium, the concentrations of the various defects have to *simultaneously* satisfy Eqs. (6.23) to (6.26), together with one further condition, namely, that the crystal as a whole remain electrically neutral or

$$\sum \text{positive charges}\,(m^{-3}) = \sum \text{negative charges}\,(m^{-3})$$

Note that in writing the neutrality condition, it is the *number of defects per unit volume that is important rather than their mole fractions.* For the example chosen, if one assumes that the only defects present in any appreciable quantities are $h^{\bullet}$, e', $V_O^{\bullet\bullet}$, and V_M''; the neutrality condition reads

$$p + 2V_O^{\bullet\bullet} = 2V_M'' + n \tag{6.27}$$

At this point, all the necessary information needed to relate the concentrations of the various defects to the oxygen potential or partial pressure surrounding the crystal is available. In Eqs. (6.23) to (6.27), there are

---

[72] Note that in general the reactions that are occurring are usually not known. In practice, various defect models are proposed to explain the experimental results. For an a priori prediction, a complete set of all the relevant thermodynamic data for all possible defect reactions (a difficult task indeed, which has only been accomplished for a few oxides) would have to be known to solve the puzzle completely. This comment notwithstanding, as discussed in Computational Materials Science 6.1, DFT calculations are getting quite accurate and can now, in principle, be used to generate Kroger–Vink diagrams.

[73] To see that more clearly, reaction (III) can be rewritten as:

$$\tfrac{1}{2}O_2(g) + V_O^{\bullet\bullet} \Leftrightarrow O_O^x + 2h^{\bullet}$$

In other words, the gas phase O atoms are incorporated into the crystal by filling the oxygen vacancies present, which naturally decreases their concentration. This, in turn, *must* increase the cation vacancy concentration to maintain the Schottky equilibrium, viz. Eq. (6.9).

four unknowns [n, p, $V_O^{\bullet\bullet}$, V_M''] and five equations. Thus, in principle, these equations can be solved simultaneously, provided, of course, that all the Δg values for the various reactions are known. Whereas this is not necessarily a trivial exercise, fortunately the problem can be greatly simplified by appreciating that under various P_{O_2} regimes, one defect pair will dominate at the expense of all other pairs such that only two terms remain in the neutrality condition. How this **so-called Brouwer approximation** is used to solve the problem is as follows.

At sufficiently low P_{O_2}, the driving force to lose oxygen to the atmosphere is quite high [i.e., reaction (I) is shifted to the right], and consequently the number of oxygen vacancies in the crystal increases. If the oxygen vacancies are doubly ionized, it follows that for every oxygen that leaves the crystal, two electrons are left behind in the conduction band (see Fig. 6.4*c*). In that case it is not unreasonable to assume that at sufficiently low P_{O_2}

$$n \approx 2V_O^{\bullet\bullet} >>> \sum \text{(all other defects)} \qquad (6.28)$$

Combining Eq. (6.28) with (6.23) and solving for n or $V_O^{\bullet\bullet}$ yields

$$n = 2V_O^{\bullet\bullet} = [2K'_{red}]^{1/3} P_{O_2}^{-1/6} = [2K_{red} N_{an} N_c^2]^{1/3} P_{O_2}^{-1/6} \qquad (6.29)$$

where $K'_{red} \approx K_{red} N_{an} N_c^2$. According to this relationship, a plot of log n (or $V_O^{\bullet\bullet}$) versus log P_{O_2} should yield a straight line with a slope of −1/6 (Fig. 6.6*a*, range I), i.e., both n and $V_O^{\bullet\bullet}$ decrease with increasing P_{O_2}. The physical picture is simple: Upon reduction, the oxygen ions are being "pulled" out of the crystal, leaving electrons and oxygen vacancies behind.

By similar arguments, in the high-P_{O_2} regime, the electroneutrality condition can be assumed to be $p \approx 2V_M''$, which, when combined with Eq. (6.25), results in

$$p \approx 2V_M'' = [2K'_{oxid}]^{1/3} P_{O_2}^{-1/6} = [2K_{oxid} N_{cat} N_v^2]^{1/3} P_{O_2}^{-1/6} \qquad (6.30)$$

where $K'_{oxid} \approx K_{oxid} N_{cat} N_v^2$. In this region, a plot of the defect concentration vs. log P_{O_2} yields a straight line with a *positive* slope of 1/6 (Fig. 6.6*a*, range III).

In the intermediate P_{O_2} regime, two possibilities exist:

1. $K_s >> K_i$, in which case the neutrality condition becomes

$$V_O^{\bullet\bullet} = V_M'' = \sqrt{K'_s} \qquad (6.31)$$

where $K'_s = N_{cat} N_{an} K_S$, and the point defect concentrations become P_{O_2} independent (Fig. 6.6*a*, range II).

By combining the three regimes, the functional dependence of the defect concentrations over a wide range of P_{O_2} can be succinctly graphed in what is known as a **Kroger–Vink diagram**, shown in Fig. 6.6*a*.

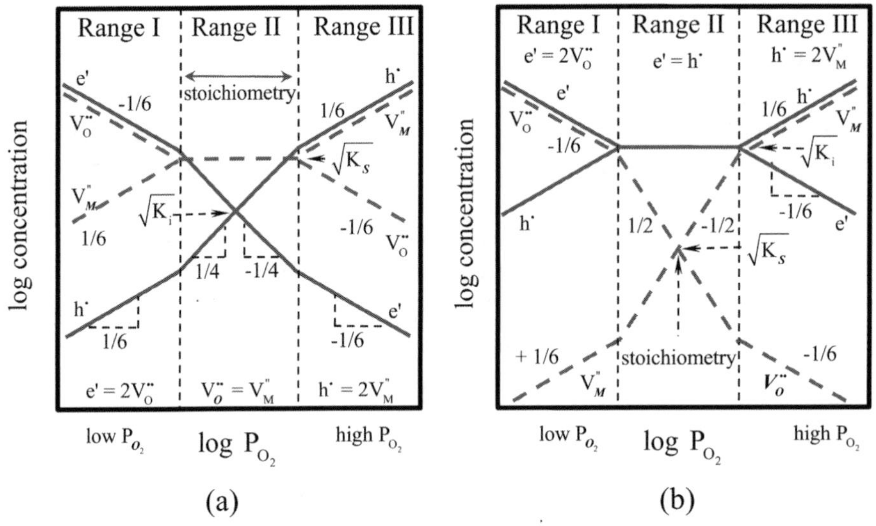

FIGURE 6.6 Variation in defect concentration in an MO oxide as a function of P_{O_2} for: (a) $K_S >> K_i$. Oxide is stoichiometric over a reasonably large P_{O_2} range, (*b*) $K_i >> K_S$. Here the "stoichiometry" reduces to a point where $V''_M = V_O^{\bullet\bullet}$.

2. $K_i >> K_s$, in which case the neutrality condition reads

$$n = p = \sqrt{K'_i}$$

where $K'_i = N_c N_v K_i$. It is left as an exercise to the reader to show that the corresponding Kroger–Vink diagram is the one shown in Fig. 6.6*b*.

Up to this point, the focus has been on the effect of P_{O_2} on the **majority defects**, that is, $V_O^{\bullet\bullet}$ and n under reducing conditions, (V''_M and p under oxidizing conditions and so forth). What about the electron holes and the metal vacancies in the low P_{O_2} region, the so-called **minority defects**?

To answer this question, it is important to appreciate that, at equilibrium Eqs. (6.23) to (6.26) have to be satisfied at all times. For example, equilibrium dictates that, at all times and under all circumstances, the product $[V_O^{\bullet\bullet}][V''_M]$ has to remain a constant equal to K_S. And since it was just established that in the low oxygen pressure region [Eq. (6.29)]:

$$V_O^{\bullet\bullet} = (\text{const.})\left[P_{O_2}\right]^{-1/6}$$

it follows that for Schottky equilibrium to be satisfied, V''_M has to increase by the same power law, or

$$V''_M = (\text{const.})\left[P_{O_2}\right]^{1/6}$$

Similarly, since $n = (\text{const.})\,[P_{O_2}]^{-1/6}$, it follows that to satisfy Eq. (6.26), $p = (\text{const.})\,[P_{O_2}]^{1/6}$. The behavior of the minority defects in region I is plotted in Fig. 6.6*a* and *b* (lower lines).

In the intermediate region when $K_s >> K_i$, Eq. (6.31) holds and $V_O^{\bullet\bullet} = \sqrt{K'_s}$ and is thus independent of P_{O_2}. Substituting this result in Eq. (6.23) yields

$$n = \left[\frac{K'_{red} P_{O_2}^{-1/2}}{V_O^{\bullet\bullet}}\right]^{1/2} = \sqrt{\frac{K'_{red}}{\sqrt{K'_s}}}\,P_{O_2}^{-1/4} \tag{6.32}$$

In other words, n decreases with a 1/4 power with increasing P_{O_2}, which implies that p is increasing with the same power in that range.

Diagrams such as Fig. 6.6 are very useful when trying to understand the relationship between externally imposed thermodynamic parameters, such as the partial pressure of one of the components of a crystal, and any property that is related to the crystal's defect concentration. For instance, as described in detail in Chap. 7, the diffusivity of oxygen is proportional to its vacancy concentration. Thus it follows that if the Kroger–Vink diagram of that given oxide is the one shown in Fig. 6.6*a*, then oxygen diffusivity will be highest at the extreme left, i.e., under reducing conditions, and will decrease with a slope of −1/6 with increasing P_{O_2}, will become constant over an intermediate P_{O_2} regime and finally start to drop again with a slope of −1/6 at higher P_{O_2} values. Similarly, if the oxide is an electronic conductor and its conductivity is measured over a wide enough P_{O_2} range, the conductivity is expected to change from n type at low P_{O_2} to p type at higher oxygen potentials.

Worked Example 6.2

The following information for NaCl is given:

At 600 K: $K'_S = 3.74 \times 10^{35}$ cm^{-6} and K'_F (on cation sublattice) $= 5.8 \times 10^{34}$ cm^{-6}.
At 800 K: $K'_S = 7.06 \times 10^{37}$ cm^{-6} and $K'_F = 1.7 \times 10^{37}$ cm^{-6}.

Calculate the equilibrium number of defects at 600 K and 800 K.

ANSWER

The three pertinent equations are

$$(V'_{Na})(V^{\bullet}_{Cl}) = K'_S$$

$$(V'_{Na})(Na^{\bullet}_i) = K'_F$$

$$V'_{Na} = Na^{\bullet}_i + V^{\bullet}_{Cl} \quad \text{electroneutrality condition}$$

In this problem we have three equations and three unknowns that we need to solve for. Adding the first two equations, one obtains

$$V'_{Na}(Na^{\bullet}_i + V^{\bullet}_{Cl}) = K'_S + K'_F$$

which when combined with the electroneutrality condition, yields

$$(V'_{Na})^2 = K'_S + K'_F$$

Solving for the various concentrations at 600 K one obtains

$$V'_{Na} = 6.6\times10^{17}\ cm^{-3}\ \ Na_i^{\bullet} = 8.8\times10^{16}\ cm^{-3}\ \ V_{Cl}^{\bullet} = 5.7\times10^{17}\ cm^{-3}$$

At 800 K

$$V'_{Na} = 9.4\times10^{18}\ cm^{-3}\ \ Na_i^{\bullet} = 1.8\times10^{18}\ cm^{-3}\ \ V_{Cl}^{\bullet} = 7.5\times10^{18}\ cm^{-3}$$

Note, the final concentration values are given in units of cm^{-3} instead of mole fractions because the equilibrium constants were given in those units. Once such a problem is solved it is always a good idea to go back and substitute the values obtained into the equilibrium expressions above and double check their accuracy.

Worked Example 6.3

Work on a pure MO oxide has shown that:

$$K_o\,(cm^{-9}\,atm^{-1/2}) = \frac{(V''_M)p^2}{P_{O_2}^{1/2}} = 7.7\times10^{63}\exp\left(-\frac{390\ kJ/mol}{RT}\right)$$

(*a*) Calculate the M vacancy concentration at 1327°C in air. State all assumptions.

(*b*) If MO is doped with 10 ppm of Al_2O_3, write down a proper defect reaction that generates M vacancies and calculate the vacancy concentration.

(*c*) Compare the vacancy concentrations you calculate in parts (*a*) and *(b)*. Do you think the properties that depend on the M-vacancy concentration of the doped oxide will be intrinsically or extrinsically dominated. Explain.

(*d*) Estimate the hole concentrations in parts (a) and (b). Which crystal do you expext to be more conductive?

Information you may find useful: Density of MO = 5.4 g/cm³; atomic wt. of M = 25 g/mol. Unit cell = rock salt. Unit cell parameter = 370 pm.

ANSWER

(*a*) Given that the values given for K_o are not dimensionless, we have to keep track of them. At 1327° C, then

$$K_o\,(cm^{-9}\,atm^{-1/2}) = \frac{(V''_M)p^2}{P_{O_2}^{1/2}} = 7.7\times10^{63}\exp\left(-\frac{390000}{8.314(1327+273)}\right) \approx 1.42\times10^{51}$$

To solve for V''_M one has to assume that $2V''_M = p$. Recalling that in air, $P_{O_2} = 0.21$ atm. Putting it all together we get:

$$(V_M'')(2V_M'')^2 = 1.42\times 10^{51}(0.21)^{1/2}$$

$$\left(V_M''\right) = \left[\frac{1}{4}1.42\times 10^{51}(0.21)^{1/2}\right]^{1/3} = 5.5\times 10^{16}\,\text{cm}^{-3} = 5.5\times 10^{22}\,\text{m}^{-3}$$

(*b*) A defect reaction that generates V_M'' upon doping with Al_2O_3 is

$$Al_2O_3 \Rightarrow 2Al_M^{\bullet} + V_M'' + 3O_o^x$$

Hence every mole of Al_2O_3 generates one mole of V_M''. It follows that 10 ppm of Al_2O_3 will generate 10×10^{-6} moles of vacancies. To convert to a number per m^{-3}, we need to calculate the number of cations per m^{-3}. In the rock salt structure there are four cations per unit cell, thus:

$$V_M'' = \frac{4}{\left(370\times 10^{-12}\right)^3}\times 10\times 10^{-6} = 7.9\times 10^{23}\,\text{m}^{-3} = 7.9\times 10^{17}\,\text{cm}^{-3}$$

(*c*) Since the number of extrinsic defects is greater than an order of magnitude of the ones generated in air at 1327°C, it follows that the properties will be dominated by the doping.

(*d*) In part (a), it was assumed that $2V_M'' = p$ and thus $p = 2\times 5.5\times 10^{23} = 1.1\times 10^{23}\ \text{m}^{-3}$. In part (b), since equilibrium is assumed, then all we have to do is insert the value of V_M'' calculated in (b) in the mass action expression, thus:

$$K_o(\text{cm}^{-9}\,\text{atm}^{-1/2}) = \frac{(7.9\times 10^{17})p^2}{P_{O_2}^{1/2}} \approx 1.42\times 10^{51}$$

Solving for p yields $2.9\times 10^{16}\ \text{cm}^{-3}$ or $2.9\times 10^{22}\ \text{m}^{-3}$. This value is roughly a factor of 4 smaller than that calculated above. It follows that the conductivity of the doped oxide – if all our assumptions are correct - should be roughly 4 times less conducting than in the intrinsic case.

Note in this problem, the equilibrium constant was given in cm^{-9} and not in m^{-3} to make the point that you have to keep track of the units to obtain the correct answers. Another important point to be made is that mass action expressions do *not* apply to extrinsic defect reactions.

6.2.9 STOICHIOMETRIC VERSUS NONSTOICHIOMETRIC COMPOUNDS

Based on the foregoing analysis, stoichiometry (defined as the point at which the numbers of anions and cations equal a simple ratio based on the chemistry of the crystal) is a **singular** point that occurs at a very specific P_{O_2}. This immediately begs the question: If stoichiometry is a singular point in a partial pressure domain, then why are some oxides labeled stoichiometric and others nonstoichiometric? To answer the question, examine Table 6.1 in which a range of stoichiometries and chemical stability domains for a number of oxides are listed. The deviation from stoichiometry, defined by Δx, where Δx is the difference between the maximum and minimum values of b/a in a $MO_{b/a}$ oxide, varies from oxide to oxide. Note that FeO and MnO exhibit only positive deviations from stoichiometry, i.e., they are always oxygen-rich, whereas TiO exhibits both negative and positive deviations.

From Table 6.1, one can conclude that an oxide is labeled stoichiometric if Δx is a weak function of P_{O_2}. Conversely, an oxide is considered nonstoichiometric if the effect of P_{O_2} on the composition is significant. This concept can be better appreciated graphically as shown in Fig. 6.7.

As a typical example of a nonstoichiometric compound, consider the variations in composition in MnO as the oxygen partial pressure is varied at 1000 K (Fig. 6.8*a*). MnO is stable between a P_{O_2} of $10^{-34.5}$ atm (below which Mn is the stable phase) and $10^{-10.7}$ atm (above which Mn_3O_4 is the stable phase and O/M = 1.18). The range of stoichiometry is depicted by the dotted lines normal to the x-axis. Such a variation is quite large and consequently MnO is considered a nonstoichiometric oxide.

TABLE 6.1 Range of stoichiometry and existence domains of select oxides at 1000 K

Oxides		Deviation from stoichiometry			Stability regime[a] ($-\log P_{O_2}$)	
		x_{min}	x_{max}	Δx	Min.	Max.
Nonstoichiometric oxides						
TiO_x	TiO	0.65	1.25	0.65	44.2[b]	41.5
TiO_x	Ti_2O_3	1.501	1.512	0.011	41.5	30.1
TiO_x	TiO_2	1.998	2.00	0.008	25.7	–
VO_x	VO	0.8	1.3	0.50	35.9	33.2
Mn_xO	MnO	0.848	1.0	0.152	34.5[b]	10.7
Fe_xO	FeO	0.833	0.975	0.155	21.6[b]	17.9
FeO_x	Fe_3O_4	1.336	1.381	0.045	17.9	10.9
Co_xO	CoO	0.998	1.000	0.012	17.1[b]	2.5
CeO_x	Ce_2O_3	1.50	1.52	0.02		
CeO_{2-x}	CeO_2	0	0.50	0.50		
Ni_xO	NiO	0.999	1.000	0.001	16.5[b]	–
CuO_x	Cu_2O	0.50	0.5016	0.0016	9.97[b]	7.0
UO_x	UO_2	1.65	2.25	0.60		
ZrO_x	ZrO_2	1.70	2.0	0.30		
Li_xWO_3		0	0.50			
$Li_xV_2O_5$		0.2	0.33			
Stoichiometric oxides						
AlO_x	Al_2O_3	1.5000	1.5000		71.3[b]	–
MgO		1.0000	1.0000		51.5[b]	–

[a] See Sec. 5.4 for more details.
[b] In equilibrium with the parent metal.
Source: Adapted from T. B. Reed, *The Chemistry of Extended Defects in Non-Metallic Solids*, L. Eyring and M. O'Keeffe, eds., North-Holland, Amsterdam, 1970.

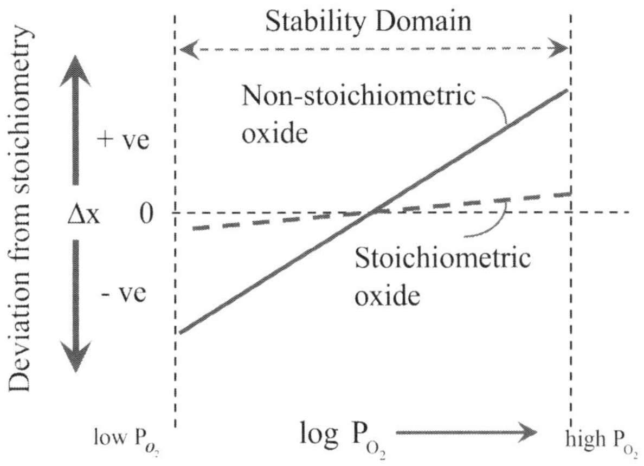

FIGURE 6.7 Distinction between a stoichiometric and a nonstoichiometric $MO_{b/a\pm\delta}$ oxide, where the functional dependence of the changes in stoichiometry, Δx, on P_{O_2} for two hypothetical compounds having the same range of chemical stability is compared. An oxide for which Δx varies widely over its stability domain will be labeled nonstoichiometric, and vice versa.

Similarly, the phase diagram of the Fe-O system is shown in Fig. 6.8*b* and *c*. Note that while FeO and Fe_3O_4 are nonstoichiometric, Fe_2O_3 is stoichiometric.

It is worth noting that as a class, the transition metal oxides are more likely to be nonstoichiometric than stoichiometric. The reason is simple: the loss of oxygen to the environment and the corresponding adjustments in the crystal are much easier when the cations can readily change their oxidation states.

Worked Example 6.4

(*a*) Given the expense and difficulty of obtaining powders that contain much less than 10 ppm of aliovalent impurities, estimate the formation enthalpy for intrinsic defect formation above which the properties would be dominated by the impurities. State all assumptions.

(*b*) Repeat part (*a*) for MnO in equilibrium with Mn_3O_4 at 1000 K.

ANSWER

Make the following assumptions:

- ∞ A fraction (1 ppm) of the impurities create vacancies on one of the sublattices in rough proportion to their concentration. In other words, it is assumed that the mole fraction of extrinsic vacancies on one of the sublattices is on the order of 10^{-6}.
- ∞ Defects are stoichimetric Schottky defects for which formation enthalpy is Δh_S.
- ∞ Ignore ΔS_S.
- ∞ Temperature is 1000°C.

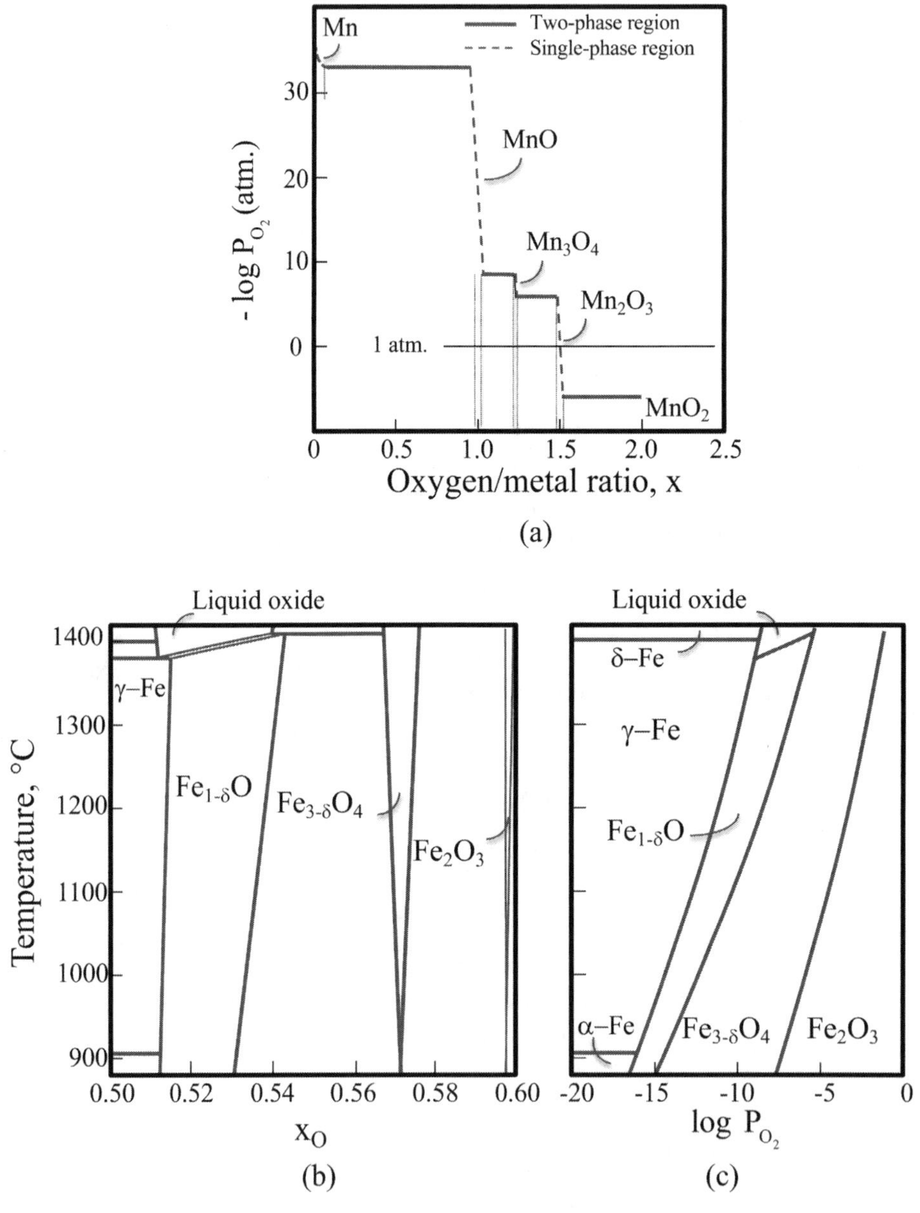

FIGURE 6.8 (*a*) Stability domains of various phases in the Mn-O system and the corresponding deviations in stoichiometry. (T. B. Reed, *The Chemistry of Extended Defects in Non-Metallic Solids*, L. Eyring and M. O'Keeffe, eds., North-Holland, Amsterdam, 1970.) (*b*) Phase diagram of Fe-O system, x_o is mole fraction of oxygen. (*c*) Stability domains of the various phases in Fe-O system. (R. Dieckmann, J. Electrochem. Soc., **116**, 1409, 1969.)

For the solid to be dominated by the intrinsic defects their mole fraction has to exceed the mole fraction of defects created by the impurities (i.e., 10^{-6}). Thus

$$\left[V_M''\right]\left[V_O^{\bullet\bullet}\right] \approx 10^{-12} = \exp\left(-\frac{96{,}500\Delta h_S}{1273 \times 8.314}\right)$$

Solving for Δh_S, one obtains ≈3.03 eV. *Note*: 1 eV/particle = 96 500 J/mol.

This is an important result because it implies that defect concentrations in stoichiometric oxides, or compounds, for which the Schottky or Frenkel defect formation energies are much greater than 3 eV will most likely be dominated by impurities.

(*b*) According to Table 6.1, at 1000 K, MnO in equilibrium with Mn_3O_4 has the composition $MnO_{1.18}$. It follows that for this oxide to be dominated by impurities, the dopant would have to generate a mole fraction of vacancies in excess of 0.18!

EXPERIMENTAL DETAILS: MEASURING NONSTOICHIOMETRY

Probably the easiest and fastest method to find out whether an oxide is stoichiometric is to carry out thermogravimetric measurements as a function of temperature and P_{O_2}. In such experiments, a crystal is suspended from a sensitive balance into a furnace. The furnace is then heated, and the sample is allowed to equilibrate in a gas of known P_{O_2}. Once equilibrium is established (weight change is zero), the P_{O_2} in the furnace is changed suddenly, and the corresponding weight changes are recorded as a function of time.

Typical curves are shown in Fig. 6.9. From the weight gain the new stoichiometry can be easily calculated (see Prob. 6.9). If the same experiment were repeated on, say, MgO or Al_2O_3, the weight changes, over a wide range of P_{O_2} would be below the detectability limit of the most sensitive balances, which is why they are considered stoichiometric oxides.

At this stage it is not a bad idea to think of a nonstoichiometric crystal as some kind of oxygen "sponge" that responds to P_{O_2} in the same way as a sponge responds to humidity. How the oxygen is incorporated and diffuses into the crystal is discussed in the next chapter.

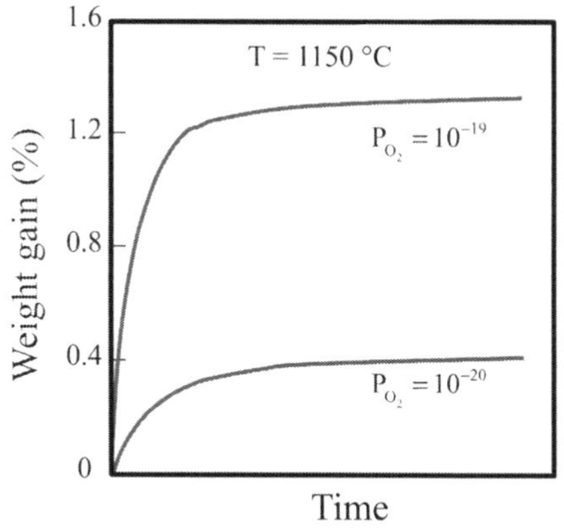

FIGURE 6.9 Typical thermogravimetric results for the oxidation of FeO_x.

6.2.10 ENERGETICS OF POINT DEFECTS FORMATION

Clearly, a knowledge of the free-energy changes associated with the various defect reactions described above is needed to be able to calculate their equilibrium concentrations. Unfortunately, much of that information is lacking for many oxides and compounds.

As noted at the outset of this chapter, the creation of a vacancy can be visualized by removing an ion from the bulk of the solid to infinity, which costs $\approx E_{bond}$, and bringing it back to the crystal surface recovering $\approx E_{bond}/2$. Similarly, the formation of a Schottky vacancy *pair* of a binary MX compound costs $\approx 2E_{bond}/2 \cong E_{bond}$. In general, the lattice energies for the alkali halides fall in the range of 650 to 850 kJ/mol, and hence one would expect Schottky defect formation energies to be of the same order. Experimentally, however, the enthalpies of formation of Schottky and Frenkel defects in alkali halides fall in the range of 100 to 250 kJ/mol (Table 6.2). The discrepancy arises from neglecting the (1) long-range polarization of the lattice as a result of the formation of a charged defect and (2) relaxation of the ions surrounding the defect. When these effects are taken into account, better agreement between theory and experiment is usually obtained.

At the end of this chapter, how defect energies can be calculated from first principles is described briefly.

6.3 LINEAR DEFECTS

6.3.1 DISLOCATIONS

Dislocations were originally postulated to account for the large discrepancy between the theoretical and actual strengths observed during the plastic deformation of metals. For plastic deformation to occur, some part of the crystal has to move, or shear, with respect to another part. If whole planes had to move simultaneously, i.e., all the bonds in that plane had to break and move at the same time, then plastic deformation would require stresses on the order of Y/15 as estimated in Chap. 4. Instead, it is a well-established fact that metals deform at significantly lower stresses. The defect that is responsible for the ease of plastic deformation is known as a *dislocation*. There are essentially two types of dislocations: edge (shown in Fig. 6.10*a*) and screw (not shown). Every dislocation is characterized by the **Burgers vector**, **b**, which is defined as the unit slip distance for a dislocation shown in Fig. 6.10*a*. For an edge dislocation, the Burgers vector is always perpendicular to the dislocation line; for a screw dislocation, it is parallel.

In ionic solids, the structure of dislocations can be quite complex because of the need to maintain charge neutrality. For example, for an edge dislocation to form in an NaCl crystal, it is not possible to simply insert one row of ions as one would do in a metallic crystal. Here two half planes have to be inserted, as shown in Fig. 6.10*c*. The plane shown here is the (010) plane in NaCl and slip would occur along the $(10\bar{1})$ plane.

The structure of dislocations in diamond lattices, which, as discussed in Chap. 3, is adopted quite frequently by elements that have tetrahedral covalent bonding, has to conform to the comparatively rigid tetrahedral bonds, as shown in Fig. 6.10*d*. This, as discussed in Chap. 11, makes them highly resistant to shear, which is why solids such as Si, SiC and diamond are brittle at room temperature.

TABLE 6.2 Defect formation and migration energies for select halides and oxides

Crystal	Defect type	Δh_{form}, kJ/mol	Δs_{form}, in units of R	ΔH_{mig}, kJ/mol	ΔS_{mig}, in units of R
AgCl	Frenkel	140	9.4R	28 (V'_{Ag})	−1.0 (V'_{Ag})
				1–10 ($Ag_i^{\bullet}$)	−3.0 ($Ag_i^{\bullet}$)
AgBr	Frenkel	116	6.6R	30 (V'_{Ag})	
				5–20 ($Ag_i^{\bullet}$)	
β-AgI	Frenkel	67			
BaF_2	Frenkel	190		40–70 ($V_F^{\bullet}$)	
				60–80 (F'_i)	
CaF_2	Frenkel	270	5.5R	40–70 ($V_F^{\bullet}$)	1–2 ($V_F^{\bullet}$)
				80–100 (F'_i)	5 (F'_i)
CsCl	Schottky	180	10.0R	60 (V'_{Cs})	
KCl	Schottky	250	9.0R	70 (V'_K)	2.4 (V'_K)
LiBr	Schottky	180		40 (V'_{Li})	
LiCl	Schottky	210		40 (V'_{Li})	
LiF	Schottky	226	9.6R	70 (V'_{Li})	1 (V'_{Li})
LiI	Schottky	110		40 (V'_{Li})	
NaCl	Schottky	240	10.0R	70 (V'_{Na})	1–3 (V'_{Na})
SrF_2	Frenkel	67		50–100 (V'_F)	
MgO	Schottky	637			
CaO	Schottky	588			
UO_2	Frenkel	328		O_i	
ZrO_2	Frenkel	396		O_i	

Most of these values were taken from J. Maier *Physical Chemistry of Ionic Materials*, Wiley, 2004.

6.3.2 RIPPLOCATIONS

It has long been assumed—implicitly or explicitly—that basal dislocations were the operative micromechanism in the deformation of layered solids. In early 2015, evidence was presented for a new micromechanism termed a *ripplocation*. The latter were defined as surface ripples operative during the deformation of 2D, weakly bonded, van der Waals solids. In 2016, we extended this idea to *all* layered solids and showed—by a combined modeling on graphite with nanoindentation experiments on Ti_3SiC_2—a MAX phase—that ripplocations are essentially an atomic-scale buckling phenomenon. A ripplocation is best described as an atomic-scale ripple (Fig. 6.10*b*) that occurs when a layered solid is loaded in such a way as to cause the layers to buckle. In other words, we showed that ripplocations are not restricted to van der Waals solids, as initially postulated, occur in bulk and thus are *not* necessarily restricted to surfaces, or near surfaces

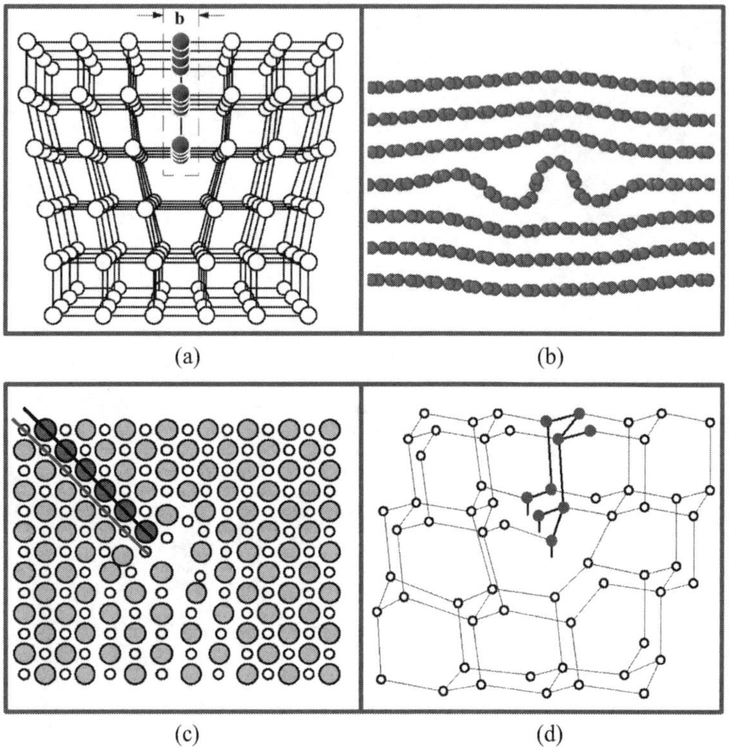

FIGURE 6.10 (*a*) Edge dislocation, the width of which is characterized by a Burgers vector b. (*b*) Schematic of a ripplocation in graphite. This ripplocation was created by adding C atoms shown in gray to the atomic plane that ultimately formed the ripplocation. (*c*) Edge dislocation in NaCl produced by the insertion of two extra half planes of ions (red solid lines and circles). (*d*) A 60° dislocation along <110> in the diamond cubic structure. The glide plane is (111); the extra half plane is depicted by the heavier lines and red circles.

as originally postulated. We also showed that unlike dislocations, ripplocations have no Burgers vector or polarities. Their energies only depend on the amount of "excess" material that needs to be accommodated, or degree of in-plane compressive strain. In graphite, ripplocations are attracted to other ripplocations, both within the same and on adjacent layers, the latter resulting in kink boundaries. Our atomistic calculations on graphite showed them to be quite mobile indeed even at 10 K. Lastly, we made the case that ripplocations are a topological imperative, as they allow atomic layers to glide relative to each other *without* breaking the all-important in-plane bonds.

6.4 PLANAR DEFECTS

Grain boundaries and free surfaces are considered to be planar defects. Free surfaces were discussed in Chap. 4. This section deals with grain boundary structure and grain boundary segregation.

6.4.1 GRAIN BOUNDARY STRUCTURE

A grain boundary, GB, is simply the interface between two grains. The two grains can be of the same material, in which case it is known as a **homophase boundary**, or of two different materials, in which case it is referred to as a **heterophase boundary**. The situation in ceramics is further complicated because often other phases that are only a few nanometers thick can be present between the grains (see below), in which case the grain boundary represents three phases. These phases usually form during processing (see Chap. 10) and can be crystalline or amorphous. In general, the presence or absence of these films has important ramifications on processing, electrical properties and creep, hence their importance.

Typically GBs are distinguished according to their structure as low-angle (<15°), special or random. The easiest to envision is the **low-angle grain boundary**, which can be described as an array of dislocations separated by areas of strained lattice. An example of such a grain boundary is shown in Fig. 6.11*a*, where two dislocations are highlighted with red atoms. The angle of the grain boundary, θ, is determined from the dislocation spacing λ_d and **b**. From Fig. 6.11*a* it is easy to appreciate that the tilt angle or misorientation is given by

$$\sin\theta = \frac{\mathbf{b}}{\lambda_d} \tag{6.33}$$

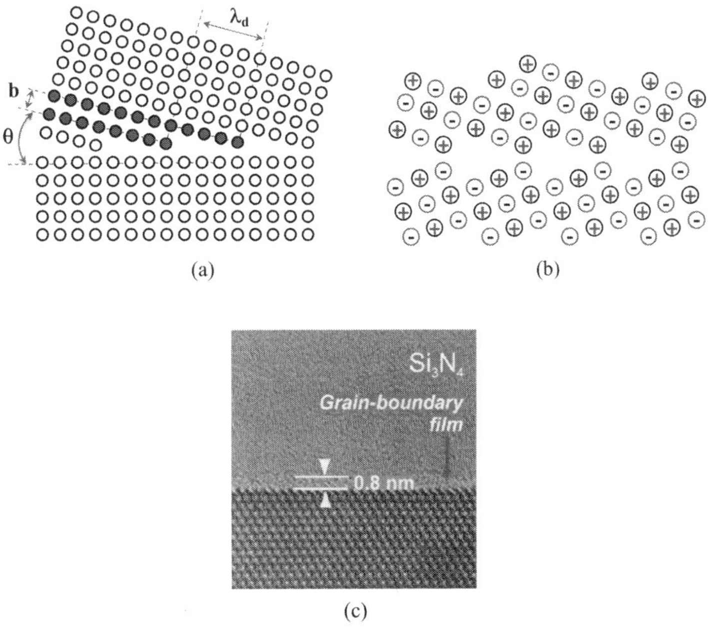

FIGURE 6.11 (*a*) Schematic representation of a low-angle tilt grain boundary made up of a series of dislocations (red circles) with Burgers vectors b spaced λ_d apart. (*b*) Structure of special or coincident boundary in NiO. (*c*) High-resolution TEM image of a 0.8 nm thick grain boundary film in Si_3N_4. (H.-J. Kleebe, J. *Amer. Cer. Soc.*, 85, 43–48, 2002.)

Special or coincident GBs are those in which a special orientational relationship exists between the two grains on either side of the GB. In these boundaries, a fraction of the total lattice sites between the two grains coincide. For example, at 36.87° the Ni and O ions in NiO coincide periodically, as shown in Fig. 6.11*b*. These special GBs have lower energies, diminished diffusivities and higher mobilities relative to general boundaries. They are, however, quite rare and thus their impact on properties is typically not very pronounced.

The vast majority of GBs, however, are neither low-angle nor special, but are believed to be composed of islands of disordered material where the fit is bad, separated by regions where the fit is better. This so-called island model was first proposed by Mott[74] and appears to qualitatively describe one view of GB structures.

6.4.2 IMPURITY SEGREGATION AT GRAIN BOUNDARIES

The role of GB chemistry on properties cannot be overemphasized. For many ceramics, the presence of small amounts of impurities in the starting material can vastly influence their mechanical, optical, electrical and dielectric properties. The effect of impurities is further compounded since they have a tendency to segregate at grain boundaries. If the concentration of solute is not too large, then the ratio of the GB concentration C_{gb} to bulk concentration C_{bulk} depends on the free-energy change due to segregation ΔG_{seg} and is given by

$$\frac{C_{gb}}{C_{bulk}} = \exp\frac{\Delta G_{seg}}{kT} \tag{6.34}$$

One of the contributions to the decrease in free energy comes from the decrease in strain energy resulting from solute misfit in the lattice. It can be shown that this decrease in strain energy scales as $[(r_2 - r_1)/r_1]^2$, where r_1 and r_2 are the ionic radii of the solvent and solute ions, respectively. Hence the larger the radii differences, the greater the driving force for segregation, which has been experimentally verified. Note that it is the absolute size difference that is important; i.e., both smaller and larger ions will segregate to a GB. The reason is simple: The GB is a region of disorder that can easily accommodate different-sized ions as compared to the bulk. Consequently, if ΔG_{seg} is large, the GB chemistry can be quite different from that of the bulk, magnifying the effect of the impurities.

6.4.3 GRAIN BOUNDARY FILMS

As discussed in Chap. 10, in many cases, densification is mediated by liquids because diffusivity in liquids in significantly faster than in solids among other reasons. Upon cooling, these films can end up at the GBs that, as already noted a few times, can affect properties. For example, in Si_3N_4 and Al_2O_3 the presence of such films, especially if they are amorphous, are detrimental to their high-temperature mechanical

[74] N. F. Mott, *Proc. Phys. Soc.*, 60: 391 (1948).

properties in general, and creep in particular. In some cases, the films are so thin that they can only be imaged by a high-resolution transmission electron microscope. An example of an 0.8 nm thick film in Si_3N_4 is shown in Fig. 6.11*c*.

COMPUTATIONAL MATERIALS SCIENCE 6.1: POINT DEFECT ENERGIES

Measuring point defect energies experimentally is quite challenging indeed for several reasons, chief among them is that their concentration, in many cases, is quite small and thus not easy to quantify. It also takes much work to just identify the nature of a given defect. The problem is further compounded by the fact that in many cases the defect concentrations are not intrinsic, but are present because of small concentrations of aliovalent cations. For example, the defect chemistry of rutile is not determined by stoichiometric defects, but rather by trace impurities, that can very from batch to batch.

Luckily, today, DFT calculations can shed much light on the subject. Before delving into the details it is important to appreciate that there are at least two ways by which defects can be modeled: a periodic and a cluster approach. In the former, a supercell—repeated infinitely in all three directions—is built and a defect is introduced in the supercell. However, because the total number of atoms that can be included in a supercell is relatively small—the larger the number the higher the computational cost—the defect concentrations calculated are typically much higher than experimental ones. The opposite is true of the cluster approach. Here a cluster of atoms containing a certain defect is created and then embedded in a perfect crystal. Thus implicity, infinite dilution is assumed.

For a given defect, d, with net charge z_i its formation energy is given by:

$$E_d = E_{tot} - E_{perf} - \sum_i n_i m_i \tag{6.35}$$

where E_{tot} is the energy of a supercell with one defect and E_{perf} is the energy of the perfect crystal. If a cluster approach is used instead, then the energy difference would be between a perfect crystal and one with an imperfect cluster. The last term represents the chemical potentials of the so-called reservoirs (see below).

To better understand the various terms in Eq. (6.35) consider the following simple reaction:

$$MO_2 = n_o O + n_o V_O^x \tag{6.36}$$

To calculate this reaction's energy, we need to calculate the energy of a perfect MO_2 crystal and one in which there are n_o oxygen vacancies formed by the loss of n_o *neutral* O atoms. It is crucial to note here that what is calculated is the formation of V_O^x, whose effective charge is zero. In other words, the calculations assume two electrons are left behind to yield the arrangement shown in Fig. 6.4*a*.

When we extract the O atoms, however, we need to worry about their reference state. If each O atom were placed in an isolated reservoir and prevented from reacting with others, then one would obtain $E_d(O)$. However, O is not a noble gas; its reference state is O_2 and not O. When two O atoms combine, they release energy that has to be accounted for. The last term in Eq. (6.35) thus represents the chemical potentials of the reservoirs. In other words, the energy change that has to be calculated is that for the following reaction:

$$MO_2 = MO_{2-n} + n_o V_O^x + \frac{1}{2} n_o O_2 \tag{6.37}$$

In this case, the sought-after result is: $E_d = E_d(O) - \frac{1}{2}E_{diss}$, where E_{diss} is the dissociation energy of a O_2 molecule (see Sec. 2.5.2 in Chap. 2). Since E_{diss} can be calculated by DFT, all is well as long as it is accounted for.

Transition metal oxides are important in heterogeneous catalysis, photoelectrochemistry and other applications. Depending on their electronic conductivities they can also be used as mixed conductors in fuel cells (see Chap. 7) and sensors. In some of these applications the number of oxygen vacancies is key to their reactivity/usefulness, etc. It is thus important to understand/quantify the energies of oxygen vacancy formation in these oxides.

The energies for formation, E_d of oxygen vacancies [i.e., Eq. (6.37)], calculated by DFT for TiO_2, ZrO_2, V_2O_5 and CeO_2 are summarized in Table 6.3. The calculations first assume V_O^x forms at the surface or in the bulk and subsequently, the fate of the two electrons remaining at the vacant site (e.g., Fig. 6.4*a*) is decided as well.

Lets discuss each oxide separately, with reference to Table 6.3.

i) Rutile and anatase: In TiO_2, upon formation of a bridging oxygen defect on a rutile TiO_2 (110) surface in particular, the excess electrons occupy Ti 3d states localized at two Ti ions close to the defect (but not necessarily both nearest neighbors). The defect reaction can be written as:

$$2Ti_{Ti}^x + O_O^x = 2Ti'_{Ti} + V_O^{\bullet\bullet} + \frac{1}{2}O_2 \tag{6.38}$$

In other words, once the O atoms escape, the two electrons left behind associate with neighboring Ti^{4+} ions changing their oxidation state to +3. In this case, when the vacancies are formed, the atoms around it rearrange. Not shown in Table 6.3, is the higher cost of forming V_O^x in anatase a result that is consistent with the fact that, under similar reducing conditions, it is more difficult to release O from anatase than rutile.

TABLE 6.3 Vacancy formation energies (eV/defect) in bulk and surfaces of select oxides

Oxide	Bulk	Surface	Comments
TiO_2 rutile	4.4±0.25	3.8±0.6 (110)	Sizable rearrangement of atoms
ZrO_2	5.8±0.3	5.5 [(101) t-ZrO_2	Fairly small and localized rearrangement
V_2O_5		1.9±0.2 (001)	Large displacements and bond formation
CeO_2		1.6-5.5 (110)	Small displacements

Source: Ganduglia-Pirovano et al. *Surface Science Reports*, 62 (2007).

Heating single crystals of TiO_2 leads to visible color changes from transparent to first light and then dark blue. The defect reaction is

$$Ti_{Ti}^{x} + O_{o}^{x} = Ti_{Ti}^{'} + V_{O}^{\bullet} + \frac{1}{2}O_2 \tag{6.39}$$

In this case, the combination of the vacancy and the electron associated with it is labeled a color center (see Chap. 16). Note this color center introduces localized levels in the band gap.

ii) Zirconia: DFT calculations of V_O^x formation in ZrO_2 suggest that after the O atom leaves, the remaining electrons are localized at the vacant sites thus again acting as color centers. In this case the defect reaction is:

$$O_o^x = V_O^x + \frac{1}{2}O_2 \tag{6.40}$$

The exact ZrO_2 polymorph has a small effect on E_d. The atomic rearrangements are small and localized.

iii) Vanadia: V_2O_5 is a layered solid comprised of VO_5 tetrahedra, where every V^{5+} cation is surrounded by five oxygens (Fig. 6.12). This solid is considered layered because the interlayer bonds are relatively weak. There are, however, three different O sites. Referring to Fig. 6.12, they are O^1, O^2 and O^3 bonded to one, two and three V^{5+} tetrahedra, respectively. Upon removal of a surface O, the end result depends on whether the resulting structure is allowed to relax or not. Without relaxation, one V^{3+} cation forms and the electrons are localized. However, if the system is allowed to relax, two V^{4+} cations and a V-O-V bond between them form. This bond, in turn, results in a significant reduction in the energy of defect formation (see Table 6.3). X-ray photoelectron spectroscopy of V_2O_5 surfaces have shown evidence for the presence of V^{4+} cations upon reduction.

iv) Ceria: CeO_2 is an insulator with a band gap of about 6 eV. The calculated energies needed to form V_O^x in this oxide vary over a wide range for the simple reason that DFT cannot model f-electrons as well as p- or d-electrons. To obtain more accurate results we have to wait for DFT to do a better job with f-electrons.

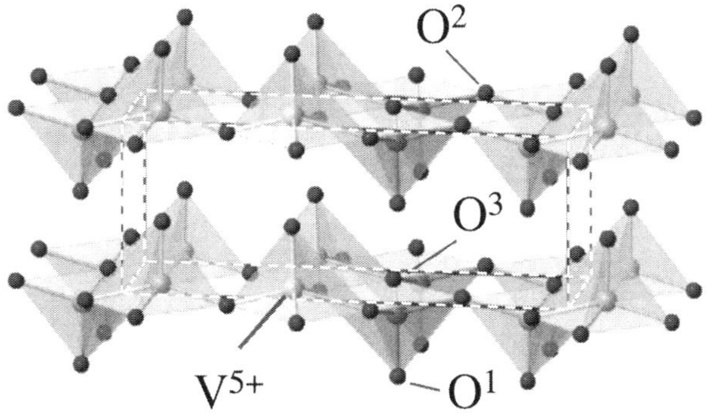

FIGURE 6.12 Schematic of V_2O_5 structure. The tetrahedra in this case are pyramids with square bases. The dashed line represents the unit cell.

The point of this section is not to list the energies needed to form certain defects in select oxides, but rather to highlight the fact that DFT today can not only predict defect energies, but as importantly also say something about the fate of any electrons left behind and, as importantly, any concomitant atomic relaxations.

In other calculations, not discussed here, DFT can in essence also be used to plot Kroger–Vink diagrams. These calculations can predict what defect species are the most likely in an oxide for a given oxygen partial pressure. If done properly, these calculations are quite useful and can shed important light on observed phenomena.

6.5 SUMMARY

Point and electronic defects reduce the free energy of a solid by increasing its entropy. The concentration of defects increases exponentially with temperature and is a function of their free energy of formation.

In compound crystals, balanced-defect reactions must conserve mass, charge neutrality and the ratio of the regular lattice sites. In *pure* compounds, the point defects that form can be classified as either stoichiometric or nonstoichiometric. By definition, stoichiometric defects do not result in a change in chemistry of the crystal. Examples are Schottky (simultaneous formation of vacancies on the cation and anion sublattices) and Frenkel (vacancy–interstitial pairs).

Nonstoichiometric defects form when a compound loses one (or more) of its constituents selectively. Mass is transferred across the crystal boundary, and compensating defects have to form to maintain charge neutrality. For instance, when exposed to severe enough reducing conditions, most oxides will lose oxygen, which in turn results in the simultaneous formation of oxygen vacancies and free electrons. An oxide is labeled nonstoichiometric when its composition is susceptible to changes in its surroundings and is usually correlated to the ease with which the cations (or anions) can change their oxidation states.

Extrinsic defects form as a result of the introduction of impurities. The incorporation of aliovalent impurities in any host compound results in the formation of defects on one of the sublattices, in order to preserve the lattice site ratio.

To relate the concentrations of point and electronic defects to temperature and externally imposed thermodynamic conditions such as oxygen partial pressures, the defects are treated as chemical species and their equilibrium concentrations are calculated from mass action expressions. If the free-energy changes associated with all defect reactions were known, then in principle diagrams, known as Kroger–Vink diagrams, relating the defect concentrations to the externally imposed thermodynamic parameters, impurity levels, etc. can be constructed.

With the notable exception of transition metal oxides some of which exhibit wide deviations from stoichiometry, the concentration of intrinsic or nonstoichiometric defects in most ceramic compounds is so low that their defect concentrations are usually dominated by the presence of impurities.

In addition to point and electronic defects, ceramic crystals can contain dislocations and grain boundaries.

PROBLEMS

6.1. (*a*) Starting with Eq. (6.6), derive Eq. (6.7).

(*b*) On the same graph, plot Eq. (6.6) for two different values of h_d for the same temperature and compare the equilibrium number of vacancies. Which will have the higher number of defects at equilibrium? Why?

(*c*) Following the same steps taken to get to Eqs. (6.6) and (6.7), derive Eq. (6.9).

6.2. (*a*) A crystal of ferrous oxide, Fe_yO, is found to have a lattice parameter $a = 0.43$ nm and a density of 5.72 g/cm³. What is the composition of the crystal (i.e., the value of y in Fe_yO)? Clearly state all assumptions.

Answer: $y = 0.939$.

(*b*) For $Fe_{0.98}O$, the density is 5.7 g/cm³. Calculate the site fraction of iron vacancies and the number of iron vacancies per cubic centimeter.

Answer: Site fraction = 0.02; $V''_{Fe} = 9.7 \times 10^{20}$ cm^{-3}.

6.3. (*a*) Write two possible defect reactions that would lead to the formation of a metal-deficient compound. From your knowledge of the structures and chemistry of the various oxides, cite an example of an oxide that you think would likely form each of the defect reactions you have chosen.

(*b*) Write possible defect reactions and corresponding mass action expressions when possible for:

(i) Oxygen from atmosphere going interstitial
(ii) Schottky defect in M_2O_3
(iii) Metal loss from ZnO
(iv) Frenkel defect in Al_2O_3
(v) Dissolution of MgO in Al_2O_3
(vi) Dissolution of Li_2O in NiO

6.4. Calculate the equilibrium number of Schottky defects n_{eq} in an MO oxide at 1000 K in a solid for which the enthalpy for defect formation is 2 eV. Assume that the vibrational contribution to the entropy can be neglected. Calculate ΔG for this system as a function of the number of Schottky defects for three concentrations, namely, n_{eq}, $2n_{eq}$ and $0.5\, n_{eq}$. State all assumptions. Plot the resulting data as a function of n.

1. Answer: $\Delta G_{n_{eq}} = -0.15\,\text{J}$, $\Delta G_{2n_{eq}} = -0.095\,\text{J}$, and $\Delta G_{0.5n_{eq}} = -0.129\,\text{J}$.

6.5. Compare the concentration of positive ion vacancies in a NaCl crystal due to the presence of 10^{-4} mol fraction of $CaCl_2$ impurity with the intrinsic concentration present in equilibrium in a pure NaCl crystal at 400°C. The formation energy Δh_S of a Schottky defect is 2.12 eV, and the mole fraction of Schottky defects near the melting point of 800°C is 2.8×10^{-4}.

2. Answer: $V'_{Na(extrinsic)} / V'_{Na(intrinsic)} = 6.02 \times 10^{19} / 1.85 \times 10^{17} = 324$

6.6. (*a*) Using the data given in Worked Example 6.2, estimate the free energies of formation of the Schottky and Frenkel defects in NaCl.

Answer: $\Delta g_S = 104.5$ kJ/mol; $\Delta g_F = 114$ kJ/mol.

(*b*) Repeat Worked Example 6.2, assuming Δg_F is double that calculated in part (*a*). What implications does it have for the final concentrations of defects?

6.7. The crystal structure of cubic yttria is shown in Fig. 3.7*b*.

(*a*) What structure does yttria resemble most?

(*b*) Estimate its lattice parameter, a. *Hint*: make use of touching ions to relate the radii to a.

(*c*) Calculate its theoretical density. Compare your result with the actual value of 5.03 g/cm³. Why do you think the two values are different?

(*d*) What stoichiometric defect do you think such a structure would favor? Why?

(*e*) The experimentally determined density changes as a function of the addition of ZrO_2 to Y_2O_3 are as follows:

Composition, mol% ZrO_2	0.0	2.5	5.2	10.0
Density, g/cm^3	5.03	5.04	5.057	5.082

Propose a defect model that would be consistent with these observations.

6.8. (*a*) If the lattice parameter of ZrO_2 is 0.513 nm, calculate its theoretical density.

Answer: 6.06 g/cm³.

(*b*) Write down two possible defect reactions for the dissolution of CaO in ZrO_2. For each of your defect models, calculate the density of a 10 mol% CaO–ZrO_2 solid solution. Assume the lattice parameter of the solid solution is the same as that of pure ZrO_2, namely, 0.513 nm.

Answer: Interstitial 6.03 g/cm³, vacancy 5.733 g/cm³.

6.9. The O/Fe ratio for FeO in equilibrium with Fe is quite insensitive to temperature in the 750 to 1250°C temperature range and is fixed at ≈1.06. When this oxide is subjected to various P_{O_2}s in a thermobalance at 1150°C, the results obtained are shown in Fig. 6.9.

(*a*) Explain in your own words why the higher P_{O_2} resulted in a greater weight gain.

(*b*) From these results determine the O/Fe ratio at the two P_{O_2} values indicated.

a. *Answer*: 1.08 and 1.12.

(*c*) Describe atomistically what you think happens to the crystal that was equilibrated at the higher P_{O_2} if the partial pressure were suddenly changed to 10^{-20} atm.

6.10. (*a*) Show that for any material $\Delta h_{red} + \Delta h_{ox} = 2E_g$. Discuss the implications of this result in terms of what is required for a material to be stoichiometric.

3. (*b*) Carrying out a calculation similar to the one shown in Worked Example 6.4(*a*), show that an oxide can be considered fairly resistant (i.e., stoichiometric) to moderately reducing atmospheres (≈10^{-12} atm) at 1000°C as long as the activation energy for reduction Δh_{red} is greater than ≈ 6 eV. What would happen if Δh_{red} were lower? State all assumptions.

6.11. Using the relevant thermodynamic data, calculate the chemical stability domain (in terms of P_{O_2}) of FeO and NiO. Plot to scale a figure such as Fig. 6.7 for each compound, using the data given in Table 6.1. Which of these two oxides would you consider the more stoichiometric at 1000°C? Why? Information you may find useful can be found in Fig. 5.4.

6.12. As depicted in Fig. 6.11*a*, the grain boundaries in polycrystalline ceramics can be considered to be made up of a large accumulation of edge dislocations. Describe a simple mechanism by which such grains can grow when the material is annealed at an elevated temperature for a long time.

6.13. (*a*) If the initial concentration of Y^{3+} in Al_2O_3 is uniform at C_0, estimate its grain boundary concentration in terms of C_0 at 1200°C if the segregation energy is 0.2 eV.

Answer: 4.85 C_0.

(*b*) Repeat part (*a*) for a temperature of 1300°C. Can you rationalize your result in terms of thermodynamics.

Answer: 4.84 C_0.

ADDITIONAL READING

1. F. A. Kroger and H. J. Vink, *Solid State Physics*, vol. 3, F. Seitz and D. Turnbull, eds., Academic Press, New York, 1956, Chap. 5.
2. J. Maier, *Physical Chemistry of Ionic Materials*, Wiley, 2004.
3. R. J. D. Tilley, *Defects in Solids*, Wiley, Hoboken, NJ, 2008.
4. L. A. Girifalco, *Statistical Physics of Materials*, Wiley-Interscience, New York, 1973.
5. F. A. Kroger, *The Chemistry of Imperfect Crystals*, North-Holland, Amsterdam, 1964.
6. I.Kaur and W. Gust, *Fundamentals of Grain and Interphase Boundary Diffusion*, 2nd ed., Zeigler Press, Stuttgart, 1989.
7. P. Kofstad, *Nonstoichiometry, Diffusion and Electrical Conductivity in Binary Metal Oxides*, Wiley, New York, 1972.
8. O. T. Sorensen, ed., *Nonstoichiometric Oxides*, Academic Press, New York, 1981.
9. W. D. Kingery, Plausible concepts necessary and sufficient for interpretation of ceramic grain-boundary phenomena, I and II, *J. Amer. Cer. Soc.*, **57**, 74–83 (1974).
10. D. Hull, *Introduction to Dislocations*, Pergamon Press, New York, 1965.
11. F. R. N. Nabarro, *Theory of Crystal Dislocations*, Clarendon Press, Oxford, UK, 1967.
12. N. Tallan, ed., *Electrical Conduction in Ceramics and Glasses, Parts A and B*, Marcel Decker, New York, 1974.
13. D. M. Smyth, *The Defect Chemistry of Metal Oxides*, Oxford University Press, 2000.
14. C. B. Carter and M. G. Norton, *Ceramic Materials*, 2nd ed., Springer, New York, 2013.

7

DIFFUSION AND ELECTRICAL CONDUCTIVITY

Electroconductivity depends
On Free Electrons: in Germanium
A touch of Arsenic liberates; in blends
Like Nickel Oxide, Ohms thwart Current. From
Pure Copper threads to wads of Chewing Gum
Resistance varies hugely. Cold and Light
as well as "doping" modify the sum
of Fermi levels, Ion scatter, site
Proximity, and other Factors recondite.

John Updike, The Dance of the Solids*

7.1 INTRODUCTION

The solid state is far from static. Thermal energy keeps the atoms vibrating vigorously about their lattice positions; they continually bump into each other and exchange energy with their neighbors and surroundings. Every now and then, an atom will gain sufficient energy to leave its mooring and migrate. This motion is termed **diffusion**, without which the sintering of ceramics, oxidation of metals, tempering of steels, precipitation hardening of alloys and doping of semiconductors, to name a few phenomena, would not be possible. Furthermore, diffusion is critical in determining the creep and grain growth rates in ceramics, hence its importance.

* J. Updike, Midpoint and Other Poems, A. Knopf, Inc., New York, 1969. Reprinted with permission.

For reasons that will become clear shortly, a prerequisite for diffusion and electrical conductivity is the presence of point and electronic defects. Consequently, this chapter and the preceding one are intimately related, and one goal of this chapter is to make that relationship clear.

In many ceramics, diffusion and electrical conductivity are inextricably linked for two reasons. The first is that *ionic species* can be induced to migrate under the influence of a chemical potential gradient (diffusion) or an electric potential gradient (electrical conductivity). In either case, the basic atomic mechanism is the same, and one of the major conclusions of this chapter is that the diffusivity of a given species is directly related to its conductivity. The second important link is that the defects required for diffusion and electrical conductivity are often created in tandem. For example, as discussed in Chap. 6, the reduction of an oxide can result in the formation of oxygen vacancies and free electrons in the conduction band. Thus the oxide not only becomes more electronically conductive but also the diffusivity of oxygen ions increase.

This chapter is structured as follows: Sec. 7.2.2 deals with the atomistics of diffusion. The relationship between atom diffusivities and activation energies, temperature and defect concentrations responsible for their motion is developed. In Sec. 7.2.3, the diffusion of ions and defects subjected to a chemical potential gradient is dealt with in detail, without reference to their electrical conductivity. But since the two cannot be separated (after all, the diffusion of charged defects is nothing but a current!), Sec. 7.2.4 makes the connection between them. Section 7.2.5 goes a step further, where it is shown that in essence the true driving force that gives rise to a flux of any charged species is the gradient in its electrochemical potential.

In Sec. 7.3.1, the concept of electrical conductivity is introduced. Now in addition to ionic conductivity, the contribution of mobile electronic defects has to be factored into the total conductance. How these electronic defects are introduced in a crystal was first encountered in the previous chapter and is further elaborated on in Sec. 7.3.2.

In Sec. 7.4, situations where the fluxes of the different species are *coupled* are considered. This coupled, or ambipolar, diffusion is of paramount importance since it is responsible for such diverse phenomena as creep, sintering and high-temperature oxidation of metals among others.

In Sec. 7.5, the relationships between the various diffusion coefficients introduced throughout this chapter are elucidated.

7.2 DIFFUSION

There are a number of atomic mechanisms by which atoms or ions can diffuse. Herein the two most important are discussed in detail and a third is simply mentioned for the sake of completion. The first, is the **vacancy mechanism**, that involves the jump of an atom or ion from a regular site into an adjacent vacant site (Fig. 7.1*a*). The second, **interstitial diffusion**—shown schematically in Fig. 7.1*b*—requires the presence of interstitial atoms or ions. The third, less common mechanism is the **interstitialcy mechanism**, shown in Fig. 7.1*c*, where an interstitial atom replaces an atom from a regular site into an interstitial site and takes its place.

In all cases, to make the jump, the atom has to squeeze through a narrow passage, which implies it has overcome an activation or energy barrier. This barrier is known as the *energy of migration* and is shown schematically in Fig. 7.1*d* for the diffusing interstitial ion shown in Fig. 7.1*b*.

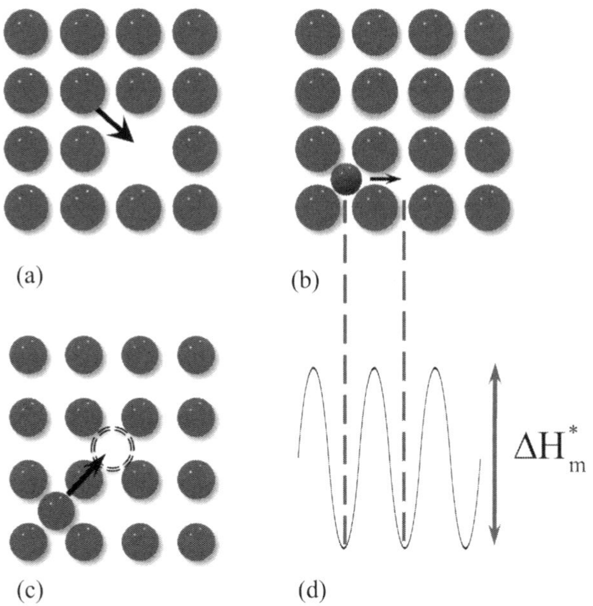

FIGURE 7.1 *(a)* Diffusion of atoms by vacancy mechanism. *(b)* Interstitial diffusion mechanism. *(c)* Interstitialcy mechanism. *(d)* For the interstitial atom shown in *(b)*, to make a jump, it must overcome an energy barrier ΔH_m^*.

7.2.1 PHENOMENOLOGICAL EQUATIONS

Toward the end of Chap. 5 it was noted that for many physical phenomena that entail transport—whether it is charge, mass, or momentum—the assumption is usually made that the resulting flux J is linearly proportional to the driving force F, or

$$J = \beta F \tag{7.1}$$

where β is a system property. In case of diffusion, the relationship between the flux J and the concentration gradient dc/dx is given by Fick's first law,[75] namely,

$$J_A^B \left(\frac{\text{mol}}{\text{m}^2 \text{s}} \right) = -D_A^B \left(\frac{\partial c_A}{\partial x} \right) \left(\frac{\text{m}^2}{\text{s}} \frac{\text{mol}}{\text{m}^3 \cdot \text{m}} \right) \tag{7.2}$$

where D_A^B is the chemical diffusion coefficient of species A in matrix B. The units of D are m^2/s; c_A is the concentration, that can be expressed in a number of units, such as moles or kilograms per cubic meter, etc. The resulting flux is then expressed in units consistent with those chosen for c_A.

The **self-diffusivity D** of an atom or ion is a *measure of the ease and frequency with which that atom or ion jumps around in a crystal lattice in the absence of external forces*, i.e., in a totally

[75] The reason why Fick's first law does not have the same form as Eq. (7.1)—after all, a concentration gradient is not a force—is discussed in greater detail later on.

random fashion. Experimentally, it has long been appreciated that D is thermally activated and could be expressed as

$$D = D_0 \exp\left(-\frac{Q}{kT}\right) \tag{7.3}$$

where Q is the activation energy for diffusion which is not a function of temperature, whereas the pre-exponential term D_0 is a weak function of temperature. It also has been long appreciated that diffusivity depends critically on the stoichiometry and purity level of a ceramic. To understand how these variables affect D, the phenomenon of diffusion has to be considered at the atomic level. Before doing so, however, it is useful to briefly explore how one measures D.

EXPERIMENTAL DETAILS: MEASURING DIFFUSIVITIES

There are many techniques by which diffusion coefficients can be measured. The most common is to anneal a solid in an environment with a well-defined and known activity or concentration of the diffusing species at a given location, for a given time, and then to measure the resulting concentration profile, that is, c(x), of the diffusing species. The profile will clearly depend on the diffusivity (the larger D, the faster and deeper the diffusing species will penetrate into the material), time, and temperature of the diffusion anneal. To determine the diffusivity, Fick's second law

$$\frac{\partial c}{\partial t} = \frac{\partial}{\partial x}\left(D\frac{\partial c}{\partial x}\right) \tag{7.4}$$

has to be solved using the appropriate boundary and initial conditions. The derivation of this equation can be found in most textbooks on diffusion, and it is nothing but a conservation of mass expression. If D is not a function of position, which implies it is also not a function of concentration, then Eq. (7.4) simplifies to read

$$\frac{\partial c}{\partial t} = D\frac{\partial^2 c}{\partial x^2} \tag{7.5}$$

where c is a function of both x and t.

Once Eq. 7.5 is solved for the appropriate initial and boundary conditions employed during the experiment, the value of D that best fits the experimental profile is taken to be the diffusivity of that species at the anneal temperature.

One convenient method to measure c(x) is to use radioactive isotopes of the atom or ion for which the D is to be measured. For instance, if one is interested in the diffusivity of Mn in MnO, a layer of radioactive ^{54}MnO is applied as a thin film on one end of a long rod of nonradioactive MnO. After an appropriate anneal time, t, at a given temperature, the rod is quenched and sectioned normal to the direction of the diffusing species, and the experimental concentration profile is evaluated by measuring the radioactivity of each section. The solution of Fick's second law for these conditions is given by[76]

$$c(x,t) = \frac{\beta}{2\sqrt{\pi Dt}} \exp\left(-\frac{x^2}{4Dt}\right) \tag{7.6}$$

where β is the total quantity per unit cross-sectional area of solute present initially that has to satisfy the condition

$$\int_0^\infty c(x)dx = \beta$$

According to Eq. (7.6), a plot of ln c(x) versus x^2 should result in a straight line with a slope equal to 1/(4Dt). Given t, D is readily calculated.

It is important to note that what one measures in such an experiment is known as a **tracer diffusion coefficient D_{tr}**, which is *not* the same as the self-diffusion coefficient defined above. The two are related, however, by a **correlation coefficient**, f_{cor}, the physics of which is discussed in greater detail in Sec. 7.5.

7.2.2 ATOMISTICS OF SOLID-STATE DIFFUSION

The fundamental relationship relating the self-diffusion coefficient D of an atom or ion to the atomistic processes occurring in a solid[77] is

$$D = \alpha \Omega \lambda^2 \tag{7.7}$$

where Ω is the frequency of *successful* jumps, i.e., number of successful jumps per second; λ is the elementary jump distance which is on the order of the atomic spacing; and α is a geometric constant that depends

---

[76] The methods of solution will not be dealt with here. The interested reader can consult J. Crank, *Mathematics of Diffusion*, 2nd ed., Clarendon Press, Oxford, 1975, or H. S. Carslaw and J. C. Jaeger, *Conduction of Heat in Solids*, Clarendon Press, Oxford, 1959. See also R. Ghez, *A Primer of Diffusion Problems*, Wiley, New York, 1988.

[77] Equation (7.7) can be derived from random walk theory considerations. A particle after n random jumps will, on average, have traveled a distance proportional to $\sqrt{n}$ times the elementary jump distance λ. It can be easily shown that, in general, the characteristic diffusion length is related to the diffusion coefficient D and time t through the equation

$x^2 \approx Dt$

from which it follows that:

$(\sqrt{n}\lambda)^2 \propto Dt$

Rearranging yields

$D \propto \lambda^2 n/t \propto \lambda^2 \Omega$

where Ω is defined as n/t, or the number of successful jumps per second. For further details see P. G. Shewmon, *Diffusion in Solids*, McGraw-Hill, New York, 1963, Chap. 2.

on the crystal structure and whose physical significance will become clearer later on. Here we only remark that for cubic lattices, where only nearest-neighbor jumps are allowed and diffusion is by a vacancy mechanism, $\alpha = 1/\zeta$, where ζ is the coordination number, or number of nearest atoms, of the vacancy.

The frequency Ω is the product of the probability of an atom having the requisite energy to make a jump ν and the probability θ that the site adjacent to the diffusing entity is available for the jump, or

$$\Omega = \nu\theta \tag{7.8}$$

From this relationship it follows that to understand diffusion and its dependence on temperature, stoichiometry and atmosphere requires an understanding of how ν and θ vary under the same conditions. Each is dealt with separately in the following subsections.

JUMP FREQUENCY, ν

For an atom to jump from one site to another, it has to be able to break the bonds attaching it to its original site and to squeeze between adjacent atoms, as shown schematically in Fig. 7.1*d*. This process requires an energy, ΔH_m^*, which is usually much higher than the average thermal energy available to the atoms, viz. $\approx kT$. Hence at any instant only a fraction of the atoms will have sufficient energy to make the jump. Therefore, to understand diffusion, one must first answer the question: at any given temperature, what fraction of the atoms has an energy $\geq \Delta H_m^*$ and are thus capable of making the jump? Or to ask a slightly different question: how often, or for what fraction of time, does an atom have sufficient energy to overcome the diffusion barrier?

To answer this question, the ***Boltzmann distribution law*** is invoked, which states that the probability P of a particle having an energy ΔH_m^* or greater is given by:[78]

$$P(E > \Delta H_m^*) = (\text{const.})\exp\left(-\frac{\Delta H_m^*}{kT}\right) \tag{7.9}$$

where k is Boltzmann's constant and T is the temperature in Kelvin.

It follows that the υ with which a particle can jump, provided that an adjacent site is vacant, is equal to the probability that it is found in a state of sufficient energy to cross the barrier multiplied by the frequency ν_0 at which that barrier is being approached. In other words,

$$\nu = \nu_0 \exp\left(-\frac{\Delta H_m^*}{kT}\right) \tag{7.10}$$

where ν_0 is the natural vibration of the atoms,[79] which is on the order of 10^{13} s^{-1} (see Worked Example 5.3*b*). For low temperatures or large values of ΔH_m^*, the frequency of successful jumps becomes vanishingly small, which is why, for the most part, solid-state diffusion occurs readily only at higher temperatures.

[78] Equation (7.9) is only valid ΔH_m^* is >>kT, the average energy of the atoms in the system. In most cases for solids, that is the case. For example, the average energy of the atoms in a solid is on the order of kT, which at room temperature is ≈0.025 eV and at 1000°C is ~0.11 eV. Typical activation energies for diffusion, vacancy formation, etc. are on the order of a few eV, so all is well.

[79] This is easily arrived at by equating the vibrational energy hv to the thermal energy kT. At 1000 K, $\nu_0 \cong 2 \times 10^{13}$ s^{-1}.

Conversely, at sufficiently high temperatures, that is, $kT \gg \Delta H_m^*$, the barrier ceases to be one and every vibration could, in principle, result in a jump.

PROBABILITY Θ OF SITE ADJACENT TO DIFFUSING SPECIES BEING VACANT

The probability of a site being available for the diffusing species to make the jump will depend on whether one considers the motion of the defects or of the ions themselves. Consider each separately.

Defect diffusivity. As noted above, the two major defects responsible for the atom mobility are vacancies and interstitials. For both, at low concentrations (which is true for the vast majority of solids) the site adjacent to the defect will almost always be available for it to make the jump and thus $\theta \approx 1$.

There is a slight difference between vacancies and interstitials, however. An interstitial can, and will, make a jump with a rate that depends solely on its frequency of successful jumps ν_{int}. By combining Eqs. (7.7), (7.8) and (7.10), with $\theta_{int} = 1$, the interstitial diffusivity D_{int} is given by

$$D_{int} = \alpha_{int}\lambda^2 \nu_0 \exp\left(-\frac{\Delta H^*_{m,int}}{kT}\right) \tag{7.11}$$

where $\Delta H^*_{m,int}$ is the activation energy needed by the interstitial to make the jump.

For a vacancy, however, the probability of a successful jump is increased ζ-fold, where ζ is the number of atoms adjacent to that vacancy (i.e., the coordination number of the atoms), since if *any* of the ζ neighboring atoms attains the requisite energy to make a jump, the vacancy will jump. Thus, for vacancy diffusion

$$\nu_{vac} = \zeta \nu_0 \exp\left(-\frac{\Delta H^*_m}{kT}\right)$$

Combining this equation with Eqs. (7.7) and (7.8), assuming $\theta_{vac} \approx 1$, yields

$$D_{vac} = \alpha\zeta\lambda^2 \nu_0 \exp\left(-\frac{\Delta H^*_m}{kT}\right) \tag{7.12}$$

Atomic or ionic diffusivity. In contrast to the defects, for an atom, or ion, in a regular site, $\theta \ll 1$, because most of its nearest neighbors are occupied by other atoms. In that situation, the *probability of a site being vacant is simply equal to the mole or site fraction, denoted by Λ (lambda), of vacancies in that solid.* Thus, the frequency of successful jumps for diffusion of atoms by a *vacancy mechanism* is given by

$$\Omega = \theta \nu_{vac} = \Lambda \zeta \nu_0 \exp\left(-\frac{\Delta H^*_m}{kT}\right)$$

The factor ζ appears here because the probability of a site next to a diffusing atom being vacant is increased ζ-fold. The diffusion coefficient is given by

$$D_{ion} = \alpha\lambda^2 \Lambda \zeta \nu_0 \exp\left(-\frac{\Delta H^*_m}{kT}\right) \tag{7.13}$$

Comparing Eqs. (7.12) and (7.13) reveals an important relationship between vacancy and ion diffusivity, namely,

$$D_{ion} = \Lambda D_{vac} \tag{7.14}$$

Given that usually $\Lambda \ll 1$, it follows that $D_{ion} \ll D_{vac}$, a result that at first sight appears paradoxical—after all one is dealing with the same species.[80] Going one step further, however, and noting that $\Lambda \approx c_{vac}/c_{ion}$ [Eq. (6.11)], where c_{vac} and c_{ion} are the concentrations of vacancies and ions, respectively, we see that

$$\boxed{D_{ion} c_{ion} = D_{vac} c_{vac}} \tag{7.15}$$

Now the physical picture is a little easier to grasp: the defects move often (high D) but are not that numerous—the atoms move less frequently, but there are lots of them. The full implication of this crucial result will become obvious shortly.

For the sake of simplicity, in deriving Eq. (7.13), the effect of the jump on the vibrational entropy was ignored. This is taken into account by postulating the existence of an excited equilibrium state (Fig. 7.4*a*), albeit of very short duration, that affects the frequency of vibration of its neighbors and is associated with an entropy change given by $\Delta S_m^* \approx kT \ln(\nu'/\nu)$, where ν and ν' are the frequencies of vibration of the ions in their ground and activated states, respectively. A more accurate expression for D_{ion} thus reads

$$D_{ion} = \alpha \lambda^2 \Lambda \zeta \nu_o \exp\left(-\frac{\Delta G_m^*}{kT}\right) \tag{7.16}$$

where ΔG_m^* is defined as

$$\Delta G_m^* = \Delta H_m^* - T\Delta S_m^* \tag{7.17}$$

Putting all the pieces together, one obtains a final expression that most resembles Eq. (7.3), viz,

$$\boxed{D_{ion} = \nu_o \lambda^2 \alpha \zeta \Lambda \exp\frac{\Delta S_m^*}{k} \exp\left(-\frac{\Delta H_m^*}{kT}\right) = D_o \exp\left(-\frac{Q}{kT}\right)} \tag{7.18}$$

except that now the physics of why diffusivity takes that form should be clearer. The temperature dependence of diffusivity of some common ceramics is shown in Fig. 7.2.

The values of the activation energies Q and their variation with temperature are quite useful in deciphering the nature of the diffusional processes occurring. For instance, if Λ is thermally activated, as in the case of intrinsic point defects, then the energy needed for the defect formation will appear in the final expression for Q. If, however, the vacancy or defect concentration is fixed by impurities, then Λ is no longer thermally activated but is proportional to the concentration of the dopant. The following worked examples

80 The implication here is that the vacancies are jumping around much more frequently than the atoms, which indeed is the case. For example, in a simulation of a diffusion process, if one were to focus on a vacancy, its hopping frequency would be quite high since it does not have to wait for a vacant site to appear before making a jump. If, however, one were to focus on a given atom, its average hopping frequency would be much lower because that atom will hop if, and only if, a vacancy appears next to it.

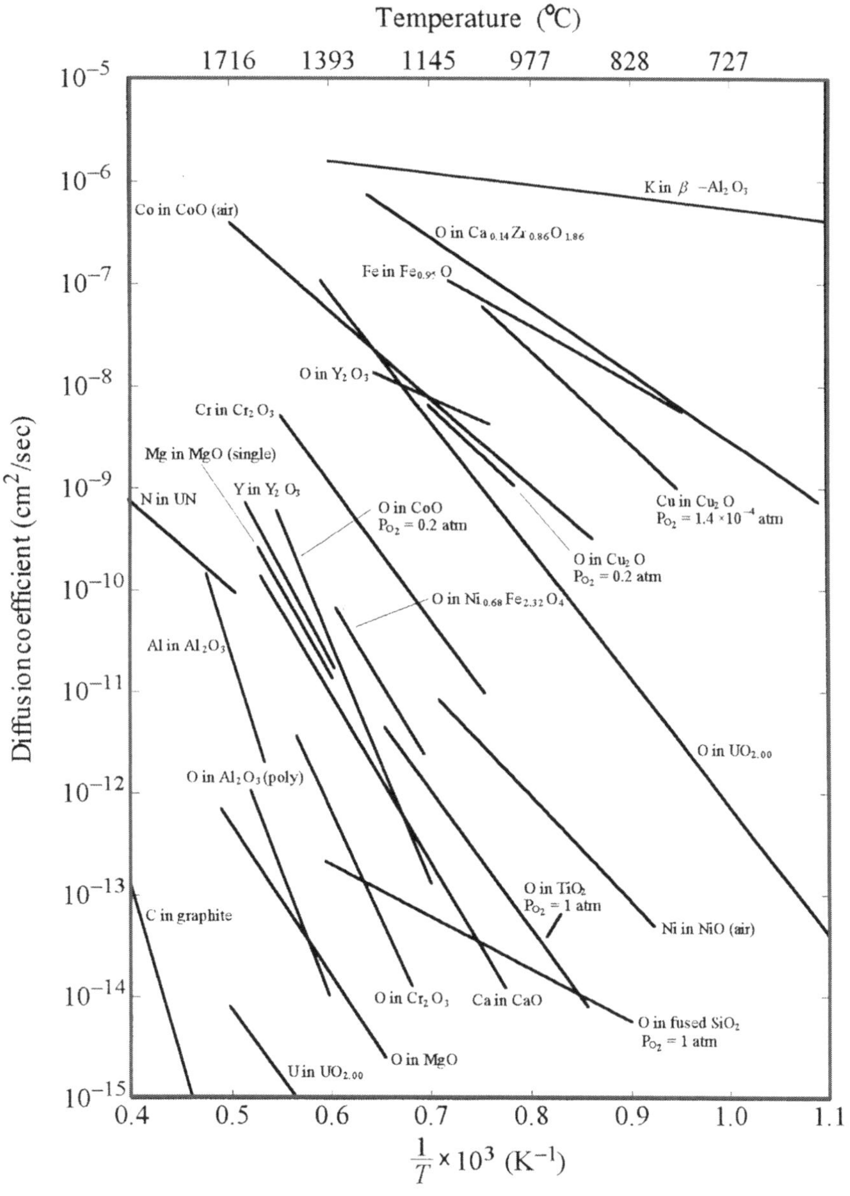

FIGURE 7.2 Temperature dependence of diffusion coefficients for some common ceramic oxides. (Adapted from W. D. Kingery, H. K. Bowen, and D. R. Uhlmann, *Introduction to Ceramics*, 2nd ed., Wiley, New York, 1976. Reprinted with permission.)

should make that point clearer. Finally, it is worth noting here that the pre-exponential term D_0, calculated from first principles, i.e., from Eq. (7.18), does not, in general, agree with experimental data, for reasons that are not entirely clear at this time.

Worked Example 7.1

For Na^+ ion migration in NaCl, ΔH_m^* is 77 kJ/mol, while the enthalpy and entropy associated with the formation of a Schottky defect are, respectively, 240 kJ/mol and 10 R (see Table 6.2).

(*a*) At approximately what temperature does the diffusion change from extrinsic (i.e., impurity-controlled) to intrinsic in a NaCl–$CaCl_2$ solid solution containing 0.01% $CaCl_2$? You can ignore ΔS_m^*.

(*b*) At 800 K, what mole percent of $CaCl_2$ must be dissolved in pure NaCl to increase D_{Na^+} by an order of magnitude?

ANSWER

(*a*) To solve this problem, an expression for D_{Na^+} as a function of temperature in both the extrinsic and intrinsic regions has to be derived. Once derived, the two expressions are equated, and T is solved for. In the intrinsic region, the vacancy concentration is determined by the Schottky equilibrium, or

$$\left[V'_{Na}\right]\left[V^{\bullet}_{Cl}\right] = \exp\frac{\Delta S_S}{k}\exp\left(-\frac{\Delta H_S}{kT}\right)$$

where ΔH_S and ΔS_S are, respectively, the Schottky formation energy and entropy. Assuming that the Schottky defects dominate, that is, $[V'_{Na}] = [V^{\bullet}_{Cl}]$ then

$$[V'_{Na}] = [V^{\bullet}_{Cl}] = \exp\frac{\Delta S_S}{2k}\exp\left(-\frac{\Delta H_S}{2kT}\right) \tag{7.19}$$

which when combined with Eq. (7.18), and noting that $\Lambda = [V'_{Na}]$ yields the desired expression in the intrinsic regime

$$D_{Na^+} = \lambda^2\alpha\zeta\nu_0\exp\frac{\Delta S_S}{2k}\exp\left(-\frac{\Delta H_S}{2kT}\right)\exp\left(-\frac{\Delta H_m^*}{kT}\right) \tag{7.20}$$

If the following incorporation reaction (see Chap. 6)

$$CaCl_2 \Rightarrow V'_{Na} + Ca^{\bullet}_{Na} + 2Cl^{x}_{Cl}$$

is assumed, it follows that in the extrinsic region, for every 1 mol of $CaCl_2$ dissolved in NaCl, 1 mol of Na vacancies is created. In other words, $\Lambda = [V'_{Na}] = [Ca^{\bullet}_{Na}] = 0.0001$. Thus in the extrinsic regime, the vacancy concentration is fixed and independent of temperature. In other words,

$$D_{Na^+} = [Ca^{\bullet}_{Na}]\lambda^2 \alpha \zeta \nu_0 \exp\left(-\frac{\Delta H^*_m}{kT}\right) \tag{7.21}$$

Equating Eqs. (7.20) and (7.21) and solving for T yields a temperature of 743°C.

(*b*) At 800 K, the intrinsic mole fraction of vacancies [Eq. (7.19)] is 2.2×10^{-6} or 1.3×10^{18} vacancies per mole. To increase D_{Na^+} by an order of magnitude, the doping must create 10 times the number of intrinsic vacancies. It follows that the addition of 2.2×10^{-5} mol fraction of $CaCl_2$ to a mole of NaCl should do the trick.

Note that when the defect concentrations are intrinsically controlled, the activation energy for their formation appears in the final expression for D [i.e., Eq. (7.20)], whereas when the defect concentration was extrinsically controlled, the final expression includes only migration energy terms. How this fact is used to experimentally determine both ΔH^*_m and ΔH_S is discussed in Worked Example 7.3. Before doing so, we discuss how to interpret the effects of P_{O_2} on D.

Worked Example 7.2

The functional dependence of the diffusion of ^{54}Mn in MnO on P_{O_2} is shown in Fig 7.3. Explain the origin of the slope. Would increasing the temperature alter the slope?

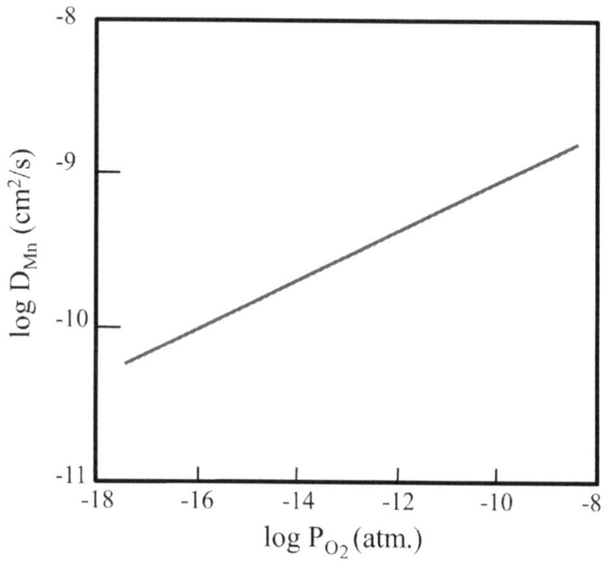

FIGURE 7.3 Functional dependence (on log–log plot) of the diffusion coefficient of ^{54}Mn in MnO on oxygen partial pressure.

ANSWER

Since D_{Mn} clearly depends on the oxygen partial pressure, P_{O_2} the first step in solving the problem is to relate the diffusivity to P_{O_2}. Assuming the diffusivity of Mn in MnO occurs by a vacancy diffusion mechanism, i.e., $\Lambda = [V''_{Mn}]$, and replacing D_o in Eq. (7.18) by $[V''_{Mn}]D'_o$, one sees that

$$D_{Mn} = [V''_{Mn}]D'_o \exp(-\Delta H^*_m/kT)$$

The next step is to relate $[V''_{Mn}]$ to P_{O_2}. Given that the slope of the curve is +1/6, that is, increasing P_{O_2} increases the diffusivity of Mn, the most likely defect reaction occurring is reaction (III) in Chap. 6. If now one assumes the neutrality condition to be $p = 2[V''_{Mn}]$, then from the mass action expression given by Eq. (6.30), it follows that

$$[V''_{Mn}] = (\text{const.})P_{O_2}^{+1/6}$$

Combining this result with the expression for D_{Mn} reveals the observed behavior. The physics of the situation can be summed up as follows: Increasing P_{O_2} (i.e., going from left to right in Fig. 7.3) decreases the concentration of oxygen vacancies which, in order to maintain the Schottky equilibrium, results in an increase in $[V_{Mn}]$ and a concomitant increase in D_{Mn}.

Note that had $[V_{Mn}]$ been fixed by extrinsic impurities, then D_{Mn} would not be a function of P_{O_2}. Finally, increasing the temperature should, in principle, only shift the lines to higher values but not alter their slopes.

7.2.3 DIFFUSION IN A CHEMICAL POTENTIAL GRADIENT

In the foregoing discussion, the implicit assumption was that diffusion was totally *random*, a randomness that was assumed in defining D by Eq. (7.7). This self-diffusion, however, is of no practical use and is not easily measured. Diffusion is important inasmuch as it can be used to effect compositional and microstructural changes. In such situations, atoms diffuse from areas of higher free energy, or chemical potential, to areas of lower free energy, in which case the process is no longer random, but is now biased in the direction of decreasing free energy.

Consider Fig. 7.4*b*, where an ion is diffusing in the presence of a chemical potential gradient $d\mu/dx$. If the chemical potential is given per mole, then the gradient or *force per atom*, f, is given by

$$f = \frac{\Xi}{\lambda} = -\frac{1}{N_{Av}}\frac{d\mu}{dx} \tag{7.22}$$

where N_{Av} is Avogadro's number.[81] Consequently (see Fig. 7.4*b*), the difference Ξ between the energy barrier in the forward and backward directions is

$$\Xi = \lambda f = -\lambda \frac{d[\mu/N_{Av}]}{dx} \tag{7.23}$$

[81] Note that f is defined here, and throughout this book, as *positive*.

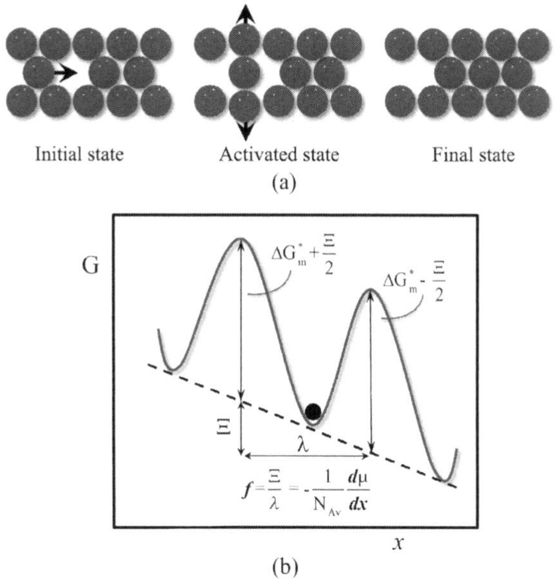

FIGURE 7.4 (*a*) Schematic of activated state during diffusion. (*b*) Diffusion of an ion down a chemical potential gradient.

The forward rate for the atom to jump is thus proportional to

$$\nu_{\text{forward}} = \Lambda\alpha\zeta\nu_0 \exp\left(-\frac{\Delta G_m^* - \Xi/2}{kT}\right) \tag{7.24}$$

while the backward jump rate is

$$\nu_{\text{back}} = \Lambda\alpha\zeta\nu_0 \exp\left(-\frac{\Delta G_m^* + \Xi/2}{kT}\right) \tag{7.25}$$

and Λ appears here because, as noted earlier, for a jump to be successful, the site into which the atom is jumping must be vacant. Also α is the same constant that appears in Eq. (7.7) and whose physical meaning becomes a little more transparent here—it is a factor that takes into account that only a fraction of the total $\zeta\nu$ hops are hops in the x-direction or direction of interest. For instance, in simple cubic lattices for which $\zeta = 6$, only one-sixth of successful jumps are in the forward x-direction, that is, $\alpha = 1/6$ and thus $\alpha\zeta = 1$.

The existence of a chemical potential gradient will bias the jumps in the forward direction, and the net rate will be given by

$$\begin{aligned} \nu_{\text{net}} = \nu_{\text{forward}} - \nu_{\text{back}} &= \Lambda\alpha\zeta\nu_0 \left\{ \exp\left(\frac{\Delta G_m^* - \Xi/2}{kT}\right) - \exp\left(-\frac{\Delta G_m^* + \Xi/2}{kT}\right) \right\} \\ &= \Lambda\alpha\zeta\nu_0 \exp\left(-\frac{\Delta G_m^*}{kT}\right)\left\{1 - \exp\left(-\frac{\Xi}{kT}\right)\right\} \end{aligned} \tag{7.26}$$

In general, chemical potential gradients are small compared to the thermal energy, i.e., $\Xi/(kT) \ll 1$, and Eq. (7.26) reduces to ($e^{-x} \cong 1 - x$ for small x)

$$\nu_{net} = \Lambda\alpha\zeta\nu_o\left(\frac{\Xi}{kT}\right)\exp\left(-\frac{\Delta G_m^*}{kT}\right) \tag{7.27}$$

The average drift velocity v_{drift} is given by $\lambda\nu_{net}$, which, when combined with Eqs. (7.27) and (7.23), gives

$$v_{drift} = \lambda\nu_{net} = \frac{\alpha\lambda^2\zeta\nu_o\Lambda}{kT}\exp\left(-\frac{\Delta G_m^*}{kT}\right)f \tag{7.28}$$

and the resulting flux

$$J_i = c_i v_{drift} = \frac{c_i}{kT}\left\{\alpha\lambda^2\Lambda\zeta\nu_o\exp\left(-\frac{\Delta G_m^*}{kT}\right)\right\}f \tag{7.29}$$

where c_i is the total concentration (atoms/ions per cubic meter) of atoms or ions diffusing through the solid. Since the term in brackets is nothing but D_{ion} [Eq. (7.16)] it follows that

$$\boxed{J_{ion} = \frac{c_{ion}D_{ion}}{kT}f} \tag{7.30}$$

This equation is of fundamental importance because

1. It relates the flux to the *product* $c_{ion}D_{ion}$. The full implication of Eq. (7.15), namely, that $D_{ion}c_{ion} = c_{vac}D_{vac}$, should now be obvious. When one is considering the diffusion of a given species, it is immaterial whether one considers the ions themselves or the defects responsible for their motion: *the two fluxes have to be, and are, equal.*
2. It relates the flux to a *driving force*, f. Given that f has the dimensions of force, Eq. (7.29) can be considered a true flux equation in that it is identical in form to Eq. (7.1). Note also that this relationship has general validity and is not restricted to chemical potential gradients. For instance, as discussed in the next two sections, f can be related to gradients in electrical or electrochemical potentials as well.
3. It can be shown (see App. 7A) that for ideal and dilute solutions Eq. (7.30) is identical to Fick's first law, that is, Eq. (7.2).

7.2.4 DIFFUSION IN AN ELECTRIC POTENTIAL GRADIENT

The situation where the driving force is a chemical potential gradient has just been addressed. If, however, the driving force is an electric potential gradient, then the force on the ions is given by[82]

[82] As defined here, f is positive for positive charges and negative for negative charges. This implies that positive charges will flow down the potential gradient, whereas negative charges flow "uphill," so to speak.

$$f_i = -z_i e \frac{d\phi}{dx} \tag{7.31}$$

where ϕ is the electric potential in volts and z_i is the net charge on the moving ion or defect. The current density $I_i(A/m^2) = C/(m^2 \cdot s)$ is related to the ionic flux J_{ion}[atoms/$m^2 \cdot s$)], by

$$I_i = z_i e J_{ion} \tag{7.32}$$

Substituting Eqs. (7.31) and (7.32) in Eq. (7.30) shows that

$$I_i = z_i e J_{ion} = \frac{z_i e c_{ion} D_{ion}}{kT} f = -\frac{z_i e c_{ion} D_{ion}}{kT}\left[z_i e \frac{d\phi}{dx}\right] \tag{7.33}$$

which, when compared to Ohm's law viz. $I = -\sigma_{ion}\, d\phi/dx$ [see Eq. (7.39)], yields

$$\boxed{\sigma_{ion} = \frac{z_i^2 e^2 c_{ion} D_{ion}}{kT} = \frac{z_i^2 e^2 c_{def} D_{def}}{kT}} \tag{7.34}$$

where σ_{ion} is the ionic conductivity. This relationship is known as the **Nernst–Einstein relationship**, and it relates the self-diffusion coefficient to the ionic conductivity. The reason for the connection is obvious: in both cases, one is dealing with the jump of an ion or a defect from one site to an adjacent site. The driving forces may vary, but the basic atomic mechanism remains the same.

In applying Eq. (7.34), the following should be kept in mind:

1. The conductivity σ_{ion} refers to only the ionic component of the total conductivity (see next section for more details).
2. This relationship is valid only as long as θ for the defects is ≈1 (i.e., at high dilution).
3. The variable c_i introduced in Eq. (7.29) and now appearing in Eq. (7.34) *is* the *total concentration of the diffusing ions in the crystal.*[83] For example, in calcia-stabilized zirconia, which is an oxygen ion conductor, c_{ion} is the total number of oxygen ions in the crystal and not the total number of defects (see Worked Example 7.4). On the other hand, in a solid in which the diffusion or conductivity occurs by an interstitial mechanism, c_{ion} represents the total number of interstitial ions in the crystal, which is identical to the number of defects.

[83] It may be argued that since only a few ions are moving at one time, the use of c—the concentration of *all* the ions in the system—is not warranted. After all, most of the ions are not migrating down the chemical potential gradient simultaneously. The way out of this apparent dilemma is to appreciate that if given enough time, indeed *all* the ions would eventually migrate down the gradient. To illustrate, assume that a single crystal of a binary oxide, in which diffusion of the cations is much faster than that of the anions and occurs via a vacancy mechanism, separates two compartments of differing oxygen partial pressures. At the *high oxygen partial pressure side, oxygen atoms will adsorb on the surface*, creating cation vacancies and holes. These defects will in turn diffuse ambipolarly (see Sec. 7.4), i.e., together, toward the low oxygen partial pressure side, where they will be eliminated (i.e., combine with oxygen vacancies) in order to maintain the local Schottky equilibrium. But since the movement of cation vacancies toward the low oxygen pressure side is tantamount to cations moving toward the *high oxygen partial pressure side, the net result is that the entire crystal will be growing at the high oxygen partial pressure side and shrinking at the low oxygen pressure side.* Thus, the solid is actually *moving* with respect to the laboratory frame of reference in very much the same way as a fluid flows in a pipe—the pipe in this case is an imaginary external frame!

Worked Example 7.3

The tracer diffusivity of Na cations in NaCl as well as their electrical conductivities are plotted in Fig. 7.5*a*. In another study, the electrical conductivities of pure and $CdCl_2$-doped NaCl were measured (Fig. 7.5*b*).

(*a*) From these results calculate the migration enthalpy for Na ion migration and the enthalpy of Schottky defect formation. Discuss all assumptions.
(*b*) Are the conductivity and diffusion results consistent?

ANSWER

(*a*) The behavior shown in Fig. 7.5*a* is typical of many ceramics and indicates a transition from intrinsic behavior at higher temperatures to extrinsic behavior at lower temperatures. In other words, at higher temperatures Eq. (7.20) applies and the slope of the line equals $\Delta H_m^*/k + \Delta H_S/2k$. At lower temperatures, Eq. (7.21) applies and the slope of the line is simply equal to $\Delta H_m^*/k$. Calculating the corresponding slopes from the figure and carrying out some simple algebra, one obtains $\Delta H_m^* \approx 74$ kJ/mol and $\Delta H_S = 199$ kJ/mol.

It is worth noting here that experiments leading to the results shown in Fig. 7.5 are one technique by which the data reported in Table 6.2 are obtained.

(*b*) From Fig. 7.5*a*, at 727°C (1000 K), $D \approx 4.5 \times 10^{-9}$ cm²/s. Converting all values to SI units (a very useful habit to get into) and applying Eq. (7.34) one obtains:

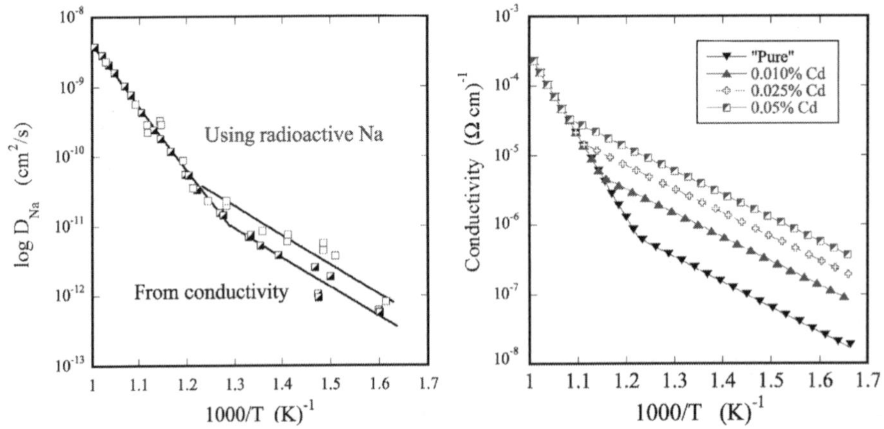

(a) (b)

FIGURE 7.5 (*a*) Arrhenian plot of diffusivity of Na cations in NaCl. The red squares were obtained from tracer experiments, the black ones from conductivity measurements. Mapother et al. J. Chem. Phys. 18, 1231 (1950). (*b*) Electrical conductivity of "pure" and $CdCl_2$ doped NaCl. With increasing $CdCl_2$ concentration the conductivity increases. (P. G. Shewmon, *Diffusion in Solids*, McGraw Hill, NY, 1963.)

$$\sigma = \frac{(1.6\times10^{-19})^2 4.5\times10^{-13}\times2.23\times10^{28}}{1000\times1.38\times10^{-23}} = 0.0186\,S/m = 1.86\times10^{-4}\,S/cm$$

Note that to solve this problem we need c_{ion}. There are several ways to do this. The simplest is probably to find the lattice parameter of NaCl (564 pm). Since every unit cell contains 4 Na cations, then $c_{ion} = 4/(564\times10^{-12})^3 = 2.23\times10^{-28}\ m^{-3}$.

Thus, based on D, the conductivity of NaCl at 727°C should $\approx 1.86\times10^{-4}$ S/cm. From Fig. 7.5*b*, the ionic conductivity at 1000 K is $\approx 3\times10^{-4}$ S/cm. Given that these were two different studies using different techniques, the agreement is acceptable.

7.2.5 DIFFUSION IN AN ELECTROCHEMICAL POTENTIAL GRADIENT

In some situations, the driving force is neither purely chemical nor electrical but rather electrochemical, in which case the flux equation has to be modified to reflect the influence of both driving forces. This is taken into account as follows: expressing Eq. (7.30) as a current by use of Eq. (7.32) and combining the results with Eqs. (7.34) and (7.22) one obtains

$$I'_k = -\frac{z_k e c_k D_k}{kT}\frac{d\tilde{\mu}_k}{dx} = -\frac{\sigma_k}{z_k e}\frac{d\tilde{\mu}_k}{dx} \tag{7.35}$$

where $d\tilde{\mu}_k/dx$ is now the driving force *per ion*,[84] that is, $\tilde{\mu}_k = \mu_k/N_{Av}$. Assuming that the total current due to an ion subjected to both a chemical and an electrical potential gradient is simply the sum of Eq. (7.35) and Ohm's law, or $I''_k = -\sigma_k\, d\phi/dx$ one obtains the following fundamental equation

$$\boxed{I_k = I'_k + I''_k = -\frac{\sigma_k}{z_k e}\left(\frac{d\tilde{\mu}_k}{dx} + z_k e\frac{d\phi}{dx}\right) = -\frac{\sigma_k}{z_k e}\left(\frac{d\tilde{\eta}_k}{dx}\right)} \tag{7.36}$$

where $\tilde{\eta}_k = \tilde{\mu}_k + z_k e\phi$ is the electrochemical potential and $d\tilde{\eta}_k/dx$ its gradient–which has the dimension of Newtons. Equation (7.36) related the latter to the current density. The corresponding flux equation (in particles per square meter per second) is

$$\boxed{J_k = -\frac{D_k c_k}{kT}\frac{d\tilde{\eta}_k}{dx} = -\frac{\sigma_k}{(z_k e)^2}\left(\frac{d\tilde{\eta}_k}{dx}\right)} \tag{7.37}$$

Equations (7.36) and (7.37) are of fundamental importance and general validity *since they describe the flux of all charged species—including electrons and holes—under all conditions*. The following should be clear at this point:

1. The driving force acting on a charged species is the gradient in its electrochemical potential.
2. For neutral species, the electric potential does not play a role, and the driving force is simply the gradient in the chemical potential, $d\mu/dx$.

---

[84] For the remainder of this chapter, the tilde over μ or η will denote energy per ion or atom.

3. In the absence of an electric field, Eq. (7.37) reverts to Eq. (7.30).
4. If the driving force is simply an electric field, i.e., $d\mu/dx = 0$, then Eq. (7.36) degenerates to Ohm's law [Eq. (7.39)].
5. Equilibrium is achieved only when $d\tilde{\eta}/dx$ vanishes, as discussed in Chap. 5 and below.
6. In all equations dealing with flux, the product $D_i c_i$ has to appear. That ensures the identical flux is obtained whether one focuses on the defects and their diffusivity or on the ions and their diffusivity.

7.3 ELECTRICAL CONDUCTIVITY

Historically technical ceramics were exploited for their electric insulation properties, which together with their chemical and thermal stability rendered them ideal insulating materials in applications ranging from power lines to cores bearing wire-wound resistors. Today their use is much more ubiquitous—in addition to their traditional role as insulators, they are used as electrodes, catalysts, fuel cells, photoelectrodes, varistors, sensors and substrates, among many other applications.

This section deals solely with the response of ceramics subjected to a constant electric field and the nature and magnitude of the *steady-state current* that results. As discussed below, the ratio of this current to the applied electric field is proportional to a material property known as *conductivity*, which is the focus of this section. The displacement currents or nonsteady-state response of solids which gives rise to capacitive properties are dealt with separately in Chaps. 14 and 15 which deal with linear and nonlinear dielectric properties, respectively.

In metals, free electrons are solely responsible for conduction. In semiconductors, the conducting species are electrons and/or electron holes. In ceramics, however, because of the presence of ions, the application of an electric field can induce these ions to migrate. Therefore, when dealing with conduction in ceramics, one must consider *both* the ionic and the electronic contributions to the overall conductivity.

Before one makes that distinction, however, it is important to develop the concept of conductivity. A good starting point is *Ohm's law*, which states that

$$V = iR \tag{7.38}$$

where V is the applied voltage (V) across a sample, R its resistance in ohms (Ω) and i the current (C/s) passing through the solid. Rearranging Eq. (7.38), dividing both sides by the cross-sectional area through which the current flows A, and multiplying the right-hand side by d/d, where d is the thickness of the sample, one gets

$$I = \frac{i}{A} = \frac{d}{RA}\frac{V}{d}$$

where $I = i/A$ is the **current density** passing through the sample. Given that V/d is nothing but the electric potential gradient, $d\phi/dx$, *Ohm's law* can be rewritten as[85]

[85] The minus sign appears for the same reason it appears in Fick's first law: A current of positive charges is positive when it flows *down* an electric potential gradient.

$$\boxed{I_i = -\sigma_i \frac{d\phi}{dx}} \tag{7.39}$$

where

$$\sigma = \frac{d}{RA}$$

Equation (7.39) states that the *flux* I is proportional to $d\phi/dx$. The proportionality constant σ is the **conductivity** of the material, which is the conductance of a cube of material of unit cross section. The units of conductivity are **Siemens** per meter or S m^{-1}, where S is the reciprocal of an ohm, Ω. Said otherwise, $S = 1/\Omega$.

The range of electronic conductivity (Fig. 7.6, right-hand side) in ceramics is phenomenal—it varies over 24 orders of magnitude, and that does not even include superconductivity! Few, if any, other physical

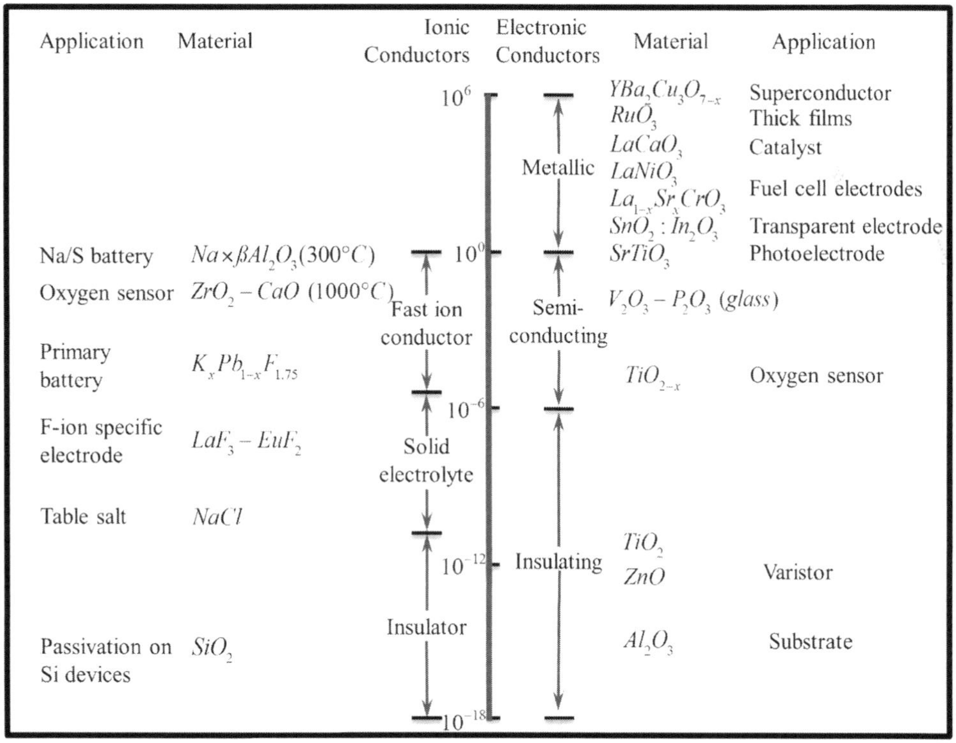

FIGURE 7.6 Range of electronic (right-hand side) and ionic (left-hand side) conductivities in Ω^{-1} cm^{-1} exhibited by ceramics and some of their uses. (H. Tuller, in *Glasses and Ceramics for Electronics*, 2nd ed., Buchanan, Ed., Marcel Dekker, New York, 1991. Reprinted with permission.)

properties vary over such a wide range. In addition to electronic conductivity, some ceramics are known to be ionic conductors (Fig. 7.6, left-hand side). In order to understand the reason behind this phenomenal range and why some ceramics are ionic conductors while others are electronic conductors, it is necessary to delve into the microscopic domain and relate the macroscopically measurable σ to more fundamental parameters, such as carrier mobilities and concentrations. This is carried out in the following subsections.

7.3.1 GENERALIZED EQUATIONS

If one assumes there are c_m mobile carriers per cubic meter drifting with an average drift velocity v_d, it follows that their flux is given by

$$I_i = |Z_i| e v_{d,i} c_{m,i} \tag{7.40}$$

The **electric mobility**, μ_d ($m^2/V{\cdot}s$), is defined as the average drift velocity per electric field, or

$$\mu_{d,i} = \frac{-v_{d,i}}{d\phi/dx} \tag{7.41}$$

Combining Eqs. (7.39) to (7.41) yields the important relationship

$$\boxed{\sigma_i = c_{m,i} e |Z_i| \mu_{d,i}} \tag{7.42}$$

between the macroscopically measurable quantity σ and the microscopic parameters μ_d and c_m.

In deriving this equation, it was assumed that only one type of charge carrier was present. However, in principle, any mobile charged species can and will contribute to the overall conductivity. Thus, the total conductivity is given by

$$\boxed{\sigma_{tot} = \sum_i c_{m,i} |z_i| e \mu_{d,i}} \tag{7.43}$$

The absolute value sign about z_i ensures that the conductivities are always positive and additive, regardless of the sign of the carrier.

The total conductivity is sometimes expressed in terms of the **transference** or **transport number**, defined as

$$t_i = \frac{\sigma_i}{\sigma_{tot}} \tag{7.44}$$

from which it follows that $\sigma_{tot} = t_{elec}\sigma_{tot} + t_{ion}\sigma_{tot}$, where t_{ion} is the ionic transference number and includes both anions and cations and t_{elec} is the electronic transference number which includes both electrons and electron holes. For any material, $t_{ion} + t_e = 1$.

From Eq. (7.42) it follows that an understanding of the factors that affect conductivity boils down to understanding how both the mobility and the concentration of mobile carriers, be they ionic or electronic, vary with temperature, doping, surrounding atmosphere, etc.

7.3.2 IONIC CONDUCTIVITY

By definition, t_{ion} for an ionic conductor should be ≈1, that is, $\sigma_{elec} << \sigma_{ion} \approx \sigma_{tot}$. In these solids, the mobile carriers are the charged ionic defects, or $c_{m,i} = c_{def}$, where c_{def} represents the concentration of vacancies and/or interstitials.[86] Replacing $c_{m,i}$ by c_{def} in Eq. (7.42) and comparing the resulting expression with Eq. (7.34), one sees immediately that

$$\mu_{d,i} = \frac{|z_i| e_i D_{def}}{kT} = \frac{|z_i| e_i D_{ion}}{kT\Lambda} \tag{7.45}$$

This is an important result because it implies that the mobility of a charged species is directly related to its defect diffusivity, a not-too-surprising result since the mobility of an ion must reflect the ease by which the defects jump around in a lattice.

Note that if diffusion is occurring by a vacancy mechanism, $\Lambda \approx c_{def}/c_{ion} << 1.0$, whereas if diffusion is occurring by an interstitial mechanism, then $\Lambda \cong 1.0$ and

$$\mu_{int} = |z_i| e_i D_{int}/(kT)$$

EXPERIMENTAL DETAILS: MEASURING IONIC CONDUCTIVITY

There are several techniques by which the ionic conductivity of a solid can be measured. One of the simpler setups is shown schematically in Fig. 7.7. Here two compartments of, say, molten Na are separated by a solid electrolyte, or membrane, that is known to be a Na^+ ion conductor (i.e., $t_e << t_{ion} \approx 1.0$). The application of a dc voltage V will result in the flow of an ionic current, I_{ion}, from the anode to the cathode. If one assumes that electrode polarization effects can be ignored, then the ratio V/I_{ion} is a measure of the ionic resistance of the solid, which is easily converted to a conductivity if the cross-sectional area through which the current is flowing, and the thickness of the solid membrane are known.

For this experiment to work, the following reaction

$$\text{Na (electrode)} \Rightarrow \text{Na}^+ \text{(in solid)} + e^{-1} \text{(in external circuit)}$$

has to occur at the anode. Simultaneously, the reverse reaction

$$\text{Na}^+ \text{(solid)} + e^{-1} \text{(from molten Na)} \Rightarrow \text{(electrode)}$$

has to occur at the cathode. (Given the polarity shown in Fig. 7.7, all the Na will end up on the right-hand side.) Thus, to measure ionic conductivity:

[86] The rationale for using the number of mobile carriers rather than the total number of ions involved is similar to the one made for Eq. (7.29). Here it can be assumed that one sublattice is the "pipe" through which the conducting ions are flowing. Referring to Fig. 7.7, the ions or defects enter the solid on one side and leave at the other. In contrast to the case where the crystal as a whole is placed in a chemical potential gradient, here the crystal itself does *not* move relative to an external frame of reference.

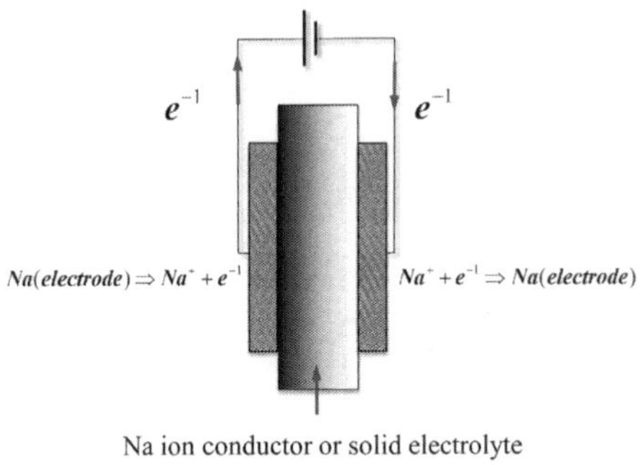

FIGURE 7.7 Experimental setup for measuring ionic conductivity. If the electrodes are nonblocking, then the ionic conductivity is simply $R_{ion} = V/I_{ion}$.

1. The solid has to conduct ions rather than electrons or holes. If that were not the case, the current would simply be carried by electronic defects or a combination of ions and electronic defects. If the latter occurs the solid would be considered a mixed conductor.
2. The electrodes have to be nonblocking to the electroactive species (Na^+ in this case), i.e., the Na^+ ions had to be able to cross, unhindered, from the solid electrolyte into the liquid electrode and vice versa. None of this is very surprising to anyone familiar with rudimentary electrochemistry; the only difference here is that the ions actually pass through a solid.

For the most part, ceramics, if they conduct at all, are electronic conductors. Sometimes if the band gap is large (see Worked Example 7.4) and the ceramic is exceptionally pure, it is possible to measure its ionic conductivity. In general, however, these conductivities are quite low. There is a certain class of solids, however, that exhibit ionic conductivities that are exceptionally high and can even approach those of molten salts (i.e., $\sigma > 10^{-2}$ S cm^{-1}). These solids are known as **fast ion conductors** (FICs), sometimes also referred to as **solid electrolytes**. Figure 7.8 shows the temperature dependence of a number of these solids, both crystalline and amorphous.

FICs fall in roughly three groups. The first is based on the halides and chalcogenides of silver and copper. The most notable of these is α-AgI, which is a silver ion conductor. The second are alkali metal conductors based on nonstoichiometric aluminates, the most important of which is β-Al_2O_3, with the approximate formula $Na_2O{\cdot}11Al_2O_3$. The third is based on oxides with the fluorite structure that have been doped with aliovalent oxides to create a large number of vacancies on the oxygen sublattice and are hence oxygen ion conductors. In all these structures, the concentration of defects is quite large and, depending on the class of FIC, is accomplished either intrinsically or extrinsically. The halides and the β-aluminas are good examples of "intrinsic" FIC where, as a consequence of their structures (see Fig. 7.9, for example), the conduction planes contain a large number of vacant sites. Furthermore, the activation energy needed for migration ΔH_m^* between sites is quite small and is in the range of 0.01 to 0.2 eV, which results in quite large defect mobilities and hence conductivities.

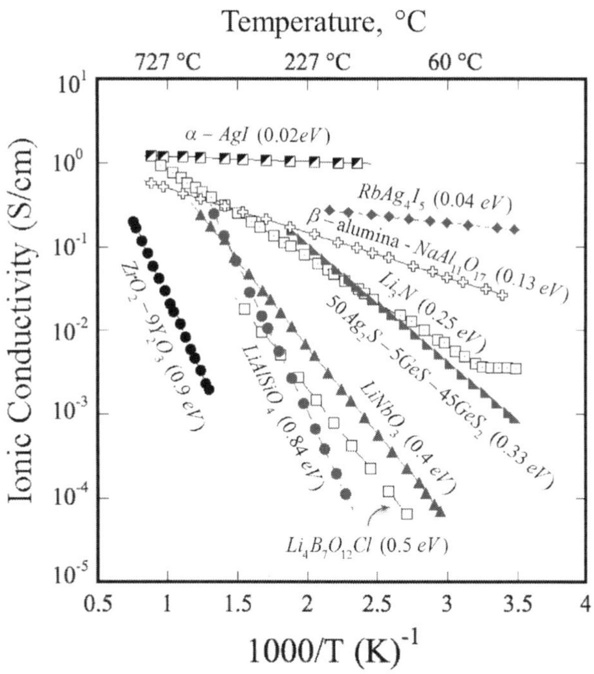

FIGURE 7.8 Arrhenius plot of ionic conductivities for a number of fast Ag, Na, Li and O ion conductors. (H. Tuller, D. Button, and D. Uhlmann, *J. Non-Cryst. Solids*, 40, 93, 1980. Reprinted with permission.)

Calcia-stabilized zirconia is an example where the number of defects is extrinsically controlled by aliovalent doping. For every mole of CaO added to ZrO_2, 1 mole of oxygen vacancies is created according to

$$CaO \underset{ZrO_2}{\Rightarrow} Ca''_{Zr} + V_O^{\bullet\bullet} + O_O^x$$

Consequently, one would expect the conductivity to be a linear function of doping, which is only true at small doping concentration as shown in Fig. 7.10. The conductivity does not increase monotonically, however, but rather goes through a maximum at higher doping levels, a fact that has been attributed to defect–defect interactions and the breakdown of the dilute approximation.

In general, the following two common structural features characterize most FICs. First, is the presence of a highly ordered, immobile sublattice, which provides a framework and defines continuous open channels for ion transport. Second, a highly disordered complementary mobile carrier sublattice with an excess of total equipotential sites n_o, compared to the number of available mobile ions n_{mob} that fill them. Under these conditions, it can be shown (see Prob. 7.2*b*) that the conductivity of FICs can be expressed as[87]

$$\sigma_{FIC} = (\text{const})\frac{\beta[1-\beta]}{kT}\exp\left(-\frac{\Delta H_m^*}{kT}\right) \tag{7.46}$$

[87] This expression is valid only if defect–defect interactions are ignored.

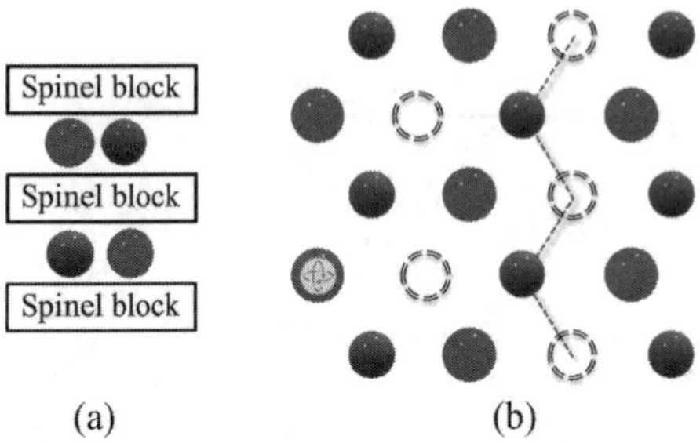

FIGURE 7.9 Structure of β-alumina. (*a*) Plane parallel to c axis. (*b*) Arrangement of atoms in conduction plane (i.e., top view of conduction planes). Empty circles denote equivalent Na ion sites that are *vacant*. Note zigzag Na^+ chains that alternate between full and occupied sites.

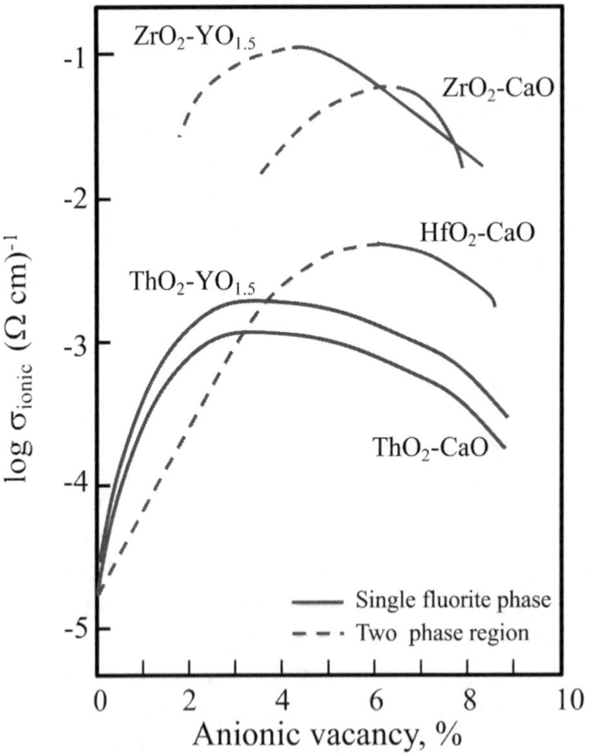

FIGURE 7.10 Effect of doping on the ionic conductivity of a number of oxygen ion conductors. (Adapted from W. H. Flygare and R. A. Huggins, *J. Phys. Chem. Solids*, 34, 1199, 1973.)

where $\beta = n_{mob}/n_o$. This function reaches a maximum when $\beta = 1/2$. Refer once again to Fig. 7.9*b*. The reason why this expression takes this form becomes obvious: Maximum conductivity will occur when the number of ions in the plane equals the number of vacant sites. The flow of traffic offers a useful analogy here. The maximum flux of cars does not occur very late at night when n_o is small, nor during rush hour where $1 - n_o$ is small, but rather sometime in between.

7.3.3 ELECTRONIC CONDUCTIVITY

Electronic conductivity, like its ionic counterpart, is governed by Eq. (7.42), and is proportional to the concentration of mobile *electronic* carriers, both electrons and holes, and their mobility. In general, there are three ways by which these mobile electronic carriers are generated in ceramics, namely (1) by excitation across the band gap (intrinsic), (2) due to impurities (extrinsic) or (3) as a result of departures from stoichiometry (nonstoichiometric). Each is considered in some detail below.

7.3.3.1 INTRINSIC SEMICONDUCTORS

In this case the electrons and holes are generated by excitation across the band gap of the material. For every electron that is excited into the conduction band, a hole is left behind in the valence band; consequently, for an intrinsic semiconductor, $n = p$.

To predict the number of electrons that are excited across the band gap at any given temperature, both the density-of-states function and the probability of occupancy of each state must be known (see App. 7B for more details). The **density of states** is defined as the number of states per unit energy interval. The probability of their occupancy is given by the Fermi–Dirac energy function, namely,

$$f(E) = \frac{1}{1 + \exp\left[(E - E_f)/kT\right]} \tag{7.47}$$

This equation is plotted as a function of temperature in Fig. 7.11. Here E is the energy of interest, and E_f is the Fermi energy, defined as the energy for which the probability of finding an electron is 0.5; k and T have their usual meanings. It can be shown (see App. 7B) that for this energy function, the number of electrons per cubic meter, n, in the conduction band is given by:

$$n = N_c \exp\left(-\frac{E_c - E_f}{kT}\right) \tag{7.48}$$

and the number of holes, p, by

$$p = N_v \exp\left(-\frac{E_f - E_v}{kT}\right) \tag{7.49}$$

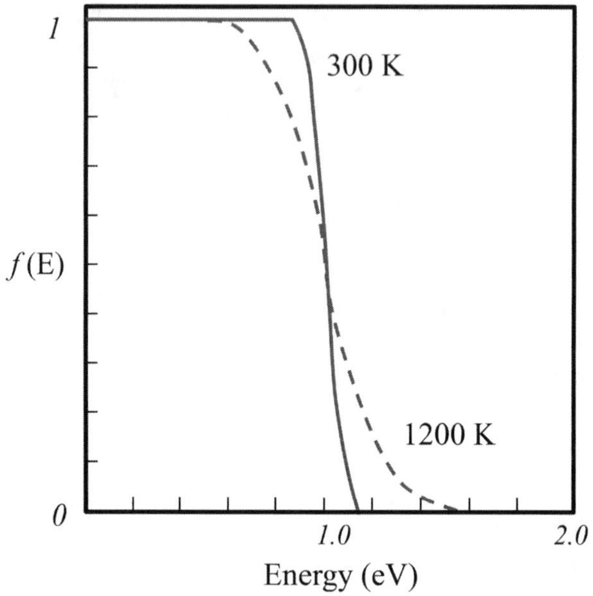

FIGURE 7.11 Fermi–Dirac distribution function for two different temperatures; E_f was assumed to be 1 eV. Note that whereas the distribution shifts to higher energies as the temperature increases, E_f, defined as the energy at which the probability of finding an electron is 0.5, does not change.

where E_c and E_v refer to the energy of the lowest and highest levels in the conduction and valence bands, respectively (see Fig. 7.12); N_c and N_v were defined in Chap. 6 and are given by (see App. 7B)

$$N_c = 2\left(\frac{2\pi m_e^* kT}{h^2}\right)^{3/2} \text{ and } N_v = 2\left(\frac{2\pi m_h^* kT}{h^2}\right)^{3/2}$$

where m_i^* is the effective mass of the migrating species. The product of Eqs. (7.48) and (7.49) yields a fundamental equation in semiconducting physics

$$np = N_c N_v \exp\left(-\frac{E_g}{kT}\right) = N_c N_v K_i \tag{7.50}$$

The np product is thus not a function of E_f. Note this equation is identical to Eq. (6.20).

As noted above for an intrinsic semiconductor, $n = p$, that is,

$$n = p = \sqrt{N_c N_v} \exp\left(-\frac{E_g}{2kT}\right) \tag{7.51}$$

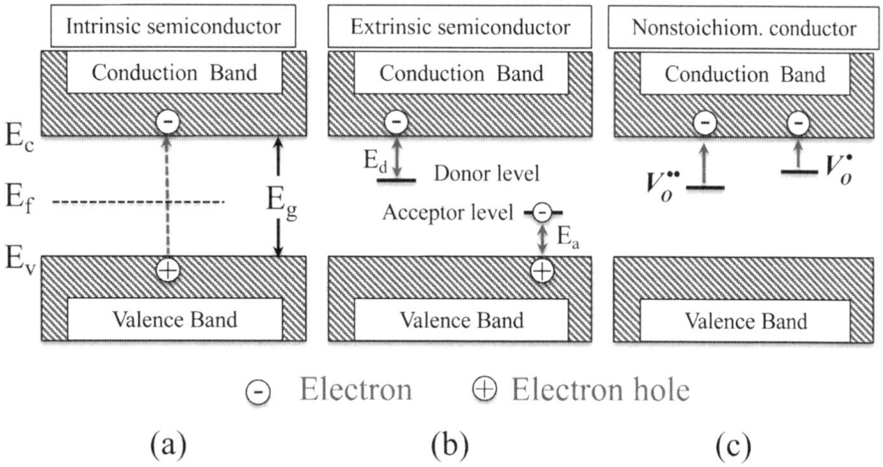

FIGURE 7.12 Schematic of energy levels for (*a*) intrinsic semiconductors, (*b*) extrinsic semiconductors and (*c*) nonstoichiometric semiconductors.

and the conductivity is given by

$$\sigma = e\mu_n n + e\mu_p p = ne(\mu_n + \mu_p) \tag{7.52}$$

Up to this point, the mobility of the electronic carriers and their temperature dependencies were not discussed. The effect of temperature on the mobility of electrons and holes will depend on several factors, with the width of the conduction and/or valence bands being the most important. For wideband materials (not to be confused with wide-band-gap materials), the mobility of the electronic carriers decreases with increasing temperature as a result of lattice or phonon scattering, not unlike what happens in metallic conductors. It can be shown that in this case both μ_n and μ_p are proportional to $T^{-3/2}$. Thus the temperature dependence of $\sigma_{el,int}$ is given by

$$\sigma_{el,int} = K''T^{-3/2}T^{3/2}\exp\left(-\frac{E_g}{2kT}\right) = K''\exp\left(-\frac{E_g}{2kT}\right) \tag{7.53}$$

Note the $T^{3/2}$ term comes from the pre-exponential term in Eq. (7.51), i.e., from the density of states. This result is applicable to intrinsic semiconductors in which phonon scattering is responsible for the temperature dependence of the electronic mobilities, i.e., one in which the mobility *decreases* with *increasing* temperature.

Other possibilities exist, however; two of the more important ones are the **small** and **large polaron** mechanisms. A *polaron* is a defect in an ionic crystal that is formed when an excess of charge at a point polarizes or distorts the lattice in its immediate vicinity. For example, if an oxygen vacancy captures an electron[88] (Fig. 6.4*b*), the cations surrounding it will be attracted to the defect and move toward it, whereas the anions will move away. This polarization essentially traps or slows down the electronic defect as it moves through the lattice.

Small polaron. In this mechanism, conduction occurs by the "hopping" of electronic defects between adjacent ions of, usually but not necessarily, the same type but with varying oxidation states. Because of the ease by which transition-metal ions can vary their oxidation states, this type of conduction is most often observed in transition-metal oxides. For example, if the charge carrier is an electron, the process can be envisioned as

$$M^{+n} + e^{-1} \Rightarrow M^{n-1}$$

Polarization of the lattice results in a reduction of the energy of the system, and the carrier is then assumed to be localized in a potential energy well of effective depth E_B. It follows that for migration to occur, the carrier has to be supplied with at least that much energy, and consequently the mobility becomes thermally activated. It can be shown (see Prob. 7.2*c*) that polaron conductivity can be described by an expression very similar to that for FICs, that is,

$$\sigma_{hop} = (\text{const})\frac{x_i[1-x_i]}{kT}\exp\left(-\frac{E_B}{kT}\right) \tag{7.54}$$

where x_i is the fraction of sites occupied by one charge and $[1-x_i]$ the fraction of sites occupied by the other charge. For example, for polaron hopping between Fe cations, x and $(1-x)$ would represent the concentrations of Fe^{2+} and Fe^{3+} cations, respectively. Based on this simple model, one would expect a conduction maximum at $x \approx 0.5$. Experimentally, this is not always observed, however.

Large polaron. If the distortion is not large enough to totally trap the electron, but is still large enough to slow it down, the term *large polaron* is applicable. Large polarons behave as free carriers except that they have a higher effective mass than a free electron. It can be shown that the large polaron mobility is proportional to $T^{-1/2}$, and consequently, the temperature dependence of an intrinsic semiconductor for which conductivity occurs by large polarons is

$$\sigma = (\text{const})T^{3/2}T^{-1/2}\exp\left(-\frac{E_g}{2kT}\right) = (\text{const})T\exp\left(-\frac{E_g}{2kT}\right) \tag{7.55}$$

Worked Example 7.4

A good solid electrolyte should have an ionic conductivity of at least 0.01 $(\Omega\cdot cm)^{-1}$ with an electronic transference number that should not exceed 10^{-4}. At 1000 K, show that the minimum band gap for

[88] Note that the combination of a trapped electron or hole at an impurity or defect is called a *color center*, which can have interesting optical and magnetic properties (see Chap. 16).

such a solid would have to be ≈ 4 eV. Assume that the electronic and hole mobilities are equal and that each is 100 $cm^2/V{\cdot}s$. State all other assumptions.

ANSWER

Based on the figures of merit stated, the electronic conductivity should not exceed

$$\sigma_{elec} = t_e\sigma_{tot} \approx t_e\sigma_{ion} = 0.01\times10^{-4} = 1\times10^{-6}(\Omega\cdot cm)^{-1}$$
$$= 1\times10^{-4}(\Omega\cdot m)^{-1}$$

Inserting this value for σ_{elec} in Eq. (7.52), assuming $\mu_e = \mu_p = 0.01$ $m^2/(V{\cdot}s)$ at 1000 K, and solving for n yields

$$n = \frac{1\times10^{-4}}{1.6\times10^{-19}(0.01+0.01)} = 3.12\times10^{16}\ \text{electrons}/m^3$$

Furthermore, by assuming that $m_e = m_e^*$, it follows that

$$N_c = 2\left(\frac{2\pi m_e^* kT}{h^2}\right)^{3/2} = 2\left(\frac{2\pi\times9.1\times10^{-31}\times1.38\times10^{-23}\times1000}{(6.63\times10^{-34})^2}\right)^{3/2}$$
$$= 1.5\times10^{26}\,m^{-3}$$

Finally, assuming $N_c = N_v$, substituting the appropriate values in Eq. (7.51) and solving for E_g give

$$E_g = -2kT\ln\frac{n}{N_c} = -2\times1000\times8.62\times10^{-5}\ln\frac{3.12\times10^{16}}{1.5\times10^{26}} = 3.84\,eV$$

7.3.3.2 EXTRINSIC SEMICONDUCTORS

The conductivity of extrinsic semiconductors is mainly determined by the presence of foreign impurities. The best way to illustrate the notion of an extrinsic semiconductor is to take the specific example of elemental Si. Consider the addition of a known amount of a group V element, such as phosphorus. If a P atom substitutionally replaces a Si atom, four electrons will be used up to bond with adjacent Si atoms, leaving behind an extra electron (Fig. 7.13*a*). Given that this electron will not be as tightly bound as the others, it is easily promoted into the conduction band. In other words, each P atom *donates* one electron to the conduction band and in so doing increases the conductivity of the Si host, which is now termed an ***n*-type semiconductor**. In terms of band theory, a donor is depicted by a localized level within the band gap, at an energy E_d lower than E_c (Fig. 7.12*b*), where E_d is a measure of the energy binding the electron to its site.

Conversely, if Si is doped with a group III atom—such as B (Fig. 7.13*b*)—that only has three electrons in its outer shell, a hole will be created in the valence. Upon application of an electric field, the hole acts as a vacancy into which electrons from neighboring atoms can jump. Such a material is called a ***p*-type semiconductor,** and the corresponding energy diagram is also shown in Fig. 7.12*b* (bottom of diagram).

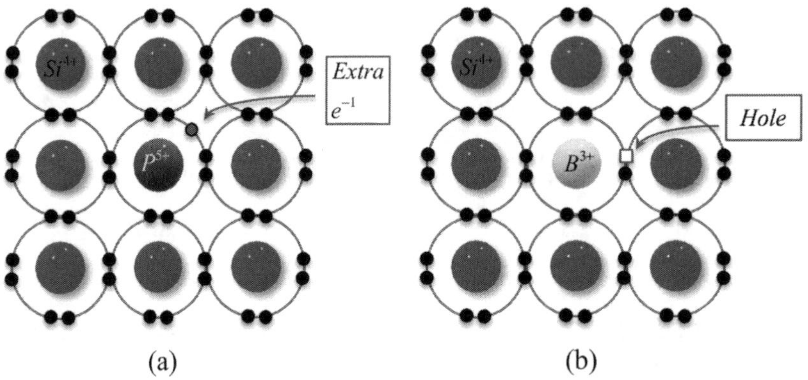

FIGURE 7.13 Schematic of a (*a*) n-type semiconductor and (*b*) p-type semiconductor. The missing electron (definition of a hole) is formed when B is added to Si. With the input of energy E_a, an electron from the valence band is promoted into the site labeled *electron hole,* and a hole is created in the valence band.

To quantify the conductivity of an extrinsic semiconductor, consider an n-type semiconductor doped with a concentration N_D of dopant atoms. The ionization reaction of the donor can be written as

$$D \Rightarrow D^{\bullet} + e'$$

Mass balance dictates that

$$N_D = D + D^{\bullet}$$

where D and $D^{\bullet}$ denote, respectively, the concentrations of un-ionized and ionized donors. The corresponding mass action expression is

$$\frac{[D^{\bullet}][n]}{[D]} = \exp\left(-\frac{E_d}{kT}\right) \tag{7.56}$$

where E_d is the energy required to ionize the donor (see Fig. 7.12*b*). Recall once again that the square brackets denote mole fractions and not concentrations. Thus [D] and $[D^{\bullet}]$ denote, respectively, the mole fractions of un-ionized and ionized donors, whereas $[n] = n/N_c$ and $[D^{\bullet}] = D^{\bullet}/N_D$.

To make the problem more tractable, consider the following three temperature regimes:

Low-temperature region (region 1 in Fig. 7.14*).* At low temperatures, when only a few donors are ionized, the neutrality condition can be written as $n = D^{\bullet} << N_D$. In other words, $[D] \approx 1$ and $n = D^{\bullet}$. Making the appropriate substitutions in Eq. (7.56), one obtains

$$\sigma_{elec} = \sigma_n = e\mu_n\sqrt{N_D N_c}\exp\left(-\frac{E_d}{2kT}\right) \tag{7.57}$$

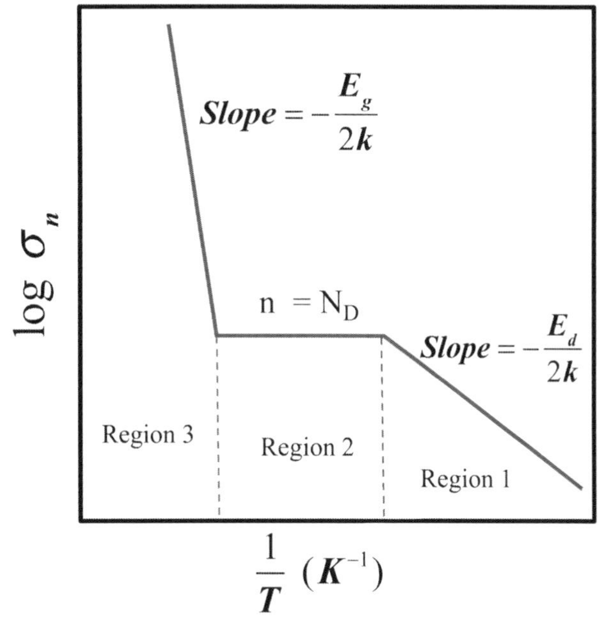

FIGURE 7.14 Temperature dependence of the electronic conductivity of an extrinsic semiconductor. Region 2 is sometimes referred to as the *exhaustion region.*

Thus, in this region, a plot of log σ_n versus reciprocal temperature should yield a straight line with slope $E_d/(2k)$.

Intermediate-temperature region (region 2 in Fig. 7.14*).* In this region, $kT \approx E_d$, and one can assume that most of the donor atoms are ionized. Hence the total number of mobile carriers will simply equal the dopant concentration N_D, in which case

$$\sigma_{elec} = \sigma_n = eN_D\mu_n \tag{7.58}$$

In this region, the conductivity will be a weak function of temperature and may even decrease with increasing temperature if phonon scattering becomes important.

High-temperature region (*region 3 in* Fig. 7.14). Here it is assumed that the temperature is high enough that the number of electrons excited from the valence band into the conduction band dominate, in which case the semiconductor behaves intrinsically, and its conductivity is given by Eq. (7.53). In this region, a plot of log σ_n versus reciprocal temperature yields a straight line with slope $E_g/(2k)$ (Fig. 7.14).

To reiterate, at low temperatures, the conductivity is low because of the paucity of mobile carriers—most are trapped. As the temperature increases, the defects start to ionize and the conductivity increases with an activation energy needed to release the electrons from their traps. At intermediate temperatures, when $kT \approx E_d$, most of the impurities will have donated their electrons to the conduction band, and a saturation in the conductivity sets in. With further increases in temperature, however, it is now possible (provided the crystal does not melt beforehand) to excite electrons clear across the band gap, and the conductivity starts increasing again, but this time with a slope that is proportional to $E_g/2k$.

7.3.3.3 NONSTOICHIOMETRIC SEMICONDUCTORS

In the preceding subsection, n and p were fixed by the doping level, especially at lower temperatures, and the concepts of donor and acceptor localized levels were discussed. The band picture for nonstoichiometric electronic semiconductors is very similar to that of extrinsic semiconductors, except that the electronic defects form not as a result of doping, but rather by varying the stoichiometry of the crystal.

To appreciate the similarities between an extrinsic semiconductor and a nonstoichiometric oxide, compare Fig. 7.13*a* with Fig. 6.4*a* or *b*. In both cases the electrons are loosely bound to their moorings and are easily excited into the conduction band. The corresponding energy diagrams for the singly and doubly ionized oxygen vacancies are shown in Fig. 7.12*c*. In essence, a nonstoichiometric semiconductor is one where the electrons and holes excited in the conduction and valence bands are a result of reduction or oxidation. For example, the reduction of an oxide entails the removal of oxygen atoms, which have to leave their electrons behind to maintain electroneutrality. These electrons, in turn, are responsible for conduction.

The starting point for understanding the behavior of nonstoichiometric oxides involves constructing their Kroger–Vink diagram, as discussed in Sec. 6.2.8. To illustrate, consider the following examples:

ZnO: Experimentally, it has been found that the electrical conductivity of ZnO decreases with increasing oxygen partial pressure, as shown schematically in Fig. 7.15*a*. When plotted on a log–log plot, the resulting slope is measured to be ¼, which is explained as follows. The defect incorporation reaction is presumed to be

$$O_O^x + Zn_{Zn}^x \Leftrightarrow \tfrac{1}{2}O_2(g) + Zn_i^{\bullet} + e^{-1}$$

for which the mass action expression is

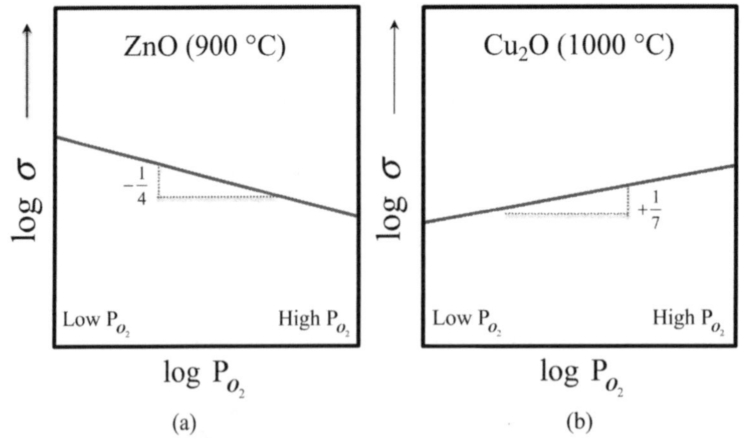

FIGURE 7.15 Schematic of changes in conductivity as a function of oxygen partial pressure for (*a*) ZnO and (*b*) Cu_2O. These curves are based on actual experimental results.

$$P_{O_2}^{1/2}[Zn_i^{\bullet}][n] = \text{const}$$

Combining this expression with the electroneutrality condition, $Zn_i^{\bullet} \approx n$, results in

$$\sigma \propto n = (\text{const})P_{O_2}^{-1/4}$$

as observed. Note that had one assumed the Zn interstitials to be doubly ionized, the P_{O_2} dependence predicted would not have been consistent with the experimental results.

Cu₂O: In contradistinction to ZnO, the conductivity of Cu_2O increases with increasing oxygen partial pressure with a slope of ≈+1/7. To explain this result, the following incorporation reaction is assumed:

$$\tfrac{1}{2}O_2(g) \Leftrightarrow O_O^x + 2V_{Cu}' + 2h^{\bullet}$$

which, when combined with the neutrality condition $V_{Cu}' = p$ and the mass action expression, results in

$$\sigma \propto p = (\text{const})P_{O_2}^{1/8}$$

which is not too far off the measured value of 1/7 shown in Fig. 7.15*b*.

CoO: CoO is a metal-deficient oxide[89] $Co_{1-x}O$ (see Table 6.1), where the conductivity is known to *be p-type and thermally* activated, i.e., occurs by polaron hopping such that $\mu_p = (\text{const})[\exp - E_B/kT]$. At high P_{O_2}, the conductivity changes are steeper with a slope of ≈+1/4, whereas at lower P_{O_2} the slope changes to ≈ +1/6. This suggests that at high P_{O_2} the defect reaction is given by

$$\tfrac{1}{2}O_2(g) \Leftrightarrow O_O^x + V_{Co}' + h^{\bullet} \quad \Delta G_{V_{Co}}$$

with an equilibrium constant $K_{V_{Co}'} = \exp[-\Delta G_{V_{Co}}/(kT)]$. Making use of the corresponding mass action expression and the electroneutrality condition $p = V_{Co}'$, one obtains

$$\sigma = \sigma_p = pe\mu_p = (\text{const})(K_{V_{Co}}')^{1/2} P_{O_2}^{1/4} \exp\left(-\frac{E_B}{kT}\right)$$

which is the pressure dependence observed. Furthermore, since

$$K_{V_{Co}}' = \exp\frac{\Delta S_{V_{Co}'}}{k}\exp\left(-\frac{\Delta H_{V_{Co}'}}{kT}\right)$$

the final expression for the electrical conductivity is given by

[89] The actual situation for CoO is not as simple as put forth here. For a good interpretation, see H.-I. Yoo, et al., *Solid State Ion.*, 67, 317–322 (1994).

$$\sigma = \sigma_p = pe\mu_p = (\text{const})\exp\left(-\frac{\Delta H_{V_{Co}'}}{2kT}\right)\exp\left(-\frac{E_B}{kT}\right)P_{O_2}^{1/4} \tag{7.59}$$

Similarly, at lower P_{O_2}, the data suggest that the corresponding reaction is one where doubly ionized cobalt vacancies form, namely,

$$\tfrac{1}{2}O_2(g) \Leftrightarrow O_O^x + V_{Co}'' + 2h^\bullet \quad \Delta G_{V_{Co}''}$$

It is left as an exercise for the reader to show that in the low P_{O_2} regime

$$\sigma = \sigma_p = (\text{const})\exp\left(-\frac{\Delta H_{V_{Co}''}}{3kT}\right)\exp\left(-\frac{E_B}{kT}\right)P_{O_2}^{1/6} \tag{7.60}$$

where $\Delta H_{V_{Co}''}$ is the enthalpy of formation of the doubly ionized cobalt vacancies.

It is worth noting that when the conductivity is dominated by redox reactions such as the ones discussed here, the final expression does not depend on the band gap or on doping but rather depends on the ease with which an oxide is oxidized or reduced, i.e., on Δg_{red} and Δg_{oxid}. Generally, this is directly related to the ability of the cations to exist in more than one oxidation state—and consequently it is intimately related to the range of nonstoichiometry discussed in Sec. 6.2.6.

ZrO_2: The Kroger–Vink diagram for yttria-doped zirconia is shown in Fig. 7.16*a*, the construction of which is left as an exercise to the reader. In pure zirconia, the concentration of oxygen vacancies is simply $\sqrt{K_F}$. However, as noted earlier, that value can be dramatically increased by doping with aliovalent cations such as Ca^{2+} or Y^{3+}. Based on this diagram, in the range where the conductivity is ionic, the minority carriers are electronic defects.

Actual data for yttria-doped zirconia are shown in Fig. 7.16*b* as a function of temperature, and generally confirm the results shown in Fig. 7.16*a*. At 1000°C and a P_{O_2} of 10^{-10} atm, $t_{ion} \approx 10^{-1}/(10^{-1}+1\times10^{-6}) \approx 1.0$, whereas at $\approx 10^{-30}$ atm, $t_{ion} \approx t_{elec} \approx 0.5$.

Note that, in contrast to electronic conduction, the ionic conductivity is P_{O_2} independent because the concentration of ionic defects is fixed extrinsically, i.e., by doping. Such an independence of conductivity on P_{O_2} is usually taken to be strong evidence that a solid is indeed an ionic conductor. There are exceptions, however, as discussed below.

Worked Example 7.5

In a classic paper,[90] Kingery et al. measured the conductivity of $Zr_{0.85}Ca_{0.15}O_{1.85}$ as a function of P_{O_2} and temperature. They found that the conductivity (S/m) was independent of P_{O_2} and obeyed the relation

$$\sigma = 1.5\times10^5 \exp\left(-\frac{1.26\text{eV}}{kT}\right)$$

[90] W. D. Kingery, J. Pappis, M. E. Doty, and D. C. Hill, Oxygen mobility in $Zr_{0:85}Ca_{0:15}O_{1:85}$, *J. Amer. Cer. Soc.*, 42(8), 393–398 (1959).

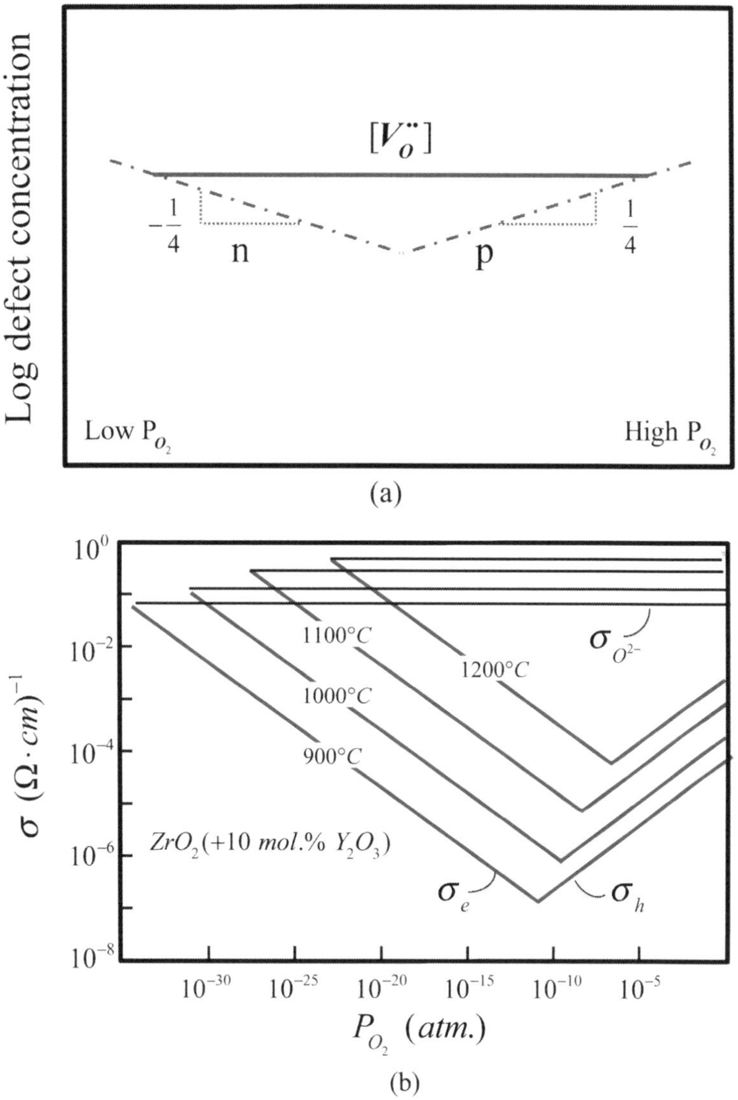

FIGURE 7.16 (*a*) Schematic of defect concentration dependence on P_{O_2} for yttria-doped zirconia. (*b*) Functional dependence of ionic and electronic conductivities on P_{O_2} and temperature for ZrO_2 (+10 mol.% Y_2O_3). Note that except at very low P_{O_2}, where $\sigma_e \approx \sigma_{ion}$, the conductivity is ionic and independent of P_{O_2}. (L. D. Burke, H. Rickert, and R. Steiner; *Z. Physik. Chem. N.F.*, 74, 146, 1971.)

The diffusion coefficient of the oxygen ions was also measured in a *separate* experiment on the same material and was found to obey

$$D_O(m^2/s) = 1\times10^{-6}\exp\left(-\frac{1.22\,eV}{kT}\right)$$

What conclusions can be reached regarding the conduction mechanisms in this oxide and its defect structure? Information you may find useful: density of zirconia ≈6.1 g/cm³; molecular wt. of Zr is 91.22 g/mol; molecular wt. of O_2 is 32 g/mol.

ANSWER

Since the total conductivity was not a function of P_{O_2}, one can assume that the conductivity is ionic.[91] The concentration of oxygen ions per cubic meter is

$$c_{ions} = \frac{6.1\times1.85\times6.02\times10^{23}}{(91.22+32)\times10^{-6}} = 5.52\times10^{28}\ \text{ions/m}^3$$

Using the Nernst–Einstein relationship, and converting all units to SI units, one obtains the following expression for the conductivity:

$$\sigma_{O\,ions} = \frac{(2^2)(1.6\times10^{-19})^2(5.5\times10^{28})(1\times10^{-6})}{1.38\times10^{-23}\times T}\exp\left(-\frac{1.22\,eV}{kT}\right)$$

$$= \frac{4.07\times10^8}{T}\exp\left(-\frac{1.22\,eV}{kT}\right)\text{S/m}$$

At 1000 K, the pre-exponential term yields a value of 4.07×10^5 S/m, which is in fairly close agreement with the pre-exponential term in the conductivity expression shown above. The fit is even better at higher temperatures. These results unambiguously prove that conduction in calcia-stabilized zirconia occurs by the movement of oxygen ions.

7.4 AMBIPOLAR DIFFUSION

In the discussion so far, the diffusional and electrical fluxes of the ionic and electronic carriers were treated separately. However, as will become amply clear in this section and was briefly touched upon in Sec. 5.6, in the absence of an external circuit such as the one shown in Fig. 7.7, the diffusion of a charged species by itself is very rapidly halted by the electric field it creates and thus *cannot* lead to steady-state conditions.

[91] This is not always the case. There are situations where the conductivity can be electronic and yet oxygen partial-pressure-independent (e.g., see Fig. 7.21c).

For steady state, the fluxes of the diffusing species have to be coupled such that electroneutrality is maintained. Hence, in most situations of great practical importance, such as creep, sintering, oxidation of metals, efficiency of fuel cells and solid-state sensors, to name a few, it is the *coupled diffusion*, or *ambipolar diffusion*, of two fluxes that is critical. To illustrate, four phenomena that are reasonably well understood and that are related to this coupled diffusion are discussed in some detail in the next subsections. The first deals with the oxidation of metals, the second with ambipolar diffusion in general in a binary oxide, the third with the interdiffusion of two ionic compounds to form a solid solution. The last subsection explores the conditions for which a solid can be used as a potentiometric sensor.

7.4.1 OXIDATION OF METALS

To best illustrate the notion of ambipolar diffusion, the oxidation of metals will be used as an example following the elegant treatment first developed by C. Wagner.[92] Another reason to go into this model is to appreciate that it is usually the electrochemical potential, rather than the chemical or electric potential, that is responsible for the mobility of charged species in solids. It also allows a link to be made between the notions of chemical stability and nonstoichiometry. However, before proceeding much further, it is instructive to briefly review how oxidation rates are measured and to introduce the parabolic rate constant.

EXPERIMENTAL DETAILS: MEASURING OXIDATION RATES

Oxidation rates can be measured by a variety of methods. One of the simplest is to expose the material for which the oxidation resistance is to be measured (typically metal foils) to an oxidizing atmosphere of a given P_{O_2} (most commonly air) for a given time, cool, and measure the thickness of the oxide layer that forms as a function of time, t. Long before any atomistic models were put forth, it was empirically fairly well established that for many metals the oxidation rate was parabolic. In other words, the increase in thickness Δx of the oxide layer was related to t by

$$\Delta x^2 = 2K_x t \tag{7.61}$$

with the proportionality constant K_x (m^2/s), known as the **parabolic rate constant**. The latter is a function of both temperature and oxygen partial pressure.

An alternate technique is to carry out a thermogravimetric experiment and measure the weight gain of the material for which the oxidation resistance is to be measured as a function of time (Fig. 7.17). In this case, the weight change per *unit area* Δw is related to time by[93]

$$\Delta w^2 = K_w t \tag{7.62}$$

[92] C. Wagner, *Z. Physikal. Chem.*, B21, 25 (1933).

[93] Sometimes Eq. (7.62) is written with a factor of 2 in analogy to Eq. (7.61). In this book K_w is defined as shown.

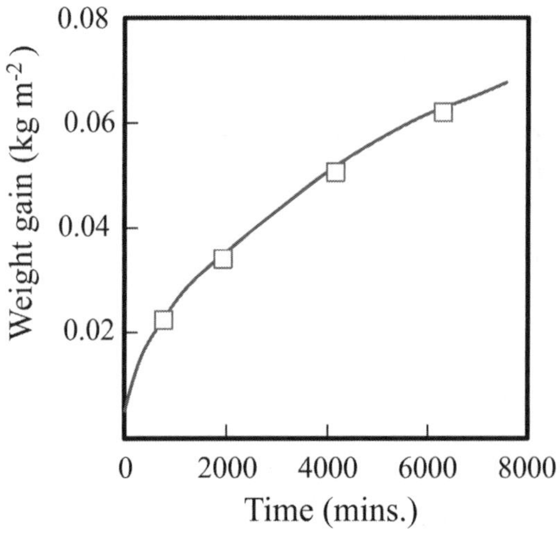

FIGURE 7.17 Weight gain during the oxidation of Ti_3SiC_2 in air at 1000°C.

where K_w (kg²/m⁴·s) is also a constant that depends on temperature and P_{O_2}. Needless to say, K_x and K_w are related (see Prob. 7.17). K_w is also known as a gravimetric parabolic rate constant.

In the remainder of this subsection, the goal is to relate these phenomenological rate constants to more fundamental parameters of the growing oxide layer. Table 7.1 lists the parabolic rate constants for a number of metals oxidized in pure oxygen at 1000°C.

For the sake of illustration, consider one of the simplest possible cases depicted schematically in Fig. 7.18*a*. Here a metal is exposed to oxygen at elevated temperatures, and an oxide layer is formed by the *outward* diffusion of *cation interstitials,* henceforth denoted by the subscript "def" for defect, together with electrons. Note that both the cations and electrons are diffusing in the same direction. ZnO is a good example of such an oxide.

At the metal/oxide interface (Fig. 7.18*a*), the incorporation reaction is

$$M^* \leftrightarrow M_{def}^{z+} + ze^{-1} \tag{7.63}$$

where z is the valence on the cation and M* denotes the neutral or metallic species. From here on, the superscript asterisk * denotes neutral species.

TABLE 7.1 Parabolic rate constants K_w, for various metals oxidized in 1 atm oxygen at 1000°C

Metal-oxide	Fe–FeO	Ni–NiO	Co–CoO	$Cr–Cr_2O_3$	$Si–SiO_2$
K_w (kg²/m⁴ · s)	4.8×10^{-5}	2.9×10^{-6}	2.1×10^{-8}	1.0×10^{-8}	1.4×10^{-14}

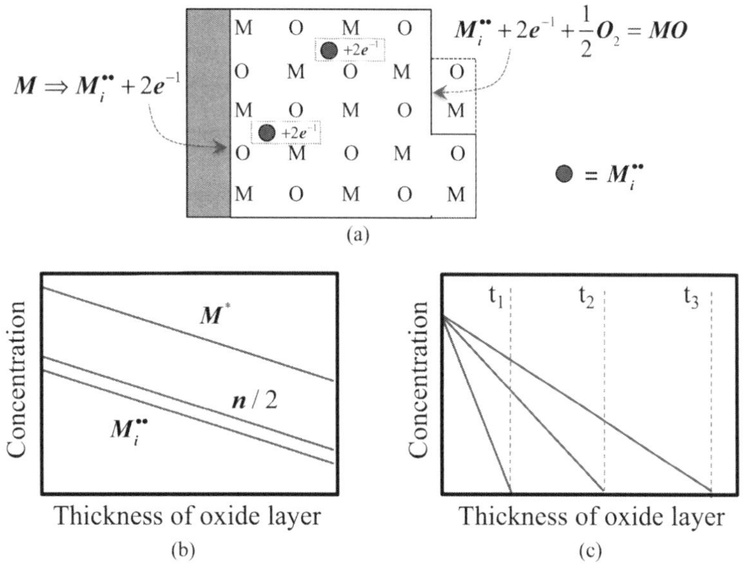

FIGURE 7.18 (*a*) Growth of an oxide layer by the outward diffusion of metal cations and electrons. (*b*) Concentration profile of defects and neutral species. (*c*) Schematic of the quasi-steady-state approximation. The concentration profile, which is proportional to the flux at any time t, is always linear and thus not a function of x. Nevertheless, as x increases, the flux decreases (slope decreases). In both (*b*) and (*c*) the profile is assumed to be linear for simplicity.

Conversely, at the oxide/gas interface, the cation and electron will combine with oxygen according to

$$\tfrac{z}{4}O_2(g) + M_{def}^{z+} + ze^{-1} \leftrightarrow MO_{z/2} \tag{7.64}$$

forming a reaction layer that grows with time with a velocity that should be proportional to the flux of the defects through that layer. The net reaction is simply the sum of Eqs. (7.63) and (7.64), or

$$\tfrac{z}{4}O_2(g) + M^* \leftrightarrow MO_{z/2} \quad \Delta G_{MO_{z/2}} \tag{7.65}$$

where $\Delta G_{MO_{z/2}}$ is the free energy of formation of the growing oxide layer.

Wagner made the following assumptions:

1. The process is diffusion-limited through a scale that is compact, fully dense and crack-free. Such layers will occur only if the volume change upon oxidation is not too great. Otherwise, the growing oxide layer cannot accommodate the mismatch strain that develops and will tend to crack or buckle.

   A good indicator of whether an oxide layer is protective is given by the **Pilling–Bedworth ratio**

$$\text{P–B ratio} = \frac{V_{MO_{z/2}}}{V_M} = \frac{MW_{MO_{z/2}}\rho_M}{MW_M \rho_{MO_{z/2}}} \tag{7.66}$$

that compares the molar volume of the metal V_M to that of the oxide, $V_{MO_{z/2}}$. MW and ρ denote the molecular weights and densities of the metal and oxide, respectively.

Metals with a P–B ratio less than unity, tend to form porous and nonprotective oxides because the volume of the latter is insufficient to cover the underlying metal surface. For ratios greater than unity, compressive stresses develop in the film. If the mismatch is too great (P–B ratios > 2), the oxide coating tends to buckle and flake off, continually exposing fresh metal, and is thus again nonprotective. The ideal P–B ratio is 1, but protective coatings normally form for metals having P–B ratios between 1 and 2.

2. Diffusion of charged species is by independent paths. In other words, it is assumed that the flux of species i is proportional to its electrochemical potential gradient solely and is independent of the gradient in the electrochemical potential of the other components.
3. Charge neutrality is maintained, and there is no *macroscopic* separation of charge. For instance, in the example shown above, the electronic defects—as a result of their higher mobility—will attempt to move along faster than the ions. But as they do so, they will create a *local electrical potential gradient*, $d\phi/dx$, which will hold them back. That same field, however, will enhance the flux of the ionic defects. As discussed in greater detail shortly, it is this *coupling* of the two fluxes that gives rise to an effective or ambipolar diffusion coefficient.
4. There is local equilibrium both at the phase boundaries and throughout the scale. This implies that at every point, reaction (7.63) holds, or

$$\tilde{\mu}_{M^*} = z\tilde{\eta}_e + \tilde{\eta}_{def} = z(\tilde{\mu}_e - e\phi) + (\tilde{\mu}_{def} + ze\phi) \tag{7.67}$$

where ϕ is the local electric potential acting on the defects.[94] It thus follows that across the layer

$$\frac{d\tilde{\mu}_{M^*}}{dx} = \frac{d(z\tilde{\mu}_e)}{dx} + \frac{d(\tilde{\mu}_{def})}{dx} \tag{7.68}$$

Given these assumptions, the flux of the defects can be related to the rate of growth of the layer. Assuming one-dimensional diffusion, the defect and electronic flux densities (particles per square meter per second) subject to an electrochemical potential gradient $d\tilde{\eta}_i/dx$ are given by Eq. (7.37), or

$$J_{def} = -\frac{\sigma_{def}}{(ze)^2}\frac{d\tilde{\eta}_{def}}{dx} = -\frac{\sigma_{def}}{(ze)^2}\left(\frac{d\tilde{\mu}_{def}}{dx} + ez\frac{d\phi}{dx}\right) \tag{7.69}$$

$$J_e = -\frac{\sigma_e}{e^2}\frac{d\tilde{\eta}_e}{dx} = -\frac{\sigma_e}{e^2}\left(\frac{d\tilde{\mu}_e}{dx} - e\frac{d\phi}{dx}\right) \tag{7.70}$$

To maintain electroneutrality, these two fluxes have to be equal, and mass balance dictates that they in turn must be equal to the flux of the neutral metal species J_{M^*}. In other words,

94 As noted earlier, the tilde denotes that the quantities are expressed per defect or per electron rather than per mole.

$$\frac{J_e}{z} = J_{def} = J_{M^*} \tag{7.71}$$

Using this condition to solve for $d\phi/dx$ results in

$$\frac{d\phi}{dx} = \frac{t_e}{e}\frac{d\tilde{\mu}_e}{dx} - \frac{t_{def}}{ze}\frac{d\tilde{\mu}_{def}}{dx} \tag{7.72}$$

which, when substituted back in Eq. (7.69) or (7.70) while use is made of Eq. (7.68), yields

$$\boxed{J_{M^*} = \frac{-\sigma_{def} t_e}{(ze)^2}\frac{d\tilde{\mu}_{M^*}}{dx} = \frac{-\sigma_{def}\sigma_e}{(ze)^2(\sigma_{def} + \sigma_e)}\frac{d\tilde{\mu}_{M^*}}{dx}} \tag{7.73}$$

Here J_{M^*} is the *number of neutral metal atoms passing through a unit area of the oxide per second* and $d\mu_{M^*}/dx$ is the chemical potential gradient of the *neutral* metal species.

This is an important result because it implies that for oxidation to occur, the oxide layer must conduct *both* ions and electrons—if either σ_e *or* σ_{def} vanishes, the permeation flux also vanishes. This comes about because it was assumed early on that both the electronic and ionic defects must diffuse together to maintain charge neutrality. It also follows from Eq. (7.73) that it is the *slower* of the two diffusing species that is rate-limiting. If an oxide is predominantly an electronic conductor, then $t_e \approx 1$, and the permeation flux will be determined by the ionic conductivity. Conversely, if $t_{ion} \approx 1$, then the permeation rate will be determined by the rate at which the electronic defects move through the oxide layer.

Equation (7.73) also implies that the driving force for the growth of the layer is nothing but the gradient in the chemical potential of the *neutral* species, which, in turn, is nothing but the free-energy change associated with reaction (7.65). In other words, the more stable the oxide, the higher the driving force for its formation.

To relate the permeation flux to the parabolic rate constant, Wagner further assumed quasi-steady-state growth conditions. This assumption implies that the flux into the reaction layer is equal to the flux out of it and that there is no accumulation of material in the film. In other words, at any time, the flux is *not* a function of x, but only a function of time. This condition is shown schematically in Fig. 7.18c for various times during scale growth. Mathematically it implies that the flux is inversely proportional to Δx, and hence dx in Eq. (7.73) can be replaced by Δx. Making that substitution, and noting that the rate at which the oxide layer is growing is given by

$$\frac{d(\Delta x)}{dt} = J_{M^*}\Omega_{MO} = \frac{-\sigma_{def}\sigma_e\Omega_{MO}}{(ze)^2(\sigma_{def} + \sigma_e)}\frac{1}{\Delta x}d\tilde{\mu}_{M^*} \tag{7.74}$$

where Ω_{MO} is the atomic volume of an MO molecule.[95] Rearranging terms and integrating, one obtains

[95] To see how this comes about, multiply the flux J_{M^*} by the area A of the layer, which gives the total number of metal atoms per second (AJ_{M^*}) reaching the surface and reacting with oxygen. If the volume of an MO molecule is Ω_{MO}, it follows that its thickness is simply Ω_{MO}/A. Hence, the rate at which the layer grows is simply $\Omega_{MO}J_{M^*}$.

$$\Delta x^2 = 2\left[\frac{\Omega_{MO}}{(ze)^2}\int_{\mu_{M^*}\text{ at metal/oxide interface}}^{\mu_{M^*}\text{ at oxide/gas interface}} \frac{-\sigma_{def}\sigma_e}{\sigma_{def}+\sigma_e}\, d\tilde{\mu}_{M^*}\right]t \tag{7.75}$$

Comparing this result to Eq. (7.61), one sees that the parabolic rate constant is given by[96]

$$K_x = \frac{\Omega_{MO}}{(ze)^2}\int_{\mu_{M^*}\text{ at metal/oxide interface}}^{\mu_{M^*}\text{ at oxide/gas interface}} \frac{-\sigma_{def}\sigma_e}{\sigma_{def}+\sigma_e}\, d\tilde{\mu}_{M^*} \tag{7.76}$$

By making use of the Gibbs–Duhem relation [Eq. (5.40)], that is, $d\tilde{\mu}_{M^*} = -d\tilde{\mu}_{O_2} = -(zkT/4)d\ln P_{O_2}$, this relationship can be recast as

$$K_x = \frac{zkT\Omega_{MO}}{4(ze)^2}\int_{P_{O_2}^{I}}^{P_{O_2}^{II}} \frac{\sigma_{def}\sigma_e}{\sigma_{def}+\sigma_e}\, d\ln P_{O_2} \tag{7.77}$$

where $P_{O_2}^{I}$ and $P_{O_2}^{II}$ are the oxygen partial pressures at the metal/oxide and oxide/gas interfaces, respectively. Integrating Eq. (7.77) is nontrivial since both the electronic and ionic defect concentrations are functions of P_{O_2}. What is customarily done, however, is to assume average values for the conductivities across the layer, with the final result being

$$K_x = \frac{zkT\Omega_{MO}}{4(ze)^2}\frac{\bar{\sigma}_{def}\bar{\sigma}_e}{\bar{\sigma}_{def}+\bar{\sigma}_e}\ln\frac{P_{O_2}^{II}}{P_{O_2}^{I}} \tag{7.78}$$

At this point, it is a useful exercise to recast Eq. (7.73) in terms of Fick's first law, in order to get a different perspective on the so-called ambipolar diffusion coefficient. It can be shown (see App. 7C) that

$$\frac{d\mu_{M^*}}{dx} = kT\left(\frac{z}{n}+\frac{1}{c_{def}}\right)\frac{dc_{M^*}}{dx} \tag{7.79}$$

which when combined with Eq. (7.73) results in

$$J_{M^*} = J_{def} = -\frac{J_e}{z} = \frac{-\sigma_{def}\sigma_e kT}{(ze)^2(\sigma_{def}+\sigma_e)}\left(\frac{z}{n}+\frac{1}{c_{def}}\right)\frac{dc_{M^*}}{dx} \tag{7.80}$$

Further, by making use of the Nernst–Einstein equation it can be shown that

[96] Wagner defined a **rational scaling rate constant** K_r as the time rate of formation of an oxide expressed as equivalents per unit scale thickness. *An* **equivalent** *is defined as the fraction of the compound that transports one positive and one negative unit charge.* In general, for an M_aX_b oxide, the number of equivalents $\phi = \left(b|Z^-|\right)^{-1} = \left(a|Z^+|\right)^{-1}$. For example, for $Ni_{0:5}O_{0:5}$, $\phi = 0.5$, while for $Al_{1/3}O_{1/2}, \phi = 1/6$. It can be shown that $K_r = K_x/(\phi\Omega_{MO})$ (see Prob. 7.17).

$$J_{M^*} = -(t_e D_{def} + t_{def} D_e)\frac{dc_{M^*}}{dx} \tag{7.81}$$

Since this equation is in the form of Fick's first law, it follows that the chemical or ambipolar diffusion coefficient responsible for oxidation is

$$\boxed{D_{ambi}^{oxid} = t_e D_{def} + t_{ion} D_e} \tag{7.82}$$

which is an alternate way of saying that the rate at which permeation will occur depends on the diffusivities or conductivities of both the ionic and electronic carriers. If either of the two vanishes, then D_{ambi}^{oxid} vanishes and the oxide layer behaves as a passivating layer, protecting the metal from further oxidation. Aluminum provides an excellent example—the aluminum oxide that grows on Al is quite insulating, water-insoluble and adherent, which is why aluminum need not be protected from the elements in the way, say, Fe is.

Worked Example 7.6

The self-diffusion coefficient of Ni in NiO was measured at 1000° C to be 2.8×10^{-14} cm²/s. At the same temperature in air, K_x was measured to 2.9×10^{-13} cm²/s. NiO is known to be a predominantly electronic conductor. What conclusions can be drawn concerning the rate-limiting step during the oxidation of Ni? The lattice parameter of NiO is 0.418 nm. The free energy of formation of NiO at 1000°C is –126 kJ/mol.

ANSWER

Given that the oxide is predominantly an electronic conductor, and $z=2$, Eq. (7.77) simplifies to

$$K_x = \frac{kT\Omega_{MO}}{2(ze)^2}\int_{P_{O_2}^{I}}^{P_{O_2}^{II}} \sigma_{def}\, d\ln P_{O_2}$$

Substituting for σ_{def}, using the Nernst–Einstein relationship [Eq. (7.34)], and integrating gives

$$K_x = \frac{\Omega_{NiO} c_{Ni} D_{Ni}}{2}\ln\frac{P_{O_2}^{II}}{P_{O_2}^{I}}$$

Note that in this case $\Omega_{NiO} c_{Ni} = 1$. The limits of integration are $P_{O_2}^{II}$ in air (0.21 atm) and $P_{O_2}^{I}$ at the Ni/NiO interface. The latter is calculated as follows: For the reaction $Ni + \frac{1}{2}O_2 \Rightarrow NiO$, the equilibrium $P_{O_2}^{I}$ is given by (see Worked Example 5.4 for details)

$$P_{O_2}^{I} = \exp\left(\frac{-2\times126{,}000}{8.314\times1273}\right) = 4.56\times10^{-11}\ \text{atm}$$

Thus

$$\ln \frac{0.21}{P_{O_2}^{I}} = 22.24$$

Thus if Ni diffusion is the rate-limiting step, then the theoretically calculated K_x would be

$$K_x^{theo} = \frac{2.8 \times 10^{-14} \times 22.24}{2} = 3.1 \times 10^{-13}\ cm^2/s$$

which is in excellent agreement with the experimentally determined value of 2.9×10^{-13} cm²/s, indicating that the oxidation of Ni is indeed rate-limited by the diffusion of Ni ions from the Ni side to the oxygen side.

7.4.2 AMBIPOLAR DIFFUSION IN A BINARY OXIDE

The problem considered here is slightly different from the one just examined. Consider, for simplicity, an MO oxide subjected to an electrochemical potential gradient $d\eta_{MO}/dx$ which in turn must result in the mass transport of MO "units" from one region to another. Typically, this occurs during sintering or creep where, as a result of curvature or externally imposed pressures, the oxide diffuses down its electrochemical potential gradient (see Chaps. 10 and 12). To preserve electroneutrality and mass balance, the fluxes of the M and O ions have to be *equal and in the same direction*.

Applying Eq. (7.37) to the fluxes of the M^{2+} and O^{2-} ions, one obtains

$$J_{M^{2+}} = -\frac{D_M c_{M^{2+}}}{kT}\frac{d\tilde{\eta}_{M^{2+}}}{dx} = -\frac{D_M c_{M^{2+}}}{kT}\left(\frac{d\tilde{\mu}_{M^{2+}}}{dx} + 2e\frac{d\phi}{dx}\right) \quad (7.83)$$

$$J_{O^{2-}} = -\frac{D_O c_{O^{2-}}}{kT}\frac{d\tilde{\eta}_{O^{2-}}}{dx} = -\frac{D_O c_{O^{2-}}}{kT}\left(\frac{d\tilde{\mu}_{O^{2-}}}{dx} - 2e\frac{d\phi}{dx}\right) \quad (7.84)$$

where c_i and D_i represent the concentration and diffusivity of species i, respectively. Making use of the following three conditions that reflect electroneutrality, local equilibrium and mass balance, respectively, one obtains

$$J_{M^{2+}} = J_{O^{2-}},\ \mu_{MO} = \mu_{O^{2-}} + \mu_{M^{2+}} \text{ and } c_{O^{2-}} = c_{M^{2+}} = c_{MO}$$

where c_{MO} is the molar concentration of MO "molecules" per unit volume (that is, $c_{MO} = 1/V_{MO}$, where V_{MO} is the molar volume). In complete analogy to how Eq. (7.73) was derived, it is a lengthy but not difficult task to show that the flux of MO units or molecules is given by

$$J_{MO} = \frac{D_M D_O}{D_M + D_O}\frac{c_{MO}}{kT}\frac{d\tilde{\mu}_{MO}}{dx} \quad (7.85)$$

The driving force here is the electro chemical potential gradient in MO. By reformulating this expression in terms of Fick's first law (see App. 7A) it can be shown that[97]

$$D_{ambi} = \frac{D_M D_O}{D_M + D_O} \tag{7.86}$$

This expression is only valid for an MO oxide; for the more general case of an $M_k O_\beta$ oxide, however, the appropriate expression is

$$\boxed{D_{ambi} = \frac{D_M D_O}{\beta D_M + \kappa D_O}} \tag{7.87}$$

Equation (7.87) has far-reaching ramifications and basically predicts that in binary ionic compounds, diffusion-controlled processes are determined by D_{ambi}, which in turn is a function of the individual component diffusivities. For most oxides, however, there are typically orders-of-magnitude differences between the diffusivities on the different sublattices (see, e.g., Fig. 7.2). Consequently, with little loss in accuracy, D_{ambi} can be equated to the slower diffusing species. For instance, in MgO, $D_{Mg^{2+}} \gg D_{O^{2-}}$ and $D_{ambi} \approx D_{O^{2-}}$.

Before proceeding further, it is important to be cognizant of the underlying assumptions made in deriving Eq. (7.87). They are:

1. The oxide is a pure intrinsic oxide, where the dominant defects are *Schottky* defects. This was implied when it was assumed that $c_{O^{2-}} = c_{M^{2+}} = c_{MO}$.
2. The vacancy concentrations are everywhere at equilibrium.
3. Local electroneutrality holds everywhere.
4. The ionic transport number is unity.

7.4.3 REACTION BETWEEN SOLIDS—INTERDIFFUSION

The reactions that can occur between solids are quite diverse and for the most part are quite complicated. In this subsection, the focus is on one very simple case, namely, that of interdiffusion of two ionic crystals in which the cations have the same charge, e.g., AO and BO. To simplify the problem even further, the following assumptions are made:

- The anion sublattice is immobile.
- Cations A and B counter diffuse independently, with self-diffusion coefficients D_{A+} and D_{B+}, respectively, that are not functions of composition.
- Electroneutrality is maintained by having the counterdiffusing cation fluxes coupled. Note that for this to happen, the system must be predominantly an ionic conductor, that is, $t_e \ll t_i$—if not, decoupling of the fluxes will occur (see below).

[97] The term D_{ambi} is used here to differentiate it from D_{chem}. It should be emphasized, however, that in the literature, in most complex expressions for diffusivity, i.e., for any process where there is some coupling between fluxes, the term *chemical diffusion* is used almost exclusively.

∞ Within the interdiffusion layer the system behaves ideally.

With these assumptions, the flux equations for the two cations are given by

$$J_{A^+} = -\frac{D_{A^+}c_{A^+}}{kT}\frac{d\tilde{\eta}_{A^+}}{dx} = -\frac{D_A c_{A^+}}{kT}\left(\frac{d\tilde{\mu}_{A^+}}{dx} + ez_{A+}\frac{d\phi}{dx}\right) \tag{7.88}$$

$$J_{B^+} = -\frac{D_{B+}c_{B^+}}{kT}\frac{d\tilde{\eta}_{B^+}}{dx} = -\frac{D_{B+}c_{B^+}}{kT}\left(\frac{d\tilde{\mu}_{B^+}}{dx} - ez_{B+}\frac{d\phi}{dx}\right) \tag{7.89}$$

Mass conservation requires that the sum of the mole fractions $X_A + X_B = 1$, from which it follows that if the solution is ideal as assumed then, $dc_{A+}/dx = -dc_{B+}/dx$. Making use of this result together with the fact that electroneutrality requires $J_{A+} = J_{B+}$, and following a derivation similar to the one that led to Eq. (7.86), one can show that the interdiffusion coefficient D_{AB} is given by

$$\boxed{D_{AB} = \frac{D_{A^+}D_{B^+}}{X_{AO}D_{A^+} + X_{BO}D_{B^+}}} \tag{7.90}$$

This expression is sometimes referred to as the **Nernst–Planck** expression, and given the many simplifying assumptions made in deriving it, should be used with care.

Now D_{AB} is not to be confused with D_{ambi}; comparing Eqs. (7.90) and (7.86) or (7.87) makes it obvious that the two expressions are not equivalent. When two charged carriers are moving in opposite directions, D_{AB} results. It is the appropriate coefficient to use whenever the two constituents of the same charge migrate in opposite directions—to be used for, e.g., analyzing ion-exchange experiments.

Note that Eq. (7.90) is valid only if the system is predominantly an ionic conductor, since only under these conditions can a diffusion, or so-called Nernst potential, be built up. For a predominantly electronic conductor no coupling occurs between the fluxes, in which case D_{AB} is given by an equation of the type[98]

$$D_{AB} = X_{AO}D_B + X_{BO}D_A \tag{7.91}$$

For example, MgO–NiO interdiffusion has been interpreted using such an expression.

7.4.4 EMF OF SOLID-STATE GALVANIC CELLS

Technologically, one important use of ionic ceramics is for potentiometric sensors. These are solids that, by virtue of being predominantly ionic conductors, are capable of measuring the absolute thermodynamic activities of various species.

It can be shown that (see App. 7D) when a solid is placed between two electrodes with chemical potentials μ_I and μ_{II}, as shown schematically in Fig. 7.19*a*, the open-cell voltage V of such a cell is given by

[98] In the metallurgical literature this expression is known as a Darken-type expression.

$$V = -\frac{1}{z\chi e}\int_{\mu_I}^{\mu_{II}} t_{ion} d\mu_{M^*} \tag{7.92}$$

where z and χ are, respectively, the charge and stoichiometry of the electroactive ion in its standard state.[99] It follows that if the solid separating the electrodes is an ionic conductor, that is, $t_{ion} \approx 1.0$, then V of such a cell is simply

$$V = -\frac{1}{z\chi e}(\mu_{II} - \mu_I) = -\frac{kT}{z\chi e}\ln\frac{a_{II}}{a_I} \tag{7.93}$$

where the activities a_I and a_{II} correspond to the chemical potentials μ_I and μ_{II}, respectively. It follows directly from this expression that if one of the electrodes is in its standard state, say, $a_I = 1$, then V is a direct measure of the activity a_{II} in the second electrode. Needless to say, this is a powerful and elegant technique to measure thermodynamic parameters such as activities, activity coefficients, heats of solutions, solubility limits and extents of nonstoichiometry.

To understand how the observed voltage develops, it is instructive to go through the following thought experiment, depicted schematically in Fig. 7.19. To simplify the problem, the following assumptions are made:

1. The solid electrolyte, SE, is a perfect Na ion conductor; that is, $t_i \approx 1.0$.
2. One of the electrodes is pure Na and $a_I = 1.0$; the other is an electrode in which $a_{II} < 1.0$.
3. Initially the electrodes are separated from the SE, and the initial conditions are such that $\eta^{I}_{Na} > \eta^{SE}_{Na} > \eta^{II}_{Na}$ (Fig. 7.19*a*).
4. The electrode is perfectly blocking to electrons. In other words, there are no surface states in the electrolyte that electrons can jump into.

To illustrate how a voltage develops in such a system, consider what happens when the electrodes are brought into intimate contact with the SE (Fig. 7.19*b*). Upon contact, at the interface where $\eta^{I}_{Na} > \eta^{SE}_{Na}$, the Na ions will jump across the interface from the electrode into the electrolyte, *leaving their electrons behind* (Fig. 7.19*b*). This has two consequences. The first is to increase the Na ion concentration in the SE in the vicinity of the interface. These ions will tend to diffuse into the bulk (i.e., a diffusional flux of Na ions into the SE bulk results, depicted by the arrow labeled *diffusion* in Fig. 7.19*c*). The second is the creation of a space charge at the interface. This space charge, or voltage, creates an electric field that tends to attract the same ions back to the electrode (see arrow labeled *migration* in Fig. 7.19*c*). The voltage developed thus increases up to a point where the electric field, ϕ_I responsible for the migrational flux is exactly balanced by the diffusional flux into the bulk. At this point the system is said to have reached a state of *local* dynamic equilibrium. Note that another way to look at the process is to appreciate that it will continue until $\eta^{I}_{Na} = \eta^{SE}_{Na}$—which is the definition of equilibrium [Eq. (5.35)].

[99] For all metals, $\chi = 1$, whereas for O_2, Cl_2, etc., $\chi = 2$.

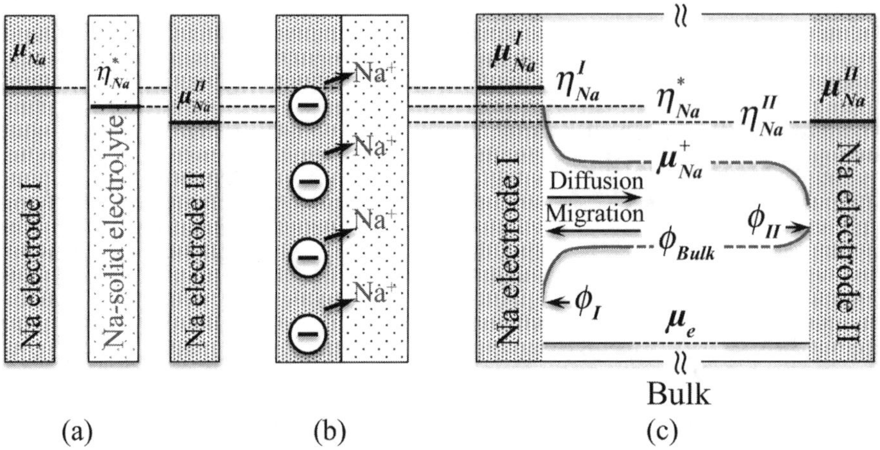

FIGURE 7.19 Development of a space charge and corresponding potential difference $\phi_I - \phi_{II}$ across a solid electrolyte subjected to electrodes of different chemical potentials. (*a*) Initially, before contact is made, it is assumed that $\mu^{I}_{Na} > \eta^{SE}_{Na} > \mu^{II}_{Na}$. (*b*) Upon contact because $\mu^{I}_{Na} > \eta^{*}_{Na}$, Na ions will jump across the interface, leaving their electrons behind and creating a space charge. (*c*) Local dynamic equilibrium is established at each electrode when the migrational and diffusional fluxes become equal.

Conversely, at the electrode at which it was assumed that $\eta^{SE}_{Na} > \eta^{II}_{Na}$, the Na ions will jump from the SE into the electrode. Once again, the process will proceed until a voltage ϕ_{II} (opposite in polarity to the one developed at interface I) develops that is sufficient to equate the electrochemical potentials $\eta^{II}_{Na} = \eta^{SE}_{Na}$ across that interface. To summarize, the space charge that forms at the electrode/electrolyte interface gives rise to a measurable voltage difference, $V = \phi_{II} - \phi_I$, which is in turn related to the activities of the electroactive species in the electrodes—a fact that is embodied in Eq. (7.93).

7.5 RELATIONSHIPS BETWEEN SELF-, TRACER, CHEMICAL, AMBIPOLAR AND DEFECT DIFFUSION COEFFICIENTS

Up to this point, quite a number of diffusion coefficients, listed below, have been introduced. For the reader not well versed in the field, this could lead, understandably enough, to some confusion. The purpose of this section is to shed some more light on the subject.

In this chapter, the following diffusion coefficients were introduced and discussed:

- ∞ *Self-diffusion coefficient*, D_{ion}, is a measure of the ease and frequency with which A atoms hop in pure A, but it applies equally to compounds $M_\kappa X_\beta$ where the M and X species form two independent sublattices.
- ∞ *Tracer diffusion coefficient*, D^*_{tr}, is a measure of the ease and frequency with which radioactive, or tagged atoms, are diffusing in a matrix. It can be shown that $D^*_{tr} = f_{cor} D_{ion}$ where f_{cor} is a correlation coefficient that depends on the crystal structure and diffusion mechanism.

- ∞ *Defect diffusion coefficient*, D_{def}, is a measure of the ease and frequency with which the defects are hopping in a solid; $D_{ion} = \Lambda D_{def}$, where Λ is the fraction of sites available for the diffusing atom or ions to make a jump.
- ∞ *Chemical diffusion coefficient*, D_{chem}, is formally defined as

$$D_{chem} = -\frac{J_i}{dc_i/dx}$$

  where J_i is the flux of species i and dc_i/dx is the gradient in its concentration.[100] In essence, D_{chem} represents the phenomenological coefficient that describes the effective rate at which a given diffusional process is occurring. As discussed below, to relate D_{chem} to more fundamental parameters in a given system, such as the diffusion of component ions or the diffusivities of defects, more information about the latter must be available.
- ∞ *Ambipolar diffusion coefficient* D_{ambi}^{oxid} and D_{ambi} are special cases of D_{chem}, and reflect the fact that in ionic compounds, the fluxes of the ions and defects are by necessity coupled, in order to maintain charge neutrality.
- ∞ *Interdiffusion diffusion coefficient*, D_{AB}, are a measure of the rate at which a diffusional process will occur when ions are interdiffusing.

To illustrate the subtle differences and nuances between the various diffusion coefficients, it is instructive to revisit NiO, which was considered earlier in Worked Example 7.6. To obtain a measure of how fast Ni diffuses into NiO, one can carry out a tracer diffusion experiment, as described earlier. By analyzing the concentration profile of the radioactive tracer, it is possible to determine the so-called tracer diffusivity D_{tr} of Ni in NiO. To measure D_{tr} of oxygen, a crystal is typically exposed to a gas in which the oxygen atoms are radioactive. The D_{tr} is then related to the self-diffusivity, D_{Ni}, by a correlation coefficient f_{cor}. The coefficient f_{cor} has been calculated for many structures and can be looked up.[101]

To relate D_{ion} to defect diffusivities, however, more information about the system is needed. For starters, it is imperative to know the diffusion mechanism—if diffusion is by vacancies, their concentration or mole fraction, Λ, has to be known to relate the two since, according to Eq. (7.14), $D_{vac} = D_{ion}/\Lambda$. Needless to say, if the number of defects is not known, the two cannot be related. If, however, the D_{tr} by an interstitial mechanism, then it can be shown that $D_{int} = D_{tr} = D_{ion}$. In this case the correlation coefficient is unity.[102]

The next level of sophistication involves the determination of the nature of the rate-limiting step in a given process, i.e., relating D_{chem} to D_{tr} or D_{ion}. For example, it was concluded, in Worked Example 7.6, that the rate-limiting step during the oxidation of Ni was the diffusion of Ni ions through the oxide scale.

[100] This is why one strives to cast flux equations in the form of Fick's first law [Eq. (7.81)]. Once in that form, the ratio of J to dc/dx is, by definition, a chemical D.

[101] See, e.g., J. Philibert, *Atom Movements, Diffusion and Mass Transport in Solids*, Les Editions de Physique, in English, trans. S. J. Rothman, 1991.

[102] Only when diffusion occurs by uncorrelated elementary steps are D_{ion} and D_{tr} equal. A case in point is interstitial diffusion where, after every successful jump, the diffusing particle finds itself in the *identical* geometric situation as before the diffusional step. However, for diffusion by a vacancy mechanism, this is not true. After every successful jump, the tracer ion has exchanged places with a vacancy, and thus the probability of the ions jumping back into the vacant site, and canceling the effect of the original jump, is >0.

This conclusion was only reached, however, because the nature of the conductivity in the oxide layer was known. To generalize, consider the following two limiting cases:

1. The oxide layer that forms is predominantly an electronic conductor, that is, $t_e >> t_{ion}$. Hence, according to Eq. (7.82), $D_{ambi}^{oxid} \approx D_{def}$ and the permeation is rate-limited by the diffusivity of the ionic defects. Furthermore, it can be shown that under these conditions

$$K_x = D_{fast} \frac{\left|\Delta \tilde{G}_{MO_{z/2}}\right|}{kT} \quad (7.94)$$

where D_{fast} is the diffusion coefficient of the *faster* of the two *ionic* species and $\Delta \tilde{G}_{MO_{z/2}}$ is the free energy of formation of the $MO_{z/2}$ oxide [Eq. (7.65)]. For most transition-metal oxides, $D_{O^{2-}} \ll D_{M^+}$ and Eq. (7.94) reads

$$K_x = D_{M+} \frac{\left|\Delta \tilde{G}_{MO_{z/2}}\right|}{kT}$$

which, not surprisingly, is the conclusion reached in Worked Example 7.6.

Note that since $t_e >> t_{ion}$, measuring the electrical conductivity of NiO yields no information about the conductivity of the ions, only about the electronic defects.

2. The oxide layer that forms is predominantly an ionic conductor, so, according to Eq. (7.82), $D_{ambi}^{oxid} \approx D_e$ or D_h (note that in order to determine which is the case, even more information about whether the oxide was p- or n-type is required). In this situation Eq. (7.78) reads

$$K_x = \frac{\bar{\sigma}_e \Omega_{MO}}{(ze)^2} \left|\Delta \tilde{G}_{MO_{z/2}}\right| \quad (7.95)$$

where $\bar{\sigma}_{elec}$ is the average partial electronic conductivity across the growing layer. Note that in contrast to the aforementioned situation, measuring K_x yields information not about the diffusivity of the ions, but rather about the electronic carriers, which is not too surprising since in these circumstances the rate-limiting step is the diffusion of the electronic defects. Information about the ionic diffusivity, however, can be deduced from conductivity experiments by relating the conductivity to the diffusivity via the Nernst–Einstein relationship (see, e.g., Worked Examples 7.3 and 7.5).

CASE STUDY 7.1: MODELING OXIDATION KINETICS OF ALUMINA FORMERS

As desirable as they are for protecting metals at high temperatures, unfortunately, parabolic kinetics are still too rapid, in that at longer times, the oxides formed eventually fail to protect. In practice the oxidation kinetics have to be subparabolic. In other words

$$x = K'\left(\frac{t}{t_o}\right)^n \tag{7.96}$$

where $n < 0.5$, i.e., slower than the parabolic exponent of 0.5; t_o defines the time dimension.

As it turns out, only three common elements oxidize with subparabolic kinetics: Si, Al and Cr. Not surprisingly the vast majority of high-temperature applications are possible because of the oxidation of one of these three elements. It is fair to claim that in the absence of these elements and their oxides jet engines and many other high-temperature applications would simply not be possible. Commercially, alumina, Al_2O_3, forming alloys are of paramount importance because the scale is quite protective indeed. It this behooves us thus to try and understand their oxidation kinetics.

Equation (7.74) can be recast to read

$$\frac{dx}{dt} = D_{eff}\frac{\Delta\mu}{kT} \times \frac{1}{x} \tag{7.97}$$

where D_{eff} is an effective diffusion coefficient. If one assumes that the oxidation kinetics are controlled by the grain-boundary diffusion of oxygen through the Al_2O_3 scale (Fig. 7.20), then

$$D_{eff} = D_{GB}f_{GB} = D_{GB}\frac{2\delta_{GB}}{d} \tag{7.98}$$

where D_{GB} and f_{GB} are the oxygen GB diffusion coefficient and the GB density, respectively. The latter can be estimated assuming the oxide grains are cubes of a grain size d and grain-boundary width, δ_{GB} as shown in Fig. 7.20.

In general, as discussed in Chap. 10, grain-coarsening kinetics can usually be described by the following relationship:

$$d^m = d_o^m + K_d t \tag{7.99}$$

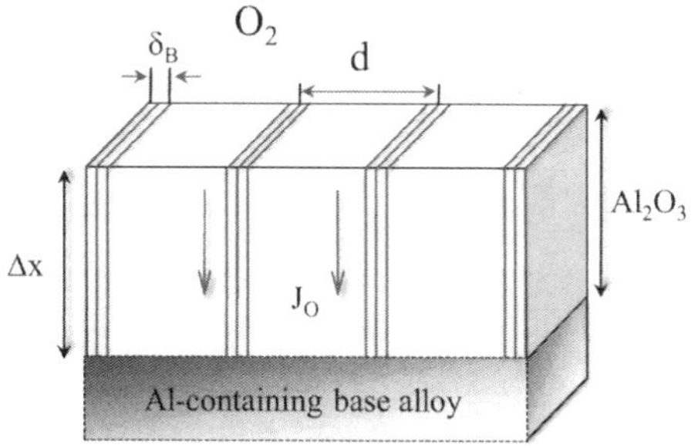

FIGURE 7.20 Schematic of a growing alumina layer by diffusion through grain boundaries.

where K_d is a constant, m is a grain growth exponent [in Eq. (10.36) it is 2] and d_0 is the initial grain size. Combining Eqs. (7.97) to (7.99) it can be shown that at longer times, that are of interest here,[103]

$$x^2 \approx K''\left(\frac{t}{t_0}\right)^{(m-1)/m} \tag{7.100}$$

where K″ is a constant and t_0 is 1 s. If the assumptions made above are correct, then by comparing this equation to Eq. (7.96), it follows that

$$n \approx (m-1)/2m \tag{7.101}$$

In other words, the oxidation kinetics exponent, n, is related to the grain growth exponent, m. This is not too surprising given that it was assumed that the oxygen diffuses through the GBs and if the oxidation is studied over a long enough time interval that the oxide grains grow, then the flux through the GBs should decrease concomitantly.

To test this idea, we measured the grain sizes and oxidation kinetics of Al_2O_3 films that formed on a Ti_2AlC sample that was oxidized for almost 3000 h. From the experimental results we found $m \approx 3.44$. According to Eq. (7.101), then n should be ≈0.35, which, coincidentally or not, was in excellent agreement with the value of 0.36 derived from our oxidation experiments.

CASE STUDY 7.2: SOLID OXIDE FUEL CELLS

In contrast to batteries where the amount of active material is fixed, the fuel in solid oxide fuel cells, SOFC, is provided continuously from an external source, usually as a gas. There are many types of SOFCs; herein we are only interested in ones where the electrolyte is an oxygen ion conductor. More specifically we are interested in SOFC in which the electrolyte is stabilized zirconia. A schematic of a typical cell is shown in Fig. 7.21. Like all electrochemical devices it is comprised of an anode, cathode and electrolyte.

The following reaction occurs at the anode,

$$\frac{1}{2}O_2(g) + V_O^{\bullet\bullet} + 2e' \Rightarrow O_O^x \tag{7.102}$$

The oxygen then traverses through the solid electrolyte to the fuel on the anode side at which point the following reaction:

$$H_2(g) + O_O^x \Rightarrow V_O^{\bullet\bullet} + 2e' + H_2O(g) \tag{7.103}$$

occurs. For an overall reaction of:

$$H_2(g) + \frac{1}{2}O_2(g) \Rightarrow H_2O(g)$$

103 Z. Liu, W. Gao, and Y. He, *Oxid. Met.*, 53, 341–350 (2000).

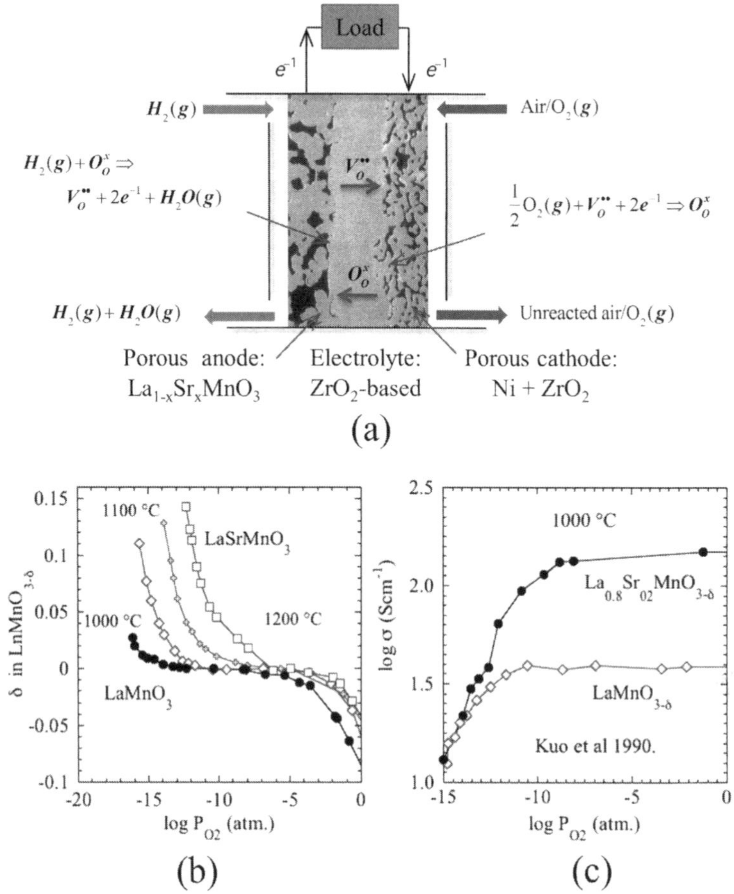

FIGURE 7.21 (*a*) Schematic of a SOFC. Effect of temperature and oxygen partial pressure on (b) non stoichiometry and (c) conductivity of $LaSrMnO_3$.

The driving force is the ΔG change of the last reaction, which in this case—237 kJ/mol. The corresponding equilibrium open-cell voltage is:

$$\Delta V_{ocv} = -\frac{\Delta G}{nF} = 1.23\,V$$

The electrochemical—not to be confused with thermodynamic—efficiency of a SOFC is given by:

$$\varepsilon_V = -\frac{\Delta V_{cell}}{\Delta V_{ocv}} = 1 - \frac{(\eta_c(i) + \eta_a(i) + R_e i)}{\Delta V_{ocv}} \quad (7.104)$$

where η_c and η_a are the overpotentials at the cathode and anode, respectively. R_e is the resistance of the electrolyte. These three terms result in energy losses and have to be minimized.

We now consider the various components of a SOFC separately.

Solid Electrolyte: SOFC commercial systems use either Ca- or Y-stabilized ZrO_2 as the electrolyte. The latter has a $t_i \approx 1$ at elevated temperatures under a wide range of oxygen partial pressures (see Fig. 7.16). The cells operate around 900°C to ensure both rapid electrode kinetics and high ionic conductivities.

Cathode: The most widely used cathode material is a La-Sr mangante, more specifically: $La_{0.8}Sr_{0.2}MnO_{3\pm\Delta}$. The cathode, like the anode, must be a mixed ionic/electronic conductor. More specifically, it must be a p-type conductor since it has to be electronically conducting under high P_{O_2} conditions.

The defect chemistry of $LaMnO_{3\pm\Delta}$ and its doped relatives $La_{1-x}A_xMnO_{3\pm\Delta}$, where A is an alkaline earth cation—typically Ca, Sr or Mg—substituting the La^{3+} cations is important here. The defect incorporation reaction is given in Eq. (15.45). One of the reasons these oxides are as useful as they are is that their stoichiometry, and hence transport properties, can be readily tuned by changing the P_{O_2} and/or doping. From a plot of Δ as a function of P_{O_2} (Fig. 7.21*b*) it is clear that the former can assume both positive and negative values of the order of 0.1 at 1000°C.

As discussed in Chap. 15, the extent of doping, x, is a direct measure of the hole concentration. Said otherwise the Mn^{4+}/Mn^{3+} ratio for the most part is fixed by the dopant and is *not* a function of P_{O_2} unless the latter is low. Transport in these oxides is believed to be due to hole hopping from Mn^{3+} to Mn^{4+}. It follows that the fact that their ratios do *not* change in the high P_{O_2} regime explains why the conductivity is P_{O_2} independent (Fig. 7.21*c*). This is a good example of a why a P_{O_2} independent conductivity cannot always be ascribed to ions.

Intriguingly, but consistent with this picture, is the fact that at low is P_{O_2} the conductivity decreases (Fig. 7.21*c*). To understand why, combine the following two defect reactions:

$$O_O^x \xrightarrow[LaMnO_3]{} V_O^{\bullet\bullet} + 0.5O_2 + 2e' \tag{7.105}$$

$$2e' + 2Mn_{Mn}^{\bullet} \xrightarrow[LaMnO_3]{} 2Mn_{Mn}^x \tag{7.106}$$

for a net reaction of:

$$O_O^x + 2Mn_{Mn}^{\bullet} \xrightarrow[LaMnO_3]{} 2Mn_{Mn}^x + V_O^{\bullet\bullet} + 0.5O_2 \tag{7.107}$$

The first reaction is the classic oxygen reduction reaction, whereas the second emphasizes that the electrons generated change the oxidation state of the Mn ions from Mn^{4+} to Mn^{3+}. It follows that under reducing conditions, the Mn^{4+} concentration is reduced, which in turn reduces the conductivity (Fig. 7.21*c*).

Anode: The anode, like the cathode, should be a good mixed ionic and electronic conductor under reducing conditions. The material of choice is a Ni/ZrO_2 cermet or metal-ceramic composite. At about 30 vol.% Ni, the conductivity of the composite increases by about 4 orders of magnitude as the percolation threshold for Ni is reached. In addition to being mixed conductors, the anodes must also be resistant to sudden increases in P_{O_2} that will occur if the fuel runs out.

The role of the ZrO_2 is to (i) delocalize the electrochemical active zone, (ii) inhibit coarsening of Ni at cell operating temperatures, (iii) lower the CTE of the otherwise metallic anode to better match that of the electrolyte and (iv) provide ionic conductivity as a complementary to the electronic conductivity of the metal particle.

7.6 SUMMARY

1. Ions and atoms will move about in a lattice if they have the requisite energy to make the jump across an energy barrier ΔG_m^* *and* if the site adjacent to them is vacant.
2. In general, diffusivity in ceramics can be expressed as follows:

$$D_k = \Lambda D_0 \exp\left(\frac{-\Delta H_m^*}{RT}\right)$$

   where Λ is the probability that the site adjacent to the diffusing ion is vacant. For defects, $\Lambda \approx 1.0$, and D_0 is a temperature-independent term that includes the vibrational frequency of the diffusing ions, the jump distance and the entropic effect of the atomic jumps. For atoms diffusing by a vacancy mechanism, $\Lambda << 1$ and is equal to the mole fraction of vacancies.
3. The temperature dependence of D will depend on the diffusion mechanism. If diffusion occurs interstitially, the temperature dependence of D will include only the migration energy term, ΔH_m^*, since the probability of the site adjacent to an interstitial atom being vacant is ≈ 1.
4. For diffusion by a vacancy mechanism, the temperature dependence of diffusivity will depend on both the migration enthalpy ΔH_m^* and the energy required to form the vacancies if the latter are thermally activated; i.e., the concentration of intrinsic defects is much greater than the concentration of extrinsic defects. If, however, Λ is fixed by doping, it becomes a constant independent of temperature and proportional to the doping level. The activation energy for diffusion in the latter case will thus only depend on ΔH_m^*.
5. The diffusivity will also depend on the chemical environment surrounding a crystal. This is especially true of nonstoichiometric oxides in which the stoichiometry, and consequently the concentration of the defects, is a relatively strong function of the partial pressure of one of the components of the compounds. In these instances, the partial pressure dependence of the diffusivity is the same as that of the defects responsible for the diffusivity.
6. The presence of a potential gradient, whether chemical or electrical, will result in an ionic flux down the potential gradient. Consequently, the ionic conductivity is directly proportional to the ionic diffusivity through the Nernst–Einstein relationship [Eq. (7.34)].
7. The most fundamental, and general, equation relating the flux of any charged species—by they ions of electric defects—to the gradient in its electrochemical potential is

$$J_k = -\frac{D_k c_k}{kT}\frac{d\tilde{\eta}_k}{dx} = -\frac{\sigma_k}{(z_k e)^2}\frac{d\tilde{\eta}_k}{dx}$$

This expression embodies both Fick's first law and Ohm's law.

8. The total electrical conductivity is governed by

$$\sigma = \Sigma \left| z_i \right| e c_{m,i} \mu_{d,i}$$

where $c_{m,i}$ and $\mu_{d,i}$ are, respectively, the concentrations and mobilities of the mobile species. The total conductivity is the sum of the partial electronic and ionic conductivities.

9. Since ionic conductivities and diffusivities are related by the Nernst–Einstein relationship, what governs one, governs the other. Fast ionic conductors are a class of solids in which their ionic conductivities are much larger than their electronic conductivities. For a solid to exhibit fast ion conduction, the concentration of ionic defects must be much higher than the concentration of electronic defects. The ionic mobilities have to be high as well. The band gap and purity of the material must be quite high to minimize the electronic contribution to the overall conductivity.

10. The electronic conductivity depends critically on the concentration of free electrons and holes. There are essentially three mechanisms by which mobile electronic carriers can be generated in a solid:
    a. Intrinsically by having the electrons excited across the band gap of the material. In this case, the conductivity is determined by the size of the band gap, E_g, and is a strong function of temperature.
    b. Extrinsically by doping the solid with aliovalent impurities that result in the generation of holes or electrons. In this case, if the dopant is fully ionized, the conductivity is fixed by the concentration of the dopant and is almost temperature-independent.
    c. As a result of departures from stoichiometry. The oxidation or reduction of an oxide can generate electrons and holes.

11. Exposing a binary compound to a chemical potential gradient of one of its components results in a flux of that component through the binary compound as a neutral species. The process, termed *ambipolar diffusion*, is characterized by a chemical diffusion coefficient, D_{chem}, that is related to the defect and electronic diffusivities by

$$D_{chem} = t_i D_{elec} + t_e D_{def}$$

Since this process involves the simultaneous, coupled diffusion of ionic and electronic defects, it is the slower of the two that is rate-limiting. To maximize oxidation resistance, electrochemical sensing capabilities and the successful use of ceramics as solid electrolytes, D_{chem}, should be minimized.

12. Exposing a binary MX compound as a whole to a chemical potential, i.e., for $d\mu_{MX}/dx \neq 0$, results in the ambipolar migration of both constituents of that compound down that gradient. The resulting ambipolar diffusion coefficient for an MX is given by

$$D_{ambi} = \frac{D_M D_X}{D_M + D_X}$$

In this case D_{ambi} is determined by the slower of the two components.

13. In a quasi-binary system, interdiffusion of ions also results in a so-called interdiffusion diffusion that is also rate-limited by the diffusivity of the slower of the two ions. This process occurs, e.g., when solid-state reactions between ceramics or ion-exchange experiments are carried out.

14. Solid electrolytes can be used as sensors to measure thermodynamic data, such as activities and activity coefficients. The voltage generated across these solids is directly related to the activities of the electroactive species at each electrode.

APPENDIX 7A: RELATIONSHIP BETWEEN FICK'S FIRST LAW AND EQ. (7.30)

The chemical potential and concentration are related [see Eq. (5.23)] by

$$\mu_i = \mu_i^\circ + RT\ln a_i = \mu^\circ + RT\ln c_i\gamma_i \tag{7A.1}$$

where γ_i is the activity coefficient.[104] It follows that Eq. (7.22) can be written as

$$f = -\frac{1}{N_{Av}}\frac{d\mu}{dx} = -\frac{RT}{N_{Av}}\left[\frac{d\ln c}{dx} + \frac{d\ln\gamma}{dx}\right] = -kT\left[\frac{1}{c}\frac{dc}{dx} + \frac{1}{\gamma}\frac{d\gamma}{dx}\right] \tag{7A.2}$$

For ideal or dilute solutions, γ is a constant and the second term inside the brackets drops out. Substituting Eq. (7A.2) in (7.30) yields

$$J_{ion} = -\frac{c_{ion}D_{ion}}{kT}\left\{kT\left[\frac{1}{c_{ion}}\frac{dc_{ion}}{dx}\right]\right\} = -D_{ion}\frac{dc_{ion}}{dx} \tag{7A.3}$$

which is nothing but Fick's first law. This is an important result since it indicates that whenever $\gamma \neq f(x)$, that is, it is not a function of c, the generalized flux equation [Eq. (7.30)] degenerates into Fick's first law.

If γ_i is a function of concentration, then the second term in Eq. (7A.2) cannot be ignored. Noting that

$$\frac{\partial\ln\gamma}{dx} = \frac{\partial\ln\gamma}{\partial\ln c}\frac{\partial\ln c}{dx}$$

and carrying out the same procedure to obtain to Eq. (7A.3), one obtains

$$D_{chem} = D_{atom}\left(1 + \frac{\partial\ln\gamma_{atom}}{\partial\ln c_{atom}}\right)$$

---

104 Strictly speaking, Eq. (7A.1) should read $\mu_i = \mu_i^\circ + RT\ln x_i\gamma_i$, where x_i is the mole fraction of species i. However, c_i and x_i are related by

$$x_i = \frac{c_i V_i}{\Sigma_i c_i V_i}$$

where V_i is the molar volume of species i. Assuming there are only two species, it follows that for dilute concentrations $X_i \approx c_1V_1/(c_2V_2)$. But since $V_1/(c_2V_2)$ is approximately a constant, it can be incorporated into μ_i°.

where the term in parentheses is known as the **thermodynamic factor**. In other words, the self-diffusivity of the atoms is modified by a factor that takes into account that the diffusing particles now interact with one another. Note that since one cannot define an activity coefficient for a charged species (see Chap. 5), this expression for D_{chem} is valid only for neutral species, hence the subscript "atom."

APPENDIX 7B: EFFECTIVE MASS AND DENSITY OF STATES

For free electrons in a metal, it can be shown that the E–k relationship in three dimensions is

$$E = \frac{h^2}{8\pi^2 m_e}(k_x^2 + k_y^2 + k_z^2) \tag{7B.1}$$

where k_i is the electron wave number in the three principal directions and m_e is the rest mass of an electron. An almost identical relationship for the density of states in a semiconductor or insulator (see Chap. 2) is

$$E = \frac{h^2}{8\pi^2 m_e^*}\left(k_x^2 + k_y^2 + k_z^2\right) \tag{7B.2}$$

where the effective electron mass m_e^* replaces the rest mass. The **effective electron mass** is defined as

$$m_e^* \frac{h}{2\pi}\left(\frac{\partial^2 E}{\partial k^2}\right)^{-1} \tag{7B.3}$$

This is another way of saying that as the electron energy approaches that of a band edge, its effective mass, or the force needed to accelerate is, becomes very large.[105]

To calculate the total number of electrons in the conduction band at any temperature T, the following integral

$$n = \int_{\text{bottom of cond. band}}^{\text{top of cond. band}} (\text{density of states}) \times (\text{prob. of electron occupying given state}) dE$$

Or more succinctly,

$$n = \int_{E_g}^{\infty} f(E) Z(E) dE \tag{7B.4}$$

105 In the limit that the electron energy satisfies the Bragg diffraction condition (i.e., at the top of the valence band), the electron forms a standing wave, and even though it may be experiencing a force, it is "going nowhere." In other words, it behaves as an infinitely heavy object.

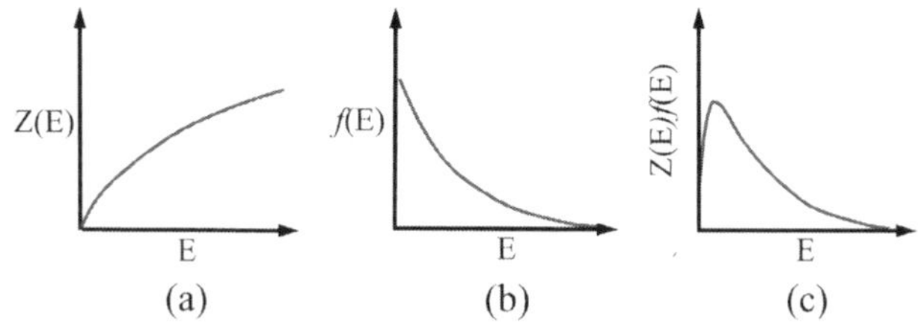

FIGURE 7.22 (*a*) Dependence of the density of states near the bottom of the conduction band on energy. (*b*) Probability of finding electron at energy E, that is, f (E) or Eq. (7B.6). (*c*) A plot of f(E)Z(E) versus E showing that most of the electrons are clustered near the bottom of the conduction band. Conversely, holes would be clustered near the top of the valence band.

has to be evaluated. Here f(E) is the density of states, given by Eq. (7.47), and Z(E)dE is the density or number of electronic states per unit volume having energies between E and E + dE. Taking zero energy to be at the top of the valence band, one can show that for the electrons[106]

$$Z(E)dE = \chi_e (E - E_g)^{1/2} dE \text{ where } \chi_e = \frac{4\pi(2m_e^*)^{3/2}}{h^3} \tag{7B.5}$$

For $E - E_F \gg kT$ the Fermi–Dirac function [Eq. (7.47)] may be approximated by

$$f(E) = \exp\left(-\frac{E - E_f}{kT}\right) \tag{7B.6}$$

Substituting Eqs. (7B.5) and (7B.6) in Eq. (7B.4) and integrating leads to sought after result, Eq. (7.48),

$$n = N_c \exp\left(-\frac{E_c - E_f}{kT}\right) \tag{7B.7}$$

where

$$N_c = 2\left(\frac{2\pi m_e^* kT}{h^2}\right)^{3/2} \tag{7B.8}$$

Similarly, for holes

$$p = N_\upsilon \exp\left(-\frac{E_f - E_\upsilon}{kT}\right) \tag{7B.9}$$

where

$$N_\upsilon = 2\left(\frac{2\pi m_h^* kT}{h^2}\right)^{3/2} \tag{7B.10}$$

[106] L. Solymar and D. Walsh, *Lectures on the Electrical Properties of Materials*, 4th ed., Oxford University Press, New York, 1988.

and m_h^* is the effective mass of a hole.

It is important to note that while the density of states increases monotonically with energy [Eq. (7B.5)], as shown in Fig. 7.20*a*, the probability of occupancy of the higher levels drops rapidly (Fig. 7.20*b*), such that in the end the *filled* electron states are all clustered together near the bottom of the conduction band (Fig. 7.20*c*). Finally, note that for many ceramic materials, the effective masses of the electrons and holes are not known, and the assumption that $m_e = m_e^* = m_h^*$ is oftentimes made.

APPENDIX 7C: DERIVATION OF EQ. (7.79)

In the dilute approximation regime where defect–defect interaction can be ignored, it is possible to express their chemical potential as

$$\mu_{def} = \mu_{def}^{o} + kT \ln c_{def} \tag{7C.1}$$

and

$$\mu_e = \mu_e^{o} + kT \ln n \tag{7C.2}$$

Electroneutrality and mass balance dictate that $c_{def} = n/z = c_{M^*}$ from which

$$\frac{dc_{M^*}}{dx} = \frac{1}{z}\frac{dn}{dx} = \frac{dc_{def}}{dx} \tag{7C.3}$$

Combining this result with Eqs. (7C.1) and (7C.2) together with Eq. (7.68), one can show that

$$\frac{d\mu_{M^*}}{dx} = kT\left[\frac{z}{n} + \frac{1}{c_{def}}\right]\frac{d\mu_{M^*}}{dx}$$

which is Eq. (7.79).

APPENDIX 7D: DERIVATION OF EQ. (7.92)

The situations depicted in Figs. 7.18*a* and 7.19*c* are quite similar in that in both cases a driving force exists for mass transport. The origin of this force is the chemical potential gradient, $d\mu/dx$, that exists across the growing oxide layer in one case and the solid electrolyte or sensor in the other.

The magnitudes of Na ion and electronic fluxes are given, respectively, by [Eq. (7.37)]

$$J_{Na^+} = -\frac{\sigma_{Na^+}}{e^2}\frac{d\bar{\eta}_{Na^+}}{dx} \tag{7D.1}$$

$$J_e = -\frac{\sigma_e}{e^2}\frac{d\bar{\eta}_e}{dx} \tag{7D.2}$$

Since electroneutrality requires that $J_{Na^+} = J_e$, it follows that

$$\frac{d\bar{\eta}_{Na^+}}{dx} = \frac{\sigma_e}{\sigma_{Na^+}} = \frac{d\bar{\eta}_e}{dx} \tag{7D.3}$$

The assumption of local equilibrium implies that

$$d\eta_{Na}\big|_{electrode} = \big|d\bar{\eta}_e + d\bar{\eta}_{Na^+}\big|_{SE} \tag{7D.4}$$

where SE refers to the solid electrolyte. Combining this equation with (7D.3), one obtains

$$\frac{d\tilde{\mu}_{Na^+}}{dx} = \frac{\sigma_e}{\sigma_{Na^+}}\frac{d\tilde{\eta}_e}{dx} + \frac{d\tilde{\eta}_e}{dx} = \frac{1}{t_{ion}}\frac{d\tilde{\eta}_e}{dx} \tag{7D.5}$$

which upon rearrangement and integration and by noting that

$$\Delta\tilde{\eta}_e = \eta_e^{II} - \eta_e^{I} = -eV$$

it follows that

$$V = -\frac{1}{e}\int_{\mu_{Na}^{I}}^{\mu_{Na}^{II}} t_{ion}\, d\tilde{\mu}_{Na}$$

In the more generalized case, where the charge on the cation is not 1 but z and the stoichiometry of the electroactive species is not χ but the more general result given in Eq. (7.92) holds.

PROBLEMS

7.1. (*a*) Calculate the activation energy at 300 K given that $D_0 = 10^{-3}$ m²/s and $D = 10^{-17}$ m²/s.
(*b*) Estimate the value of D_0 for the diffusion of Na ions in NaCl, and compare with the experimental value of ≈0.0032 m²/s. State all assumptions. (Consult Table 6.2 for most of requisite information.)
Answer: $D_0 = 8.8 \times 10^{-4}$ m²/s.
(*c*) Referring to Fig. 7.4a, confirm that $\Delta H_m^* \approx 74$ kJ/mol and $\Delta H_S = 199$ kJ/mol.
(*d*) The ionic conductivity of NaCl as a function of Cd concentration is plotted in Fig. 7.5b. Suggest of defect model that explains these results.
(*e*) Calculate the mobility of interstitial oxygen ions in UO_2 at 700°C. The diffusion coefficient of the oxygen ions at that temperature is 10^{-17} m²/s. State all assumptions. Compare this mobility with electron and hole mobilities in semiconductors.
Answer: 2.4×10^{-16} m²/(V·s).

7.2. (*a*) Explain why for a solid to exhibit predominantly ionic conductivity, the concentration of mobile ions must be much greater than the concentration of electronic defects.

(*b*) Derive Eq. (7.46) and show that the conductivity should reach a maximum when one-half the sites are occupied.

(*c*) Derive Eq. (7.54) and comment on the similarity of this expression to that derived in part (*b*).

7.3. Estimate the number of vacant sites in an ionic conductor at room temperature in which the cations are the predominant charge carriers. Assume that at room temperature the conductivity is $10^{-17}(\Omega \cdot m)^{-1}$ and the ionic mobility is 10^{-17} m²/(V·s). State all assumptions.

7.4. (*a*) What determines whether the conductivity of a ceramic is ionic or electronic?

(*b*) It is often said that reducing a ceramic will increase its electronic conductivity. Do you agree with this statement? Explain.

(*c*) Distinguish between p- and n-type oxides with respect to their oxygen partial pressure dependence of their majority carriers. Describe an experiment by which you could distinguish between the two.

7.5. Consider a NiO unit cell (a = 212 pm) with one V_{Ni} in its center (i.e., assume the atom in the center of the unit cell is missing).

(*a*) If the Ni ion diffusivity at 1500 K is 9×10^{-12} m²/s, estimate σ_{ion}.

Answer: 14.02 S/m.

(*b*) Calculate the Ni vacancy diffusivity. State all assumptions.

Answer: 2.7×10^{-11} m²/s.

7.6. (*a*) A stoichiometric oxide M_2O_3 has a band gap of 5 eV. The enthalpy of Frenkel defect formation is 2 eV, while that for Schottky defect formation is 7 eV. Further experiments have shown that the only ionic mobile species are cation interstitials, with a diffusion coefficient $D_{M,int}$ at 1000 K of 1.42×10^{-10} cm²/s. The mobilities of the holes and electrons were found to be 2000 and 8000 cm²/(V·s), respectively. At 1000 K would you expect this oxide to be an ionic, electronic or mixed conductor? Why? You may find this information useful: Molecular weight of oxide = 40 g/mol, density = 4 g/cm³. Assume the density of states for holes and electrons to be on the order of 10^{22} cm⁻³.

Answer: $\sigma_{ion} = 2.6\times10^{-9}$ S/cm, $\sigma_p = 8.0\times10^{-7}$ S/cm, $\sigma_n = 3.2\times10^{-6}$ S/cm.

(*b*) If the oxide in part (*a*) is doped with 5 mol.% of an MbO oxide [final composition: $(MbO)_{0.05}(M_2O_3)_{0.095}$, write two possible defect reactions for the incorporation of MbO in M_2O_3 oxide. Calculate the molar fraction of each defect formed.

(*c*) Assume one of the defect reactions in part (*b*) involves the creation of Mb_i. Recalculate the ionic conductivity, given that the diffusivity of Mb interstitials D_{Mbi} in M_2O_3 at 1000 K is measured to be 10^{-9} cm²/s. *Hint*: Start with 1 mol of final composition, and calculate the fraction of Mb ions that go interstitially. Make sure you take the effective charge into account.

Answer: $\sigma_{Mbi} = 7.44\times10^{-6}$ S/cm.

7.7. (*a*) Construct the Kroger–Vink diagrams for pure zirconia.

Answer: See Fig. 7.16*a*.

(*b*) Repeat part (*a*) for calcia-doped zirconia and compare to Fig. 7.16*a*. State all assumptions. On the same diagram, explain what happens if the dopant concentration is increased.

7.8. Refer to Worked Example 7.3.

(*a*) What do you think will happen in part (*c*), if the Al_2O_3 doping is increased to 20 ppm. Explain your reasoning, again stating all assumptions.
Answer: Extrinsic.
(*b*) Plot the log conductivity vs. log P_{O_2} you would expect for the oxide doped with 0.1 ppm Al_2O_3 and the one doped with 20 ppm Al_2O_3? Explain your reasoning.
Answer: 0.1 ppm: straight line with +1/6 slope; 20 ppm no P_{O_2} dependence.
(*c*) Clearly explain whether V_M, D_v or D_M will change with doping or not. Explain your reasoning, again stating all assumptions, etc.
(*d*) Calculate the electronic conductivity at 1700 K of the sample doped with 20 ppm Al_2O_3. Explain your reasoning, again stating all assumptions, etc.
Answer: 5.92 S/m.

7.9. (*a*) To increase the electron (n-type) conductivity of ZnO, which would you add, Al_2O_3 or Li_2O? Explain.
Answer: Al_2O_3 (not a typo).
(*b*) The resistivity of ZnO was found to decrease from 4.5 to 1.5Ω·cm as the doping level was increased from 0.23 to 0.7 mol.%. Which dopant do you think was used (Al_2O_3 or Li_2O) and why? Derive an expression for the conductivity of this oxide in terms of dopant concentration that takes into account the latter. Is the expression derived consistent with the changes in conductivity observed?
(*c*) ZnO is a semiconductor. Except at very high temperatures the carrier concentration is related to the excess Zn which dissolves into the structure as interstitials. The first ionization energy of the Zn interstitial is 0.04 eV, the second is ≈1.5 eV. The electronic structure of the Zn atom is . . . $3d^{10}4s^2$, of an oxygen atom is . . . $2s^2 2p^4$. The band gap is 3.2 eV.
(*a*) Sketch schematically the band structure of ZnO. Label the valence band, conduction band, and the two zinc interstitial defect levels.
(*b*) What is the defect reaction for the incorporation of Zn vapor into ZnO and what kind of electronic carriers result?
(*c*) Samples of ZnO were annealed at 1300°C, in a Zn vapor and quenched to room temperature. Their electrical transport properties were then measured in the range −200 to +300°C. The results are shown in Fig. 7.23. What is the reason for the decrease in the carrier concentration below room temperature?
(*d*) Is the mobility due to free carriers or to "hopping"? Explain.

7.10. Positive and negative vacancies are attracted to one another coulombically. Show that if the energy of attraction of such pairs is E_p, then the fraction of such pairs at equilibrium is given by

1. $$\frac{V_p}{(V - V_p)^2} = K = 6e^{E_p/RT}$$

where V_p is the number of pairs and $V = V_{cat} = V_{an}$. State all assumptions.

7.11. Thomas and Lander[107] measured the solubility and conductivity of hydrogen in ZnO and found that they varied as $P_{H_2}^{1/4}$. Derive a model that explains their observation. *Hint*: Hydrogen dissolves interstitially and ionizes.

[107] D. G. Thomas and J. J. Lander, *J. Chem. Phys.*, 25, 1136–1142 (1956).

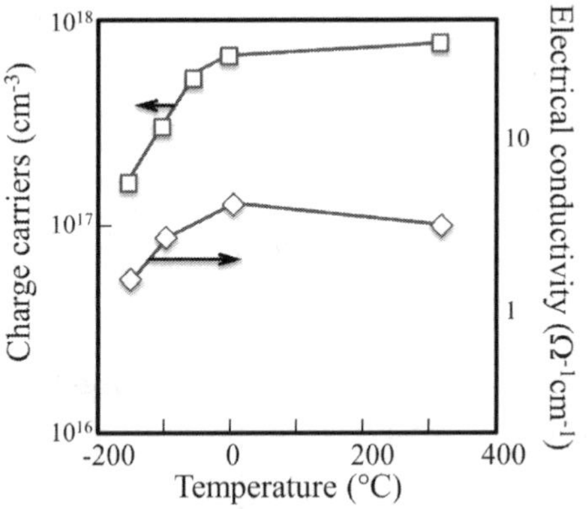

FIGURE 7.23 Temperature dependence of electrical conductivity of ZnO (right axis) and charge carrier concentration (left axis). Note log scale on y-axes.

7.12. Undoped PbS is an n-type semiconductor and crystallizes in the NaCl structure. The predominant defects are Frenkel defect pairs on the Pb sublattice.

(*a*) Would you expect the diffusivity of Pb or S to be greater in PbS? Why?

(*b*) Explain what would happen to the diffusional flux of Pb upon the addition of Ag_2S to PbS.

Answer: Pb ion flux will increase with additions (not a typo).

(*c*) How would Bi_2S_3 additions affect the diffusion of Pb?

7.13. Cuprous chloride is an electronic p-type conductor at high Cl partial pressures. As the chlorine pressure decreases, ionic conductivity takes over.

(*a*) Suggest a mechanism or combination of mechanisms to explain this behavior.

(*b*) Obtain a relationship between conductivity and the defect population for your proposed mechanism(s) that is consistent with the experimental observations. *Hint*: Consider two mechanisms, one stoichiometric and the other nonstoichiometric.

7.14. The electrical conductivity σ of a solid is predicted to vary as

$$\sigma = \frac{C}{T}\exp\left(-\frac{Q}{kT}\right)$$

where C is a constant and k is Boltzmann's constant. Measurements of σ, in arbitrary units, for ice as a function of T were as follows:

σ	31	135	230	630
T, K	200	220	230	250

Based on these results, what do you expect the conduction mechanism in ice to be if: (i) band gap of ice is 0.1 eV, (ii) proton transport involving the breaking of a H-bond is 0.25 eV and (iii) transfer of complex ions requiring simultaneous breaking of four hydrogen bonds is 1 eV?

Answer: Activation energy for conduction = 0.3 eV, thus operative mechanism is (ii).

7.15. The functional dependence of the electrical conductivity of an oxide on the P_{O_2} and temperature is shown in Fig. 7.24*a*.
The temperature dependence of the conductivity is shown in Fig. 7.24*b*.
Answer the following questions.
(*a*) Is this oxide stoichiometric or nonstoichiometric? Explain.
(*b*) Develop the defect reaction or reactions that would explain this behavior. Pay special attention to the slopes.
(*c*) What type of conductor (ionic, p-type, n-type, etc.) do you expect this oxide to be? Elaborate, using appropriate equations.
(*d*) To which energy does the slope of the line in Fig. 7.24*b* correspond? Explain, stating all assumptions.
(*e*) Label the curves in Fig. 7.24*a* in terms of increasing temperature. Explain.
(*f*) Do these figures assume equilibrium of any kind? Elaborate briefly.
(*g*) Describe what changes, if any, would occur to the defects in this crystal if the temperature were suddenly changed from, say, T_1 to T_2 (assuming $T_1 > T_2$). Elaborate on the atomic mechanisms that would be occurring to affect the changes, if any.

7.16. (*a*) The tracer diffusion of coefficient of oxygen in calcia-stabilized zirconia (CSZ) was measured to fit the relationship

$$D = 1.8 \times 10^{-6} \exp\left(-\frac{1.35\text{eV}}{kT}\right) m^2/s$$

Assuming the transport number of oxygen is unity, estimate the electrical conductivity at 1000°C. Assume the unit cell side of 513 pm. State all assumptions.
Answer: ≈ 2.81 S/m.
(*b*) CSZ membranes are currently used as solid electrolytes in fuel cells. To get maximum efficiency of the fuel cell, however, it is imperative to reduce the permeation of oxygen across the membrane. Assume that CSZ in part (*a*) has a hole transport number of 10^{-3} at 1000°C. If the thickness of the CSZ is 1 mm, estimate the molar flux of oxygen permeating through, if the fuel cell is operating between a P_{O_2} of 10^{-10} atm and air. State all assumptions.
Answer: 8.33×10^{-6} mol of O/m²s.

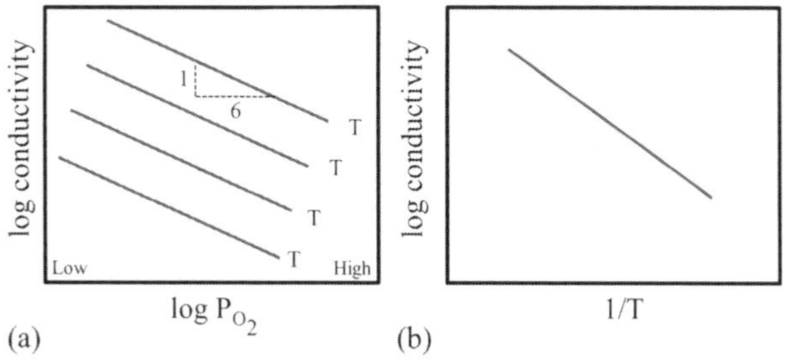

FIGURE 7.24 (*a*) Functional dependence of log σ on log P_{O_2}. (*b*) Effect of temperature on conductivity, at a fixed P_{O_2}.

7.17. One side of an oxygen sensor is exposed to *air* and the other to an equilibrium mixture of Ni and NiO. The following results were obtained at the various temperatures indicated:

T, K	1200	1300	1400
Emf, V	0.644	0.6	0.55

(*a*) Calculate P_{O_2} on the Ni/NiO side at 1300 K.

(*b*) Calculate the standard free energy of enthalpy and entropy of formation of NiO at 1300 K.

Answer: P_{O_2} at 1300 K = 1×10^{-10} atm; $\Delta G_{1300} = -124$ kJ/mol.

7.18. (*a*) Do the results shown in Fig. 7.17, obey Eq. (7.62)? Explain.

(*b*) If your answer is yes, calculate K_w, and compare to those listed in Table 7.1.

7.19. (*a*) Show that in the case of oxidation, the rational rate constant is related to the parabolic rate constant, K_x, by

$$K_r = \frac{K_x}{\Omega_{M_aO_b}\phi}$$

where ϕ is the number of equivalents (see footnote 108).

(*b*) Further show that K_r and K_w are related by

$$K_r = \frac{1}{2}\frac{|z|^2 V_{MO}}{M_O^2}\frac{\Delta w^2}{t} = \frac{1}{2}\frac{|z|^2 \phi V_{MO}}{M_O^2} K_w$$

where M_O is the atomic weight of oxygen. In this case, z is the valence on the anion.

(*c*) The time dependence of the weight gain of 6×3 mm² Gd metal foils at two different temperatures is shown in Fig. 7.25*a*. Calculate K_w and K_r for this oxide.

(*d*) How long would it take to grow an oxide layer 25 μm thick at 1027°C?

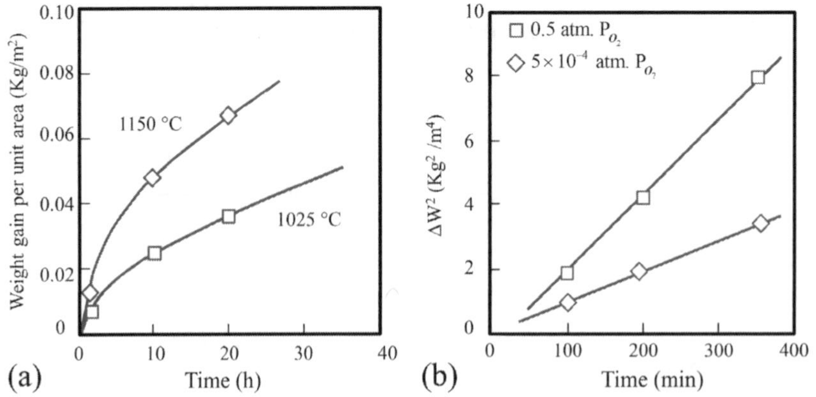

FIGURE 7.25 (*a*) Weight gain during the oxidation of Gd metal at two different temperatures. (*b*) Weight gain during the oxidation of Gd metal at 1027°C at two different P_{O_2}. (Adapted from D. B. Basler and M. F. Berard, *J. Amer. Cer. Soc.*, 57, 447, 1974.)

(*e*) In your own words explain the oxygen partial pressure dependence of the data shown in Fig. 7.25*b*. In other words, why does the sample gain more weight at higher partial pressures? Information you may find useful: molecular weight of Gd_2O_3 = 362.5 mol, density = 7.41 g/cm^3.

7.20. Based on their P-B ratio, predict which of the following metals will form a protective oxide layer and which will not: Be, Nb, Ni, Pd, Pb, Li and Na.

7.21. Derive Eqs. (7.94) and (7.95).

ADDITIONAL READING

J. Maier, *Physical Chemistry of Ionic Materials*, Wiley, Chichester, 2004.

R. Bube, *Electrons in Solids*, Academic Press, New York, 1988.

L. Solymar and D. Walsh, *Lectures on the Electrical Properties of Materials*, 4th ed., Oxford University Press, Oxford, 1988.

F. A. Kröger, *The Chemistry of Imperfect Crystals*, 2nd ed. rev., vols. 1 to 3, North-Holland, Amsterdam, 1973.

L. Heyne, in *Topics in Applied Physics*, S. Geller, Ed., Springer-Verlag, Berlin, 1977.

P. Kofstad, *Nonstoichiometry, Diffusion and Electrical Conductivity in Binary Metal Oxides*, Wiley, New York, 1972.

O. Johannesen and P. Kofstad, Electrical conductivity in binary metal oxides, Parts 1 and 2, *J. Mater. Ed.*, 7, 910–1005 (1985).

C. Kittel, *Introduction to Solid State Physics*, 6th ed., Wiley, New York, 1986.

P. Kofstad, *High Temperature Oxidation of Metals*, Wiley, New York, 1966.

H. Tuller in *Nonstoichiometric Oxides*, O. T. Sorensen, Ed., Academic Press, New York, 1981, p. 271.

R. E. Hummel, *Electronic Properties of Materials*, Springer-Verlag, Berlin, 1985.

L. Azaroff and J. J. Brophy, *Electronic Processes in Materials*, McGraw-Hill, New York, 1963.

H. Rickert, *Electrochemistry of Solids*, Springer-Verlag, Heidelberg, 1982.

H. Schmalzreid, *Solid State Reactions*, 2nd rev. ed., Verlag Chemie, Weinheim, 1981.

P. G. Shewmon, *Diffusion in Solids*, 2nd ed., TMS, Warrendale, PA, 1989.

R. Allnatt and A. B. Lidiard, *Atomic Transport in Solids*, Cambridge University Press, Cambridge, MA, 1993.

J. Philibert, *Atom Movements, Diffusion and Mass Transport in Solids*, Les Editions de Physique, in English, trans. S. J. Rothman, 1991.

R. J. Borg and G. J. Dienes, *An Introduction to Solid State Diffusion*, Academic Press, New York, 1988.

Diffusion Data, 1967–1973, vols. 1–7, Diffusion Information Center, Cleveland, OH.

F. H. Wohlbier and D. J. Fisher, Eds., *Diffusion and Defect Data (DDD)*, vol. 8, Trans. Tech. Pub., Aerdermannsdorf, Switzerland, 1974.

J. C. Bachman et al. Inorganic solid-state electrolytes for lithium batteries: Mechanisms and properties governing ion conduction, *Chem. Rev.*, 116, 140–162 (2016).

OTHER REFERENCES

1. Animation of diffusion in solids: https://www.youtube.com/watch?v=toNKd6YKm2U.
2. Explanation of drift velocity: https://www.youtube.com/watch?v=qgoJY4GNKow.
3. Animation of an electron flowing in a metal: https://www.youtube.com/watch?v=07qqC85Qcpg.

8

PHASE EQUILIBRIA

Like harmony in music; there is a dark inscrutable workmanship that reconciles discordant elements, makes them cling together in one society.

William Wordsworth

8.1 INTRODUCTION

Phase diagrams are graphical representations of what equilibrium phases are present in a material system at various temperatures, compositions and pressures. A **phase** is defined as a region in a system in which the properties and composition are spatially uniform. The condition for equilibrium is one where the electrochemical gradients of all the components of a system vanish. A system is said to be at equilibrium when there are no observable changes in either properties or microstructure with the passing of time, provided, of course, that no changes occur in the external conditions during that time.

The importance of knowing the phase diagram in a particular system cannot be overemphasized. It is the roadmap without which it is quite difficult to interpret and predict microstructure and its evolution, that in turn can have a profound effect on the ultimate properties of a solid.

In principle, phase diagrams provide the following information:

1. The phases present at equilibrium
2. The composition of the phases at equilibrium

3. The fraction of each phase present
4. The range of solid solubility of one element or compound in another

Like Chaps. 2 and 5, this chapter is not intended to be a comprehensive treatise on phase equilibria and phase diagrams. It is included more for the sake of completeness and is to be used as a reminder to the reader of some of the more important concepts invoked. For more information, the reader is referred to the references listed at the end of this chapter.

The subject matter is introduced by a short exposition of the Gibbs phase rule in Sec. 8.2. Unary component systems are discussed in Sec. 8.3. Binary and ternary systems are addressed in Secs. 8.4 and 8.5, respectively. Section 8.6 makes the connection between free energy, temperature and composition, on one hand, and phase diagrams, on the other.

8.2 PHASE RULE

As noted above, phase diagrams are equilibrium diagrams. J. W. Gibbs showed that the condition for equilibrium places constraints on the degrees of freedom, F, that a system may possess. This constraint is embodied in the **phase rule** that relates F to the number of phases P present and the number of components C that states

$$F = C + 2 - P \tag{8.1}$$

where the 2 on the right-hand side denotes that two external variables are being considered, usually taken to be the temperature and pressure of the system.

The number of phases P is the number of physically distinct and, in principle, mechanically separable portions of the system. One of the easiest and least ambiguous methods to identify a phase is by analyzing its X-ray diffraction pattern—every phase has a unique pattern with peaks that occur at very well-defined angles (see Chap. 4). For solid solutions and nonstoichiometric compounds, the situation is more complicated; the phases still have a unique X-ray diffraction pattern, but the angles at which the peaks appear will depend on composition.

In the liquid state, the number of phases is much more limited than in the solid state, since for the most part liquid solutions are single-phase (alcohol and water are a common example). However, in some systems, most notably the silicates, liquid–liquid immiscibility results in the presence of two or more phases (e.g., oil and water). The gaseous state is always considered one phase because gases are miscible in all proportions.

The number of components C is the minimum number of constituents needed to fully describe the compositions of all the phases present. When one is dealing with binary systems, then perforce the number of components is identical to the number of elements present. Similarly, in ternary systems, one would expect C to be 3. There are situations, however, when C is only 2. For example, for any binary join in a ternary-phase diagram the number of components is 2, since one element is common.

The number of degrees of freedom F represents the number of variables, which include temperature, pressure and composition, that have to be specified to completely define a system at equilibrium.

8.3 ONE-COMPONENT SYSTEMS

For a one-component system C = 1, and the degrees of freedom F = 2. In other words, to completely define the system, both temperature and pressure must be specified. If two phases are present, F = 1, and either pressure or temperature needs to be specified, but not both. For example, at 1 atm pressure, water and ice can coexist at only one temperature (0 °C). At the triple point, three phases coexist, and there are *no* degrees of freedom left—the three phases must coexist at a unique temperature and pressure.

As single-phase substances are heated or cooled, they can undergo a number of polymorphic transformations. **Polymorphs** are different crystalline modifications of the same chemical substance. These transformations are quite common and include crystallization of glasses, melting, and many solid-solid phase transformations, some of which are described below. In general, there are two types of polymorphic transformations, displacive and reconstructive.

8.3.1 RECONSTRUCTIVE TRANSFORMATIONS

As shown schematically in Fig. 8.1*a*, reconstructive transformations involve the breaking and rearrangement of bonds. Such transformations usually occur by nucleation and growth, which in turn usually depend on the rate at which atoms diffuse and consequently are relatively sluggish and easily suppressed (see Chap. 9). The reconstructive transformations that occur in quartz, specifically the α-β transformation (see below), are good examples.

8.3.2 DISPLACIVE TRANSFORMATIONS

In contrast to reconstructive transformations, displacive transformations do not involve the breaking of bonds, but rather occur by the displacement of atomic planes relative to one another, as illustrated in Fig. 8.1*b*. These reactions occur quite rapidly, and the resulting microstructures are usually heavily twinned. In these transformations, the role of thermal entropy is important since the enthalpies of the phases on either side of the transformation temperature are quite comparable. It follows that the transformation *usually* results in the formation of more open (less dense) structures at higher temperatures, for reasons touched upon in Chap. 5, namely that the more open structures have higher thermal entropies.[108]

Martensitic transformations in steel are probably the most studied of these transformations. Examples in ceramic systems of technological importance include the tetragonal-to-monoclinic transformation in zirconia, ZrO_2, the cubic-to-tetragonal transformation in $BaTiO_3$, and numerous transformations in silica. In the remainder of this section, for the sake of illustration, each is discussed in greater detail.

[108] Note that there are exceptions; for example, the tetragonal-to-monoclinic transformation of ZrO_2 is one where the more "open" structure is more stable at lower temperatures.

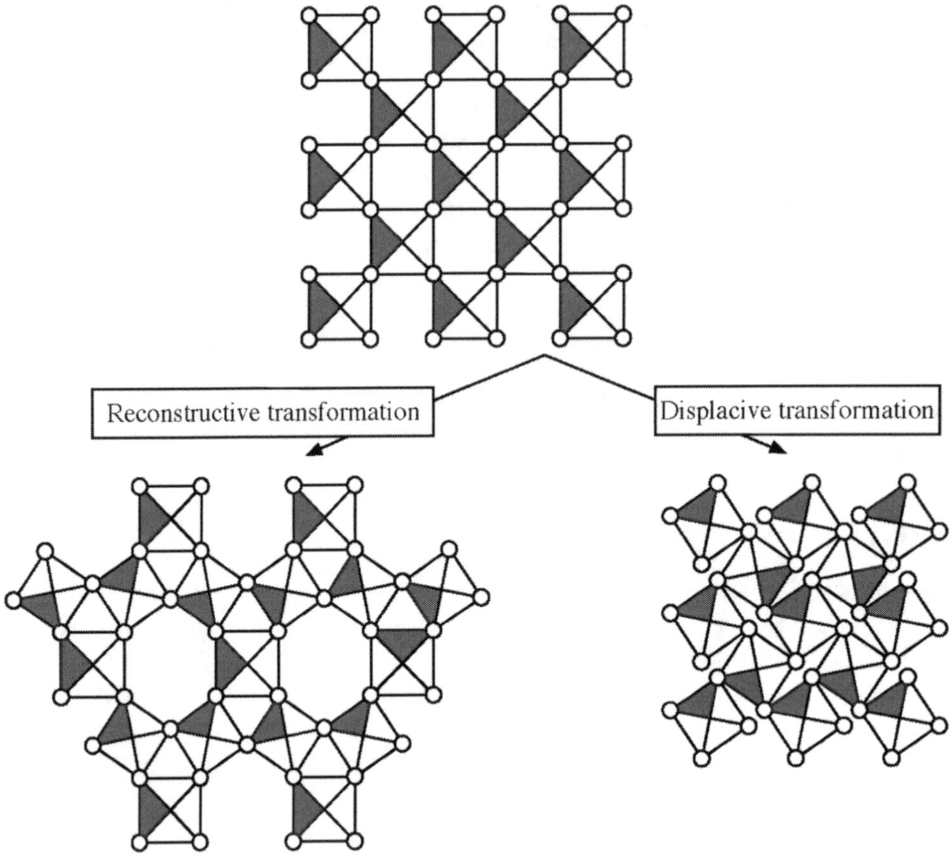

FIGURE 8.1 Schematic of reconstructive (left) and displacive transformations (right).

BARIUM TITANATE

Barium titanate goes through the following phase transitions upon heating:

$$\text{Rhombohedral} \underset{-90^\circ\text{C}}{\rightarrow} \text{orthorhombic} \underset{0^\circ\text{C}}{\rightarrow} \text{tetragonal} \underset{130^\circ\text{C}}{\rightarrow} \text{cubic}$$

Above 130°C, the unit cell is cubic, and the Ti ions are centered in the unit cell. Between 0°C and 130°C, however, $BaTiO_3$ has a distorted perovskite structure with an eccentricity of the Ti ions. As discussed in greater detail in Chaps. 14 and 15, it is this eccentricity that is the origin of this compounds' high dielectric constants that in turn renders it the material of choice for capacitors.

SILICA

Silica has a multitude of polymorphs that undergo a number of both displacive and reconstructive transformations, the most important of which are summarized in Fig. 8.2a. The displacive transformation from

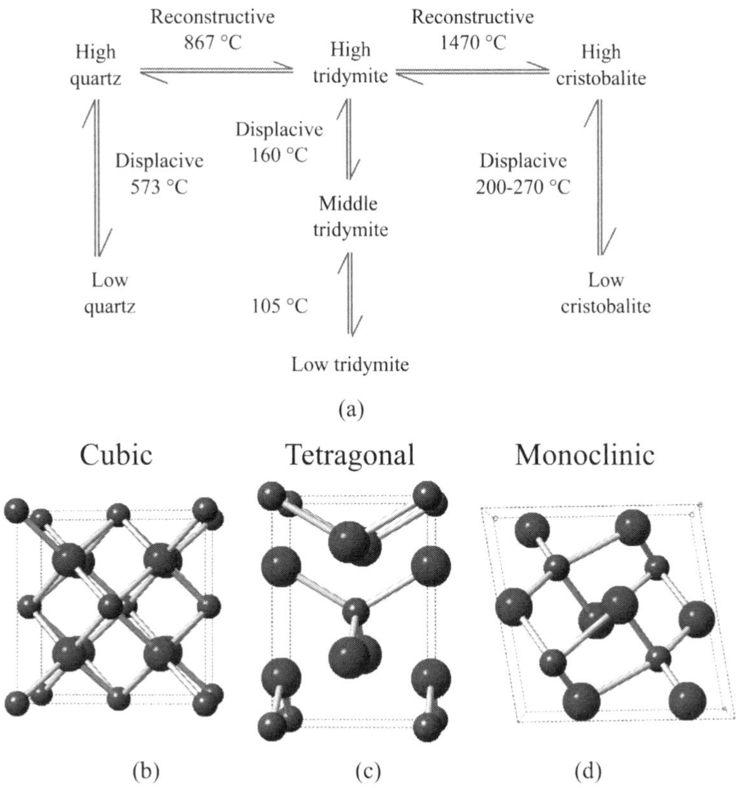

FIGURE 8.2 (a) Polymorphic transformations in silica. (b) Cubic, (c) tetragonal and (d) monoclinic unit cells of zirconia.

high to low quartz is associated with a large volume change (see Fig. 4.5) that, upon cooling, can create large residual stresses and result in a loss of strength (see Chap. 13). The best way to avoid the problem is to ensure that during processing all the quartz is converted to cristobalite, which because of its sluggish reconstructive transformation is metastable at room temperature. The volume change from high to low cristobalite is not as severe as that for quartz.

ZIRCONIA

Upon heating under 1 atm pressure, ZrO_2 goes through the following transformations:

$$\text{Monoclinic} \underset{1170^{\circ}\text{C}}{\rightarrow} \text{tetragonal} \underset{2370^{\circ}\text{C}}{\rightarrow} \text{cubic} \underset{2680^{\circ}\text{C}}{\rightarrow} \text{liquid}$$

It exhibits three well-defined polymorphs: a monoclinic phase, a tetragonal phase, and a cubic phase shown in Fig. 8.2*b–d*, respectively. The low-temperature phase is monoclinic (Fig. 8.2*b*), stable to 1170°C at which temperature it changes reversibly to the tetragonal phase (Fig. 8.2*c*), which in turn is stable to 2370°C. Above that temperature the cubic phase (Fig. 8.2*d*) becomes stable up to the melting point of 2680°C. The tetragonal-to-monoclinic (t ⇒ m) transformation is believed to occur by a diffusionless shear process that is similar to the formation of martensite in steels. This transformation is associated with a large volume change and undergoes extensive shear that is the basis for transformation toughening of zirconia, addressed in greater detail in Chap. 11.

8.4 BINARY SYSTEMS

A binary system consists of two components and is influenced by three variables: temperature, pressure and composition. When two components are mixed together and allowed to equilibrate, three outcomes are possible:

1. Mutual solubility and solid solution formation over the *entire* composition range, also known as *complete solid solubility.*
2. Partial solid solubility *without* the formation of intermediate phases.
3. Partial solid solubility with the formation of intermediate phases.

One objective of this section is to qualitatively describe the relationship between these various outcomes and the resulting phase diagrams. First, however, it is important to appreciate what is meant by a solid solution in a ceramic system and the types of solid solutions that occur—a topic that was dealt with indirectly, and briefly, in Chap. 6. The two main types of solid solutions, described below, are substitutional and interstitial.

In a **substitutional solid solution**, the solute ion directly substitutes for the host ion nearest to it in electronegativity, which implies, as noted in Chap. 6, that cations substitute for cations and anions for anions. Needless to say, the rules for defect incorporation reactions (see Chap. 6) have to be satisfied at all times. For instance, the incorporation reaction of NiO in MgO would be written as

$$\mathrm{NiO} \underset{\mathrm{MgO}}{\rightarrow} \mathrm{O_O^x} + \mathrm{Ni_{Mg}^x}$$

where the Ni^{2+} ions substitute for Mg^{2+} ions. The resulting substitutional solid solution is denoted by $(Ni_{1-x},Mg_x)O$ and shown schematically in Fig. 8.3*a*. The factors that determine the extent of solid solubility are discussed below.

If the solute atoms are small, they may dissolve interstitially in the host crystal which results in a **interstitial solid solution**. The ease with which interstitial solid solutions form depends on the size of the interstitial sites in the host lattice relative to that of the solute ions. For example, in a close-packed structure, such as rock salt, the only available interstitial sites are small tetrahedral sites and thus interstitial solid solubility is not very likely. In contrast, in thoria, or ThO_2, with its fluorite structure where the interstitial sites are relatively large, interstitial solid solutions can form more easily. For example, it has been established that when YF_3 is dissolved in CaF_2, the incorporation reaction is

$$\mathrm{YF_3} \underset{\mathrm{CaF_2}}{\rightarrow} \mathrm{Y_{Ca}^{\bullet}} + \mathrm{F_i'} + 2\mathrm{F_F^x}$$

In other words, to maintain charge neutrality, the extra F^- anions reside on interstitial sites located in the center of the fluorite unit cell, shown as a black circle in the center of unit cell in Fig. 8.3*b*. Another

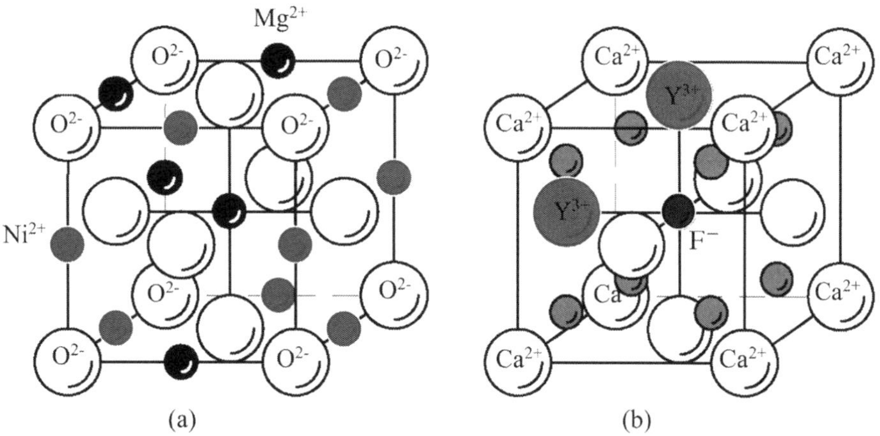

FIGURE 8.3 Representative unit cell of, (*a*) $(Mg_{0.5}Ni_{0.5})O$ and (*b*) 0.25 mol fraction of YF_3 in CaF_2 solid solutions. Note in both unit cells, only the corner atoms are labeled.

example involves the dissolution of ZrO_2 in Y_2O_3, where it has been established that the appropriate defect reaction is

$$2ZrO_2 \underset{Y_2O_3}{\rightarrow} 2Zr_Y^{\bullet} + 3O_O^x + O_i''$$

Worked Example 8.1

Draw a representative unit cell for a, a) $(Mg_{0.5},Ni_{0.5})O$ solid solution and, (*b*) 0.25 mol fraction of YF_3 in CaF_2.

ANSWER

(*a*) A representative unit cell for this solid solution must contain 2 Ni^{2+} and 2 Mg^{2+} cations. It is left as an exercise to the reader to show that the unit cell in Fig. 8.3*a* fulfills that condition.

(*b*) A representative unit cell must reflect the composition of the solid solution, that is, $Y_1Ca_3F_9$.Such a unit cell is shown in Fig. 8.3*b*. Note excess F ion occupies the large central interstitial octahedral site that is normally vacant in the fluorite structure.

After this brief introduction to solid solutions, it is instructive to consider the type of phase diagrams expected for each of the three possible outcomes mentioned above.

8.4.1 COMPLETE SOLID SOLUBILITY

For *complete* solid solubility to occur between two end members, the following conditions have to be satisfied:

1. *Structure type*. The two end members must have the same structure type. For instance, SiO_2 and TiO_2 would not be expected to form a complete solid solution.
2. *Valency factor*. The two end members must have the same valence. If this condition is not satisfied, compensating defects must form in the host crystal in order to maintain charge neutrality. Given that the entropy increase associated with defect formation is not likely to be compensated for by the energy required to form them over the entire composition range, complete solid solubility is unlikely.
3. *Size factor*. As a result of the mismatch in size of the solvent and solute ions, strain energy will develop as one is substituted for the other. For complete solid solubility to occur, that excess strain energy has to be low. Hence, in general, the size difference between the solvent and solute ions has to be less than 15%.
4. *Chemical affinity*. The two end members cannot have too high a chemical affinity for each other. Otherwise the free energy of the system will be lowered by the formation of an intermediate compound.

A typical phase diagram for two compounds that form a complete solid solubility over their entire composition range is shown in Fig. 8.4. Both NiO and MgO crystallize in the rock salt structure, and their cationic radii are quite comparable.

To illustrate the use and usefulness of phase diagrams, consider the 60 mol.% MgO composition depicted by a vertical dashed line labeled Y in Fig. 8.4, and let us examine what happens as it is cooled from the melt. At T_1, a solid solution of MgO and NiO (roughly 80 mol.% Mg^{2+}) will start solidifying. At T_2 or ≈2500°C, two phases coexist: a solid solution of composition Z (see top of Fig. 8.4) and a liquid solution of composition X. The relative amounts of each phase are given by the **lever rule:**

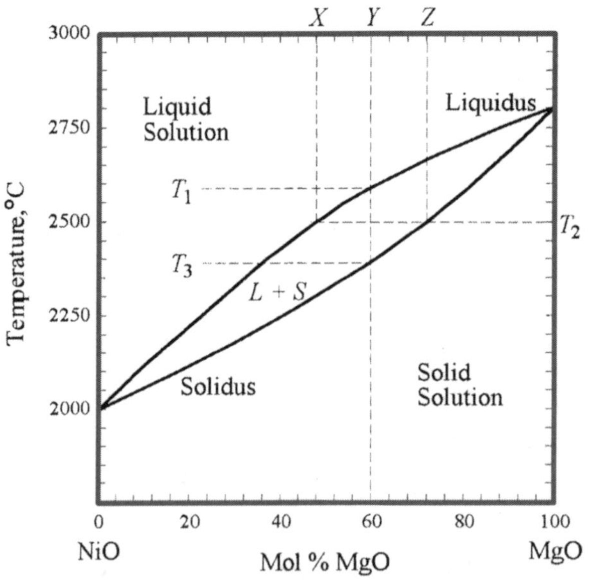

FIGURE 8.4 MgO-NiO phase diagram exhibiting solid solubility over entire composition range. Note liquidus and solidus lines.

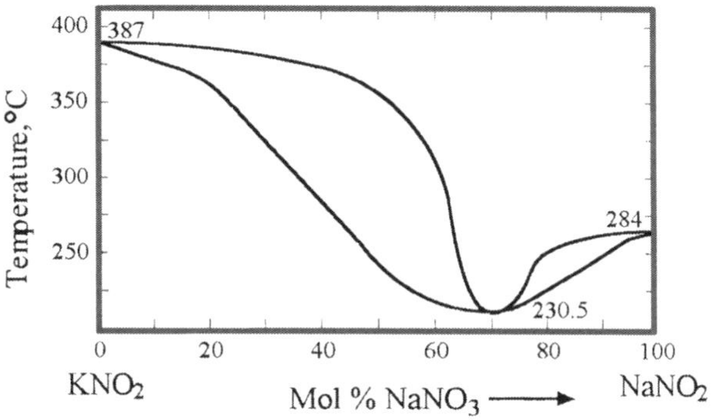

FIGURE 8.5 System with complete solid solubility with a minimum in temperature.

$$\text{Mole fraction liquid} = \frac{yz}{xz} \quad \text{and} \quad \text{mole fraction solid} = \frac{xy}{xz}$$

Note that as the temperature is lowered, the composition of the solid solution moves along the **solidus** line toward NiO, while that of the liquid moves along the **liquidus** line.[109] At T_3 or ≈2400°C, the final liquid solidifies, and the composition of the solid solution is now the same as the initial composition.

Sometimes systems that exhibit complete solid solubility will also exhibit either a maximum (rare in ceramic systems) or a minimum, as shown in Fig. 8.5.

8.4.2 EUTECTIC DIAGRAMS WITH PARTIAL SOLID SOLUBILITY AND NO INTERMEDIATE COMPOUNDS

Given the numerous restrictions needed for the formation of complete solid solutions, they are the exception rather than the rule—most ceramic binary-phase diagrams exhibit partial solubility instead. Furthermore, the addition of one component to another will lower a mixture's freezing point relative to the melting point of the end members. The end result is lowered liquidus curves for both end members that intersect at a point. The point of intersection defines the lowest temperature at which a liquid can exist and is known as the **eutectic temperature** T_E. This type of diagram is well illustrated by the MgO-CaO system shown in Fig. 8.6, where MgO dissolves some CaO and vice versa. The limited solubility comes about mostly because the size difference between the Ca and Mg ions is too large for complete solid solubility to occur. Beyond a certain composition, the increase in strain energy associated with increasing solute content can no longer be compensated for by the increase in configuration entropy.

To illustrate the changes that occur upon cooling consider what happens when a 40 mol.% CaO composition, depicted by the dotted vertical line in Fig. 8.6, is cooled from the melt. Above 2600°C the liquid phase is stable. Just below 2600°C, a MgO solid solution (≈95 mol. % MgO saturated with CaO) will

109 Lines separating single-phase *liquid* regions from two-phase (S + L) regions are known as liquidus lines. Similarly, lines separating single-phase *solid* regions from two-phase regions are known as solidus lines.

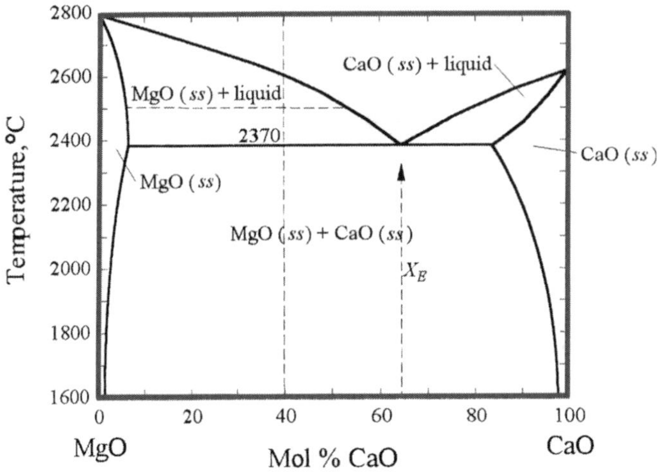

FIGURE 8.6 MgO-CaO phase diagram, which exhibits partial solid solubility of the end members for each other and a single eutectic.

start to precipitate out. At 2500°C, two phases will coexist: a MgO-CaO solid solution and a liquid that is now richer in CaO (≈55 mol % CaO) than the initial composition. Upon further cooling, the composition of the liquid follows the liquidus line toward the eutectic composition, whereas the composition of the precipitating solid follows the solidus line toward the point of maximum solubility. Just above T_E, that is, at $T_E + \delta$, *a* solid solution of CaO in MgO and a liquid with the eutectic composition $X_E \approx 65$ mol.% CaO coexist.

Just below T_E, however, the following reaction

$$L \Rightarrow S_1 + S_2 \tag{8.2}$$

known as a **eutectic reaction**, occurs, and the liquid disproportionates into two phases of quite different compositions[110]—a calcia-rich and a magnesia-rich solid solution.

It is important to note here that the solution of one compound in another is unavoidable—a perfectly pure crystal is a thermodynamic impossibility for the same reason that a defect-free crystal is impossible.[111] The only legitimate question therefore is: How much solubility is there? In many binary systems, the regions of solid solution that are necessarily present do not appear on the phase diagrams. For example, according to Fig. 8.7*a* or 8.8, one could conclude, incorrectly, that neither Na_2O nor Al_2O_3 dissolves in SiO_2. This is simply a reflection of the scale over which the results are plotted—expanding the x-axis will

[110] At the eutectic temperature, three phases coexist and there are no degrees of freedom left. In other words, the coexistence of three phases in a two-component system can occur only at a *unique* temperature, pressure *and* composition.

[111] The decrease in free energy due to increase in entropy associated with the mixing process is infinitely steep as the concentration of solute goes to zero; that is, $\partial \Delta G/\partial n$ goes to $-\infty$ as $n \rightarrow 0$ [see Eq. (6.6)].

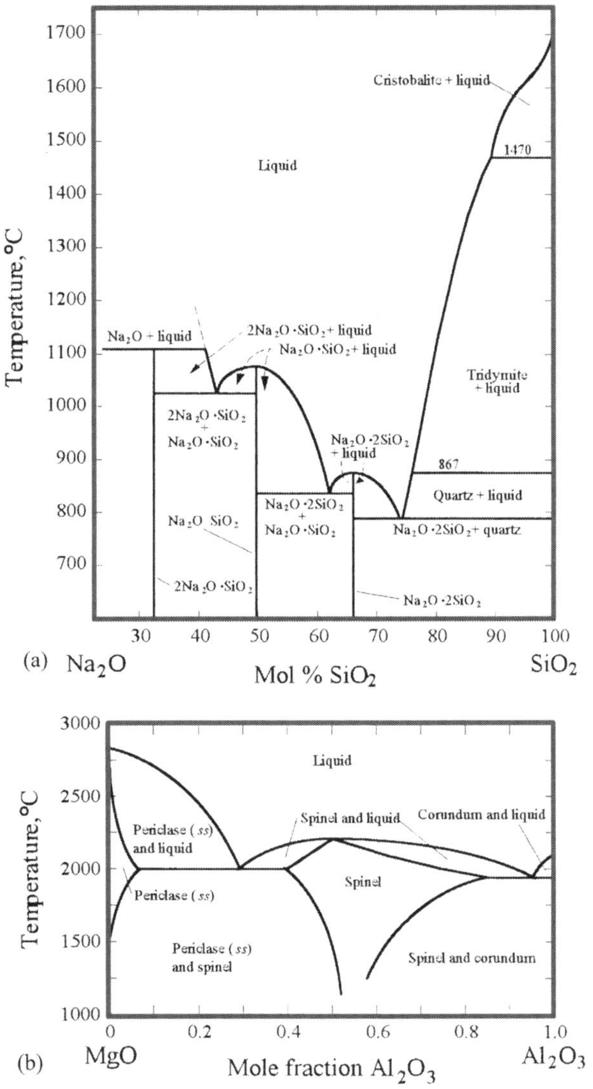

FIGURE 8.7 Phase diagrams of (*a*) Na_2O-SiO_2 and (*b*) MgO–Al_2O_3 systems.

indicate the range of solubility that must be present. Note that in many applications and processes, this is far from being a purely academic question. For example, as noted in the previous chapter, the electrical and electronic properties of a compound can be dramatically altered by the addition of a few parts per million of impurities. Optical properties and sintering kinetics are also strongly influenced by small amounts of impurities. This is especially true when, as noted in Chap. 6, these impurities tend to segregate at grain boundaries.

8.4.3 PARTIAL SOLID SOLUBILITY WITH FORMATION OF INTERMEDIATE COMPOUNDS

As noted above, one of the conditions for the existence of a wide solid solution domain is the absence of a strong affinity of the end members for one another. That is not always the case—in many instances, the two end members react to form intermediate ternary compounds. For instance, the compound $A_xB_yO_2$ can be formed by the reaction

$$xAO_{1/x} + yBO_{1/y} \Rightarrow A_xB_yO_2$$

where the free energy change for the reaction exceeds that for the simple mixing of the two end members to form a solid solution. Under these conditions, intermediate ternary compounds appear in the phase diagram, which in analogy to the end members can either be line compounds (i.e., solubility of end members in the intermediate compound is small) or can have a wide range of stoichiometry. Furthermore, these intermediate phases can melt either congruently or incongruently.

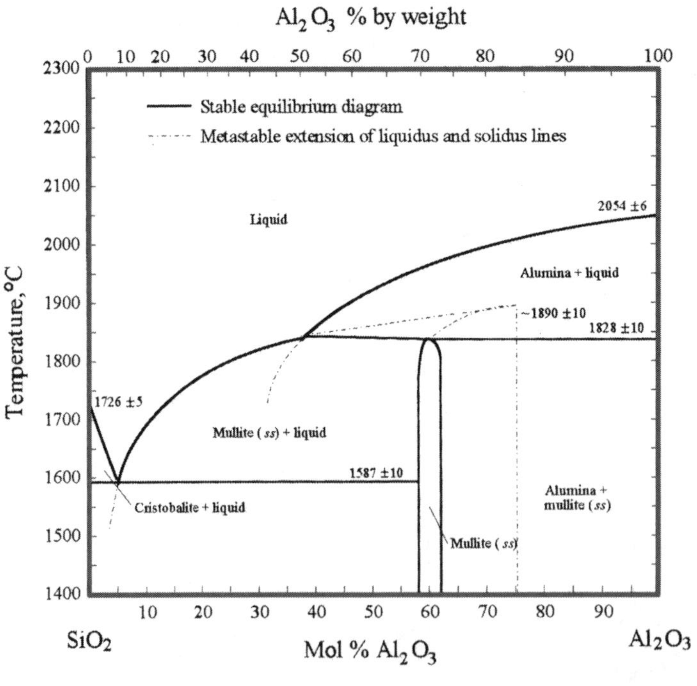

FIGURE 8.8 SiO_2-Al_2O_3 phase diagram. (I. Aksay and J. Pask, *Science*, 183, 69, 1974. See also October issue of *J. Amer. Cer. Soc.*, 74, 2341, 1991, dedicated to the processing, structure, and properties of mullite.)

CONGRUENTLY MELTING INTERMEDIATE PHASES

The ternary compounds $Na_2O{\cdot}2SiO_2$ and $Na_2O{\cdot}SiO_2$, shown in Fig. 8.7*a*, are examples of line compounds that melt **congruently**, i.e., without a change in composition. Note that in this case the resulting phase diagram is simply split into a series of smaller simple eutectic systems (e.g., Fig. 8.7*a*, for compositions greater than 50 mol.% silica).

Spinel, $MgO{\cdot}Al_2O_3$, however, which also melts congruently and splits the phase diagram into two simple eutectic systems (Fig. 8.7*b*), is not a line compound but readily dissolves significant amounts of both MgO and Al_2O_3.

INCONGRUENTLY MELTING INTERMEDIATE PHASES

If the intermediate compound melts **incongruently**, i.e., the compound dissociates before melting into a liquid and another solid, then the phase diagram becomes slightly more complicated. A typical example of such a system is the SiO_2–Al_2O_3 system (Fig. 8.8) where mullite, $2SiO_2{\cdot}3Al_2O_3$, melts at ≈1828° C, by the formation of a liquid containing ≈40 mol.% Al_2O_3 and "pure" alumina according to the reaction

$$S_1 \Rightarrow L + S_2 \tag{8.3}$$

This reaction is known as a **peritectic** reaction and is quite common in ceramic systems. Other examples of incongruently melting ternary compounds are $2Na_2O{\cdot}SiO_2$ (Fig. 8.7*a*) and $3Li_2O{\cdot}B_2O_3$ (Fig. 8.9).

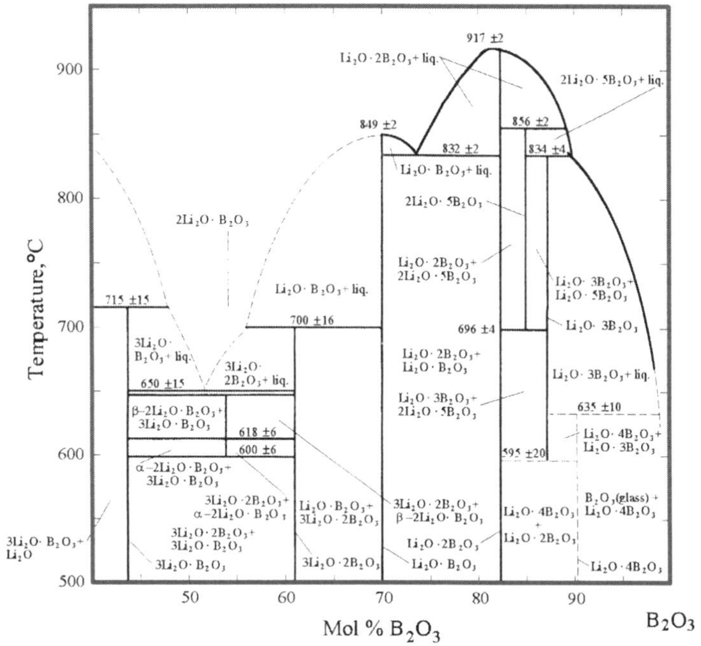

FIGURE 8.9 Li_2O-B_2O_3 phase diagram.

A variation of the aforementioned ideas, are phase diagrams in which a ternary compound will dissociate into two other solid phases upon either cooling or heating. For example, according to Fig. 8.9, at about 700°C, $2Li_2O{\cdot}5B_2O_3$ will dissociate into the 1:2 and 1:3 compounds.

8.5 TERNARY SYSTEMS

Ternary-phase diagrams relate the phases to temperature in a three-component system, and the four variables to be considered are temperature, pressure and the concentration of two components (the composition of the third is fixed by the other two). A graphical representation is possible if the three components are represented by an equilateral triangle, where the apexes of the triangle represent the pure components, and temperature on a vertical axis as shown in Fig. 8.10*a*. The two-dimensional representation of the same diagram is shown in Fig. 8.10*c*, where the intersection of two surfaces is a line, the intersection of

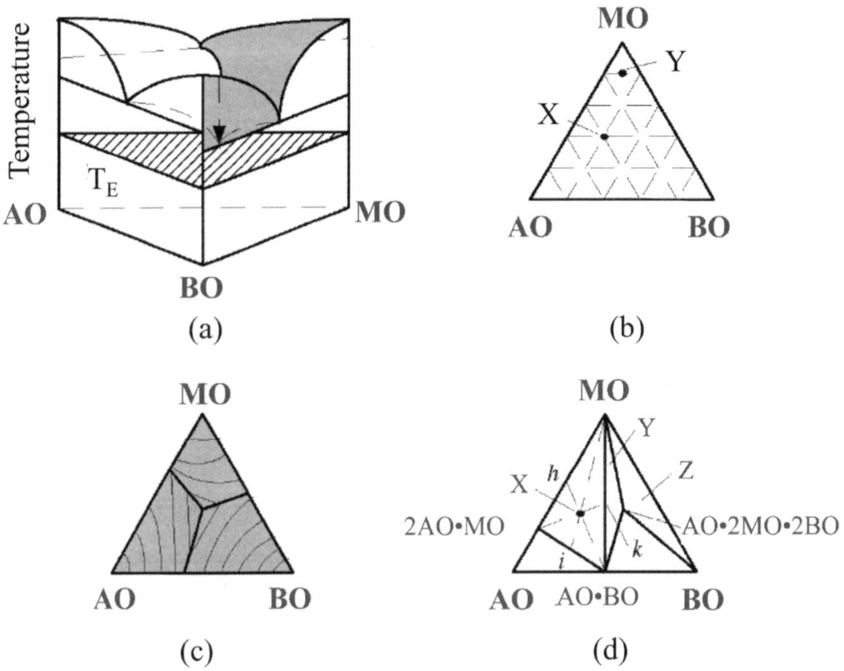

FIGURE 8.10 (*a*) Three-dimensional representation of a ternary-phase diagram. (*b*) Triangular grid for representing compositions in a three-component system. (*c*) Two-dimensional representation of part (*a*) where boundary curves between two surfaces are drawn as heavy lines and temperature is represented by a series of lines corresponding to various isotherms. (*d*) Isothermal section of a ternary system that includes two ternary phases, AO·BO and 2AO MO and a quaternary phase AO·2MO·2BO. Compatibility triangles are drawn with solid lines. (To those familiar with topographical maps, height and temperature are analogous.)

three surfaces is a point, and the temperature is represented by isotherms. The boundary curves represent equilibrium between two solids and the liquid, and the intersection of the boundary curves represents four phases in equilibrium (three solid phases and a liquid). This point is the lowest temperature at which a liquid can exist and, in complete analogy to the binary case, is called the **ternary eutectic**. The composition at any point is found by drawing lines parallel to the three sides of the triangle. For example, the composition of point X in Fig. 8.10*b* is 40 mol.% AO, 20 mol.% BO and 40 mol.% MO, while that at point Y is 80 mol.% MO, 10 mol.% AO, with a balance of BO. The temperature of the liquidus surface is depicted by isothermal contours, as shown in Fig. 8.10*c*.

Any ternary-phase diagram that does not include a solid solubility region will consist of a number of compatibility triangles, shown as solid lines in Fig. 8.10*d*, with the apexes of the triangles representing the solid phases that would be present at equilibrium. For example, if the starting mixture is point X in Fig. 8.10*d*, then at equilibrium the phases present will be MO, 2AO·MO and AO·BO. Similarly, for an initial composition at point Z, MO, BO and the quaternary phase AO·2BO·2MO ($AB_2M_2O_5$) are the equilibrium phases, and so forth. This does not mean that the composition of any of the phases that appear or disappear during cooling are restricted to the triangle, but simply that at equilibrium, any phases that are not within the boundaries of the original triangle have to disappear.

Once the compatibility triangles are known, both the phases present at equilibrium and their relative amounts can be determined. Refer once more to Fig. 8.10*d*. At equilibrium, composition X would comprise the phases MO, 2AO·MO and AO·BO in the following proportions:

$$\text{Mol fraction MO} = \frac{X_i}{i - MO}$$

$$\text{Mol fraction of 2AO} \cdot \text{MO} = \frac{Xk}{k - 2AO \cdot MO}$$

$$\text{Mol fraction of AO} \cdot \text{BO} = \frac{Xh}{h - AO \cdot BO}$$

Note that in going from a ternary to a binary representation, a dimension is lost; planes become lines and lines become points. Thus a ternary phase is a point, and the edges of the triangles represent the corresponding binary-phase diagrams (compare Fig. 8.10*a* and *c*).

8.6 FREE-ENERGY COMPOSITION AND TEMPERATURE DIAGRAMS

The previous sections dealt with various types of phase diagrams and their interpretations. What has been glossed over, however, is what determines their shape. In principle, the answer is simple: The phase, or combination of phases, for which the free energy of the system is lowest is by definition the equilibrium state. However, to say that a phase transformation occurs because it lowers a system's free energy is a tautology, since it would not be observed otherwise—thermodynamics forbids it. The more germane question, and one that is much more difficult to answer, asks: Why does any given phase have the lower free energy at any given temperature, composition, or pressure? The difficulty lies in the fact that to answer the questions, precise knowledge of all the subtle interactions between all the atoms that make up the solid and their vibrational characteristics, etc. is required. It is a many-body problem that is quite sensitive to many variables, the least of which is the nature of the interatomic potentials one chooses to carry out the calculations. It is worth noting that today DFT theory has advanced to the extent that theoretical phase diagrams can be predicted.

The objective of this section is much less ambitious and can be formulated as follows: If the free-energy function for all phases in a given system were known as a function of temperature and composition, how could one construct the corresponding phase diagram? In other words, what is the relationship between free energies and phase diagrams? Two examples are considered below: polymorphic transformations in unary systems and complete solid solubility.

8.6.1 POLYMORPHIC TRANSFORMATIONS IN UNARY SYSTEMS

Congruent melting of a compound, or any of the polymorphic transformations discussed earlier, is a good example of this type of transformation. To illustrate, consider the melting of a compound. The temperature dependence of the free-energy functions for the liquid is

$$G_{T,liq} = H_{T,liq} - TS_{liq}$$

while that for the solid phase is

$$G_{T,s} = H_{T,s} - T_s S_s$$

where H and S are the enthalpies and entropies of the solid and liquid phases, respectively. The two functions are plotted in Fig. 9.1, assuming they are linear functions of temperature. The latter is only valid as long as (1) the heat capacities are not strong functions of temperature and (2) the temperature range considered is not too large.

Here $G_{T,liq}$ is steeper than $G_{T,s}$ because the entropy content of the liquid is larger (more disorder) than that of the solid. The salient point here is that at the temperature above which the lines intersect, the liquid has the lower energy and thus is the more stable phase, whereas below that temperature the solid is. Not surprisingly, the intersection temperature is the melting point of the solid.

8.6.2 COMPLETE SOLID SOLUTIONS

The free energy versus composition diagram for a system that exhibits complete solid solubility is shown in Fig. 8.11. The components of the diagram are the two vertical axes that represent pure AO (left) and pure BO (right). The point labeled μ^o_{AO} represents the molar free energy of formation ΔG_{form} of AO from its elements, and similarly, μ^o_{BO} for BO. In this case AO has a lower free energy of formation than BO. If, for simplicity's sake, the solution is assumed to be ideal, that is, $\Delta H_{mix} = 0$, then the free energy of mixing of AO and BO is given by

$$\Delta G_{mix} = X_{AO}\mu^o_{AO} + X_{BO}\mu^o_{BO} - T\Delta S_{mix} \tag{8.4}$$

where X_i represents the mole fraction of phase i. The entropy of mixing ΔS_{mix} [see Eq. (5.11)] is given by

$$\Delta S_{mix} = -R(X_{AO} \ln X_{AO} + X_{BO} \ln X_{BO}) \tag{8.5}$$

Combining these two equations and plotting ΔG_{mix} versus composition yields the curve $\mu^o_{AO} - M - \mu^o_{BO}$ shown in Fig. 8.11.

Using the same arguments, a free-energy versus composition function for the liquid solution can be determined. Superimposing the two functions as a function of temperature results in the curves depicted in Fig. 8.12*b–d*. It is, in principle, from these types of curves that the corresponding phase diagram shown in Fig. 8.12*a* can be plotted. At T_1 the free energy of the liquid solution is lowest at all compositions (Fig. 8.12*b*) and is the only phase that exists at that temperature. Conversely, at T_3, the solid solution is the most stable phase (Fig. 8.12*d*). At some intermediate temperatureT_2, the free-energy versus composition curves have to intersect (Fig. 8.12*c*), from which it is obvious that, as depicted in Fig. 8.12*a*:

- Between AO and point M, the lowest energy of the system is that of the liquid solution.
- Between BO and N, the solid solution has the lowest energy.
- Between compositions M and N, the system's lowest energy state is given by the common tangent construction. In other words, the system's lowest free energy occurs when two phases (a solid phase of composition N and a liquid phase of composition M) coexist.

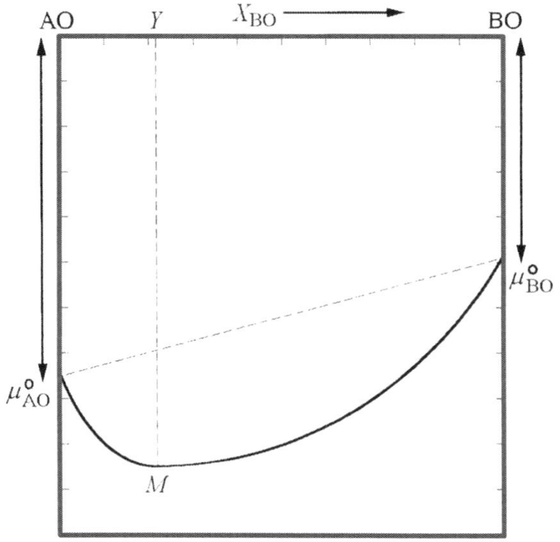

FIGURE 8.11 Free-energy versus composition diagram for an AO-BO mixture exhibiting complete solid solubility at a temperature that is lower than the solidus line.

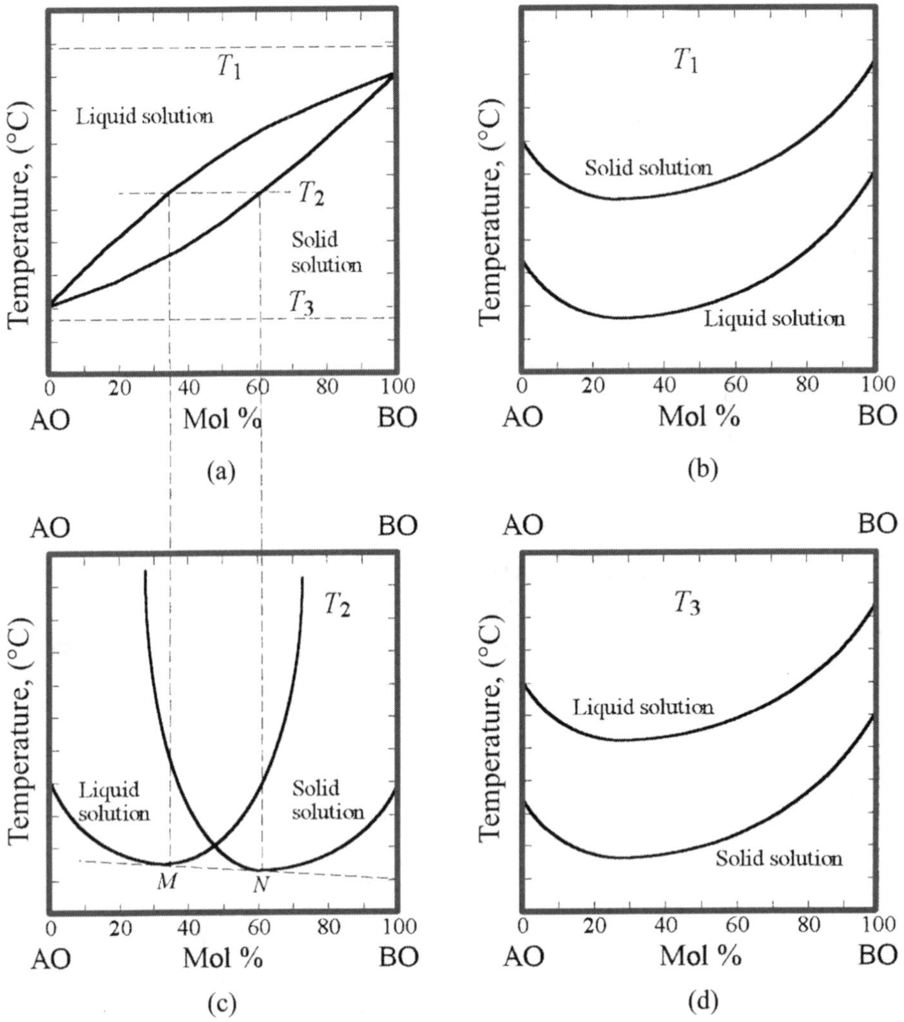

FIGURE 8.12 Temperature versus composition phase diagram and corresponding free-energy composition diagrams at various temperatures. In the two-phase region, a mixture of the solid and liquid solution is the lowest energy configuration.

8.6.3 STOICHIOMETRIC AND NONSTOICHIOMETRIC COMPOUNDS REVISITED

In Chap. 6, the notions of stoichiometry and nonstoichiometry were discussed at some length, and it was noted that a nonstoichiometric compound was one in which the composition range over which the compound was stable was not negligible. In the context of this chapter, the pertinent question is: How does one represent such a compound on a free-energy versus composition diagram? To answer the question, consider Fig. 8.13, where a nonstoichiometric compound $A_{1/2}B_{1/2}O$ is presumed to exist between two stoichiometric compounds, namely,

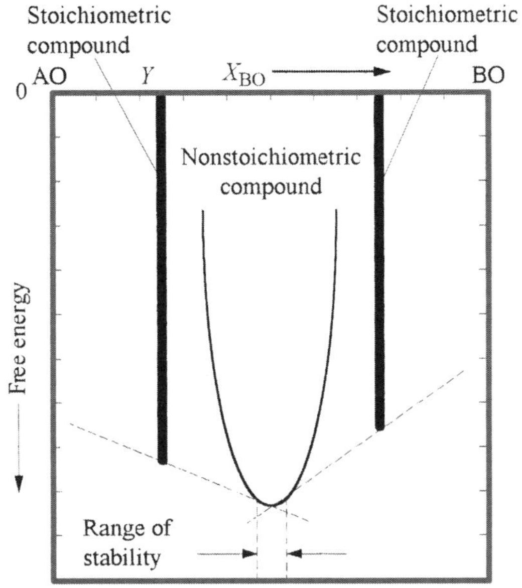

FIGURE 8.13 Free-energy versus composition curves of a nonstoichiometric compound $A_{1/2}B_{1/2}O$ that exists between two stoichiometric, or line compounds, namely, $A_{3/4}B_{1/4}O$ and $A_{1/4}B_{3/4}O$.

$A_{3/4}B_{1/4}O$ and $A_{1/4}B_{3/4}O$. The latter are drawn as straight vertical lines, to emphasize that they exist only over a very narrow composition range; i.e., they are stoichiometric or **line compounds**. Note that the two tangents to the nonstoichiometric phase from the adjacent phases do not meet at a point, implying that there is a range of compositions over which the nonstoichiometric phase has the lowest free energy and thus exists.

In comparing Figs. 8.11 and 8.13, the similarities between the free-energy versus composition curves for a nonstoichiometric compound and a solid solution should be obvious. It follows that an useful way to look at the nonstoichiometric phase $A_{1/2}B_{1/2}O$ is to consider it to be for $X_{BO} < 1/2$ a solid solution between $A_{3/4}B_{1/4}O$ and $A_{1/2}B_{1/2}O$, and for $X_{BO} > 1/2$ a solid solution between $A_{1/4}B_{3/4}O$ and $A_{1/2}B_{1/2}O$. Note that for this to occur, the cations in the nonstoichiometric phase must exist in *more* than one oxidation state.

EXPERIMENTAL DETAILS: DETERMINING PHASE DIAGRAMS

Probably the simplest method to determine phase diagrams is to hold a carefully prepared mixture of known composition isothermally at elevated temperatures until equilibrium is achieved, quench the sample to room temperature rapidly enough to prevent phase changes during cooling and then examine the samples to determine the phases present. The latter is usually carried out by using a combination of X-ray diffraction and microscopy techniques.

And while in principle the procedure seems straightforward enough, the challenge encountered in all phase diagram determinations is to ensure that equilibrium has actually been achieved. The most extensive and up-to-date set of phase diagrams for ceramists is published by the American Ceramic Society.

8.7 SUMMARY

Equilibrium between phases occurs at specific conditions of temperature, composition and pressure. Gibbs' phase rule provides the relationship between the number of phases that exist at equilibrium, the degrees of freedom available to the system and the number of components making up the system.

Phase diagrams are the roadmaps from which the number of phases, their compositions and their fractions can be determined as a function of temperature. In general, binary-phase diagrams can be characterized as exhibiting complete or partial solid solubility between the end members. In case of the latter, they will contain one, or both, of the following reactions depending on the species present. The first is the eutectic reaction in which a liquid becomes saturated with respect to the end members, such that at the eutectic temperature two solids precipitate out of the liquid simultaneously. The second reaction is known as the peritectic reaction in which a solid dissociates into a liquid and a second solid of a different composition at the peritectic temperature. The eutectic and peritectic transformations also have their solid-state analogues, which are called eutectoid and peritectoid reactions, respectively.

Ternary-phase diagrams are roadmaps for three-component systems, where the major difference between them and binary-phase diagrams lies in how the results are presented. In ternary diagrams, the apexes of an equilateral triangle represent the compositions of the pure components, and the temperatures appear as contour lines.

In principle, were one to know the dependence of the free energy of each phase as a function of temperature and composition, it would be possible to predict the corresponding phase diagram. The number of phases present at any temperature are simply the ones for which the total free energy of the system is at a minimum. Given that the free-energy versus composition information is, more often than not, lacking it follows that to date most phase diagrams are determined experimentally. This comment notwithstanding, there are powerful computational methods by which one can determine phase diagrams. The quality of these diagrams, however, are directly linked to the quality of the input results.

PROBLEMS

8.1. Which of the following transformations can be considered displacive and which can be considered reconstructive? Explain.
   (*a*) Melting.
   (*b*) Crystallization.
   (*c*) Tetragonal-to-monoclinic transformation in zirconia.

8.2. (*a*) Explain why complete solid solubility can occur between two components of a substitutional solid solution but not an interstitial solid solution.
   (*b*) Can NaCl and CsCl form an extensive solid solution? Explain.
   (*c*) What type of solid solutions—substitutional or interstitial—would you expect to be more likely in yttria? magnesia? Explain.

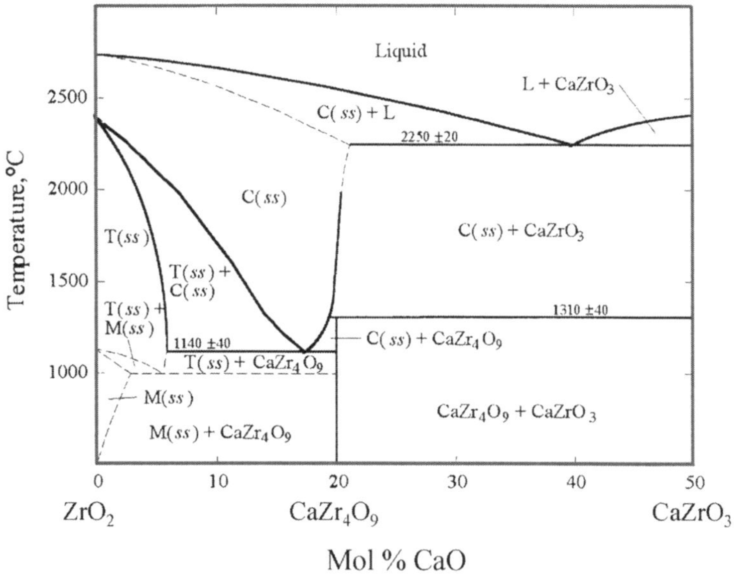

FIGURE 8.14 ZrO_2-$CaZrO_3$ phase diagram.

8.3. The CaO–ZrO_2 phase diagram is shown in Fig. 8.14. What range of initial compositions can be used to manufacture toughened zirconia? Explain. See Chap. 11.

8.4. Starting with stoichiometric spinel, write the incorporation reactions for both alumina and magnesia. Why do you consider spinel to have a wider range of solubility for its end members as compared to, say, $Na_2O{\cdot}SiO_2$?

8.5. (*a*) Show geometrically that the range of nonstoichiometry of a compound is related to the sharpness of the free-energy versus composition diagram; i.e., show that as $\partial G/\partial X$ approaches infinity, one obtains a line compound.

(*b*) From a structural point of view, what factors do you think are likely to determine $\partial G/\partial X$? Consider strain effects and defect chemistry.

ADDITIONAL READING

1. W. D. Kingery, H. K. Bowen, and D. R. Uhlmann, *Introduction to Ceramics*, 2nd ed., Wiley, New York, 1976.
2. P. Gordon, *Principles of Phase Diagrams in Materials Systems*, McGraw-Hill, New York, 1968.
3. J. E. Ricci, *The Phase Rule and Heterogeneous Equilibrium*, Dover, New York, 1968.
4. A. Muan and E. F. Osborne, *Phase Equilibria among Oxides in Steelmaking*, Addison-Wesley, Reading, MA, 1965.
5. M. Alper, Ed., *Phase Diagrams: Materials Science and Technology*, vols. 1 to 3, Academic Press, New York, 1970.
6. L. S. Darken and R. W. Gurry, *Physical Chemistry of Metals*, McGraw-Hill, New York, 1953.

7. G. Bergeron and S. H. Risbud, *Introduction to Phase Equilibria in Ceramics*, American Ceramic Society, Columbus, OH, 1984.
8. M. F. Berard and D. R. Wilder, *Fundamentals of Phase Equilibria in Ceramic Systems*, R. A. N. Publications, Marietta, OH, 1990.
9. F. A. Hummel, *Introduction to Phase Equilibria in Ceramic Systems*, Marcel Dekker, New York, 1984.
10. E. M. Levin, C. R. Robbins, and H. F. McMurdie, *Phase Diagrams for Ceramists*, vol. 1, 1964; vol. 2, 1969.
11. E. Levin and H. McMurdie, Eds., *Phase Diagrams for Ceramists*, vol. 3, 1975.
12. R. Roth, T. Negas, and L. Cook, Eds., *Phase Diagrams for Ceramists*, vol. 4, 1981.
13. R. Roth, M. Clevinger, and D. McKenna, Eds., *Phase Diagrams for Ceramists*, vol. 5, 1983, and cumulative index, 1984.
14. R. Roth, J. Dennis, and H. McMurdie, Eds., *Phase Diagrams for Ceramists*, vol. 6, 1987.

PHASE DIAGRAM INFORMATION

The most comprehensive compilation of ceramic phase diagrams is published by the American Ceramic Society, Columbus, Ohio. Additional volumes are anticipated.

9

FORMATION, STRUCTURE AND PROPERTIES OF GLASSES

Prince Glass, Ceramic's son though crystal-clear
Is no wise crystalline. The fond Voyeur
And Narcissist alike devoutly peer
Into Disorder, the Disorderer
Being Covalent Bondings that prefer
Prolonged Viscosity and spread loose nets
Photons slip through. The average Polymer
Enjoys a Glassy state, but cools, forgets
To slump, and clouds in closely patterned Minuets.

John Updike, The Dance of the Solids*

9.1 INTRODUCTION

From the time of its discovery, thousands of years ago (perhaps on a beach somewhere in ancient Egypt after a campfire was put out) to this day, glass has held a special fascination. Originally, the pleasure was purely aesthetic—glasses, unlike gems and precious stones for which the colors were predetermined by nature, could be fabricated in a multitude of shapes and vivid, extraordinary colors. Today that aesthetic appeal is further enhanced, scientifically speaking, by the challenge of trying to understand their structures and properties.

* J. Updike, *Midpoint and Other Poems*, A. Knopf, Inc., New York, 1969. Reprinted with permission.

Numerous diffraction studies of glasses have shown that while glasses possess short-range order, they clearly lack long-range order and can therefore be classified as solids in which the atomic arrangement is more characteristic of liquids. This observation suggests that if a liquid is cooled rapidly enough such that the atoms do not have enough time to rearrange themselves in a crystalline pattern, before their motion is arrested, a glass is formed. As a consequence of their structure, glasses exhibit many properties that crystalline solids do not; most notably, glasses do not have unique melting points but rather soften over a temperature range. Similarly, their viscosity increases gradually as the temperature is lowered.

This chapter focuses on why glasses form, their structure and the properties that make them unique, such as their glass transition temperature and viscosity. In Sec. 9.2 the question of how rapidly a melt would have to be cooled to form a glass is addressed. Section 9.3 briefly describes glass structure. In Sec. 9.4, the focus is on trying to understand the origin of the glass transition temperature and the temperature and composition dependence of viscosity. The first case study deals with another technologically important class of materials, namely glass-ceramics, their processing, advantages and properties. The second describes how ultrathin and ultra-strong glasses are made. Other properties such as mechanical, optical and dielectric, that show similarities to those of crystalline solids, are dealt with in the appropriate chapters.

9.2 GLASS FORMATION

Most liquids, when cooled from the melt, will, at a well-defined temperature, namely, their melting point, abruptly solidify into crystalline solids. There are some liquids, however, for which this is not the case; when cooled, they form amorphous solids instead. Typically, the transformation of a liquid to a crystalline solid occurs by the formation of nuclei and their subsequent growth—two processes that require time. Consequently, if the rate of removal of the thermal energy is faster than the time needed for crystallization, the latter will not occur and a glass will form. It follows that it is only by understanding the nucleation and growth kinetics that the critical question concerning glass formation, namely, how fast must a melt be cooled to result in a glass, can be answered.

9.2.1 NUCLEATION

The two main mechanisms by which a liquid crystallizes are homogeneous and heterogeneous nucleation. *Homogeneous nucleation* refers to nucleation that occurs without the benefit of preexisting heterogeneities. It is considered first, and in some detail, because of its relative simplicity. *Heterogeneous nucleation* occurs at heterogeneities in the melt, such as container walls, insoluble inclusions and free surfaces. And even though the vast majority of nucleation occurs heterogeneously, it is not as well understood or amenable to analysis as homogeneous nucleation, a fact reflected in the following discussion.

HOMOGENEOUS NUCLEATION

Consider the crystallization of a melt with a melting point T_m. At T_m the free-energy change per mole associated with the solid-to-liquid transformation, ΔG_f, is zero [see Eq. (4.1)], and

$$\Delta S_f = \frac{\Delta H_f}{T_m}$$

where ΔH_f and ΔS_f are the molar enthalpies and entropies of fusion, respectively.

For temperatures $T < T_m$, the solid phase with its lower free energy will be more stable and will tend to form. The free-energy change, ΔG_v, for the transformation is the energy difference between the undercooled liquid and the solid (Fig. 9.1). Assuming that for small undercooling, ΔH_f and ΔS_f remain essentially unchanged, ΔG_v at T is given by

$$\Delta G = \Delta H_f - T\Delta S_f \approx \Delta H_f - \frac{T\Delta H_f}{T_m} = \Delta H_f\left(\frac{\Delta T}{T_m}\right) \tag{9.1}$$

where, the undercooling, $\Delta T = T_m - T$. This equation implies that the driving force for crystallization increases linearly with increasing ΔT.

The energy changes that need be considered during homogeneous nucleation include:

- Bulk free energy released as a result of the liquid-to-solid transformation at $T < T_m$.
- Surface energy required to form new solid surfaces. This term is endothermic.
- Strain energy associated with any volume changes resulting from the transformation.

If one assumes spherical nuclei of radii, r, with a solid–liquid interfacial energy between the growing nucleus and the melt, γ_{sl}, and if one ignores strain effects, the energy changes that accompany their formation are

$$\text{Volume free energy} = -\frac{4}{3}\pi r^3 \frac{\Delta G_v}{V_m} = -\frac{4}{3}\pi r^3 \frac{\Delta H_f}{V_m}\left(\frac{\Delta T}{T_m}\right) \tag{9.2}$$

$$\text{Surface energy} = 4\pi r^2 \gamma_{sl} \tag{9.3}$$

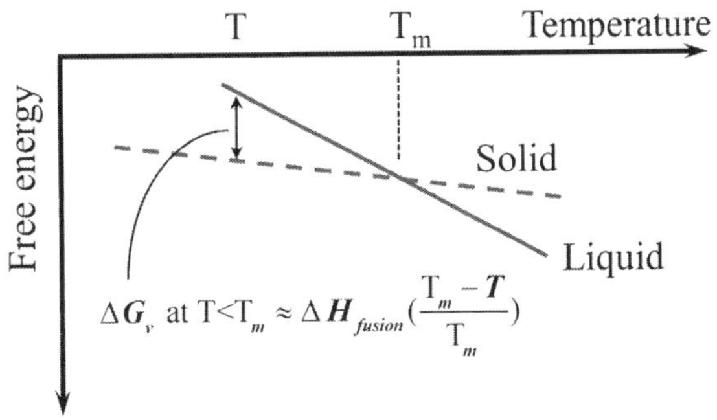

FIGURE 9.1 Schematic of free-energy changes as temperature is lowered below the equilibrium melting point temperature, T_m.

where V_m is the molar volume of the crystal phase. The sum of Eqs. (9.2) and (9.3) represents the excess free-energy change, ΔG_{exc}, resulting from the formation of a nucleus, or

$$\Delta G_{exc} = 4\pi r^2 \gamma_{sv} - \frac{4}{3}\pi r^3 \frac{\Delta H_f}{V_m}\left(\frac{\Delta T}{T_m}\right) \tag{9.4}$$

The functional dependence of ΔG_{exc} on r is plotted in Fig. 9.2. Since the surface energy term (top curve in Fig. 9.2*a*) scales with r^2, whereas the volume energy term (bottom curve in Fig. 9.2*a*) scales with r^3, this function has to go through a maximum at a critical radius, r_c. Said otherwise, the formation of small clusters, with $r < r_c$, *locally* increases the free energy of the system.[112] Differentiating Eq. (9.4), equating to zero, and solving for r_c gives

$$r_c = \frac{2\gamma_{sv} V_m}{\Delta H_f (1 - T/T_m)} \tag{9.5}$$

which, when substituted back into Eq. (9.4), yields the height of the energy barrier ΔG_c (Fig. 9.2*a*)

$$\Delta G_c = \frac{16\pi\gamma_{sl}^3 V_m^2}{3\Delta H_f^2 (1 - T/T_m)^2} \tag{9.6}$$

Small clusters, with $r < r_c$ are called embryos, and are more likely to redissolve than grow. Occasionally, however, an embryo becomes large enough ($r \approx r_c$) and is transformed into a *nucleus*, with an equal probability

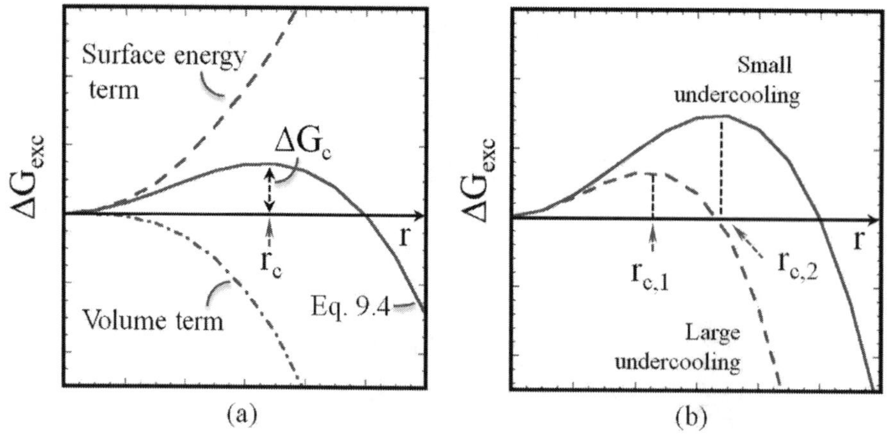

FIGURE 9.2 (*a*) Free energy versus embryo radius for a given $T < T_m$. Note that ΔG_{exc} goes through a maximum at $r = r_c$. (b) Effect of undercooling on ΔG_c. With increased undercooling both ΔG_c and r_c decrease.

[112] Equation (9.4) represents the local increase in free energy due to the formation of a nucleus, and *not* the total free energy of the system. The latter must include the configurational entropy of the mixing of n nuclei in the liquid. When that term is included, the total free energy of the system decreases, as it must (see App. 9A).

of growing or decaying. It is important to note that both ΔG_c and r_c are strong functions of undercooling as shown in Fig. 9.2*b*.

By minimizing the free energy of a system containing N_v total number of molecules or formula units of nucleating phase per unit volume, it can be shown (see App. 9A) that the metastable equilibrium concentration (per unit volume) of nuclei N_n^{eq} is related to ΔG_c by

$$N_n^{eq} = N_v \left[\exp\left(-\frac{\Delta G_c}{kT} \right) \right] \tag{9.7}$$

The rate of nucleation per second per unit volume (# of nuclei per cubic meter) can be expressed by

$$I_v = \nu N_n^{eq} \tag{9.8}$$

where ν is the frequency of successful atom jumps across the nucleus-liquid interface or the rate at which atoms are added onto the critical nucleus, given by

$$\nu = \nu_0 \exp\left(-\frac{\Delta G_m^*}{kT} \right) \tag{9.9}$$

where ν_0 is the vibrational frequency of an atom and ΔG_m^* is the free energy of activation needed for an atom to jump across the nucleus-liquid interface[113] (Fig. 9.3). Combining Eqs. (9.7)–(9.9) yields the final expression for the rate of homogeneous nucleation

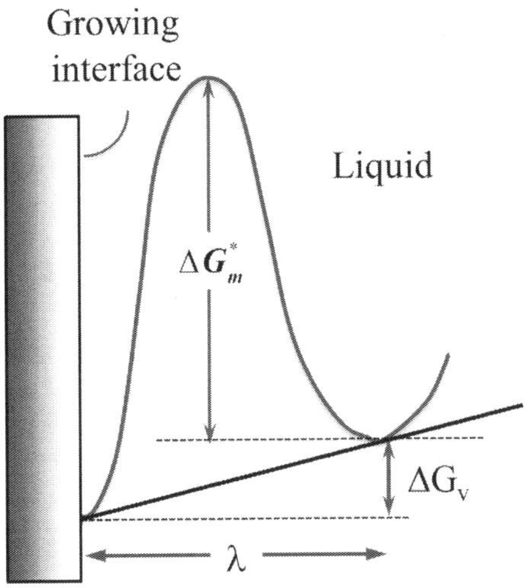

FIGURE 9.3 Schematic of a growing nucleus or crystal interface. Atoms or molecules that result in growth jump a distance, λ, across an energy barrier ΔG_m^* down a chemical potential gradient $\Delta G_v/\lambda$, to the growing interface.

[113] The implicit assumption here is that every atom that makes the jump sticks to the interface and contributes to the growth of the nucleus. In other words, a sticking coefficient of 1 is assumed.

$$I_v = \nu_0 N_v \left[\exp\left(-\frac{\Delta G_m^*}{kT} \right) \right] \left[\exp\left(-\frac{\Delta G_c}{kT} \right) \right] \quad (9.10)$$

The first exponential term is sometimes referred to as the *kinetic barrier to nucleation*, whereas the second exponential term is known as the *thermodynamic barrier to nucleation*. And although it is not immediately obvious from Eq. (9.10), I_v goes through a maximum as a function of ΔT for the following reason: Increased ΔT reduces both r_c and ΔG_c (Fig. 9.2*b*), which in turn strongly *enhances* the nucleation rate, *but* simultaneously severely *reduces* atomic mobility and the rate of attachment of atoms to the growing embryo (see Prob. 9.2*a*). The net effect is that a maximum is expected and is well established experimentally (see Fig. 9.4*a*).

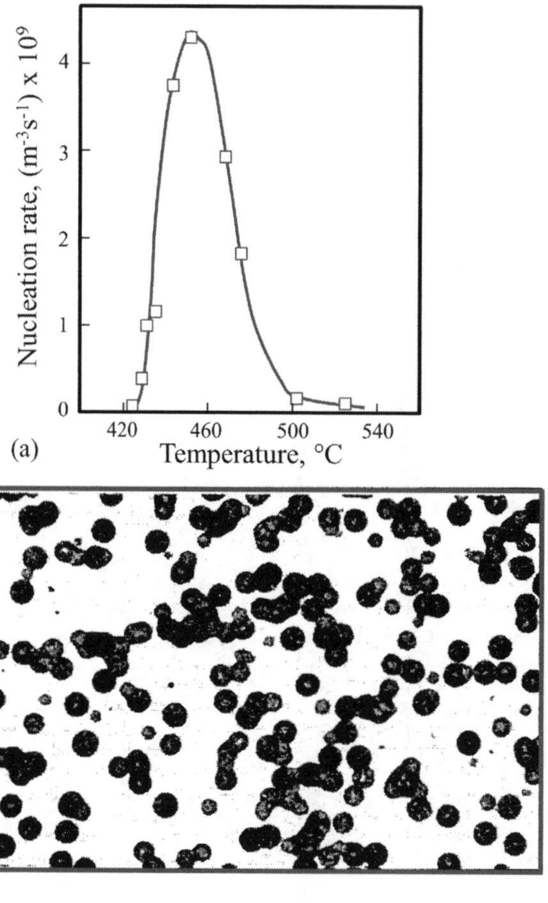

FIGURE 9.4 (*a*) Steady-state nucleation rate as a function of temperature for a glass close to the $Li_2O{\cdot}2SiO_2$ composition. (*b*) Reflection optical micrograph of a $BaO{\cdot}2SiO_2$ glass after heat treatment. (P. James, Chap. 3 in *Glasses and Glass-Ceramics*, M. H. Lewis, Ed., Chapman & Hall, New York, 1989. Reprinted with permission.)

Experimentally, it is much easier to measure the viscosity, η, of an undercooled liquid than it is to measure ν. It is therefore useful to relate the nucleation rate to η, which is done as follows: Given the similarity between the elementary jump shown in Fig. 9.3 and a diffusional jump (see Fig. 7.5*b*), the two can be assumed to be related by

$$D_{liq} = \text{const.}\,\nu\lambda^2 = \text{const.}\,\nu_0\lambda^2 \exp\left(-\frac{\Delta G_m^*}{kT}\right) \tag{9.11}$$

where D_{liq} is the diffusion coefficient of the "formula units" in the liquid and λ is the distance advanced by the growing interface (Fig. 9.3) in a unit kinetic process usually taken as that of a molecular, or formula unit, diameter. Usually λ is taken to equal $(V_m/N_{Av})^{1/3}$, where V_m is the molar volume of the crystallizing phase. If it is further assumed that D_{liq} is related η of the melt by the **Stokes–Einstein** relationship, namely,

$$D_{liq} = \frac{kT}{3\pi\lambda\eta} \tag{9.12}$$

then by combining Eqs. (9.10) to (9.12), one obtains

$$\boxed{I_v = (\text{const})\frac{N_v kT}{3\pi\lambda^3\eta}\exp\left(-\frac{\Delta G_c}{kT}\right)} \tag{9.13}$$

With both η (see below) and the exponential term in Eq. (9.13) increasing at differing rates with decreasing temperature, it should once again be apparent why I_v goes through a maximum as a function of undercooling.

In deriving Eqs. (9.10) and (9.13), the following assumptions were made:

- Nucleation is homogeneous. This is seldom the case. As discussed below, crystal nucleation occurs almost always heterogeneously on impurity particles, or container walls, at relatively low ΔTs.
- The rate given in Eq. (9.7) is a steady-state rate. In other words, density fluctuations develop and maintain an equilibrium distribution of subcritically sized embryos. As nucleation drains off critically sized embryos, new ones are produced at a rate sufficiently fast to maintain the equilibrium distribution.
- Nucleation occurs without a change in composition. If a change in composition accompanies the formation of a nucleus, the expression for the energy gained [i.e., Eq. (9.2)] and is no longer simply ΔH_f, but must now include the free-energy change associated with the formation of the new phase(s) (see Chap. 8).
- Nucleation occurs without a change in volume. In other words, strain energy was ignored. If there is a volume change, and the associated strain energy is known, an extra term is added to Eq. (9.3).

Worked Example 9.1

1. (*a*) If ΔH_f of a glass-forming silicate liquid is 53 kJ/mol, its molar volume is 60 cm³/mol, the solid liquid interfacial energy, γ_{sl}, is 150 mJ/m², and its melting point is 1034°C, calculate the size

of r_c, the height of the nucleation barrier, and the steady-state, metastable equilibrium concentration of nuclei at a $\Delta T = 500°C$.

(*b*) Repeat part (*a*) for a solid–liquid interfacial energy of 130 mJ/m².

(*c*) Explain why the lower limit of γ_{sl} for this glass-forming liquid must be ≈50 mJ/m².

ANSWER

(*a*) If ΔT is 500°C, then the nucleation temperature is 1034 − 500 = 534°C or 807 K. Converting to SI units and substituting in Eq. (9.5), one obtains

$$r_c = \frac{2\gamma_{sl} V_m}{\Delta H_f(1-T/T_m)} = \frac{2(150\times10^{-3})(60\times10^{-6})}{53{,}000\left(1-\frac{534+273}{1034+273}\right)} = 8.9\times10^{-10}\,\text{m}$$

Given that the Si–Si distance in silicates is on the order of 3×10^{-10} m, this result is not unreasonable—the critical radius would appear to be comprised of a few SiO_4 tetrahedra [see part (*c*)]. The corresponding ΔG_c [(Eq. (9.6)] is

$$\Delta G_c = \frac{16\pi\gamma_{sl}^3 V_m^2}{3\Delta H_f^2(1-T/T_m)^2} = \frac{16\pi(150\times10^{-3})^3(60\times10^{-6})^2}{3(53{,}000)^2(0.38)^2} \approx 5\times10^{-19}\,\text{J}$$

With 1 mole as a basis, N_v is simply N_{Av}/V_m, or 1×10^{22} molecular units per cm^{-3}. It follows [Eq. (9.7)] that

$$N_n^{eq} = N_v\left[\exp\left(-\frac{\Delta G_c}{kT}\right)\right] = 1\times10^{22}\exp\left(-\frac{5\times10^{-19}}{1.38\times10^{-23}\times807}\right)$$

$$= 317\ \text{nuclei/cm}^3$$

(*b*) Repeating the calculations using the slightly lower γ_{sl} of 130 mJ/m² yields: $N_n^{eq} \approx 2.7\times10^9$ nuclei/cm^{-3}, which is >6 *orders of magnitude* higher than that calculated in part (*a*)! This simple calculation makes it amply clear the paramount importance of the γ_{sl} term during nucleation. It is worth noting that surface energies, in general, and solid–liquid interfacial energies, in particular, are fiendishly difficult to measure experimentally.

(*c*) Since an embryo cannot be smaller than a single SiO_4 tetrahedron, where the O-O distance is ≈0.26 nm, then the minimum value of $\gamma_{sl,}$ for conditions listed above, is

$$\gamma_{sl} = r_c\frac{\Delta H_f(1-T/T_m)}{2V_m} = 0.26\times10^{-9}\frac{53{,}000\times0.38}{2\times60\times10^{-6}} = 43.6\ \text{mJ/m}^2$$

If a more realistic embryo size—comprised of say 50 near close-packed tetrahedra—is assumed, then $r_c \approx 1$ nm and the minimum value of γ_{sl} would be ≈168 mJ/m².

HETEROGENEOUS NUCLEATION

As noted above, technologically, the vast majority of nucleation occurs heterogeneously at defects such as dislocations, interfaces, pores, grain boundaries, GBs, and especially free surfaces. These sites present preferred nucleation sites for three reasons. First, they are regions of higher free energy, and that excess energy becomes available to the system upon nucleation. Second, and more importantly, the heterogeneities tend to reduce γ, which allows nucleation to occur at relatively smaller undercoolings, where homogeneous nucleation is unlikely. Third, the presence of pores, or free surfaces, reduces any strain energy contributions that may suppress the nucleation and/or growth process.

It can be shown that the steady-state heterogeneous rate of nucleation of a supercooled liquid on a flat substrate is given by[114]

$$I_v = \nu_0 N_s \left[\exp\left(-\frac{\Delta G_m^*}{kT}\right)\right]\left[\exp\left(-\frac{\Delta G_{het}}{kT}\right)\right]$$

where N_s is the number of atoms, or formula units, of the liquid in contact with the substrate per unit area, and $\Delta G_{het} = (1/2 - 3/4 \cos\theta + ¼ \cos^3\theta) \times \Delta G_c$, where θ is the contact angle between the crystalline nucleus and the substrate (see Chap. 10). Note that in the limit of a complete wetting, that is, $\theta = 0$, the thermodynamic barrier to nucleation vanishes.

EXPERIMENTAL DETAILS: MEASURING NUCLEATION RATES

Since the overwhelming majority of glasses usually nucleate heterogeneously, homogeneous or volume nucleation is rarely observed. There are a few glass systems, however, that nucleate homogeneously, and have been studied in order to test the validity of Eqs. (9.10) or (9.13). Of these, probably lithium disilicate, $Li_2O{\cdot}2SiO_2$, has been one of the more intensively studied. In a typical nucleation experiment, the glass is heat-treated to a certain temperature for a given time, cooled and sectioned. The number of nuclei is then counted using optical, or electron, microscopy and, assuming steady-state nucleation, the nucleation rate is calculated.[115] When the nucleation rate is plotted versus temperature, the typical bell-shaped curve (Fig. 9.4*a*), predicted from nucleation theory, is obtained. A typical reflection optical microscope micrograph of a glass after heat treatment to induce nucleation is shown in Fig. 9.4*b*.

It should be pointed out, however, that while Eq. (9.13) correctly represents the temperature dependence of the nucleation rate, the measured rates are typically 20 or more *orders of magnitude* larger than predicted! The reason for this huge discrepancy is not entirely clear. One explanation has been to allow the surface energy term to be weakly temperature-dependent. Many other possibilities have been proposed; the problem to this day remains unresolved, however.[116]

---

114 For more details, see J. W. Cahn, *Acta Met.*, 4, 449 (1956) and 5, 168 (1957). See also J. W. Christian, *The Theory of Transformations in Metals and Alloys*, 2nd ed., Pergamon Press, London, 1975.

115 Sometimes if the size of the nuclei that form is too small to observe, a second heat treatment at a higher temperature is carried out to grow the nuclei to an observable size. Implicit in this approach is that the nuclei formed at the lower temperatures do not dissolve during the second heat treatment.

116 For a comprehensive review of homogeneous nucleation theory in glasses and attendant mysteries see V. M. Fokin et al., *J. Non-Cryst. Solid*, 352, 2681 (2006).

9.2.2 CRYSTAL GROWTH

Once the nuclei are formed, they will tend to grow until they start to impinge upon each other. The growth of these crystals depends on the nature of the growing interface which has been related to the entropy of fusion, ΔS_f.[117] It can be shown that for crystallization processes in which ΔS_f is small, i.e., <2R, the interface will be rough and the growth rate will be, more or less, isotropic. In contrast, when ΔS_f is large, (>4R), the most closely packed faces should be smooth and the less closely packed faces should be rough, resulting in large growth-rate anisotropies. Based on these notions, various models of crystal growth have been developed, most notably:

STANDARD GROWTH, $\Delta S_f < 2R$

In this model, the interface is assumed to be rough on the atomic scale, and a sizable fraction of the interface sites are available for growth to take place. Under these circumstances, the rate of growth is solely determined by the rate of atoms jumping across the interface (i.e., it is assumed that the process is controlled by the surface reaction rate and not diffusion). Using an analysis that is almost identical to the one carried out in Sec. 7.2.3, where the net rate of atom movement down a chemical potential gradient was shown to be [Eq. (7.26)]

$$\nu_{net} = \nu_0 \exp\left(-\frac{\Delta G_m^*}{kT}\right)\left\{1-\exp\left(-\frac{\Xi}{kT}\right)\right\}$$

where Ξ was defined by Eq. (7.23), it is possible to derive an expression for the growth rate as follows. Comparing Figs. 7.5 and 9.3, the equivalence of Ξ and ΔG_v is obvious. Hence, under these conditions, the growth rate, u, of the interface is given by

$$u = \lambda \nu_{net} = \lambda \nu_0 \left\{\exp\left(-\frac{\Delta G_m^*}{kT}\right)\right\}\left\{1-\exp\left(-\frac{\Delta G_v}{RT}\right)\right\} \tag{9.14}$$

where ΔG_v is given by Eq. (9.1). It is left as an exercise to the reader to show that in terms of viscosity η this equation can be rewritten as

$$u = (\text{const.})\frac{kT}{3\pi\eta\lambda^2}\left[1-\exp\left(-\frac{\Delta H_f}{RT}\frac{\Delta T}{T_m}\right)\right] \tag{9.15}$$

when the growth is occurring at temperature, T, with an undercooling of ΔT.

This is an important result because it predicts that the growth rate, like the nucleation rate, should also go through a maximum as a function of undercooling. The reason, once more, is that with *increasing ΔT, the driving force for growth, ΔG_v, increases, while the atomic mobility, expressed in terms of η, decreases. Note both do so exponentially.* It is important to note that the temperatures at which the maximum growth rates occur are usually higher compared to those at which the nucleation rates peak.

For small ΔT values ($e^x \approx 1-x$), a linear relation exists between u and ΔT (see Prob. 9.2*b*). Conversely, for large undercoolings the limiting growth rate is predicted to be

[117] K. A. Jackson in *Progress in Solid State Chemistry*, vol. 3, Pergamon Press, NY, 1967.

$$u = (\text{const.})\frac{kT}{3\pi\eta\lambda^2} = (\text{const.})\frac{D_{liq}}{\lambda} \tag{9.16}$$

Once a stable nucleus has formed, it will grow until it encounters other crystals or until the molecular mobility is sufficiently reduced that further growth is cut off.

SURFACE NUCLEATION GROWTH, $\Delta S_F > 4R$

In the normal growth model, all atoms that arrive at the growing interface are assumed to be incorporated in the growing crystal. This only occurs when the interface is rough on an atomic scale. If, however, the interface is smooth, growth will take place only at preferred sites such as ledges or steps. In other words, growth will occur by the spreading of a monolayer across the surface.

SCREW DISLOCATION GROWTH

Here the interface is viewed as being smooth, but imperfect on an atomic scale. Growth is assumed to occur at step sites provided by screw dislocations intersecting the interface. The growth rate is given by

$$u = f_g \lambda \nu \left[1 - \exp\left(-\Delta H_f \frac{\Delta T}{T_m RT}\right)\right] \tag{9.17}$$

where f_g is the fraction of preferred growth sites. It can be shown that the fraction of such sites is related to the undercooling by[118]

$$f_g \approx \frac{\Delta T}{2\pi T_m}$$

Hence in this model, at small undercoolings, u is expected to be proportional to ΔT^2.

Worked Example 9.2

Consider a glass with a melting point of 1300°C and an entropy of fusion of 8 J/(mol·K). If the constant in Eq. (9.15) is 0.1, and the temperature dependence of η is given in the table below, how long can a 1 cm³ sample be held at 1000°C without sensible bulk crystallization? What is the volume of the fraction crystallized? You can assume the concentration of nucleation sites to be constant at 2×10^6 cm^{-3}, and a molar volume of $\approx$10 cm³/mol. State all assumptions.

T, °C	1400	1300	1200	1000
η, Pa·s	10	250	1000	100,000

ANSWER

Since the number of nuclei is fixed and constant, the growth rate, u, of the nuclei will determine the extent of crystallization. Once u is calculated, the size of the nuclei after a given time can be calculated. Rewriting Eq. (9.15) in terms of ΔS_f and making use of the fact that the constant is 0.1 one obtains

118 W. B. Hillig and D. Turnbull, *J. Chem.* Phys., **24**, 914 (1956).

$$u = \frac{0.1\,kT}{3\pi\eta\lambda^2}\left[1 - \exp\left(-\frac{\Delta S_f \Delta T}{RT}\right)\right]$$

The jump distance can be approximated by

$$\lambda = \left[\frac{10}{(6.02\times10^{23})}\right]^{1/3} = 2.55\times10^{-10}\text{ cm}$$

Inserting the appropriate values gives a linear growth rate of

$$u = \frac{0.1(1.38\times10^{-23})(1273)}{3\pi(10^5)(2.55\times10^{-10})^2}\left[1 - \exp\left(-\frac{8\times300}{8.314\times1273}\right)\right] = 5.8\times10^{-9}\text{ m/s}$$

If we assume the nuclei are detectable when they reach a diameter of 1 μm, then based on their growth rate, the time to reach that size would be: $(0.5\times10^{-6})/(5.8\times10^{-9}) \approx 86$ s.

If one starts with a volume, V_0 of 1 cm³, then the volume fraction crystallized after converting the radius of each nucleus to cm, is:

$$\frac{V_t}{V_0} = \frac{4\pi}{3}r^3 N_c = \frac{4\pi}{3}(0.5\times10^{-4})^3(2\times10^6) \approx 1\times10^{-6}$$

As discussed below, this volume fraction transformed is typically used as an upper limit below which a solid is considered a glass.

9.2.3 KINETICS OF GLASS FORMATION

At this point, the fundamental question, posed at the outset of this section—How fast must a melt be cooled to avoid the formation of a detectable volume fraction of the crystallized phase?—can be addressed somewhat more quantitatively. The first step entails the construction of a **time–temperature–transformation** (TTT) curve for a given system. Such a curve defines the time required, at any temperature, for a given volume fraction to crystallize.

If at any time t, in a total volume, V, the nucleation rate is I_v, it follows that the number, N_t, of new particles formed in time interval, $d\tau$, is

$$N_t = I_v V d\tau$$

For a time-independent constant growth rate, u, and assuming isotropic growth (i.e., spheres), the radius of the sphere after time *t* will be

$$r = \begin{cases} u(t-\tau) & \text{for } t > \tau \\ 0 & \text{for } t < \tau \end{cases}$$

and its volume will be

$$V_\tau = \frac{4}{3}\pi u^3 (t-\tau)^3$$

where τ is the time at which a given nucleus appears. Hence the total volume transformed after time t, denoted by V_t, is given by the number of nuclei at time t multiplied by their volume at that time, or

$$V_t = V_\tau N_t = \int V_\tau I_v V d\tau = \int_{\tau=0}^{\tau=t} V I_v \left(\frac{4}{3}\pi u^3\right)(t-\tau)^3 d\tau$$

Upon integration and rearranging, this gives

$$\frac{V_t}{V} = \frac{\pi}{3} I_v u^3 t^4 \tag{9.18}$$

An implicit assumption made in deriving this expression is that the transformed regions do not interfere or impinge on one another. In other words, this expression is valid only for the *initial* stages of the transformation. A more exact, and general analysis, which takes impingement into account, but which will not be derived here, yields

$$\boxed{\frac{V_t}{V} = 1 - \exp\left(-\frac{\pi}{3} I_v u^3 t^4\right)} \tag{9.19}$$

This relationship is known as the *Johnson–Mehl–Avrami equation.*[119] This equation reduces to Eq. (9.18) at short times. The assumptions made in deriving this equation are:

1. Both the nucleation and growth rates follow Boltzmann distributions.
2. The growth rate is isotropic and linear (i.e., surface reaction rate controlled) and three-dimensional (3D) with time. If the growth is diffusion-limited, the growth rate would not be linear with time, but parabolic.
3. Nucleation rate is random and continuous.

Given the nucleation and growth rates at any given temperature, the fraction crystallized can be calculated as a function of time from Eq. (9.19). Repeating the process for other temperatures and joining the loci of points having the same volume fraction transformed yields a TTT diagram, shown schematically in Fig. 9.5.

Once constructed, an estimate of the **critical cooling rate** (CCR) is given by

$$CCR \approx \frac{T_L - T_n}{t_n}$$

[119] For an excellent derivation, see K. Tu, J. Mayer, and L. Feldman, *Electronic Thin Film Science for Electrical and Materials Engineers*, Macmillan, New York, 1992, Chap. 10. For the original references, see W. L. Johnson and R. F. Mehl, *Trans. AIME*, 135, 416 (1936), and M. Avrami, *J. Chem. Phys.*, 7, 1103 (1937), 8, 221 (1940), 9, 177 (1941).

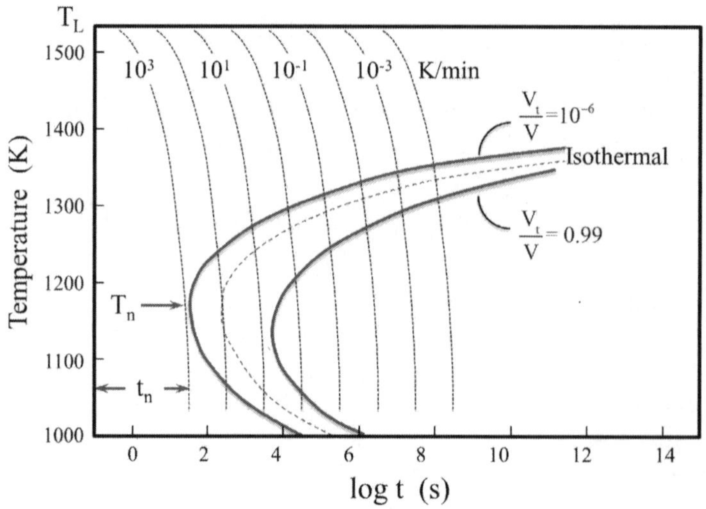

FIGURE 9.5 Isothermal Time–Temperature–Transformation or TTT diagram.

where T_L is the temperature of the melt and T_n and t_n are the temperature and time corresponding to the nose of the TTT curve, respectively (see Fig. 9.5). The CCRs in degrees Celsius per second for a number of silicate glasses are shown in Fig. 9.6, where the salient feature is the strong (note log scale on y-axis) functionality of CCR on glass composition. In principle, if the requisite data were available, the TTT diagram for any material could be generated, and the CCR required to keep it from crystallizing can be calculated.

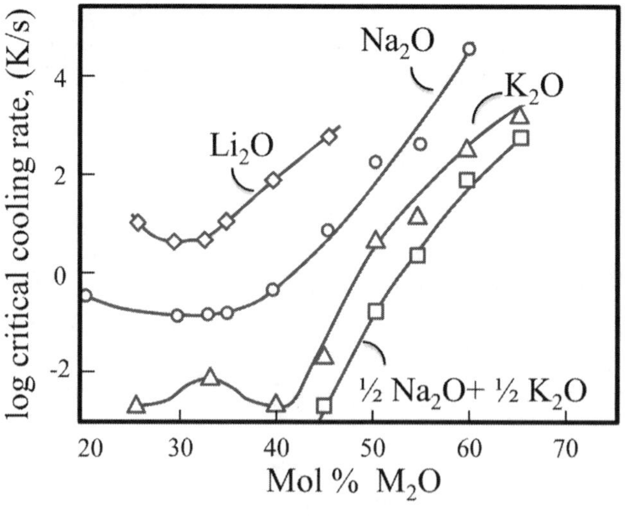

FIGURE 9.6 Critical cooling rates as a function of glass composition. (A. C. Havermans, H. N. Stein and J. M. Stevels, *J. Non-Cryst. Solids*, **5**, 66–69, 1970.)

TABLE 9.1 Summary of glass-forming ability of various compounds

Compound	Melting point °C	ΔS_f J/(mol·K)	η_m, Pa·s	$\Delta S_f/\eta_m$	Comments
B_2O_3	450	33.2	5,000	0.0066	Excellent glass former
SiO_2	1423	4.6	230,000	2.0×10^{-5}	Excellent glass former
$Na_2Si_2O_5$	874	31.0	200	0.155	Good glass former
Na_2SiO_3	1088	38.5	20	1.9	Poor glass former
GeO_2	1116	10.8	71,428	2×10^{-4}	Excellent glass former
P_2O_5	427	73.22	5×10^{6}	1.5×10^{-5}	Glass former
$NaAlSi_3O_8$			44,668		Glass former
$CaSiO_3$	1544	31.0	1.0	31.0	Difficult to form glass
NaCl	800	25.9	0.002	1.3×10^{4}	Difficult to form glass

Since that information is much more likely to be lacking than known, it is useful to develop a simpler criterion.

9.2.4 CRITERIA FOR GLASS FORMATION

The question of glass formation can now be restated: why do some liquids readily form glasses while others do not? Based on the foregoing discussion, for a glass to form, the following conditions must exist:

1. Low nucleation rate. This occurs if ΔS_f is small or the crystal/liquid interfacial energy, γ_{sl}, is high. In either case, ΔG_c is large and consequently nucleation is rendered more difficult.
2. High viscosity, at or near the melting point, η_m. This translates to low growth rates.
3. Absence of heterogeneities. The latter can act as potent nucleating agents and reduce the critical nuclei size and thus greatly enhance the nucleation kinetics; they have to be avoided.

Based on 1 and 2, a useful criterion for the formation of a glass is the ratio $\Delta S_f/\eta_m$. The smaller this ratio, the more likely a melt will form a glass and vice versa. Table 9.1 shows that indeed to be the case. Further inspection reveals that atom mobility, as reflected in η_m is by far the dominant factor. In short *the higher the viscosity of a liquid at its liquidus temperature, or freezing point, the higher the likelihood that a glass will form.*

9.3 GLASS STRUCTURE

As noted above, if cooled rapidly enough, any liquid will form a glass, and indeed glasses have been formed from ionic, organic and even metallic melts. What is of interest here, however, are inorganic glasses formed from covalently bonded—for the most part silicate-based—oxide melts. These glass-forming oxides are characterized by having a continuous three-dimensional network of linked polyhedra and are known as **network formers.** They include silica, boron oxide, B_2O_3, phosphorous pentoxide, P_2O_5, and germania,

GeO_2. Commercially, silicate-based glasses are by far the most important and the most studied and consequently are the only ones discussed here.[120]

Since glasses possess only short-range order, the idea of a repeating unit cell is inapplicable. Thus the best way to describe a glass is to first describe the building block that possesses the short-range order (e.g., the coordination number of each atom) and then describe how these blocks are connected. The simplest of the silicates is fused or **vitreous silica**, v-SiO_2, and understanding its structure is fundamental to understanding the structure of other silicates.

9.3.1 VITREOUS OR SILICA, SIO_2

The basic building block for all crystalline silicates is the SiO_4 tetrahedron (see Chap. 3). In the case of quartz, every silica tetrahedron is attached to four other tetrahedra, and a three-dimensional, 3D, *periodic* network results (see top of Table 3.4). The structure of **fused silica**, v-SiO_2, is quite similar, except that the network lacks long-range periodicity. This so-called random network model, first proposed by Zachariasen,[121] is generally accepted as the best description of the structure of v-SiO_2 and is shown schematically in two dimensions in Fig. 1.1*b*. Quantitatively it has been shown that the Si–O–Si bond angle in v-SiO_2—while centered on 144°, which is the angle for quartz (see Prob. 9.5)— has a distribution of roughly ±10%. In other words, most of the Si–O–Si bond angles fall between 130° and 160°. To summarize, the structure of fused silica is quite uniform at a short range, but that order does not persist beyond a few tetrahedra.

9.3.2 MULTICOMPONENT SILICATES

In Sec. 3.6, the formation of nonbridging oxygens (NBOs) upon the addition of alkali or alkaline earth oxides to silicate melts was discussed in some detail. Because, as discussed shortly, these oxides usually strongly modify the properties of a glass, they are referred to as **network modifiers.** The resulting structure is not unlike that of pure silica, except that now the continuous three-dimensional network is broken up due to the presence of NBOs, as shown in Fig. 9.7.

Table 9.2 lists typical compositions of some of the more common commercial glasses and their softening points (see below). Most of these glasses are silica-based. Alumina is interesting in that it sometimes behaves as a glass network and sometimes as a glass modifier. If the Al^{3+} ion substitutes for a Si^{4+} ion, it becomes part of the network. If, on the other hand, it results in the formation of a NBO it is considered a modifier (see Sec. 3.6). Which role alumina plays is usually a complex function of glass chemistry.

It should be pointed out that whereas this network model of silicate structures was very useful during the earlier stages of the development of glass theory, it does not fully explain several experimental facts. For example, it has been observed that significant structural changes occur at about 10 mol% alkali content, and it is well documented that the molar volume of silicate melts remains fairly constant over a wide range of alkali concentrations—both observations at odds with the simple network model described above. As a result, other models have been proposed, such as the discrete polyanions or "iceberg" model, which

[120] See Kingery et al. for descriptions of the structure of other glasses or many of the references listed at the end of this chapter.
[121] W. H. Zachariasen, *J. Amer. Chem. Soc.*, 54, 3841 (1932).

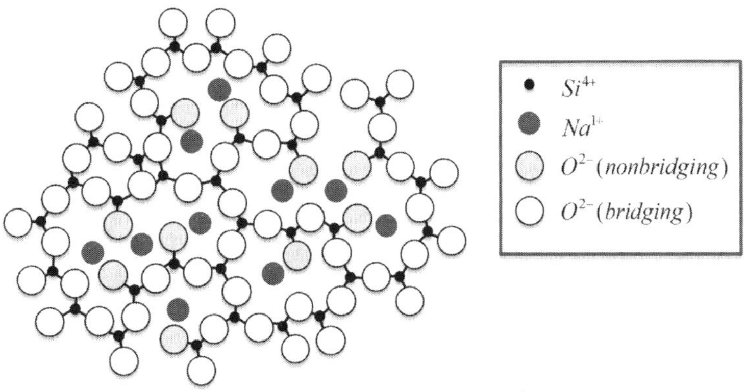

FIGURE 9.7 Two-dimensional schematic of the silicate glass structure in the presence of modifier ions such as Na^+ and the resulting formation of nonbridging oxygens. Recall every NBO is negatively charged.

seems to fit the experimental results better. In this model, it is assumed that between 0 and 0.1 mol fraction alkali oxide, bonds are broken by the formation of NBOs, whereas between 0.1 and 0.33 mol fraction M_2O, discrete six-member rings $(Si_6O_{15})^{6-}$ exist. Between 0.33 and 0.5, a mixture of $(Si_6O_{15})^{6-}$ and $(Si_3O_9)^{6-}$ or $(Si_4O_{12})^{8-}$ and $(Si_6O_{20})^{8-}$ rings are presumed to exist.

A further complication, that is beyond the scope of this book but is mentioned for the sake of completeness, is the fact that in the composition range between 12 and 33 mol%, M_2O and SiO_2 are not completely miscible in the liquid state.

9.3.3 MULTICOMPONENT BORATES

In pure B_2O_3, each B atom is bonded to three O atoms and each O to two B atoms. When network modifiers, such as Na_2O are added, however, they do not immediately form NBOs. Instead, rigid BO_4 tetrahedra form. Beyond ≈0.2 mole fraction of the modifier oxide, NBOs start appearing. This is why when the glass transition, T_g, (see below) and Vickers hardness of a series of borate glasses are plotted as a function of say Na_2O mole fraction (see Prob. 9.12 and Fig. 9.19), both properties peak around 0.2 mole fraction. In the following section, the relationships between glass structure and various properties are explored.

Commercially, there are few pure borate glasses. Borosilicate glasses, on the other hand, in which B_2O_3 and SiO_2 are mixed together, with some alkali and alkali earth metal oxides are much more common (see Table 9.2).

9.4 GLASS PROPERTIES

The noncrystalline nature of glasses endows them with unique characteristics as compared to their crystalline counterparts. Once formed, the changes that occur in a glass upon further cooling are quite subtle and different from those that occur during other phase transitions such as solidification or crystallization. The

TABLE 9.2 Approximate compositions (wt.%) and softening temperatures, T_s, of common glasses

	Network formers			Network modifiers					
	SiO_2	B_2O_3	Al_2O_3	Na_2O	K_2O	MgO	CaO	PbO	T_s, °C
Fused silica	99.8					0.1	0.1		1600
Vycor	96.0	3	1						
Pyrex	81.0	13	2	3.5	0.5				830
Soda silica	72.0		1	20.0		3.0	4.0		
Lead silica	63.0		1	8.0	6.0		1.0	21	
Window	72.0	1	2	15.0	1.0	4.0	5.0		700
E glass	55.0	7	15	1.0	1.0		21.0		830

change is not from disorder to order, but, rather, from disorder to disorder with *less* empty space. In this section, the implication of this statement on glass properties is discussed.

9.4.1 THE GLASS TRANSITION TEMPERATURE

The temperature dependencies of several properties of crystalline solids and glasses are compared schematically in Fig. 9.8. Figures 9.8*a–d* compare, respectively, the temperature dependencies of the specific volume, V_s, configurational entropy, S_{cofig}, heat capacity, c_p and thermal expansivities, α, of crystalline solids and glasses. Typical crystalline solids will normally crystallize at T_m with an abrupt and significant decrease in V_s and S_{cofig} (Fig. 9.8*a* and *b*). For glasses the changes are much more gradual; there is no abrupt change at T_m, but rather the properties follow the liquid line up to a temperature where the slopes of V_s, or S_{cofig} vs. T curves are markedly decreased. The point at which the break in slope occurs is known as the ***glass transition temperature*** T_g and denotes the temperature at which a glass-forming liquid transforms from a rubbery, soft, plastic state to a rigid, brittle, glassy state. In other words, the temperature at which a supercooled liquid becomes a glass, i.e., a rigid, amorphous, brittle body, is referred to as T_g. Between T_m and T_g, the material is usually referred to as a **supercooled liquid, SCL.**

THERMODYNAMIC CONSIDERATIONS

At first glance, given that (see Fig. 9.8) at T_g, V_s and S_{conf} are continuous, whereas c_p and α are discontinuous, it is not unreasonable to characterize the transformation occurring at T_g as a second-order phase transformation. After all, by definition, second-order phase transitions require that the properties that depend on the first derivative of the free energy G such as

$$V_s = \left(\frac{\partial G}{\partial P}\right)_T \text{ and } S = -\left(\frac{\partial G}{\partial P}\right)_P$$

be continuous at the transformation temperatures, but that the ones that depend on the second derivative of G, such as

$$\alpha = \frac{1}{V}\left(\frac{\partial V}{\partial T}\right)_P = \frac{1}{V}\left(\frac{\partial^2 G}{\partial P \partial T}\right) \text{ and } c_p = \left(\frac{\partial H}{\partial T}\right)_P = -T\left(\frac{\partial^2 G}{\partial T^2}\right)_P$$

be discontinuous.

What is occurring at T_g, however, is more complex, because it is experimentally well established that T_g is a function of cooling rate. As shown in Fig. 9.8*a*, with decreasing cooling rates, T_g shifts to lower temperatures. This implies that with more time for the atoms to rearrange, a denser glass results and is strong evidence that T_g is not a thermodynamic quantity, but rather a kinetic one.

Further evidence for this conclusion includes the changes in α and c_p. The abrupt decrease in these properties at T_g has to be related to a sudden inability of some molecular degrees of freedom to contribute to these thermodynamic quantities. It is thus this "freezing out" of molecular degrees of freedom (see below) that is responsible for the observed behavior. As discussed below, glass viscosity, η at T_g is quite large and on the order of 10^{15} Pa·s, which in turn implies that atomic mobility is quite low. It follows that if the timescale of the experiment is less than the average time for an atom to move permanently, then that atom will not contribute to the property being measured and, for all intents and purposes, T_g would appear as a relatively abrupt phenomenon, as observed.

Interestingly enough, if a glass-forming liquid were cooled slowly enough (at several times the age of the universe!) such that it follows the dotted line shown in Fig. 9.8*b* at a temperature, T_{KAU}, the entropy of the

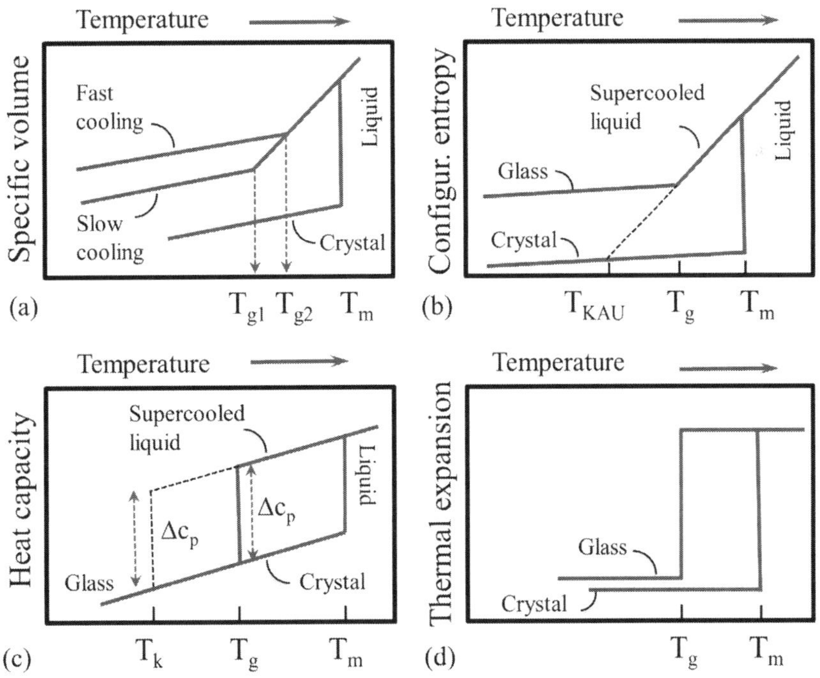

FIGURE 9.8 Schematic of property changes observed as a glass or a crystal is cooled through T_g. (*a*) Specific volume; (*b*) configurational entropy; (*c*) heat capacity; and (*d*) thermal expansion coefficient.

supercooled liquid would become lower than that of the crystal—a clearly untenable situation first pointed out by Kauzmann and referred to since as the Kauzmann paradox. This paradox is addressed in more detail in Sec. 9.4.2.

EFFECT OF COMPOSITION ON T_g

In a very real sense, T_g is a measure of the rigidity of the glass network. In general, the addition of network modifiers tends to reduce T_g, while the addition of network formers increases it. This observation is so universal that experimentally one of the techniques of determining whether an oxide goes into the network or forms NBOs is to follow the effect of its addition on T_g. An example for multicomponent borates (see Prob. 9.12) is a good illustration of that concept.

EXPERIMENTAL DETAILS: MEASURING THE GLASS TRANSITION TEMPERATURE

The glass transition temperature, T_g can be determined by measuring any of the properties shown in Fig. 9.8 as a function of cooling rate. The temperature at which the property changes slope whether continuously, or abruptly, defines T_g.

In principle, differential scanning calorimetry, DSC, and differential thermal analysis, DTA—not to be confused with TGA—ca ben used. These two techniques are quite similar except that in DSC instead of simply measuring temperature differences (see Fig. 4.8), heat flow is also measured. Upon heating of a glass, T_g, being a weak endothermic process, appears as a small step in the baseline of the DSC curves, as shown in Fig. 9.9. If a glass does not crystallize before melting then this anomaly is the only signature of the glass. If, however, the glass crystallizes at some temperature, T_{cryst} between T_g and T_m, then two extra peaks emerge (Fig. 9.9); the first is exothermic and is due to the crystallization, or devitrification, of the supercooled liquid; the second is endothermic and due to the melting of these same crystals.

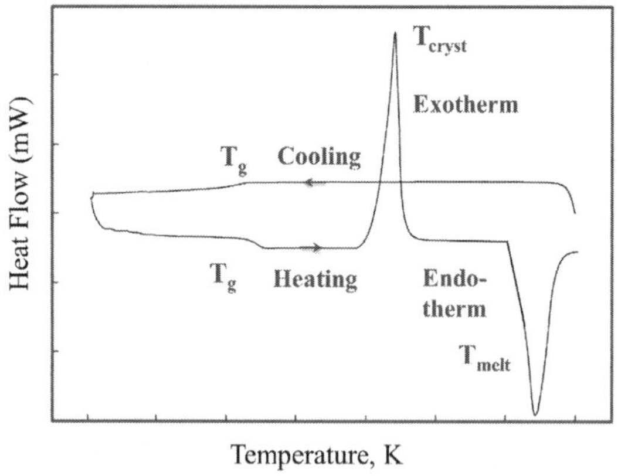

FIGURE 9.9 Typical DSC traces obtained upon heating and cooling of a glass. Upon heating the first endotherm occurs at T_g, which is followed by an exothermic crystallization peak at T_{cryst}, which is, in turn, followed by melting at T_{melt}. If the cooling is fast enough, the only observable feature is a small step at T_g.

9.4.2 VISCOSITY

Technologically, the viscosity, η, of a glass and its temperature dependence, are key because they determine the melting times and temperatures required to homogenize a melt, the working and annealing temperatures (see below), the rate of devitrification and thus the CCR, as well as the temperatures for annealing of residual stresses.

Viscosity, η, is the ratio of the applied shear stress to the rate of flow, v, of a liquid. If a liquid contained between two parallel plates of area, A, a distance, d, apart is subjected to a shear force, F, then

$$\eta = \frac{Fd}{Av} = \frac{\tau_s}{\dot{\varepsilon}} \tag{9.20}$$

where $\dot{\varepsilon}$ is the strain rate, in s^{-1}, and τ_s the applied shear stress in Pa. The units of η are thus Pa·s.

As noted above, upon solidification, η of a crystalline solid will vary abruptly and, over an extremely narrow temperature range, increase by orders of magnitude. The viscosities of glass-forming liquids, however, change in a more gradual fashion. A schematic of the effect of temperature on η of a glass-forming liquid is shown in Fig. 9.10, where in addition to T_g, four other temperatures of practical importance are defined. The **strain point** is defined as the temperature at which $\eta = 10^{15.5}$ Pa·s. At this temperature, any internal strain is reduced to an acceptable level within 4 h.[122] The **annealing point** is the temperature at which $\eta = 10^{14}$ Pa·s and any internal strains are reduced sufficiently within about 15 min. The **softening point** is the temperature at which $\eta = 10^{8.6}$ Pa·s. At that temperature, a glass article elongates at a strain rate of 0.03 s^{-1}. Finally, the **working point** is the temperature at which $\eta = 10^5$ Pa·s, and glass can be readily shaped, formed or sealed.

EFFECT OF TEMPERATURE ON VISCOSITY

The functional dependence of η on temperature has been measured for a large number of glass-forming liquids. Phenomenologically it has been determined that the most accurate three-parameter fit for the data over a wide temperature range is given by the **Vogel–Fulcher–Tammann** (V–F–T) equation[123]

$$\ln \eta = A + \frac{B}{T - T_0} \tag{9.21}$$

where A, B and T_0 are temperature-independent adjustable parameters.

A number of theories have been proposed to explain this behavior, most notable among them being the free volume theory[124] and the configuration entropy theory.[125] The former predicts that the transition

[122] For the sake of comparison, water, motor oil no. 10, chocolate syrup and caulking paste have viscosities of 0.001, 0.5, 50 and >1000 Pa·s, respectively. In cgs units, viscosity is given in poise (abbreviated P). 1 centipoise (cP) = 0.01 P = 0.1 Pa·s.

[123] G. Fulcher, *J. Amer. Cer. Soc.*, **75**, 1043–1059 (May 1992). Commemorative reprint.

[124] M. Cohen and D. Turnbull, *J. Chem. Phys.*, 31, 1164169 (1959), and D. Turnbull and M. Cohen, *J. Chem. Phys.*, 34, 120125 (1961).

[125] G. Adams and J. Gibbs, *J. Chem. Phys.*, 43, 139 (1965).

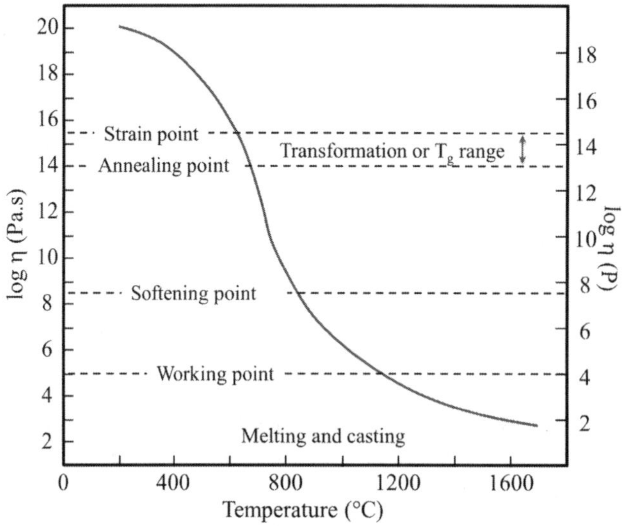

FIGURE 9.10 Functional dependence of viscosity on temperature. Note log scale on y-axis.

occurring at T_g is a first-order phase transition and cannot account for the fact that in some systems T_g has been found to increase with increasing pressure, and thus will not be discussed further. The remainder of this section is devoted to the configuration entropy model, which while not perfect, succeeds rather well in explaining many experimental observations and at this time appears to be the most promising.

In Chaps. 5 and 6, it was established that the entropy content of a crystal is the sum of its vibrational and configurational entropies—the latter due to the introduction of either defects and/or impurities into the crystal. The entropy of a liquid or glass contains, in addition, a term reflecting its ability to change configurations. In the configuration entropy model, a simplified version of which is presented here, the liquid is divided into N_c blocks, each containing $n = N/N_c$ atoms, where N is the total number of atoms in the system. These blocks - termed *cooperatively rearranging regions* - are defined as the smallest region that can undergo a transition to a new configuration without a requisite simultaneous configuration change at its boundaries. It is further assumed, for the sake of simplicity here, that for each block only two configurations exist. The total entropy of each block is thus k ln 2, and the total configuration entropy of the supercooled liquid (SCL) is simply $\Delta S_{config} = N_c k \ln 2$. Replacing N_c by N/n and rearranging results in

$$n = \frac{Nk \ln 2}{\Delta S_{config}} \tag{9.22}$$

If it is further assumed that at $T = T_k$, where T_k is an adjustable parameter, the entropies of the SCL and the glass are identical and that the heat capacity differences between the glass and the SCL, denoted by Δc_p, is a constant and independent of temperature (that is, Δc_p at T_g is equal to its value at T_k, see Fig. 9.8c), then it can be shown that for any temperature $T > T_k$ (see Prob. 9.7)

$$\Delta S_{config} = S_{config}^{SCL} - S_{config}^{glass} = \Delta c_p \ln \frac{T}{T_k} \tag{9.23}$$

Combining Eqs. (9.22) and (9.23) yields

$$n = \frac{Nk \ln 2}{\Delta c_p \ln(T/T_k)} \tag{9.24}$$

This expression predicts that n—the number of atoms in each *cooperatively rearranging region*—increases with decreasing temperature. The situation is depicted schematically in Fig. 9.11, for three different temperatures $T > T_g$, $T \approx T_g$, and $T = T_k$. As the temperature is lowered and various configurations are frozen out, the cooperatively rearranging regions decrease in number and increase in volume. At T_k only one configuration remains and there is a total loss of configurational entropy.

The major tenet of this model is that there is a direct relationship between configurational entropy and the rate of molecular transport. In other words, it is postulated that as the blocks become larger, it takes more time for them to switch configurations. It follows that the relaxation time τ can be assumed to be proportional to n,[126] and thus

$$\tau = (\text{const.}) \exp[\zeta n] \tag{9.25}$$

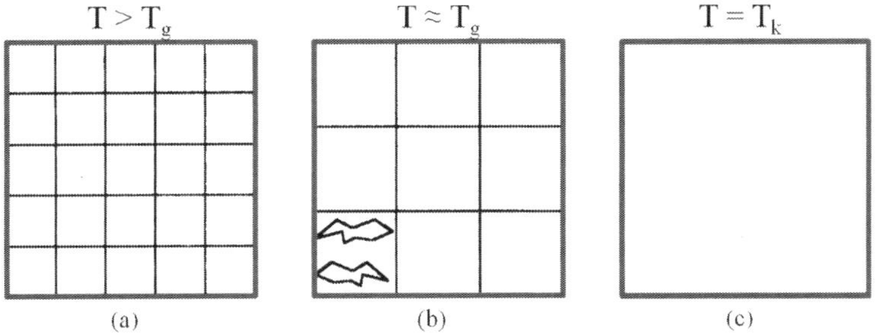

FIGURE 9.11 Schematic of effect of temperature on number of blocks and their size. As temperature decreases from left to right, the number of blocks decreases but the number of atoms in each block increases. Shown in the corner of the middle diagram is what is meant by the two configurations that the atoms in a given cell block can have. That number was chosen to be two for simplicity; the original model does not make that simplifying assumption.

[126] The barrier to rearrangement increases in proportion to n because the potential energy increase of the barrier ΔE scales as $n\Delta\mu$, where $\Delta\mu$ is the potential barrier per molecule hindering rearrangement.

where ζ is an undetermined constant. Since τ represents a characteristic time for structural relaxation, it is not unreasonable to assume that it is proportional to η. Putting all the pieces together, it can be shown that

$$\ln\eta = \hat{K}\exp - \frac{\zeta Nk\ln 2}{\Delta c_p \ln(T/T_k)} \tag{9.26}$$

where $\hat{K}$ is a constant. Taking the natural logarithm of both sides yields

$$\ln\eta = \ln\hat{K} + \frac{\zeta Nk\ln 2}{\Delta c_p \ln(T/T_k)} = A' + \frac{B'}{\ln(T/T_k)} \tag{9.27}$$

where A′ and B′ are constants. And while at first glance this equation does not appear to have the same temperature dependence as Eq. (9.21), it can be shown that when $T_0 \approx T_k$, that dependence is indeed recovered (see Prob. 9.7).

Regardless of whether or not the two expressions are mathematically equivalent, both are equally good at describing the temperature dependencies of η. This is clearly shown in Fig. 9.12, where the temperature

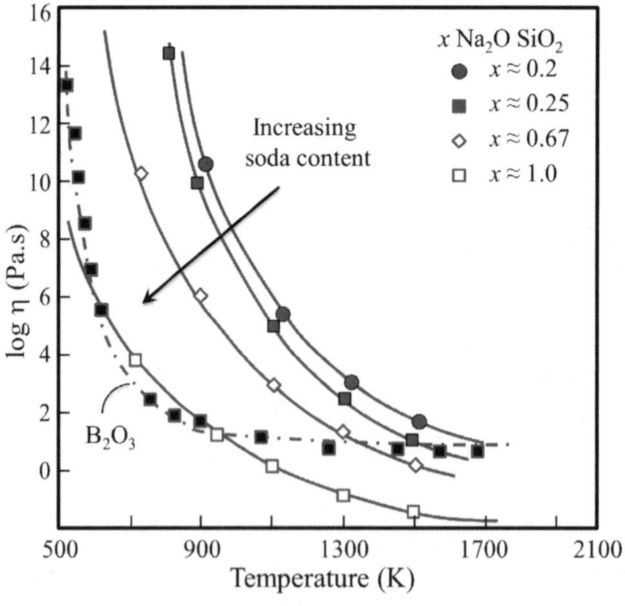

FIGURE 9.12 Temperature dependence of viscosity of select glasses. (Data for the silicates are from Fulcher; those for B_2O_3 from Macedo and Napolitano *J. Chem. Phys.*, 49, 1887–1895, 1968.) The solid lines are fit according to Eq. (9.27). For all except the highest alkali concentration (for which T_k was chosen to be 136 K, compared to 0 K by Fulcher), T_k in Eq. (9.27) was identical to that used by Fulcher. For B_2O_3, the best fit was obtained for $T_k = 445$ K. (G. S. Fulcher, *J. Amer. Cer. Soc.*, 8(6), 339355, 1925. Reprinted in *J. Amer. Cer. Soc.*, 75(5), 1043–1059, 1992.)

dependence of η is plotted for a number of sodium silicate melts and B_2O_3 glass. The data points for the silicates were generated using Eq. (9.21), which in turn were best fits of experimental results. The lines are plotted using Eq. (9.27). In the case of B_2O_3, the points are experimental points, and the line is again plotted using Eq. (9.27). The fit in all cases is excellent, using a single adjustable parameter, namely, T_k.

Finally, it is worth noting that the values of T_0 or T_k needed to fit the η results are close to the temperature at which the Kauzmann temperature, T_{KAU}, is estimated from extrapolations of other properties such as those shown in Fig. 9.8, lending credence to the model. This model also provides a natural way out of the Kauzmann paradox, since not only do the relaxation times go to infinity as T approaches T_k, but also the configuration entropy vanishes since at $T = T_k$ only one configuration is possible.

EFFECT OF COMPOSITION ON VISCOSITY

For pure SiO_2 to flow, the high-energy directional Si–O–Si bonds have to be broken. The activation energy for this process is quite large (565 kJ/mol), and consequently, the viscosity of pure liquid SiO_2 at 1940°C is 1.5×10^4 Pa·s, which, considering the temperature, is quite high.

As discussed earlier, the addition of **basic oxides** to a melt (a basic oxide is an oxide which when dissolved in a melt contributes an oxygen ion to the melt, such as Na_2O and CaO) will break up the silicate network—by the formation of NBOs—resulting in the breakdown of the original 3D network into progressively smaller discrete units. Consequently, the number of Si–O bonds needed to be broken during viscous flow decreases, and the shear process becomes easier. The dramatic effects of basic oxide additions on the viscosity of fused silica are shown in Fig. 9.13. (Note log scale of y-axis.)

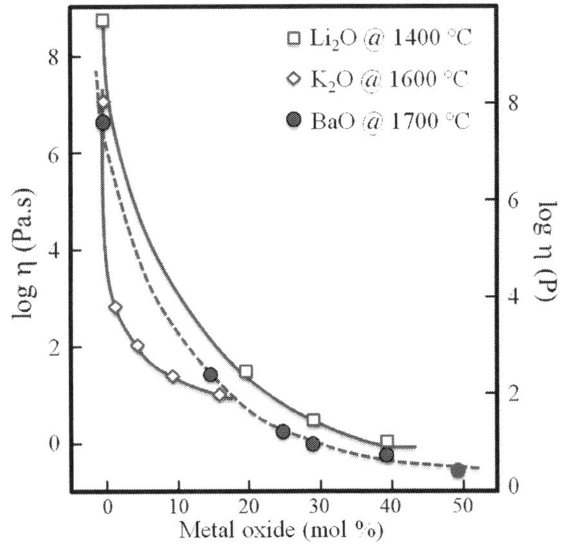

FIGURE 9.13 Dramatic effect of modifier metal oxide content on the viscosities of fused silica at various temperatures. The addition of about 10 mol % of modifier oxide results in a 5 or more orders-of-magnitude viscosity drop. (J. O'M. Bockris, J. D. Mackenzie, and J. A. Kitchener, *Faraday Soc.*, 51, 1734, 1955.)

EXPERIMENTAL DETAILS: MEASURING VISCOSITY

The technique used to measure η usually depends on the viscosity range of the glass. Up to $\approx 10^7$ Pa·s, a *viscometer,* shown schematically in Fig. 9.14*a*, is used. Here, the fluid for which η is to be measured is placed between two concentric cylinders of length L that are rotated relative to each other. Usually, one of the cylinders, say, the inner one, is rotated at an angular velocity, ω_a, while the outer cylinder is held stationary by a spring that measures the torque, T, acting upon it. If that is the case, it can be shown that[127]

$$\eta = \frac{(b^2 - a^2)T}{4\pi a^2 b^2 L \omega_a} \tag{9.28}$$

where a and b are defined in Fig. 9.14*a*.

As noted above, viscometers are usually good up to about 10^7 Pa·s. For higher values the fiber elongation method is sometimes used. In this method (Fig. 9.14*b*), a load is attached to the material for which η is to be measured and the material is heated to a given temperature. The strain rate at which the fiber elongates is then measured. It can be shown[128] that the rate of energy dissipation, $\dot{E}_v$, as a result of viscous flow of a cylinder of height L_c and radius R is given by

$$\dot{E}_v = \frac{3\pi\eta R^2}{L_c}\left(\frac{dL_c}{dt}\right)^2 \tag{9.29}$$

Integrating this equation with respect to time, assuming a constant strain rate, and equating it to the decrease in system's potential energy, while ignoring surface energy changes, one can show that

$$\eta = \frac{mg}{3\pi R^2 \dot{\varepsilon}} = \frac{mgL_0 t}{3\pi R^2 \Delta L} \tag{9.30}$$

where ΔL_c is the elongation in a given time t and L_0 is the original fiber length. Here it was also assumed that $\Delta L << L_0$.

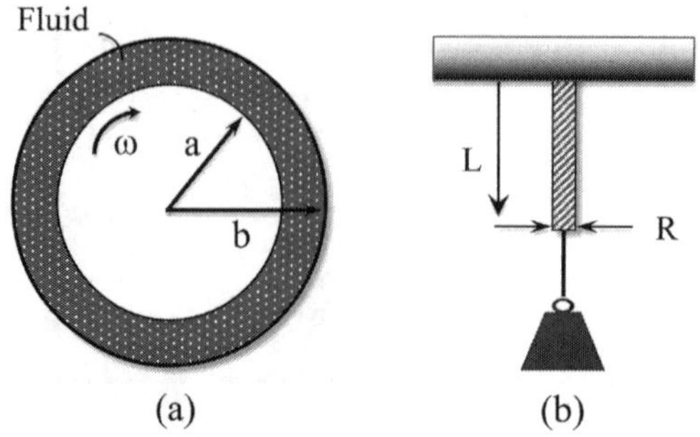

FIGURE 9.14 Measuring viscosity using a (*a*) viscometer, and (*b*) fiber elongation method.

[127] R. Feynman et al. *The Feynman Lecture on Physics*, vol. 2, Addison-Wesley, Reading, MA, 1964, pp. 41–43.
[128] J. Frenkel, *J. Phys. (Moscow)*, 9(5), 385–391 (1945).

9.4.3 OTHER PROPERTIES

As any phase diagram will show, nonstoichiometry notwithstanding, most crystalline phases exist over a very narrow range of compositions, with limited solubility for other compounds. All this tends to limit the possibilities for property tailoring. Glasses are not subject to this constraint—they can be thought of as "universal solvents" or "garbage cans" for other compounds. Needless to say, it is this degree of freedom that has rendered glasses quite useful and fascinating materials.

CASE STUDY 9.1: GLASS-CERAMICS

Glass-ceramics are an important class of materials that have been commercially quite successful. They are polycrystalline materials produced by the controlled crystallization of glass and are composed of randomly oriented crystals with some residual glass, typically between 2 and 5%, with no voids or porosity.

A typical temperature versus time cycle for the processing of a glass-ceramic is shown in Fig. 9.15*a*, and consists of four steps.

1. *Mixing and melting.* Raw materials such as quartz, feldspar, dolomite and spodumene are mixed with potent nucleating agents, such as TiO_2 or ZrO_2, and melted.
2. *Forming.* As noted below, one of the major advantages of glass-ceramics lies in the fact that they can be formed using conventional glass-forming techniques such as spinning, rolling, blowing and casting. Complex-shaped, pore-free articles can thus be easily and economically manufactured. The cooling rate during the formation process, however, has to be rapid enough to avoid crystallization or growth of any nuclei if they form.
3. *Nucleation.* Once formed, the glass body is heated to a temperature high enough to obtain a large nucleation rate. Efficient nucleation is the key to success of the process. The nucleation is heterogeneous, and the crystals grow on the particles of the nucleating agents, typically TiO_2 or ZrO_2, that are added to the melt. To obtain crystals, on the order of 1 μm, the nucleating agent density has to be on the order of 10^{12} to 10^{15} cm^{-3}.
4. *Growth.* Following nucleation, the temperature is raised to a point where growth of the crystallites occurs readily. Once the desired microstructure is achieved, the parts are cooled. During this stage the body usually shrinks slightly—by about 1 to 5%.

Glass-ceramics offer several advantages over both glassy and crystalline phases, including:

1. The most important advantage of glass-ceramics, over their crystalline counterparts, is the ease of processing inherent to glasses, to shape and form complex shapes, followed by transforming the glass phase to a more refractory solid in which the properties can be tailored by judicious crystallization. Unlike ceramic bodies made by conventional pressing and sintering, glass-ceramics tend to be pore-free. This is because during crystallization the glass can flow and accommodate changes in volume.

2. Usually the presence of the crystalline phase results in much higher deformation temperatures than the corresponding glasses of the same composition. For example, many oxides have T_g values of 400 to 450°C and soften readily at temperatures above 600°C. A glass-ceramic of the same composition, however, can retain its mechanical integrity and rigidity to temperatures as high as 1000 to 1200°C.
3. The strength and toughness of glass-ceramics are usually higher than those of glasses. For example, the strength of a typical glass plate is on the order of 100 MPa, while that its glass-ceramics can be several times higher. The reason, as discussed in greater detail in Chap. 11, is that the crystals present in the glass-ceramics tend to limit the size of the flaws present in the material, increasing its strength. Furthermore, the presence of the crystalline phase enhances toughness (see Chap. 11).
4. As with glasses, the properties—most notably the thermal expansion coefficients—of glass-ceramics can be controlled by adjusting the composition. In many applications, such as glass-metal seals and the joining of materials, it is important to match the thermal expansion coefficients to avoid the generation of thermal stresses.
5. One of the major advantages of some glass-ceramics is the ease by which they can be machined. Coming from a dental—father and brother—family, so to speak, I am more familiar than many with dental materials. In the twentieth century, when a tooth decayed and had to be removed, it was usually replaced by a crown that was comprised of a porcelain fused to metal. The procedure was lengthy, where the dentist made a mold of the cavity which was sent to a dental lab, where the crown was fabricated and a second and sometimes a third appointment was needed. At the end of the twentieth century, all-ceramic solutions were sought and found. On my last visit to the dentist, a digital image of the cavity where the crown was to be fabricated was taken, sent to a computer in another room. The latter controlled a three-axis mill to machine an all glass-ceramic block into the required shape in about 15 mins. This was followed by a quick firing of the crown and in about one hour I walked out of the dentist's office with a full restoration. A typical microstructure of a lithium disilicate-based glass-ceramic used for dental restorations is shown in Fig. 9.15*b*, where the brighter areas are crystalline and the darker areas are regions where the glass was etched.

The most important glass-ceramic compositions are lithium silicate based. The phase diagram of the Li_2O–SiO_2 system is shown in Fig. 9.16. Commercial compositions typically contain more than about 30 mol% lithia, that upon crystallization yield $Li_2Si_2O_5$ as the major crystalline phase, with some SiO_2 and Li_2SiO_3 as minor phases.

CASE STUDY 9.2: ULTRATHIN AND ULTRA-STRONG GLASS

Glass manufacturing is a multibillion business for good reasons. Our mobile devices, skyscrapers, cars and trains, drinks make extensive use of this versatile material. The fact that the raw materials are "dirt-cheap" is another huge advantage. In sharp contrast to organic glasses—also known as polymers that are ubiquitous, transparent and quite a bit cheaper to shape as well—inorganic glasses are resistant to scratching, wear and environmental degradation in general. My first convertible car had a plastic back window, which

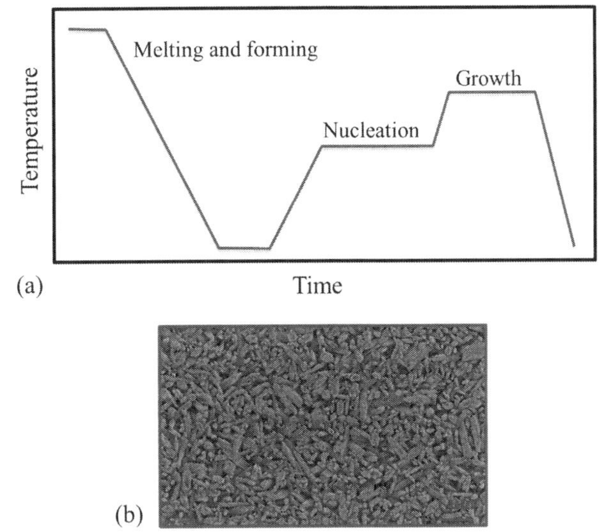

FIGURE 9.15 (*a*) Temperature versus time cycle for controlled crystallization of a glass-ceramic body. (*b*) Typical SEM micrograph of a lithium disilicate-based glass-ceramic after etching with 40% HF vapor for 30 s. (Adapted from Ritzberger et al. *Materials, 3*, 3700, 2010.)

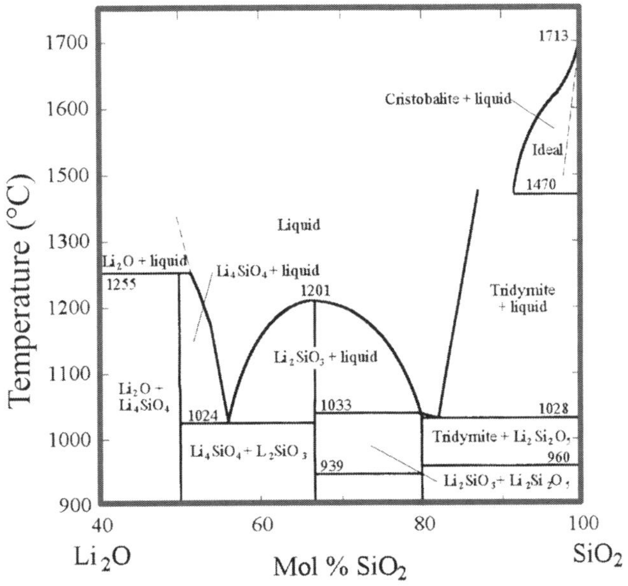

FIGURE 9.16 The LiO_2-SiO_2 phase diagram. (F. C. Kracek, *J. Phys. Chem.*, 34, 2645, Part II, 1930.)

within a few years turned an ugly brown and became quite brittle; I had to replace it. The back window in my new convertible, on the other hand, is glass and unless it breaks, I will not have to ever replace it.

For hundreds of years mankind was perfectly happy with window glass. However, quite recently, humanity decided it could not live without mobile devices. As those devices got smaller and lighter, new thinner, hence stronger, glass was required. Necessity being the mother of invention, glass scientists rose to the challenge. Today thin glass plates are routinely used in a wide variety of high-tech applications, not the least of which is in cell phones. Such glasses are also used as substrates for flexible displays, hard disk covers among many other high-tech applications. All of a sudden glass is "sexy" again.

How the thermal tempering of a glass can introduce surface compressive stresses that, in turn, increases its strength and renders it less dangerous is discussed in Chap. 13. Here we are interested in chemical strengthening by ion exchange. This technology is not new, but has been around since the 1940s and 1950s. The idea is simple: Immerse a glass containing cations with radii, r_g, in a molten salt bath containing cations with radii, r_m, and allow them to exchange (Figs. 9.17*a* and *b*). As long as $r_m > r_g$, surface compressive stresses—balanced by bulk tensile stresses (Fig. 9.17*c*)— will accrue. Since the glass can only fail when that compressive stress is exceeded, its strength is greatly increased. The ion exchange is typically carried out at a $T <$ strain point of the glass (400 to 600°C) before it is rapidly cooled. Since $r_{Li+} < r_{Na+} < r_{K+}$, Na-containing glasses, typically alkali aluminosilicates can be strengthened by immersing them in K^+-containing melts such as KNO_3.

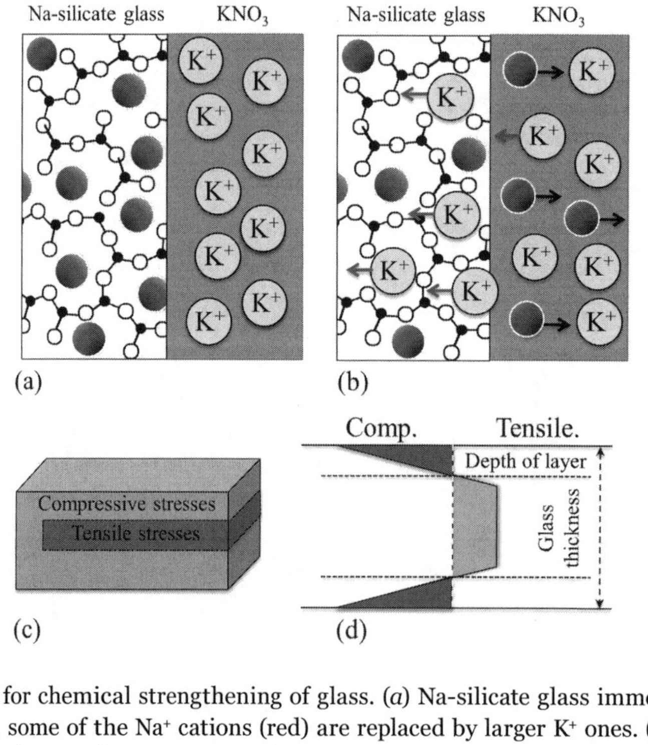

FIGURE 9.17 Ion exchange for chemical strengthening of glass. (*a*) Na-silicate glass immersed in KNO_3 molten salt solution. (*b*) After a time t, some of the Na^+ cations (red) are replaced by larger K^+ ones. (*c*) State of stress after ion exchange. (*d*) Cross-sectional state of stress. Near surface in compression (red) while the bulk is in tension (gray).

The process kinetics are determined by the ambipolar diffusion of the two cations in opposite direction—a problem tackled in Sec. 7.4.3. And while the situation here is slightly more complicated since the state of stress changes with ion exchange, and has to be taken into account, the fundamental idea remains the same. The total flux of ions into the molten salt has to equal the opposite flux of the ions going the other direction. The depth of the compressive layer thus depends on the nature of the cations being exchanged, glass structure, temperature and time of the process. Research has shown that a depth of 20 to 30 μm is optimum. Among other factors, a glass chemical composition determines the sustainable depth. For example, Na-lime glass cannot sustain the high compressive stresses because of rapid stress relaxation. Today alkali-aluminosilicate glass plates—0.5 to 0.7 mm thick—exhibit strengths in the 800 to 1000 MPa range! Such strong glasses, which are, by their thin nature, also quite flexible, will play an important role in the development of flexible displays, solar cells and next-generation touch-screen devices among many others.

9.5 SUMMARY

Glasses are supercooled liquids that solidify without crystallizing. They are characterized by having only short-range order. Glass structure is best described by a random network model and is determined by the relative amounts of network formers to network modifiers.

To form a glass, a melt has to be cooled rapidly enough such that there is insufficient time for crystalline phases to nucleate and grow. Low atom mobility at, or around, the melting point, together with the absence of potent nucleating agents, is a necessary condition for glasses to form at moderate cooling rates. By far the most important predictor of whether a glass will form is its viscosity at, or near, the melting point.

Upon cooling of a glass melt, the driving force for nucleation increases, but atom mobility decreases. These two counteracting forces result in maxima for both the nucleation and growth rates. The convolution of the two functions results in temperature–time–transformation diagrams, from which one can, in principle, quantify the critical cooling rate, that would yield a glass.

At the glass transition temperature T_g the supercooled liquid transforms to a solid. The transformation is kinetic in origin and reflects the fact that, on the timescale of the observation, the translational and rotational motions of atoms or molecules that contribute to various properties "freeze out." In other words, they cease to contribute to the properties measured. Below T_g, a glass behaves as a brittle, elastic solid.

Glass viscosity is a strong, but smoothly changing, function of temperature. The relationship between and temperature, for the most part, cannot be described by a simple Arrhenian equation. The gradual change in viscosity with temperature is important from a processing point of view and allows glasses to be processed rapidly and relatively easily into pore-free complex shapes.

The introduction of nonbridging oxygens in a melt will reduce both T_g and η.

Glass-ceramics are processed in the same way as glasses, but are then given a further heat treatment to nucleate and grow a crystalline phase, such that the final microstructure is composed of small polycrystals, with a glass phase between them. The possibility of tailoring both the initial composition and size and the volume fraction of the crystalline phase allows for precise tailoring of properties. Lastly, chemical strengthening of glasses by ion exchange—that results in large surface compressive stresses—is now routinely being used to mass produce very thin, but ultrastrong glasses.

APPENDIX 9A: DERIVATION OF EQ. (9.7)

Consider a homogeneous phase containing N_v atoms per unit volume in which in a smaller volume, containing n atoms, the density fluctuates to form a new phase. As discussed in Sec. 9.2, the formation of these embryos results in a *local* increase in the free energy ΔG_c. So the reason the nuclei form must be related to an increase in the system's entropy. This increase is configurational and comes about because once the nuclei form, it is possible to distribute N_n embryos on any of N_v possible sites. The free energy of the system can be expressed as

$$\Delta G_{sys} = N_n \delta G_c - kT \ln \Omega \tag{9A.1}$$

where Ω is the number of independent configurations of embryos and host atoms, with each configuration having the same energy; k and T have their usual meaning. The number of configurations of distributing N_n embryos on N_v sites is (see Eq. 5.9)

$$\Omega = \frac{N_v!}{N_n!(N_v - N_n)!} \tag{9A.2}$$

By combining Eqs. (9A.1) and (9A.2) it can be shown that at equilibrium (i.e., *when* $\partial G_{sys}/\partial N_n = 0$) the number of nuclei is

$$N_n^{eq} \approx N_v \exp\left(-\frac{\Delta G_c}{kT}\right)$$

The similarity between this problem and that of determining the equilibrium number of defects [i.e., Eq. (6.7)] should be obvious at this point.

PROBLEMS

9.1. (*a*) The water-ice interface tension was measured to be 2.2×10^3 J/m². If the water is exceptionally clean, it is possible to undercool it by 40°C before it crystallizes. Estimate the size of the critical nucleus if the enthalpy of fusion of ice is 6 kJ/mol.
Answer: 0.9 Å.
(*b*) Discuss what you think would happen if the water is not clean. Would the undercooling increase or decrease? Explain.

9.2. (*a*) Making use of the values used in Worked Example 9.1 and assuming $\Delta G_m^* = 50$ kJ/mol, plot I_v as a function of temperature.
(*b*) Starting with Eq. (9.14) or (9.15), show that for small ΔTs, a linear relationship should exist between growth rate and degree of undercooling.

9.3. (*a*) Take ΔS to be on the order of 2R, and assume growth is occurring at 1000°C. For a liquid that melts at 1500°C, estimate what would represent a small undercooling. State all assumptions.
Answer: $T < 100$°C.

(*b*) Based on the results listed in Table 9.1, repeat part (*a*) for NaCl. Based on your answer, decide whether it would be easy or difficult to obtain amorphous NaCl. Explain. State all assumptions.

9.4. Based on your knowledge of the silicate structure, would you expect the viscosity of $Na_2Si_2O_5$ to be greater or smaller than that of Na_2SiO_3 at their respective melting points? Which would you expect to be the better glass former? Explain.

9.5. (*a*) Show that for quartz the Si–O–Si bond angle is 144°.

(*b*) Classify the following elements as modifiers, intermediates or network formers in connection to their role in oxide glasses:

Si———; Na———; P———; Ca———; Al———.

9.6. The nucleation rate of an amorphous solid [i.e., Eq. (9.13)] can be expressed as $I_v = I_{v,0} \exp(-\Delta H_N/kT)$. If for the same solid, $I_{v,0} = 8 \times 10^4$ m^{-3}·s^{-1} and the nucleation rate was measured to be 16.7 m^{-3}·s^{-1} at 140°C. The growth rates of the crystals were measured to be 7×10^{-7} and 3×10^{-6} m/s at 140 and 160°C, respectively. How long would it take to crystallize 95% of this solid at 165°C? Assume that the growth rate is isotropic and linear with time and that nucleation is random and continuous.

Answer: 1.7 h.

9.7. Show that the entropy change between a supercooled liquid and the glass can be expressed by Eq. (9.23). Also show that when $T \approx T_k$, Eq. (9.21) is recovered from Eq. (9.27). *Hint*: When $T \approx T_k$, $\ln (T/T_k) \approx (T - T_k)/T_k$.

9.8. The activation energy for of pure silica drops from 565 to 163 kJ/mol upon the addition of 0.5 mol fraction of MgO or CaO. The addition of alkali metal oxides has an even more dramatic effect, lowering the activation energy to 96 kJ/mol for 0.5 mol fraction additions. Explain, using sketches, why this is so.

9.9. The results of a two-dimensional simulation of nucleation and growth are shown in Fig. 9.18. The nucleation rate was assumed to be constant and equal to 0.0015 per mm²/s; the growth rate was assumed to be 1 mm/s. Repeat the experiment on your computer and compare the surface fraction

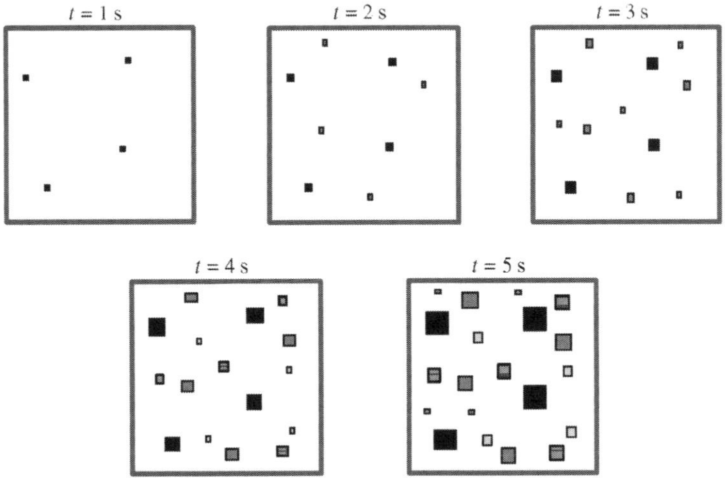

FIGURE 9.18 Two-dimensional simulation of nucleation and growth of a second phase as a function of time.

crystallized that you obtain from your simulation to what you would derive analytically for this particular problem.

9.10. Material X has two allotropic forms: a high temperature β-phase that transforms, at 1000 K, to a lower temperature α-phase. At $T > 1000$ K the solid exists purely in the β-*phase*. Consider two separate cases.

Case 1: Sample A contains inclusions in the amount of 10^8 cm^{-3}. These inclusions act as heterogeneous nuclei in the phase transformation. The sample is quenched to 980 K and held at that temperature while the transformation to the α-phase occurs.

Case 2: Sample B is sufficiently clean that nucleation of α is homogeneous. The sample is quenched to 800 K and held at that temperature.

(*a*) In which case will the sample take longer to transform, i.e., go to 99% completion? You can assume thermal conductivity is high and normal growth occurs.

(*b*) If growth is not normal, would this change your ranking in part (*a*)? Explain.

Information you may find useful:

Enthalpy of transformation	$\Delta H = 0.5\, RT_e$ where T_e is transform. T
Molar volume of α	$V_m = 10$ cm^3/mol
Lattice parameter in α	$a = 3 \times 10^{-8}$ cm
Surface energy along α-β	$\gamma_{\alpha\text{-}\beta} = 0.05$ J/m^2
Lattice diffusivities	$D = 10^{-8}$ cm^2/s at 980 K
	$D = 10^{-10}$ cm^2/s at 800 K
Diffusivity across α-β interface	$D = 10^{-7}$ cm^2/s at 980 K
	$D = 10^{-9}$ cm^2/s at 800 K

9.11. Loehman[129] reported on the formation of oxynitride glasses in the Y–Si–Al–O–N system. The silica-rich glasses contained up to 7 at.% nitrogen. The changes in thermal expansion coefficients and glass transition temperatures are tabulated below.

(*a*) Discuss the incorporation of nitrogen in the oxide glass. Where is it located, and how does it affect the properties?

(*b*) Do these results have any technological implications?

Atomic % Nitrogen	0.0	1.5	7.0
Thermal expansion ($\times 10^{-6}$°C^{-1})	7.5	6.8	4.5
Glass transition, °C	830	900	920
Microhardness (Vickers)	1000	1050	1100

9.12. Smedskjaer et al. (*Phys. Rev. Lett.* 105 115503 (2010)) measured the hardness values of a series of Na-borate glasses where they systematically varied the Na_2O content at the expense of the B_2O_3. The results are shown in Fig. 9.19. The authors developed an elegant model to explain these results based on the average coordination number of the B atoms. Find the paper and in your own words explain the model and discuss how well the model and experimental results agree or disagree.

[129] R. Loehman, *J. Amer. Cer. Soc.*, 62, 491 (1979).

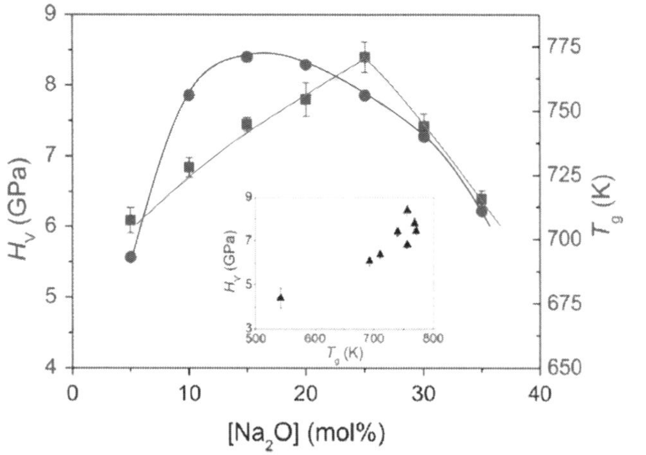

FIGURE 9.19 Compositional dependence of the Vickers hardness, H, and T_g of sodium borosilicate glasses. Inset shows relationship between Vickers hardness H_v and T_g. (Adapted from Smedskjaer et al. *PRL* 105, 2010.)

ADDITIONAL READING

1. H. Rawson, Properties and Applications of Glass, vol. 3, Glass Science and Technology, Elsevier, Amsterdam, 1980.
2. R. Doremus, Glass Science, Wiley, New York, 1973.
3. G. Beall, Synthesis and design of glass-ceramics, J. Mater. Ed., 14, 315 (1992).
4. G. Scherer, Glass formation and relaxation, in Materials Science and Technology, Glasses and Amorphous Materials, vol. 9, R. Cahn, P. Haasen, and E. Kramer, Eds., VCH, New York, 1991.
5. G. O. Jones, Glass, 2nd ed., Chapman & Hall, London, 1971.
6. J. Jackle, Models of the glass transition, Rep. Prog. Phys., 49, 171 (1986).
7. G. W. Morey, The Properties of Glass, Reinhold, New York, 1954.
8. J. W. Christian, The Theory of Transformations in Metals and Alloys, 2nd ed., Pergamon Press, London, 1975.
9. M. H. Lewis, ed., Glasses and Glass-Ceramics, Chapman & Hall, London, 1989.
10. J. Zarzycki, Glassses and the Vitrous State, Cambridge University Press, Cambridge, UK, 1975.
11. The December 2016 of the International Journal of Applied Glass Science is a special issue in which the first paper argues that our age should be referred to as The Glass Age.

OTHER REFERENCES

Videos/animations were chosen based on how entertaining/illuminating they are. They are not an endorsement of any product or company.

1. The Glass Age, Part 1: Flexible, Bendable Glass: https://www.youtube.com/watch?v=12OSBJwogFc.
2. The Glass Age, Part 2: Strong, Durable Glass: https://www.youtube.com/watch?v=13B5K_lAabw.
3. Crystallization of sodium acetate: https://www.youtube.com/watch?v=BLq5NibwV5g.
4. Movie of spherulite formation in polymers; a beautiful example of nucleation and growth. https://www.youtube.com/watch?v=9qbCyJ-nF5A.
5. MD of water freezing to form ice Part I: https://www.youtube.com/watch?v=gmjLXrMaFTg.
6. MD of water freezing to form ice Part II https://www.youtube.com/watch?v=RIW65QLWsjE.

10

SINTERING AND GRAIN GROWTH

The best way to have a good idea is to have lots of ideas.

Linus Pauling

10.1 INTRODUCTION

Up to this point, the effect of microstructure on properties has been neglected, mainly because the properties discussed so far, such as Young's modului, thermal expansions, electrical conductivities, melting points, and densities, among others, are to a large extent microstructure-insensitive. In the remainder of this book, however, it will become apparent that microstructure can and does play a significant role in determining properties. For example, as shown in Table 10.1, the optimization of various properties requires various microstructures.

And while metals and polymers are usually molten, cast, and, when necessary, machined or forged into the final desired shape, ceramic processing poses more of a challenge on account of their refractoriness and brittleness. With the notable exception of glasses, few ceramics are processed from the melt—the fusion temperatures are simply too high. Instead, the starting point is usually fine powders that are milled, mixed and molded into the desired shape by a variety of processes and subsequently heat-treated, or fired, to convert them to dense solids. Despite the fact that the details of shaping and forming of the green (unfired) bodies can have a profound influence on the final microstructure, they are not directly addressed here, but will be touched upon later. The interested reader is referred to a number of excellent books and monographs

TABLE 10.1 Desired microstructures for optimizing properties	
Property	**Desired microstructure**
High strength	Small grain size, uniform microstructure and flaw-free
High toughness	Duplex microstructure with high aspect ratios
High creep resistance	Large grains and absence of amorphous GB phases
Optical transparency	Pore-free microstructure with grains that are either much smaller or much larger than the wavelength of light being transmitted
Varistor behavior	Control of GB chemistry
Catalyst	Very large surface area that is resistant to densification/coarsening

that have been written on the subject, some of which are listed in Additional Reading at the end of this chapter.

As noted above, once shaped, the parts are fired or sintered. Sintering is the process by which a powder compact is transformed to a strong, dense ceramic body upon heating. In an alternate definition given by Herring,[130] **sintering** is "understood to mean any changes in shape which a small particle or a cluster of particles of uniform composition undergoes when held at high temperature." As will become clear in this chapter, sintering is a complex phenomenon in which several processes are occurring simultaneously.

Sintering can occur in the presence or absence of a liquid phase. In the former case, it is called *liquid-phase sintering,* where the compositions and firing temperatures are chosen such that some a liquid forms during processing, as shown schematically in Fig. 10.1*a*. In the absence of a liquid phase, the process is referred to as *solid-state sintering* (Fig. 10.1*b*).

There are many papers in the ceramic literature devoted to understanding and modeling of the sintering process; and if there were such a thing as the holy grail of ceramic processing science, it probably would be

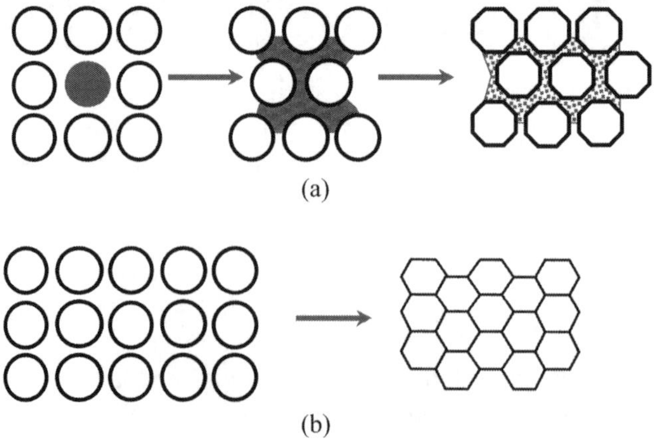

FIGURE 10.1 Schematics of (*a*) Liquid-phase sintering; (*b*) solid-state sintering.

[130] C. Herring, *J. Appl. Phys.*, 21, 301–303 (1950).

how to consistently obtain theoretical or full density at the lowest possible temperature. The main difficulty in achieving this goal, however, lies in the fact that the driving force for sintering is quite small, usually on the order of a few joules per mole, compared to a few kilojoules per mole in the case of chemical reactions like oxidation (see Worked Example 10.1). Consequently, unless great care is taken during *solid state* sintering, full density is difficult to achieve. Liquid state sintering, on the other hand, is much more forgiving and is technologically the process of choice.

This chapter is mainly devoted to understanding the science behind the sintering process. In the next section, the driving forces and atomic mechanisms responsible for solid state sintering are examined. Section 10.3 deals with sintering kinetics, and the factors affecting solid-state sintering are elucidated on the basis of the ideas presented in Sec. 10.2. In Sec. 10.4, liquid-phase sintering is discussed, and in Sec. 10.5 hot pressing and hot isostatic pressing are briefly touched upon. Lastly two case studies of relatively new densification techniques are outlined and discussed.

10.2 SOLID-STATE SINTERING

The macroscopic driving force operative during sintering is the reduction of the excess energy associated with surfaces. This can happen by (1) reduction of the total surface area by an increase in the average size of the particles, which leads to coarsening (Fig. 10.2*b*) and/or (2) the elimination of solid/vapor interfaces and the creation of grain boundaries, GBs, instead, followed by grain growth, which leads to densification (Fig. 10.2*a*). These two mechanisms are usually in competition. If the atomic processes that lead to densification dominate, the pores get smaller and disappear with time and the compact shrinks. But if the atomic processes that lead to coarsening are faster, both the pores and grains get larger with time.

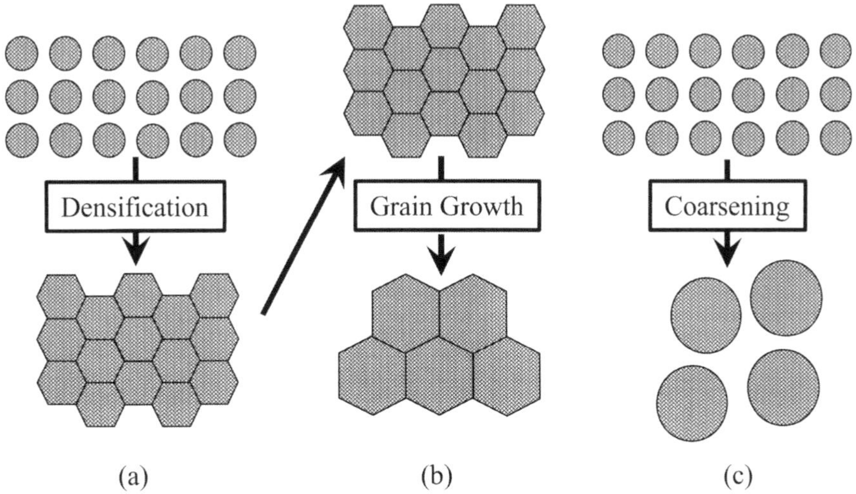

FIGURE 10.2 Schematics of two possible paths by which a collection of particles can lower its energy. (*a*) Densification followed by (*b*) grain growth. In this case, the compact shrinks. (*c*) Coarsening, where large grains grow at the expense of smaller ones.

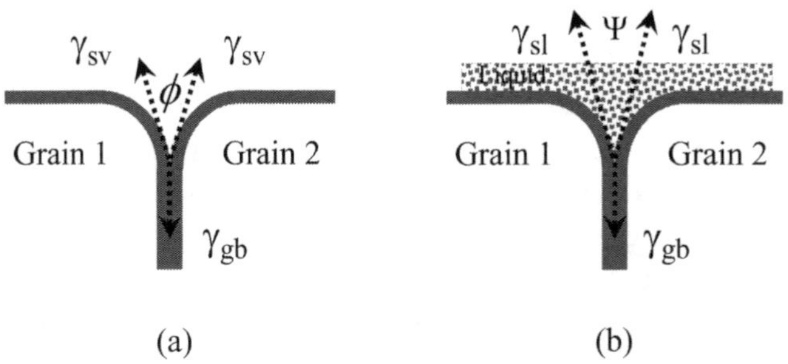

FIGURE 10.3 Equilibrium dihedral angle between (*a*) grain boundary, GB, and solid/vapor interfaces and, (*b*) GB and liquid phase.

A necessary condition for densification to occur is that the GB energy γ_{gb} be less than twice the solid/vapor surface energy γ_{sv}. This implies that the equilibrium dihedral angle ϕ shown in Fig. 10.3*a* and defined as

$$\gamma_{gb} = 2\gamma_{sv}\cos\frac{\phi}{2} \tag{10.1}$$

has to be less than 180°. For many oxide systems,[131] the dihedral angle is around 120°, implying that $\gamma_{gb}/\gamma_{sv} \approx 1.0$, in contrast to metallic systems where that ratio is closer to between 0.25 and 0.5.

Worked Example 10.1

(*a*) Calculate the enthalpy change for an oxide as the average particle diameter increases from 0.5 to 10 μm. Assume a molar volume of 10 cm³/mol and a surface energy of 1 J/m². (*b*) Repeat part *(a)* if, instead of coarsening, the 0.5 μm spheres are sintered together into cubes, if the dihedral angle for this system was measured to be 100°.

ANSWER

(*a*) Take 1 mol or 10 cm³ to be the basis and assume monosize spheres 0.5 μm in diameter. Their total number N is given by

$$N = \frac{10\times10^{-6}\,m^3}{(4/3\pi)(0.25\times10^{-6})^3} = 1.5\times10^{14}$$

Their corresponding surface area *S* is

$$S = 4\pi r^2(1.5\times10^{14}) = 120\,m^2$$

The total energy of the 0.5 μm spheres:

[131] See, e.g., C. Handwerker, J. Dynys, R. Cannon, and R. Coble, *J. Amer. Cer. Soc.*, 73, 1371 (1990).

$$E_{0.5} = N\gamma_{sv} = (120\ m^2)(1\ J/m^2) = 120\ J$$

Similarly, it can be shown that the total surface energy of the 10 μm spheres is $\approx$ 6 J. The change in enthalpy associated with the coarsening is

$$6 - 120 = -114\ J/mol$$

In other words, the process is exothermic because surfaces were eliminated. This energy is the driving force for coarsening.

(*b*) Given that the dihedral angle is 100°, applying Eq. (10.1), one obtains

$$\gamma_{gb} = 2\gamma_{sv}\cos\frac{\phi}{2} = 2\times 1\times \cos 50 = 1.28\ J/m^2$$

If a is the length of the side of the cubes, then mass conservation requires that the volume of the cube be equal to that of the sphere, or $a^3 \approx 4/3\pi r^3$. In other words, a $\approx$ 0.4 μm. The total GB area of the cubes (neglecting free surfaces) is

$$S_{gb} \cong \frac{6}{2}(0.4\times 10^{-6})^2(1.5\times 10^{14}) = 72\ m^2$$

Thus, the energy of the system after sintering is $1.28\times 72 \approx 92.2$ J, which is less than the original of 120 J. The difference between the two is the driving force for densification.

EXPERIMENTAL DETAILS: SINTERING KINETICS

Based on the foregoing discussion, to understand what is occurring during sintering, one needs to measure the grain and pore sizes as well as the overall shrinkage of a powder compact as a function of the sintering variables, such as time, temperature and initial particle size. If the powder compact shrinks, its density will increase with time. Hence, densification is best followed by measuring the density of the compact (almost always reported as a percentage of the theoretical density) as a function of sintering time. This is usually carried out dilatometrically (see Fig. 4.9), where the length of a powder compact is measured as a function of time at a given temperature. Typical shrinkage curves are shown in Fig. 10.4 for two different temperatures $T_2 > T_1$. For reasons that will become clear shortly, the densification rate is a strong function of temperature, as shown in the figure.

In contradistinction, if a powder compact coarsens, no shrinkage is expected in a dilatometric experiment. In that case, the coarsening kinetics are best followed by measuring the average particle size and pores as a function of time via optical or scanning electron microscopy.

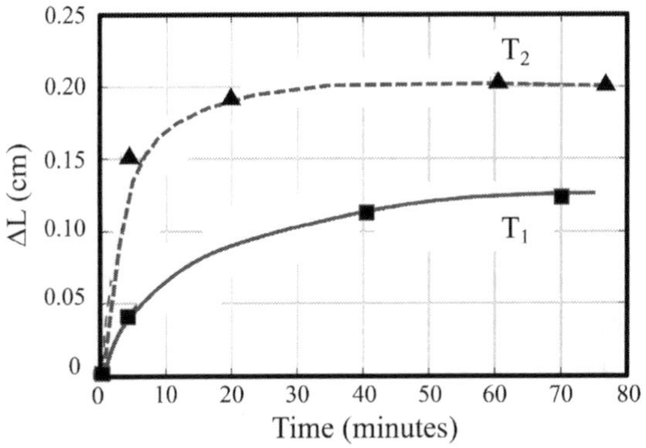

FIGURE 10.4 Typical axial shrinkage curves during sintering as a function of temperature, where $T_2 > T_1$.

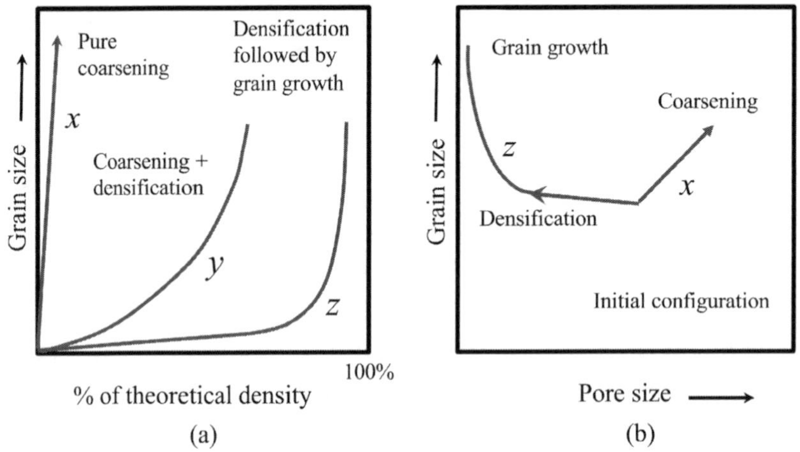

FIGURE 10.5 (*a*) Grain size versus density trajectories for densification (curve z) and coarsening (curve x). Curve y shows a powder for which both coarsening and densification are occurring simultaneously. (*b*) Alternate scheme to represent data in terms of grain and pore size trajectories.

It is useful to plot the resultant behavior in what is known as **grain size versus density trajectories**, such as those shown in Fig. 10.5*a* and *b*. Typically, a material will follow the path denoted by curve y, where both densification and coarsening occur simultaneously. However, to obtain near-theoretical densities, coarsening has to be suppressed until most of the shrinkage has occurred; i.e., the system should follow the trajectory denoted by curve z. A powder that follows trajectory x, however, is doomed to remain porous—the free energy has been expended, large grains have formed, but more importantly so have large pores. Once formed, these pores are kinetically very difficult to remove, and as discussed below, they may even be thermodynamically stable, in which case they would be impossible to remove.

An alternate method of presenting the sintering data is shown in Fig. 10.5*b*, where the time evolution of the grain and pore sizes is plotted; coarsening leads to an increase in both, whereas densification

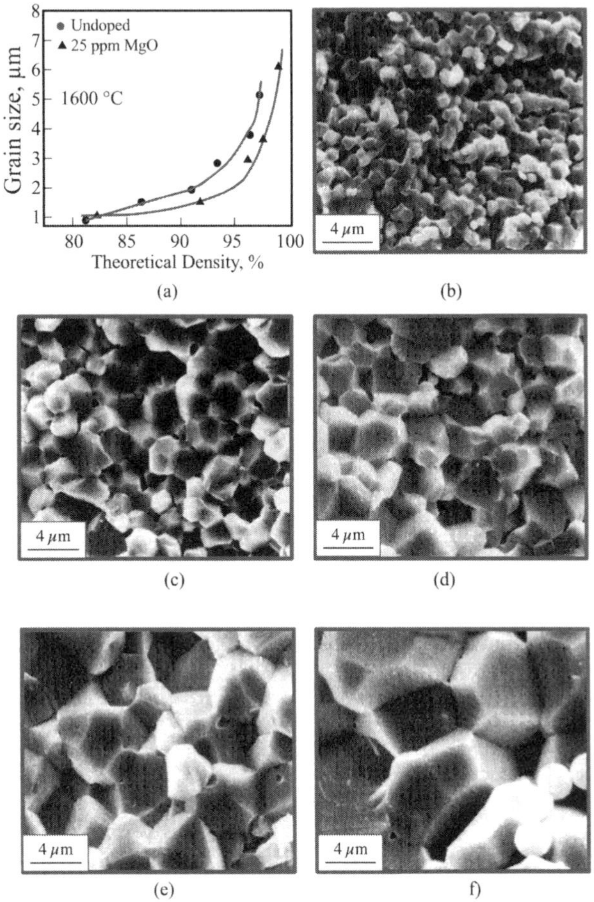

FIGURE 10.6 (*a*) Grain size-densification trajectory of nominally pure Al_2O_3 and Al_2O_3 doped with 500 ppm MgO sintered at 1600°C in air. (*b–f*) Microstructure development of the MgO-doped compact as a function of time. (K. A. Berry and M. P. Harmer, *J. Amer. Cer. Soc*, **69**:143–149, 1986. Reprinted with permission.)

eliminates pores. Figure 10.6*a* compares the grain size vs. density trajectory of undoped and MgO-doped Al_2O_3 sintered for various times at 1600°C. Such trajectories are typical of many ceramics that sinter to full density. Clearly the minuscule doping has a dramatic effect on the trajectory, allowing the compact to achieve almost 100% density. The corresponding time dependence of the microstructural development is shown in Figs. 10.6*b–f*. As time progresses, the average grain size increases whereas the average pore size decreases.

A good example of a powder that coarsens without densification is Fe_2O_3 sintered in HCl-containing atmospheres. The final microstructures after firing for 5 h at 1200°C in air and Ar-10% HCl are shown in

Fig. 10.7*a* and *b*, respectively. The corresponding time dependence of the relative density as a function of HCl content in the gas phase is shown Fig. 10.7*c*. These results clearly indicate that whereas Fe_2O_3 readily sinters to high density in air, in an HCl atmosphere it coarsens instead.

Full density is thus obtained only when the atomic processes associated with coarsening are suppressed, while those associated with densification are enhanced. It follows that in order to understand and control what occurs during sintering, the various atomic processes responsible for each of these outcomes need to identified and described. Before doing so, however, it is imperative to understand the effect of curvature on the chemical potential of the ions or atoms in a solid.

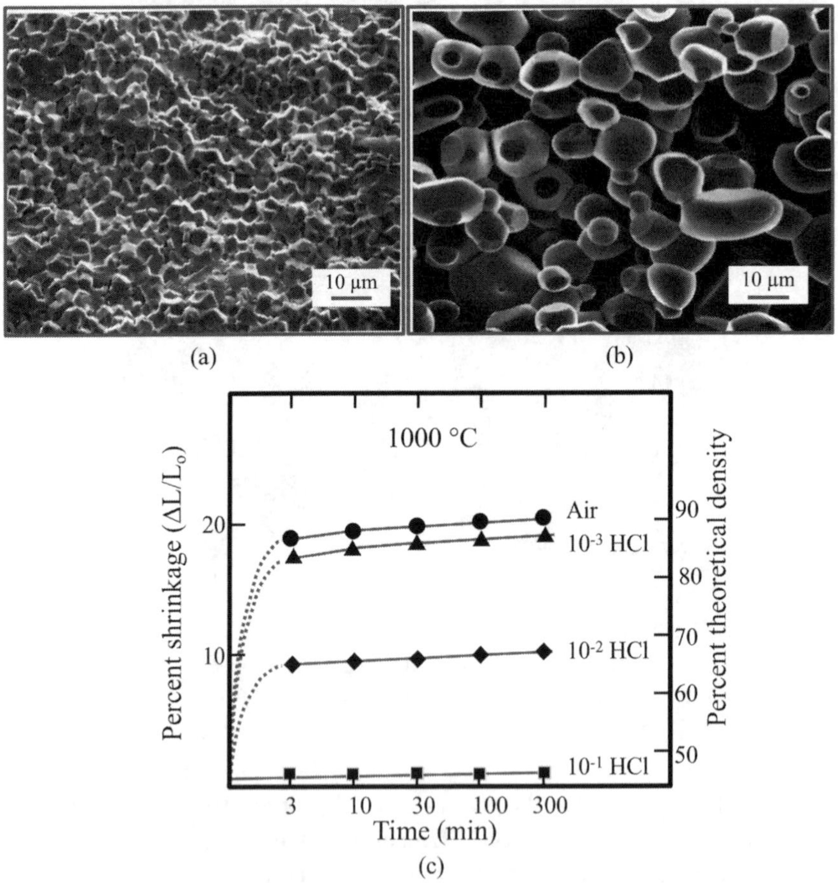

FIGURE 10.7 (*a*) Microstructure of air-sintered Fe_2O_3. (*b*) Same as *(a)* but sintered in HCl-containing atmospheres. Note significant coarsening of microstructure. (*c*) Effect of atmosphere on relative density versus time for Fe_2O_3 sintered at 1000°C. (D. Ready, in *Sintering of Advanced Ceramics, Ceramic Trans.*, vol. 7, C. A. Handwerker, J. E. Blendell, and W. Kayser, Eds., American Ceramic Society, Westerville, OH, 1990, p. 86. Reprinted with permission.)

10.2.1 LOCAL DRIVING FORCE FOR SINTERING

As mentioned above, the global driving force operating during sintering is the reduction in surface energy, which manifests itself locally as curvature differences. From the Gibbs–Thompson equation (see App. 10A for derivation), it can be shown that the chemical potential difference *per formula unit, $\Delta\mu$,* between atoms on a flat surface and under a surface of curvature κ is given by

$$\Delta\mu = \mu_{curv} - \mu_{flat} = \gamma_{sv}\Omega_{MX}\kappa \tag{10.2}$$

where Ω_{MX} is the volume of a MX formula unit. For simplicity in the following discussion, it will be assumed that one is dealing with an MX compound. The curvature κ depends on geometry; e.g., for a sphere of radius ρ, $\kappa = 2/\rho$ (see App. 10B for further details). Equation (10.2) has two important ramifications that are critical to understanding the sintering process. The first is related to the partial pressure of a material above a curved surface; the second involves the effect of curvature of vacancy concentration.

EFFECT OF CURVATURE ON PARTIAL PRESSURE

At equilibrium, this chemical potential difference translates to a difference in partial pressure above the curved surface, i.e.,

$$\Delta\mu = kT \ln \frac{P_{curv}}{P_{flat}} \tag{10.3}$$

Combining the two equations,

$$\ln \frac{P_{curv}}{P_{fat}} = \kappa \frac{\Omega_{MX}\gamma_{sv}}{kT} \tag{10.4}$$

If $P_{curv} \approx P_{flat}$, then Eq. (10.4) simplifies to the more common expression, namely,

$$\frac{\Delta P}{P_{flat}} = \frac{P_{curv} - P_{flat}}{P_{fat}} = \kappa \frac{\Omega_{MX}\gamma_{sv}}{kT} \tag{10.5}$$

As noted above, for a sphere of radius ρ, $\kappa = 2/\rho$, and Eq. (10.5) can be written as

$$\boxed{P_{curv} = P_{flat} = \left(1 + \frac{2\Omega_{MX}\gamma_{sv}}{\rho kT}\right)} \tag{10.6}$$

Given that the *radius of curvature is defined as negative for a concave surface and positive for a convex* one, this expression is of fundamental importance because it predicts that the pressure of a material above a convex surface is *greater* than that over a flat surface, and vice versa for a concave surface. For example, the pressure inside a pore of radius ρ is *less* than that over a flat surface; conversely,

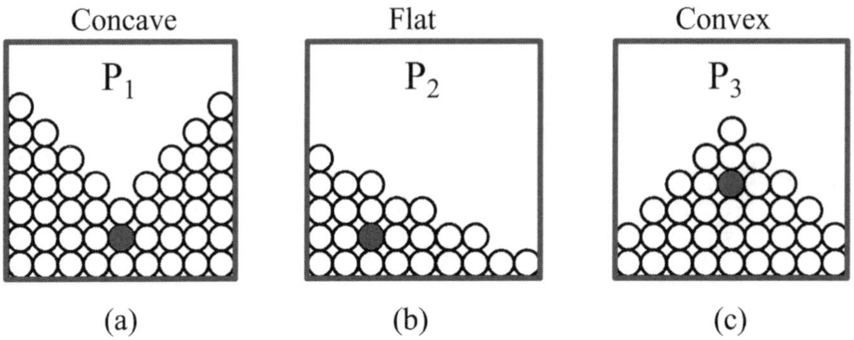

FIGURE 10.8 Effect of surface curvature on equilibrium vapor pressure. At this scale it is easy to appreciate why $P_1 < P_2 < P_3$.

the pressure surrounding a collection of fine spherical particles will be *greater* than that over a flat surface.[132]

It is only by appreciating this fact that coarsening can be understood. Given the importance of this conclusion, it is instructive to explore what occurs on the atomic level that allows this to happen. To do so, consider the following thought experiment: Place each of three different-shaped surfaces, of the same solid, in a sealed and evacuated chamber, as shown in Fig. 10.8, and heat until an equilibrium vapor pressure is established. Examining the figures shows that $P_1 < P_2 < P_3$, since, on average, the atoms on a convex surface are less tightly bound to their neighbors than atoms on a concave surface and will thus more likely escape into the gas phase, resulting in a higher partial pressure.

EFFECT OF CURVATURE ON VACANCY CONCENTRATIONS

The other important ramification of Eq. (10.2) is that the equilibrium vacancy concentration is also a function of curvature. In Chap. 6, the relationship between the equilibrium concentration of vacancies over a flat, stress-free surface, c_v, their enthalpy of formation Q and temperature was given by Eq. 6.7. This equation can be re-expressed as

$$c_0 = K' \exp\left(-\frac{Q}{kT}\right) \tag{10.7}$$

where c_0 is the equilibrium concentration of vacancies, under a *flat and stress-free* surface, h_d is replaced by Q, and the entropy of vacancy formation, and all pre-exponential terms are included in the constant K′.

[132] Implicit in this result is that the MX compound is evaporating as MX molecules.

Note that for clarity's sake c_v is replaced by c_o in this chapter. An implicit assumption made in deriving this expression is that the vacancies are formed under a flat, stress-free surface.

Since the chemical potential of an atom under a curved surface is either greater or smaller than that over a flat surface by $\Delta\mu$, this energy has to be accounted for when one is considering the formation of a vacancy under a curved surface, c_{curv}. Hence it follows that

$$c_{curv} = K' \exp\left(-\frac{Q + \Delta\mu}{kT}\right) = c_o \exp\left(-\frac{\kappa \Omega_{MX} \gamma_{sv}}{kT}\right) \tag{10.8}$$

And since for the most part $\gamma_{sv} \Omega_{MX} \kappa \ll kT$ then with little loss in accuracy

$$\boxed{c_{curv} = c_o \left(1 - \frac{\kappa \Omega_{MX} \gamma_{sv}}{kT}\right)} \tag{10.9}$$

which is identical to

$$\boxed{\Delta c_{vac} = c_{curv} - c_o = -c_o \left(\frac{\kappa \Omega_{MX} \gamma_{sv}}{kT}\right)} \tag{10.10}$$

It is left as an exercise for the reader to show that the vacancy concentration under a concave surface is greater than that under a flat surface, which, in turn, is greater than that under a convex surface.[133] Referring once again to Fig. 10.8 and focusing on the shaded atoms, the physics can be explained as follows: A good measure of the enthalpy of formation of a vacancy is the difference in bonding between an atom in the bulk versus one on the surface. Now since an atom forming under a convex surface will, on average, be bonded to fewer atoms than it would have under a concave surface, it follows that it costs more energy to create a vacancy in the vicinity of a convex surface than a concave one. Said otherwise in the convex situation you get *less* energy back when you place the atom on the surface than you would if you placed it on a concave surface.

To recap, curvature causes local variations in partial pressures and vacancy concentrations. *The partial pressure over a convex surface is higher than that over a concave surface. Conversely, the vacancy concentration under a concave surface is higher than that below a convex surface.* In either case, a local driving force is present that induces the atoms to migrate from the convex to the concave areas, i.e., from

[133] An important assumption made in deriving Eq. (10.10) is that under the curved surface, the defects form in their stoichiometric ratios; i.e., the defect concentrations are dictated by the Schottky equilibrium $[V_M] = [V_X]$.

the mountaintops to the valleys. Armed with these important conclusions, it is now possible to explore the various atomic mechanisms taking place during sintering.

10.2.2 ATOMIC MECHANISMS OCCURRING DURING SINTERING

There are five atomic mechanisms by which mass can be transferred in a powder compact:

1. Evaporation-condensation depicted as path 1 in Fig. 10.9*a*.
2. Surface diffusion, or path 2 in Fig. 10.9*a*.
3. Volume diffusion. Here there are two paths. The mass can be transferred from the surface to the neck area—path 3 in Fig. 10.9*a*—or from the GB area to the neck area—path 5 in Fig. 10.9*b*.
4. Grain boundary diffusion from the GB area to the neck area—path 4 in Fig. 10.9*b*.
5. Viscous or creep flow. This mechanism entails either the plastic deformation or viscous flow of particles from areas of high stress to low stress and can lead to densification. However, since it is essentially the same process that occurs during creep, it is dealt with in Chap. 12.

Consider now which of these mechanisms leads to coarsening and which to densification.

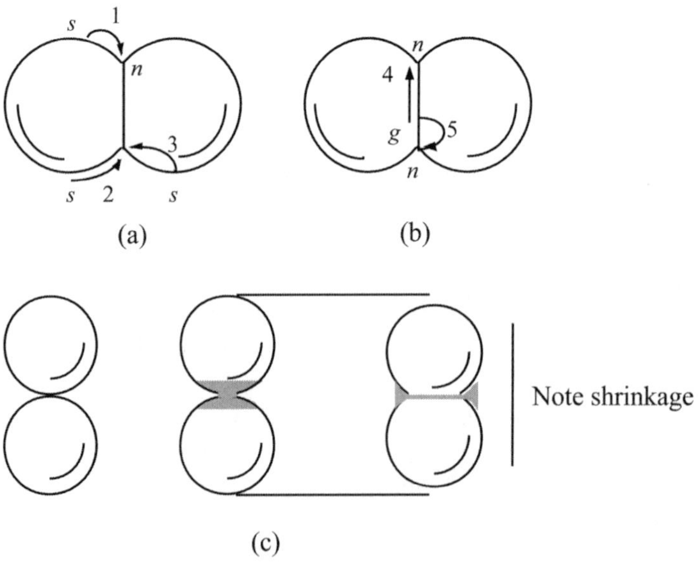

FIGURE 10.9 Basic atomic mechanisms that can lead to, (*a*) coarsening and change in pore shape and (*b*) densification. (*c*) Thought experiment illustrating how removal of material from the area between particles into the pore leads to shrinkage and densification.

COARSENING

At the outset, it is important to appreciate that any mechanism in which the source of material is the surface of the particles and the sink is the neck area *cannot* lead to densification, because such a mechanism does *not* allow the particle centers to move closer together. Consequently, evaporation-condensation, surface diffusion and lattice diffusion from the surface to the neck area *cannot* lead to densification. They can, however, result in a change in pore shapes, a growth in neck size and a concomitant increase in compact strength. Moreover, the smaller grains, with their smaller radii of curvature, will tend to "evaporate" away and plate out on the larger particles, resulting in a general coarsening of the microstructure.

The driving force in all cases is the partial pressure differential associated with the local variations in curvatures. For instance, the partial pressure at point s in Fig. 10.9*a* is greater than that at point n, which in turn results in mass transfer from the convex to the concave surfaces. The actual path taken will depend on the kinetics of the various paths, a topic that will be dealt with shortly. At this point, it suffices to say that since the atomic processes are occurring in parallel, at any given temperature, it is the fastest mechanism that will dominate.

DENSIFICATION

If mass transfer from the surface to the neck area, or from the surface of smaller to larger grains, does not lead to densification, other mechanisms have to be invoked to explain the latter. *For densification to occur the source of material has to be the GB or region between powder particles and the sink has to be the neck or pore region.* To illustrate why this is the case, consider the thought experiment illustrated in Fig. 10.9*c*: Cut a volume (shaded area in Fig. 10.9*c*) from between two spheres, bring the two spheres closer together and then place the extra volume removed in the pore area. Clearly such a process leads to shrinkage and the elimination of pores. Consequently, the only mechanisms, apart from viscous or plastic deformation, that can lead to densification are GB and bulk diffusion from the GB area to the neck area (Fig. 10.9*b*).

Atomistically both mechanisms entail the diffusion of ions from the GB region toward the neck area, for which the driving force is the curvature-induced vacancy concentration. Because there are more vacancies in the neck area than in the region between the grains, a vacancy flux develops away from the pore surface into the GB area, where the vacancies are eventually annihilated. Needless to say, an equal atomic flux will diffuse in the opposite direction, filling the pores.

10.3 SOLID-STATE SINTERING KINETICS

Based on the foregoing discussion, a powder compact can reduce its energy by following various paths, some of which can lead to coarsening, others to densification. This brings up a central and critical question:

What governs whether a collection of particles will densify or coarsen? To answer the question, models for each of the paths considered above must be developed and compared with the fastest path determining the behavior of the compact. For instance, a compact in which surface diffusion is much faster than bulk diffusivity would tend to coarsen rather than densify.

In practice, the question is much more difficult to answer, however, because sintering kinetics are dependent on so many variables, including particle size and packing, sintering atmosphere, degree of agglomeration, temperature, presence of impurities, among others. The difficulty of the problem is best illustrated by comparing the densification kinetics of an "as-received" yttria-stabilized zirconia powder compact to that of the same powder that was rid of agglomerates before sintering (Fig. 10.10). The marked reduction in the temperature, by about 300°C, needed to fully densify the compact which was prepared from an agglomerate-free compact is obvious.

In this section, some of the many sintering models proposed over the years to model this complex process are outlined. Because of the problem's complex geometry, analytic solutions are only possible by making considerable geometric and diffusion flow field approximations, which are rarely realized in practice. Consequently, the models discussed below have limited validity and should be used with extreme care when one is trying to predict/understand the sintering behavior of real powders. Much of the usefulness of these sintering models thus lies more in appreciating the general trends that are to be expected and identifying the critical parameters than in their predictive capabilities.

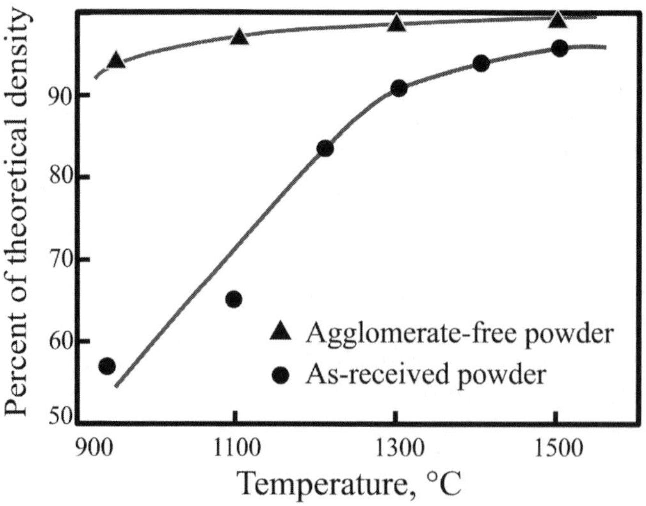

FIGURE 10.10 Temperature dependence of sintered density for an agglomerated or "as-received" and agglomerate-free yttria-stabilized zirconia powder. Eliminating the agglomerates in the green body resulted in a powder compact that densified much more readily. (W. H. Rhodes, *J. Amer. Cer. Soc.*, 64, 19, 1981. Reprinted with permission.)

Sintering stages

Coble[134] described a sintering stage as an "interval of geometric change in which pore shape is totally defined (such as rounding of necks during the initial-stage sintering) or an interval of time during which the pore remains constant in shape while decreasing in size." Based on that definition, three stages have been identified: an initial, an intermediate and a final stage.

During the **initial stage**, the interparticle contact area increases by neck growth from 0 (Fig. 10.11*a*) to ≈0.2 (Fig. 10.11*b*), and the relative density increases from about 60% to 65%.

The **intermediate stage** is characterized by continuous pore channels that are coincident with three-grain edges (Fig. 10.11*c*). During this stage, the relative density increases from 65% to about 90% by having matter diffuse toward, and vacancies away from the long cylindrical channels.

The **final stage** begins when the pore phase is eventually pinched off and is characterized by the absence of a continuous pore channel (Fig. 10.11*d*). Individual pores are either of lenticular shape, if they reside on

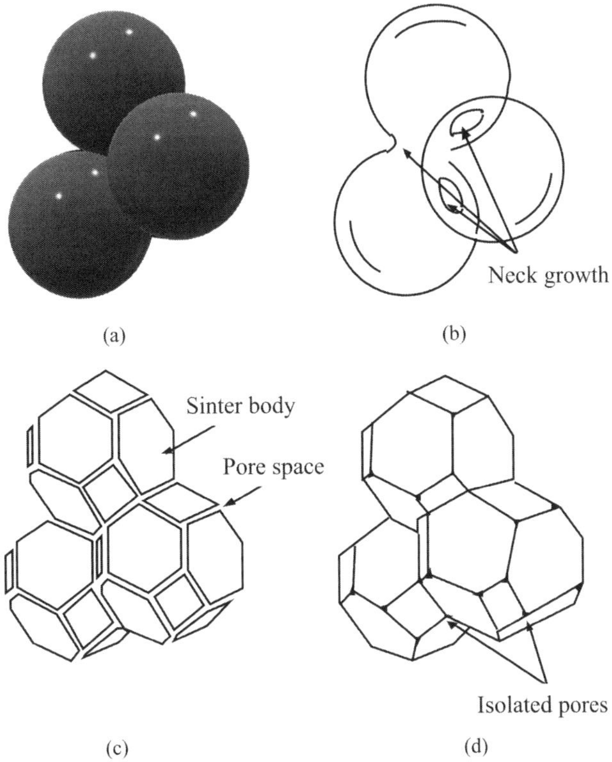

FIGURE 10.11 (*a*) Initial stage of sintering model represented by spheres in tangential contact. (*b*) Near end of initial stage; spheres have begun to coalesce. (*c*) Intermediate stage; grains adopted shape of dodecahedra, enclosing pore channels at grain edges. (*d*) Final stage; pores are tetrahedral inclusions at corners where four dodecahedra meet.

[134] R. L. Coble, *J. Appl. Phys.*, 32, 787–792 (1961); R. L. Coble, *J. Appl. Phys.*, 36, 2327 (1965).

the GBs, or rounded if they reside within a grain. Important characteristics of this stage are the increases in pore and GB mobilities, which have to be controlled if theoretical density is to be achieved.

Clearly, the sintering kinetics will be different during each of the aforementioned stages. To further complicate matters, in addition to having to treat each stage separately, the kinetics will depend on the specific atomic mechanisms operative. Despite these complications, most, if not all, sintering models share the following common approach:

1. A representative particle shape is assumed.
2. The surface curvature is calculated as a function of geometry.
3. A flux equation that describes the rate-limiting step is adopted.
4. The flux equation is integrated to predict the rate of geometry change.

In the following subsections, this approach is used to predict the rates of various processes occurring during the aforementioned stages. In particular, Sec. 10.3.1 deals with the initial stage, while Sec. 10.3.2 addresses densification kinetics. Coarsening and grain growth, because of their similarities, are discussed in Sec. 10.3.3.

10.3.1 INITIAL-STAGE SINTERING

Given the multiplicity of paths available to a powder compact during this stage, it is impossible to address them all in detail. Instead, the following approach has been adopted: The rate of neck growth by evaporation-condensation (path 1 in Fig. 10.9*a*) is worked out in detail; the final results for the other mechanisms, namely, surface, grain boundary, lattice diffusion and viscous sintering, are given without proof.[135]

EVAPORATION-CONDENSATION MODEL

In this mechanism (path 1 in Fig. 10.9a), the pressure differential between the surface of a particle and the neck area results in a net matter transport, via the gas phase from the surface to the neck. The evaporation rate (in molecules of MX per square meter per second), is given by the Langmuir expression

$$\dot{j} = \frac{\alpha \Delta P}{\sqrt{2\pi m_{MX} kT}} = K_r \Delta P \tag{10.11}$$

where α and m_{MX} are, respectively, the evaporation coefficient and the mass of the evaporating gas molecules assumed to be MX; ΔP is the pressure differential between the surface and neck areas. Starting with Eq. (10.5), it can be shown that ΔP between these two regions is

$$\Delta P = \frac{\Omega_{MX} P_{flat} \gamma_{sv}}{kT} \left[\frac{1}{\rho} - \frac{1}{r} \right] \approx \frac{\Omega_{MX} P_{flat} \gamma_{sv}}{\rho kT} \tag{10.12}$$

[135] For a summary and a critical analysis of the initial-stage sintering models, see, e.g., W. S. Coblenz, J. M. Dynys, R. M. Cannon, and R. L. Coble, in *Sintering Processes*, G. C. Kuczynski, Ed., Plenum Press, New York, 1980.

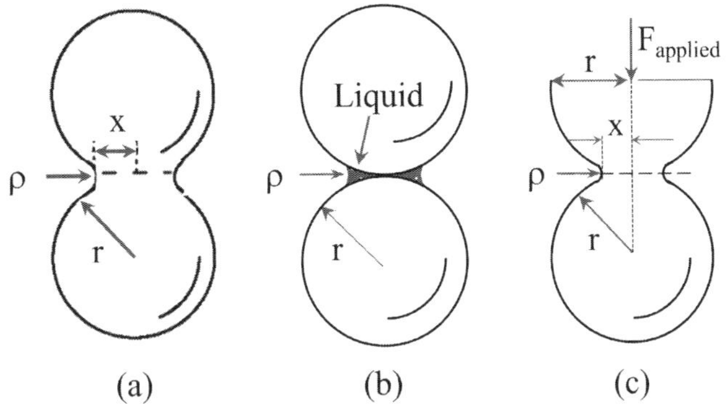

FIGURE 10.12 (*a*) Sphere tangency construction used for initial-stage sintering modeling. (*b*) Spherical particles held together by liquid capillary pressure. (*c*) Sphere tangency construction during hot pressing. Ratio of applied to boundary stress is proportional to $(x/r)^2$ (see Sec. 10.5).

where ρ and r, defined in Fig. 10.12*a*, are, respectively, the radius of curvature of the neck area and the sphere radius. Furthermore, according to this figure,

$$(r+\rho)^2 = (x+\rho)^2 + r^2$$

where x is the neck radius. For x << r, this equation simplifies to

$$\rho = \frac{x^2}{2(r-x)} \approx \frac{x^2}{2r} \tag{10.13}$$

Multiplying the flux of material arriving to the neck area by Ω_{MX} yields the rate at which the neck will grow, or

$$\frac{dx}{dt} = j\Omega_{MX} \tag{10.14}$$

Combining Eqs. (10.11) to (10.14) and integrating yields

$$\left(\frac{x}{r}\right)^3 = \left[\frac{6\alpha\gamma_{sv}\Omega_{MX}^2 P_{flat}}{kTr^2\sqrt{2\pi m_{MX} kT}}\right] t \tag{10.15}$$

This equation predicts that the rate of growth of the neck region is: (1) initially quite rapid but then flattens out, (2) a strong function of initial particle size, and (3) a function of the partial pressure P_{flat} of the compound, which in turn depends exponentially on temperature.

To recap, Eq. (10.15) was derived by assuming a representative shape (Fig. 10.12*a*), from which the surface curvature was calculated as a function of geometry [Eq. (10.13)]. A flux equation [Eq. (10.11)] was then assumed and integrated to yield the final result. By using essentially the same procedure, the following results for other models are obtained.

LATTICE DIFFUSION MODEL

If it is assumed that material diffuses away from the GB area through the bulk and plates out in an area of width 2ρ (path 5, Fig. 10.9*b*), the following expression for neck growth is obtained:[136]

$$\left(\frac{x}{r}\right)^4 = \left[\frac{64 D_{ambi} \gamma_{sv} \Omega_{MX}^2}{kTr^3}\right] t \tag{10.16}$$

where D_{ambi} is the ambipolar diffusivity, given by Eq. (7.87). It is important to note that the use of D_{ambi} in Eq. (10.16) implies that the compound is pure and stoichiometric, with the dominant defects being Schottky defects.[137] It is also implicit that the two component fluxes both diffuse through the bulk. In general, however, it is important to appreciate that it is the *slowest species diffusing along its fastest path that is rate-limiting,* a conclusion that applies to all mechanisms discussed below as well.

GRAIN BOUNDARY DIFFUSION MODEL

In this model, the mass is assumed to diffuse from the GB area radially along a GB of width δ_{gb} and plate out at the neck surface (path 4, Fig. 10.9*b*). The neck growth is given by[138]

$$\left(\frac{x}{r}\right)^6 = \left[\frac{192 \delta_{gb} D_{gb} \gamma_{sv} \Omega_{MX}^2}{kTr^4}\right] t \tag{10.17}$$

which leads to the following linear shrinkage:

$$\left(\frac{\Delta L}{L}\right)^3 = \left[\frac{3 \delta_{gb} D_{gb} \gamma_{sv} \Omega_{MX}}{kTr^2}\right] t$$

where D_{gb} is the GB diffusivity of the rate-limiting ion.

SURFACE DIFFUSION MODEL

In this model, the atoms are assumed to diffuse along the surface from an area that is near the neck region toward the neck area (path 2, Fig. 10.9*a*). The appropriate expression for the growth of the neck with time is[139]

$$\left(\frac{x}{r}\right)^5 = \left[\frac{225 \delta_s D_s \gamma_{sv} \Omega_{MX}}{kTr^4}\right] t \tag{10.18}$$

where D_s and δ_s are, respectively, the surface diffusivity and surface thickness.

---

[136] D. L. Johnson, *J. Appl. Phys.*, 40, 192 (1969).

[137] The situation gets more complicated quite rapidly if the oxide is impure. See, e.g., D. W. Ready, *J. Amer. Cer. Soc.*, 49, 366 (1966).

[138] W. S. Coblenz, J. M. Dynys, R. M. Cannon, and R. L. Coble, in *Sintering Processes*, G. C. Kuczynski, Ed., Plenum Press, NY, 1980. See also: R. L. Coble, *J. Amer. Cer. Soc.*, 41, 55 (1958).

[139] W. S. Coblenz, J. M. Dynys, R. M. Cannon, and R. L. Coble, in *Sintering Processes*, G. C. Kuczynski, Ed. Plenum Press, New York, 1980.

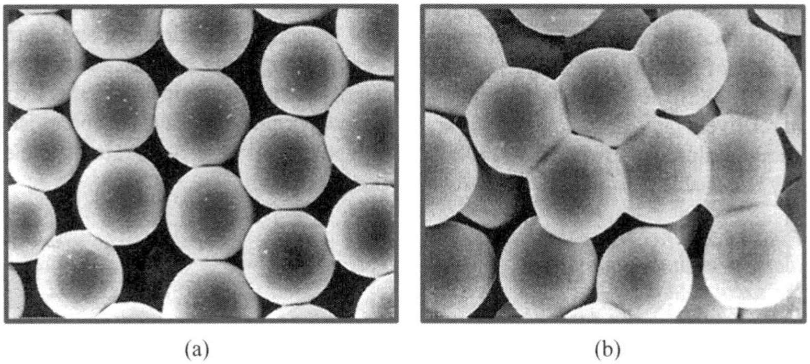

FIGURE 10.13 Micrographs of glass spheres: (*a*) before and (*b*) after sintering by heating in air in a temperature range at which viscous flow can occur. Note here no pressure was applied; the process was driven solely by surface energy considerations.

VISCOUS SINTERING

The shrinkage of two glass spheres during the initial stage is given by the Frenkel equation[140]

$$\frac{\Delta L}{L} = \frac{3\gamma_{sv}}{4\eta r} t \tag{10.19}$$

where η is the glass viscosity (see Chap. 9). A micrograph of glass spheres that sinter by viscous flow upon heating is shown in Fig. 10.13.

GENERAL REMARKS

Usually the activation energies for surface, GB, and lattice diffusivity increase in that order. Thus, surface diffusion is favored at lower temperatures and lattice diffusion at higher temperatures. By comparing Eqs. (10.16), (10.17) and (10.19), it should be obvious why GB and surface diffusion are preferred over lattice diffusion for smaller particles. Lattice diffusion, however, is favored at long sintering times, high sintering temperatures and larger particles. However, by far the most forgiving mechanism with respect to particle size is viscous sintering (see Worked Example 10.2). It is important to note that these general trends also extend through the intermediate- and final-stage sintering stages.

Worked Example 10.2

If it takes 0.2 h for the relative density of a 1 μm average diameter powder to increase from 60 to 65%, estimate the time it would take for a powder of 10 μm average diameter to achieve the same degree of densification if the rate-controlling mechanism were (*a*) lattice diffusion and (*b*) viscous flow.

[140] J. Frenkel, *J. Phys. (USSR)*, 9, 305 (1945).

ANSWER

(*a*) If one assumes that the *ratio x/r does* not vary much as the density increases from 60 to 65%, which is not a bad assumption, it follows that Eq. (10.16) for lattice diffusion can be recast to read

$$\Delta t = \left(\frac{x}{r}\right)^4 \frac{kTr^3}{64 D_{ambi} \gamma_{sv} \Omega_{MX}^2} \approx K' \cdot r^3$$

If the 0.1 μm particles densify by lattice diffusion, the time needed to go from 60 to 65% will be given by $\Delta t = K' \, r^3$ from which

$$K' = \frac{0.2}{(0.5 \times 10^{-6})^3} = 1.6 \times 10^{18} \text{ h/m}^3$$

The 10 μm particles will densify by the same amount after a time Δt given by

$$\Delta t = K'r^3 = 1.6 \times 10^{18} (5 \times 10^{-6})^3 = 200 \text{ h}$$

(*b*) In the case of viscous flow, a similar analysis shows that since t scales with r [Eq. (10.19)], Δt in this case is of about 20 h.

This calculation makes clear why viscous phase sintering is much more forgiving concerning the initial particle size. It also explains the absolute need to start with fine crystalline powders if densification is to occur in reasonable times.

10.3.2 DENSIFICATION KINETICS

INTERMEDIATE SINTERING MODEL

Most of the densification of a powder compact occurs during the intermediate stage. Unfortunately, this stage is the most difficult to tackle because it depends strongly on the details of particle packing—a variable that is quite difficult to model. To render the problem tractable, Coble made the following assumptions:

1. The powder compact is composed of ideally packed tetrakaidecahedra of length a_p separated from each other by long porous channels of radii r_c (Fig. 10.14*a*).
2. Densification occurs by the bulk diffusion of vacancies away from the cylindrical pore channels toward the GBs (curved arrows in Fig. 10.14*b*).
3. A linear, steady-state profile of the vacancy concentration is established between source and the sink.
4. The vacancies are annihilated at the GBs; i.e., the GBs act as vacancy sinks. It is also assumed that where the vacancies are annihilated, their concentration is given by c_0 [Eq. (10.7)], i.e., under a stress-free planar interface.

Making these assumptions, one can show (App. 10C) that during the intermediate sintering stage, the fractional *porosity P_c should* decrease linearly with time according to

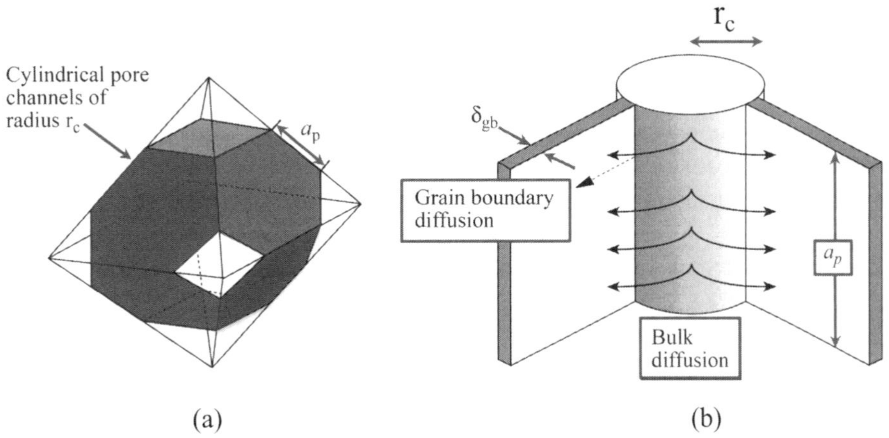

FIGURE 10.14 (*a*) Tetrakaidecahedron model of intermediate-stage sintering. (*b*) Expanded view of one of the cylindrical pore channels. Vacancies can diffuse down the GB (dashed arrow) or through the bulk (solid arrows). In both cases, vacancies are annihilated at the GBs.

$$P_c \approx (\text{const})\frac{D_{ambi}}{d^3}\frac{\gamma_{sv}\Omega_{MX}}{kT}(t_f - t) \tag{10.20}$$

where t_f is the time at which the cylindrical channels vanish and d is the average diameter of the sintering particles, which is assumed to scale with a_p at t = 0.

Repeating the procedure used to arrive at Eq. (10.20) but assuming densification takes place by GB diffusion (straight arrows in Fig. 10.14*b*), one can show that

$$P_c = (\text{const})\left[\frac{D_{gb}\delta_{gb}}{d^4}\left(\frac{\gamma_{sv}\Omega_{MX}}{kT}\right)(t_f - t)\right]^{2/3} \tag{10.21}$$

where δ_{gb} is the GB width shown in Fig. 10.14*b*.

It is rather unfortunate that of all the sintering stages the most important is also the most difficult to model. For example, any intermediate-stage model that does not take into account the details of particle packing has limited validity. A cursory examination of the results shown in Fig. 10.10 should make this point amply clear.

What is interesting about this process is that it is self-accelerating, since as the cylinders gets smaller in diameter, their curvature increase and the concomitant vacancy concentration gradient also increases. This process cannot, and does not, go on indefinitely; as the cylindrical pores get longer and thinner, at some point they become unstable and break up into smaller spherical pores along the GBs and/or at the triple points between grains (see Fig. 10.11*d*). It is at this point that the intermediate sintering stage gives way to the final-stage sintering, where both the annihilation of the last remnants of porosity and the simultaneous coarsening, i.e., grain growth, of the microstructure occur. The next subsection deals with the elimination of porosity; grain growth is dealt with in Sec. 10.3.3.

PORE ELIMINATION

When atoms diffuse toward the pores and vacancies are transported away from the pores to a sink such as GBs, dislocations or external surfaces of the crystal the pores will be eliminated. Before proceeding it is important to understand how crystal defects can act as sources and sinks of vacancies. The following examples make the case. The simplest is how free surfaces and GBs can act as sinks and sources. If an atom jumps from the bulk to the surface or GB, it leaves a vacancy behind and is essence the surface/GB is acting as a vacancy source. If, on the other hand, an atom jumps from the surface or GB into a vacant site in the bulk, the vacancy is eliminated and the surface/GB act as vacany sinks. The same arguments can be made about atoms jumping to and out of dislocation cores. For example, if an atom jumps to the dislocation core of an edge dislocation, the line of extra atoms shrinks by one, the vacancy is elimiated and the dislocation is said have climbed. Climb is quite important during creep.

Consider the two following representative mechanisms.

Volume diffusion. In this model, the vacancy source is the pore surface of radius ρ_p and the sink is the spherical surface of radius R, where $R \ll \rho_p$ (Fig. 10.15*a*). By solving the flux equation, subject to the appropriate boundary conditions, it can be shown that (see App. 10D)

$$\rho_p^3 - \rho_{p,o}^3 = \frac{6D_{ambi}\gamma_{sv}\Omega_{MX}}{kT}t \tag{10.22}$$

where $\rho_{p,o}$ is the initial size of the pores at $t = 0$. To relate the pore radius to porosity P_c, use is again made of the model shown in Fig. 10.14*a*. A pore is assumed to be present at the vertices of each tetrakaidecahedron; and since there are 24 vertices and each pore is shared by four polyhedral. It follows that the fraction porosity is given by

$$P_c = \frac{8\pi\rho_p^3}{8\sqrt{2a_p^3}} = \frac{\pi\rho_p^3}{\sqrt{2a_p^3}} \tag{10.23}$$

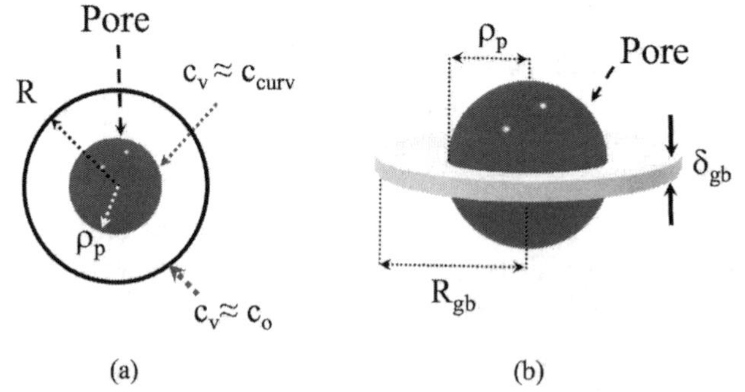

FIGURE 10.15 Geometric constructions used to model porosity elimination, assuming (*a*) bulk diffusion, where vacancies are eliminated at the grain surface, and (*b*) GB diffusion, where vacancies are restricted to diffusing along a GB of width δ_{gb}.

where the denominator represents the volume of the tetrakaidecahedron. Combining this result with Eq. (10.22), and noting that the grain size d scales with a_p at t = 0, yields

$$P_c - P_o = -(\text{const})\frac{D_{ambi}\gamma_{sv}\Omega_{MX}}{d^3kT}t \tag{10.24}$$

where P_o is the porosity at the beginning of the final stage of sintering. This model predicts that the porosity during this stage will decrease linearly with time and inversely as d^3. In other words, smaller grains should result in much faster porosity elimination, which was expected since it was assumed that the GBs act as sinks.

Grain boundary diffusion. Equation (10.24) has limited validity, however, because as discussed in greater detail below, the elimination of the last remnants of vacancies usually occurs only if they remain attached to, and are eliminated, at the GBs. The appropriate geometry is shown schematically in Fig. 10.15*b*, and by following a derivation similar to the one carried out in App. 10D, it can be shown (see Prob. 10.7) that

$$\rho_p^4 - \rho_{p,o}^4 = -\frac{8\delta_{gb}D_{gb}\gamma_{sv}\Omega_{MX}}{kT}\frac{1}{\log(R_{gb}/\rho_p)}t \tag{10.25}$$

where R_{gb} is defined in Fig. 10.15*b*. In other words, if GB diffusion is the operative mechanism, the average pore size should shrink with $t^{1/4}$ rather than the $t^{1/3}$ dependence expected if bulk diffusion were important [i.e., Eq. (10.22)]. Unfortunately, no simple analytic expression exists relating the porosity to ρ_p, and numerical methods have to be used instead.

EFFECT OF DIHEDRAL ANGLE ON PORE ELIMINATION

It should be clear by now that pore shape and volume fraction continually evolve during sintering, and understanding that evolution is critical to understanding how high theoretical densities can be achieved. An implicit, and fundamental, assumption made in the foregoing analysis is the existence of a driving force to shrink the pores at all times—an assumption that is not always valid. As discussed below, under some conditions, the pores can be thermodynamically stable.

To demonstrate the conditions for which this is the case, consider the four grains that intersect as shown in Fig. 10.16*a*. Referring to the figure, if the square pore is allowed to shrink by an amount equal to the shaded area, the excess energy eliminated is proportional to $2\gamma_{sv}\Gamma$ while the excess energy gained will be proportional to $\lambda\gamma_{gb}$, where Γ and λ are defined in the figure. It follows that the ratio of the energy gained to that lost is

$$\frac{\text{Energy gained}}{\text{Energy lost}} = \frac{2\Gamma\gamma_{sv}}{\lambda\gamma_{gb}} \tag{10.26}$$

Combining this result with Eq. (10.1) and the fact that $\cos(\varphi/2) = \Gamma/\lambda$ (see Fig. 10.16*a*), one can show that the right-hand side of Eq. (10.26) equals 1.0, when $\varphi = \phi$. In other words, when the grains around a pore meet such that $\gamma_{gb} = 2\gamma_{sv}\cos(\phi/2)$, the *driving force for GB migration and pore shrinkage goes to zero.*

This is an important conclusion since it implies that if pores are to be completely eliminated, their coordination number has to be *less* than a critical value n_c. It can be further shown (see Prob. 10.9) that n_c is

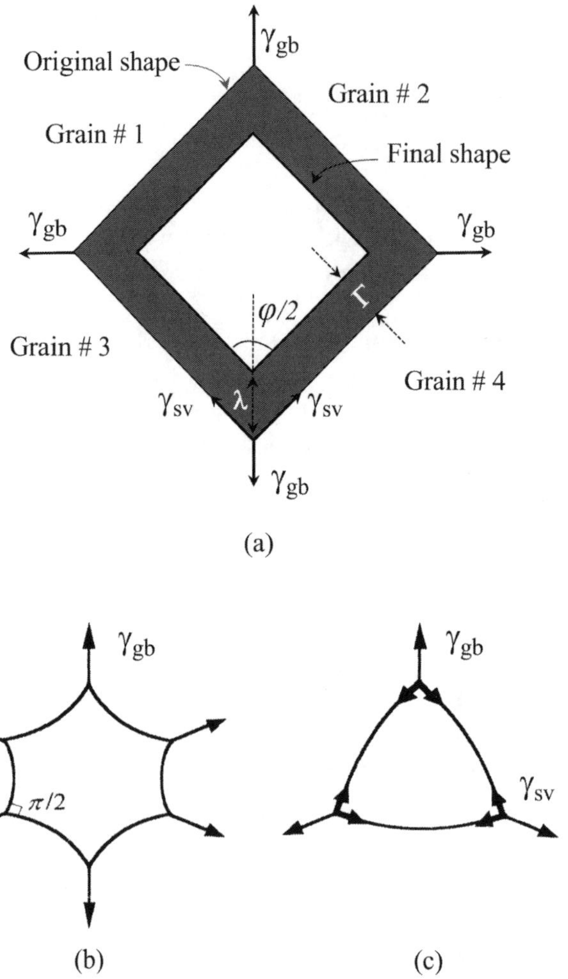

FIGURE 10.16 Effect of pore coordination on pore shrinkage for a system for which the equilibrium dihedral angle $\phi = \pi/2$. (*a*) Intersection of four grains around a pore. (*b*) Intersection of six grains; that is, $\varphi = 120°$. Note that to maintain the equilibrium dihedral angle, $\pi/2$, the surfaces of the grains surrounding the pore have to be convex. (*c*) Same system with $\varphi = 60°$. Here the pore surface has to be concave in order to maintain the equilibrium dihedral angle.

related to the dihedral angle by $n_c < 360/(180 - \phi)$. It is left as an exercise for the reader to determine which of the two pores shown in Figs. 10.16*b* and *c* is stable and which is not.

Based on these results, one may conclude that increasing the dihedral angle should, in principle, aid in the later stages of sintering. The situation is not so simple, however, since it can also be shown that the attachment of pores to GBs is stronger for lower ϕs. Now given that in order to eliminate a pore, it has to remain attached to the GB, this latter property would tend to suggest that low dihedral angles would aid in the prevention of boundary–pore breakaway and thus be beneficial. Finally, note that channel breakup at the end of the intermediate-stage sintering occurs at smaller volume fractions of pores as the dihedral angle decreases, which again is beneficial.

10.3.3 COARSENING AND GRAIN GROWTH KINETICS

Any collection of fine particles will coarsen with time, kinetics permitting, where coarsening implies an increase in the ensemble's average particle size with time. Comparing Fig. 10.2*b* and *c* shows the clear similarity between coarsening and grain growth. This section deals with the kinetics of the microstructural evolution during coarsening (Fig. 10.2*c*) and the grain growth kinetics associated with the final stages of sintering (Fig. 10.2*b*).

COARSENING

To model coarsening, consider a powder compact consisting of a distribution of particles, with an average particle radius r_{av}. Assuming all particles to be spheres, the *average* partial pressure over the ensemble is given by Eq. (10.6), or

$$P_{av} = P_{flat}\left(1 + \frac{2\Omega_{MX}\gamma_{sv}}{r_{av}kT}\right) \tag{10.27}$$

Similarly, the partial *pressure P_r over any particle of radius $r \neq r_{av}$ is also* given by Eq. (10.6), using the appropriate radius. Consequently, grains that are smaller than r_{av} will "evaporate" away, while those that are larger will grow with time (Fig. 10.17*a*).

If the interface kinetics are rate-limiting, it follows that the velocity u of the solid/gas interface is linearly dependent on the driving force, and

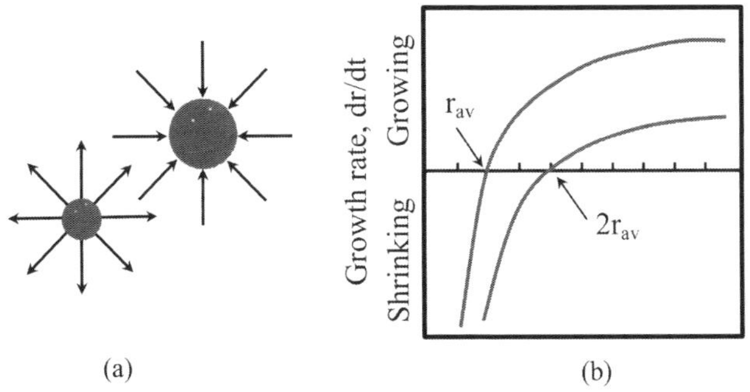

FIGURE 10.17 (*a*) Schematic of coarsening problem, showing how grains that are smaller than the average shrink at the expense of larger grains that grow. (*b*) Plot of Eq. (10.29). The lower curve has an average radius, r_{av}, that is twice that of the top curve.

$$u = \frac{dr}{dt} = K_r(P_{av} - P_r) \tag{10.28}$$

where K_r is a proportionality constant related to interface mobility.[141] Combining Eqs. (10.6), (10.27) and (10.28), one obtains

$$\frac{dr}{dt} = K_r(P_{av} - P_r) = \frac{2\Omega_{MX} P_{flat} K_r \gamma_{sv}}{kT}\left(\frac{1}{r_{av}} - \frac{1}{r}\right) \tag{10.29}$$

A plot of this equation is shown in Fig. 10.17*b* for two different average particle sizes. In addition to demonstrating that small particles shrink and larger ones grow, the figure also shows that smaller grains ($r < r_{av}$) disappear much faster than larger grains grow. Furthermore, as time progresses and r_{av} increases, the growth rate for all particles is reduced and eventually goes to zero.

The model can be taken further by making the simplifying assumption that the rate of increase of r_{av} is identical to that of particles that are twice r_{av}. In other words, by assuming

$$\frac{dr_{av}}{dt} = \frac{dr}{dt} \text{ at } r = 2r_{av}$$

Making use of this assumption, Eq. (10.29) can be integrated to yield the final result

$$\boxed{r_{av}^2 - r_{0,av}^2 = \frac{2\gamma_{sv}\Omega_{MX} P_{flat} K_r}{kT} t} \tag{10.30}$$

where $r_{0,av}$ is the average particle size at $t = 0$. A more rigorous treatment[142] gives a value of 64/81 instead of 2. Equation (10.30) predicts a parabolic increase in r_{av} with time. It also predicts coarsening kinetics are enhanced for solids with high vapor/surface interface energies and high vapor pressures, both predictions in fair agreement with experimental observations. For example, it is now well established that covalently bonded solids such as Si_3N_4, SiC, and Si coarsen, rather than densify, because they have relatively high vapor pressures at the temperatures needed to densify them.

GRAIN GROWTH

As noted above, during the final stages of sintering, in addition to pore elimination, a general coarsening of the microstructure by grain growth occurs. During this process, r_{av} increases with time as the smaller

[141] The astute reader will note that for a process that is evaporation-controlled, $K_r = \alpha/\sqrt{2\pi m_{MX} kT}$ [see Eq. (10.11)].
[142] C. Wagner, Z. *Electrochem.*, 65, 581 (1961).

grains are consumed by larger grains as shown in Fig. 10.18. Controlling and understanding the processes that lead to grain growth are important for two reasons. The first, discussed in greater detail in subsequent chapters, is related to the fact that grain size is a major factor determining many of the electrical, magnetic, optical, and especially mechanical properties of ceramics. The second is related to suppressing what is known as **abnormal grain growth**, which is the process whereby a small number of grains grow rapidly to sizes that are more than an order of magnitude larger than average in the population (Fig. 10.21*b*). In addition to the detrimental effect that such large grains have on the mechanical properties (see Chap. 11), the walls of these large grains can pull away from porosities, leaving them trapped within them, which in turn limits the possibility of obtaining theoretical densities in reasonable times.

Before one proceeds with the model, it is important to appreciate the origin of the driving force responsible for grain growth. Consider the schematic of a two dimensional microstructure composed of grains of

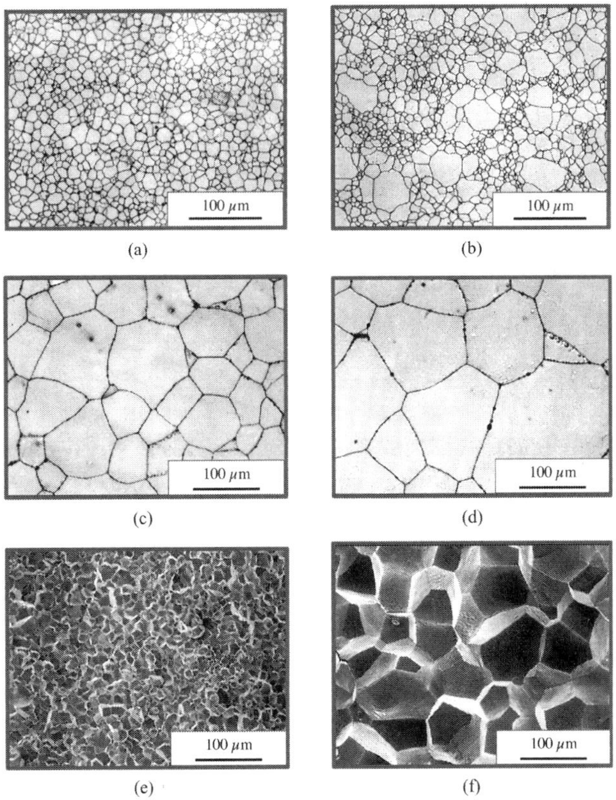

FIGURE 10.18 Time evolution of microstructure of CsI hot pressed at 103 MPa at 100°C for (*a*) 5 min, (*b*) 20 min, (*c*) 1 h and (*d*) 120 min. (*e*) Fractured surface of *a*, (*f*) fractured surface of *d*. (H.-E. Kim and A. Moorhead, *J. Amer. Cer. Soc.*, **73**, 496, 1990. Reprinted with permission.)

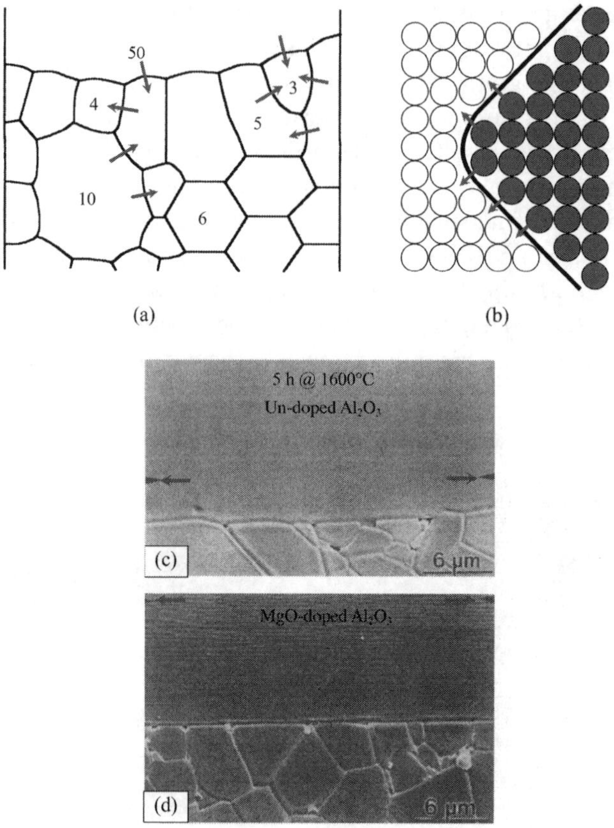

FIGURE 10.19 (*a*) Grain shape equilibrium and direction of motion of GBs in a two-dimensional sheet. Note grains with six sides are stable, while those <6 sides shrink and those with >6 sides grow. (*b*) Atomic view of a curved boundary. Atoms will jump from right to left as depicted by red arrows, and the GB will *per force* move in the opposite direction. (*c*) Interface between basal single-crystal and polycrystalline Al_2O_3 after 5 h at 1600°C. (*d*) Same as c, but starting with MgO-doped Al_2O_3. Arrows indicate position of original interface. Clearly, doping enhanced boundary mobility. (Adapted from J. Rodel and A. M. Glaeser, *J. Am. Ceram. Soc.* 73, 3292, 1990.)

varying curvatures (Fig. 10.19*a*). Since in this structure the dihedral equilibrium angle has to be 120°, it follows that grains with more than six sides will tend to grow, while those with less than six sides will tend to shrink.[143] This occurs by migration of GBs in the direction of the arrows (Fig. 10.19*a*).

To appreciate the origin of the driving force, consider the atomic-scale schematic of such a boundary (Fig. 10.19*b*). At this level, it should be obvious why an atom on the convex side of the boundary would rather be on the concave side—it would, on average, be more tightly bound, i.e., have lower potential energy. Consequently, atoms will jump from right to left as depicted by arrows, which means that the GB will move from left to right. Looking at the problem at that level, one can see why straight (i.e., no curvature)

[143] The reason a grain with less than six sides shrinks is the same as why the pore shown in Fig. 10.16*c* does.

GBs would be stable and would not move. The latter is only true when there is no anisotropy in the surface energies (see Chap. 4). If such anisotropy exists, as it must, then straight boundaries can, and will, move to increase the areas of the surfaces with the lower surface energies at the expense of the higher energy ones.

Figure 10.19*c* and *d* are micrographs of a basal plane single crystal Al_2O_3 that was placed on top of a polycrystalline Al_2O_3 sample and annealed at for 5 h at 1600°C. The heating induced the single crystal boundary downwards into the polycrystal. The original interface is depicted by red arrows. This is an elegant example of GB motion. Figure 10.19*d* is the same as c, but in this case the Al_2O_3 was doped with MgO. Note here that it appears that the straight boundary did indeed move, which in turn implies that the surface energy of the basal plane is lower than others, as is well established.

More quantitatively, the driving forces, per MX molecule, $\Delta\mu_{gb}$ across the grain boundary is given by

$$\Delta\mu_{gb} = \kappa\gamma_{gb}\Omega_{MX} \tag{10.31}$$

where γ_{gb} is the GB energy and κ its radius of curvature.

To model the process, one needs to obtain a relationship between the GB velocity u_{gb} and the driving force acting on the boundary. And since the situation is almost identical to that encountered during the growth of a solid/liquid interface (Fig. 9.3), Eq. (9.14) is directly applicable and;

$$u_{gb} = \lambda\upsilon_{net} = \lambda\nu_o \exp\left(-\frac{\Delta G_m^*}{kT}\right)\left\{1-\exp\left(-\frac{\Delta\mu_{gb}}{kT}\right)\right\} \tag{10.32}$$

where ΔG_υ is replaced by $\Delta\mu_{gb}$. Expanding the term within braces in Eq. (10.32), for the usual case when $\Delta\mu_{gb} << kT$, one obtains

$$u_{gb} = \lambda\upsilon_{net} = \frac{\lambda\nu_o}{kT}\exp\left(-\frac{\Delta G_m^*}{kT}\right)\Delta\mu_{gb} \tag{10.33}$$

This expression is usually abbreviated to

$$u_{gb} = M\Delta\mu_{gb} \tag{10.34}$$

where

$$M = \frac{\lambda\nu_o}{kT}\exp\left(-\frac{\Delta G_m^*}{kT}\right) \tag{10.35}$$

Comparing Eqs. (10.33) and (10.28) reveals the similarities between this problem and that of coarsening worked out in the previous section. So by modifying Eq. (10.30) to the problem at hand, the final result for grain growth is

$$\boxed{d_{av}^2 - d_{av,0}^2 = \frac{4M\gamma_{gb}\Omega_{MX}}{\beta}t} \tag{10.36}$$

where $d_{av,0}$ is the average grain size at $t = 0$, and β is a geometric factor that depends on the curvature of the boundary. For example, for a solid that is made up entirely of straight, noncurved grain boundaries, no grain growth would occur and β would be infinite. Recall here again that this conclusion is only valid if the GB energies are isotropic. If not, then the GBs with the lowest energies will grow at the expense of others. The process just described is sometimes referred to as **Ostwald ripening** and is characterized by a parabolic dependence of grain size on time.

EFFECT OF MICROSTRUCTURE AND GRAIN BOUNDARY CHEMISTRY ON BOUNDARY MOBILITY

In deriving Eq. (10.36), the implicit assumption that the GBs were pore-, inclusion-, and essentially solute-free—a very rare occurrence, indeed—was made, and as such Eq. (10.36) predicts the so-called intrinsic grain growth kinetics. Needless to say, the presence of "second phases" or solutes at the GBs can have a dramatic effect on their mobility, and from a practical point of view it is usually the mobility of these phases that is rate-limiting. To illustrate the problem's complexity, consider just a few possible rate-limiting processes:

1. Intrinsic GB mobility discussed above.
2. Extrinsic or solute drag. If the diffusion of the solute segregated at the GBs is slower than the intrinsic GB mobility, it becomes rate-limiting. In other words, if the moving GB must drag the solute along, it will slow it down.
3. The presence of inclusions, viz. second phases, at the GBs. It can be shown that larger inclusions have lower mobilities than smaller ones, and that the higher the volume fraction of a given inclusion at the GBs, the larger the resistance to their migration.
4. Material transfer across a continuous boundary phase. For instance, in Si_3N_4 boundary movement can occur only if both silicon and oxygen diffuse through the thin, glassy film that usually exists between grains.
5. In some cases, the re-dissolution of the boundary-anchoring second phase inclusions into the matrix can be rate limiting.

In addition to these, the following interactions, between pores and GBs can also occur:

1. What is true of second phases is also true of pores. Pores *cannot* enhance boundary mobility; they only leave it unaffected or reduce it. During the final stages of sintering as the pores shrink, the boundary mobility will increase (see below).
2. The pores do not always shrink—they can also coarsen as they move along or intersect a moving GB.
3. Pores can grow by the Ostwald ripening mechanism.
4. Pores can grow by reactive gas evolution and sample bloating.

To discuss even a fraction of these possibilities in any detail is clearly not within the scope of this book. What is attempted instead is to consider, in some detail, one of the more important GB interactions namely that between GBs and pores.

As the grains get larger and the pores fewer, the grain mobility increases accordingly. In some cases, at a combination of grain size and density, the GB mobility becomes large enough that the pores can no longer keep up with them; the boundaries simply move too fast for the pores to follow and consequently unpin themselves. This region is depicted on the grain size versus density trajectory in the upper right-hand corner of Fig. 10.20*a*.

If theoretical density is to be achieved, it is important that the GB versus density trajectory not intersect this separation region. The importance of having the pores near GBs is illustrated in Fig. 10.20*b*; in this the

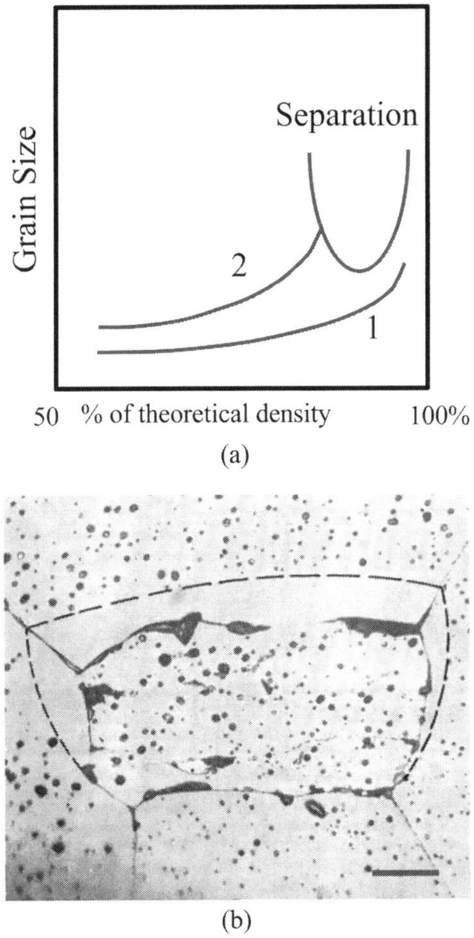

FIGURE 10.20 (*a*) Grain size versus densification trajectory including region where separation of boundaries and pores occurs. In that region, the GBs break away from the pores, entrapping them within the grains. To achieve full density, path 1 has to be followed. Path 2 will result in entrapped pores. (*b*) Micrograph of sweeping out of pores by migration of GBs. Original position of the boundary is depicted by the dotted line (X250). (J. E. Burke, *J. Amer. Cer. Soc.*, 40, 80, 1957. Reprinted with permission.)

migration of the GB downwards has swept and eliminated all the pores in its wake. (In this micrograph the original GB location is given by dashed line and the final position by the solid lines.) Pores that are trapped within grains tend to remain there because the diffusion distances between sources and sinks, viz. the GBs in this case, become too large.

There are essentially two strategies that can be employed to prevent pore breakaway, namely, reduce GB mobility and/or enhance pore mobility. An example of how slowing GB mobility enhances the final density is shown in Fig. 10.6*a*, where the grain size versus density trajectories for two aluminas, one pure and the other doped with 250 ppm MgO, are compared. It is obvious from the results that the doped alumina achieves higher density—the reason is believed to be due to impurity drag on the boundary by the MgO.

ABNORMAL GRAIN GROWTH

In some systems, a small number of grains in the population grow rapidly to very large sizes relative to the average size of the population (see Fig. 10.21*b*). This phenomenon is referred to as *abnormal grain growth* (AGG). AGG is to be avoided for the same reason as pore-grain boundary unpinning.

Although it is not entirely clear as to what results in AGG, there is mounting evidence that it is in most cases, likely associated with the formation of liquid phases, or very thin liquid films at the GBs. These can result from dopants intentionally added or simply from impurities in the starting powders. The effect of having small amounts of liquid during solid-state sintering and its effect on the sintering and grain growth kinetics are discussed in the next section. There is little doubt, however, that small amounts of liquid can result in substantial coarsening of the microstructure, as shown in Fig. 10.24.

10.3.4 FACTORS AFFECTING SOLID-STATE SINTERING

Typically, a solid-state sintered ceramic is an opaque material containing some residual porosity and grains that are much larger than the starting particle sizes. On the basis of the discussion and models just presented, it is useful to summarize the more important factors that control solid state sintering. Implicit in the following arguments is that theoretical density is desired.

1. *Temperature.* Since diffusion is responsible for sintering, clearly increasing temperature will greatly enhance the sintering kinetics, because D is thermally activated. As noted earlier, the activation energies for bulk diffusion are usually higher than those for surface and GB diffusion. Therefore, increasing the temperature usually enhances the bulk diffusion mechanisms, which in turn can lead to densification.
2. *Green density.* Usually a correlation exists between the green (prior to sintering) and final density, since with higher green densities, less pore volume has to be eliminated.
3. *Uniformity of green microstructure.* As important as the green density is the uniformity of the green microstructure and the lack of agglomerates (see Fig. 10.10). The importance of eliminating agglomerates is discussed in greater detail below.
4. *Atmosphere.* The effect of atmosphere can be critical to the densification of a powder compact. In some cases, the atmosphere can enhance the diffusivity of a rate-controlling species, e.g., by influencing the underlying defect structure. In other cases, the presence of a certain gas can promote coarsening

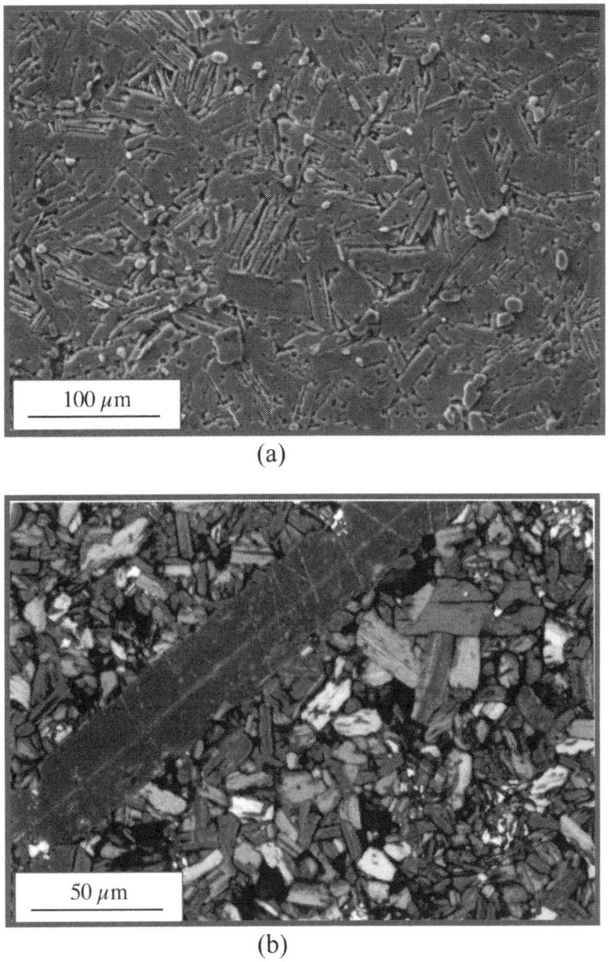

FIGURE 10.21 Microstructures showing (*a*) normal and (*b*) abnormal grain growth in Ti_3SiC_2.

by enhancing the vapor pressure and totally suppressing densification. An excellent example of the effect of atmosphere was shown in Fig. 10.7: Fe_2O_3 will readily densify in air but not in HCl-containing atmospheres. Another important consideration is the solubility of the gas in the solid. Because the gas pressure within the pores increases as they shrink, it is important to choose a sintering atmosphere gas that readily dissolves in the solid.

5. *Impurities*. The role of impurities cannot be overemphasized. The key to many successful commercial products has been the identification of the right pinch of magic dust. The role of impurities has been extensively studied, and to date their effect can be summarized as follows:
   a. *Sintering aids*. They are purposefully added to form a liquid phase (discussed in the next section). It is also important to note that the role of impurities is not always appreciated. The presence of

impurities can form low-temperature eutectics and result in enhanced sintering kinetics, even in very small concentrations.

b. Suppress coarsening by reducing the evaporation rate and lowering surface diffusion. A classic example is boron additions to SiC, without which it is quite difficult to densify.
c. Suppress grain growth and lower GB mobility (Fig. 10.6).
d. Enhance diffusion rate. Once the rate-limiting ion during sintering is identified, the addition of the proper dopant that will go into solution and create vacancies on that sublattice should, in principle, enhance the densification kinetics.

6. *Size distribution.* Narrow grain size distributions decrease the propensity for AGG.
7. *Particle size.* Since the driving force for densification is the reduction in surface area, the larger the initial surface area, the greater the driving force. Thus, it would seem that one should use the finest initial particle size possible, and while in principle this is good advice, in practice very fine particles pose serious problems. As the surface/volume ratio of the particles increases, electrostatic and other surface forces become dominant, which leads to agglomeration. Upon heating, the agglomerates have a tendency to sinter together into larger particles, which not only dissipates the driving force for densification but also create large pores between the partially sintered agglomerates that are subsequently difficult to eliminate. The dramatic effect of ridding a powder of agglomerates on the densification kinetics is well illustrated in Fig. 10.10.

The solution lies in working with nature instead of against it. In other words, make use of surface forces to colloidally deflocculate the powders and keep them from agglomerating.[144] However, once dispersed, the powders should *not* be dried, but piped directly into a mold, or a device, that gives them the desired shape. The reason is simple: In many cases, drying reintroduces agglomerates and defeats the purpose of colloidal processing.

To avoid excessive shrinkage during fluid removal requires a pourable slurry with a high-volume fraction of particles. Once the slurry has been molded, its rheological properties must be dramatically altered to allow shape retention during unmolding. What is required in this stage is to change the viscous slurry to an elastic body without fluid-phase removal. The basic idea is to avoid, at all costs, passing through a stage where a liquid/vapor interface exists, the presence of which can result in strong capillary forces that can cause particle rearrangement and agglomeration. And whereas this is desirable during liquid-phase sintering (see next section), it is undesirable when a slurry is dried because it is uncontrollable and can result in shrinkage stresses that, in turn, can result in the formation of either agglomerates or large cracks between areas that shrink at different rates. The crazing of mud upon drying is a good example of this phenomenon.

Another possible source of flaws can be introduced during the cold pressing of agglomerated powders as a result of density differences between the agglomerates and the matrix. When the pressure is removed, the elastic dilation of the agglomerates and the matrix may be sufficiently different to cause cracks to form. Said otherwise the springback of the agglomerates will be different from the matrix as a result of their density differences.

144 See, e.g., F. Lange, *J. Amer. Cer. Soc.*, 72, 3 (1989).

10.4 LIQUID-PHASE SINTERING

The term *liquid-phase sintering* is used to describe the sintering process when a proportion of the material being sintered is in the liquid state (Fig. 10.1*a*). Liquid-phase sintering of ceramics is of major commercial importance since a majority of ceramic products are fabricated via this route and include ferrite magnets, covalent ceramics such as silicon nitride, ferroelectric capacitors, and abrasives, among others. Even in products that are believed to be solid-state sintered, it has been demonstrated that in many cases the presence of small amounts of GB liquid phases probably play a significant role.

Liquid-phase sintering offers two significant advantages over solid-state sintering. First, it is much more rapid; second, it results in more uniform densification. As discussed below, the presence of a liquid reduces the friction between particles and introduces capillary forces that result in the dissolution of sharp edges and the rapid rearrangement of the solid particles.

During liquid-phase sintering, the compositions of the starting solids are such as to result in the formation of a liquid phase upon heating. The liquid formed has to have an appreciable solubility for the solid phase, as well as, wet the solid. In the next sections, the reasons these two requirements have to be met and the origin of the forces at play during liquid-phase sintering are elucidated.

10.4.1 SURFACE ENERGY CONSIDERATIONS

As noted a few times already, the driving force during sintering is the overall reduction in surface energy of the system. In solid-state sintering, the lower-surface-energy GBs replace the higher energy solid/vapor surfaces. The presence of a liquid phase introduces a few more surface energies that have to be considered, namely, the liquid/vapor, γ_{lv}, and the liquid/solid, γ_{ls}, interfacial energies.

When a liquid is placed on a solid surface, it will either spread and wet that surface (Fig. 10.22*a*) or it will bead up (Fig. 10.22*b*). The degree of wetting and whether a system is wetting or nonwetting are quantified by the equilibrium contact angle θ that forms between liquid and solid and is defined in Fig. 10.22*a* and *b*. A simple balance of forces indicates that at equilibrium

$$\gamma_{sv} = \gamma_{ls} + \gamma_{lv} \cos\theta \tag{10.37}$$

from which it is clear that high values of γ_{sv} and low values of γ_{ls} and/or γ_{lv} promote wetting. By using an argument not unlike the one used to derive Eq. (10.26), it can be shown that a necessary condition for liquid-phase sintering to occur is that the contact angle lie between 0 and $\pi/2$, that is, the system must be wetting. For nonwetting systems, the liquid will simply bead up in the pores, and sintering can only occur by one of the solid-state mechanisms discussed previously.

The complete penetration of GBs with a liquid is also important for microstructural development (e.g., it could lead to the breakup of agglomerates). It can be shown (Prob. 10.12) that a necessary condition for the penetration and separation of the grains with a continuous liquid film is $\gamma_{gb} > 2\gamma_{ls}$. This implies that the equilibrium dihedral angle Ψ, shown in Fig. 10.3*b*, and defined as

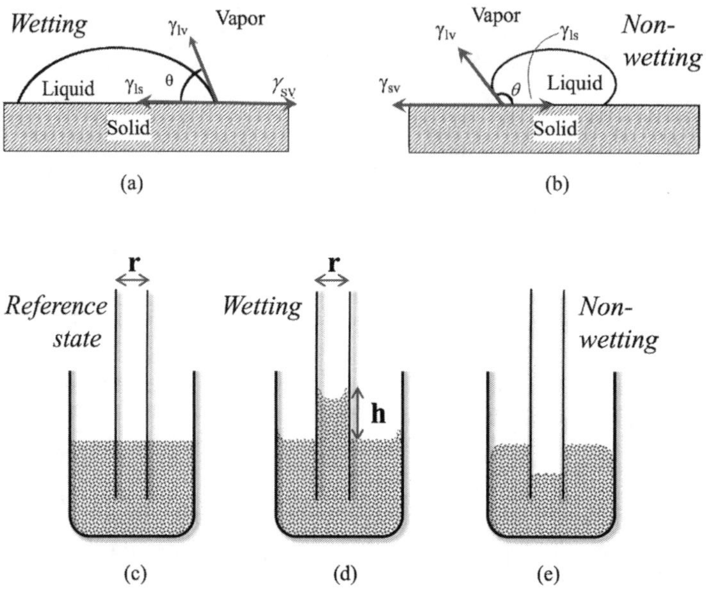

FIGURE 10.22 (*a*) Wetting system showing forces acting on the a liquid drop. (*b*) No-wetting system with $\theta > 90°$. (*c*) Initial system before equilibrium is established; (*d*) after equilibration for $\gamma_{sv} > \gamma_{sl}$, and (*e*) after equilibration for $\gamma_{sv} < \gamma_{sl}$.

$$\gamma_{gb} = 2\gamma_{sl} \cos\frac{\Psi}{2} \tag{10.38}$$

must be zero. In this context, it follows that high γ_{gb} and low γ_{ls} values are desirable.

10.4.2 CAPILLARY FORCES

When solids and liquids coexist, capillary forces are generated. These forces, as discussed below, can give rise to strong attractive forces between neighboring particles which, when combined with a liquid's lubricating potential , can lead to quite rapid and significant particle rearrangement and densification. Before discussing the effect of capillary forces on sintering and densification it is important to understand the atomic origins of capillarity.

ATOMIC ORIGINS OF CAPILLARITY

Capillarity is typically discussed in terms of γ_{lv}. However, as discussed below, the atomic origin of capillarity as it relates to liquid-phase sintering has less do with γ_{lv} and more to do with the difference between γ_{sv} and γ_{sl}. Consider the system shown in Fig. 10.22*c*. Before the liquid is allowed into the long thin tube, with radius r, the energy of the system is given as:

$$E_{Initial} = \pi r^2 \gamma_{lv} + 2\pi r h \gamma_{sv} \tag{10.39}$$

where h is the ultimate height of the liquid column in the tube. When the liquid is in the tube, the system's energy is given by

$$E_{Final} = \frac{4\pi r^2}{2}\gamma_{lv} + 2\pi r h \gamma_{sl} + \rho_{den} g h \pi r^2 h \quad (10.40)$$

where ρ_{den} is the density of the liquid and g is the gravitational constant. (Note ρ_{den}gh is an energy per unit volume, which when multiplied by a volume results in energy.) The energy difference between the initial and final state is thus:

$$\Delta E = \pi r^2 \gamma_{lv} + 2\pi r h\left[\gamma_{ls} - \gamma_{sv}\right] + \rho_{den} g \pi r^2 h^2 \quad (10.41)$$

At equilibrium, the energy is at a minimum and thus

$$\frac{dE}{dh} = 2\pi r\left[\gamma_{sl} - \gamma_{sv}\right] + 2\rho g \pi r^2 h = 0 \quad (10.42)$$

which upon rearrangement yields:

$$\boxed{h = \frac{\gamma_{sv} - \gamma_{sv}}{\rho_{den} g r}} \quad (10.43)$$

This is an important result with far-reaching implications, among them:

1) The driving force for liquid penetration into a thin tube is the difference between γ_{sv} and γ_{sl}. If $\gamma_{sv} > \gamma_{sl}$, then h is positive, and the liquid will rise into the column to a height h as shown in Fig. 10.22*d*. However, if $\gamma_{sv} < \gamma_{sl}$ then the liquid column will be depressed by a height h as shown in Fig. 10.22*e*. The former occurs when a thin glass tube is placed in water; the latter when it is placed in mercury. Note this result is valid irrespective of the value of γ_{lv}.
2) The height, h is inversely proportional to r. This is because thinner tubes have larger surface to volume ratios than thicker ones.
3) Interestingly, γ_{lv} plays no role whatsoever, and yet in almost all explanations of capillarity, γ_{lv} is invoked. This comes about because $\cos\theta$ is proportional to $(\gamma_{sv} - \gamma_{sl})$. When γ_{lv} is invoked, the argument is that the pressure above a concave surface is less than over a flat surface and that pressure difference, ΔP "pulls" the liquid up, which is, by the way, what happens every time water is sucked through a straw. By combining Eqs. (10.37) and (10.43) one can show that:

$$\Delta P = \rho_{den} g h = \frac{\gamma_{lv}\cos\theta}{r} \quad (10.44)$$

This comes about because like ρ_{den}gh, ΔP is also a measure of energy per unit volume. And while there is nothing wrong with this approach, it is less transparent than the one based on the relative values of γ_{sv} and γ_{ls}, especially when it comes to liquid-phase sintering.

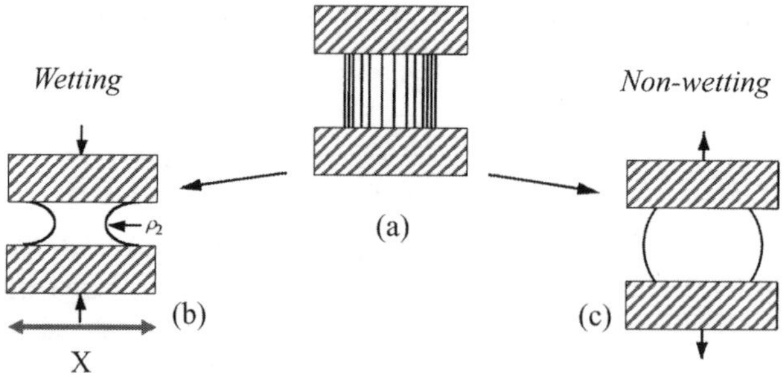

FIGURE 10.23 Thought experiment illustrating origin of main force acting on a liquid drop held between two plates for, (*a*) $\theta = 90°$, (*b*) wetting, $\theta < 90°$, and (*c*) nonwetting or $\theta > 90°$.

To better appreciate what happens during liquid-phase sintering, consider the following thought experiment: A solid cylinder of radius X is placed in between two plates (Fig. 10.23*a*), and the system is heated so as to melt the solid. If $\gamma_{sl} > \gamma_{sv}$, $\theta > 90°$ and the liquid will not wet the solid but will push the plates apart are shown in Fig. 10.23*c*, and no densification can occur. If $\gamma_{sv} > \gamma_{sl}$, the liquid will wet the solid (Fig. 10.23*b*). For a fixed volume of liquid, the only way the liquid can spread is if the two plates are brought closer together. If the plates are prevented from doing so, a compressive stress normal to the plates will develop, the magnitude of which is simply

$$F_{att} = \pi X^2 \Delta P \qquad (10.45)$$

where ΔP is the pressure across the curved surface given by Eq. (10.44)

$$\Delta P = \frac{\gamma_{lv}\cos\theta}{\rho_2} = \frac{\gamma_{sv} - \gamma_{sl}}{\rho_2} \qquad (10.46)$$

and ρ_2 is defined in Fig. 10.23*b*.[145] Note the maximum ΔP possible is when θ is 0. In all cases, the driving force is proportional to $(\gamma_{sv} - \gamma_{sl})$.

Another way to think of the problem is to assume that the negative pressure in the pores is pulling the liquid into them, which in turn pulls the particles together.[146] Typically, the pores during liquid-phase sintering are on the order of 0.1 to 1 mm. For a liquid with a γ_{lv} on the order of 1 J/m², this translates to compressive stresses [Eq. (10.46)] between particles on the order of 1 to 10 MP*a*. These stresses, together with the greatly enhanced diffusion rates in the liquid (see below), are key to the process.

10.4.3 LIQUID-PHASE SINTERING MECHANISMS

Upon melting, a wetting liquid will penetrate between grains, as shown schematically in Fig. 10.12*b*, and exert an attractive force, pulling them together. The combination of these forces and the lubricating effect

[145] In general, ΔP is related to the two principal radii of curvature of the saddle, as shown in Fig. 10.26*b* in App. 10B. In most cases, however, $_1 >> _2$ and Eq. (10.41) is a good approximation.

[146] For example, it is this force that results in the strong cohesion of two glass slides when a thin layer of water is inserted between them.

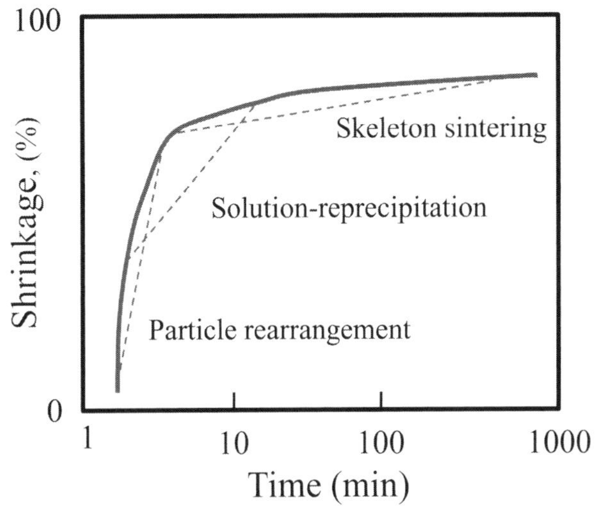

FIGURE 10.24 Three liquid phase sintering stages occurring when a powder is heated in the presence of a wetting liquid. Note log scale on x-axis.

of the liquid, as it penetrates between grains, leads to the following three mechanisms that operate in succession. A schematic of a typical shrinkage curve for the three sintering regimes is shown in Fig. 10.24 (note *log* scale on the x-axis). These regimes are:

PARTICLE REARRANGEMENT

Here densification results from particle rearrangement under the influence of capillary forces and the filling of pores by the liquid phase. This process is quite rapid and if, during the early stages of sintering, the liquid flows and completely fills the finer pores between particles, 100% densification can result almost instantaneously.

SOLUTION-REPRECIPITATION

At points where the particles touch, the capillary forces generated will increase the chemical potential of the atoms at the point of contact relative to areas that are not in contact. The chemical potential difference between the atoms at the two sites is given by (see Chap. 12)

$$\mu - \mu_0 = kT\ln\frac{a}{a_0} = \Delta P\Omega_{MX} \tag{10.47}$$

where ΔP is given by Eq. (10.46), a is the activity of the solid in the liquid at the point of contact, and a_0 is the same activity under no stress, i.e., at, or near, the pore surfaces. This chemical potential gradient induces the dissolution of atoms at the contact points and their reprecipitation away from the area between the two particles, which in turn leads to shrinkage and densification. As importantly, the densification kinetics will be much faster than in the case of solid-state sintering because diffusion is now occurring in the liquid where the diffusivities are orders of magnitude higher than those in the solid state. Obviously, for this process to occur, there must be some solubility of the solid in the liquid—but not vice versa—and wetting. It is worth noting here that in addition to densification, coarsening or Ostwald ripening will occur simultaneously by the dissolution of the finer particles in the liquid phase and their reprecipitation on the larger particles. The dramatic effect of having a small amount of liquid on the final microstructure is shown in Fig. 10.25.

SOLID-STATE SINTERING

Once a rigid skeleton is formed, liquid-phase sintering stops and solid-state sintering takes over, and the overall shrinkage or densification rates are significantly reduced.

Relative to the other two processes, particle rearrangement is the fastest, occurring in the time scale of minutes. The other two processes take longer since they depend on diffusion through the liquid or solid (Fig. 10.24)

Based on the discussion so far and numerous studies, it is now appreciated that for rapid densification to occur during liquid-phase sintering, the following conditions have to be met:

- There must be an appreciable solubility of the solid in the liquid in order for material transfer away from the contact areas to occur.
- Since the capillary pressure is proportional to $1/\rho_2$, which in turn scales with the particle diameter, the finer the solid phase, the higher the capillary pressures that can develop and the faster the densification rate.
- Wetting of the solid phase by the liquid is crucial.
- A sufficient amount of liquid to wet the solid phase must also be present.

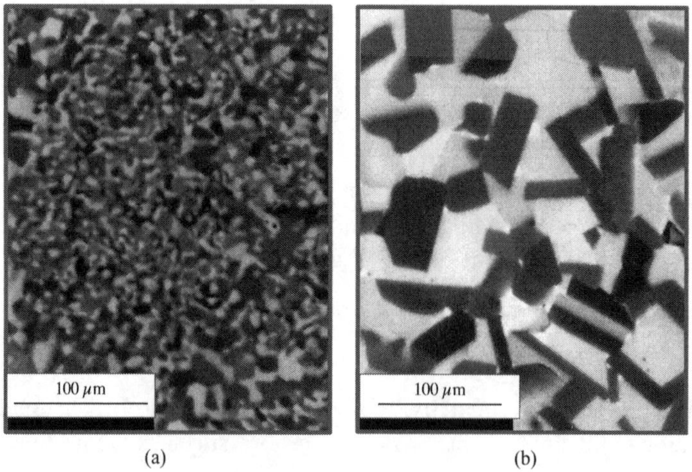

FIGURE 10.25 Effect of small amounts of liquid on the coarsening of a TiB_2/TiC microstructure at a temperature at which (*a*) there is no liquid; (*b*) a small amount of liquid is present. (T. Lien, MSc. thesis, Drexel University, 1992.)

Clearly liquid-phase sintering of ceramics is more forgiving in terms of powder packing, more rapid, and hence more economical, than the solid-state version. Indeed, as noted above, most commercial ceramics are liquid-phase sintered. Some of the more advanced materials such as Si_3N_4 and B_4C cannot be easily densified without the presence of a liquid phase. If the properties required of the part are not adversely affected by the presence of a liquid, then it is the preferred route. However, for some applications, the presence of a glassy film at the GBs can have a detrimental effect on properties. For example, creep resistance can be significantly compromised by the presence of a glassy phase that softens upon heating. Another application in which the presence of an ionically conducting glassy phase cannot be tolerated is in the area of ceramic insulators. Consequently, in the processing of today's electronic ceramics, liquid phases and residual porosities are avoided as much as possible.

10.5 HOT PRESSING AND HOT ISOSTATIC PRESSING

By now it should be clear that the driving force for densification is the chemical potential gradient between the atoms in the neck region and those adjacent to the pores. This can be accomplished by the application of a compressive pressure to the compact during sintering (Fig. 10.12*c*); in other words, the simultaneous application of heat and pressure. If the applied pressure is uniaxial, the process is termed hot pressing, HP, whereas if it is hydrostatic, the process is termed hot isostatic pressing or HIP. Figure 10.26*a* demonstrates how pressure accelerates densification of Al, alumina, magnesia and zirconia powders.

The effect of applied stress on chemical potential and vacancy concentrations is discussed in detail in Chap. 12. Here we simply note that the concentration of vacancies c_b in an area subjected to a stress σ_b is related to c_0 (i.e., concentration in absence of a stress) by an equation similar to Eq. (10.9), that is,

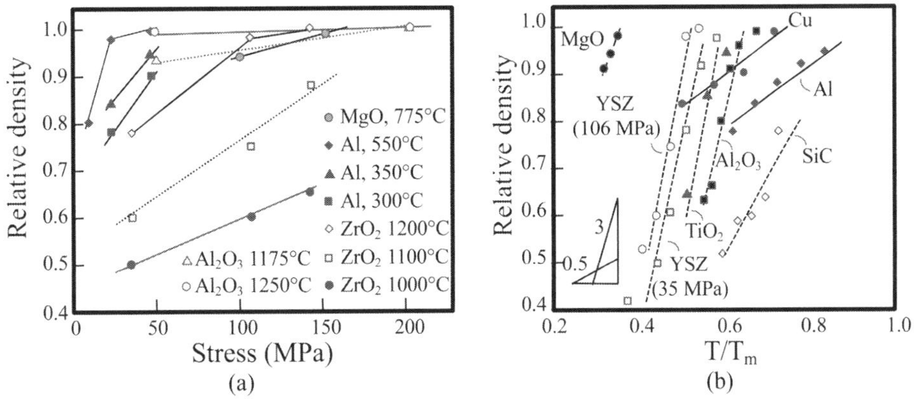

FIGURE 10.26 (*a*) Relative density vs. applied pressure for various materials as a function of temperature. (*b*) Relative density vs. homologous temperature for several metals and ceramic powders. Ceramic materials have slopes of ≈3, while metals have slopes of ≈0.5. Ionic solids fall on the left, while covalent solids, such as SiC, fall on the right. Both sets of results were obtained using either the FAST or PECS process. (Adapted from J. E. Garay, *Annu. Rev. Mater. Res.* 40, 445, 2010.)

$$c_b = c_0\left(1+\frac{\Omega_{fu}\sigma_b}{kT}\right) \tag{10.48}$$

where σ_b is the effective stress at the boundary due to the applied force (see Worked Example 10.3). By convention, if the applied stress is compressive, σ_b is negative[147] and thus according to Eq. (10.48), c_b (i.e., between the particles) is less than that at the edges, which results in a net vacancy flux from the neck into the boundary areas which, in turn, leads to densification.

The major advantage of HP and HIP is the fact that the densification occurs quite readily and rapidly, minimizing the time for grain growth, resulting in finer and more uniform microstructures. The major disadvantages, however, are the costs associated with tooling and dies and the fact that the process does not lend itself to continuous production, since the pressing is usually carried out in a vacuum or an inert atmosphere.

Worked Example 10.3

Calculate the effect of applying a load corresponding to a stress of 50 MPa on the vacancy concentration in a powder compact when $x/r=0.2$, during the initial sintering stage at 1200°C. Information you may find useful: Surface energy is 1.2 J/m^2; molar volume = 10 cm^3; average particle radius is 2 μm.

ANSWER

If $x/r=0.2$ and $r=2\times10^{-6}$ m, then $x=4\times10^{-7}$ m. From Eq. (10.13), $\rho\approx x^2/(2r)=4\times10^{-8}$ m. In the neck region $\kappa\approx-1/\rho$ and the vacancy concentration in the neck area due to curvature is given by Eq. (10.9)

$$c_{neck}^{curv} = c_0\left(1+\frac{V_m\gamma_{sv}}{\rho RT}\right) = c_0\left(1+\frac{10\times10^{-6}\times1.2}{4\times10^{-8}\times8.314\times1473}\right) = 1.0245c_0$$

Assuming the particle arrangement shown in Fig. 10.12c and a perfect cubic array, one can use a simple balance-of-forces argument, namely,

$$F_{app} = \sigma_a(2r)^2 = F_{boundary} = \pi\sigma_b x^2$$

where σ_b is the stress on the neck or boundary. Rearranging this equation, it follows that

$$\sigma_b \approx -(4\sigma_a/\pi)(r/x)^2 \approx -1.6\,\text{GPa}$$

Consequently, the vacancy concentration due to σ_b between grains is given by Eq. (10.48):

$$c_{bound}^{stress} = \left(1+\frac{V_m\sigma_b}{RT}\right)c_0 = \left(1-\frac{10\times10^{-6}\times1.6\times10^9}{8.314\times1473}\right)c_o = -0.3c_o$$

147 Here and throughout the book, the chosen convention is that applied tensile forces are positive and compressive stresses are negative.

The total vacancy concentration *gradient* acting over a distance of $\approx x/2$, is thus roughly

$$\Delta c/(x/2) \approx c_0\left[1.0245-(-0.3)\right]/(x/2) \approx 2.65c_0/x$$

which is greater than either alone. The effect of stress, however, is more significant. From these calculations it should be obvious why the application of moderate pressures can result in significant enhancements in the densification rates during all sintering stages.

CASE STUDY 10.1: ELECTRIC FIELD ASSISTED SINTERING

It has long been known that passing a current through metal powders while they are being hot pressed aided in their densification. Since initially it was thought that a spark and a plasma were generated, the process was referred to as spark plasma sintering or SPS. Over the past couple of decades, the same has been shown to be the case with ceramic powders. However, since SPS was coined, it has, more or less, been established that neither a spark nor a plasma is generated and yet many in the field still refer to the process as SPS. That is clearly incorrect and should be abandoned, which begs the question: Replace it with what? The leading contenders are: PECS for pulsed electric current sintering, FAST for field-activated sintering technique and CAPHAD current activated, pressure assisted densification. Which brings up a philosophical problem: do you describe a process by what is done, e.g., passing a large pulsed current through a graphite die or try and describe the underlying physics. If it is the latter, then the only acronym that is anywhere close to being accurate for insulating powders is FAST for the simple reason that in a typical experiment a ceramic powder is loaded in a graphite die and a large DC current is pulsed through the system. Given that the conductivity of a typical ceramic oxide is orders of magnitude lower than your typical graphite die it is not difficult to ascertain that most of the current actually passes through the graphite die. More specifically, it can be shown that unless the conductivity of the ceramic powder is >100 S/m, very little, if any, current passes through it. The acronym FAST has the added advantage that it reinforces the facts that (i) in many cases densification rates are faster than if the dies are conventionally furnace heated, and (ii) at 100 to 600°C/min, the typical heating rates are also quite fast. That said, if the powders that are to be sintered, or reacted, are conductive then PECS is the best acronym. It follows that in the remaining discussion FAST will be used for insulating powders and PECS for conductive ones.

At the outset, it is important to concede that how, or why, FAST works remains a mystery. None of the arguments put forth are compelling enough and therefore instead of professing and pontificating, let me leave it at that. However, it is useful, in this context to present at least some results as food for thought.

Figure 10.26*b* plots the relative densities of a number of materials densified either using the FAST or PECS process on the homologous temperature defined as the ratio of the actual sintering temperature to a material's melting point, T_M in degrees Kelvin. It is an intriguing graph to say the least. At ≈ 3, the slopes for the ceramic powders are significantly higher than the ≈ 0.5 values for the metals powders. Furthermore, the homologous temperatures needed for ionic ceramics are lower than their more covalent counterparts, dramatically exemplified when MgO and SiC are compared (Fig. 10.26*b*).

CASE STUDY 10.2: MICROWAVE SINTERING

When you place a potato in a microwave oven to cook it, the microwaves couple with the water dipoles in the potato resulting in energy dissipation that, in turn, increases its temperature and cooks the potato from the inside out. The process is faster and uses less energy than if you had placed the potato in a conventional oven. So it is with some ceramic powders; some of them can be placed in a microwave oven and densified faster and/or at lower temperatures than when they are placed in a furnace. A good example is shown in Fig. 10.27*a* for the sintering of alumina powders co-doped with MgO and Y_2O_3. This set of results is illuminating

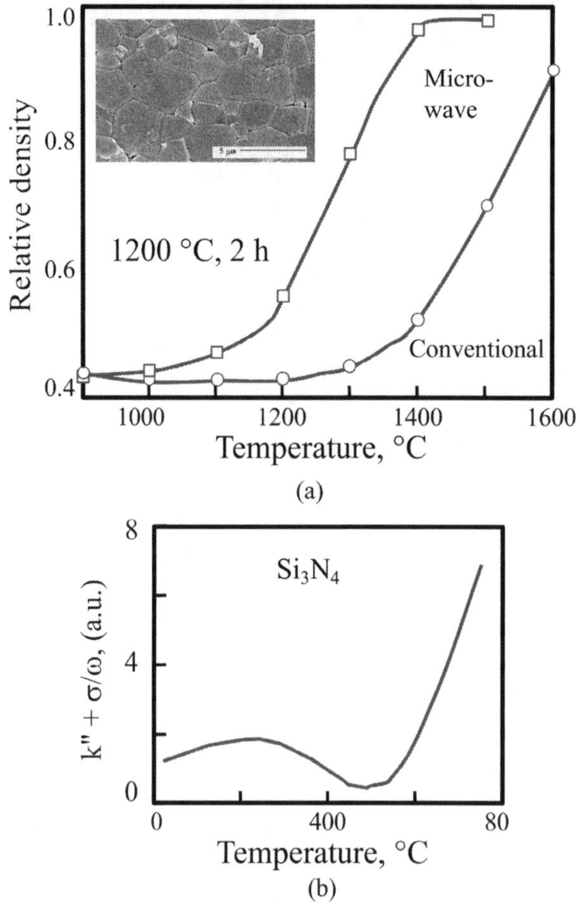

FIGURE 10.27 (*a*) Density of microwave- and conventionally sintered Al_2O_3 samples as a function of sintering temperature. In this experiment, the heating rates were the same. Inset shows microstructure after microwave sintering at 1200°C for 2 h. (Adapted from Brosnan et al. J. Amer. Ceram. Soc. 86, 1307 (2003).) (*b*) Temperature dependence of dielectric loss in Si_3N_4. (Adapted from Peng et al. *J. Microw. Power Electromag. Energ.*, 38, 243, 2003.)

because the researchers used the *same* heating rate during conventional sintering and in the presence of microwaves. That the microwaves aid in densification is pretty convincing at least for this composition.

To start understanding why this is the case one needs to understand what happens when a ceramic material is exposed to microwaves. When a dielectric material is placed in an AC electric field, with magnitude, E_o, oscillating at a frequency ω, it can be shown [see Eq. (14.25)] that the power dissipated per unit volume, P_V, is given by:

$$P_V = \frac{1}{2}\{\sigma_{dc} + \omega\varepsilon_o k''\} E_o^2 \tag{10.49}$$

where σ_{dc} is the low frequency or DC conductivity of the material, ε_o the permittivity of free space and k″ is the imaginary part of the dielectric constant. The first term in brackets is typically referred to as Joule heating and is equal to I^2R or IV—where I and V are the applied current and voltage, respectively—normalized by the sample volume. The second term depends on k″ and is related to the various dissipative polarization mechanisms operative at microwave frequencies. As discussed in Chap. 14, the only mechanism that is dissipative at microwave frequencies is dipolar polarization. However, the maximum energy dissipation in the presence of dipoles usually occurs at temperatures, that are significantly lower than those needed for densification. Said otherwise, at typical densification temperatures, $\alpha_{dc} >> \omega\varepsilon_o k''$ and the power dissipated is simply due to the Joule heating term. Furthermore, since as discussed in Chap. 7, electrical conductivity is a thermally activated process then:

$$P_V \approx \sigma_{dc}\frac{E_o^2}{2} \approx \sigma_o \frac{E_o^2}{2} e^{-Q/kT} \tag{10.50}$$

where Q is the activation energy for the process and σ_o is a pre-exponential temperature independent term. However, the results that exist—and they are not many—show that the increase in dielectric loss is more linear than exponential with temperature (see Fig. 10.27*b*).

These comments notwithstanding, the reason why microwaves can be used to densify some ceramic powders at lower temperatures is, like FAST and PECS, basically not understood. A few hypotheses have been put forward. The one with the most merit at this time is that dielectric loss at the GBs—with their higher disorder—is higher than in the bulk of the grains. This enhancement, in turn, could increase the grain boundary temperature locally leading to faster diffusional transport along the GBs. And while this conjecture sounds reasonable, it needs to be experimentally verified.

10.6 SUMMARY

1. Local variations in curvature κ result in mass transfer from areas of positive curvature (convex) to areas of negative curvature (concave). Quantitatively, on a molar basis, this chemical potential differential is given by

$$\Delta\mu = \mu_{conv} - \mu_{conc} = \gamma_{sv} V_m \kappa$$

which has to be positive if sintering is to occur.

2. On the atomic scale, this chemical potential gradient results in *a local increase in the partial pressure of the solid and a local decrease in the vacancy concentration under convex areas relative to those under concave areas.* Looked at from another perspective, matter will always be displaced from the peaks into the valleys.
3. High vapor pressures and small particles will tend to favor gas transport mechanisms, which lead to coarsening, whereas low vapor pressure and fast bulk or GB diffusivities will tend to favor densification. If the atomic flux is from the surface of the particles to the neck region, or from the surface of smaller to larger particles, this leads to, respectively, neck growth and coarsening. However, if atoms diffuse from the GB area, to the neck region, densification results. Hence, all models that invoke shrinkage invariably assume that the GB areas or free surfaces are vacancy sinks whereas neck surfaces are vacancy sources.
4. Sintering kinetics are dependent on particle size and the relative values of the transport coefficients, with smaller particles favoring GB and surface diffusion and larger particles favoring bulk diffusion.
5. During the intermediate stage of sintering, the porosity is eliminated by the diffusion of vacancies from porous areas to GBs, free surfaces or dislocations. The uniformity of particle packing and lack of agglomerates are important for the achievement of rapid and full densification.
6. In the final stages of sintering, the goal is usually to eliminate the last remnants of porosity. This can only be accomplished, however, if the pores remain attached to the GBs. One way to do this is to slow down GB mobility by doping or by the addition of inclusions or second phases at the boundaries.
7. During liquid-phase sintering, the capillary forces that develop can be quite large. These result in the rearrangement of the particles as well as enhance the dissolution of matter between them, resulting in fast shrinkage and densification. Most commercial ceramics are manufactured via some form of liquid-phase sintering.
8. The application of an external force to a powder compact during sintering can greatly enhance the densification kinetics by increasing the chemical potential gradients of the atoms between the particles, inducing them to migrate away from these areas.

APPENDIX 10A: DERIVATION OF THE GIBBS–THOMPSON EQUATION

The work of expansion of a bubble must equal the increase in surface energy, or

$$\Delta PdV = \gamma dA$$

For a sphere of radius ρ, $dA/dV = 8\pi\rho/(4\pi\rho^2) = 2/\rho$. It follows that $\Delta P = 2\gamma/\rho$.

The Gibbs free energy change is given by

$$dG = VdP - SdT$$

For an isothermal process, $dT = 0$ and $dG = V\,dP$. Integrating yields $\Delta G = V\Delta P$. Substituting for the value of ΔP given above and assuming 1 mol, one obtains

$$\Delta\mu = 2\gamma V_m/\rho$$

where V_m is the molar volume. On a per formula unit, MX basis,

$$\Delta\mu = 2\gamma V_m / N_{Av}\rho = 2\gamma\Omega_{MX}/\rho$$

This comes about because $\Omega_{MX}N_{Av} = V_m$.

APPENDIX 10B: RADII OF CURVATURE

In order to understand the various forces that arise as a result of curvature it is imperative to understand how curvature is defined and its implication. Any surface can be defined by its radius of curvature, κ, that in turn can be defined by two orthogonal radii of curvature, ρ_1 and ρ_2 where

$$\kappa = \frac{1}{\rho_1} + \frac{1}{\rho_2} \tag{10B.1}$$

For a spherical particle, the two radii of curvature are defined as positive, equal to each other and equal to the radius of the sphere ρ_{sphere} (Fig. 10.28*a*). Thus, for a sphere, $\kappa = 2/\rho_{sphere}$ and ΔP is positive.

Conversely, for a spherical pore, the two radii are equal, but since in this case the surface of the pore is concave, it follows that $\kappa = -2/\rho_{pore}$ and ΔP is negative, which renders Δc_{vac} [Eq. (10.10)] positive. That is, the vacancy concentration just below a concave surface is less than that under a flat surface.

In sintering, the geometry of the surface separating particles is modeled to be a saddle with two radii of curvature, as shown in Fig. 10.28*b*. It follows that

$$\kappa = \frac{1}{\rho_1} + \frac{1}{\rho_2} = \frac{2}{d} - \frac{1}{\rho_{neck}} \tag{10B.2}$$

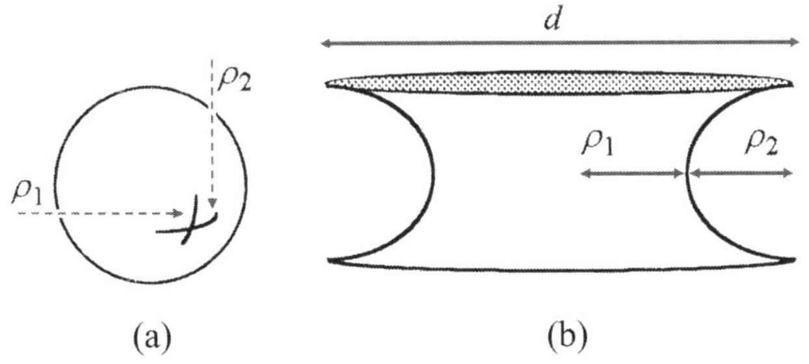

FIGURE 10.28 Definition of radii of radii of curvature for a, (*a*) sphere and, (*b*) saddle.

In most sintering problems, ρ_1 is on the same order as the particle diameter d, which is usually much larger than the radius of curvature of the neck ρ_{neck} and thus $\kappa \approx -1/\rho_{neck}$.

APPENDIX 10C: DERIVATION OF EQ. (10.20)

Refer to Fig. 10.14*a*. The volume of each polyhedron is

$$V = 8\sqrt{2}a_p^3$$

and the volume of the porous channel per polyhedron is given by

$$V_P = \tfrac{1}{3}(36 a_p \pi r_c^2)$$

where r_c is the radius of the channel. The factor 1/3 comes about because each cylinder is shared by three polyhedra. The fraction of pores is thus simply

$$P_c = \frac{V_P}{V} = \frac{12\pi a_p r_c^2}{8\sqrt{2}a_p^3} \approx 3.33\frac{r_c^2}{a_p^2} \quad (10C.1)$$

The total flux of *anion* vacancies, $c_{o,X}$, diffusing away from the cylinder of surface area, S is given by[148]

$$J_X \cdot S = -\frac{D_{\upsilon,X}\Delta c_{\upsilon,X}}{a_p}(2\pi r_c a_p) = 2\pi D_{\upsilon,X}\Delta c_{\upsilon,X} r_c \quad (10.C2)$$

where $D_{v,X}$ is the diffusivity of the anion vacancies, assumed here to be rate limiting. Since for a cylindrical pore, $\kappa = -1/r_c$, $\Delta c_{v,X}$ in this case is given by Eq. (10.10):

$$\Delta c_{v,X} = c_{v,X}\left(\frac{\Omega_X \gamma_{sv}}{r_c kT}\right) \quad (10C.3)$$

[148] In cylindrical coordinates, Fick's second law is given by

$$\frac{\partial c}{\partial t} = \frac{D}{r}\frac{\partial}{\partial r}\left(r - \frac{\partial c}{\partial r}\right)$$

which at steady state, i.e., $\partial c/\partial t = 0$, has solution of the form A + B log r, where A and B are constants that are determined from the boundary conditions. Thus strictly speaking, Eq. (10C.2), is incorrect since it assumes a planar geometry. It is used here for the sake of simplicity. Using the more exact expression, namely

$$J = D_\upsilon \frac{\Delta C}{\log(r/a_p)}\frac{1}{r}$$

does not greatly affect the final result.

where Ω_X is the volume of the anion vacancy. Combining Eqs. (10C.2) and (10C.3) and noting that $D_{v,X}$ $\Omega_X c_v = D_X$ where D_X is the diffusivity of the anions, it follows that

$$J_X \cdot S = -2\pi D_{v,X} c_v \left(\frac{\Omega_X \gamma_{sv}}{kT} \right) = -\frac{2\pi D_X \gamma_{sv}}{kT} \tag{10C.4}$$

Since it was assumed that the rate-limiting step was the diffusion of the anions, it follows that the ambipolar diffusion coefficient, $D_{ambi} \approx D_X$ (see Sec. 7.5). Make that substitution in Eq. (10C.4) and note that the total volume of matter transported per unit time is $J \cdot S \cdot \Omega_{MX}$, or

$$\frac{dV}{dt} = 2\pi D_{ambi} \left(\frac{\gamma_{sv} \Omega_{MX}}{kT} \right)$$

For a cylinder of radius r and length a_p, $dV = 2\pi a_p r dr$. Substituting for dV, and integrating from t to t_f yields

$$\frac{r_c^2}{a_p^2} = \frac{2D_{ambi}}{a_p^3} \left(\frac{\Omega_{MX} \gamma_{sv}}{kT} \right) (t_f - t) \tag{10C.5}$$

where t_f is the time at which the pore vanishes. Experimentally, it is much easier to measure the average porosity than the actual radii of the pores. By combining Eqs. (10C.1) and (10C.5) and noting that the particle diameter d scales with a_p, this model predicts that

$$P_c - P_o = (\text{const.}) \frac{D_{ambi} \gamma_{sv} \Omega_{MX}}{d^3 kT} t$$

where P_o is the porosity at the end of the intermediate stage of sintering.

APPENDIX 10D: DERIVATION OF EQ. (10.22)

Given the spherical symmetry of the problem, it can be shown that

$$c_{vac}(r) = B + \frac{A}{r} \tag{10D.1}$$

is a solution to Fick's first law when the latter is expressed in spherical coordinates,[149] and $c_v(r)$ is the vacancy concentration at any r. Referring to Fig. 10.15*a*, and assuming that the following boundary conditions apply—at

[149] Fick's second law in spherical coordinates is:

$$\frac{\partial c}{\partial t} = \frac{D}{r^2} \frac{\partial}{\partial r} \left(r^2 \frac{\partial c}{\partial r} \right)$$

which at steady state becomes

$$\frac{d}{dr} \left(r^2 \frac{dc}{dr} \right) = 0$$

$r = \rho_p$, $c_v(r) = c_{curv}$ given by Eq. (10.9), and at $r = R$, $c_o(r) = c_o$, given by Eq. 10.7—it can be shown (see Prob. 10.7*a*) that

$$\left(\frac{dc_{vac}}{dr}\right)_{r=\rho_p} = -\frac{1}{\rho_p}\Delta c_{vac}\frac{R}{R-\rho_p}$$

satisfies the boundary and steady-state conditions. Here Δc_{vac} is given by Eq. (10.10).

Consequently, the total flux of vacancies moving radially away from the pore is

$$J_X \cdot S = -4\pi\rho_p^2 D_v\left(\frac{dc_v}{dr}\right)_{r=\rho_p} = -4\pi\rho_p D_v \Delta c_{vac}\left(\frac{R}{R-\rho_p}\right) \tag{10D.3}$$

where S is the surface area of the pore. The total volume eliminated per unit time is thus

$$\frac{dV}{dt} = -J_a S\Omega_{MX} = -D_a\frac{8\pi\gamma_{sv}\Omega_{MX}}{kT}\left(\frac{R}{R-\rho_p}\right) \tag{10D.4}$$

where the product $D_v\Omega_a c_{vac}$ was replaced by the diffusivity of the *rate-limiting ion* D_a in the bulk. For a spherical pore, $dV = 4\rho_p^2 d\rho_p$; when this is combined with Eq. (10D.4), integrating the latter and neglecting ρ_p with respect to R, one obtains the final solution, namely

$$\rho_p^3 - \rho_{p,0}^3 = -\frac{6D_a\gamma_{sv}\Omega_{MX}}{kT}t \tag{10D.5}$$

PROBLEMS

10.1. (*a*) Explain, in your own words, why a necessary condition for sintering to occur is that $\gamma_{gb} < 2\gamma_{sv}$. Furthermore, show why this condition implies that $\phi > 0°$.

(*b*) The dihedral angles ϕ for three oxides were measured to be 150°, 120° and 60°. If the three oxides have comparable surface energies, which of the three would you expect to densify most readily? Explain.

(*c*) At 1850°C, the surface energy of the interface between alumina and its vapor is approximately 0.9 J/m². The average dihedral angle for the GBs intersecting the free surface was measured as 115°. In an attempt to toughen the alumina, it is dispersed with ZrO_2 particles that end up at the GBs. Prolonged heating at elevated temperatures gives the particles their equilibrium shape. If the average dihedral angle between the particles at the GBs was measured to be 150°, estimate the interface energy of the alumina/zirconia interface. What conclusions, if any, could be reached concerning the interfacial energy if the particles had remained spherical?

Answer: 1.87 J/m².

10.2. When 8.5 g of ZnO powder were pressed into a cylindrical pellet the diameter was 2 cm and height was 1 cm. The pellet was then placed in a dilatometer and rapidly heated to temperature T_2. The isothermal axial shrinkage was monitored as a function of time and the results are plotted in Fig. 10.4. Calculate the relative theoretical density of the pellet at the end of the run, i.e., after 80 min, at T_2. State all assumptions. Hint, radial shrinkage cannot be ignored.

Answer: 0.94.

10.3. (*a*) Estimate the value of P/P_{flat} for a sphere of radius 1 nm at 300 K if the surface tension is 1.6 J/m^2 and the atomic volume is 20×10^{-30} m^3.

Answer: $P/P_{flat} = 5.2 \times 10^6$!

(*b*) Calculate the relative change in average partial pressure at 1300 K as the average particle size increases from 0.5 to 10 μm.

Answer: $P_{0.5\mu m}/P_{10\mu m} = 1.0068$.

(*c*) Would your answer in part (*b*) have changed much if the final diameter of the particle were that of a 10 cm^3 sphere?

Answer: $P_{0.5\mu m}/P_{10cm} \approx 1.0072$.

10.4. (*a*) Calculate the volume of a single SiO_2 molecule in a solid with a density of 2.3 g/cm^3.

Answer: 43 Å^3.

(*b*) Calculate the vapor pressure of liquid silica over a flat surface at 2000 K if $\Delta G = 253$ kJ/mol at 2000 K, for the reaction: SiO_2 (1iq) $\Rightarrow$ SiO_2 (g). *Hint*: Start with the mass action expression.

Answer: $P_{flat} = 2.5 \times 10^{-7}$ atm.

(*c*) Compare value calculated in part b to the equilibrium vapor pressure inside a 0.5 μm diameter SiO_2 bubble suspended in liquid SiO_2 at the same temperature. Assume γ_{sv} of liquid SiO_2 is 1.2 J/m^2.

Answer: 0.985 P_{flat}.

10.5. Derive an expression for the ΔP between points s and n in Fig. 10.9.

10.6. A 1 cm^2 surface is covered by a bimodal distribution of hemispherical clusters, one-half of which are 10 nm and the other half are 30 nm in diameter.

(*a*) Describe, in your own words, what you believe will happen as the system is held at an elevated temperature for a given time.

(*b*) Assuming that the large clusters consume all the small clusters, at equilibrium, what will be the reduction in energy of the system, given that γ_{sv} of the material making up the clusters is 2.89 J/m^2 and the initial total number of clusters is 4×10^{10}? Hints: Ignore changes occurring to substrate and work with radii, not diameters.

Answer: −6.8 μJ.

10.7. (*a*) Using an approach similar to that used to arrive at Eq. (10.20), derive Eq. (10.21).

(*b*) In the intermediate-phase sintering model, the grains were assumed to be tetrakaidecahedra (Fig. 10.14*a*). Repeat the analysis, using cubic particles. Does the final dependence of shrinkage on the dimensions of the cube that you obtain differ from those of Eq. (10.20) or (10.21)?

10.8. Derive Eq. (10.25). Plot this equation for various values of R_{gb} and ρ_p, and comment on the importance of the log term.

10.9. (*a*) Develop an expression relating the equilibrium coordination number of a pore *n* to the dihedral angle ϕ of the system.

Answer: $n = 360/(180 - \phi)$.

(*b*) For a given packing of particles, will increasing the dihedral angle aid or retard pore elimination? Explain.

(*c*) Which pore(s) is thermodynamically stable in Fig. 10.16? Explain.

10.10. Given an oxygen-deficient MO oxide that is difficult to densify, for which it is known that $D_{cat} < D_{an}$. Detail a strategy for enhancing the densification kinetics.

10.11. The sintering of a compound MO is governed by diffusion of oxygen ions. If this compound is cation-deficient, propose a method by which the sintering rate may be enhanced.

10.12. (*a*) What is the range of contact angles for the following conditions?

1. $\gamma_{lv} = \gamma_{sv} < \gamma_{ls}$
2. $\gamma_{lv} > \gamma_{sv} > \gamma_{ls}$
3. $\gamma_{lv} > \gamma_{sv} = \gamma_{ls}$

(*b*) Redraw Fig. 10.3*b* for the following cases: $\gamma_{gb} = \gamma_{sl}$, $\gamma_{gb} = 2\gamma_{sl}$, and $\gamma_{gb} = 0.1\gamma_{sl}$. Make sure that you pay special attention to how far the liquid penetrates into the GB as well as the dihedral angle. Which of the three cases would result in faster densification? Explain.

10.13. (*a*) Referring to Fig. 10.23*b*, qualitatively describe, using a series of sketches, what would happen to the shape of the cylinder as it melted such that the liquid could not spread on the solid surface (that is, X is fixed in Fig. 10.23*b*), but the plates were free to move vertically.

(*b*) Assuming that an equilibrium shape is reached in part (*a*) and that the radius of curvature of the resulting shape is 1.5X, calculate the wetting angle.

Answer: 160°.

(*c*) Refer once again to Fig. 10.23*b*. If one assumes that upon melting the resulting shape is a cube of volume V, derive an expression for the resulting force. Is it attractive or repulsive? Explain.

Answer: $4\gamma_{lv}V^{1/3}$.

(*d*) Repeat part (*c*), assuming the liquid between the plates takes the shape of a tetrakaidecahedron (Fig. 10.14*a*). Does the answer depend on which face of the tetrakaidecahedron is wetting? Explain.

10.14. (*a*) Show that the total surface energy of the "spherical cap" shown in Fig. 10.29*a* is given by.

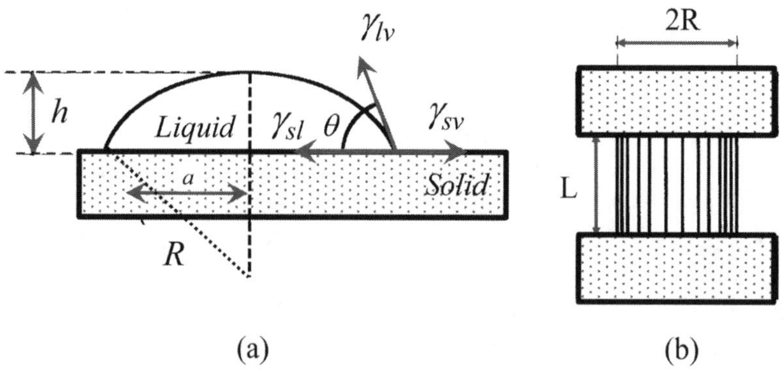

FIGURE 10.29 (*a*) Schematic of sessile drop assuming it is a spherical cap. (*b*) Liquid cylinder between two plates.

$$E_{tot} = \gamma_{sv}A + \gamma_{lv}\left[\frac{2V}{h} + \frac{2\pi h^2}{3}\right] + (\gamma_{sl} - \gamma_{sv})\left[\frac{6V - \pi h^3}{3h}\right]$$

where V is the volume of the droplet, given by $(\pi/6)(h^3 + 3h^2a)$, and A is the total area of the slab. *Hint*: Surface area of spherical cap, $S = 2\pi Rh$, and $R = (a^2 + h^2)/2h$.

(*b*) Show that the surface energy is a minimum when

$$h^3 = \frac{3V}{\pi} \frac{\gamma_{sl} + \gamma_{lv} - \gamma_{sv}}{\gamma_{sv} + 2\gamma_{lv} - \gamma_{sl}}$$

(*c*) Show this expression is consistent with Eq. (10.37); that is, show that for complete wetting, $\theta = 0°$, $h = 0$ and for $\theta = 90°$, $h = R$.

10.15. (*a*) Long thin cylinders of any material and long thin cylindrical pores are inherently unstable and tend to break up into one or more spheres. This is termed a Rayleigh instability. Explain why this occurs?

(*b*) For the liquid cylinder shown in Fig. 10.29*b*, show that the length beyond which it will become mechanically unstable is given by $L = 2\pi R$. Is this the same value at which the cylinder becomes thermodynamically unstable? Explain. You may ignore solid–liquid interactions and gravity.

10.16. (*a*) If a thin wire of radius r and length l is heated, it will tend to shorten. Explain.

(*b*) To keep it from shrinking, a force F must be applied to the wire. Derive an expression for F in terms of the wire's γ_{sv} and its dimensions.

(*c*) Write an expression for the change in energy associated with an incremental increase in length of the wire in terms of γ_{sv} and the wire dimensions. Relate this result to that obtained in part (b) and show that $\sigma = \gamma_{sv}$.

(*d*) Explain how this technique can be used to measure γ_{sv} of solids.

10.17. (*a*) Discuss the wetting phenomena that are generally relevant to liquid-phase sintering.

(*b*) If γ_{sv} of Al_2O_3 (s) is 0.9 J/m² and the surface tension of liquid Cr is 2.3 J/m² at its melting point of 1875°C, can liquid Cr completely wet Al_2O_3? Explain.

(*c*) Calculations for the interfacial energy between liquid Cr and Al_2O_3 yield a value of 0.32 J/m². Assuming this value is correct, discuss its implications to the liquid-phase sintering of this cermet (i.e. metal ceramic) system.

10.18. Show that Eq. (10.2) for a cube of side h becomes: $\gamma_{sv}\Omega/h$.

ADDITIONAL READING

1. R. M. German, *Liquid Phase Sintering*, Plenum, New York, 1985.
2. V. N. Eremenko, Y. V. Naidich, and I. A. Lavrinenko, *Liquid Phase Sintering*, Consultants Bureau, New York, 1970.
3. W. D. Kingery, Densification and sintering in the presence of liquid phase, I. Theory, *J. Appl. Phys.*, 30, 301–306 (1959).
4. J. Philibert, *Atom Movements, Diffusion and Mass Transport in Solids*, in English, S. J. Rothman, trans., Les Editions de Physique, Courtabeouf, France, 1991.
5. J. W. Martin, B. Cantor, and R. D. Doherty, *Stability of Microstructures in Metallic Systems*, 2nd ed., Cambridge University Press, Cambridge, UK, 1996.

6. J. McColm and N. J. Clark, *High Performance Ceramics*, Blackie, Glasgow, Scotland, 1988.
7. R. L. Coble, Diffusion models for hot pressing with surface energy and pressure effects as driving forces, *J. Appl. Phys.*, 41, 4798 (1970).
8. D. L. Johnson, A general method for the intermediate stage of sintering, *J. Amer. Cer. Soc.*, 53, 574–577 (1970).
9. W. D. Kingery and B. Francois, The sintering of crystalline oxides. I. Interactions between grain boundaries and pores, in *Sintering and Related Phenomena*, G. C. Kuczynski, N. A. Hooten, and C. F. Gibbon, Eds., Gordon and Breach, New York, 1967, pp. 471–498.
10. S. Somiya and Y. Moriyoshi, Eds., *Sintering Key Papers*, Elsevier, New York, 1990.[150]
11. J. Reed, *Principles of Ceramic Processing*, 2nd ed., Wiley, New York, 1995.
12. The August 2012 issue of the J. Amer. Ceram. Soc. was devoted to sintering.

OTHER REFERENCES

1. In this video, soap bubbles are generated on water—so-called bubble rafts—and then allowed to interact freely. The bubbles act as atoms would if restricted to 2D. The results are fascinating. https://www.youtube.com/watch?v=ah1Q6yqTdpA.
2. Sintering simulation in 3D: https://www.youtube.com/watch?v=48Is5ENhkGE&index=1&list=PLXY9aXdWOPapbz4QEeoikL3WJ6ojGm8Z1.
3. MD simulation of nano-capillary: https://www.youtube.com/watch?v=Z8n3m36GVQo.
4. MD of bilayer formation by self-assembly: https://www.youtube.com/watch?v=lm-dAvbl33o.
5. Grain Growth in 2D https://www.youtube.com/watch?v=J_2FdkRqmCA&list=PL23652608CEF19DB9&index=7.
6. Grain Growth in 3D https://www.youtube.com/watch?v=Ac_ca_NeRnw.

[150] This compilation of the most important papers in the field of sintering is noteworthy. In addition to reproducing many of the classic papers in the field, there are several more recent papers that critically assess the validity of the various models and summarize current sintering paradigms.

11

MECHANICAL PROPERTIES

Fast Fracture

The careful text-books measure
(Let all who build beware!)
The load, the shock, the pressure
Material can bear.
So when the buckled girder
Lets down the grinding span.
The blame of loss, or murder,
Is laid upon the man.
Not on the stuff—the Man!

R. Kipling, "Hymn of the Breaking Strain"

11.1 INTRODUCTION

Sometime, before the dawn of civilization, some hominid discovered that the edge of a broken stone was quite useful for killing prey and warding off predators. This seminal juncture in human history has been recognized by archeologists and anthropologists who refer to it as the Stone Age. C. Smith[151] goes further by stating, "Man probably owes his very existence to a basic property of inorganic matter, the brittleness of certain ionic compounds." In this context, Kipling's hymn and Gordon's statement[152] that "The worst sin in an engineering material is not lack of strength or lack of stiffness, desirable as these properties are, but lack

[151] C. S. Smith, *Science*, 148, 908 (1965).
[152] J. E. Gordon, *The New Science of Engineering Materials*, 2nd ed., Princeton University Press, Princeton, NJ, 1976.

of toughness, that is to say, lack of resistance to the propagation of cracks" stand in sharp contrast. But it is this contrast that in a very real sense summarizes the short history of structural ceramics; what was good enough for eons and millenia now falls short. After all, the consequences of a broken mirror or pot are not as dire as those of, say, a ruptured turbine blade. It could be argued, with some justification, that were it not for their brittleness, the use of ceramics for structural applications, especially at elevated temperatures, would be much more widespread since at least some of them possess other quite attractive properties such as low densities, high specific stiffness values. As importantly some are also quite oxidation and creep resistant.

The application of a load, or stress, to any solid will initially result in a reversible elastic strain, that is followed by either fracture without much plastic deformation (Fig. 11.1*a*) or fracture that is preceded by plastic deformation (Fig. 11.1*b*). For the most part, ceramics and glasses fall in the former category which is why they are considered brittle solids, whereas most metals and polymers, above their glass transition temperatures, fall into the latter category.

The theoretical stress level at which a material is expected to fracture by bond rupture was discussed in Chap. 4 and estimated to be on the order of $Y/15$, where Y is Young's modulus. Given that Y for ceramics (see Table 11.1) ranges between 100 and 500 GPa, the expected "ideal" fracture stress is quite high—on the order of 10 to 50 GPa. For reasons that will become apparent shortly, the presence of flaws, such as shown in Fig. 11.2, in brittle solids greatly reduce their failure stresses. Conversely, it is well established that extraordinary strengths can be achieved if they are flaw-free. For example, as discussed in Case Study 12.1, defect-free, silica glass fibers can be elastically deformed to stresses that exceed 10 GPa! Thus, it may be concluded that certain flaws within a material serve to promote fracture at stress levels that are well below their ideal fracture stress.

The stochastic nature of flaws present in brittle solids, together with their flaw sensitivity, has important design ramifications as well. Strength variations of ±25% from the mean are not uncommon and are quite large when compared to, say, the spread of flow stresses in metals, which are typically within just a few percent. Needless to say, such variability, together with the sudden nature of brittle failure, pose a veritable challenge for design engineers considering using ceramics for structural and other critical applications.

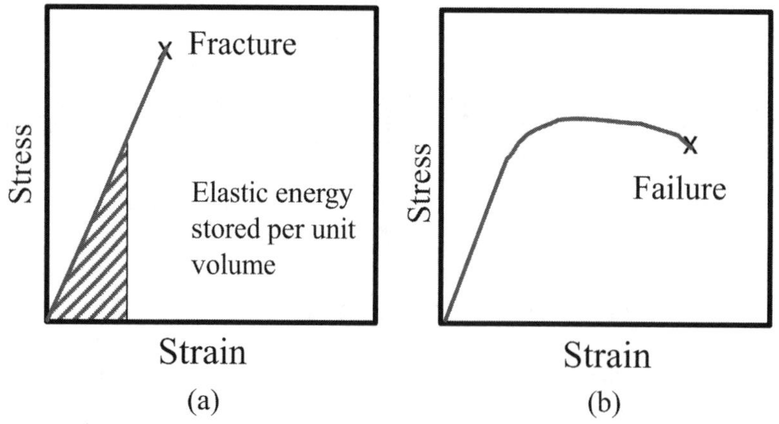

FIGURE 11.1 Typical stress–strain curve for a (*a*) brittle, and (*b*) ductile solid.

TABLE 11.1 Young's moduli (Y), Poisson's ratios (ν), fracture toughness (K_{Ic}) and Vickers hardness, H, values of select ceramics at ambient temperatures[a]

	Y (GPa)	Poisson's Ratio, ν	K_{Ic}, MPa·m$^{1/2}$	Vickers hard-ness, GPa
Oxides				
Al_2O_3	390	0.20–0.25	2.0–6.0	19–26
Al_2O_3 (single crystal, $(10\bar{1}2)$)	340		2.2	
Al_2O_3 (single crystal, (0001))	460			
$BaTiO_3$	125			
BeO	386	0.34		0.8–1.2
HfO_2 (monoclinic)	240			
MgO	250–300	0.18	2.5	6–10
$MgTi_2O_5$	250			
$MgAl_2O_4$	248–270		1.9–2.4	14–18
Mullite (fully dense)	230	0.24	2.0–4.0	15
Nb_2O_5	180			
$PbTiO_3$	81			
SiO_2 (quartz)	94	0.17		12 (011)
SnO_2	263	0.29		
TiO_2	282–300			10 ± 1
ThO_2	250		1.6	10
Y_2O_3	175		1.5	7–9
$Y_3Al_5O_{12}$				18 ± 1
ZnO	124			2.3 ± 1
$ZrSiO_4$ (zircon)	195	0.25		≈15.0
ZrO_2 (cubic)	220	0.31	3.0–3.6	12–15
ZrO_2 (partially stabilized)	190	0.30	3.0–15.0	13
Carbides, Borides, Nitrides and Silicides				
AlN	308	0.25		12
B_4C	417–450	0.17		30–38
BN	675			
Diamond	1000			
$MoSi_2$	400	0.15		
Si	107	0.27		10
SiC (hot pressed)	440 ± 10	0.19	3–6	26–36
SiC (single crystal)	460		3.7	
Si_3N_4 (hot pressed, dense)	300–330	0.22	3–10	17–30
TiB_2	500–570	0.11		18–34
TiC	456	0.18	3.0–5.0	16–28
Ti_3SiC_2	340	0.20	7–15	2–4
WC	450–650		6–20	

(Continued)

TABLE 11.1 (Continued) Young's moduli (Y), Poisson's ratios (ν), fracture toughness (K_{Ic}) and Vickers hardness, H, values of select ceramics at ambient temperatures[a]

	Y (GPa)	Poisson's Ratio, ν	K_{Ic}, $MPa \cdot m^{1/2}$	Vickers hard-ness, GPa
ZrB_2	440	0.14		22
Halides				
CaF_2	76	0.26	0.8	1.8
NaCl	≈ 33	0.2	≈ 0.2 (SC)	≈ 0.2
KCl (forged single crystal)	24		≈ 0.35	0.15
MgF_2	138	0.28	1.0	6.0
SrF_2	88		1.0	1.4
Glasses and Glass Ceramics				
Aluminosilicate (Corning 1720)	89	0.24	0.96	6.6
Borosilicate (Corning 7740)	63	0.20	0.75	6.5
Borosilicate (Corning 7052)	57	0.22		
LAS (glass-ceramic)	100	0.30	2.00	
Silica (fused)	72	0.16	0.80	6.0–9.0
Silica (96% dense)	66		0.70	
Soda lime silica glass	69	0.25	0.82	5.5

[a] Fracture toughness is a function of microstructure; values listed are mostly for comparison's sake.

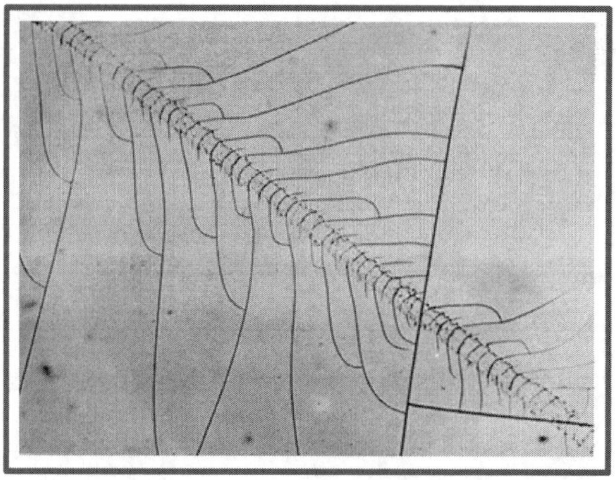

FIGURE 11.2 Surface cracks caused by the accidental contact of a glass surface with dust particles or another solid surface. These flaws result in significant reductions in strength.

Flaws, their shape and their propagation are central themes of this chapter. The various aspects of brittle failure are discussed from several viewpoints. The concepts of fracture toughness and flaw sensitivity are discussed first. The factors influencing the strengths of ceramics are dealt with in Sec. 11.3.[153] Toughening mechanisms are dealt with in Sec. 11.4. Section 11.5 introduces the statistics of brittle failure and a methodology for design. Two case studies end the chapter.

11.2 FRACTURE TOUGHNESS

11.2.1 FLAW SENSITIVITY

To illustrate what is meant by flaw or notch sensitivity, refer to Fig. 11.3. Upon the application of a load F_{app}, to a crack-free sample (Fig. 11.3*a*), each chain of atoms will carry its share of the load or F/n, where n is the number of chains. Under such conditions the applied load or stress, σ_{app}, is said to be uniformly distributed. The introduction of a surface crack, however, results in a stress redistribution such that the load that was supported by the severed bonds is now being carried by only a few bonds at the crack tip (Fig. 11.3*b*). Said otherwise, the presence of a flaw will *locally amplify the applied stress at the crack tip*, σ_{tip}.

In Sec. 4.4, the general shape of the force vs. interatomic distance, as well as the fact that a bond can only be pulled in tension to a maximum force before rupturing, were discussed in detail. Replacing the force in Fig. 4.6, to stress, one recovers Fig. 11.3*c*. Here the maximum stress at which the bond will rupture is labeled σ_{max}. Going back to our crack tip: as σ_{app} is increased, σ_{tip} increases accordingly and moves up the stress versus interatomic distance curve as shown by the short black arrow in Fig. 11.3*c*. As long as $\sigma_{tip} < \sigma_{max}$, the situation is stable and the flaw will not propagate. However, if at any time σ_{tip} exceeds σ_{max}, the situation becomes catastrophically unstable (not unlike the bursting of a dam). Based on this simple picture, the reason why brittle fracture occurs rapidly and without warning, with cracks propagating at velocities approaching the speed of sound, should now be obvious. Furthermore, it should also be obvious why ceramics are significantly stronger in compression than in tension.

To be a little more quantitative in predicting when σ_{app} would lead to failure, σ_{tip} would have to be calculated and equated to σ_{max} or $\approx Y/15$. Calculating σ_{tip} is rather complicated (only the final result is given here) and is a function of the type of loading, sample, crack geometry, etc.[154] However, for a thin sheet, it can be shown that σ_{tip} is related to the applied stress by

$$\sigma_{tip} = 2\sigma_{app}\sqrt{\frac{c}{\rho}} \tag{11.1}$$

where c and ρ are, respectively, the crack length and its radius of curvature,[155] shown schematically in Fig. 11.4.

[153] Time-dependent mechanical properties such as creep and subcritical crack growth are dealt with separately in the next chapter.

[154] C. E. Inglis, *Trans. Inst. Naval Archit.*, 55, 219 (1913).

[155] This equation strictly applies to a surface crack of length c, or an interior crack of length 2c, in a thin sheet. Since the surface of the material cannot support a stress normal to it, this condition corresponds to what is known as a plane stress condition (viz. the stress is two-dimensional). In thick components, the situation is more complicated, but for brittle materials the two expressions are not too different.

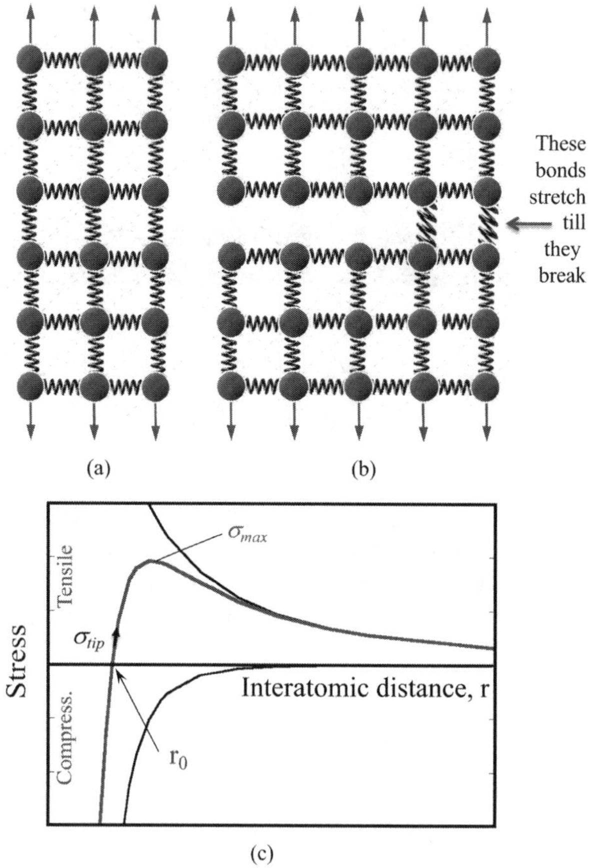

FIGURE 11.3 (*a*) Depiction of a uniform stress. (*b*) Stress redistribution as a result of the presence of a crack. (*c*) For a given applied load, as the crack grows and the bonds are sequentially ruptured, σ_{tip} moves up the stress versus displacement curve toward σ_{max}. When $\sigma_{tip} = \sigma_{max}$ catastrophic failure occurs. This figure is identical to Fig. 4.6, except that here the y-axis represents the stress on the bond rather than the applied load.

Since, as noted above, fracture can be reasonably assumed to occur when $\sigma_{tip} = \sigma_{max} \approx Y/15$, it follows that

$$\sigma_f \approx \frac{Y}{30}\sqrt{\frac{\rho}{c}} \tag{11.2}$$

where σ_f is the stress at fracture. This equation predicts that (i) σ_f is inversely proportional to the square root of the flaw size, c, and (ii) sharp cracks, i.e., those with small values of ρ are more deleterious than blunt cracks. Both predictions are in good agreement with numerous experimental observations.

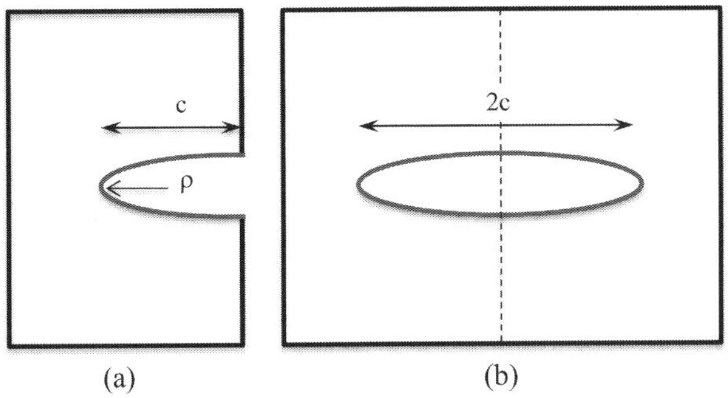

FIGURE 11.4 (*a*) Surface crack of length c and radius of curvature, ρ; (*b*) interior crack of length 2c. From a fracture point of view, the two are equivalent.

11.2.2 ENERGY CRITERIA FOR FRACTURE—THE GRIFFITH CRITERION

An alternate, and ultimately more versatile, approach to the problem of fracture was developed in the early 1920s by Griffith.[156] His basic idea was to balance the energy consumed in forming new surfaces as a crack propagates against the elastic energy released. The critical condition for fracture, then, occurs when the rate at which energy is released is greater than the rate at which it is consumed. The approach taken here is a simplified version of the original approach, and it entails deriving an expression for the energy changes resulting from the introduction of a flaw of length, c, in a material subjected to a uniform stress σ_{app}.

STRAIN ENERGY

When a solid is elastically stressed uniformly, all bonds in the material elongate and the work done by the applied stress is converted to elastic energy, which is stored in the stretched bonds. The magnitude of the elastic energy stored per unit volume is given by the area under the stress–strain curve[157] (Fig. 11.1*a*), or

$$U_{elas} = \frac{1}{2}\varepsilon\sigma_{app} = \frac{1}{2}\frac{\sigma_{app}^2}{Y} \tag{11.3}$$

It follows that the total energy of the parallelopiped of volume (V_0) subjected to a uniform stress σ_{app} (Fig. 11.5*a*) increases to

$$U = U_0 + U_{elas}V_0 = U_0 + \frac{1}{2}\frac{V_0\sigma_{app}^2}{Y} \tag{11.4}$$

where U_0 is free energy in the absence of stress.

[156] A. A. Griffith, *Philos. Trans. R. Acad.*, A221, 163 (1921).

[157] When a bond is stretched, energy is stored in that bond in the form of elastic energy. This energy can be converted to other forms of energy as any schoolboy with a slingshot can attest; the elastic energy stored in the rubber band is converted into kinetic energy of the projectile. If, by chance, a pane of glass comes in the way of the projectile, that energy will, in turn, be converted to other forms of energy such as thermal, acoustic and surface energy. In other words, the glass will shatter and some of the kinetic energy will have created new surfaces.

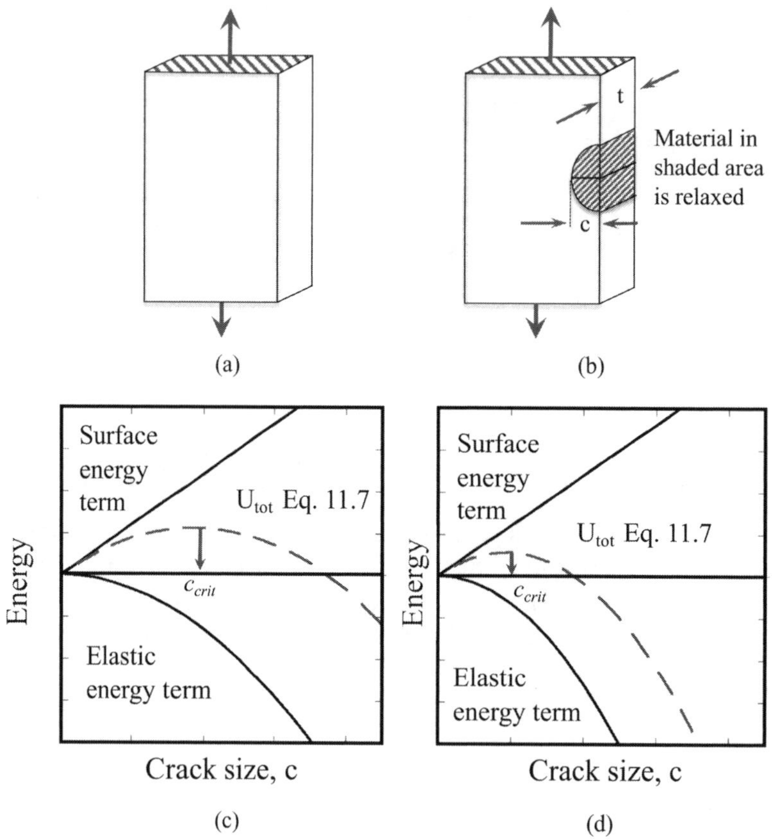

FIGURE 11.5 (*a*) Uniformly stressed solid, (*b*) relaxed volume in vicinity of crack of length c. (*c*) Plot of Eq. (11.7) as a function of c. Top curve represents surface energy term, and lower curve represents the strain energy release term. Curve labeled U_{tot} is sum of the two curves. The critical crack length c_{crit} at which fast fracture will occur corresponds to the maximum. (*d*) Plot of Eq. (11.7) on the same scale as in (*c*) but for $\sqrt{2}$ times the applied stress applied in (*c*). Increasing the applied stress by that factor reduces c_{crit} by a factor of 2.

In the presence of a surface crack of length, c (Fig. 11.5*b*), it is fair to assume that some volume around that crack will relax (i.e., the bonds in that volume will relax and lose their strain energy). Assuming—it is not a bad assumption, as will become clear shortly—that the relaxed volume is given by the shaded area in Fig. 11.5*b*, it follows that the strain energy of the system in the presence of the crack is given by

$$U_{strain} = U_o + \frac{V_o \sigma_{app}^2}{2Y} - \frac{\sigma_{app}^2}{2Y}\left[\frac{\pi c^2 t}{2}\right] \tag{11.5}$$

where t is the thickness of the plate. The third term represents the *strain energy released* in the relaxed volume.

SURFACE ENERGY

To form a crack of length *c*, an energy expenditure of

$$U_{surf} = 2\gamma ct \tag{11.6}$$

is required, where γ is the intrinsic surface energy of the material (see Chap. 4). The factor 2 arises because two - bottom and top - new surfaces are created by the fracture event.

The total energy change of the system upon introduction of the crack is simply the sum of Eqs. (11.5) and (11.6), or

$$U_{tot} = U_0 + \frac{1}{2}\frac{V_0\sigma_{app}^2}{Y} - \frac{1}{2}\frac{\sigma_{app}^2}{Y}\left(\frac{\pi c^2 t}{2}\right) + 2\gamma ct \tag{11.7}$$

Since the surface energy term scales with c and the strain energy term scales with c^2, U_{tot} has to go through a *maximum* at a critical crack size c_{crit} as shown in Fig. 11.5*c*. This is a crucial result since it says that extending a crack that is smaller than c_{crit} would *consume rather than liberate energy and is thus stable.* In contrast, a *flaw that are longer than c_{crit} is unstable since extending it releases more energy than is consumed.* Note that increasing σ_{app} (Fig. 11.5*d*) will result in failure at smaller c_{crit}. For instance, a solid—for which the size of the largest[158] flaw lies somewhere between those shown in Figs. 11.5*c* and *d*—will *not* fail at the stress shown in Fig. 11.5*c*, but will fail if that stress is increased (Fig. 11.5*d*). Note that increasing σ_{app} only affects the strain energy term; the suface energy term remains the same.

The location of the maximum is determined by differentiating Eq. (11.7) and equating it to zero. Carrying out the differentiation, replacing σ_{app} by σ_f and rearranging terms, one can show that the condition for failure is

$$\sigma_f\sqrt{\pi c_{crit}} = 2\sqrt{\gamma Y} \tag{11.8}$$

A more exact calculation yields

$$\boxed{\sigma_f\sqrt{\pi c_{crit}} \geq \sqrt{2\gamma Y}} \tag{11.9}$$

and is the expression used in subsequent discussions.[159] This equation predicts that a critical combination of *applied stress* <u>*and*</u> *flaw size is required to cause failure.* The combination $\sigma\sqrt{\pi c}$ occurs so often in discussing fast fracture that it is abbreviated to a single symbol K_I the units of which are $MPa{\cdot}m^{1/2}$, and is referred to as the **stress intensity factor**. Similarly, the combination of terms on the right-hand side of Eq. (11.9), sometimes referred to as the **critical stress intensity factor**, or more commonly the

---

[158] The largest flaw is typically the one that will cause failure, since it becomes critical before other smaller flaws (see Fig. 11.8*a*).

[159] Comparing Eqs. (11.8) and (11.9) shows that the estimate of the volume over which the stress is being relieved in Fig. 11.5*b* was off by a factor of √2, which is not too bad.

fracture toughness, is abbreviated by the symbol K_{Ic}. It follows that the condition for fracture can be succinctly rewritten as

$$\boxed{K_I \geq K_{Ic}} \tag{11.10}$$

Equations (11.9) and (11.10) were derived with the implicit assumption that the only factor keeping the crack from extending was the creation of new surface. This is only true, however, for extremely brittle systems such as inorganic glasses. In general, when other energy dissipating mechanisms—such a plastic deformation at the crack tip or friction, are operative—K_{Ic} is defined more generally as

$$\boxed{K_{Ic} = \sqrt{YG_c}} \tag{11.11}$$

where G_c is the **toughness** of the material. The units of G_c are the same as surface energy, namely J/m^2. For purely brittle solids,[160] G_c approaches the lower limit of 2γ. Table 11.1 lists the Young's moduli, Poisson's ratios and K_{Ic} values of a number of ceramic materials. It should be pointed out that since (see below) K_{Ic} is a material property that is also microstructure-dependent, the values listed in Table 11.1 are to be used with care.

Finally, it is worth noting that the Griffith approach, Eq. (11.10), can be reconciled with Eq. (11.2) by assuming that ρ is on the order of $10r_0$, where r_0 is the equilibrium interionic distance (see Prob. 11.3). In other words, the Griffith approach implicitly assumes that the flaws are atomically sharp, a fact that must be borne in mind when determining K_{Ic} experimentally.

To summarize, fast fracture will occur in a material when the product of σ_{app} and the square root of the flaw dimension is greater than a material's fracture toughness.

Worked Example 11.1

(*a*) A sharp edge notch 120 μm deep is introduced in a thin MgO plate. The plate is then loaded in tension normal to the plane of the notch. If $\sigma_{app} = 150$ MPa, will the plate survive? (*b*) Would your answer change if the notch were the same length but an internal one (Fig. 11.4*b*) instead of an edge notch?

ANSWER

(*a*) To determine whether the plate will survive σ_{app}, K_I at the crack tip needs to be calculated and compared to the fracture toughness of MgO, which, according to Table 11.1, is 2.5 $MPa{\cdot}m^{1/2}$. Making use of the definition of K_I,

$$K_I = \sigma\sqrt{\pi c} = 150\sqrt{\pi \times 120 \times 10^{-6}} = 2.91\ \text{MPa}\cdot\text{m}^{1/2}$$

Since this value is > K_{Ic} of MgO the plate will fail.

(*b*) In this case, because the notch is an internal one (Fig. 11.4*b*), it is not as detrimental as a surface or edge notch (Fig. 11.4*a*) and

160 Under these conditions, one may calculate the surface energy of a solid from a measurement of K_{Ic}.

$$K_I = \sigma\sqrt{\pi\frac{c}{2}} = 150\sqrt{\pi \times 60 \times 10^{-6}} = 2.06\ \text{MPa} \cdot \text{m}^{1/2}$$

Since this value is < K_{Ic}, the plate would survive the applied load. The reason only c/2 is used for an internal crack is easily appreciated by simply slicing Fig. 11.4b along the dotted line and comparing the result with Fig. 11.4b.

Before exploring the various strategies used to increase the fracture toughness of ceramics, it is important to understand how K_{Ic} is measured.

EXPERIMENTAL DETAILS: MEASURING K_{IC}

There are several techniques by which K_{Ic} can be measured. The two most common methods entail (i) measuring the fracture stress for a given geometry and known c, and (ii) measuring the lengths of cracks emanating from the corners of hardness indentations.

FRACTURE STRESS

Equation (11.9) can be recast in its most general form

$$\Psi \sigma_f \sqrt{\pi c} \geq K_{Ic} \tag{11.12}$$

where Ψ is a dimensionless constant on the order of unity that depends on sample shape, crack geometry and its relative size to the dimension of the sample. This relationship suggests that to measure K_{Ic}, one would start with an *atomically* sharp crack [an implicit assumption made in deriving Eq. (11.10)—see Prob. 11.3] of length c and measure the stress at which fracture, σ_f, occurs. Given a sample and the crack geometries, Ψ can be looked up in various fracture mechanics handbooks and K_{Ic} is calculated from Eq. (11.12). Thus, in principle, it would appear that measuring K_{Ic} is fairly straightforward; experimentally, however, the difficulty lies in introducing an atomically sharp crack.

Two of the more common test configurations are shown in Fig. 11.6. A third geometry, not shown here, is the **double torsion test**, which in addition to measuring K_{Ic} can be used to measure crack velocity versus K curves. The latter is described in Chap. 12.

SINGLE-EDGE NOTCHED BEAM (SENB) TEST

In this test, a notch of initial depth c_i is introduced, usually using a diamond wheel with a thin blade, on the tensile side of a flexure specimen (Fig. 11.6*a*). The sample is loaded until failure and c is taken as the initial crack length. K_{Ic} is calculated assuming

$$K_{Ic} = \frac{3\sqrt{c_i}(S_1 - S_2)\xi F_{fail}}{2BW^2}$$

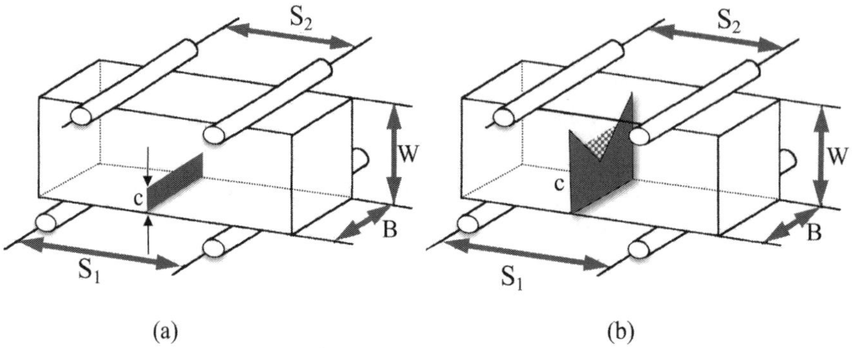

FIGURE 11.6 Schematic of (*a*) single-edge notched beam, and (*b*) Chevron notch specimens.

where F_{fail} is the load at which the specimen failed and ξ is a calibration factor. The other symbols are defined in Fig. 11.6*a*. The advantage of this test lies in its simplicity—its major drawback, however, is that the condition that the crack be atomically sharp is, more often than not, unfulfilled, which causes one to overestimate K_{Ic}.

CHEVRON NOTCH SPECIMEN[161]

The chevron notch (CN) specimen configuration looks quite similar to the SENB except for the vital difference that the shape of the initial crack is not flat but V or chevron-shaped, as shown by the shaded area in Fig. 11.6*b*. The constant widening of the crack front as it advances causes crack growth to be stable *prior* to failure. Since an increased load is required to continue crack extension, it is possible to create an atomically sharp crack in the specimen *before* final failure, which eliminates the need to precrack the specimen. K_{1c} is then related to F_{fail} and the minimum of a compliance function,[162] ξ^* by,

$$K_{Ic} = \frac{(S_1 - S_2)\xi^* F_{fail}}{BW^{3/2}}$$

GENERAL REMARKS

Unless care is taken in carrying out these measurements, different tests will result in different K_{Ic} values. Among the more common problems are: (1) sample dimensions are too small, compared to the process zone (viz. the zone ahead of the propagating crack tip); (2) internal stresses generated during machining of the specimens were not sufficiently relaxed before the measurements were made; (3) crack tip was not atomically sharp. As noted above, if the fracture initiating the flaw is not atomically sharp, apparently

[161] A *chevron* is a figure, or a pattern, having the shape of a V.

[162] For more information, see J. Sung and P. Nicholson, *J. Amer. Cer. Soc.*, 72 (6), 1033–1036 (1989).

higher K_{Ic} values will be obtained. Thus, although simple in principle, the measurement is fraught with pitfalls, and care must be taken if reliable and accurate data are to be obtained.

VICKERS HARDNESS AND RELATED INDENTATION METHOD

Due to its simplicity, its nondestructive nature and the fact that minimal machining is required to prepare samples, the use of the Vickers hardness, H, indentations to measure K_{1c} has become quite popular. Before describing this technique, it is important to describe the Vickers hardness test. In the latter, a square diamond indenter is thrust into a polished surface to a pre-chosen load, F, held at that load for a few seconds and then retracted. The Vickers hardenss is defined as the load divided by the actual contact area of the indeter (i.e. the surfaces of the 4 sides of the pyramidal indetation. If F is in Newtons, then:

$$H(GPa) = 1.854\, F/(2a)^2$$

where 2a is the indentation diagonal side in mm (see Fig. 11.7a).

Coming back to measuring K_{1c} using a diamond indenter. This test starts off as a Vickers test. Upon removal, the sizes of the cracks that emanate (sometimes) from the edges of the indent are measured (see Fig. 11.7). A number of empirical and semiempirical relationships have been proposed relating K_{Ic}, c, Y and H, and in general the expressions take the form

$$K_{Ic} = \Phi\sqrt{aH}\left(\frac{Y}{H}\right)^{0.4} f\left(\frac{c}{a}\right) \tag{11.13}$$

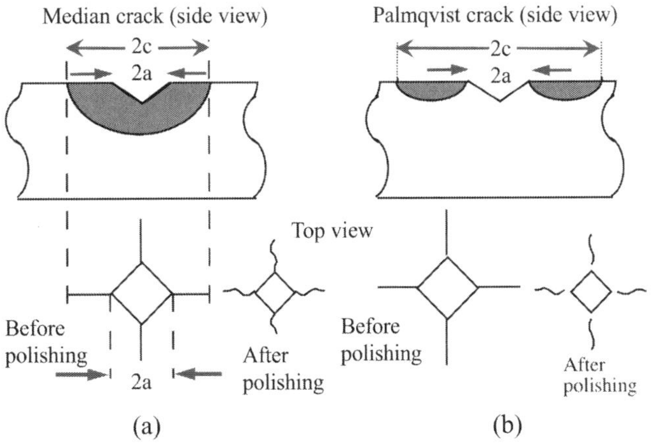

FIGURE 11.7 Crack systems developed from Vickers indents. (*a*) Side and top views of a median crack. (*b*) Top and side views of a Palmqvist crack.

where Φ is a geometric constraint factor and c and a are defined in Fig. 11.7. The exact form of the expression used depends on the type of crack that emanates from the indent.[163] A cross-sectional view and a top view of the two most common types of cracks of interest are shown in Fig. 11.7. At low loads, Palmqvist cracks are favored, while at high loads fully developed median cracks result. A way to differentiate between the two types is to polish the surface layers away; the median crack system will always remain connected to the inverted pyramid of the indent while the Palmqvist will become detached, as shown in Fig. 11.8*b*.

It should be emphasized that the K_{Ic} values measured using this technique are at best approximate values, in many cases they are simply wrong. When no cracks emanate from the corners an infinite K_{Ic} value is predicted! It is only included here for the sake of completion since it has been widely adopted. It is not recommended and should only be used when all other avenues are blocked.

Before moving to the next topic, in general there are three modes of failure, known as modes I, II and III. Mode I (Fig. 11.8*a*) is the one that we have been dealing with so far. Modes II and III are shown in Fig. 11.18*b* and *c*, respectively. The same energy concepts that apply to mode I also apply to modes II and III. Mode I, however, is by far the more pertinent to crack propagation in brittle solids. In many practical problems the modes are mixed.

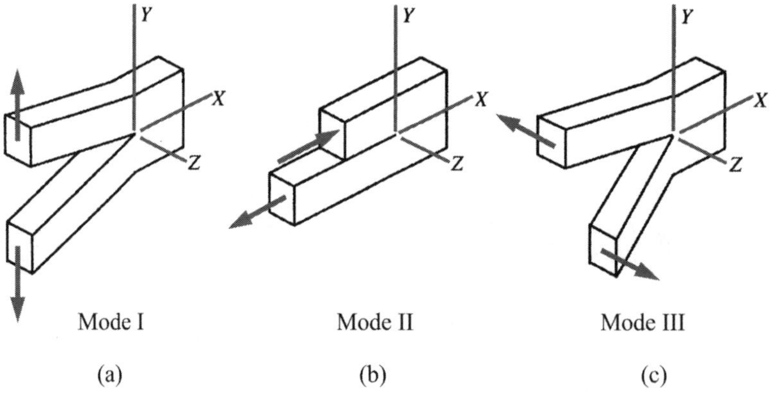

FIGURE 11.8 Three failure modes: (*a*) opening mode, or mode I, characterized by K_{Ic}; (*b*) sliding mode, or mode II, characterized by K_{IIc}; (*c*) tearing mode, or mode III, characterized by K_{IIIc}.

[163] For more information, see G. R. Anstis, P. Chantikul, B. R. Lawn, & D. B. Marshall, *J. Amer. Cer. Soc.*, 64, 533 (1981), & R. Matsumoto, *J. Amer. Cer. Soc.*, 70(C), 366 (1987). See also Prob. 11.9.

11.3 ATOMISTIC ASPECTS OF FRACTURE

Up to this point, the discussion has been mostly couched in macroscopic terms. Flaws were shown to concentrate the applied stress at their tip that ultimately led to failure. No distinction was made between brittle and ductile materials, and yet experience clearly indicates that the different classes of materials behave quite differently—after all, the consequences of scribing a glass plate are quite different from those of a metal one. Thus, the question is, what renders brittle solids notch-sensitive, or more directly, why are ceramics brittle?

The answer is related to **crack-tip plasticity,** or lack thereof in the case of ceramics. In the foregoing discussion, it was assumed that intrinsically brittle fracture was free of crack-tip plasticity, i.e., dislocation generation and motion. Given that dislocations are generated and move under the influence of *shear stresses*, two limiting cases can be considered:

1. The theoretical tensile stress, $\approx Y/15$, is *smaller* than the cohesive strength in shear, in which case the solid can sustain a sharp crack and the Griffith approach is valid.
2. $Y/15$ is *greater* than the cohesive strength in shear, in which case shear breakdown will occur—i.e., dislocations will move away from the crack tip—and the crack will lose its atomic sharpness. In other words, the emission of dislocations from the crack tip, as shown in Fig. 11.9*a*, will move material away from the crack tip, absorbing energy and causing crack blunting, as shown in Fig. 11.9*b*.

Theoretical calculations have shown that the ratio of theoretical shear strength to tensile strength diminishes as one proceeds from covalent to ionic to metallic bonds. For metals, the intrinsic shear strength is

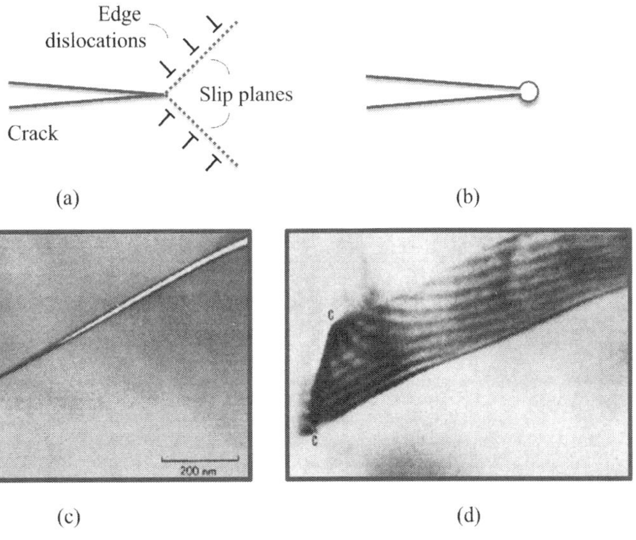

FIGURE 11.9 (*a*) Emission of dislocations from crack tip. (*b*) Blunting of crack tip due to dislocation motion. (*c*) Transmission electron microscope micrograph of a crack in Si at 25°C. (*d*) Another crack in Si formed at 500°C, where dislocation activity in vicinity of crack tip is evident. (B. R. Lawn, B. J. Hockey, and S. M. Wiederhorn, J. Mater. Sci., 15, 1207, 1980. Reprinted with permission.)

so low that plastic flow by dislocation motion at ambient temperatures is almost inevitable. Conversely, for covalent materials such as diamond and SiC, the opposite is true: the exceptionally rigid tetrahedral bonds would rather extend in a mode I type of crack than shear!

Theoretically, the situation for ionic solids is less straightforward, but direct observations of crack tips in transmission electron microscopy tend to support the notion that most covalent and ionic solids are truly brittle at room temperature (see Fig. 11.9*c*). Note that the roughly order-of-magnitude difference between the fracture toughness of metals (20 to 100 $MPa{\cdot}m^{1/2}$) and ceramics (1 to 10 $MPa{\cdot}m^{1/2}$) is directly related to the lack of crack-tip plasticity in the latter—moving dislocations consumes quite a bit of energy and also can blunt cracks. Simply put this is why metals are as useful and tough as they are.

These comments notwithstanding it is important to note that ceramic single crystals loaded in the right orientation vis-a-vis an operative slip systems can be quite ductile. For example if KCl single crystals are loaded so that the resolved shear stress on the (110) planes is non-zero then dislocations can nucleate and move on that system at shear stresses in the order of 1 to 30 MPa (Fig. 11.10). It thus follows that the reason some ceramics are brittle is not lack of dislocations per se, but lack of the 5 slip systems needed for ductility.

The situation is different at higher temperatures. Since dislocation mobility is thermally activated, increasing the temperature in some cases, will tend to favor dislocation activity, as shown in Fig. 11.9*d*, which in turn can increase a material's ductility. Thus, the condition for brittleness can be restated as follows: Solids are brittle when the energy barrier for dislocation motion is large relative to the thermal energy kT available to the system.

Given the large flow stresses required to move dislocations at elevated temperatures in oxide single crystals it is once again not surprising that ceramics are brittle at room temperatures. Finally, note that dislocation activity is not the only mechanism for crack blunting. At temperatures above the glass transition temperature, viscous flow can also effectively blunt cracks.

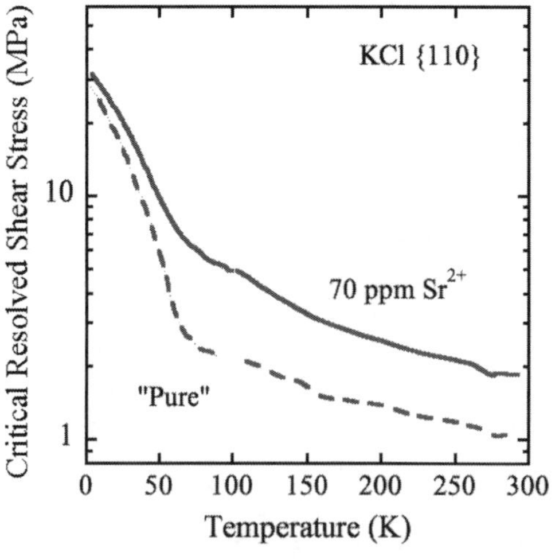

FIGURE 11.10 Temperature dependence of flow stress for {110} slip in KCl for a pure and impure crystal. (W. Skrotzki and P. Haasen, Hardening Mechanisms of Ionic Crystals on 110 and 100 Slip Planes, *J. De Physique Colloques*, 1981, 42, pp. C3-119.)

11.4 STRENGTH OF CERAMICS

It has very long been appreciated that the strength of ceramics loaded in compression are significantly higher than when they are loaded in tension. It is this property that allows us to build tall buildings with essentially a dried paste also known as Portland cement. In the next sections some of the factors that come into play in determining the strengths of ceramics are discussed.

11.4.1 COMPRESSIVE FAILURE

To this point implicity or explicity, the applied load was tensile and perpendicular to the plane of the flaw. If a brittle material containing a distribution of crack sizes, such as shown in Fig. 11.11*a*, is loaded in tension, the longest and worst oriented flaw will cause failure as shown in Fig. 11.11*b*. This occurs because σ_{tip} for the longest crack reaches σ_{max} before all others. In contradistinction to tension, in theory, the theoretical strength of a bond under compression is infinite (Fig. 11.3c). This is clearly manifested in high-pressure diamond anvil cell experiments where the pressure is hydrostatic. When the load is hydrostatic, stresses in excess of 300 GPa can be applied to the most brittle of solids, and not result in mechanical failure! At worst, such stresses can cause phase transformations, a topic that is quite important for geologist trying to understand how the earth's crust deforms.

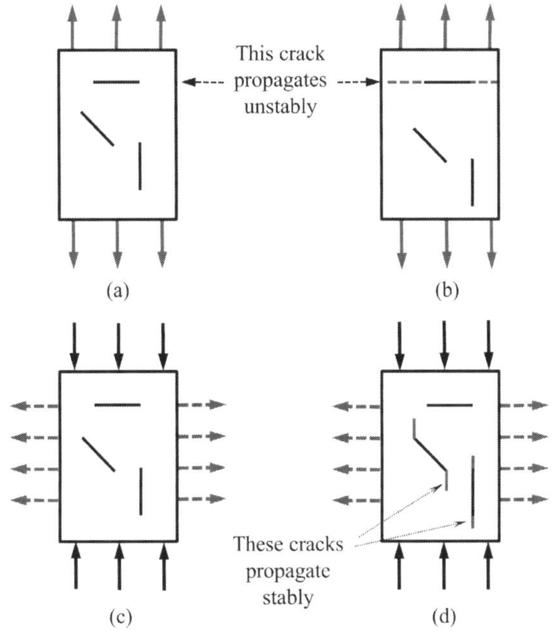

FIGURE 11.11 Fracture in ceramics due to preexisting flaws (*a*) tested in tension; (*b*) failure occurs by the unstable propagation of the longest crack that is also most favorably oriented. (*c*) tested in compression. (*d*) failure occurs by *stable* propagation of cracks due to Poisson's expansion. Cracks eventually link up creating a crush zone. (G. Quinn and R. Morrell, *J. Am. Cer. Soc.*, 74, 2037–2066, 1991.)

This begs the question; why and how do ceramics fail in compression? The answer to this question lies in appreciating that when a uniaxial compressive load is applied to most solids—an expansion in the direction normal to the applied stress—as shown by red arrows in Fig. 11.11*c*—results. This is known as Poisson's effect and is quantified by **Poisson's ratio,** ν, that is defined as the negative of the ratios of the transverse to axial strains, or

$$\nu = -\frac{\varepsilon_{\text{transverse}}}{\varepsilon_{\text{axial}}}$$

Table 11.1 lists the Poisson's ratios for a number of ceramic materials.

Getting back to the question at hand: This *tensile* stress, denoted by red arrows, now can act on all cracks that are not normal to the applied stress and cause them to grow in a direction that is parallel to the applied load (depicted in Fig. 11.11*d* by red tips to the cracks). So even though we are applying a compressive stress, the driving force for crack extension is still tensile in nature. With this preamble, it should now be clear why cracks in compression tend to propagate stably, twist out of their original orientation and propagate parallel to the compression axis. Fracture is thus caused not by the unstable propagation of a single crack, as would be the case in tension (Fig. 11.11*b*), but by the slow extension and linking up of many cracks to form a crushed zone. Hence it is not the size of the largest crack that counts, but the average one, c_{av}. The compressive stress to failure is still given by an expression similar in spirit to Eq. 11.12, except that

$$\sigma_f \geq Z\frac{K_{Ic}}{\sqrt{\pi c_{av}}} \tag{11.14}$$

Z is a constant on the order of 15. This, combined with the fact that the c_{av} is not the longest crack, but the average, is why compressive strengths of ceramics are in many cases about an order of magnitude, or higher, as compared to their tensile strengths.

11.4.2 HARDNESS

In a hardness experiment, a small region in the ceramic is placed under high pressure by the indenter. The last column in Table 11.1 lists the H values of select ceramics. How H is measured was discussed above. What is immediately obvious from these results is the very wide range of values—from lows of ≈ 0.1 GPa, to highs of over 35 GPa—observed. The reason for this extremely wide range is related to the fact that what can happen under an indenter in ceramic materials can be quite complex indeed. Generally speaking, H depends on whether dislocations and/or twins are nucleated below the indenter or not. In the halides, ZnO and MgO, dislocations are nucleated and H is on the low side. In Al_2O_3, twins are nucleated and H is quite high. In layered solids, ripplocations and concomitant kink bands form, which in turn can lead to massive delaminations and large pileups at the edges of the indentation marks. The latter are typical of layered solids such as mica, graphite and the MAX phases.

11.4.3 MODULI OF RUPTURE

As discussed in Chap. 10, most forming methods that are commonly used in the metal and polymer industries, such as melting and casting, are not applicable for ceramics. Their brittleness precludes deformation methods, and their high melting points, and in some cases (e.g., Si_3N_4, SiC) decomposition prior to melting, preclude casting. Consequently, as discussed in Chap. 10, most polycrystalline ceramics are fabricated by either solid- or liquid-phase sintering, that unless great care is taken (see Case Study 12.1) will inevitably result in the presence of flaws. For example, how agglomeration and inhomogeneous packing during powder preparation often lead to the development of flaws in the sintered body was discussed in Chap. 10. In this section, the various types of flaws that form during processing and their effect on strength are discussed. Subsequent sections deal with the effect of grain size on strength and how ceramics can be strengthened by the introduction of compressive surface layers (see also Case Study 9.2). Before proceeding further, however, it is important to briefly review how the strength of a ceramic is measured.

EXPERIMENTAL DETAILS: MODULUS OF RUPTURE

Tensile testing of ceramics is time-consuming and expensive because of the difficulty in machining test specimens with the proper geometry. Instead, the simpler transverse bending, or flexure test, is used, where the specimen is loaded to failure in either three- or four-point bending. The maximum stress, or stress at fracture, is commonly referred to as the *modulus of rupture* (MOR). For rectangular cross sections, the MOR in *four-point* bending is given by

$$\sigma_{MOR} = \frac{3(S_1 - S_2)F_{fail}}{2BW^2} \tag{11.15}$$

where F_{fail} is the load at fracture and all the other symbols are defined in Fig. 11.6*a*. Note that the MOR specimen—unlike what is shown in this figure—is un-notched and fails as a result of preexisting surface or interior flaws.

Once again, a word of caution: Although the MOR test appears straightforward, it is also fraught with pitfalls.[164] For example, in some materials the edges of the samples have to be carefully beveled before testing since sharp corners can act as stress concentrators that, in turn, can significantly reduce the measured strengths. Also the skill of the machinist can be a factor. If you get into this field make sure you find a good machinist and pay them well.

PROCESSING AND SURFACE FLAWS

The flaws in ceramics can be either internal or surface flaws generated during processing or introduced later, during machining or service. Here we consider the following flaws.

INCLUSIONS

Impurities in the starting powders can react with the matrix and form inclusions that can have different mechanical, elastic and thermal properties from the matrix in which they are embedded. Consequently, as a result of the mismatch in the thermal expansion coefficients of the matrix, α_m, and the inclusions, α_i, large residual stresses can develop as the part is cooled from the processing temperature (see Chap. 13). For

[164] For a comprehensive review of the MOR test, see G. Quinn and R. Morrell, *J. Am. Cer. Soc.*, **74**, 2037–2066 (1991).

example, a spherical inclusion of radius R in an infinite matrix will result in both radial, σ_{rad}, and tangential, σ_{tan}, residual stresses at a radial distance, r, away from the inclusion/matrix interface given by

$$\sigma_{rad} = -2\sigma_{tan} = \frac{(\alpha_m - \alpha_i)\Delta T}{\left[(1-2v_i)/Y_i + (1-v_m)/2Y_m\right]}\left(\frac{R}{r+R}\right)^3 \tag{11.16}$$

where ν is Poisson's ratio; m and i refer to the matrix and inclusion, respectively. ΔT is the difference between the initial and final temperatures. Note ΔT is defined as positive during cooling and negative during heating. On cooling, the initial temperature is the maximum temperature below which stresses are not relaxed. (See Chap. 13 for more details.)

It follows from Eq. (11.16) that upon cooling, if $\alpha_i < \alpha_m$, large tangential tensile stresses develop that, in turn, could result in the formation of radial matrix cracks. Conversely, if $\alpha_i > \alpha_m$, the inclusion will tend to detach itself from the matrix and produce a porelike flaw.

PORES

Pores are usually quite deleterious to the strength of ceramics not only because they reduce the cross-sectional area over which the load is applied, but more importantly because they can act as stress concentrators. Typically, the strength and porosity have been related by the following empirical relationship:

$$\sigma_p = \sigma_0 e^{-BP} \tag{11.17}$$

where P, σ_p and σ_0 are, respectively, the volume fraction porosity and the strength of the specimen with and without porosity; B is a constant that depends on the distribution and morphology of the pores. The strong

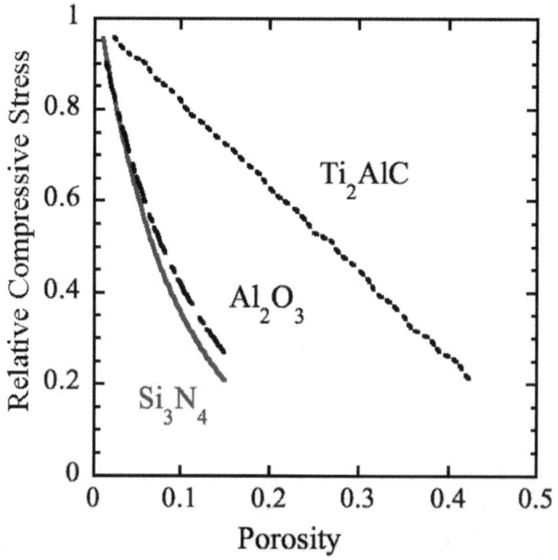

FIGURE 11.12 Functional dependence of strength on porosity for Si_3N_4, alumina and the MAX phase Ti_2AlC. (Data taken from Hu et al. Acta Mater. **60**, 6266, 2012.)

dependence of various brittle ceramics such Si_3N_4 and alumina to porosity is demonstrated in Fig. 11.12. The effect of porosity on the strengths of Ti_2AlC is different because in this case ripplocations are nucleated Fig. 11.12.

Usually, the stress intensities associated with the pores themselves are insufficient to cause failure, and as such the role of pores can be indirect. Fracture from pores is typically dictated by the presence of other defects in their immediate vicinity. For example, if the pore is much larger than the surrounding grains, atomically sharp cusps around the surface of the former can result. The critical flaw thus becomes comparable to the pore dimension. If the pores are spherical, as in glasses, they are less detrimental to the strength. Thus, both the largest dimension of the pore and the smallest radius of curvature at the pore surface are what determine their effect on strength. A typical micrograph of a pore that resulted in failure is shown in Fig. 11.13*a*.

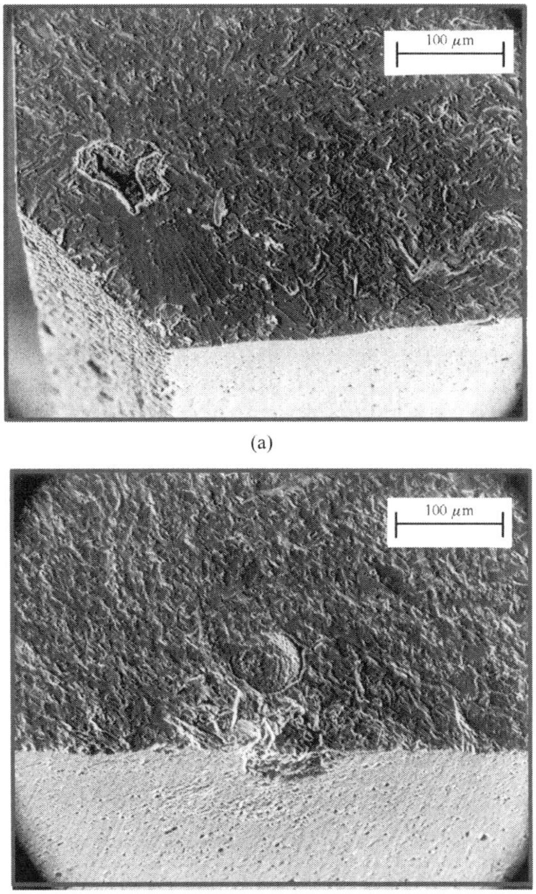

(a)

(b)

FIGURE 11.13 (*a*) Large pore associated with a large grain in sintered α-SiC. (*b*) An agglomerate with associated porosity in a sintered α-SiC. (G. Quinn and R. Morrell, *J. Am. Cer. Soc.*, 74, 2037–2066, 1991. Reprinted with permission.)

AGGLOMERATES AND LARGE GRAINS

Voids and cracks usually tend to form around agglomerates, as shown in Fig. 11.13*b*. These voids form as a result of the rapid and large differential shrinkage of the agglomerates during the early stages of sintering. Since these agglomerates form during the fabrication of the green bodies (see Chap. 10), care must be taken to avoid them.

Similarly, large grains caused by exaggerated grain growth during sintering (see Chap. 10) can also often result in strength degradation. Large grains, if noncubic, will be anisotropic with respect to such properties as thermal expansion and elastic modulus, and their presence in a fine-grained matrix essentially can act as inclusions in an otherwise homogeneous matrix. The degradation in strength is also believed to be partly due to residual stresses at grain boundaries that result from thermal expansion mismatches between the large grains and their surrounding matrices. The magnitude of the residual stresses will depend on the grain shape factor and the grain size, but can be approximated by Eq. (11.16). The effect of grain size on the residual stresses and spontaneous microcracking will be dealt with in greater detail in Chap. 13.

SURFACE FLAWS

Surface flaws can be introduced in a ceramic as a result of high-temperature grain boundary grooving, post-fabrication machining operations or accidental damage to the surface during use, among others. During grinding, polishing or machining, the grinding particles act as indenters that introduce flaws into the surface. These cracks can propagate through a grain along cleavage planes or along the grain boundaries, as shown in Fig. 11.14. In either case, the cracks do not extend much farther than one grain diameter before they are usually arrested. The machining damage thus penetrates approximately one grain diameter from the surface. Consequently, according to the Griffith criterion, the fracture stress is expected to decrease with increasing grain size—a common observation. This brings up the next important topic, which relates ceramic strengths to their grain size.

EFFECT OF GRAIN SIZE ON STRENGTH

Typically, the strengths of ceramics show an inverse correlation to the average grain size, d_{gr}. A schematic of the dependence is shown in Fig. 11.15*a*, where the fracture strength is plotted versus $d_{gr}^{-1/2}$. The simplest

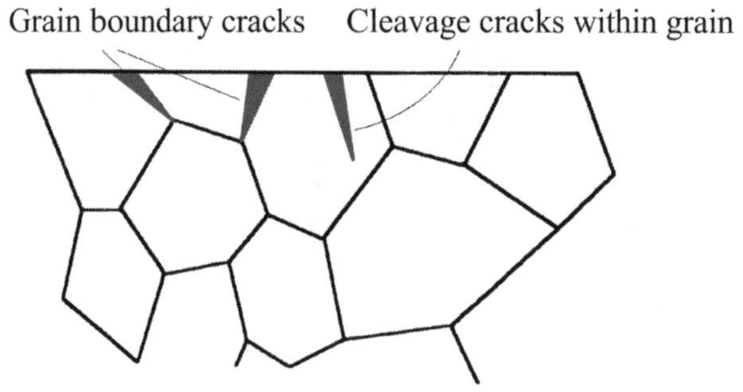

FIGURE 11.14 Schematic of cleavage and grain boundary cracks that can form on the surface of ceramics as a result of machining. The flaws are usually limited to one grain diameter because they are deflected or arrested at GBs.

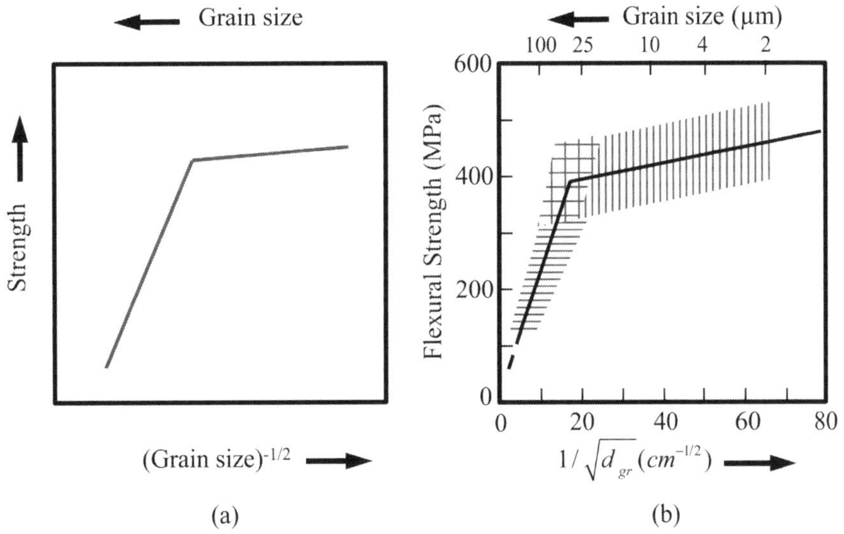

FIGURE 11.15 (*a*) Schematic relationship between grain size and strength for a number of ceramics. (*b*) Actual data for $MgAl_2O_4$. (Courtesy of R. W. Rice.)

explanation for this behavior is that the intrinsic flaw size scales with d_{gr}, a situation not unlike the one shown in Fig. 11.14. The flaws form at the grain boundaries, that are weak areas to begin with, and propagate up to about one grain diameter. Thus, once again invoking the Griffith criterion, one expects the strength to be proportional to $d_{gr}^{-1/2}$, as observed (Fig. 11.15). It is worth noting that the strength does not keep on increasing with decreasing grain size. For very fine-grained ceramics, fracture usually occurs from preexistent processing, or surface, flaws that are unrelated to grain size, and thus the strength becomes relatively grain-size-insensitive. In other words, the line shown in Fig. 11.15*b* becomes less steep for finer grain sizes.

EFFECT OF COMPRESSIVE SURFACE RESIDUAL STRESSES

The introduction of surface compressive layers can strengthen ceramics and is a well-established technique for glasses (see Sec. 13.5 for more details). The underlying principle is to introduce a state of compressive surface residual stress, the presence of which would inhibit failure from surface flaws since these compressive stresses would have to be overcome before a surface crack can propagate. These compressive stresses have also been shown to enhance thermal shock resistance and contact damage resistance.

There are several approaches to introducing a state of compressive residual stress, but in all cases the basic principle is to generate a surface layer with a higher volume than the original matrix. This can be accomplished in a variety of ways:

∞ Incorporation of an outer layer having a lower coefficient of thermal expansion, as in glazing of ceramics or tempering of glass. These are discussed in more detail in Chap. 13.

- ∞ Using transformation stresses in certain zirconia ceramics (see next section).
- ∞ Physically stuffing the outer layer with atoms or ions such as by ion implantation.
- ∞ Ion-exchanging smaller ions for larger ions. The larger ions that go into the matrix place it in a state of compression. This is similar to physical stuffing and is most commonly used in glasses (see Case Study 9.2).

One aspect of this technique is that to balance the compressive surface stresses, a tensile stress develops in the center of the part. Thus, if a flaw actually propagates through the compressive layer, the material is then weaker than in the absence of the compressive layer, and the release of the residual stresses can actually cause the glass to shatter. This is the principle at work in the manufacture of tempered glass for car windshields (See Sec. 13.5). The latter are designed to shatter on impact into a large number of small pieces that are much less dangerous than larger shards of glass, which can be quite lethal.

EFFECT OF TEMPERATURE ON STRENGTH

The effect of temperature on the strength of a ceramic depends on many factors the most important of which is whether the atmosphere in which the testing is being carried out heals or exacerbates preexisting surface flaws. In general, when a ceramic is exposed to a corrosive atmosphere at elevated temperatures, one of two scenarios is possible: (1) A protective, usually oxide, layer forms on the surface, which tends to blunt and partially heal preexisting flaws and can result in an increase in the strength. (2) The atmosphere attacks the surface, either forming pits on the surface or simply etching the surface away at selective areas; in either case, a drop in strength is observed.

As importantly, for ceramics containing glassy grain boundary phases, at high enough temperatures, the drop in strength is most often related to the softening of these phases.

11.5 TOUGHENING MECHANISMS

Despite the fact that ceramics are inherently brittle, a variety of approaches have been used to enhance their fracture toughness and resistance to fracture. The essential idea behind all toughening mechanisms is to increase the energy needed to extend a crack, that is, G_c in Eq. (11.11). The basic approaches are crack deflection, crack bridging, and transformation toughening.

11.5.1 CRACK DEFLECTION

It is experimentally well established that the fracture toughness of a polycrystalline ceramic is appreciably higher than that of single crystals of the same composition. For example, K_{Ic} of single-crystal alumina is about 2.2 MPa·m$^{1/2}$, whereas K_{Ic} for polycrystalline alumina is closer to 4 MPa·m$^{1/2}$. Similarly, the K_{Ic} of glass is ≈0.8 MPa·m$^{1/2}$, whereas K_{Ic} of a glass-ceramic, of the same composition, is closer to 2 MPa·m$^{1/2}$. One of the reasons invoked to explain this effect is crack deflection at the grain boundaries, a process illustrated in Fig. 11.17*a*. In a polycrystalline material, as the crack is deflected along the weak grain boundaries, the average stress intensity at its tip, K_{tip}, is reduced, because the stress is no longer always *normal* to the crack plane [an implicit assumption made in deriving Eq. (11.9)]. In general, it can be shown that K_{tip} is related to the applied stress intensity K_{app} and the angle of deflection, θ (defined in Fig. 11.17*a*), by

$$K_{tip} = \left(\cos^3 \frac{\theta}{2}\right) K_{app} \tag{11.18}$$

Based on this equation, and assuming an average θ value of, say, 45°, the increase in fracture toughness expected should be about 1.25 above the single-crystal value. By comparing this conclusion with the experimental results listed above, it is clear that crack deflection by itself accounts for some, but not all, of the enhanced toughening. Case Study 11.1 goes into more detail about how crack deflection enhances fracture toughness.

In polycrystalline materials, crack bifurcation around grains can also lead to a much more potent toughening mechanism, namely, crack bridging—the topic tackled next.

11.5.2 CRACK BRIDGING

In this mechanism, the toughening results from bridging of the crack surfaces behind the crack tip by strong bridging ligaments. The latter (Fig. 11.16*b* and *c*) generate closure forces on the crack face that

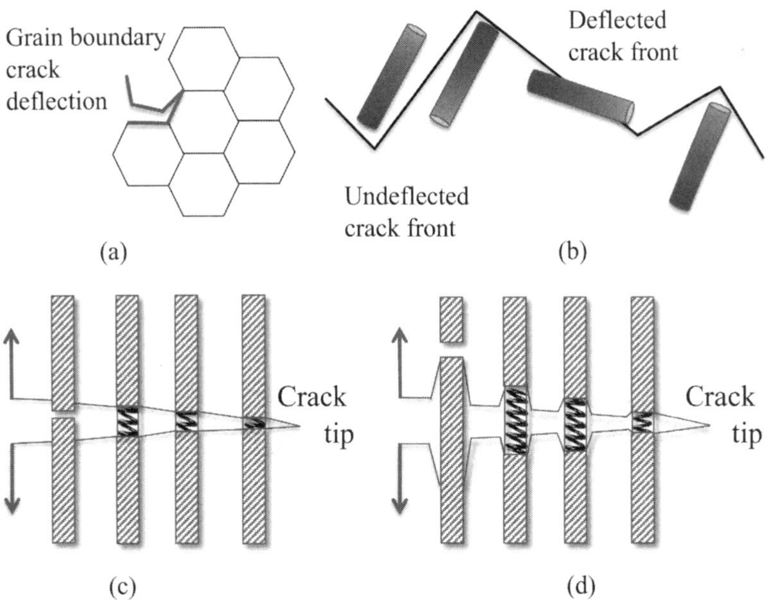

FIGURE 11.16 Schematic of (a) crack deflection mechanism at grain boundaries; (*b*) deflection of crack front around rod-shaped particles; (*c*) ligament bridging mechanism with no interfacial debonding, and (*d*) with debonding. Note that in the latter case the strain on the ligaments is delocalized, and the toughening effect is enhanced.

reduce K_{tip}. In other words, by providing some partial support of the applied load, the bridging constituent reduces the crack-tip stress intensity. The nature of the ligaments varies but they can be whiskers , elongated grains (Fig. 11.17*b*) or continuous fibers (Fig. 11.17*c*). Figure 11.17*c* is a schematic of how such elastic ligaments can result in a closure force. A useful way to think of the problem is to imagine the unbroken ligaments in the crack wake as tiny springs that have to be stretched, and hence consume energy, for the crack front to advance.

It can be shown that the fracture toughness of a composite due to elastic stretching of a partially debonded reinforcing phase at the crack tip, with no interfacial friction, is given by[165]

$$K_{Ic} = \sqrt{Y_c G_m + \sigma_f^2 \left(\frac{r V_f Y_c \gamma_f}{12 Y_f \gamma_i} \right)} \tag{11.19}$$

where the subscripts c, m and f represent the composite, matrix and reinforcement, respectively; Y, V and σ_f are the Young's modulus, volume fraction and strength of the reinforcement phases, respectively; r is the radius of the bridging ligament and G_m is the toughness of the unreinforced matrix. The ratio γ_f/γ_i represents the ratio of the fracture energy of the bridging ligaments to that of the reinforcement/matrix interface. Equation (11.19) predicts that the fracture toughness increases with

∞ Increasing fiber volume fraction of reinforcing phase.
∞ Increasing Y_c/Y_f ratio.
∞ Increasing γ_f/γ_i ratio (i.e., the toughness is enhanced for weak fiber/matrix interfaces).

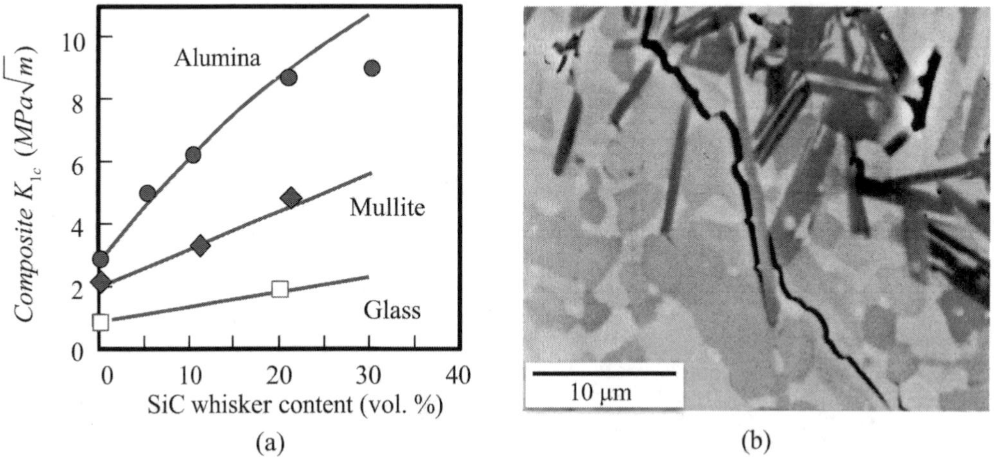

FIGURE 11.17 (*a*) Effect of SiC whisker content on toughness enhancement in different matrices. (*b*) Toughening is associated with crack bridging and grain pullout of elongated matrix grains in Si_3N_4. (P. Becher, *J. Amer. Cer. Soc.*, 74, 255–269, 1991.)

[165] See, e.g., P. Becher, *J. Amer. Cer. Soc.*, 74, 255–269 (1991).

Comparing Fig. 11.16*c* and *d* reveals how the formation of a debonded interface spreads the strain displacement imposed on the bridging reinforcing ligament over a longer gauge length. As a result, the stress supported by the ligaments increases more slowly with distance behind the crack tip, and greater crack-opening displacements are achieved in the bridging zone, which in turn significantly enhances the composite's fracture resistance. An essential ingredient of persistent bridge activity is that substantial pullout can occur well after whisker or fiber rupture. The fiber bridging mechanism is thus usually supplemented by a contribution from fibers and fail away *from* the crack plane (Fig. 11.16*c*). As the ligaments pull out of the matrix, they consume energy that has to be supplied to the advancing crack, further enhancing the composite's toughness.

That toughening contributions obtained by crack bridging and pullout can yield substantially increased fracture toughness is demonstrated in Fig. 11.17*a* for a number of whisker-reinforced ceramics. The solid lines are predicted curves and the data points are experimental results; the agreement is quite good. A similar mechanism accounts for the high toughness values achieved in Si_3N_4 with acicular grains (Fig. 11.17*b*), coarser grain-sized aluminas and other ceramics.

11.5.3 TRANSFORMATION TOUGHENING

Transformation-toughened materials owe their very large toughness to stress-induced transformations of a metastable phase in the vicinity of a propagating crack. Since the original discovery[166] that the tetragonal-to-monoclinic ($t \Rightarrow m$) transformation of zirconia (see Chap. 8) has the potential for increasing both the fracture stress and the toughness of zirconia and zirconia-containing ceramics, a large effort has been dedicated to understanding the phenomenon.[167]

To understand this phenomenon, it is useful to refer to Fig. 11.18, where fine tetragonal zirconia grains are dispersed in a matrix. If these tetragonal particles are fine enough, then upon cooling from the processing temperatures, they can be constrained from transforming by the surrounding matrix and consequently can be retained in a *metastable tetragonal phase*. If, for any reason, that constraint is lost, the transformation—which is accompanied by a relatively large volume expansion or dilatation of ≈4% and shear strains of ≈7%—is induced. In transformation toughening, the approaching crack front, being a free surface, is the catalyst that triggers the transformation, which in turn places the zone ahead of the crack tip in compression. Given that the transformation occurs in the vicinity of the crack tip, extra energy is required to extend the crack through that compressive layer, which increases both the toughness and the strength of the ceramic.

The effect of the dilation strains is to reduce the stress intensity at the crack tip K_{tip} by a shielding factor K_s such that

$$K_{tip} = K_a - K_s \tag{11.20}$$

[166] R. Garvie, R. Hannick, and R. Pascoe, *Nature*, 258, 703 (1975).

[167] See, e.g., A. G. Evans and R. M. Cannon, *Acta. Metall.*, 34, 761–800 (1986) and D. Marshall, M. Shaw, R. Dauskardt, R. Ritchie, M. Ready, and A. Heuer, *J. Amer. Cer. Soc.*, 73, 2659–2666 (1990).

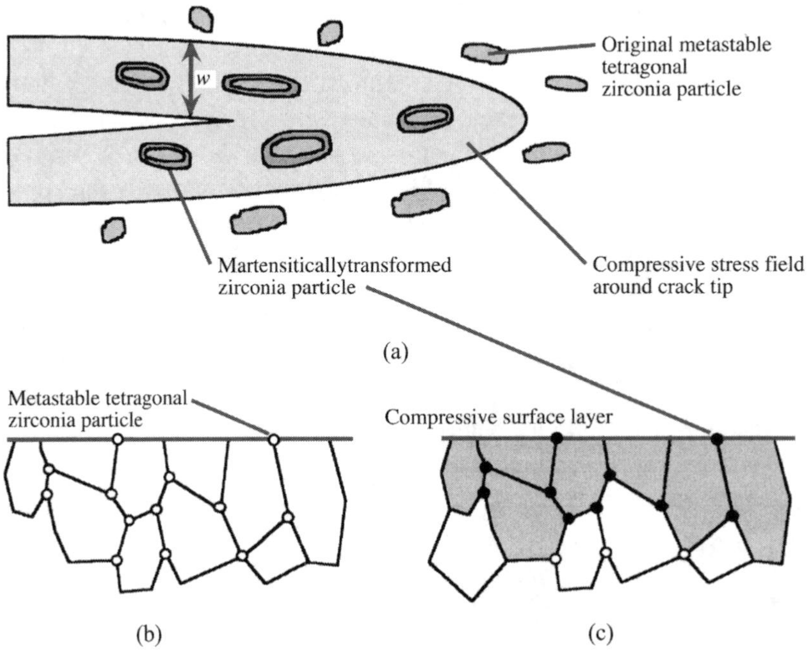

FIGURE 11.18 Schematics of *(a)* transformation zone ahead and around crack tip and *(b)* near surface containing untransformed particles. *(c)* same as *(b)* but after surface grinding that induces the martensitic transformation, that in turn creates compressive surface layers and a concomitant increase in strength.

It can be shown that if the zone ahead of the crack tip contains a uniform volume fraction V_f of transformable phase that transforms in a zone of width w, shown in Fig. 11.18*a*, from the crack surface, then the shielding crack intensity factor is given by[168]

$$K_s = A'YV_f\varepsilon^T\sqrt{w} \tag{11.21}$$

where A′ is a dimensionless constant on the order of unity that depends on the shape of the zone ahead of the crack tip and ε^T is the transformation strain. A methodology to calculate ε^T is discussed in Chap. 13.

Fracture will still occur when $K_{tip} = K_{1c}$ of the matrix in the absence of shielding; however, now the enhanced fracture toughness comes about by the shielding of K_{tip} by K_s. Careful microstructural characterization of crack-tip zones in various zirconias has revealed that the enhancement in fracture toughness does in fact scale with the product $V_f\sqrt{w}$, consistent with Eq. (11.21).

Unfortunately, the reason transformation toughening works so well at ambient temperatures—mainly the tetragonal phase's metastability—is the same reason it is ineffective at elevated temperatures. Increasing the temperature reduces the driving force for transformation and, consequently, the extent of

[168] R. M. McMeeking and A. G. Evans, *J. Amer. Cer. Soc.*, 63, 242–246 (1982).

the transformed zone, leading to less tough materials. Humididty also tends to trigger the transformation and thus these materials cannot be used in humid envrionments.

It is worth noting that the transformation can be induced any time the hydrostatic constraint of the matrix on the metastable particles is relaxed. For example, it is now well established that compressive surface layers can develop as a result of this spontaneous transformation. The process is shown schematically in Fig. 11.18*b* and c. The fracture strength can be almost doubled by simply abrading the surface, since surface grinding has been shown to be an effective method for inducing the transformation. Practically this is important, because we now have a ceramic that, in principle, becomes stronger as it is handled and small scratches are introduced on its surface.

At this stage, three classes of toughened zirconia-containing ceramics have been identified:

- *Partially stabilized zirconia* (PSZ). In this material, the cubic phase is less than totally stabilized by the addition of MgO, CaO or Y_2O_3. The cubic phase is then heat-treated to form coherent tetragonal precipitates. The heat treatment is such as to keep the precipitates small enough so they do not spontaneously transform within the cubic zirconia matrix but only as a result of stress.
- *Tetragonal zirconia polycrystals* (TZPs). The zirconia in TZPs is 100% tetragonal, with small amounts of yttria and other rare-earth additives. With bend strength exceeding 2 GPa, these ceramics are among the strongest known.
- *Zirconia-toughened ceramics* (ZTCs). These consist of tetragonal or monoclinic zirconia particles finely dispersed in other ceramic matrices such as alumina, mullite and spinel.

11.5.4 R CURVE BEHAVIOR

One of the important consequences of the toughening mechanisms described above is that they result in what is known as *R curve behavior.* In contrast to a typical Griffith solid, where the fracture toughness is independent of crack size, R curve behavior refers to a fracture toughness, which increases as the crack grows, as shown schematically in Fig. 11.19*a*. The main mechanisms responsible for this type of behavior are the same as those operative during crack bridging or transformation toughening, i.e., the closure forces imposed by either the transformed zone or the bridging ligaments. For example, referring once again to Fig. 11.16*c*, one sees that as the number of bridging ligaments increases in the crack wake, so will the energy required to extend the crack. The fracture toughness does not increase indefinitely, however, but reaches a plateau, when the number of ligaments in the crack wake reach a steady-state with increasing crack extension. Further away from the crack tip, the ligaments tend to break and pull out completely and thus become ineffective.

Both fine- and coarse-grained MAX phases, such as Ti_3SiC_2, exhibit strong R-curve behavior. As shown in Fig. 11.19*c*, for a coarse-grained sample, K_{Ic}, increases from ≈10–11 initially to values in some cases that exceed 15 MPa$\sqrt{m}$. These values are record K_{Ic} values for layered non-transforming ceramics. Figure 11.19*d* shows a scanning electron microscope micrograph of a crack in Ti_3SiC_2 extending from left to right. As the crack intersects a large grain, it bifurcates and appears on the other side of the grain. With increased crack opening, this configuration leads to the unique wood-like ligaments shown in Fig. 1.4a. These liagaments form by the nucleation and propagation of ripplocations.

There are four important implications for ceramics that exhibit R curve behavior:

1. The degradation in strength with increasing flaw size is less severe than for ceramics *with R curve* behavior. This is shown schematically in Fig. 11.19*b*.
2. The reliability of the ceramic increases. This will be discussed in detail in Sec. 11.6.
3. On the down side, there is now an increasing body of evidence that seems to indicate that ceramics that exhibit R curve behavior are more susceptible to fatigue than ceramics that do not exhibit R curve behavior. This is discussed in more detail in Chap. 12.
4. There is some recent evidence to suggest that R curve behavior enhances the thermal shock resistance of some ceramics. The evidence at this point is not conclusive, however, and more work is needed in this area.

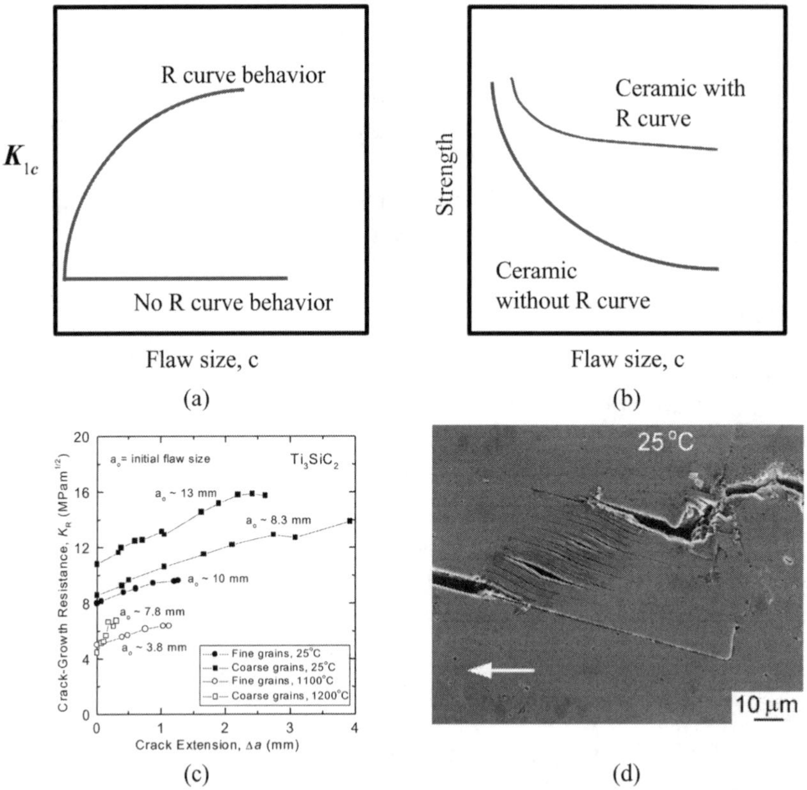

FIGURE 11.19 Functional dependence of fracture toughness on flaw size for a ceramic exhibiting R curve behavior (top curve) and one that does not (lower curve). (*b*) Effect of R curve behavior on strength degradation as flaw size increases. Ceramics exhibiting R curve behavior should be more flaw-tolerant than those that do not. (c) R-curve behavior in Ti_3SiC_2, (d) SEM micrograph of a crack in Ti_3SiC_2 extending from left to right. (D. Chen et al., *J. Amer. Cer. Soc.* 84, 2914, 2001.)

To summarize, fracture toughness is related to the work required to extend a crack and is determined by the details of the crack propagation process. Only for the fracture of the most brittle solids is the fracture toughness simply related to surface energy. The fracture toughness can be enhanced by increasing the energy required to extend the crack. Crack bridging and martensitic transformations are two mechanisms that have been shown to result in R curve behavior and thus increase K_{1c}.

11.6 DESIGNING WITH CERAMICS

In light of the preceding discussion, one expects that the failure stress, being as sensitive as it is to flaw sizes and their distributions, will exhibit considerable variability or scatter. This begs the question: Given this variability, is it still possible to design critical, load-bearing parts with ceramics? In theory, if the flaws in a part were fully characterized (i.e., their size and orientation with respect to the applied stresses) and the stress concentration at each crack tip could be calculated, then given K_{1c}, the exact stress at which a component would fail could be determined, and the answer to the question would be yes. Needless to say, such a procedure is quite impractical for several reasons, least among them the difficulty of characterizing all the flaws inside a material and the time and effort that would entail.

An alternative approach, described below, is to characterize the behavior of a large number of samples of the same material and to use a statistical approach to design. Having to treat the problem statistically has far-reaching implications since now the best that can be hoped for in designing with brittle solids is to state the *probability* of survival of a part at a given stress. The design engineer must then assess an acceptable risk factor and estimate the appropriate design stress.

Other approaches being taken to increase the reliability of ceramics are nondestructive testing and proof testing. The latter approach is briefly discussed in Sec. 11.6.2.

11.6.1 WEIBULL DISTRIBUTIONS

One can describe the strength distribution of a ceramic in a variety of formalisms. The one most widely used today is the *Weibull distribution*.[169] This two-parameter semiempirical distribution is given by

$$f(x) = m(x)^{m-1} \exp(-x^m) \tag{11.22}$$

where f(x) is the frequency distribution of the random variable x and m is a shape factor, usually referred to as the **Weibull modulus**. When Eq. (11.22) is plotted (see Fig. 11.20*a*), a bell-shaped curve results, the width of which depends on *m*; as *m* gets larger, the distribution narrows.

Since one is dealing with a strength distribution, here the random variable, x, is defined as σ/σ_0, where σ is the failure stress and σ_0 is a normalizing parameter, required to render x dimensionless and whose physical significance will be discussed shortly.

[169] W. Weibull, *J. Appl. Mech.*, 18, 293–297 (1951); *Mater. Res. Std.*, May 1962, 405–411.

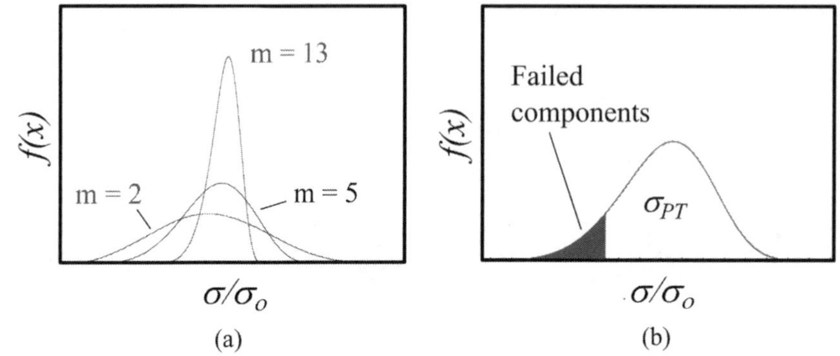

FIGURE 11.20 (*a*) Effect of *m* on the shape of the Weibull distribution. As *m* increases, the distribution narrows. (*b*) Truncation of Weibull distribution as a result of proof testing.

Replacing x by σ/σ_0 in Eq. (11.22), the survival probability, i.e., the fraction of samples that would survive a given stress level, is simply given by

$$S = \int_{\sigma/\sigma_0}^{\infty} f\left(\frac{\sigma}{\sigma_0}\right) d\left(\frac{\sigma}{\sigma_0}\right)$$

or

$$S = \exp\left[-\left(\frac{\sigma}{\sigma_0}\right)^m\right] \tag{11.23}$$

Rewriting Eq. (11.23) as $1/S = \exp(\sigma/\sigma_0)^m$ and taking the natural log of both sides twice yields

$$\ln\ln\frac{1}{S} = m\ln\frac{\sigma}{\sigma_0} = m\ln\sigma - m\ln\sigma_0 \tag{11.24}$$

Multiplying both sides of Eq. (11.24) by – 1 and plotting – ln ln (1/*S*) vs. ln σ yield a straight line with slope *m*. The physical significance of σ_0 is now also obvious: It is the stress level at S equal to 1/e, or 0.37. (When S = 1/e the RHS is zero) Once m and σ_0 are determined from a set of experimental results, then S at any stress can be easily calculated from Eq. (11.23) (see Worked Example 11.2).

The use of Weibull plots for design purposes has to be handled with care. As with all extrapolations, a small uncertainty in m can result in large uncertainties in S. Hence to increase the confidence level, the data sample has to be sufficiently large (N > 100). Furthermore, in the Weibull model, it is implicitly assumed that the material is homogeneous, with a single flaw population that does not change with time. It further assumes that only one failure mechanism is operative and that the defects are randomly distributed and are small, relative to the specimen or component size. Needless to say, whenever any of these assumptions is invalid, Eq. (11.23) has to be modified. For instance, bimodal distributions that lead to strong deviations from a linear Weibull plot are not uncommon.

Worked Example 11.2

The strengths of 10 nominally identical ceramic bars were measured and found to be 387, 350, 300, 420, 400, 367, 410, 340, 345 and 310 MPa. (*a*) Determine m and σ_0 for this material. (*b*) Calculate the design stress that would ensure a survival probability higher than 0.999.

ANSWER

(*a*) To determine m and σ_0, the Weibull plot for this set of data is generated as follows:

- Rank the specimens in order of increasing strength, 1, 2, 3, ..., j, j + 1, ..., N, where N is the total number of samples.
- Determine the survival probability for the jth specimen. As a first approximation, the probability of survival of the first specimen is $1 - 1/(N+1)$; for the second, $1 - 2/(N+1)$, for the jth specimen $1 - j/(N+1)$, etc. This expression is adequate for most applications. However, a more accurate expression deduced from a more detailed statistical analysis yields

$$S_j = 1 - \frac{j - 0.3}{N + 0.4} \tag{11.25}$$

- Plot $-\ln \ln (1/S)$ vs. $\ln \sigma$. The least-squares fit to the resulting line is m.

The last two columns in Table 11.2 are plotted in Fig. 11.21. A least-squares fit of the data yields a slope of 10.5, which is typical of many conventional, as-finished ceramics. From the table, $\sigma_0 \approx$ 385 MPa (i.e., it the failure stress when $-\ln \ln 1/S = 0$).

(*b*) To calculate the stress at which the survival probability is 0.999, use Eq. (11.23), or

$$0.999 = \exp\left\{-\left(\frac{\sigma}{385}\right)^{10.5}\right\}$$

from which $\sigma = 200$ MPa. Note that the error in using the average stress of 366 MPa instead of σ_0 changes the end result for the design stress only slightly. For most applications, it is sufficient to simply replace σ_0 by the average stress.

FACTORS AFFECTING THE WEIBULL MODULUS

Clearly, from a design point of view, the higher the m values the better. Note that m should not be confused with strength, since it is possible to have a weak solid with a high m and vice versa. For instance, a solid with large defects that are all identical in size would be weak but, in principle, would exhibit a large m. It is the *uniformity* of the microstructure, including flaws, grain size and inclusions, that is critical for obtaining high m values.

Interestingly enough, increasing the fracture toughness for a truly brittle material will not increase m. This can be shown as follows: By recasting Eq. (11.*24*), *m* can be rewritten as

TABLE 11.2 Summary of data needed to find *m* from a set of experimental results

Rank j	S_j	σ_j	$\ln \sigma_j$	$-\ln \ln(1/S)$
1	0.932	300	5.70	2.65
2	0.837	310	5.73	1.73
3	0.740	340	5.82	1.20
4	0.644	345	5.84	0.82
5	0.548	350	5.86	0.51
6	0.452	367	5.90	0.23
7	0.356	387	5.96	−0.03
8	0.260	400	5.99	−0.30
9	0.160	410	6.02	−0.61
10	0.070	420	6.04	−0.98

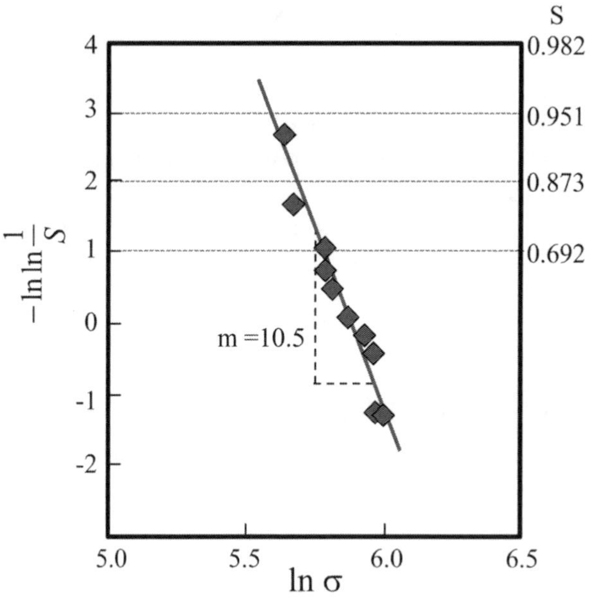

FIGURE 11.21 Weibull plot of data listed in Table 11.2. Slope of the line is the Weibull modulus, m. Actual survival probability is shown on the right-hand side y-axis. At low stresses, S is large (left-hand corner of figure). (The reason that $-\ln \ln(1/S)$ is plotted rather than $\ln \ln(1/S)$ is aesthetic, such that the high survival probabilities appear on the upper left-hand sides of the plots.)

$$m = \frac{\ln\ln(1/S_{max}) - \ln\ln(1/S_{min})}{\ln(\sigma_{max}/\sigma_{min})} \tag{11.26}$$

Thus, for any set of samples, the numerator will be a constant that depends only on the total number of samples tested [i.e., N in Eq. (11.25)]. The denominator depends on the ratio $\sigma_{max}/\sigma_{min}$, which is proportional to the ratio c_{min}/c_{max}, which is clearly independent of K_{Ic}—absent R curve effects. Thus, toughening of a solid *per se* will often not increase m. However, it can be shown that if a solid exhibits R curve behavior, then an increase in *m* should, in principle, follow (see Prob. 11.12).

EFFECT OF SIZE AND TEST GEOMETRY ON STRENGTH

One of the important ramifications of brittle failure, or weak-link statistics, as it is sometimes referred to, is the fact that strength becomes a function of volume: *larger specimens will have a higher probability of containing a larger defect, which in turn will cause lower strengths.* In other words, the larger the specimen, the weaker it is likely to be. Clearly, this is an important consideration when data obtained on test specimens, which are usually small, are to be used for the design of larger components.

Implicit in the analysis so far has been that the volumes of all the samples tested were the same size and shape. The probability of a sample of volume V_o surviving a stress σ is given by

$$S(V_o) = \exp\left\{-\left[\frac{\sigma}{\sigma_o}\right]^m\right\} \tag{11.27}$$

The probability that a batch of n such samples will all survive the same stress is lower and is given by[170]

$$S_{batch} = [S(V_o)]^n \tag{11.28}$$

Placing n pieces together to create a larger body of volume V, where $V = nV_o$, one sees that the probability S(V) of the larger volume surviving a stress σ is identical to Eq. (11.28), or

$$S(V) = S_{batch} = [S(V_o)]^n = [S(V_o)]^{V/V_o} \tag{11.29}$$

which is mathematically equivalent to

$$\boxed{S = \exp\left\{-\left(\frac{V}{V_o}\right)\left(\frac{\sigma}{\sigma_o}\right)^m\right\}} \tag{11.30}$$

This is an important result because it implies S of a ceramic depends on both the volume subjected to the stress and m. Equation (11.30) states that as the volume increases, the stress level needed to maintain a given S has to be reduced. This can be seen more clearly by equating the survival probabilities of two types

[170] An analogy here is useful: the probability of rolling a given number with a six-sided die is 1/6. The probability that the same number will appear on n dice rolled simultaneously is $(1/6)^n$.

of specimens—test specimens with a volume V_{test} and component specimens, with volume V_{comp}. Equating the survival probabilities of the two types of samples and rearranging Eq. (11.30), one can show that

$$\frac{\sigma_{comp}}{\sigma_{test}} = \left(\frac{V_{test}}{V_{comp}}\right)^{1/m} \tag{11.31}$$

A plot of this equation is shown in Fig. 11.22, where the relationship between strength and volume is plotted. The salient point here is that as either V_{comp} *increases* or m *decreases,* the more severe the downgrading of the design stress required to maintain a given S.

An implicit assumption made in deriving Eq. (11.31) was the presence of one flaw population (i.e., those due to processing rather than, say, machining) is controlling the strength. Different flaw populations will have different strength distributions and will scale in size differently. Also, implicit in deriving Eq. (11.31) is that volume defects are responsible for failure. If, instead, surface flaws were suspected of causing failure, by using a derivation similar to the one used to get to Eq. (11.31), it can be shown that

$$\frac{\sigma_{comp}}{\sigma_{test}} = \left(\frac{A_{test}}{A_{comp}}\right)^{1/m} \tag{11.32}$$

in which case the strength will scale with area instead of volume.

Lastly, another important ramification of the stochastic nature of brittle fracture is the effect of the stress distribution itself during testing on the results. When a batch of ceramics is tested in tension, the entire volume and surface are subjected to the stress. Thus, a critical flaw *anywhere* in the sample will propagate with equal probability. In three- or four-point flexure tests, however, only one-half of the sample is in tension,

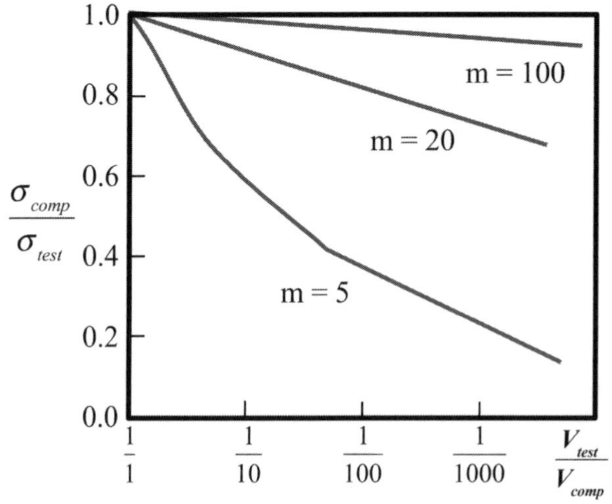

FIGURE 11.22 Effect of volume on strength degradation as a function of m. Strength decreases as V increases and is more severe for materials with low m.

and the other one-half is in compression. In other words, the effective volume tested is, in essence, reduced. It can be shown that the ratio of the tensile to flexural strength for an equal probability of survival is

$$\frac{\sigma_{3\text{-point bend}}}{\sigma_{\text{tension}}} = \left[2(m+1)^2\right]^{1/m} \tag{11.33}$$

In other words, samples of a given identical batch, subjected to flexure, will appear to be stronger, by a factor that depends on m. For example, for $m = 5$, the ratio is about 2, whereas increasing m to 20 reduces the ratio to 1.4.

11.6.2 PROOF TESTING

In proof testing, the components are briefly subjected to a stress level σ_{PT}, which is in excess of that anticipated in service. The weakest samples fail and are thereby eliminated. The resulting truncated distribution, shown in Fig. 11.20*b*, can be used with a high level of confidence at any stress that is slightly lower than σ_{PT}.

One danger associated with proof testing, however, is subcritical crack growth, discussed in the next chapter. Since moisture is usually implicated in subcritical crack growth, effective proof testing demands inert, i.e., moisture-free, testing environments and rapid loading/unloading cycles that minimize the time at maximum stress.

CASE STUDY 11.1: IMPORTANCE OF CRACK DEFLECTION

In an elegant set of experiments researches in Canada set about to better understand how deflection of cracks at interfaces can lead to enhanced toughness. Working with compact tension glass samples, they engraved defects ahead of a crack at various angles, θ, to the crack plane (Fig. 11.23*a*). They then loaded the samples in tension and calculated the critical fracture toughness, K_{1c}. From fracture mechanics, they predicted that:

$$\frac{K_c}{K_{Ic}^b} = \frac{1}{\cos^2(\theta/2)} \frac{K_{Ic}^i}{K_{Ic}^b} \tag{11.34}$$

where K_{Ic}^b and K_{Ic}^i are the bulk and interfacial critical fracture toughness values of the bulk glass and interface, respectively. The system was designed such that K_{Ic}^b/K_{Ic}^i was 0.5. In other words, the interface was deliberately made weaker than the bulk.

The introduced defects were made to lie on a plane at an angle, θ to the original crack interface (Fig. 11.23*a*). The results are shown in Fig. 11.23*b* and clearly show that up to a θ of 60°, the model given by red line in Fig. 11.23*b*, was reasonably well followed. When the angle was >60°, the crack no longer deflected and instead went straight across the glass specimen as shown in top right inset in Fig. 11.23*b*. It follows that by deflecting the cracks by 60°, the toughness was increased from 0.5 to ≈ 0.75.

The study then proceeded to create more elaborate crack surfaces such as the interlocking jigsaw interface shown schematically in inset of Fig. 11.23*a*. When these interfaces were loaded in tension, the cracks

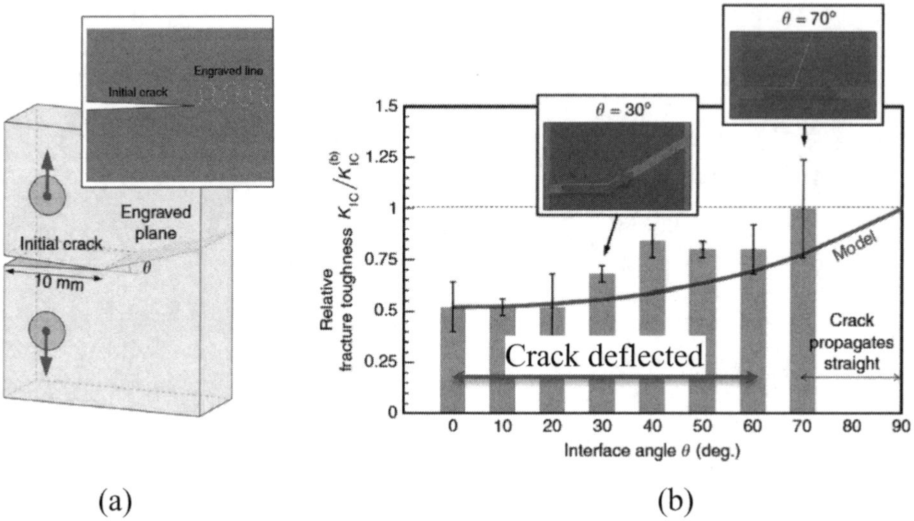

FIGURE 11.23 (*a*) Glass compact tension specimen with engraved lines at various angles θ to plane of initial crack. Inset shows complex engraved line. (*b*) Relative fracture toughness as a function of θ. Up to 60°, the crack was deflected; above that value it was not deflected. Left inset shows crack propagation when $\theta = 30°$; right inset shows crack propagation when $\theta = 70°$. (Adapted from Mirkhalaf et al. *Nat. Comm.*, 2013.)

propagated stably and the energy dissipated in their growth increased dramatically. Lastly, these cracks were filled with a thin polymer layer that further greatly enhanced the toughness under tension. This study thus confirms much that has been known in the field, some of which was discussed in this chapter.

CASE STUDY 11.2: STRONG AND TOUGH CERAMICS

Based on the ideas presented in this chapter, the requirements for high strengths—uniform and small grains—and those needed for high fracture toughness—nonhomogenous, duplex microstructures with weak interfaces (e.g., Fig. 11.17*b*)—are incompatible. It would thus appear that a choice between strong and tough ceramics would have to be made. It turns out, however, that is not the case, with the inspiration possibly coming from nature.

An ongoing theme in materials design over the past three decades, or so, has been to understand how nature designs tough materials and try to mimic it, an approach, not surprisingly, labeled biomimetics. When it comes to designing tough materials with brittle minerals, nature is difficult to beat. An excellent example is nacre from mollusk shells that is comprised of more than 95% by volume of extremely brittle calcite—$CaCO_3$—plates, bound together with very thin protein, or polymer, layers. The final design looks like a brick wall, where the bricks are held together with a thin polymer cement. Multiple studies on nacre, dental enamel and others have shown that the toughening is due to a hierarchical architecture that results in deflecting or channeling the cracks away from the mineral and keeping them in the polymer. The latter is only possible if the polymer/ceramic interfaces are weak. A weak interface, however, is only part of the puzzle. The other more important piece is the polymer; confining polymers to two-dimensions greatly enhances their toughness and strength. Some have made the case that weak interfaces and hierarchical architectures alone hold the key. Such a notion is incorrect: Heating a mollusk shell a few hundred degrees Centigrade will not affect its architecture but will render it mechanically useless.

These comments notwithstanding, it is possible to design ceramics that are both strong and tough. Consider the MAX phase, Nb_4AlC_3, where a combination of ≈17 MPa·m$^{1/2}$ and flexural strengths >1.2 GPa was reported in 2011 (Fig. 11.24*a*). This was essentially achieved by fabricating a "brick wall" with quite fine grains (Fig. 11.24*g*). By keeping the grains small, high strengths were achieved; the "brick wall" structure, greatly enhanced K_{1c}. This enhancement, in turn, can be ascribed to the tortuosity of the crack path; the crack is deflected at two scales, the milli- and micrometer as shown in Fig. 11.24*b*–*g*.

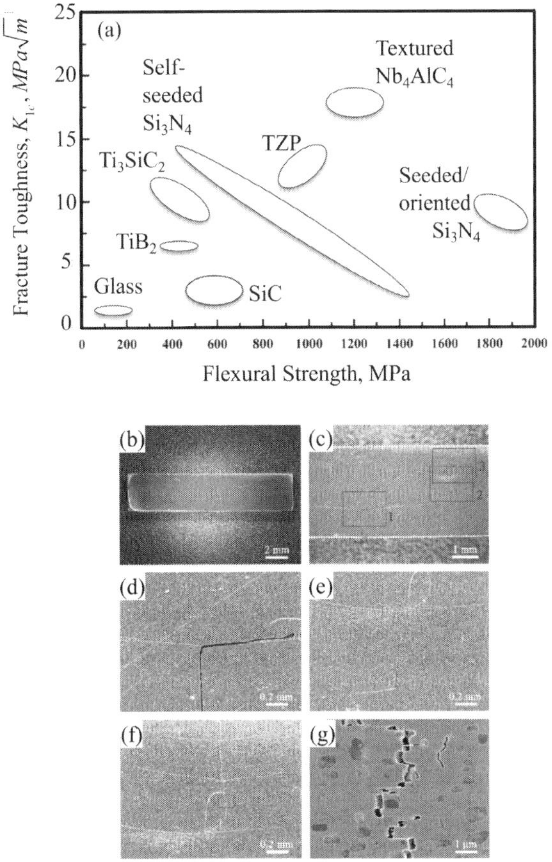

FIGURE 11.24 (*a*) Map of fracture toughness vs. flexural strength for a variety of technical ceramics. (*b*) Optical microscope, OM, micrographs of bend bar of highly textured Nb_4AlC_3 loaded to failure. (*c*) to (*f*) Higher magnification of various areas of the crack shown in (*c*). (*g*) SEM micrograph showing detail of the crack at an even higher magnification. Note tortuosity of the crack around the grains. Grains are so oriented so as to look as "bricks" in a wall. (C. Hu et al. *Scripta Mater.*, **64**, 765, 2011.)

FIGURE 11.25 Picture of various ceramic bearings.

CASE STUDY 11.3: WEAR-RESISTANT CERAMICS

Hardness and concomitant wear resistance are important attributes that render ceramics useful in many structural applications. For example, the wear resistance of ceramics is why they are used in bioimplants, including artificial hips, knees, teeth among others. Other important applications include pump seals, cutting tools, ball and roller bearings. The high hardness of SiC and B_4C, together with their low densities, render them key materials for ceramic armor applications. It should be noted here that hardness is not the only important attribute, high fracture toughness is important. Chemical inertness is also key in many applications where corrosive liquids are used. Alumina, SiC, Si_3N_4, ZrO_2 and B_4C are all widely used for these applications. Figure 11.25 is a picture of typical ceramic ball, and other, types of bearings.

11.7 SUMMARY

1. Ceramics are brittle because they lack a mechanism to relieve the stress buildup at the tips of cracks and flaws. Since one of the most effective relieve mechanisms is plastic deformation by the glide of dislocations, ceramics are brittle because they lack the five independent slip systems needed for plasticity. This makes them notch-sensitive, and consequently their strength will depend on the combination of applied stress and flaw size. The condition for failure is

$$K_I = \sigma_f \sqrt{\pi c} \geq K_{Ic}$$

where K_{Ic} is the fracture toughness of the material. Ceramic strengths be increased by either increasing K_{Ic} or decreasing the flaw size, c.

2. Processing introduces flaws in the material that are to be avoided if high strengths are to be achieved. The flaws can be pores, large grains in an otherwise fine matrix and/or inclusions, among others. Furthermore, since the strength of a ceramic component decreases with increasing grain size, it follows that to obtain a high-strength ceramic, a flaw-free, fine microstructure is desirable.
3. The high hardness and wear resistances of some ceramics render them useful in many structural and wear resistant applications.
4. It is possible to toughen ceramics by a variety of approaches. The commonality to these approaches is to render the propagation of cracks energetically expensive. This can be accomplished either by having a zone ahead of the crack martensitically transform, thus placing the crack tip in compression, or by adding whiskers, fibers or large grains (duplex microstructures) that bridge the crack faces as it propagates.

   Comparing the requirements for high strength—uniform, fine microstructures—to those needed to improve toughness—nonhomogeneous, duplex microstructure—reveals the problem in achieving both simultaneously.
5. The brittle nature of ceramics together with the stochastic nature of finding flaws of different sizes, shapes and orientations relative to the applied stress will invariably result in some scatter to their strength. According to the Weibull distribution, the survival probability is given by

$$S = \exp\left\{-\left(\frac{\sigma}{\sigma_0}\right)^m\right\}$$

   where m, known as the *Weibull modulus*, is a measure of the scatter. Large scatter is associated with low m values, and vice versa.
6. If the strength is controlled by defects randomly distributed within the volume, then the strength becomes a function of volume, with S decreasing with increasing volume. If strength is controlled by surface defects, S will scale with area instead.
7. Proof testing, in which a component is loaded to a stress level higher than the service stress, eliminates the weak samples, truncating the distribution and establishing a well-defined stress level for design.

PROBLEMS

11.1. (*a*) Following a similar analysis used to arrive at Eq. (11.7), show that an internal crack of length c is only $\sqrt{2}$ as detrimental to the strength of a ceramic as a surface crack of the same length.

(*b*) Why are ceramics usually much stronger in compression than in tension?

(*c*) Explain why the yield point of ceramics can approach the ideal strength σ_{theo}, whereas the yield point in metals is usually much less than σ_{theo}. How would you attempt to measure the yield strength of a ceramic, given that the fracture strength of ceramics in tension is usually much less than the yield strength?

11.2. (*a*) Estimate the size of the critical flaw for a glass that failed at 102 MPa if $\gamma = 1$ J/m^2 and Y = 70 GPa.

Answer: 4.3 μm.

(*b*) What is the maximum stress this glass will withstand if the largest crack is on the order of 100 μm and the smallest on the order of 7 μm?

Answer: 21 MPa.

11.3. Show that Eqs. (11.2) and (11.9) are equivalent, provided the radius of curvature of the crack $\rho \approx 14r_0$, where r_0 is the equilibrium interatomic distance; in other words, if it is assumed that the crack is atomically sharp. *Hint*: Find expressions for γ and Y in terms of n, m and r_0 defined in Chap. 4. You can assume n = 9, m = 1 and $\sigma_{th} = Y/10$.

11.4. Al_2O_3 has a fracture toughness K_{Ic} of about 4 MPa·m$^{1/2}$. A batch of Al_2O_3 samples were found to contain surface flaws about 30 μm deep. The average flaw size was more on the order of 10 μm. Estimate the (*a*) tensile, and (*b*) compressive strength of this batch.
Answer: 412 MPa, 10 GPa.

11.5. To investigate the effect of pore size on the strength of reaction-bonded Si_3N_4, Heinrich[171] introduced artificial pores (wax spheres that melt during processing) in his compacts prior to reaction bonding. The results obtained are summarized in table below. Are these data consistent with the Griffith criterion? Explain clearly, stating all assumptions.

Wax grain size, μm	Average pore size, μm	Bend strength, MPa
0–36	48	140 ± 12
63–90	66	119 ± 12
125–180	100	101 ± 14

11.6. The tensile fracture strengths of three structural ceramics—hot-pressed Si_3N_4 (HPSN), reaction-bonded Si_3N_4 (RBSN) and chemical vapor-deposited SiC (CVDSC), measured at room temperature—are listed below.
(*a*) Plot the cumulative *failure probability* as a function of fracture strength for each.
(*b*) Calculate the mean strength and standard deviation of the strength distributions, and determine the Weibull modulus for each material.
Answer: m = 8.15 (HPSN); m (RBSN) = 4.31; m(CVD) = 6.95
(*c*) Do you expect any relationship between the standard deviations and the Weibull moduli? Explain.
(*d*) Estimate the design stress for each material assuming you can only afford one part in a 1000 to fail.
Answer: σ = 197 MPa (HPSN); σ (RBSN) = 60 MPa; σ (CVD) = 37.3 MPa
(*e*) On the basis of your knowledge of these materials and how they are processed, why do you think they behave so differently? State all assumptions.

HPSN (MPa)	521, 505, 500, 490, 478, 474, 471,453, 452, 448, 444, 441, 439, 430, 428, 422, 409, 398, 394, 372, 360, 341, 279.
CVDSC	386, 351, 332, 327, 308, 296, 290, 279, 269, 260, 248, 231, 219, 199, 178, 139.
RBSN	132, 120, 108, 106, 103, 99, 97, 95, 93, 90, 89, 84, 83, 82, 80, 80, 78, 76.

11.7. When the ceramic shown in Fig. 11.26 is loaded in tension (along the length of the sample), it fractured at 20 MPa. The heavy red lines denote cracks (two internal and one surface crack). Estimate K_{Ic} for this ceramic. State all assumptions.
Answer: 3.5 MPa·m$^{1/2}$.

[171] J. Heinrich, *Ber. Dt. Keram. Ges.*, **55**, 238 (1978).

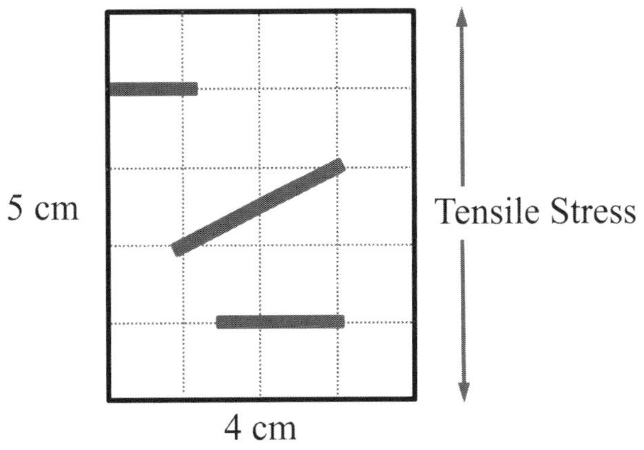

FIGURE 11.26 Cross section of ceramic part loaded in tension as shown. The heavy red lines denote flaws.

11.8. For Si_3N_4, K_{Ic} is strongly dependent on microstructure, and can vary anywhere from 3 to 10 MPa·m$^{1/2}$. Which of the following Si_3N_4 would you choose, one in which the largest flaw size is on the order of 50 μm and the fracture toughness is 8 MPa m$^{1/2}$ or one for which the largest flaw size is 25 μm, but is only half as tough? Explain.

11.9. Evans and Charles[172] proposed the following equation for the determination of fracture toughness from indentation:

$$K_{Ic} \approx 0.15(H\sqrt{a})\left(\frac{c}{a}\right)^{-1.5}$$

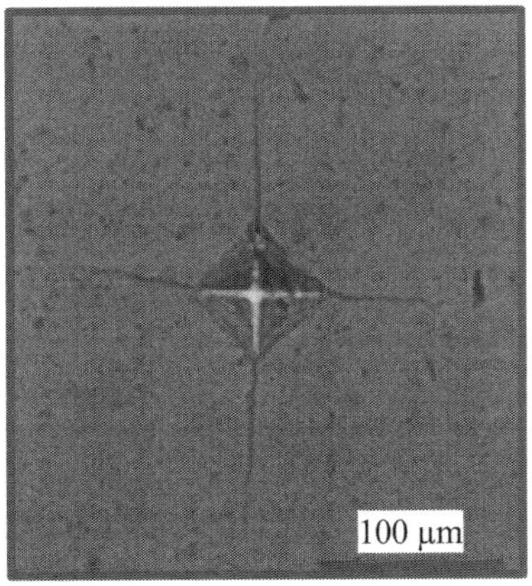

FIGURE 11.27 Optical photomicrograph of indentation in glass at a magnification of 200x.

[172] A. G. Evans and E. A. Charles, *J. Amer. Cer. Soc.*, 59, 317 (1976).

where H is the Vickers hardness in GPa and c and a are defined in Fig. 11.7. A photomicrograph of a Vickers indention in a glass slide and the cracks that emanate from it is shown in Fig. 11.27. Estimate K_{Ic} of this glass if its hardness is ≈5.5 GPa.

Answer: ≈1.2 to 1.6 MPa·m$^{1/2}$ depending on size of crack measured.

11.10. A manufacturer wishes to choose between two ceramics for a certain application. Data for the two ceramics tested under identical conditions were as follows:

Ceramic	Mean fracture stress	Weibull modulus
A	500 MPa	12
B	600 MPa	8

The service conditions are geometrically identical to the test conditions and impose a stress of 300 MPa. Based on your knowledge of Weibull statistics, which ceramic would be more reliable at 300 MPa? At what stress, if any, would the two ceramics give equal performance?

Answer: Stress for equal performance = 349 MPa.

11.11. The MORs of a set of cylindrical samples, 5 mm in diameter and 25 mm long, were tested and analyzed using Weibull statistics. The average strength was 112 MPa, with a Weibull modulus of 10.

(*a*) Estimate the stress required to obtain a S of 0.95. State all assumptions.

Answer: 74.5 MPa.

(*b*) If you now fabricate components to sell from the same materials that are 10 mm in diameter and 25 mm long, would your design stress calculated in part a change? If yes, by how much. State all assumptions.

Answer: 64.8 MPa.

11.12. Show why ceramics that exhibit R curve behavior should, in principle, also exhibit larger m values.

11.13. (*a*) In deriving Eq. (11.30), the flaw population was assumed to be identical in both volumes. However, sometimes in the manufacturing of ceramic bodies of different volumes and shapes, different flaw populations are introduced. What implications, if any, does this statement have on designing with ceramics? Be specific.

(*b*) In an attempt to address this problem, Kschinka et al.[173] measured the strength of different glass spheres in compression. Their results are summarized below, where D_0 is the diameter of the glass spheres, N is the number of samples tested, m is the Weibull modulus, σ_f is the average strength and V is the volume of the spheres.

[173] B. A. Kschinka, S. Perrella, H. Nguyen, and R. C. Bradt, *J. Amer. Cer. Soc.*, 69, 467 (1986).

D_0, cm	N	m	σ_f (50%)	V, cm³
0.368	47	6.19	143	2.6×10^{-2}
0.305	48	5.96	157	1.5×10^{-2}
0.241	53	5.34	195	7.3×10^{-3}
0.156	30	5.46	229	2.0×10^{-3}
0.127	45	5.37	252	1.1×10^{-3}
0.108	38	5.18	303	6.6×10^{-4}
0.091	47	3.72	407	3.9×10^{-4}
0.065	52	4.29	418	1.4×10^{-4}
0.051	44	6.82	435	6.9×10^{-5}

(i) On *one graph* plot the Weibull plots for spheres of 0.051, 0.108 and 0.368 cm diameters. Why are they different?
(ii) For the 0.051 cm spheres, what would be your design stress to ensure S = 0.99?
(iii) Estimate the average strength of glass spheres 1 cm in diameter.
(iv) If the effect of volume is taken into account, then it is possible to collapse all the data on a master curve. Show how that can be done. *Hint*: Normalize data to 0.156 cm spheres, for example.

ADDITIONAL READING

1. R. W. Davidge, *Mechanical Behavior of Ceramics*, Cambridge University Press, New York, 1979.
2. R. Warren, Ed., *Ceramic Matrix Composites*, Blackie, Glasgow, Scotland, 1992.
3. B. Lawn, *Fracture of Brittle Solids*, 2nd ed., Cambridge University Press, New York, 1993.
4. A. Kelly and N. H. Macmillan, *Strong Solids*, 3rd ed., Clarendon Press, Oxford, UK, 1986.
5. G. Weaver, Engineering with ceramics, Parts 1 and 2, *J. Mater. Ed.*, 5, 767 (1983); 6, 1027 (1984).
6. T. H. Courtney, *Mechanical Behavior of Materials*, McGraw-Hill, New York, 1990.
7. A. G. Evans, Engineering property requirements for high performance ceramics, *Mater. Sci. Eng.*, 71, 3 (1985).
8. S. M. Weiderhorn, A probabilistic framework for structural design, in *Fracture Mechanics of Ceramics*, vol. 5, R. C. Bradt, A. G. Evans, D. P. Hasselman, and F. F. Lange, Eds., Plenum, New York, 1978, p. 613.
9. M. F. Ashby and B. F. Dyson, in *Advances in Fracture Research*, S. R. Valluri, D. M. R. Taplin, P. Rama Rao, J. F. Knott, and R. Dubey, Eds., Pergamon Press, New York, 1984.
10. P. F. Becher, Microstructural design of toughened ceramics, *J. Amer. Cer. Soc.*, 74, 225 (1991).
11. F. Riley, *Structural Ceramics, Fundamentals and Case Studies*, Cambridge University Press, 2009.

12

CREEP, SUBCRITICAL CRACK GROWTH AND FATIGUE

The fault that leaves six thousand tons a log upon the sea.

R. Kipling, ***McAndrew's Hymn***

12.1 INTRODUCTION

As discussed in the previous chapter, at low and intermediate temperatures, failure typically emanated from a preexisting flaw formed during processing or surface finishing. The condition for failure was straightforward: Fracture occurred rapidly, and catastrophically, when $K_I > K_{Ic}$. It was tacitly implied that for conditions where $K_I < K_{Ic}$, the crack was stable, i.e., did not grow with time, and, consequently, the material would be able to sustain the load indefinitely. In reality, the situation is not that simple—preexisting cracks can, and do, grow slowly under steady and cyclic loadings, even when $K_I < K_{Ic}$. For example, it has long been appreciated that in metals, cyclic loading, even at small loads, can result in crack growth, a phenomenon referred to as **fatigue**. In contrast, it has long been accepted that ceramics, because of their lack of crack-tip plasticity or work hardening, were not susceptible to fatigue. More recently, however, this has been shown to not be the case: Some ceramics, especially those that exhibit R curve behavior, are indeed susceptible to cyclic loading.

Another phenomenon that has been well appreciated for a long time is that the exposure of a ceramic to the combined effects of a steady stress and a corrosive environment results in slow crack growth. In this mode of failure, a preexisting subcritical crack, or one that nucleates during service, grows slowly by a stress-enhanced chemical reactivity at the crack tip. This phenomena is referred to as **subcritical crack growth (SCG)**. Unfortunately, this phenomenon is also sometimes termed **static fatigue**, seemingly to differentiate it from the dynamic fatigue situation just alluded to, but more to create confusion.

Last, **creep**, or the slow deformation of a solid subjected to a stress at high temperatures, also occurs in ceramics. Sooner or later, a part experiencing creep will either fail, or undergo shape and dimensional changes that, in close-tolerance applications, render the part useless.

Despite the fact that the atomic processes and micromechanisms that are occurring during each of these phenomena are quite different, there is a commonality among them. In each case, a preexisting, or nucleated, flaw grows with time, leading to the eventual failure of the part, usually with disastrous consequences. In other words, the ceramic now has a *lifetime* that one has to contend with.

In the sections that follow, each phenomenon is dealt with separately. In Sec. 12.5, a methodology for estimating lifetimes and the choice of appropriate design criteria are developed.

12.2 CREEP

Creep is the slow and continuous deformation of a solid with time that only occurs at higher temperatures, that is, $T > 0.5T_m$, where T_m is the melting point in Kelvins. In metals, it has long been established that grain boundary sliding, and related cavity growth are the mechanisms most detrimental to creep resistance, which led to the development of single-crystal superalloy turbine blades that are quite resistant to creep. In ceramics, the situation is more complex because several mechanisms, some of which are not sufficiently well understood, can lead to creep deformation. The problem is further complicated by the fact that different mechanisms may be operative over different temperature and stress regimes. In general, creep is a convoluted function of stress, time, temperature, grain size and shape, microstructure, volume fraction and the viscosity of any glassy phases at the grain boundaries if present, dislocation mobility, etc. Before tackling the subject in greater detail, it is instructive to briefly review how creep is measured.

EXPERIMENTAL DETAILS: MEASURING CREEP

Typically, the creep response of a solid is characterized by measuring the strain as a function of applied load and time. This, most simply, can be carried out by attaching a load to a sample, heating it to a given temperature, T, and measuring its deformation as a function of time, t. The resulting strain, ε, is plotted versus t, as shown in Fig. 12.1*a*, where three regions are typically observed: (1) an initial region where, after an almost instantaneous increase in ε, is followed by a decreasing rate of increase in ε with t. This region is known as the **primary creep** region. (2) A region where the ε increases, more or less, linearly with t. This is known as the **steady-state** or **secondary creep** stage, which, from a practical point of view, is the most important stage and is of major concern here. (3) A **tertiary creep** stage, which occurs just before the specimen fails, where the strain rate increases rapidly with time.

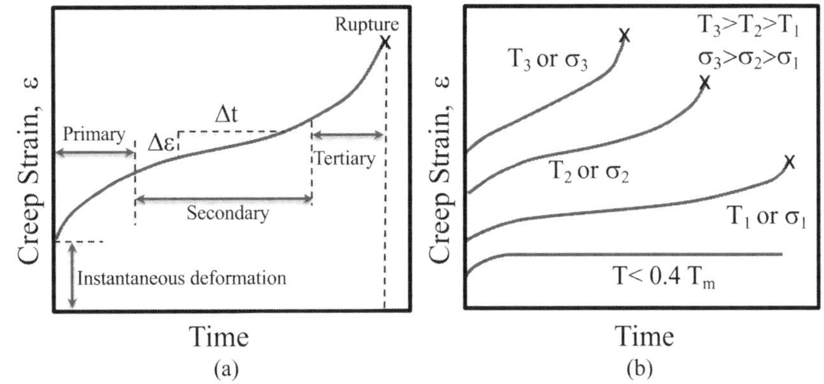

FIGURE 12.1 (*a*) Typical strain versus time creep curves. Experimentally not all three regions are always observed. (*b*) Effect of increasing stress and/or temperature on a material's creep response.

Increasing T and/or stress (Fig. 12.1*b*) results in an increase in both the instantaneous strains and the steady-state creep rates together with a decrease in the times to failure.

Data such as shown in Fig. 12.1*b* can be further reduced by plotting the log of the *steady-state creep rate*, $\dot{\varepsilon}_{ss}$, versus the log of the applied stress σ at a constant T. Such curves usually yield straight lines, which in turn implies that

$$\dot{\varepsilon}_{ss} = \frac{d\varepsilon}{dt} = \Gamma \sigma^{p} \tag{12.1}$$

where Γ is a temperature-dependent constant and p is called the **creep law exponent** and usually lies between 1 and 8. For $p > 1$, the creep is commonly referred to as *power law creep.*

Over a dozen mechanisms have been proposed to explain the functional dependence described by Eq. (12.1). In general they fall into one of four categories: *diffusion*, *viscous*, *grain boundary sliding* or *dislocation* creep. To cover even a fraction of these models in any detail is beyond the scope of this book. Instead, diffusion creep is dealt with in some detail below, followed by a brief mention of the other two important, but less well-understood, and more difficult to model, mechanisms. For more comprehensive reviews, consult the references at the end of this chapter.

12.2.1 DIFFUSION CREEP

For permanent deformation to occur, atoms have to move from one region to another, which requires a driving force of some kind. Thus, before one can even attempt to understand creep, it is imperative to appreciate the origin of the driving forces involved.

DRIVING FORCE FOR CREEP

In general, the change in the Helmholtz[174] free energy A is given as

$$dA = -SdT - pdV \tag{12.2}$$

where S is entropy, p is the pressure and V the volume. For changes occurring at constant T, as in a typical creep experiment, $dA = -p\,dV$. Upon rearrangement,

$$p = -\frac{\partial A}{\partial V}$$

Multiplying both sides by the volume of a formula unit, Ω_{fu} and noting that V/Ω_{fu} is nothing but the number of formula units, f.u., per unit volume, N, one finds that

$$p\Omega_{fu} = -\frac{\partial A}{\partial V}\Omega_{fu} = -\frac{\partial A}{\partial (V/\Omega_{fu})} = -\frac{\partial A}{\partial N} \tag{12.3}$$

Thus $\partial A/\partial N$ represents the excess (due to stress) chemical potential per f.u., or $\Delta\mu = \tilde{\mu} - \tilde{\mu}_0$ Here $\tilde{\mu}_0$ is the standard chemical potential of the f.u. in a *stress-free* solid (see Chap. 5).

By equating p with the applied stress, σ, the chemical potential of the f.u. in a stressed solid is given by

$$\tilde{\mu} = \tilde{\mu}_0 - \sigma\Omega_{fu} \tag{12.4}$$

By convention, *σ is considered positive when the applied stress is tensile and negative when it is compressive.* Note that multiplying Ω_{fu} by Avogadro's number N_{Av} yields the molar volume, V_m. By working with Ω_{fu} it is implicit that we are dealing with atoms and not moles and thus Boltzmann's constant k—and not R—is the operative constant. It is for this reason that the tilde is used in Eq. (12.4) (see also Chap. 7).

To better understand the origin of Eq. (12.4), consider the situation depicted schematically in Fig. 12.2, where four pistons are attached to four sides of a cube of material such that the pressures in the pistons are unequal with, say, $P_A > P_B$. These pressures will result in normal compressive forces $-\sigma_{11}$ and $-\sigma_{22}$ on faces A and B, respectively. If a f.u. is now removed from surface A (e.g., by having it fill in a vacancy just below the A surface), piston A will move by a volume Ω_{fu}, and the work done *on the system* is $\Omega_{fu}P_A = \Omega_{fu}\sigma_{11}$. By placing a formula unit on surface B (e.g., by having a f.u. from just below the surface diffuse to the surface), work is done by the system: $\Omega P_B = -\Omega\sigma_{22}$. The net work done is thus

$$\Delta W_{A\Rightarrow B} = \Omega_{f.u.}(P_B - P_A) = \Omega_{f.u.}(\sigma_{11} - \sigma_{22}) \tag{12.5}$$

which is a fundamental result because it implies that energy can be recovered - that is, ΔW is negative - if atoms diffuse from areas of high stress to areas of lower compressive stresses (see Worked Example 12.1*a*).

[174] The Helmholtz free energy A represents the changes in free energy of a system when they are carried out under constant *volume*. In contrast, ΔG represents the free-energy changes occurring at constant *pressure*. However, since volume changes in condensed phases, and the corresponding work against atmospheric pressure, are small, they can be neglected, and in general for solids, $\Delta G \approx \Delta A$.

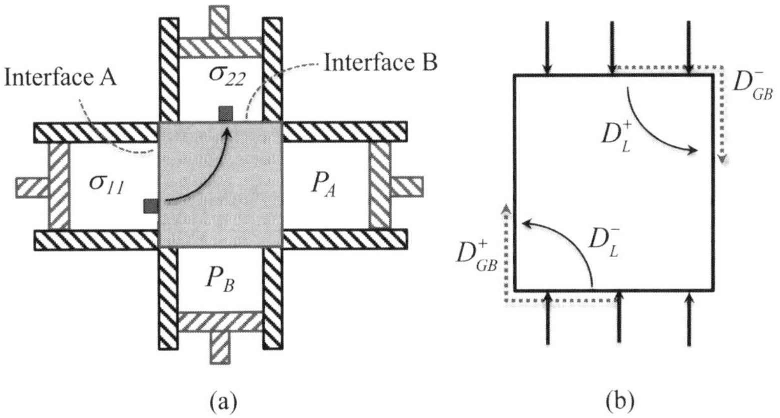

FIGURE 12.2 (a) Schematic of thought experiment invoked to arrive at Eq. (12.5). (b) Possible paths for cationic and/or anionic diffusion within and around a given grain compressed along the vertical direction.

Note that when $\sigma_{11} = -\sigma_{22} = \sigma$, the energy recovered will be

$$\Delta W = -2\Omega_{f.u.}\sigma \tag{12.6}$$

This energy is a direct measure of the driving force available for a f.u. to diffuse from a region that is subjected to a compressive stress, to an area subjected to an equal tensile stress.

Note the energy recovered when an f.u. moves from *just below* interface A to *just below* interface B is orders of magnitude lower than that given by Eq. (12.5). In other words, the strain energy contribution – see Ch. 11) to the process is *not* the driving force—it is only when the atoms or f.u. "*plate out*" onto the surface that the energy is recovered. The fundamental conclusion is thus: *Atom movements that result in shape changes are much more energetically favorable than ones that do not result in such changes.* (This last point is best made by example.

Worked Example 12.1

(*a*) Refer to Fig. 12.2a. If P_A is 25 MPa and P_B is 15 MPa, calculate the energy changes for an atom that diffuses from interface A to interface B. Compare this energy change to the average thermal energy of the atoms at room temperature and at 1000 K. (*b*) Repeat part (a) with σ_{22} = 100 MPa and σ_{11} = 0. Compare the value obtained to the elastic strain energy changes for the same system. Information you may find useful: Y = 150 GPa and V_m = 10 cm³/mol. State all assumptions.

ANSWER

(*a*) If V_m 10 cm³/mol, then $\Omega_{f.u} = 1.67 \times 10^{-29}$ m³. According to convention, $\sigma_{11} = -20$ MPa and $\sigma_{22} = -10$ MPa, and the net energy recovered is given by Eq. (12.5), or

$$\Delta W_{A\Rightarrow B} = 1.67\times10^{-29}[-25-(-15)]\times10^{6} = -1.67\times10^{-22}\ \text{J/atom}$$

$$= -100.5\ \text{J/mol} = -0.001\ \text{eV}$$

The thermal energy available to the atoms at room temperature $\approx kT = 0.026$ eV. At 1000 K, the thermal energy is ≈ 0.9 eV which is $>>0.001$.

(*b*) Since $\sigma_{22} = 100$ MPa and $\sigma_{11} = 0$, this corresponds to a simple tension experiment and

$$\Omega_{f.u.}\sigma = -1.67\times10^{-29}\times(100-0)\times10^{6} = -1.67\times10^{-21}\ \text{J/atom}$$

$$= -1005\ \text{J/mol}$$

For the second part, the elastic energy per unit volume is given by [see Eq. (11.3)]

$$U_{elas} = -\frac{1}{2}\frac{\sigma^2}{Y} = -\frac{1}{2}\frac{(100\times10^{6})^2}{150\times10^{9}} = -33{,}000\ \text{J/m}^3$$

It follows that the energy change for 1 mole, 10 cm^3 or 10^{-5} m^3 $= -0.33$ J/mol, which is roughly 3000 times *smaller* than the $\sigma\Omega$ term, which is ≈ 1005 in this case. Note that in all cases energy is recovered which is accounted for by the negative sign.

ATOMIC MECHANISMS DURING CREEP

Although Eqs. (12.4) to (12.6) elucidate the nature of the driving force operative during creep, they do not shed any light on *how* the process occurs at the *atomic level*. To do that, one has to go one step further and explore the effect of applied stresses on vacancy concentrations. For the sake of simplicity, the following discussion assumes creep is occurring in a pure elemental solid. The complications that arise from ambipolar diffusion in ionic compounds are discussed later. The equilibrium concentration of vacancies, c_0, under a *flat and stress-free* surface is given by [Eq. (6.7)], which can be rewritten as

$$c_0 = K'\exp\left(-\frac{h_d}{kT}\right) \tag{12.7}$$

where h_d is the enthalpy of vacancy formation. Here the entropy of defect formation and all pre-exponential terms are included in K'. Since the chemical potential of an atom under a surface subjected to a stress is either greater or smaller than that over a stress-free flat surface by $\Delta\mu$ [Eq. (12.4)], this energy has to be accounted for when one is considering vacancy formation. It follows that

$$c_{11} = K'\exp\left(-\frac{h_d + \Delta\mu}{kT}\right) = c_0 \exp\frac{\Omega_{fu}\sigma_{11}}{kT} \tag{12.8}$$

and similarly,

$$c_{22} = c_0 \exp\frac{\Omega_{fu}\sigma_{22}}{kT} \tag{12.9}$$

where c_{ii} is the concentration of vacancies just under a surface subjected to a normal stress σ_{ii} (refer to Fig. 12.2a). Subtracting these two equations and noting that in most situations $\sigma \Omega_{fu} \ll kT$ (see Worked Example 12.1) one obtains (recall $e^x = 1 - x$ for small x):

$$\boxed{\Delta c = c_{22} - c_{11} = \frac{c_0 \Omega_{fu}(\sigma_{22} - \sigma_{11})}{kT}} \tag{12.10}$$

which is a completely general result. In the special case where $\sigma_{11} = -\sigma_{22} = \sigma$, it simplifies to

$$\Delta c_{t-c} = c_{tens} - c_{comp} = \frac{2c_0 \Omega_{fu} \sigma}{kT} \tag{12.11}$$

Equations (12.10) and (12.11) are of fundamental importance since they predict that *the vacancy concentrations in tensile regions are higher than those in compressive regions* (Fig. 12.3*a*). In other words, stress or pressure gradients result in vacancy gradients, which in turn result in atomic fluxes carrying atoms, or

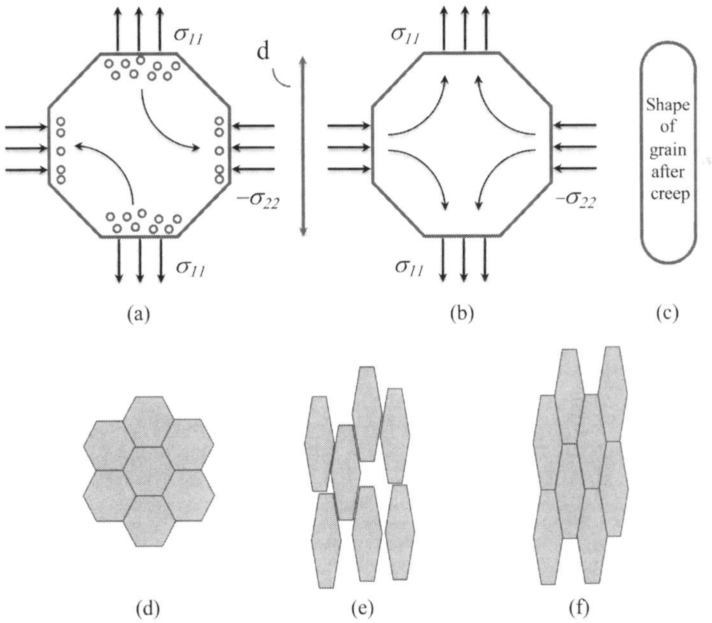

FIGURE 12.3 (*a*) Vacancy concentration gradients that develop as a result of stress gradients. The vacancy concentrations are higher below the tensile surface. Curved arrows denote direction of *vacancy* fluxes. (*b*) Schematic of a grain of diameter d subjected simultaneously to a vertical tensile and a side compressive stress. Curved arrows denote direction of *atomic* fluxes. (*c*) Shape of grain after creep has occurred; *(d)* Model microstructure prior to creep deformation; *(e)* Same as *(d)* without allowing for GB sliding. *(f)* Same as *(e)* after allowing for GB sliding.

matter, away in the opposite direction (Fig. 12.3*b*). It is only by appreciating this fact that sintering, creep and the densification occurring during hot pressing, among other phenomena, can be truly understood.

DIFFUSIONAL FLUXES

In Eq. (7.30), it was shown that the flux of atoms is related to the driving force by

$$J_i = \frac{c_i D_i}{kT} f \tag{12.12}$$

where f, c_i and D_i, are, respectively, the driving force per atom, atomic concentration and diffusivity. Once again, by assuming that $\sigma_{11} = -\sigma_{22} = \sigma$, the chemical potential *difference* per atom between the top and side faces of the grain boundary shown in Fig. 12.3*b* is simply $\Delta\tilde{\mu} = -2\sigma\Omega_{fu}$ [Eq. (12.6)]. This chemical potential difference acts over an average distance d/2, where d is the grain diameter; that is, $f = -d\tilde{\mu}/dx = 4\sigma\Omega_{fu}/d$, which when combined with Eq. (12.12), results in

$$J_i = \frac{c_i D_i}{kT}\frac{4\sigma\Omega_{fu}}{d} \tag{12.13}$$

The total number of atoms transported in a time t, crossing through an area A, is $N = J_i A_t$. Given that the volume associated with N f.u. is $\Omega_{fu}N$, it can be shown that the resulting strain from the displacement of the two opposite faces is given by

$$\varepsilon = \frac{\Delta d}{d} = \frac{2(\Omega_{fu}N/A)}{d} = \frac{2\Omega_{fu}J_i t}{d} \tag{12.14}$$

Combining Eqs. (12.13) and (12.14), the corresponding strain rate is given by

$$\boxed{\dot{\varepsilon} = \frac{8D_i\Omega_{fu}\sigma}{d^2kT}} \tag{12.15}$$

This expression is known as the **Nabarro–Herring** expression for creep, and it predicts that

1. The creep rate is inversely proportional to the square of the grain size d. Thus, solids with large grains are more resistant to creep than their fine-grained counterparts. This is well-documented experimentally.
2. The creep rate is proportional to the applied stress, which is also experimentally observed, but as discussed in greater detail below, only at lower stresses. At higher stresses, the stress exponent is usually greater than 1.
3. The slope of a plot of ln (T dε/dt) versus 1/kT should yield the activation energy for creep. If creep occurs by lattice diffusion as assumed here, that value should be the same as that measured in a diffusion experiment on the same material. This is often found to be the case, especially for metals.
4. Compressive stresses result in negative strains or shrinkage, while tensile strains result in elongation parallel to the direction of applied stress (Fig. 12.3*c*).

In deriving Eq. (12.15), the diffusion path was assumed to be through the bulk, which is usually true at higher temperatures where bulk diffusion is faster than grain boundary, GB, diffusion. However, at lower T's, or for fine-grained solids, GB diffusion may be the faster path, in which case the expression for the creep rate, known as **Coble creep**, becomes

$$\boxed{\dot{\varepsilon} = \psi \frac{\delta_{gb} D_{gb} \Omega_{fu} \sigma}{d^3\, kT}} \tag{12.16}$$

where δ_{gb} is the grain boundary width and ψ is a numerical constant $\approx 14\pi$. Here D_i in Eq. (12.15) is replaced by $D_{gb}\delta_{gb}/d$. The term 1/d represents the density or number of grain boundaries per unit area. Consequently, δ_{gb}/d can be considered to be proportional to a "grain boundary cross-sectional area" (see Fig. 7.20). The use of Ω_{fu} is justified here and in other creep expressions because under steady-state conditions an entire f.u. needs to move.

It should be emphasized that Eqs. (12.15) and (12.16) are valid under the following conditions:

- ∞ The GBs are the main sources and sinks for vacancies. If that is not the case, it is possible for creep to become interface controlled.
- ∞ Local equilibrium is established at temperature and stress levels used, i.e., the sources and sinks are sufficiently efficient.
- ∞ Cavitation does not occur either at triple junctions or at GBs. It is important to note if there are no mechanisms to eliminate cavities at the GBs, the latter will nucleate. To understand why, refer to Fig. 12.3*d* and deform each grain along its vertical axis. If that were the only mechanism operative, the crystal would devolve to that shown in Fig. 12.3*e*, viz. one full of cavities. However, if one were to slide these grains—along their GBs —while they were being deformed, it is possible to obtain a cavity-free microstructure as shown in Fig. 12.3*f*. It follows that regardless of the creep mechanism occurring, if the microstructure after creep is cavity free, then one can assume with some degree of confidence that some form of GB sliding occurs.

Since both GB and volume, or lattice diffusion can contribute independently to creep, the overall creep rate can be represented by the sum of Eqs. (12.15) and (12.16). For metals and elemental crystals, the nature of the diffusing species, *i*, is obvious. For binary ceramics, however, the nature of *i* is more complicated because the *entire* f.u. needs to diffuse. In other words, the diffusion must be ambipolar (see Chap. 7). If that is the case, it follows that the slower ion is *rate-limiting*. In binary ceramics, the rate-limiting step is always *the slower-diffusing species moving along its fastest possible path*. To illustrate refer to Fig. 12.2*b*, where four diffusion paths for moving atoms from the top surface loaded in compression to one normal to it. Two of the paths are associated with cations, D_L^+ and D_{GB}^+, and two with the anions, D_L^- and D_{GB}^- (L and GB refer to lattice and GB, respectively). Applying the aforementioned idea, consider two cases: First, a solid for which, $D_L^+ > D_{GB}^+ > D_L^- > D_{GB}^-$. Here, the rate-limiting step will be D_L^- since that is the slower species along its fastest path. The creep will thus be limited by anions diffusing through the lattice. For the second case, let us assume, $D_L^+ > D_{GB}^+ > D_{GB}^- > D_L^-$. Here the creep will, again, be limited by the anions, but in this case they would be diffusing along the GBs.

In general, for a binary or more complex compounds, a complex diffusivity, $D_{complex}$—which takes into account the various diffusion paths possible for each of the charged species in the bulk and along the GBs, as well as the effective widths of the latter—is the one to use. (See Problem 12.1 for an example of such an expression) For many practical applications, however, $D_{complex}$ simplifies to the diffusivity of the rate-limiting ion along its fastest path as just discussed.

One final note: In Chap. 7 it was stated that it was of no consequence whether the flux of atoms or defects was considered. To illustrate this important notion again, it is worthwhile to derive an expression for the creep rate based on the flux of defects. Substituting Eq. (12.11) in the appropriate flux equation for the diffusion of vacancies, i.e.,

$$J_v = -D_v \frac{\Delta C_{t-c}}{\Delta x} = -\frac{c_o D_v}{kT} \frac{4\Omega_{fu}\sigma}{d} \tag{12.17}$$

results in an expression which, but for a negative sign—which is to be expected since the atoms and vacancies are diffusing in opposite directions—is identical to Eq. (12.13) (Recall in Chap. 7, it was shown that $c_i D_i = c_v D_v = c_o D_v$. Thus, once more it is apparent that it is equivalent whether one considers the flux of the atoms or the defects; the final result has to be the same, which is comforting.

12.2.2 VISCOUS CREEP

Many structural ceramics often contain significant amounts of glassy phases at the GBs, and it is now well established that for many of them the main creep mechanism is not diffusional, but rather results from the softening and viscous flow of these glassy phases. Several mechanisms have been proposed to explain the phenomenon, most notable among them being these three:

SOLUTION REPRECIPITATION

This mechanism is similar to the one occurring during liquid-phase sintering, where the dissolution of crystalline material into the glassy phase occurs at the interfaces loaded in compression and their reprecipitation on interfaces loaded in tension. The rate-limiting step in this case can be either the dissolution kinetics or transport through the boundary phase, whichever is slower. This topic was discussed in some detail in Chap. 10, and will not be repeated here.

VISCOUS FLOW OF A GLASSY LAYER

As the temperature rises, glass viscosity decreases, and viscous flow of the glassy layer from between the grains can result in creep. The available models predict that the effective viscosity of the material is inversely proportional to the cube of the volume fraction f of the boundary phase, i.e.,

$$\eta_{eff} = (\text{const}) \frac{\eta_i}{f^3} \tag{12.18}$$

where η_i is the intrinsic or bulk viscosity of the GB phase. Since the shear strain rate and shear stress are related by

$$\dot{\varepsilon} = \frac{\tau}{\eta_{\mathrm{eff}}} \tag{12.19}$$

this model predicts a stress exponent of 1, which has sometimes been observed. One difficulty with this model is that as the grains slide past one another and the viscous fluid is squeezed out from between them, one would expect the grains to eventually interlock. Thus, unless significant volume fractions of the glassy phase exist, this process must, at best, be a transient process. One interesting way to test whether this process is operative is to compare the creep rates of the same material in both tension and compression. If this process is operative, a large difference in the creep rates should be observed.

VISCOUS CREEP CAVITATION[175]

In some ceramic materials, notably those that contain glassy phases, failure commonly occurs intergranularly by a time-dependent accumulation of creep damage in the form of GB cavities. The exact mechanism by which the damage accumulates depends on several factors such as microstructure, volume of glassy phase, temperature and applied stress, but two limiting mechanisms have been identified: distributed and localized damage.

Low stresses and long exposure times tend to favor distributed damage, where cavities are presumed to nucleate and grow throughout the GB region, with failure occurring by the coalescence of these cavities to ultimately form a critical crack. A typical intergranular failure revealing the presence of numerous cavities along two-grain facets is shown in Fig. 12.4*a*.

In ceramics that are essentially devoid of glassy phases, the cavities are believed to occur by vacancy diffusion and clustering. More commonly, however, cavitation is thought to occur by the nucleation of voids within the intergranular glassy phases.

High stresses and short exposure times, however, tend to favor a more localized type of damage, where the nucleation and growth of the cavities occur locally within the field of influence of a local stress concentrator, such as a preexisting flaw. Here two crack growth mechanisms have been identified: (1) direct extension of the creep crack along GBs by diffusive or viscous flow and/or (2) cavitation damage ahead of a crack tip and its growth by the coalescence of these cavities, followed by the nucleation and growth of fresh cavities, and so forth, as shown schematically in Fig. 12.4*b*.

The problem is further complicated by the fact that there are three timescales to worry about: sliding of grains with respect to each other, which in turn creates the negative pressure at the triple points that is responsible for the nucleation of the cavities, followed by the time needed to nucleate a cavity and finally the growth and coalescence of these cavities—any one of which can be rate-limiting.

This particular creep mechanism occurs in Si_3N_4, typically fabricated by liquid-phase sintering, and thus almost always contains some glassy phases at the GBs, which naturally renders cavitation creep important. How to best solve the problem is not entirely clear but choosing glass compositions that can be easily

[175] For a review, see K. Chan and R. Page, *J. Amer. Cer. Soc.*, **76**, 803 (1993).

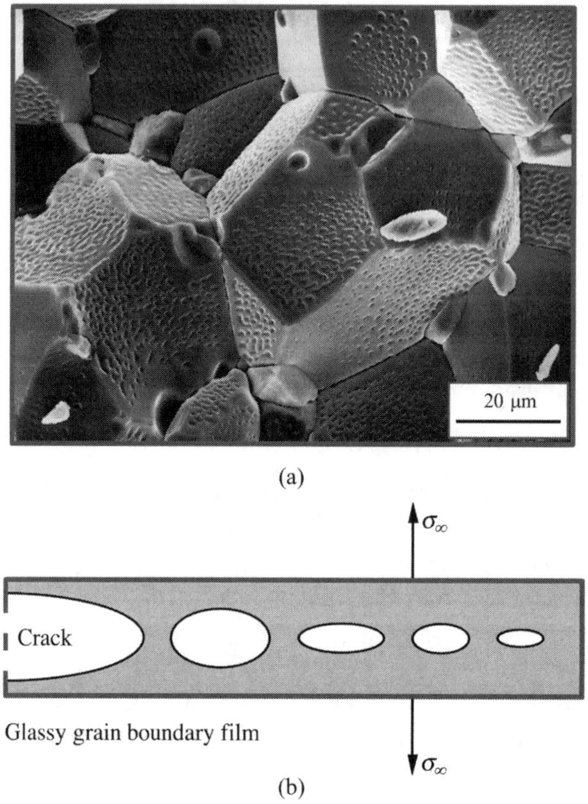

FIGURE 12.4 (*a*) Intergranular fracture of an alumina sample showing creep cavitation due to compressive creep at 1600°C. Note closely spaced cavities along the two-grain facets. (*b*) Schematic of cavity formation in a viscous grain boundary film as a result of applied tensile stress. (Micrograph courtesy of R. Page, C. Blanchard and R. Railsback, Southwest Research Institute, San Antonio, TX.)

crystallized in a post-fabrication step and/or introducing second phases such as SiC, which could control GBs sliding, have been attempted, with sometimes promising results.

12.2.3 DISLOCATION CREEP

As noted above, the experimentally observed creep power law exponents, especially at higher temperatures and applied stresses, are in the range of 3 to 8 (Fig. 12.6), which none of the aforementioned models predict. Thus, to explain the higher stress exponents, it has been proposed that the movement of atoms from regions of compression to tension occurs by the coordinated movement of "blocks" of material via dislocation glide and climb. In this mechanism, the creep rate can be formally expressed as

$$\dot{\varepsilon}_{ss} = \mathbf{b}\left\{\rho\upsilon(\sigma) + \frac{\lambda d\rho(\sigma)}{dt}\right\} \tag{12.20}$$

in which **b** is Burgers vector, ρ the dislocation density in m^{-2}, $v(\sigma)$ the average velocity of a dislocation at an applied stress σ, $d\rho/dt$ is the rate of nucleation of the dislocations at stress σ and λ is the average distance they move before they are pinned. The main difficulty in developing successful creep models and checking their validity stems primarily from the fact that many of these parameters are unknown and are quite nonlinear and interactive. Progress has been achieved for some materials, however.[176]

12.2.4 GENERALIZED CREEP EXPRESSION

In general, the steady-state creep of ceramics may be expressed in the form[177]

$$\dot{\varepsilon} = \frac{(\text{const})\, D_i G \mathbf{b}}{kT}\left(\frac{\mathbf{b}}{d}\right)^r \left(\frac{\sigma}{G}\right)^p \tag{12.21}$$

where G is the shear modulus, r is a **grain-size exponent** and p is the stress exponent defined in Eq. (12.1). It can be shown (see Prob. 12.3) that if $p=1$, then Eqs. (12.15) and (12.16) are of this form. Note that to obtain Eq. (12.21) from Eq. (12.15) or (12.16), $\Omega_{fu}c_i$ is assumed to be 1.

Based on Eq. (12.21), the creep behavior of ceramics can be divided into two regimes:

1. A low-stress, small grain-size regime, where the creep rate is a function of grain size and the stress exponent is unity. Consequently, a plot of $\log\{\dot{\varepsilon}kT/(DG\mathbf{b})\}(d/\mathbf{b})^r$ vs. $\log(\sigma/G)$ should yield a straight line with slope 1. The grain-size exponent, r, will depend on the specific creep mechanism; it is 2 for Nabarro–Herring creep and 3 for Coble creep. Figure 12.5 compares the experimental normalized

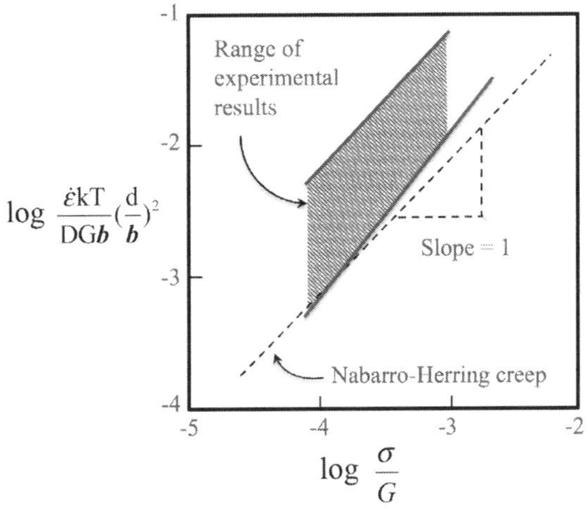

FIGURE 12.5 Summary of normalized creep rate versus normalized stress for alumina. Dotted line is what one would predict based on Eq. (12.15). (Data taken from W. R. Cannon and T. G. Langdon, *J. Mater. Sci.*, 18, 1–50, 1983.)

176 O. A. Ruano, J. Wolfenstine, J. Wadsworth, and O. Sherby, *J. Amer. Cer. Soc.*, **75**, 1737 (1992).
177 B. M. Moshtaghioun, D. G. García, A. D. Rodríguez and N. P. Padture, *J. Europ. Cer. Soc.* **35**, 1423 (2015).

creep rates of alumina to σ/G collected from several sources (shaded area) to that predicted if Nabarro–Herring creep were the operative mechanism. The agreement is rather good, considering the uncertainties in the diffusion coefficients, etc. Note that since it is d^2 that is plotted, and the slope is unity, this confirms the creep is of the Nabarro–Herring type.

2. A high stress level regime where the creep rate becomes independent of grain size, that is, $r = 0$, and p lies between 3 and 7. In this regime, plots *of log* $\{\dot{\varepsilon}kT/(DG\mathbf{b})\}$ vs. $\log(\sigma/G)$ should once again yield straight lines. In Fig. 12.6, the normalized creep rate for a number of ceramics is plotted as a function of $\log(\sigma/G)$, where it is obvious that all the data fall on straight lines with slopes between 3 and 7.

Figure 12.7 schematically summarizes the creep behavior of ceramics over a wide range of applied stresses as a function of grain size. From the figure, it is obvious that small grains are detrimental to the creep rates

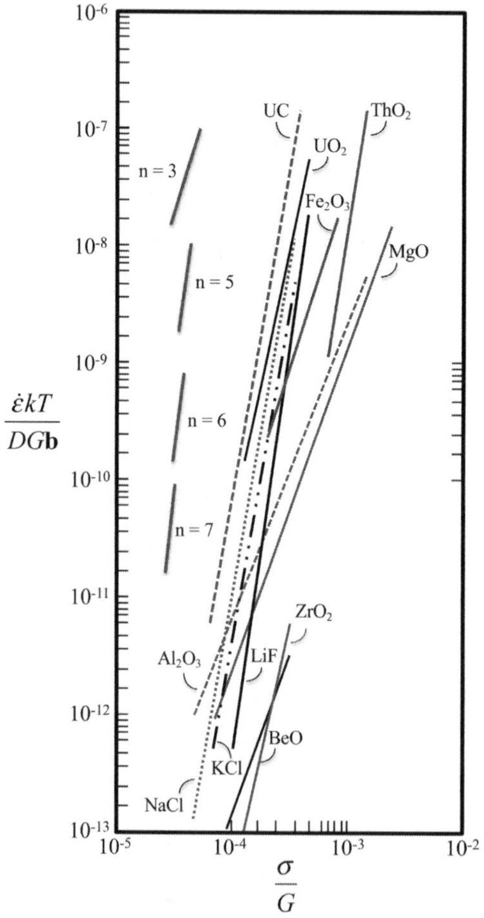

FIGURE 12.6 Summary of power law creep data for a number of ceramics. (Data taken from W. R. Cannon and T. G. Langdon, *J. Mater. Sci.*, 18, 1–50, 1983.)

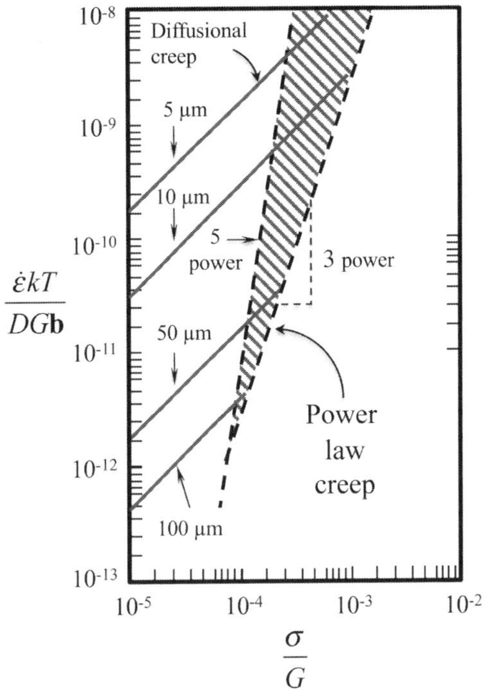

FIGURE 12.7 Effect of grain size on normalized creep rate versus normalized stress. As grains, and consequently diffusion distances, become smaller, diffusional creep becomes more important. (Data taken from W. R. Cannon and T. G. Langdon, *J. Mater. Sci.*, 18, 1–50, 1983.)

at low stresses, but that at higher stresses the intragranular movement of dislocations by climb, or glide, is the operative and more important mechanism. In Fig. 12.7, the roles of intergranular films and the formation of multigrain junction cavitation are not addressed, but the stress exponents obtained experimentally from such mechanisms also fall in the range of 2 to 7, which, needless to say, can further cloud the interpretation of creep results.

Before changing topics, it is worth noting that probably the most generalized expression for creep rate and the one recommended here is

$$\boxed{\dot{\varepsilon} = (\text{const.})\left(\frac{\Omega_{fu}}{\mathbf{b}^2}\right)(c_i D_i)\left(\frac{\mathbf{b}}{d}\right)^{n'}\left(\frac{\sigma}{G}\right)^{m''}\frac{\Omega_{fu}\sigma}{kT}} \tag{12.22}$$

The advantages of this relationship relative to Eq. (12.21) are: (i) It explicitly spells out the $c_i D_i$ product. (ii) It separates the stress into two terms: the first, with exponent m″, describes dislocation creep. The second, and last, term is nothing but the driving for creep, which applies regardless of mechanism!

12.3 SUBCRITICAL CRACK GROWTH

Subcritical crack growth, *SCG*, refers to the slow growth of a subcritical flaw as a result of its exposure to the combined effect of stress and a corrosive environment.[178] As discussed in greater detail below, the combination of a reactive atmosphere and a stress concentration can greatly enhance the rate of crack propagation. For instance, silica will dissolve in water at a rate of 10^{-17} m/s, whereas the application of a stress can cause cracks to grow at speeds greater than 10^{-3} m/s. The insidiousness of, and hence the importance of understanding, this phenomenon lies in the fact that as the crack tip advances, the material is effectively weakened and eventually can give way suddenly and catastrophically after years of service.

The objective of this section is twofold: to describe the phenomenon of SCG and to relate it to what is occurring at the atomic level at the crack tip. Before proceeding much further, however, it is important to briefly outline how this effect can be quantified.

EXPERIMENTAL DETAILS: MEASURING SCG

The techniques and test geometries that have been used to measure SCG in ceramics are several. They all, however, share a common principle, namely, the subjection of a well-defined crack to a well-defined stress intensity K_I, followed by a measurement of its velocity v. The technique considered here, the advantages of which are elaborated upon below, is the double torsion geometry shown in Fig. 12.8. For this geometry, K_I is given by

$$K_I = PW_m\sqrt{\frac{3(1+\nu)}{Wd^3d_n}}$$

where P is the applied load and ν is Poisson's ratio; all the other symbols and dimensions are defined in Fig. 12.8. Note that the crack length c does not appear in this equation, which means that for this geometry K_I is not a function of c! And since K_I is *not* a function of crack length, c, then neither is v. It follows that a constant v should be observed for any given load, which greatly simplifies the measurements and their analysis. The main disadvantage of the technique, however, is that reasonably large samples that require some machining are needed.

The measurements are carried out as follows:

1. A starter crack, of length c_0, is introduced in a specimen, and a load P is applied, as shown in Fig. 12.8. As a result, the starter crack will grow with time.
2. The rate of crack growth or velocity, v, is measured, usually optically. For instance, two marks are placed on the specimen surface and the time required for the crack to propagate that distance is

[178] A good example of this phenomenon that should be familiar to many is the slow crack growth, over time, in a car's windshield after it had been damaged.

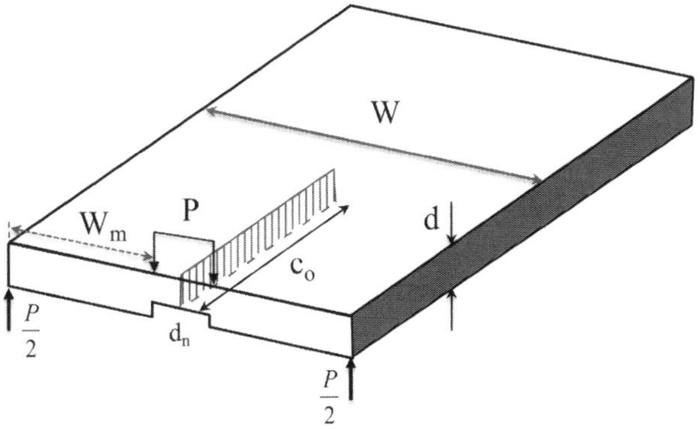

FIGURE 12.8 Schematic of double torsion specimen.

measured. The crack velocity is then simply $\upsilon = \Delta c/\Delta t$. The experiment is repeated under different loading conditions, either on the same specimen, if it is long enough, or on different specimens, if not.

If explored over a wide enough spectrum, a ln υ vs. K_I plot will exhibit four regions, shown in Fig. 12.9*a*:

- A threshold region below which no crack growth is observed
- Region I, where the crack growth is extremely sensitive to K_I and is related to it by an exponential function of the form

$$\upsilon = \mathbf{A}^* \exp \alpha K_I \tag{12.23}$$

where $\mathbf{A}^*$ and α are empirical fitting parameters.

- Region II, where υ is high appears to be independent of K_I.
- Region III, where υ increases even more rapidly with K_I than in region I.

To measure n (defined in Eq. (12.30)) directly from υ vs. K_I curves is often difficult and time consuming. Fortunately, a simpler and faster technique, referred to as a **dynamic fatigue test** (not to be confused with the normal fatigue test discussed below), is available. In this method, the strain rate dependence of the average strength to failure is measured; i.e., the samples are loaded at varying rates and their strength at failure is recorded. It can be shown (see App. 12A) that the mean failure stress σ_1 at a constant strain rate $\dot{\varepsilon}_1$ is related to the mean failure stress σ_2 at a different strain rate, $\dot{\varepsilon}_2$, by

$$\left(\frac{\sigma_1}{\sigma_2}\right)^{n+1} = \frac{\dot{\varepsilon}_1}{\dot{\varepsilon}_2} \tag{12.24}$$

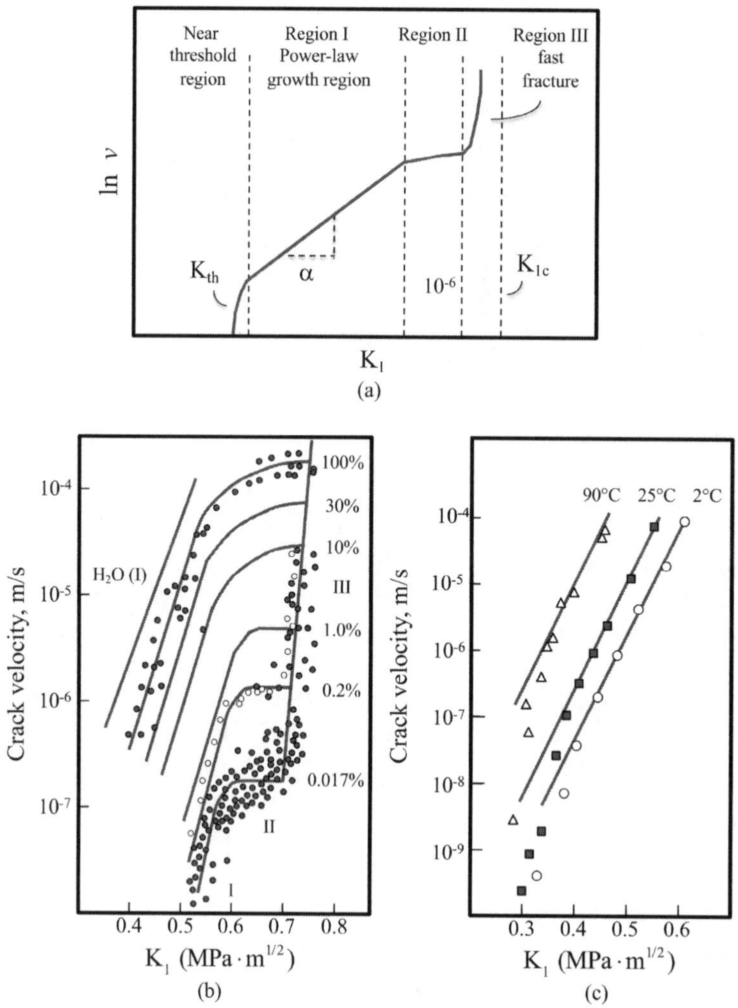

FIGURE 12.9 (*a*) Schematic of crack propagation rates as a function of K_I, where four stages are identified. (*b*) Actual data for soda-lime glass tested in N_2 gas of varying relative humidity shown on the right-hand side. (*c*) Temperature dependence of crack propagation in same glass in water. (M. Wiederhorn, *J. Amer. Cer. Soc.*, 50, 407, 1967.)

Hence by measuring the stress to failure at different strain rates, n can be directly calculated from this relationship.

Yet another variation of this test, which is also used to obtain creep information is to simply attach a load to a specimen and measure it's time to failure. The results are then plotted in a format identical to the one shown in Fig. 12.11*a* for cyclic fatigue and are referred to as **static fatigue** or **stress/life curves**.

Typical v vs. K_1 data for soda-lime silicate glasses tested as a function of humidity are shown in Fig. 12.9b, where the following points are salient:

- v is a strong function of K_1 or applied stress. Note log y-axis scale.
- Humidity has a dramatic effect on v; increasing the relative humidity in the ambient atmosphere from 0.02% to 100%, results in > 3 orders-of-magnitude increase in v.
- Clear identification of the three regions just described is possible.
- The presence of a threshold K_1 is not clearly defined because of the difficulty of measuring crack velocities that are much slower than 10^{-7} m/s.
- The dramatic effect of increasing the temperature on crack velocity is shown in Fig. 12.9c. The v increases by about 2 orders of magnitude over a temperature range of $\approx 100°C$, typical of thermally activated processes.

To understand this intriguing phenomenon, it is imperative to appreciate what is occurring at the crack tip on the atomic scale. Needless to say, the details will depend on several factors, including the chemistry of the solid in which SCG is occurring, the nature of the corrosive environment, temperature and stress levels applied. However, given the ubiquity of moisture in the atmosphere and the commercial importance of silicate glasses, the following discussion is limited to SCG in silicate-based glasses, although the ideas presented are believed to have general validity. Furthermore, since, as discussed in greater detail in Sec. 12.5.1, it is region I that determines the lifetime of a part it is dealt with in some detail below. The other regions will be briefly touched upon at the end of this section.

The models that have been suggested to explain region I for glasses in the presence of moisture can be divided into three categories: diffusional (i.e., a de-sintering of the material along the fracture plane), plastic flow and chemical reaction theories. Currently the chemical reaction approach appears to be the most consistent with experimental results and is the one developed here.

It is now generally accepted that a stressed Si–O–Si bond at a crack tip will react with water to form two Si–OH bonds according to the following chemical reaction:

$$\text{Si–O–Si} + \text{H}_2\text{O} = 2\text{Si–OH} \qquad \text{(I)}$$

In this process—illustrated in Fig. 12.10 and referred to as *dissociative chemisorption*—a water molecule is assumed to diffuse to, and chemisorb at, the crack tip (Fig. 12.10a). The molecule then rotates so as to align the lone pairs of its oxygen molecules with the unoccupied electron orbitals of the Si atoms (Fig. 12.10b). Simultaneously, the hydrogen of the water molecule is attracted to a bridging oxygen. Eventually, to relieve the strain on the Si–O–Si bond, the latter ruptures and is replaced by two Si–OH bonds (Fig. 12.10c).

Experimentally, it is now fairly well established that water is not the only agent that causes SCG in glasses; other polar molecules such as methanol and ammonia have also been found to cause SCG, provided that the molecules are small enough to fit into the extending crack (i.e., smaller than about 0.3 nm).

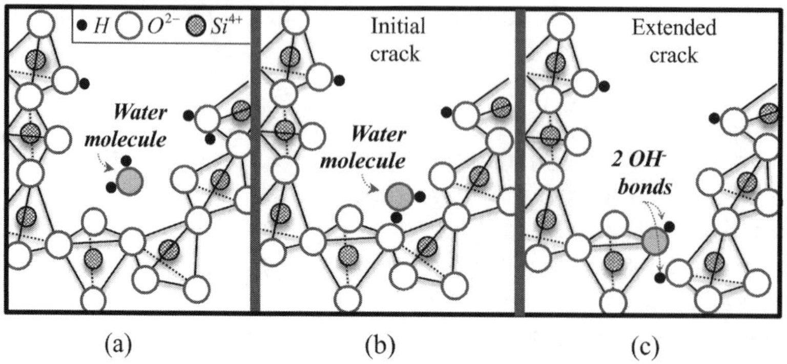

FIGURE 12.10 Steps in the dissociative chemisorption of water at the tip of a crack in silica glass. (*a*) Tip of crack with approaching water molecule; (*b*) chemisorption of water and its alignment; (*c*) breaking of Si–O–Si bond and the formation of two Si–OH bonds. (Adapted from T. Michalske and B. Bunker, *Sci. Amer.*, 257, 122, 1987.)

The basic premise of the chemical reaction theory is that the rate of reaction (I) is a strong function of K_1 at the crack tip. According to the absolute reaction rate theory,[179] the *zero-stress* reaction rate of a chemical reaction is given by

$$K_{r,o} = \kappa \exp\left(-\frac{\Delta G_P^* = 0}{RT}\right) = \kappa \exp\left(-\frac{\Delta H^* + T\Delta S^*}{RT}\right) \quad (12.25)$$

where κ is a constant and ΔH^* and ΔS^* are, respectively, the differences in enthalpy and entropy between the reactants in their ground and activated states.[180] In the presence of a hydrostatic pressure P, this expression has to be modified to read

$$K_r = \kappa \exp\left(-\frac{\Delta G^*}{RT}\right) \quad (12.26)$$

where ΔG^* is now given by

$$\Delta G^* = \Delta H^* - T\Delta S^* + P\Delta V^*$$

ΔV^* is the difference in volume between the reactants in their ground and activated states. Combining Eqs. (12.25) and (12.26), the following relationship is recovered

$$K_r = K_{r,o} \exp\left(-\frac{P\Delta V^*}{RT}\right) \quad (12.27)$$

[179] S. Glasstone, K. Laidler, and H. Eyring, *The Theory of Rate Processes*, McGraw-Hill, NY, 1941.
[180] The implicit assumption here is that the forward reaction rate is much faster than the backward rate.

This result is important because it predicts that applying a hydrostatic pressure ($P > 0$) to a reaction should slow it down, and vice versa. Physically, this can be more easily appreciated by considering a diatomic molecular bond: That a tensile stress along the bond axis must enhance the rate at which rupture occurs, and consequently its chemical reactivity, is obvious.

If now υ is assumed to be directly related to the reaction rate K_r and P is proportional to K_I at the crack tip, then Eq. (12.27) can be recast as[181] (see Prob. 12.5)

$$\boxed{\upsilon = \upsilon_0 \exp\left[\frac{-\Delta H^{o} + \beta K_I}{RT}\right]} \tag{12.28}$$

where υ_0, ΔH^* and β are empirical constants determined from $\log \upsilon$ vs. K_I types of plots.

This formalism accounts, at least qualitatively, for the principal external variables since:

1. It predicts a strong dependence of υ on K_I consistent with the behavior in region I.
2. It predicts an exponential temperature effect, as observed (Fig. 12.9*c*). Note that for constant temperatures, Eqs. (12.28) and (12.23) are of the same form.
3. Although not explicitly included in Eq. (12.28), the effect of moisture is embedded in the pre-exponential factor. This comes about by recognizing that the reaction rate [Eq. (12.26)] is proportional to the concentration of water (see Prob. 12.5*b*).

In much of the literature, an empirical power law expression of the form

$$\upsilon = A\left(\frac{K_I}{K_{Ic}}\right)^n \tag{12.29}$$

which is usually further abbreviated to[182]

$$\upsilon = A'' K_I^n \tag{12.30}$$

is often used to describe SCG, instead of Eq. (12.28). In this formalism, the temperature dependence is embedded in the A″ term. The sole advantage of using Eq. (12.29) or (12.30) over Eq. (12.28) is the ease with which the former can be integrated (see Sec. 12.5.1). Equation (12.28), however, must be considered more fundamental because, i) it has some scientific underpinnings and, ii) it explicitly predicts the exponential

[181] R. J. Charles and W. B. Hillig, pp. 511–527 in *Symposium on Mechanical Strength of Glass and Ways of Improving It*, Florence, Italy, 1961, *Union Scientifique Continentale du Verre*, Charleroi, Belgium, 1962. See also S. W. Weiderhorn, *J. Amer. Cer. Soc.*, 55, 81–85 (1972).

[182] Note that while A has the dimensions of m/s, A″ has the unwieldy dimensions of $m\ s^{-1}\ (Pa{\cdot}m^{1/2})^{-n}$.

T-dependence observed. Finally, note that the experimental results can usually be fitted equally well by either of these equations; extrapolations, however, can lead to considerable divergences.[183]

Before moving on, consider briefly the other regions observed in the v versus K_1 plots:

1. *Threshold region.* Although it is difficult to unequivocally establish experimentally that there is a threshold stress intensity, K_{th}, below which no crack growth occurs from v vs. K_1 types of curves, probably the most compelling results indicating that it indeed exists come from crack healing studies. At very low K_1 values, the driving force for crack growth is low, and it is thus not inconceivable that its growth rate of at some point would equal the rate at which it heals, the driving force for which would be a reduction in surface energy. In other words, a sort of dynamic equilibrium is established, and a threshold results.
2. *Region II.* The hypothesis explaining the weak dependence of v on K_1 in region II is that the v is limited by the rate of arrival of the corroding species at the crack tip.
3. *Region III.* This stage is not very well understood, but once again a combination of stress and chemical reaction is believed to accelerate the crack.

12.4 FATIGUE OF CERAMICS

It has long been assumed that because dislocation motion in ceramics is limited, strain hardening, and consequent crack extension during cyclic loading would not occur, and hence ceramics were not susceptible to fatigue damage. And indeed, ceramics with homogeneous microstructures such as glass or very fine-grained single-phase ceramics do not appear to be susceptible to cyclic loadings.

However, with the development of tougher ceramics that exhibit R curve behavior such as transformation-toughened zirconia and whisker- and fiber-reinforced ceramics (see Chap. 11), it is becoming clear that the situation is not as simple as first thought. R-curve behavior can be detrimental to fatigue life. Before tackling the fatigue micromechanisms, however, a brief description is warranted of what is meant by fatigue and what the relevant parameters are.

EXPERIMENTAL DETAILS: MEASURING FATIGUE

In a typical fatigue test, a sample is subjected to an alternating stress of a given amplitude and frequency. The *cyclic stress amplitude* is defined as

$$\sigma_{amp} = \frac{\sigma_{max} - \sigma_{min}}{2} \qquad (12.31)$$

---

[183] T. Michalske and B. Bunker, *J. Amer. Cer. Soc.*, **76**, 2613 (1993).

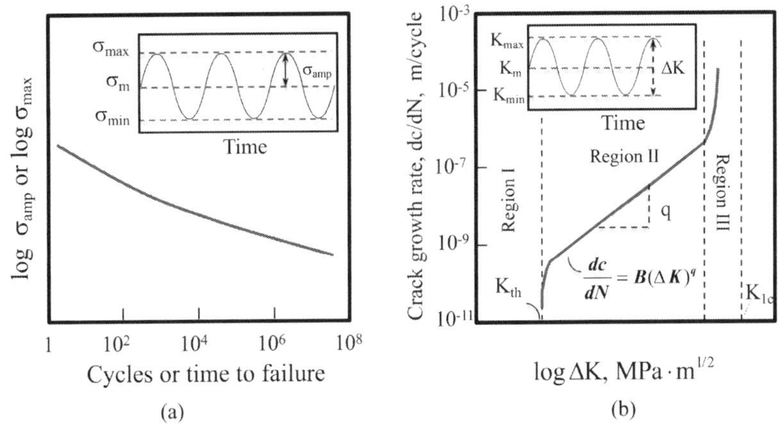

FIGURE 12.11 (a) Stress amplitude versus cycles to failure curve (S/N curve). Inset shows definition of stress amplitude σ_{amp}. (b) Curves of log (dc/dN) versus log ΔK_I. Slope of curve in region II is q. ΔK is defined in inset.

whereas the load ratio R is defined as

$$R = \frac{\sigma_{min}}{\sigma_{max}} \tag{12.32}$$

where σ_{min} and σ_{max} are, respectively, the minimum and maximum stresses to which the sample is subjected (Fig. 12.11a). The experiments can be carried out either in tension–tension, compression–compression or tension–compression, in which case R is negative.

Two types of specimens are typically used, smooth "crack-free" specimens or specimens containing long cracks, i.e., cracks of dimensions that are large with respect to the microstructural features, such as grain size, of the material.

For the smooth or crack-free specimens, the experiments are run until the sample fails. The results are then used to generate **S/N curves** where the applied stress amplitude is plotted versus the cycles to failure (which are equivalent to the time to failure if the frequency is kept constant), as shown in Fig. 12.11a.

For the specimens with long cracks, the situation is not unlike that dealt with in the previous section, except that instead of measuring v vs. K_I, the crack growth rate per cycle dc/dN is measured as a function of ΔK_I, defined as

$$\Delta K_1 = \xi(\sigma_{max} - \sigma_{min})\sqrt{\pi c} \tag{12.33}$$

where ξ is a geometric factor of the order of unity.

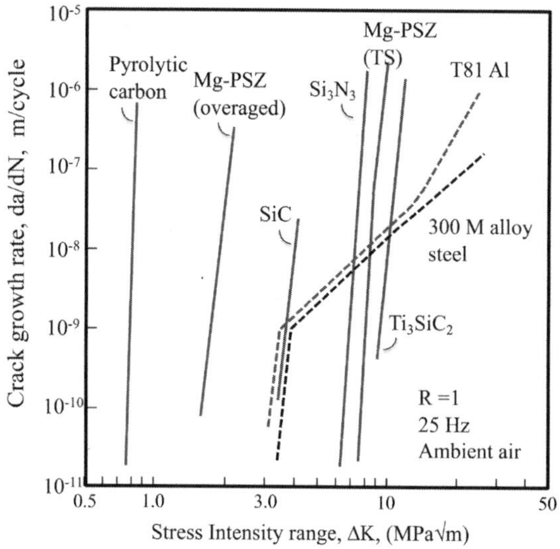

FIGURE 12.12 Cyclic fatigue long-crack propagation data for several ceramics compared to those for some typical metals. TZPs are tetragonal zirconia polycrystals (see Chap. 11). (Adapted from C. J. Gilbert et al. *Scr. Mater.*, 42, 761–767, 2000.)

Typical crack growth behavior of ceramics is represented schematically in Fig. 12.11*b* as log ΔK_I vs. log(dc/dN) curves. The resulting curve is sigmoidal and can be divided into three regions, labeled I, II and III. Below K_{th}, that is, region I, the cracks will not grow with cyclic loading. And just prior to rapid failure, the crack growth is accelerated once more (region III).

In the midrange, or region II, the growth rates are well described by

$$\frac{dc}{dN} = B(\Delta K_I)^q \qquad (12.34)$$

where B and q are empirically determined constants.

Typical long-crack data for a number of ceramics are shown in Fig. 12.12, where the log(dc/dN) vs. log ΔK_I curves are linear and quite steep, implying high q values. These studies also indicate that under cyclic loading, the thresholds for crack growth can be as low as 50% of the fracture toughness measured under monotonically increasing loads.

FATIGUE MICROMECHANISMS

At this point, the micromechanics of what is occurring at the crack tip in ceramic materials are not fully understood. The results, however, have established that (1) no one micromechanical model can successfully explain all fatigue data, (2) fatigue in ceramics appears to be fundamentally different from that in metals, where crack propagation results from dislocation activity at the crack tip, and (3) ceramics that exhibit *R*-curve behavior appear to be the most susceptible to fatigue, indicating that the cyclic nature of the loading somehow diminishes the effect of the crack-tip shielding mechanisms discussed in Chap. 11. For instance, in the case of fiber- or whisker-reinforced ceramics, it is believed that unloading induces fracture, or buckling, of the whiskers in the crack wake, that in turn reduces their shielding effect. If the toughening, on the other hand, is achieved primarily by grain bridging or interlocking, then the unloading cycle is believed to cause cracking and/or crushing of asperities between crack faces which reduces the frictional sliding at bridging interfaces.

Becuase of the similarity in behavior between fatigue and SCG [compare Figs. 12.9*a* and 12.11*b* or Eqs. (12.34) and (12.30)], one of the major experimental difficulties in carrying out fatigue experiments lies in ascertaining that the strength degradation observed is truly due to the cyclic nature of the loading and not due to SCG. Even more care must be exercised when the tests are carried out at higher temperatures since, as noted above, SCG is a thermally activated process and hence becomes more important at elevated temperatures.

The few studies on cyclic fatigue of ceramics at elevated temperatures seem to indicate that at high homologous temperatures (i.e., in the creep regime) cyclic fatigue does not appear to be as damaging as SCG. In the cases where it has been observed, the improved cyclic fatigue behavior at higher temperatures has been attributed to bridging of the crack surfaces by grain boundary glassy phases.

12.5 LIFETIME PREDICTIONS

Creep, fatigue and SCG are dangerous becuase they can result in sudden and catastrophic failure with *time*. Thus, from a design point of view, in addition to the probabilistic aspects of failure discussed in Chap. 11, the central question becomes: How long can a part serve its purpose reliably? The conservative approach, of course, would be to design with stresses that are below the threshold stresses discussed above. An alternative approach is to design a part to last for a certain lifetime, after which it would be replaced, or at least examined for damage. In the following sections, a methodology is described that can be used to calculate lifetime for each of the three phenomena dealt with so far.

12.5.1 LIFETIME PREDICTIONS DURING SCG

Replacing υ in Eq. (12.29) by dc/dt, rearranging, and integrating, one obtains

$$\int_0^{t_f} dt = \int_{c_i}^{c_f} \frac{dc}{\upsilon} = \frac{K_{Ic}^n}{A}\int_{c_i}^{c_f} \frac{1}{K_I^n} dc \qquad (12.35)$$

where c_i and c_f are the initial and final (just before failure) crack lengths, respectively; t_f is the time to failure. Recalling that $K_I = \Psi\sigma_a\sqrt{\pi c}$ [Eq. (11.12)], dc in Eq. (12.35) can be eliminated and recast in terms of K_I as

$$t_f = \frac{2K_{Ic}^n}{A\Psi^2\sigma_a^2\pi}\int_{K_i}^{K_{Ic}}\frac{dK_1}{K_1^{n-1}} \tag{12.36}$$

which upon integration yields

$$t_f = \frac{2K_{Ic}^n}{A\Psi^2\pi\sigma_a^2(n-2)}\left[\frac{1}{K_i^{n-2}} - \frac{1}{K_{Ic}^{n-2}}\right] \tag{12.37}$$

Given that for $n \approx 10$, even if K_i is as high as 0.5 K_{Ic}, the second term is less than 0.5% of the first, and thus with good accuracy[184]

$$t_f = \frac{2K_{Ic}^n}{A\Psi^2\pi\sigma_a^2(n-2)K_i^{n-2}} \tag{12.38}$$

When a similar integration is carried out on Eq. (12.28), the t_f is given by (see Prob. 12.5)

$$t_f = \frac{2}{\upsilon_0'}\left[\frac{RT}{\beta\sigma_a\Psi}\sqrt{\pi c_i} + \left(\frac{RT}{\beta\sigma_a\Psi}\right)^2\right]\exp\left(-\frac{\beta\sigma_a\Psi\sqrt{\pi c_i}}{RT}\right) \tag{12.39}$$

where $\upsilon_0' = \upsilon_0 \exp[-\Delta H^*/(RT)]$. Both Eqs. (12.38) and (12.39) predict that t_f of a component that is susceptible to SCG is a strong function of K_1 (see Worked Example 12.2).

Worked Example 12.2

For silica glass tested in ambient temperature water, $\upsilon_0 = 3\times10^{-22}$ m/s and $\beta = 0.182$ m$^{5/2}$. Estimate the effect of increasing K_1 from 0.4 to 0.5 MPa·m$^{1/2}$ on the lifetime.

ANSWER

By noting that $K_1 = \Psi\sigma_a\sqrt{\pi c}$, Eq. (12.39) can be rewritten as

$$t_f = (\text{const})e^{-\beta K_1/(RT)}$$

Here the dependence of the pre-exponential term on K_1 was ignored, which is a good approximation relative to the exponential dependence. Substituting the appropriate values for K_1 and β in this expression, one obtains

[184] Ignoring the second term in Eq. (12.37) says that the fraction of the lifetime the crack spends as its size approaches c_f is insignificant compared to the time taken by the crack to increase from c_i to $c_i + \delta c$, when its velocity is quite low.

$$\frac{t_{f@0.4}}{t_{f@0.5}} = \frac{\exp\left(-\dfrac{0.182\times 0.4\times 10^6}{8.314\times 300}\right)}{\exp\left(-\dfrac{0.182\times 0.5\times 10^6}{8.314\times 300}\right)} = 1476$$

In other words, a decrease in K_1 from 0.5 to 0.4 MPa·m$^{1/2}$ increases t_f by a factor of ≈1500!

The methodology just described—while useful in predicting t_f with a well-defined starter crack (t_f cannot be calculated without a knowledge of K_i or c_i)—is not amenable to a probabilistic analysis such as the one discussed in Sec. 11.5. However, as discussed below, the relationships derived for t_f, particularly Eq. (12.38), permit the straightforward construction of strength/probability/time (SPT) diagrams that can be used for design purposes.

SPT DIAGRAMS

For a set of identical specimens, i.e., given that A, K_1 or equivalently c_i and Ψ are constant, it follows directly from Eq. (12.38) that

$$\frac{t_2}{t_1}\left(\frac{\sigma_1}{\sigma_2}\right)^n = \text{constant} \tag{12.40}$$

where t_1 and t_2 are the lifetimes at σ_1 and σ_2, respectively. If one assumes that a sample was tested such that it failed in 1 s at a stress of σ_1 [clearly not an easy task, but how that stress can be estimated from constant strain rate experiments is outlined in App. 12A—see Eq. (12A.8)], it follows from Eq. (12.40) that σ_2 at which these same samples would survive a lifetime of $t_2 = 10^\theta$ is

$$\frac{10^\theta}{1} = \left(\frac{\sigma_1}{\sigma_2}\right)^n$$

where $t_1 = 1$ s. Taking the log of both sides and rearranging, one obtains

$$\log\sigma_2 = \log\sigma_1 - \frac{\theta}{n} \tag{12.41}$$

This is an important result, which leads to the straightforward construction of SPT diagrams from Weibull plots (Fig. 11.21). The procedure is as follows (see Worked Example 12.3):

- ∞ Convert the stress at any failure probability to the equivalent stress which would have resulted in failure in 1 s, viz. σ_{1s}, using Eq. (12A.8).
- ∞ Plot log σ_{1S} versus log log (1/S), as shown in Fig. 12.13 (line labeled 1 s). Since the Weibull modulus m is not assumed to change, the slope of the line will be that obtained from the Weibull analysis of the original dataset, viz. the one shown in Fig. 11.21.

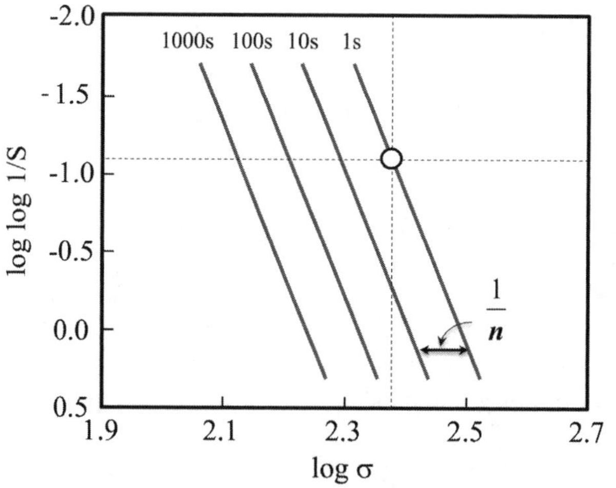

FIGURE 12.13 Effect of SCG rate exponent n on Weibull plots. Weibull plots are shifted by 1/n for every decade of life required of the part. Data are the same as plotted in Fig. 11.21, but are converted to a log scale.

∞ Draw a series of lines parallel to the original line, with a spacing between the lines equal to 1/n, as shown in Fig. 12.13. Each line represents a decade increase in t_f.

Worked Example 12.3

If the data collected in Worked Example 11.2 were measured at a strain rate of 1×10^{-3} s^{-1} and n for this material was measured to be 10, then (*a*) construct the SPT diagram for the data listed in Table 11.2. (*b*) Calculate the design stress that would result in a t_f of 10^4 s and still maintain a survival probability of 0.999. Assume Y = 350 GPa.

ANSWER

(*a*) Referring to Table 11.2, the failure stress at, say, S = 0.837 [i.e., log log (1/S) = −1.11] was 310 MPa. Using Eq. (12A.8)], convert this stress to σ_{1s}:

$$\sigma_{1s} = 310\left(\frac{310\times 10^6}{350\times 10^9 \times 10^{-3}\times (10+1)}\right)^{1/10} = 241\text{ MPa}$$

for which log σ = 2.38. The intersection of the two dotted lines in Fig. 12.13 establishes a point on the Weibull plot. A line is then drawn through this point, with the same slope m as the original one. The other lines are plotted parallel to this line but are shifted to the right by 1/n or 0.1 for every decade of lifetime, as shown in Fig. 12.13.

(*b*) In Worked Example 11.2, the stress for which S was 0.999 was calculated to be 200 MPa. The corresponding 1 s failure stress is

$$\sigma_{1s} = 200\left(\frac{200\times10^{6}}{350\times10^{9}\times1\times10^{-3}\times(10+1)}\right)^{1/10} = 149\ \text{MPa}$$

To calculate the design stress that would result in a lifetime of 10^4 s, use is made of Eq. (12.40), or

$$\frac{10^4}{1} = \left(\frac{149}{\sigma_2}\right)^{10}$$

Solving for σ_2 yields 59.3 MPa. In other words, because of SCG, the applied stress would have to be downgraded by a factor of ≈3 in order to maintain the same S of 0.999 in the absence of SCG. It is left as an exercise to the readers to convince themselves that for *higher* n value, the reduction in design stress is *less* severe.

12.5.2 LIFETIME PREDICTIONS DURING FATIGUE

Given the similarities between the shape of the curves in the intermediate region for SCG and fatigue, one can design for a given fatigue lifetime by using the aforementioned methodology (see Prob. 12.13). However, given the large values of q in Eq. (12.34), there is little gain in doing so; design based on the threshold fracture toughness ΔK_{th} alone suffices. This can be easily seen in Fig. 12.12: To avoid fatigue failure, the stress intensity during service should simply lie to the left of the lines shown.

The actual situation is more complicated, however, because the results shown in Fig. 12.12 are only applicable to long cracks. Short cracks have been shown to behave differently from long ones. Furthermore, the very high values of q imply that marginal differences in either the assumed initial crack size or component in-service stresses can lead to significant variations in projected lifetimes (see Prob. 12.13).

The more promising approach at this time appears to be to use S/N curves such as shown in Fig. 12.11*a* and simply to design at stresses below which no fatigue damage is expected, i.e., use a fatigue limit approach. The major danger of this approach, however, lies in extrapolating data that were evaluated for simple, and usually small parts, to large, complex structures where the defect population may be quite different.

12.5.3 LIFETIME PREDICTIONS DURING CREEP

The starting point for predicting t_f during creep is the **Monkman–Grant equation,** which states that the product of the time to failure t_f and the strain rate $\dot{\varepsilon}$ is a constant, or

$$\dot{\varepsilon} t_f = K_{MG} \tag{12.42}$$

What this relationship, in effect, says is that every material will fail during creep when the strain in that material reaches a certain value K_{MG}, independent of how slow or how fast that strain was reached.

That the Monkman–Grant expression is valid for Ti_3SiC_2 for two different grain sizes is shown in Fig. 12.14. On such a curve, Eq. (12.42) would appear as a straight line with a slope of 1, which appears to be the case.

If one assumes the creep rate is either grain-size-independent or for a set of samples with comparable grain sizes, Eq. (12.21) can be recast as

$$\dot{\varepsilon}_{ss} = A_o \left(\frac{\sigma_a}{\sigma_o}\right)^p \exp\left(-\frac{Q_c}{kT}\right) \tag{12.43}$$

where Q_c is activation energy for creep, A_o is a constant, σ_a is the applied stress that causes failure and σ_o is a normalizing parameter that defines the units of σ_a. Since the stresses one deals with are normally in the megapascal range, σ_o is usually taken to be 1 MPa. Combining these two equations, one obtains

$$t_f = \frac{K_{MG}}{A_o}\left(\frac{\sigma_o}{\sigma_a}\right)^p \exp\frac{Q_c}{kT} \tag{12.44}$$

In other words, t_f of a part should decrease exponentially with increasing temperature as well as with increasing applied stress; both predictions are borne out by experiments. It is worth noting that Eqs. (12.42) to (12.44) are only valid if the rate of damage generation was controlled by the bulk creep response of the material and steady-state conditions are established during the experiment.

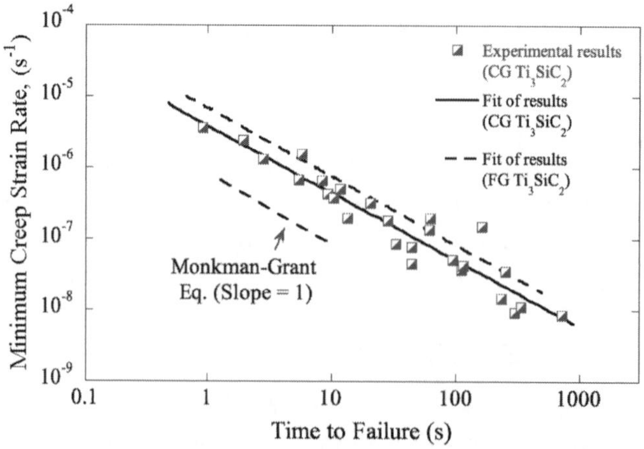

FIGURE 12.14 Plot of log tensile creep strain rate versus log t_f for coarse-grained, CG, and fine-grained, FG, Ti_3SiC_2. On such a plot the Monkton-Grant relationship (Eq. 12.42) would appear as a straight line with a slope of 1 as observed. In this plot for clarity's sake, the FG results are not shown, only their fit. (Adapted from M. Radovic et al. *J. Alloy Compds.*, **361**, 299, 2003.)

12.5.4 FRACTURE MECHANISM MAPS

High-temperature failure of ceramics typically occurs by either SCG or creep. In an attempt to summarize the data available so as to be able to identify the mechanisms responsible for failure and their relative importance, Ashby and co-workers suggested plotting the data on what is now known as *fracture mechanism maps*.[185] The starting point for constructing such a map is to recast Eqs. (12.44) and (12.38), respectively, as

$$\sigma_a = \left(\frac{K_{MG}}{A_o}\right)^{1/p} \frac{\sigma_o}{t_f^{1/p}} \exp\frac{Q_c}{pkT} \tag{12.45}$$

$$\sigma_a = B_o\sigma_o\left(\frac{t_o}{t_f}\right)^{1/n} \exp\frac{Q_{SCG}}{nkT} \tag{12.46}$$

at which point their similarities become quite obvious.[186] Variables t_o and σ_o are introduced in Eq. (12.46) to keep B_o dimensionless and define the scales—most commonly, t_o is chosen to be 1 h and σ_o is, again, 1 MPa.

To construct such a map (see Worked Example 12.4), SCG and creep rupture data must be known for various temperatures. The temperature dependence of the stress levels required to result in a given t_f are then calculated from Eqs. (12.45) and (12.46). The mechanism that results in the lowest failure stress at a given temperature defines the threshold stress or highest applicable stress for the survival of a part for a given time. In other words, t_f is determined by the fastest possible path. Such maps are best understood by actually plotting them.

Worked Example 12.4

Using the following information, construct a fracture deformation map for Si_3N_4: $K_{MG} \approx 5.4 \times 10^{-3}$, $p = 4$, $Q_c = 800$ kJ/mol, $A_o = 1.44 \times 10^{19}\ h^{-1}$, $\sigma_o = 1$ MPa (these are typical values for Si_3N_4, except p which is usually closer to 8 or 9; see Prob. 12.11 for another set of data). For SCG in Si_3N_4, assume $B_o = 80$, $n = 55$, $Q_{SCG} = 760$ kJ/mol, $\sigma_o = 1$ MPa and $t_o = 1$ h.

ANSWER

Plugging the appropriate numbers in Eq. (12.45), one obtains

$$\sigma_a(\text{MPa}) = \sigma_o\left(\frac{K_{MG}}{A_o t_f}\right)^{1/p} \exp\frac{Q_c}{pRT} = \frac{4.4 \times 10^{-6}}{t_f^{1/4}} \exp\frac{24{,}055}{T}$$

When this equation is plotted as a function of temperature, a series of steep lines emerge that are shifted toward the left with increasing lifetimes, as shown in red in Fig. 12.15.

---

[185] M. F. Ashby, C. Gandhi, and D. M. R. Taplin, *Acta Metall.*, 27, 1565 (1979).

[186] Note the temperature dependence, which was buried in A in Eq. (12.38), is now spelled out. Also note exponent of σ_a in Eq. (12.38) once K_I is replaced by $\Psi\sigma_a\sqrt{\pi c}$ is n (see Prob. 12.5).

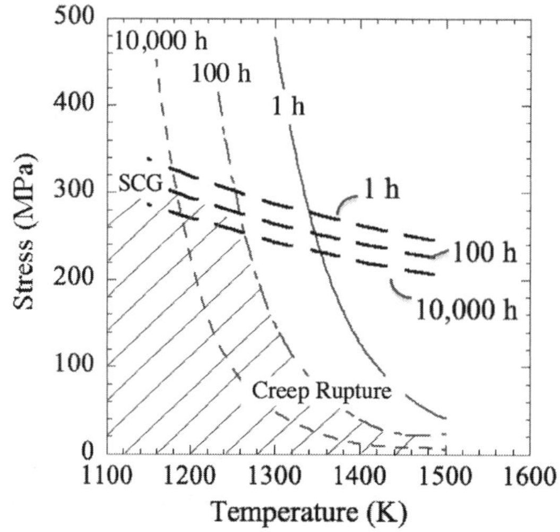

FIGURE 12.15 Fracture mechanism map for Si_3N_4 developed using data given in Worked Example 12.4. Shaded area denotes the stress-temperature regime for which the part would survive at least 100 h.

Similarly, a plot of Eq. (12.46), that is,

$$\sigma_a\,(\mathrm{MPa}) = \sigma_o B_o \left(\frac{t_o}{t_f}\right)^{1/n} \exp\frac{Q_{SCG}}{nkT} = \frac{80}{t_f^{1/55}} \exp\frac{1662}{T}$$

gives another series of almost parallel (plotted in black) lines labeled SCG in Fig. 12.15. The large n value makes Si_3N_4 much less susceptible to SCG at higher temperatures.

The advantage of such maps is that once they have been constructed, the stress–temperature regime for which a part would survive a given t_f is easily delineated. Referring to Fig. 12.15, one sees that to design a part to withstand 100 h of service, the design should be confined to the domain encompassed by the 100 h lines. In other words, a part subjected to a combination of stress and temperature that lies within the lower shaded left-hand corner of Fig. 12.15 should survive for at least 100 h—any other combination would result in a shorter lifetime.

CASE STUDY 12.1: STRENGTHS AND LIFETIMES OF OPTICAL FIBERS

An optical fiber is a single, hair-thin filament drawn from ultrapure, molten silica glass. The importance of silica-based optical fibers in today's connected world cannot be overemphasized. These fibers are the conduits by which information is carried from one side of the globe to the other at a speed close to the speed of light (see Case Study 16.2). As these fibers were being developed for their optical transparency, other questions came to the fore, most notably: How strong are these fibers and, as importantly, what is their lifetime given the propensity of silica-based glasses to suffer from SCG?

To answer the latter question, the **inert strength**, $\sigma^{\ddagger}$, of a glass fiber, defined as the strength where SCG is absent had to be measured. To do so one needs to use one of the following strategies: (i) test in a moisture-free environment, such as ultrahigh vacuum; (ii) test hermetically sealed fibers; (iii) test at very high strain rates; (iv) test at temperatures where SCG kinetics are extremely slow, such as in liquid nitrogen, N_2, or liquid helium, He.

The results of such measurements are shown in Fig. 12.16*a*, where the tensile strengths of silica glass fibers are plotted vs. temperature and test environment. From the figure, it is obvious that the strengths at liquid He temperatures (4 K) are the highest at close to 15 GPa! The fibers tested at liquid N_2 (77 K)

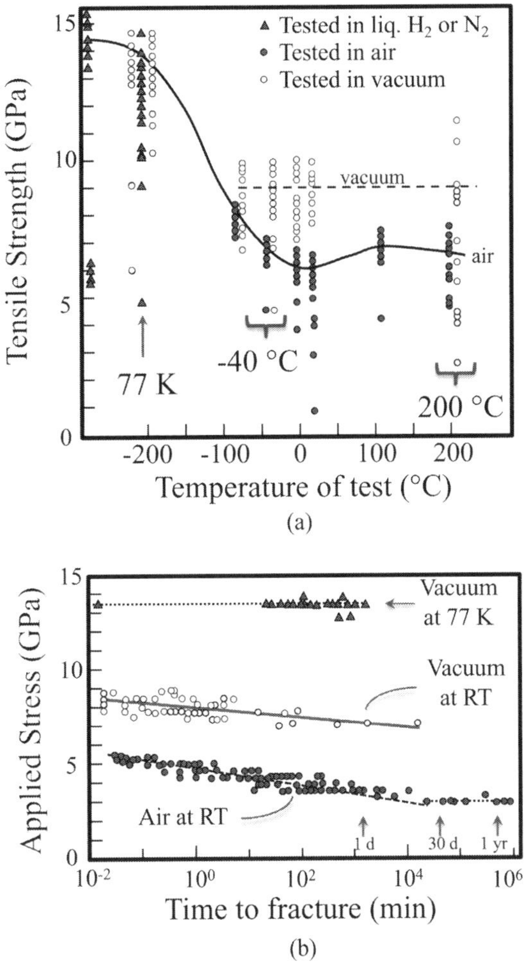

FIGURE 12.16 (*a*) Temperature dependence of tensile strengths of silica fibers tested in liquid He or liquid N_2 (filled triangles), vacuum (open circles), or air (filled circles). (*b*) Time dependence of tensile strength of silica fibers held in vacuum at 77 K (top filled triangles), in vacuum at RT (open circles) and in air at RT (filled circles). (Adapted from B. A. Proctor, I. Whitney, J. W. Johnson, *Proc. Roy. Soc. A* 297, 534, 1967.)

temperatures or those that were hermetically sealed are the next highest (≈ 12 GPa) followed by fibers tested in vacuum at T > −100°C (9 GPa). The weakest fibers are those tested at room temperature in air (≈ 7 GPa).

Figure 12.16*b* compares the times to failure as a function of applied stress for fibers tested at 77 K, in vacuum and in air at room temperature. These results elegantly illustrate the importance of both temperature and moisture on SCG in silica fibers. The tensile strengths of the fibers held at 77 K did *not* degrade with time despite being held at a stress of ≈13 GPa! When the same fibers were tested in vacuum at room temperature (open circles) a slow degradation of their strengths with time, from ≈ 9 GPa to ≈ 7 GPa after about a month was observed. The fibers held at room temperature in air, on the other hand, saw their tensile strengths drop from ≈5 GPa to ≈ 3 GPa after a year.

Commercially, optical fibers are hermetically sealed with a polymer coating just after being drawn into fibers. This approach is relatively inexpensive and protects the fibers from damage and moisture. The end result is exceedingly strong glass fibers, with tensile strengths that do not degrade with time. When their superlative transparency is added to the mix (see Chap. 16), it is no wonder that these fibers dominate the optical fibers market.

CASE STUDY 12.2: FIBER-REINFORCED CERAMIC MATRIX COMPOSITES

It has long been appreciated that because ceramics are lighter and can in general go to much higher temperatures than metals, they were attractive materials for aerospace applications. Introduction of ceramics in jet engines would lead to weight reductions and concomitant fuel savings, longer lifetimes, cleaner exhausts and cost savings. But for their brittleness, that is. Materials for aerospace applications can be subdivided into load and non-load bearing. The used of ceramics for the latter is well-established. Probably the first use of ceramics in aerospace that caught the public's imagination—and arguably forever changed the public's perception of ceramic materials, from pottery to high-tech— was as space shuttle tiles (see Case Study 13.2). The ultimate goal, however, is to fabricate an all-ceramic jet engine, or more precisely a gas turbine engine. The latter are the mainstay of commercial and military aircraft; the land-based ones produce 80% of the world's electricity.

Two or so decades ago, the idea that a ceramic could be used in jet engine was considered far-fetched. That ceramics actually fly today in some jet engines is thus a testament to the ingenuity and perseverance of the researchers working in that field. The conditions under which current jet engines operate are not for the faint of heart. Figure 12.17*a* plots the 500 h tensile rupture strengths versus temperature for Ni-based superalloys, the workhorses of today's jet engines, oxide and SiC/SiC ceramic matrix composites or CMCs. Figure 12.17*b* plots the specific (i.e., the strengths divided by density) fast-fracture strengths vs. temperature for different families of high-temperature materials. In both plots the advantages of CMCs are obvious. There are many combinations of fibers and matrices. From the results shown in Fig. 12.17 it should be clear why herein, and industrially, the focus is on SiC_f/SiC_m composites, where both the fibers and matrices are comprised of SiC. C/C composites are even more attractive (Fig. 12.17*b*), however, it is much more difficult—almost impossible—to protect them from oxidation.

The first step in manufacturing SiC_f/SiC_m composite parts is to make a porous SiC fiber preform. The latter is then infiltrated with a SiC matrix. The infiltration step can be carried out by chemical vapor

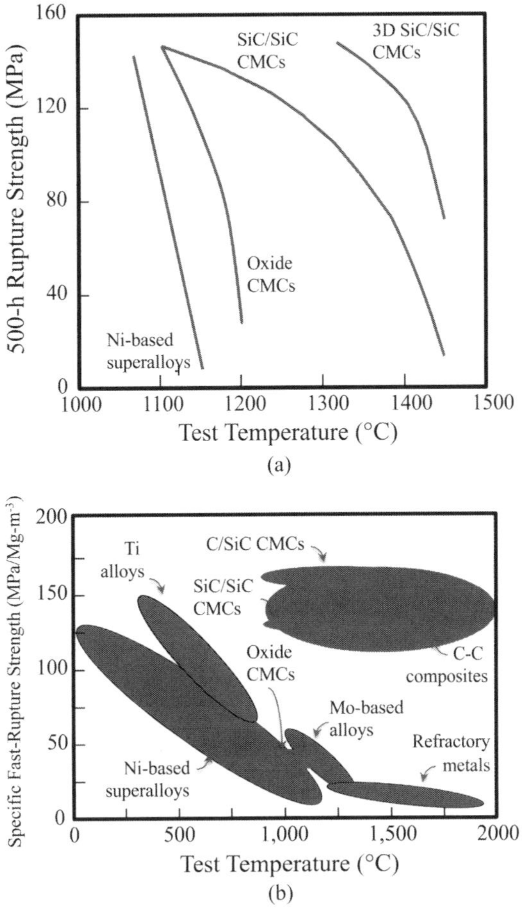

FIGURE 12.17 Temperature dependencies of a number of high-temperature structure material for, (*a*) 500 h tensile rupture strength, and (*b*) fast-rupture strengths normalized by the density of the materials. (Both figures adapted from N. P. Padture, *Nat. Mater.* **15**, 804, 2016.)

infiltration, slurry infiltration or reactive melt infiltration. Arguably the most important, but also most difficult, step in fabricating CMCs is the quality of the infiltration. What is needed is a technique that does not close surface pores before the interior of the part is fully dense.

Before infiltration, however, a thin layer of BN is deposited on the fibers to reduce the fiber/matrix interfacial strength. This interfacial layer enables extensive debonding between crack and matrix and leads to crack bridging by the fibers and frictional pull-out. Referring to Fig. 11.18*c*, if the interface is too strong, no debonding occurs, and the mechanical properties would be no different than if the material were a monolithic bulk ceramic, i.e., the failure would be brittle. The crack bridging insures that the failure is not brittle or sudden, but more graceful that allows for inspection of the parts and their replacement if the damage accumulated is high.

12.6 SUMMARY

1. The removal of atoms from regions that are in compression and placing them in regions that are in tension reduces their energy by an amount $\approx 2\Omega\sigma$, where σ is the applied stress and Ω is the atomic volume. This reduction in energy - which results in permanent shape change - is the driving force for creep.
2. Diffusional creep in binary ceramics is a thermally activated process that depends on the diffusivity of the slower species along its fastest path. When a material is subjected to relatively low stresses and/or temperatures, the creep rate typically increases linearly with applied stress, i.e., the stress exponent is unity. Because the diffusion path scales with grain size d, finer-grained solids are usually less resistant to diffusional creep than their larger-grained counterparts. If the ions diffuse through the bulk (Nabarro–Herring creep), the creep rate is proportional to d^{-2}; if they diffuse along grain boundaries (Coble creep), the creep rate scales as d^{-3}.
3. At higher stresses, the creep rate is much more sensitive to the applied stress, with stress exponents anywhere between 3 and 8, and is typically independent of grain size. The presence of glassy phases along grain boundaries can lead to cavitation and stress rupture.
4. SCG can also occur by the combined effect of stresses and corrosive environment and/or the accumulation of damage at crack tips. The basic premise at ambient and near-ambient temperatures is that SCG results from a stress-enhanced reactivity of the chemical bonds at the crack tip. This phenomenon is thus a strong function of the stress intensity at the crack tip and typically is thermally activated. At high temperatures, the phenomenon of SCG is believed to occur by the formation of cavities ahead of a crack tip. The crack then grows by the coalescence of these cavities.

1. Some ceramic materials - most notably those that exhibit R curve behavior - are prone to fatigue, as a result of the weakening of the shielding elements, such as whiskers or large grains. The results also suggest that short cracks behave differently from long ones, indicating that perhaps, from a design point of view, it is more promising/practical to use S/N curves (Fig. 12.11*a*) than crack growth rate curves (Fig. 12.11*b*). The commonality of these three time dependent deformation mechanisms lies in the fact that they endow a part with a lifetime that needs to be taken into account.

APPENDIX 12A: DERIVATION OF EQ. (12.24)

Equation (12.40), while appearing to be a useful equation to estimate the lifetime of a part, has to be used with caution since it assumes that c_i is identical in all samples. And while that is possible in a laboratory setting where well-defined cracks can be introduced in a sample, its usefulness in practice is limited. What is thus needed is a methodology to transform data (i.e., one with a distribution of failure times) to an equivalent stress that would have caused failure in, say, 1 s. In other words, a renormalization of the time-to-failure data.

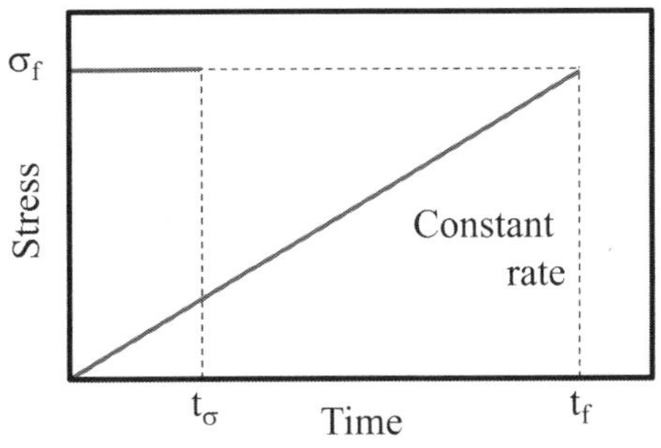

FIGURE 12.18 Comparison of stress vs. time response for constant stress and constant rate tests.

Figure 12.18 illustrates the problem. The sloping line represents the stress on a sample as a result of some increasing strain rate $\dot{\varepsilon}$. The specimen fails at stress σ_f after time t_ε. Had the stress been applied instantaneously, the sample would have failed after a shorter time t_σ because the average stress is initially higher. Equation (12.40) can be recast to read

$$t\sigma^n = \Gamma' \tag{12A.1}$$

where Γ' is a constant. Invoking the notion that the sum of the fractional times the material spends at any stress has to be unity, i.e.,

$$\sum \frac{dt}{t} = 1 \tag{12A.2}$$

implies that for a constant strain rate test

$$\int_0^{t_\varepsilon} \frac{dt}{t} = 1 \tag{12A.3}$$

where t_ε is the time for the sample to fail when the strain rate is $\dot{\varepsilon}$ (Fig. 12.18). It follows directly from Fig. 12.18 that

$$t_\varepsilon = \sigma_f \frac{dt}{d\sigma} \tag{12A.4}$$

Combining Eqs. (12A.4) and (12A.3), eliminating t by using Eq. (12A.1), and integrating Eq. (12A.3), one obtains

$$1 = \int_0^{\sigma_f} \frac{t_\varepsilon \sigma^n d\sigma}{\Gamma' \sigma_f} = \frac{t_\varepsilon \sigma_f^n}{\Gamma'(n+1)} \tag{12A.5}$$

where σ_f is the fracture stress. Making use of Eq. (12A.1) once again, one sees that

$$\frac{t_\varepsilon}{t_\sigma} = n + 1 \tag{12A.6}$$

Rewriting Hooke's law as $\sigma_f = Y\dot{\varepsilon}t_\varepsilon$ one obtains

$$\sigma_f = Y\dot{\varepsilon}_1(n+1)t\sigma \tag{12A.7}$$

Combining this equation with Eq. (12.40) yields the sought-after result

$$\sigma_{1s} = \sigma_f\left(\frac{t_a}{t_o}\right)^{1/n} = \sigma_f\left(\frac{\sigma_f}{Y\dot{\varepsilon}(n+1)}\right)^{1/n} \tag{12A.8}$$

Here t_o is 1 s. This result thus allows for the calculation of the constant stress which would have caused failure in 1 s, viz. σ_{1s}, from the stress σ_f at which the sample failed at a strain rate $\dot{\varepsilon}$, both of which are experimentally accessible.

By rewriting Eq. (12A.8) for two different strain rates, it is straightforward to show that

$$\frac{\bar{\sigma}_{\dot{\varepsilon}1}}{\bar{\sigma}_{\dot{\varepsilon}2}} = \left(\frac{\dot{\varepsilon}_1}{\dot{\varepsilon}_2}\right)^{1/n+1} \tag{12A.9}$$

where $\bar{\sigma}_{\dot{\varepsilon}1}$ is the mean value of the failure stress measured at strain $\dot{\varepsilon}_1$.

PROBLEMS

12.1. (*a*) For a M_aX_b compound it can be shown that (e.g. R. S. Gordon, *J. Amer. Cer., Soc.*, 56, 174 (1973).

$$D_{complex} \approx \frac{(D^M d + \pi\delta_{gb}^M D_{gb}^M)(D^X d + \pi\delta_{gb}^X D_{gb}^X)}{\pi\left[a(D^M d + \pi\delta_{gb}^M D_{gb}^M) + b(D^X d + \pi\delta_{gb}^X D_{gb}^X)\right]}$$

where d is the grain size, δ_{gb} is the grain boundary width and D^i and D_{gb}^i are the bulk and GB diffusivities of the appropriate species, respectively. Consider an oxide for which $dD^M \gg \delta_{gb}D_{gb}^M \gg dD^X \gg \delta_{gb}D_{gb}^X$. Which ion do you think will be rate-limiting and which path will it follow?

(*b*) Repeat part (*a*) for $dD^X \gg \delta_{gb}D_{gb}^M \gg \delta_{gb}D_{gb}^X \gg dD^M$.

12.2. If the GB diffusivity is given by

$$D_{gb} = 100\exp(-40\ \text{kJ/RT})$$

and the bulk diffusion coefficient is

$$D_{latt} = 300\exp(-50\ \text{kJ/RT})$$

at 900 K determine whether GB or lattice diffusion will dominate. How about at 1300 K? At which temperature will they be equally important?
Answer: $T = 1095$ K.

12.3. For what values of *r* and *p* does Eq. (12.21) become similar to Eqs. (12.15) and (12.16)?

12.4 (*a*) Calculate the vacancy concentration at two-thirds of the absolute melting point of Al and Cu subjected to a tensile strain of 0.2%. The enthalpies of vacancy formation for Cu and Al are, respectively, 1.28 and 0.67 eV. The Young's moduli at two-thirds of their respective melting points are 110 and 70 GPa, respectively. Densities of Al and Cu are, respectively, 2.7 and 8.96 g/cm³. State all assumptions.
Answer: $c_{Cu} \approx 1.1 \times 10^{22}\ m^{-3}$, $c_{Al} \approx 3.24 \times 10^{24}\ m^{-3}$.
(*b*) Derive an expression for the vacancy concentration difference between the side and bottom surfaces of a cylindrical wire of radius ρ subjected to a normal tensile stress σ_{nn} (see Chap. 10).
(*c*) At what radius of curvature of the wire will the surface energy contribution be comparable to the applied stress contribution? Assume the applied stress is 10 MPa. State all other assumptions. Comment on the implications of your solution to the relative importance of externally applied stresses versus those that result from curvature.
Answer: $r_c \approx 0.01\ \mu m$.

12.5. (*a*) Derive Eq. (12.28).
(*b*) Re-derive Eq. (12.28), but include a term that takes into account the concentration of moisture.
(*c*) Derive Eqs. (12.38) and (12.39).
(*d*) For $n = 10$, estimate the error in neglecting the second term within brackets in Eq. (12.37). Why can this term be safely neglected? And does that imply that t_f is independent of K_{Ic}? Explain.
(*e*) Show that the lifetime of a part subjected to SCG [Eq. (12.38)] can be equally well expressed in terms of the initial crack length c_i by

$$t_f = \frac{2K_{Ic}^n}{A(\Psi\sqrt{\pi}\sigma_a)^n(n-2)c_i^{n/2-1}}$$

12.6. Typical crack growth data for a glass placed in a humid environment are listed in the table below. Calculate the values of A″, A and n [see Eq. (12.30)], given that $K_{Ic} = 0.7\ MPa{\cdot}m^{1/2}$. What are the units of A″?

Stress intensity, $MPa{\cdot}m^{1/2}$	0.4	0.5	0.55	0.6
Crack velocity, m/s	1×10^{-6}	1×10^{-4}	1×10^{-3}	1×10^{-2}

Answer: $n = 22.5$, $A'' = 772$, $A \approx 0.25$ m/s.

12.7. If the specimen shown in Prob. 11.7 is loaded in tension under a stress of 10 MPa and all cracks but the surface crack shown on the left-hand side were ignored, calculate the lifetime for the part. Assume $n = 15$ and $A = 0.34$ m/s. State all other assumptions.
Answer: 123 s.

12.8. (*a*) The Weibull plots shown in Fig. 12.19 were generated at two different strain rates: $\dot{\varepsilon}_1 = 2 \times 10^{-6}\ s^{-1}$ and $\dot{\varepsilon}_2 = 2 \times 10^{-5}\ s^{-1}$. Which strain rate is associated with which curve? Briefly explain your choice.

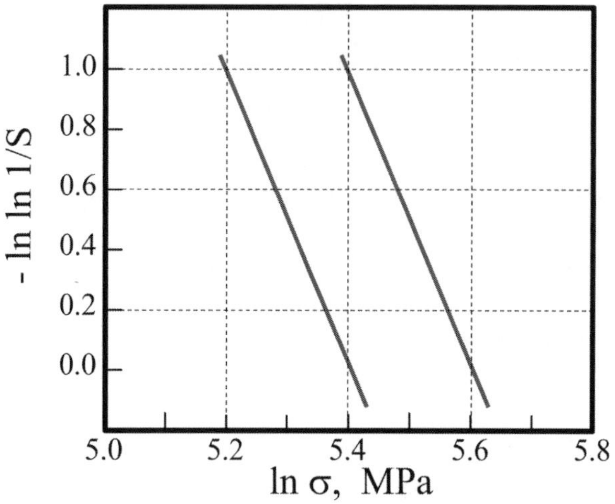

FIGURE 12.19 Effect of strain rate on Weibull plots.

(*b*) Calculate the stress needed to obtain a 0.9 survival probability and a lifetime of 10^7 s. Information you may find useful: $K_{Ic} = 3$ MPa·m$^{1/2}$, $Y = 100$ GPa.
Answer: 45 MPa.

12.9. (*a*) If you want to design an engine component such that the probability of failure will be at most 0.01, and the choice is between the following two materials:

Ceramic A	Mean strength 600 MPa	$m=25$	$n=11$
Ceramic B	Mean strength 500 MPa	$m=17$	$n=19$

Which would you recommend and why? What maximum design stress would be allowable to ensure the survival probability required? State all assumptions.
Answer: 499 MPa.
(*b*) The values shown in (*a*) were measured for samples that had the following dimensions: 1×1×8 cm^3. If the component part is 10 times that volume, would your recommendation change concerning which material to use? Would the design stress change? If yes, calculate the new stress?
Answer: 455 MPa.
(*c*) If the part were to be used in a moist environment, do you get concerned? If the lifetime of the part is to be 4 years or ≈10^9 s, what changes, if any, would you recommend for the design stress? Does your recommendation as to which material to use change? Explain. Assuming, for the sake of simplicity, that the data reported in part (*a*) were obtained in 1 s, determine which material to use by calculating the design stress needed for the lifetime noted for both.
Answer: *Ceramic A*, 76 MPa; *Ceramic B*, 128 MPa.

12.10. (*a*) To study the degradation in strength of a ceramic component, the average flexural strength of the material was measured *following* exposure under stress in a corrosive environment. The Weibull

modulus, measured to be 10, did not change with time, but the average strength was found to decrease from 350 MPa after 1 day to 330 MPa after 3 days. Qualitatively explain what is happening.
(*b*) Calculate the average strength one would expect after 10 weeks' exposure.
Answer: 279 MPa.
(*c*) Post-test examination of the samples that failed after 1 day showed that the average size of the cracks that caused failure was of the order of 120 μm. Calculate the average crack size that was responsible for failure after 3 days. Information you may find useful: $K_{Ic} = 3$ MPa·m$^{1/2}$.
Answer: 135 μm.

12.11. You are currently using the Si_3N_4 whose properties are listed in Worked Example 12.4, and a new Si_3N_4 appears on the market with the following properties: $K_{MG} = 5.4 \times 10^{-3}$, $p = 9$, $Q_c = 1350$ kJ/mol, $A_0 = 4 \times 10^{19}$ h^{-1} and $\sigma_0 = 1$ MPa. For subcritical crack growth you can assume $B_0 = 100$, $n = 50$, $Q_{SCG} = 900$ kJ/mol, $\sigma_0 = 1$ MPa and $t_0 = 1$ h. By constructing a fracture mechanism map and comparing it to Fig. 12.20, which material would you use and why?
Answer: See Fig. 12.20.

12.12. The surface energy of a glass can be measured by carrying out a zero creep experiment, where a glass fiber is suspended in a furnace and is heated. As a result of gravity, the wire will extend to a final equilibrium length l_{eq}, beyond which no further increase in length is measured.
(*a*) What do you suppose keeps the fiber from extending indefinitely?
(*b*) Show that at equilibrium $l_{eq} = 2\gamma_{sv}/\rho r_{eq} g$ where g and ρ are, respectively, the gravitational constant and the density of the glass; r_{eq} is the equilibrium radius of the wire. *Hint*: Write an expression for

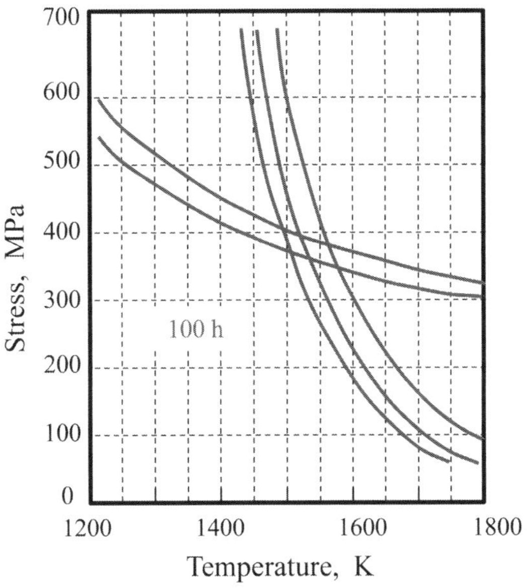

FIGURE 12.20 Fracture mechanism map for Si_3N_4, for which the properties are listed in Prob. 12.11.

the total energy of the system that includes gravitational and surface energy terms, and minimize that function with respect to l.

12.13. (*a*) Show that the number of cycles to failure during fatigue is given by

$$N_f = \frac{2}{B(\xi\sqrt{\pi}\Delta\sigma_a)^q(q-2)c_i^{q/2-1}}$$

where c_i is the initial crack size. All other terms are defined in the text. Note the similarity of this expression to the one derived in Prob. 12.5*e*.

(*b*) Estimate the values of B and q for Mg-TZP shown in Fig. 12.12. What are the units of B?

Answer: $B = 1.7 \times 10^{-48}$; $q \approx 40$.

(*c*) Estimate the effect of increasing the applied stress by a factor of 2 on N_f.

Answer: 4.3×10^{12} cycles!

(*d*) Estimate the effect of doubling the assumed initial crack size on N_f.

Answer: 5×10^5 cycles!

12.14. Rectangular glass slides were tested in three-point bending as a function of surface finish. The results obtained are listed below. Qualitatively explain the trends.

Treatment	As-received	HF etched	Abraded normal to applied stress	Abraded parallel to applied stress
Strength, MPa	87	106	42	71

12.15. From the results shown in Fig. 12.16*b*, estimate the value of n for silica glass.

ADDITIONAL READING

1. W. Cannon and T. Langdon, Creep of ceramics, Parts I and II, *J. Mater. Sci.*, 18, 1–50 (1983).
2. R. Ritchie and R. Daukardt, Cyclic fatigue of ceramics: a fracture mechanism approach to subcritical crack growth and life prediction, *J. Cer. Soc. Japan*, 99, 1047–1062 (1991).
3. R. Raj, Fundamental research in structural ceramics for service near 2000°C, *J. Am. Cer. Soc.*, 76, 2147–2174 (1993).
4. N. A. Fleck, K. J. Kang, and M. F. Ashby, The cyclic properties of engineering materials, *Acta Metall. Mater.*, 42, 365–381 (1994).
5. S. Suresh, *Fatigue of Materials*, Cambridge University Press, Cambridge, UK, 1991.
6. H. Reidel, *Fracture at High Temperatures*, Springer-Verlag, Heidelberg, Germany, 1987.
7. K. Chan and R. Page, Creep damage in structural ceramics, *J. Amer. Cer. Soc.*, 76, 803 (1993).
8. R. L. Tsai and R. Raj, Overview 18: Creep fracture in ceramics containing small amounts of liquid phase, *Acta Metall. Mater.*, 30, 1043–1058 (1982).
9. M. F. Ashby, C. Gandhi, and D. M. R. Taplin, Fracture mechanism maps for materials which cleave: FCC, BCC and HCP metals and ceramics, *Acta Metall. Mater.*, 27, 1565 (1979).
10. R. W. Davidge, *Mechanical Behavior of Ceramics*, Cambridge University Press, Cambridge, UK, 1979.
11. B. Lawn, *Fracture of Brittle Solids*, 2nd ed., Cambridge University Press, Cambridge, UK, 1993.
12. R. Warren, Ed., *Ceramic Matrix Composites*, Blackie, Glasgow, Scotland, 1992.
13. A. Kelly and N. H. Macmillan, *Strong Solids*, 3rd ed., Clarendon Press, New York, 1986.
14. G. Weaver, Engineering with ceramics, Parts 1 and 2, *J. Mater. Ed.*, 5, 767 (1983); 6, 1027 (1984).

15. A. G. Evans, Engineering property requirements for high performance ceramics, *Mater. Sci. Eng.*, 71, 3 (1985).
16. S. M. Weiderhorn, A probabilistic framework for structural design, in *Fracture Mechanics of Ceramics*, vol. 5, R. C. Bradt, A. G. Evans, D. P. Hasselman, and F. F. Lange, Eds., Plenum, New York, 1978, p. 613.
17. M. F. Ashby and B. F. Dyson, *Advances in Fracture Research*, S. R. Valluri, D. M. R. Taplin, P. Rama Rao, J. F. Knott, and R. Dubey, Eds., Pergamon Press, New York, 1984.
18. T. H. Courtney, *Mechanical Behavior of Materials*, McGraw-Hill, New York, 1990.

13

THERMAL PROPERTIES

What happens in these Lattices when Heat
Transports Vibrations through a solid mass?
T = 3Nk is much too neat;
A rigid Crystal's not a fluid Gas.
Debye in 1912 proposed Elas-
Tic Waves called phonons that obey Max Planck's
E = hν. Though amorphous Glass,
Umklapp Switchbacks, and Isotopes play pranks
Upon his Formulae, Debye deserves warm Thanks.

John Updike, *The Dance of the Solids**

13.1 INTRODUCTION

As a consequence of their brittleness and low thermal conductivities, ceramics are prone to thermal shock; i.e., they tend to shatter when subjected to large thermal gradients. This is why it is usually not advisable to pour a very hot liquid into a cold glass container, or cold water on a hot ceramic furnace tube—the rapidly cooled surface will want to contract, but will be restrained from doing so by the bulk of the body, so stresses will develop. If these stresses are large enough, the ceramic will crack.

Thermal stresses will also develop because of thermal contraction mismatches in multiphase materials or thermal expansion anisotropies in a single phase. It thus follows that thermal stresses exist in all polycrystalline ceramics that i) are noncubic, ii) undergo phase transformations and/or, iii) include second phases with differing thermal expansion characteristics as their host crystal. These stresses can result

* J. Updike, Midpoint and Other Poems, A. Knopf, Inc., New York, 1969. Reprinted with permission.

in the formation of stable microcracks and can strongly influence the strength and fracture toughness of ceramics. In a worst-case scenario, these stresses can cause the total disintegration of a ceramic body. Used properly, however, they can enhance the strength of glasses. The purpose of this chapter is to explore the problem of thermal residual stresses, why they develop and how to quantify them.

Another important thermal property, dealt with in Sec. 13.6, is thermal conductivity. It is the low thermal conductivity of ceramics, together with their chemical inertness and oxidation resistance, which renders them, as a class, uniquely qualified to play an extremely demanding, and critical role during metal smelting and refining. Many ceramics such as diaspore, alumina, fosterite and periclase are used for the fabrication of high-temperature insulative firebrick without which the refining of some metals would be impossible.

13.2 THERMAL STRESSES

ORIGIN OF THERMAL RESIDUAL STRESSES

To best illustrate the idea of how differential thermal expansion in multiphase materials leads to thermal stresses, consider the simple case shown schematically in Fig. 13.1*a*, where a solid disk is placed inside of a ring of a different material. To emphasize the similarity of this problem to that of an inclusion in a matrix, which was discussed in Chap. 11 and is one of practical significance, the disk will henceforth be referred to as the *inclusion*, and the outside ring as the *matrix*, with thermal expansion coefficients α_i and α_m, respectively.

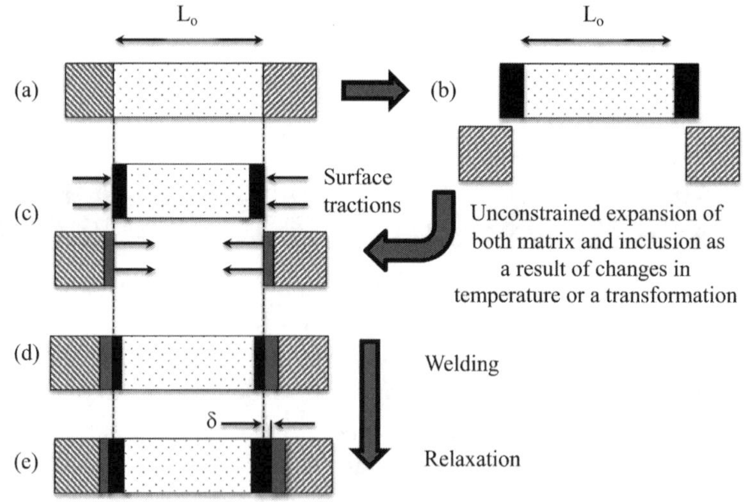

FIGURE 13.1 Steps involved in Eshellby's method. (*a*) Initial configuration. (*b*) Cutting and allowing for free expansion of both inclusion and matrix as a result of heating. Note that the radius of the outside ring *increases* upon heating. (*c*) Application of surface forces needed to restore elements to original shape. (*d*) Weld pieces together. (*e*) Allow the system to relax. Note displacement of original interface as a result of relaxation.

Before one attempts to find a quantitative answer, it is important to qualitatively understand what happens to such a system as the temperature is varied. Needless to say, the answer will depend on the relative values of α_i and α_m, and whether the system is being heated or cooled. To illustrate, consider the case where $\alpha_i > \alpha_m$ and the system is heated. Both the inclusion and the matrix will expand[187] (Fig. 13.1*b*); however, given that $\alpha_i > \alpha_m$, the inclusion will try to expand at a faster rate, but will be radially restricted from doing so by the outside ring. It follows that upon heating, both inclusion and matrix will be in radial compression. It is left as an exercise to the reader to show that if the assembly were cooled, the inclusion would develop radial tensile stresses instead. It should be noted here, and is discussed in greater detail below, that stresses, other than radial, also develop.

The quantification of the problem is nontrivial and is usually carried out today by using finite-element and other numerical techniques. However, for simple geometries, a powerful method developed by Eshellby[188] exists, which in principle is quite simple, elegant and ingenious. The problem is solved by carrying out the following series of imaginary cuts, strains and welding operations illustrated in Fig. 13.1:

1. Cut the inclusion out of the matrix.
2. Allow both the inclusion and the matrix to expand or contract as a result of either heating or cooling (or as a result of a phase transformation) (Fig. 13.1*b*).
3. Apply sufficient surface tractions to restore the elements to their original shape (Fig. 13.1*c*).
4. Weld the pieces together (Fig. 13.1*d*).
5. Allow the system to relax (Fig. 13.1*e*).

To apply this technique to the problem at hand, do the following:

1. Cut the inclusion, and allow both it and the matrix to freely expand (Fig. 13.1*b*). The thermal strain in the inclusion is given by [Eq. (4.2)]:

$$\frac{\Delta L}{L_o} = \varepsilon_i = \alpha_i \Delta T = \alpha_i \left(T_{final} - T_{init}\right)$$

$$\boxed{\varepsilon_i = \alpha_i \left(T_{final} - T_{init}\right)} \quad (13.1)$$

Similarly, for the matrix

$$\varepsilon_m = \alpha_m \Delta T \quad (13.2)$$

Note that as defined here, ΔT *is positive during heating and negative during cooling.* On cooling, T_{final} is usually taken to be room temperature; T_{init}, however, is more difficult to determine

[187] Note expansion of the matrix implies that the internal diameter of the ring *increases* with increasing temperature.
[188] J. D. Eshellby, *Proc. R. Soc.*, A241, 376–396 (1957).

unambiguously, but it is the highest temperature below which the residual stresses are *not* relieved, which, depending on the material in question, may or may not be identical to the processing or annealing temperature. At high enough temperatures, stress relaxation by diffusive, or viscous, flow will usually relieve some, if not most, of the stresses that develop. It is only below a certain temperature that these stress relaxation mechanisms become inoperative and local elastic residual stresses start to develop from the contraction mismatch.

2. Apply a stress to each element to restore it to its original shape[189] (Fig. 13.1c). For the inclusion,

$$\sigma_i = -Y_i \varepsilon_i = -Y_i \alpha_i \Delta T \tag{13.3}$$

where Y is Young's modulus. For the matrix:

$$\sigma_m = Y_m \varepsilon_m = Y_m \alpha_m \Delta T \tag{13.4}$$

Note that the applied stress needed to restore the inclusion to its original shape is compressive (see Fig. 13.1c), which accounts for the minus sign in Eq. (13.3).

3. Weld the two parts back together (Fig. 13.1*d*) and allow the stresses to relax. Since the stresses are unequal, one material will "push" into the other, and the location of the *original interface will shift by a strain* δ in the direction of the larger stress until the two stresses are equal (Fig. 13.1*e*). At equilibrium the two **radial stresses** are equal and are given by

$$\sigma_{i,eq} = Y_i\left[\varepsilon_i + \delta\right] = \sigma_{m,eq} = Y_m\left[\varepsilon_m - \delta\right] \tag{13.5}$$

Solving for δ, plugging that value back into Eq. (13.5), and making use of Eqs. (13.1) to (13.4), one can show (see Prob. 13.2) that

$$\sigma_{i,eq} = \sigma_{m,eq} = \frac{\Delta\alpha\Delta T}{1/Y_i + 1/Y_m} = \frac{(\alpha_m - \alpha_i)\Delta T}{1/Y_i + 1/Y_m} \tag{13.6}$$

This is an important result which predicts that:

- If $\Delta\alpha$ is zero, no stress develops, which makes sense since the matrix and the inclusion would be expanding at the same rate.
- For $\alpha_i > \alpha_m$, upon heating (positive ΔT), the stresses generated in the inclusion and matrix should be compressive, or negative, as anticipated.
- If the inclusion is totally *constrained from moving* (that is, $\alpha_m = 0$ and Y_m is infinite), then Eq. (13.6) simplifies to the more familiar equation

$$\sigma_{i,eq} = -Y_i \alpha_i \Delta T \tag{13.7}$$

which predicts that upon heating, the stress generated will be compressive, and vice versa upon cooling.

[189] Equations (13.2) and (13.3) are strictly true only for a one-dimensional problem. Including the other dimensions does not generally greatly affect the final result [see Eq. (13.8)].

In treating the system shown in Fig. 13.1, for simplicity's sake, only the radial stresses were considered. The situation in three dimensions is more complicated, and it is important at this stage to be able to at least qualitatively predict the nature of these stresses. Since the problem is no longer one-dimensional, in addition to the radial stresses, the **axial** and **tangential** or **hoop stresses** have to be considered.

To qualitatively predict the nature of these various stresses, a useful trick is to assume the component with the lower thermal expansion coefficient, TEC, to be *zero* and apply the Eshellby technique. To illustrate, consider the nature of the thermal residual stresses that would be generated if a fiber, with α_f, were embedded in a matrix (same problem as the one shown in Fig. 13.1, except that now the three-dimensional state of stress is of interest), densified and *cooled* from the processing temperature for the case when $\alpha_m > \alpha_f$. Given that $\alpha_m > \alpha_f$ and by making use of the aforementioned trick, i.e., assuming $\alpha_f = 0$ (which implies its dimension does not change with temperature changes), it follows that upon cooling, the matrix will shrink both axially and radially (the hole will get smaller). Consequently, the stress required to fit the matrix to the fiber will have to be axially tensile; when the matrix is welded to the fiber and allowed to relax, this will place the fiber in a state of axial residual compressive stress, which, in turn, is balanced by an axial tensile stress in the matrix. Radially, the matrix will clamp down on the fiber, resulting in radial compressive stresses in both the fiber and the matrix, in agreement with the conclusions drawn above. In addition, the system will develop tensile tangential stresses, as shown in Fig. 13.2*a*.[190] These stresses, if sufficiently high, can cause the matrix to crack radially as shown in Fig. 13.2*c*. It is left as an exercise to readers to determine the state of stress when $\alpha_m < \alpha_f$, and to compare their results with those summarized in Fig. 13.2*b*.

Finally, in this section the problem of a spherical inclusion in an infinite matrix is considered. It can be shown that the radial, σ_{rad} and tangential σ_{tan} stresses generated for a spherical inclusion of radius R at a distance r away from the interface are given by:

$$\sigma_{rad} = -2\sigma_{tan} = \frac{(\alpha_m - \alpha_i)\Delta T}{(1-2\nu_i)/Y_i + (1+\nu_m)/(2Y_m)} \left(\frac{R}{r+R} \right)^3 \tag{13.8}$$

where ν_i and ν_m are, respectively, Poisson's ratios for the inclusion and matrix. The stress is maximum at the interface, i.e., at $r = 0$, and drops rapidly with distance. Note that the final form of this expression is similar to Eq. (13.6). Incidentally, the Eshellby technique is not restricted to calculating thermal stresses; it can also be used to calculate phase transformation stresses, or any other problem where two adjacent regions change shape differently.

[190] To appreciate the nature of tangential stresses, it helps to go back to the Eshellby technique and ask: What would be required to make the hole in the matrix, which is now smaller than the fiber it surrounds, larger? The answer is, one would have to stretch the matrix in a manner similar to fitting a smaller-diameter rubber hose around a larger-diameter pipe. This naturally results in a tangential stress in the hose. Experience tells us that if the hose is too small, it will develop radial cracks similar to the ones shown in Fig. 13.2c.

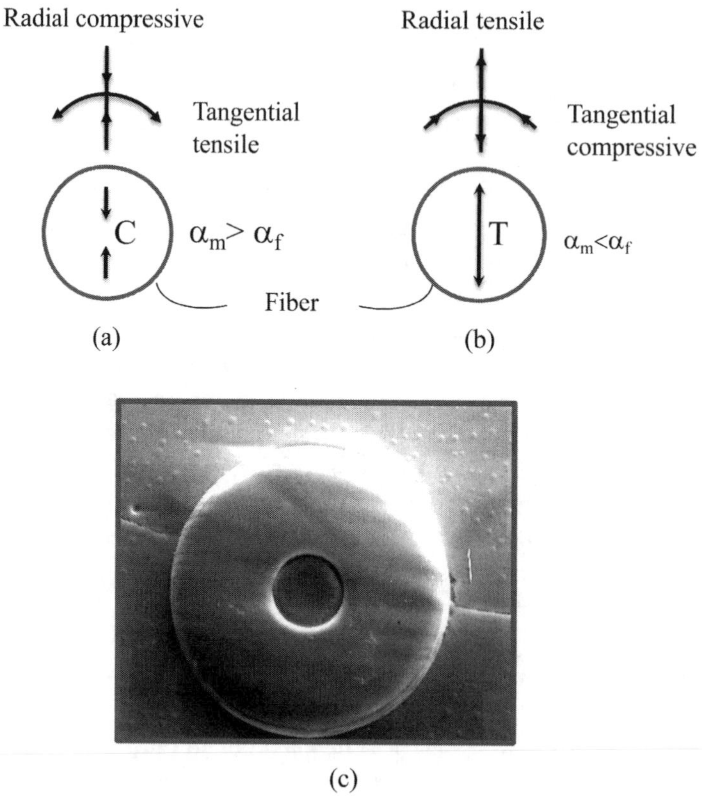

FIGURE 13.2 Radial and tangential stresses developed upon cooling of a fiber embedded in a matrix for (*a*) $\alpha_m < \alpha_f$ and (*b*) $\alpha_m > \alpha_f$. (*c*) Micrograph of radial cracks generated around a fiber upon cooling when $\alpha_m > \alpha_f$.

Thermal stresses generated when dissimilar materials are bonded together and heated can if not mitigated create serious problems. Examples include porcelain enameled metals, thin films, metal-to-ceramic bonding important in the semiconductor industry and glass-to-metal seals to name a few.

13.3 THERMAL SHOCK

Generally speaking, thermal stresses are to be avoided since they can significantly weaken a component. In extreme cases, a part can spontaneously crumble during cooling. As noted earlier, *rapid* heating, or cooling, of a ceramic will often result in its failure. This kind of failure is known as *thermal shock* and occurs when thermal gradients and corresponding thermal stresses exceed the strength of the part. For instance, as a component is rapidly cooled from a temperature T to T_0, the surface will tend to contract but will be prevented from doing so by the bulk of the component that is still at temperature T. Using arguments similar to the ones made above, it is easy to appreciate that in such a situation surface tensile stresses would be generated that have to be counterbalanced by compressive ones in the bulk.

EXPERIMENTAL DETAILS: MEASURING THERMAL SHOCK RESISTANCE

Thermal shock resistance is usually evaluated by heating samples to various temperatures, T. The samples are then rapidly cooled by quenching them from T into a medium, most commonly, ambient temperature water. The post-quench retained strengths are measured and plotted versus the severity of the quench, or $\Delta T = T - T_{ambi}$. Typical results are shown in Fig. 13.3*a*, where the salient feature is the occurrence of a rapid decrease in retained strength around a critical temperature difference ΔT_c, below which the original strength is retained. As the quench T is further increased, the strength decreases but more gradually. Actual data for single-crystal and polycrystalline alumina are shown in Fig. 13.3*b*. In both cases catastrophic losses in strength are observed at a given ΔT_c.

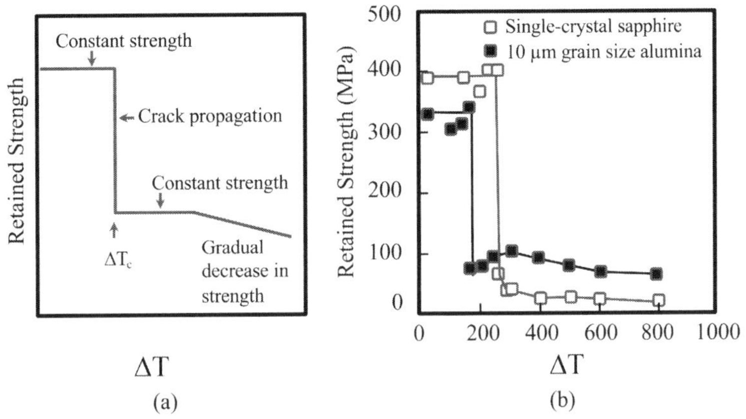

FIGURE 13.3 (*a*) Schematic of strength behavior as a function of severity of quench, ΔT. (*b*) Actual data for single-crystal and polycrystalline alumina. (Error bars were omitted for the sake of clarity.) (T. K. Gupta, *J. Amer. Cer. Soc.*, 55, 249, 1972.)

From a practical point of view, it is useful to be able to predict ΔT_c. Furthermore, it is only by understanding the various parameters that affect thermal shock that successful design of solids that are resistant to it can be developed. In the remainder of this section, a methodology is outlined for doing just that, an exercise that will by necessity highlight the important parameters that render a ceramic resistant to thermal shock.

To estimate ΔT_c, the following assumptions are made[191]

1. The material contains N identical, uniformly distributed, Griffith flaws per unit volume.
2. The flaws are circular with radii c_i.

[191] This derivation is a simplified version of one originally outlined by D. P. H. Hasselman, *J. Amer. Cer. Soc.*, **46**, 453 (1963) and 52, 600 (1969).

3. The body is uniformly cooled with the external surfaces rigidly constrained to give a well-defined triaxial tensile state of stress given by[192]

$$\sigma_{ther} = -\frac{\alpha Y \Delta T}{(1-2v)} \tag{13.9}$$

4. Crack propagation occurs by the simultaneous propagation of the N cracks, with negligible interactions between the stress fields of neighboring cracks.

The derivation is straightforward and follows the one carried out in deriving Eq. (11.9). The total energy of the system can be expressed as

$$U_{tot} = U_o - U_{strain} + U_{surf}$$

where U_o is the energy of the stress- and crack-free crystal of volume V_o; U_{surf} and U_{strain} are, respectively, the surface and strain energies of the system. Since it was assumed that the stress fields were noninteracting, in the presence of N cracks U_{tot} is modified to read

$$U_{tot} = U_o + \frac{V_o \sigma_{ther}^2}{2Y} - \frac{N\sigma_{ther}^2}{2Y}\frac{4\pi c_i^3}{3} + NG_c \pi c_i^2 \tag{13.10}$$

where the third term on the right-hand side represents the strain energy released by the existence of the cracks and the last term is the energy needed to extend them. G_c is the material's toughness [see Eq. (11.11)].

Differentiating this expression with respect to c_i, equating the resulting expression to zero and rearranging terms, one can show (see Prob. 13.6*a*) that for $\Delta T > \Delta T_c$, where ΔT_c is given by

$$\Delta T_c \geq \sqrt{\frac{G_c (1-2v)^2}{\alpha^2 Y c_i}} \tag{13.11}$$

the cracks will grow and release the strain energy. Conversely, for $\Delta T \leq \Delta T_c$, the strain energy that develops is insufficient to extend the cracks, which in turn implies that the strength remains unchanged, as experimentally observed (Fig. 13.a and b).

In contrast to the situation of a flaw propagating as a result of a constant applied stress, in which the flaw will extend indefinitely until fracture, the driving force for crack propagation during thermal shock is finite. Here, the cracks will only extend up to a certain length c_f, which is commensurate with the strain energy available to them, and then stop. To estimate c_f, one simply equates the strain energy available to the system to the increase in surface energy, or

$$\pi N G_c \left(c_f^2 - c_i^2\right) = \frac{(\alpha \Delta T_c)^2 Y}{2(1-2v)^2} \tag{13.12}$$

[192] Note the similarity of this equation to Eq. (13.7).

For short initial cracks, that is, $c_f \gg c_i$, substituting for ΔT_c from Eq. (13.11), one obtains

$$c_f \cong \sqrt{\frac{1}{\pi N c_i}} \tag{13.13}$$

which, interestingly enough, does not depend on any material parameters.

For the sake of clarity, the model used to derive Eqs. (13.11) and (13.13) was somewhat simplified here. Using a slightly more sophisticated approach, Hasselman obtained the following relationships:

$$\boxed{\Delta T_c = \sqrt{\frac{\pi G_c (1-2\nu)^2}{Y\alpha^2 (1-\nu^2) c_i}} \left[1 + \frac{16 N c_i^3 (1-\nu)^2}{9(1-2\nu)} \right]} \tag{13.14}$$

$$c_f = \sqrt{\frac{3(1-2\nu)}{8(1-\nu^2) N c_i}} \tag{13.15}$$

And while at first glance these expressions may appear different from those derived above, on closer examination, their similarity becomes obvious. For example, for small cracks of low density, the second term in brackets in Eq. (13.14) can be neglected with respect to unity, in which case, but for a few terms including Poisson's ratio and π, Eq. (13.14) is similar to Eq. (13.11). The same is true for Eqs. (13.13) and (13.15).

Before one proceeds further, it is worthwhile to summarize the physics of events occurring during thermal shock. Subjecting a solid to a rapid change in temperature results in differential dimensional changes in various parts of the solid and a buildup of stresses within it. Consequently, the system's strain energy will increase. If that strain energy increase is not too large, i.e., for small ΔT values, the preexisting cracks will not grow and the solid will not be affected by the thermal shock. However, if the thermal shock is large, the many cracks present in the solid will extend and absorb the excess strain energy. Since the available strain energy is finite, the cracks will extend only until most of the strain energy is converted to surface energy, at which point they will be arrested. The final lengths to which the cracks will grow will depend on their initial size and density. If only a few, small cracks are present, their final length will be large and strength degradation will be high. Conversely, if there are numerous small cracks, each will extend by a small amount and the corresponding degradation in strength will not be that severe. In the latter case, the solid is considered to be **thermal-shock-tolerant.**

It is this latter approach that is used in fabricating insulating firebricks for furnaces and kilns. The bricks are fabricated so as to be porous and contain many flaws. Because of the large number of flaws and pores within them, the bricks can withstand severe thermal cycles without structural failure.

Inspecting Eq. (13.11) or (13.14), it is not difficult to conclude that a good figure of merit for thermal shock resistance is

$$R_H = (\text{const})(\Delta T_c) = (\text{const})\sqrt{\frac{G_c}{\alpha^2 Y}} = \frac{K_{Ic}}{\alpha Y} \tag{13.16}$$

from which it is clear that solids with low thermal expansion coefficients, low elastic moduli, but high fracture toughnesses should be thermal shock resistant.

Kingery's[193] approach to the problem was slightly different. He postulated that failure would occur when the thermal stress, given by Eq. (13.7), was equal to the solid's tensile strength σ_t (see Prob. 13.4). Equating the two, it can be shown that the figure of merit in this case is

$$R_{TS} = (\text{const})(\Delta T_c) = (\text{const})\frac{(1-2\nu)\sigma_t}{\alpha Y} \tag{13.17}$$

However, given that σ_t is proportional to $(G_c Y/c_{max})^{1/2}$, it is an easy exercise to show that R_{TS} is proportional to $R_H / c_{max}^{1/2}$, implying that the two criteria are related.[194]

One parameter which is not included in either model, and which clearly must have an important effect on thermal shock resistance, is the thermal conductivity of the ceramic k_{th} (see Sec. 13.6). Given that thermals gradients are ultimately responsible for the buildup of stress, it stands to reason that a highly thermally conductive material would not develop large gradients and would thus be thermal shock resistant. For the same reason, the heat capacity and the heat-transfer coefficient between the solid and the environment must also play a role. Thus an even better indicator of thermal shock resistance is to multiply Eq. (13.16) or (13.17) by k_{th}. These values are calculated for a number of ceramics and listed in Table 13.1 in columns 7 and 8. Also listed in Table 13.1 are experimentally determined values. A correlation between the two sets of values is apparent, giving some validity to the aforementioned models.

Note that in general the nitrides and carbides of Si, with their lower thermal expansion coefficients, are more resistant to thermal shock than oxides. In theory, a material with zero thermal expansion would not be susceptible to thermal shock. In practice, a number of such materials do actually exist commercially, including some glass ceramics that have been developed that as a result of thermal expansion anisotropy, have extremely low αs (see Chap. 4). Another good example is fused silica, which has an extremely low α and thus is not prone to thermal shock. Fused silica is used to make high-temperature furnace tubes that are quite resistant to thermal shock.

TABLE 13.1 Comparison of thermal shock parameters for a number of ceramics. Poisson's ratio was taken to be 0.25 for all materials

Material	MOR, MPa	Y, GPa	α, 10^6 K^{-1}	k_{th}, W/(m•K)	K_{1c}, MPa•m$^{1/2}$	$k_{th}R_{TS}$, W/m	$R_H k_{th}$, W/m^2	ΔT_c, exper.
SiAlON	945	300	3.0	21	8	16 500	180	900
HP[a]–Si_3N_4	890	310	3.2	15–25	5	16 800	126	500–700
RB[b]–Si_3N_4	240	220	3.2	8–12	2	2557	28	≈500
SiC (sintered)	483	410	4.3	84	3	17 300	143	300–400
HP[a]–Al_2O_3	380	400	9.0	6–8	4	633	8	200
PSZ	610	200	10.6	2	≈10	435	9	500
Ti_3SiC_2	300	320	9.1	43	≈10		149	>1400

[a] Hot-pressed.
[b] Reaction-bonded.
[c] Partially stabilized zirconia.

[193] W. D. Kingery, *J. Amer. Cer. Soc.*, 38, 3–15 (1955).
[194] Note the Hasselman solid is a highly idealized one where all the flaws are the same size.

13.4 SPONTANEOUS MICROCRACKING OF CERAMICS

In the previous section, the emphasis was on thermal shock, where failure was initiated by a *rapid and/or severe temperature change*. This is not always the case; both single- and multiphase ceramics have been known to spontaneously microcrack upon slow cooling. Whereas thermal shock can be avoided by slow cooling, the latter phenomenon is *unavoidable* regardless of the rate at which the temperature is changed.

Spontaneous microcracking results from the buildup of residual stresses that can be caused by one, or more, of the following reasons:

- ∞ Thermal expansion anisotropy in single-phase materials
- ∞ Thermal expansion mismatches in multiphase materials
- ∞ Phase transformations and accompanying volume changes in single- or multiphase materials

In the remainder of this section each of these cases is explored in some detail.

13.4.1 SPONTANEOUS MICROCRACKING DUE TO THERMAL EXPANSION ANISOTROPY

Noncubic ceramics with high thermal expansion anisotropy have been known to spontaneously microcrack upon cooling.[195] The cracking, which occurs along the grain boundaries, becomes progressively less severe with decreasing grain size, and below a certain "critical" grain size, it is no longer observed. The phenomenon has been reported for various solids such as Al_2O_3, graphite, Nb_2O_5 and many titania-containing ceramics such as TiO_2, Al_2TiO_5, Mg_2TiO_5 and Fe_2TiO_5. Data for some anisotropic crystals are given in Table 13.2.

Before attempting to quantify the problem, it is important once again to understand the underlying physics. Consider the situation shown in Fig. 13.4*a*, where the grains, assumed to be cubes, are arranged in such a way that adjacent *grains* have different thermal expansion coefficients along their x and y axes as shown, with $\alpha_1 < \alpha_2$. To further elucidate the problem, use the aforementioned trick of equating the lower thermal expansion to zero, i.e., pretend $\alpha_1 = 0$. If during cooling, the grains are unconstrained, the shape of the assemblage would be that shown in Fig. 13.4*b*. But the cooling is *not* unconstrained, which implies boundary stresses will build up. These stresses are ultimately responsible for failure.

To estimate the critical grain size above which spontaneous microcracking would occur, various energy terms have to be considered. For the sake of simplicity, the grains are assumed to be cubes with grain size *d* in which case the total energy of the system is[196]

$$U_{tot} = U_s - NU_g d^3 + 6Nd^2 G_{c,gb} \tag{13.18}$$

[195] Thermal expansion coefficients of cubic materials are isotropic and hence do not exhibit this phenomenon.

[196] The treatment here is a slightly simplified version of that carried out by J. J. Cleveland and R. C. Bradt, *J. Amer. Cer.*, **61**, 478 (1978).

TABLE 13.2 Thermal expansion coefficients, in K^{-1}, for select ceramic single crystals with anisotropic thermal expansion behavior

Material	Normal to c-axis	Parallel to c-axis
Al_2O_3	8.3×10^{-6}	9.0×10^{-6}
Al_2TiO_5	-2.6×10^{-6}	11.5×10^{-6}
$3Al_2O_3 \cdot 2SiO_2$ (mullite)	4.5×10^{-6}	5.7×10^{-6}
$CaCO_3$	-6.0×10^{-6}	25.0×10^{-6}
C (graphite)	1.0×10^{-6}	27.0×10^{-6}
$LiAlSi_2O_6$ (β-spodumene)	6.5×10^{-6}	-2.0×10^{-6}
$LiAlSiO_4$ (β-eucryptite)	8.2×10^{-6}	-17.6×10^{-6}
$NaAlSi_3O_8$ (albite)	4.0×10^{-6}	13.0×10^{-6}
SiO_2 (quartz)	14.0×10^{-6}	9.0×10^{-6}
TiO_2	6.8×10^{-6}	8.3×10^{-6}
$ZrSiO_4$	3.7×10^{-6}	6.2×10^{-6}

where N is the number of grains relieving their stress and $G_{c,gb}$ is the grain boundary toughness, U_s is the energy of the un-microcracked body and U_g is the strain energy per unit volume stored in the grains. Differentiating Eq. (13.18) with respect to d and equating to zero yields the critical grain size

$$d_{crit} = \frac{4G_{c,gb}}{U_g} \tag{13.19}$$

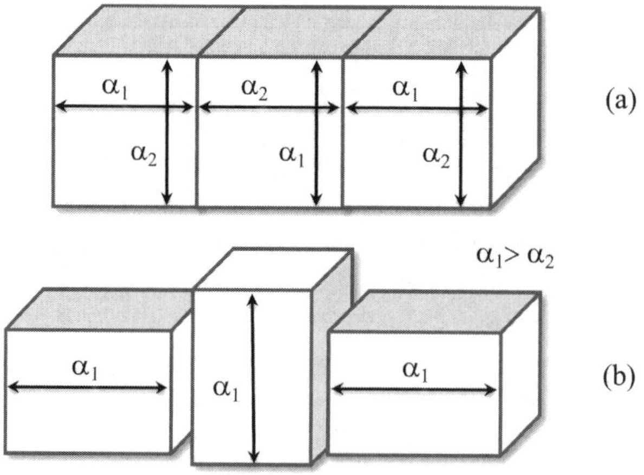

FIGURE 13.4 Schematic of how thermal expansion anisotropy can lead to the development of thermal stresses upon cooling of a polycrystalline solid. (*a*) Arrangement of grains prior to cooling shows relationship between thermal expansion coefficients and grain axis. (*b*) Unconstrained contraction of grains. Here it was assumed that $\alpha_1 = 0$.

U_g is estimated as follows: For a totally constrained grain, the stress developed is given by Eq. (13.7). Extending the argument to two adjacent grains, the residual stress can be approximated by

$$\sigma_{th} = \frac{1}{2} Y \Delta\alpha_{max} \Delta T \tag{13.20}$$

where $\Delta\alpha_{max}$ is the maximum anisotropy in thermal expansion between two crystallographic directions. Substituting Eq. (13.20) in the expression for the strain energy per unit volume, that is, $U_g = \sigma^2/(2Y)$, and combining with Eq. (13.19), one obtains

$$d_{crit} = \frac{32 G_{c,gb}}{Y \Delta\alpha_{max}^2 \Delta T^2} \tag{13.21}$$

In general, however,

$$\boxed{d_{crit} = (\text{const}) \frac{G_{c,gb}}{Y \Delta\alpha_{max}^2 \Delta T^2}} \tag{13.22}$$

where the value of the numerical constant one obtains depends on the details of the models. This model predicts that the critical grain size below which spontaneous microcracking will *not* occur is a function of the thermal expansion anisotropy, the grain boundary fracture toughness and Young's modulus. Experimentally, the functional relationship among d_{crit}, ΔT and $\Delta\alpha_{max}$, is reasonably well established (see Prob. 13.8).

EXPERIMENTAL DETAILS: DETERMINATION OF MICROCRACKING

Unless a ceramic component totally falls apart in the furnace as the sample is cooled from the sintering or processing temperature, it is experimentally difficult to directly observe grain boundary microcracks. There are, however, a number of indirect techniques to study the phenomenon. One is to fabricate ceramics of varying grain sizes and measure their flexural strengths after cooling. A dramatic decrease in strength over a narrow grain size variation is usually a good indication that spontaneous microcracking has occurred.

13.4.2 SPONTANEOUS MICROCRACKING DUE TO THERMAL EXPANSION MISMATCHES IN MULTIPHASE MATERIALS

Conceptually there is little difference between this situation and the preceding one; the similarity of the two cases is easily appreciated by simply replacing one of the grains in Fig. 13.4 by a second phase with a different TEC than its surroundings.

13.4.3 SPONTANEOUS MICROCRACKING DUE TO PHASE-TRANSFORMATIONS

Here the residual stresses do not develop as a result of thermal expansion mismatches or rapid variations in temperature, but as a result of phase transformations. Given that these transformations entail atomic rearrangements, they are always associated with a volume change (e.g., Fig. 4.5). Conceptually, the reason why such a volume change should give rise to residual stresses should at this point be obvious. Instead of using $\Delta\alpha$, however, the resultant stresses are assumed to scale with $\Delta V/V_0$, where ΔV is the volume change associated with the transformation. The stresses approximated by

$$\sigma \approx \frac{Y}{3(1-2\nu)} \frac{\Delta V}{V_0} \tag{13.23}$$

can be quite large. For example, a 3% volumetric change in a material with Y = 200 GPa and Poisson's ratio of 0.25 would result in a stress of about 4 GPa!

Residual stresses are generally deleterious to the mechanical properties and should be avoided. This is especially true if a part is to be subjected to thermal cycling. In some situations, however, residual stresses can be used to advantage. A case in point is the transformation toughening of zirconia discussed in Chap. 11, and another excellent example is the tempering of glass discussed in the next section.

13.5 THERMAL TEMPERING OF GLASS

Because of the transparency and chemical inertness of inorganic glasses, their uses in everyday life are ubiquitous. However, for many applications, especially where safety is concerned, as-manufactured, glass is deemed to be too weak and brittle. Fortunately, glass can be significantly strengthened by a process referred to as *thermal tempering*, which introduces a state of compressive residual stresses on the surface (see Sec. 11.3.3).

The appropriate thermal process, illustrated in Fig. 13.5, involves heating the glass body to a temperature above its glass transition temperature, T_g, followed by a two-step quenching process. During the first quenching stage, initially the surface layer contracts more rapidly than the interior (Fig. 13.5b) and becomes rigid while the interior is still in a viscous state. This results in a tensile state of stress at the surface, shown in Fig. 13.5c. However, since the interior is viscous these stresses will relax, as shown in Fig. 13.5*d*.

During the second quenching step, the entire glass sample is cooled to room temperature. Given that on average the glass interior will have cooled at a *slower* rate than its exterior, its final specific volume will be *smaller* than that of the exterior.[197] The situation is shown in Fig. 13.5*e* and leads directly to the desired final state of stress (Fig. 13.5*f*) in which the external surfaces are in compression and the interior is in tension.

By using this technique, the mean strength of soda-lime silicate glass can be raised to the range of 150 MPa, which is sufficient to permit its use in large doors and windows as well as safety lenses. Tempered glass is also used for the side and rear windows of automobiles. In addition to being stronger, tempered glass is preferred to untempered glass for another important reason: the release of large amounts of stored elastic energy upon fracture tends to shatter the glass into a great many fragments, which are less dangerous than larger shards. Windshields, however, are made of two sheets of tempered glass in between which a polymer layer in embedded. The function of the latter is to hold the fragments of glass together in case of fracture and to prevent them from becoming lethal projectiles.

[197] This effect was discussed briefly in Sec. 9.4.1 and illustrated in Fig. 9.8*a*. Simply put, the more time the atoms have to arrange themselves during the cooling process the denser the glass that results.

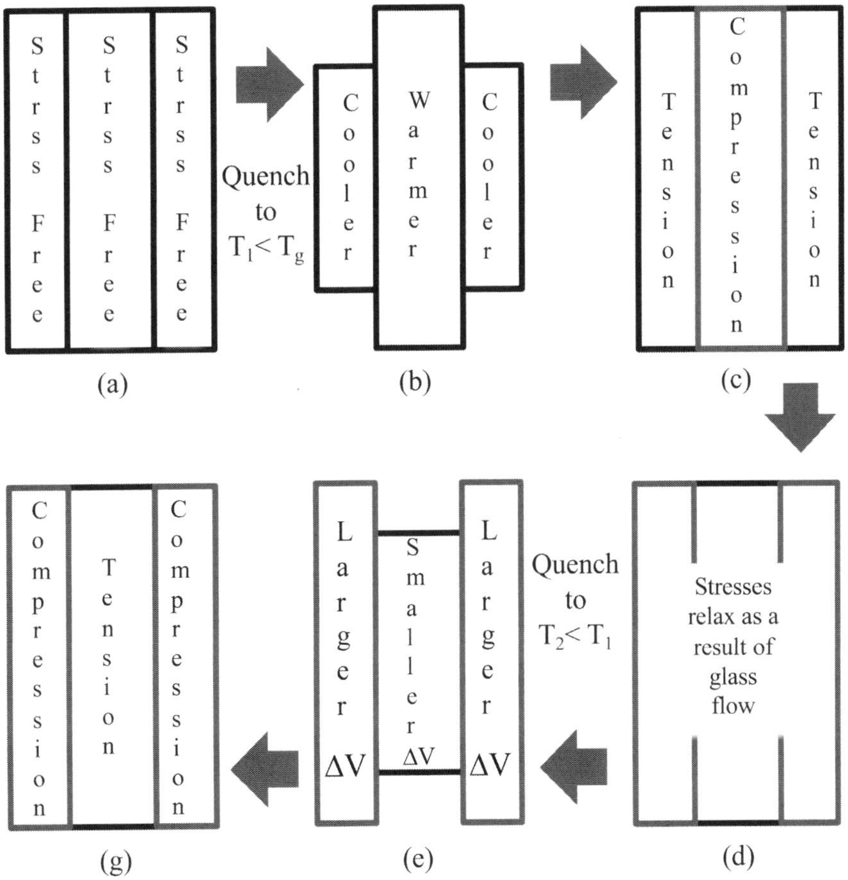

FIGURE 13.5 Thermal process that results in tempered glass. (*a*) Initial configuration. (*b*) The glass is quenched to a temperature that is below T_g, which results in the rapid contraction of the exterior. (*c*) Resulting transient state of stress. (*d*) Relaxation of these stresses occurs by the flow and deformation of the interior. (*e*) Second quenching step results in a more rapid cooling rate for the exterior than for the interior. This results in a glass with a smaller specific volume in the center than on the outside. (*f*) Final state of stress at room temperature.

13.6 THERMAL CONDUCTIVITY

The conduction of heat through solids occurs as a result of temperature gradients. In analogy to Fick's first law, the relationship between the heat flux and temperature gradients $\partial T/\partial x$ is given by

$$\frac{\partial Q}{\partial t} = k_{th} A \frac{\partial T}{\partial x} \tag{3.24}$$

TABLE 13.3 Typical values for room temperature thermal conductivities of select ceramic materials

Material	k_{th}, W/(m·K)	Material	k_{th}, W/(m·K)
Al_2O_3	30–35	Spinel ($MgAl_2O_4$)	12.0
AlN	200–280	Soda-lime silicate glass	1.7
BeO	63–216	TiB_2	40.0
MgO	37.0	Ti_3SiC_2	43.0
PSZ	2.0	Ti_2AlC	40.0
SiC	84–93	Cordierite (Mg-aluminosilicate)	4.0
SiAlON	21.0	Glasses	0.6–1.5
SiO_2	1.4	Forsterite	3.0
Si_3N_4	25.0		

where $\partial Q/\partial t$ is the heat transferred per unit time across a plane of area A normal to the flow of the thermal energy. k_{th} is a material property (analogous to diffusivity) that describes the ability of a material to transport heat. Its units are J/(s·m·K) or equivalently W/(m·K). Approximate values for k_{th} for a number of ceramics are listed in Table 13.3.

THERMAL CONDUCTION MECHANISMS

Describing thermal conduction mechanisms in solids is non-trivial. Here only a brief qualitative sketch of some of the physical phenomena is given. In general, thermal energy in solids is transported by lattice vibrations, i.e., phonons, free electrons and/or radiation. Given that the concentration of free electrons in typical ceramics is low and that most ceramics are not transparent, phonon mechanisms dominate and are the only ones discussed below.

Imagine a small region of a solid being heated. Atoms in that region will acquire larger amplitudes of vibration and will vibrate violently around their average positions. Given that these atoms are bonded to neighboring atoms, their motion will *per force* set their neighbors into oscillation. As a result, the disturbance, caused by the application of heat, propagates outward in a wavelike manner.[198] These waves, in complete analogy to electromagnetic waves, can be scattered by imperfections, grain boundaries, and pores or even reflected at other internal surfaces. In other words, every so often the disturbance will have its propagation direction altered. The average distance that the disturbance travels before being scattered is analogous to the average distance traveled by a gas molecule and is referred to as the *mean free path*, λ_{th}.

By assuming the number of these thermal energy carriers to be N_{th}, and their average velocity v_{th}, in analogy with the electrical conductivity equation viz. $\sigma = n\mu q$, k_{th} can be given by

$$k_{th} = (\text{const})(N_{th}\lambda_{th}v_{th}) \quad (3.25)$$

[198] A situation not unlike the propagation of light or sound through a solid.

In general, open, highly ordered structures made of atoms or ions of similar size and mass tend to minimize phonon scattering and result in increased k_{th} values. An excellent example is diamond, which has one of the highest thermal conductivity values of any known material. Other good examples are SiC, BeO and AlN. More complex structures, such as spinels, and ones where there is a large difference in mass between ions, such as UO_2 and ZrO_2, tend to have lower values of k_{th}. Similar arguments suggest that a solid's thermal conductivity will be decreased by the addition of a second component in solid solution. This effect is well known, as shown, e.g., by the addition of NiO to MgO or Cr_2O_3 to Al_2O_3.

Furthermore, the lack of long-range order in amorphous ceramics results in more phonon scattering than in crystalline solids and consequently leads to lower values of k_{th}.

EFFECTS OF MICROSTRUCTURE ON THERMAL CONDUCTIVITY

In addition to the factors listed above, a ceramic's microstructure can also have an important effect on its thermal conductivity. Here we explore three factors; grain size, point defects and pores.

In general, the phonon mean free path, λ_{th} is << grain size and hence, phonon scattering by grain boundaries has little effect on k_{th}. The best evidence for this claim is the fact that the values of k_{th} of single crystals and their polycrystalline counterparts are quite comparable not only at room temperature but over an extended range of temperature.

As just noted the presence of atoms of dissimilar masses and sizes can also lead to phonon scattering and a reduction in k_{th}. The effect is not very large, however. Along the same lines, point defects can also affect k_{th}. An excellent example is AlN. High purity AlN is an exceptionally good thermal conductor, with k_{th} values of the order of 300 W/(m·K). However, like most Al-containing nonoxides, AlN powders are typically surrounded by a thin native Al_2O_3 layer. When AlN powders are sintered, the Al_2O_3 dissolves in the AlN according to the following defect reaction:

$$Al_2O_3 \rightarrow 2Al_{Al}^{x} + 3O_N^{\bullet} + V_{Al}'''$$

The Al vacancies so generated, in turn, are good phonon scatterers and lead to a reduction in k_{th} to values of the order of 50–200 W/(m·K). One solution to this problem has been the addition of rare earth, or alkaline metal, oxides such as CaO or Y_2O_3 to the starting powder. At the sintering temperatures, these additives react with the native Al_2O_3 layer to form a liquid that not only aids in sintering, but also acts as a getter or oxygen reservoir. Upon cooling, this oxygen-rich liquid segregates harmlessly at triple points.

This is important because as electronic devices become increasingly smaller, lighter, thinner and more powerful, removing the heat generated rapidly and efficiently is becoming a serious hurdle. AlN is almost ideal for this application for a number of reasons, among them is its inertness and high electrical resistivity. Most important, its thermal expansion coefficient in the 24–400°C temperature range matches that of Si—the workhorse of intergrade circuits—quite closely. This minimizes the accumulation of residual stresses that can be detrimental to the devices.

Lastly, in this section, it is important to discuss porosity. Since the thermal conductivity of air is negligible compared to most ceramics, the addition of large (>0.25) volume fractions of pores can significantly reduce k_{th}. This approach is used in the fabrication of firebrick. As noted above, the addition of large-volume fractions of porosity has the added advantage of rendering the firebricks thermal-shock-tolerant. Note that heat transfer by radiation across the pores, which scales as T^3, has to be minimized. Hence, for optimal thermal resistance, the pores should be small and the pore phase should be continuous.

Another good example of the use of pores for thermal management is discussed in Case Study 13.2.

EXPERIMENTAL DETAILS: MEASURING THERMAL CONDUCTIVITY

Several techniques are used to measure k_{th}. One such method is the laser flash technique. In principle the technique attempts to measure the time evolution of the temperature on one side of a sample, as the other side is rapidly heated by a laser pulse. As it passes through the solid, the signal will be altered in two ways: There will be a time lag between the time at which the solid was pulsed and the maximum in the response. This time lag is directly proportional to the thermal diffusivity, D_{th}, of the material. The second effect will be a reduction in the temperature spike, which is directly related to the heat capacity, c_p, of the solid. The heat capacity, thermal diffusivity, and thermal conductivity and density, σ, are related by:

$$k_{th} = \rho c_p D_{th} \tag{3.26}$$

Hence k_{th} can be readily calculated if the density of the solid , ρ, is known and D_{th} and c_p are measured.

CASE STUDY 13.1: THERMAL BARRIER COATINGS

At the end of Chap. 11, how CMCs were slowly, but surely, making jet engines lighter and more efficient was discussed. Another vital technology—thermal barrier coatings or TBCs—is discussed here. TBCs provide thermal insulation to gas turbine engines allowing them to run hotter and thus be more efficient. These engines represent roughly a \$50 billion industry, with 65% of sales going to aviation, the balance used for electricity generation. More than 20% of global electricity is produced by gas turbines.

A successful TBC must: (i) provide insulation to the underlying engine part, (ii) be strain compliant to accommodate the thermal expansion mismatch between it and underlying part, (iii) reflect the radiant heat from the hot engine gas; (iv) maintain its integrity as it is cycled many times from room temperature to 1300°C and back; (v) be able to withstand thermal shock, extreme thermal gradients and energy fluxes. (vi) operate in oxidizing atmospheres, flowing at speeds close to the speed of sound and at pressures up to 10 atm.

TBCs are comprised of several layers that are deposited on top of a Ni-based superalloy component. The layers—shown in Fig. 13.6*a*—are comprised of (i) a 30 to 100 μm thick *bond coat* deposited directly on the superalloy unto which (ii) a 0.1 to 3 mm porous oxide is deposited. The most common oxide used today is a zirconia doped with 7 wt.% Y_2O_3. This oxide was chosen because of its low thermal conductivity. An actual SEM micrograph of a cross section of such a system is shown in Fig. 13.6*b*.

The bond coat—typically an Al-containing alloy—is designed to oxidize in such a way as to form a thin adherent and coherent alumina layer between it and the porous oxide layer. This oxide (shown in solid black in Fig. 13.6*a*) is usually referred to as a thermally grown oxide (TGO) and is formed by the selective oxidation of the Al that is contained in the bond coat. Figure 13.6*b* is a SEM micrograph of the schematic

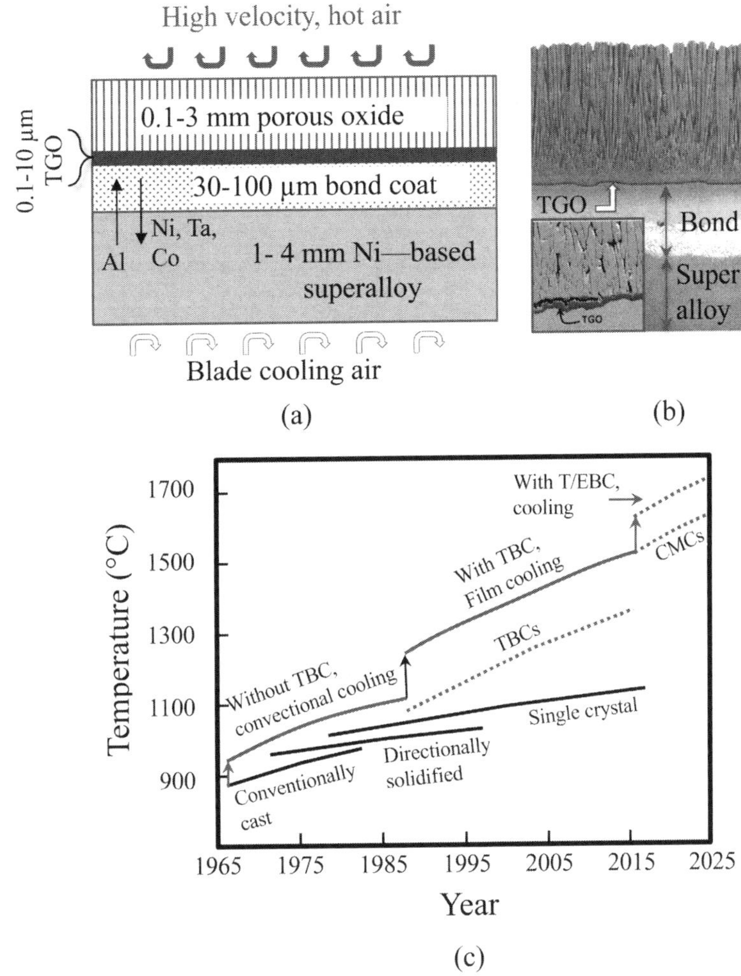

FIGURE 13.6 Thermal barriers coatings. (*a*) Schematic (not to scale) of a cross section of a Ni-based superalloy blade onto which a bond coat and a porous oxide were deposited. Vertical arrows denote direction of diffusion of various atoms; (*b*) actual SEM micrograph of microstructure; (*c*) time evolution of maximum operating temperatures of jet engines. Note large jump around 1988 when TBCs were introduced. (Adapted from D. R. Clarke, M. Oechsner and N. Padture, *MRS Bulletin*, 37, 891, 2012.)

shown in Fig. 13.6*a*. In this micrograph the TGO appears as a thin dark line labeled TGO. Inset in Fig. 13.6*b* shows a higher magnification SEM micrograph of the TGO layer.

The linchpin of this coating, and what renders it useful, is the last layer deposited, viz. a 0.1–3 mm thick oxide layer. This coating is essentially massively microcracked ZrO_2! This layer is typically deposited by vapor phase deposition, unto the blades that comprise the heart of a jet engine. In Chap. 11 microcracks were to be avoided like the plague since they are the bane of ceramics. Here the designers of these coatings took a page from the refractory ceramics world where controlled microcracks have long been used to render firebricks more resistant to spalling and thermal shock. The microcracks allow for the thermal expansion of the coating as the temperatures are rapidly increased upon engine ignition. They also accommodate any thermal expansion mismatches between the substrates and coatings. In that sense they are quite compliant. Lastly, the air-filled cracks are excellent heat insulators, which is the main purpose of the TBCs. Thermodynamics dictates that a hotter engine is a more efficient one. Figure 13.6c shows the evolution of the maximum operating temperature of gas turbine engines with time. Today's engines run roughly 400°C hotter than their counterparts 30 years ago.

CASE STUDY 13.2: SPACE SHUTTLE TILES

The space shuttles provided routine transportation of crews and cargo from the Earth's surface to space. A key component of these shuttles and the one that allowed them to be reusable was a thermal protection system that insulated the crewmembers and the orbiter from the high temperatures experienced upon reentry. This system had to withstand the high temperatures of reentry, be thermally insulating and lightweight. The idea that air is an exceptional thermal insulator and is not very dense led NASA in the 1970s to develop the space shuttle tiles (Fig. 13.7a). The tiles were 90% air, entrapped in low density, high purity, 99.8%, amorphous silica fiber tiles. With diameters between 2 to 4 μm, the silica fibers were quite thin (Fig. 13.7*b*). With a density of 0.14 g/cm^3 the tiles were quite light. Their maximum operating temperature was ≈1250°C. At room temperature and 1 atm their thermal conductivity was of the order of 0.05 W m^{-1} K^{-1}.

The top and sides of the tiles were then sprayed with a mixture of powdered tetrasilicide and borosilicate glass suspended in a liquid carrier. The coated tiles were then fired at a temperature of 1250°C in air that resulted in a black, waterproof glossy coating (inset in Fig. 13.7b). After this step, the remaining silica fibers were treated with a resin for further waterproofing.

Because they were mostly air, the tiles' thermal expansion was significantly lower than that of the airframe. Had they simply been rigidly fixed unto the load frame, the large temperature variations experienced

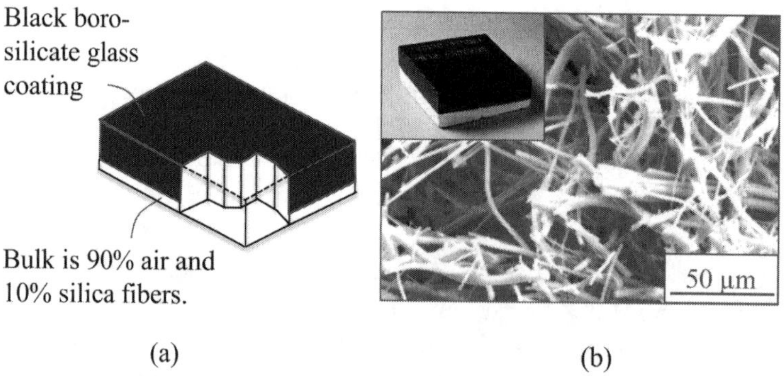

FIGURE 13.7 Space shuttle tiles. (*a*) Schematic of a typical tile; (*b*) SEM micrograph of the body of the tile comprised of thin amorphous silica fibers. Inset is a picture of an actual tile.

by the shuttle would have literally destroyed the tiles. The tiles thus had to be mechanically decoupled from the airframe. This was provided by a strain isolation pad, SIP. The SIPs were made of an aramid felt material ≈0.3 mm thick, that was bonded to the bottom of the tiles. The combination was in turn bonded to the orbiter structure using a room temperature vulcanizing silicone—aka bathroom caulk—as the glue.

Lastly to prevent tile-to-tile contact when the shuttle was in very cold temperatures of space, they needed to be assembled with gaps between them. The same felt material used for the SIP was used in the form of 2 cm strips about 1 mm thick. The strips were placed between the tiles and bonded to the structure.

Tragically upon liftoff of the space shuttle Columbia, a chunk of the insulation from the external fuel tank, punched a hole through and breached the spacecraft wing. Upon reentry in February 2003, hot atmospheric gases penetrated the heat shield and destroyed the internal wing, which resulted in its disintegration and the death of the seven on board.

13.7 SUMMARY

Temperature changes result in dimensional changes that, if not relevied, can result in thermal strains. Isotropic, unconstrained solids subjected to uniform temperatures can accommodate these strains without the generation of thermal stresses. The latter will develop, however, if one, or more, of the following situations are encountered:

- ∞ Constrained heating and cooling.
- ∞ Rapid heating or cooling. This situation is a variation of that above. By rapidly changing the temperature of a solid, its surface will usually be constrained by the bulk and will develop stresses. The magnitude of these stresses depends on the severity of thermal shock or rate of temperature change. In general, the higher the temperature from which a ceramic is quenched the more likely it is to fail or thermal shock. Thermal shock can be avoided by slow heating or cooling. Solids with high thermal conductivities and fracture toughnesses and/or low thermal expansion coefficients are less prone to thermal shock.
- ∞ Heating or cooling of multiphase ceramics in which the various constituents have differing thermal expansion coefficients. The stresses generated in this case will depend on the mismatch in thermal expansion coefficients of the various phases. These stresses cannot be avoided by slow heating or cooling.
- ∞ Heating or cooling of ceramics for which thermal expansion along the various crystallographic axis is anisotropic. The magnitude of the stresses will depend on the thermal expansion anisotropy, and can cause polycrystalline bodies to spontaneously microcrack. This damage cannot be avoided by slow cooling, but can be avoided if the grain size is kept small enough.
- ∞ Phase transformations in which there is a volume change upon transformation. In this case, the stresses will depend on the magnitude of the volume change. They can only be avoided by suppressing the transformation.

If properly introduced, thermal residual stresses can be beneficial, as in the case of tempered glass.

Finally, in the same way that solids conduct sound, they also conduct heat, i.e., by lattice vibrations. Heat conduction occurs by the excitation and interaction of neighboring atoms.

PROBLEMS

13.1. Give an example for each of, (*a*) thermal strain but no stress, (*b*) thermal stress but no strain and (*c*) a situation where both exist.

13.2. (*a*) Derive Eq. (13.6).

(*b*) A metallic rod ($\alpha = 14 \times 10^{-6}$ °C^{-1} $Y = 50$ GPa at 800°C) is machined such that it perfectly fits inside a dense Al_2O_3 tube at room temperature. The assembly is then slowly heated; at 800°C the tube cracks. Assume Poisson's ratio to be 0.25 for both materials.

(*i*) Describe the three dimensional state of stress that develops in the system as it is heated.

(*ii*) Estimate the strength of the alumina tube.
Answer: 170 MPa.

(*iii*) In order to increase the temperature at which this system can go, several strategies have been proposed (some of which are wrong): Use an Al_2O_3 with a larger grain size; use another ceramic with a higher thermal expansion coefficient; use a ceramic that does not bond well with the metal; and/or use a metal with a higher stiffness at 800°C. Explain in some detail (using calculations when possible) which of these strategies you think would work and which would not. Why?

(*iv*) If the situation were reversed (i.e., the Al_2O_3 rod were placed inside a metal tube), describe in detail the three-dimensional state of stress that would develop in that system upon heating.

(*v*) It has been suggested that one way to bond a ceramic rotor to a metal shaft is to use the assembly described in part (*iv*). If you were the engineer in charge, describe how you would do it. Note this is not a hypothetical problem but is used commercially and works quite well!

13.3. Consider a two-phase ceramic in which there are spherical inclusions B. If upon cooling, the inclusions go through a phase transformation that causes them to expand, which of the following states of stress would you expect, and why?

(*a*) Hydrostatic pressure in B; radial, compressive and tangential tensile hoop stresses.
(*b*) Debonding of the interface and zero stresses everywhere.
(c) Hydrostatic pressure in B; radial, tensile and tangential compressive hoop stresses.
(d) Hydrostatic pressure in B; radial, compressive and tangential compressive hoop stresses.
(*e*) Hydrostatic pressure in B; radial, tensile and tangential tensile hoop stresses.

13.4. (*a*) Plot the radial stress as a function of r for an inclusion in an infinite matrix, given that $\Delta\alpha = 5 \times 10^{-6}$, $\Delta T = 500$°C, $Y_i = 300$ GPa, $Y_m = 100$ GPa, and $\nu_i = \nu_m = 0.25$.

(*b*) If the size of the inclusions were 10 μm, for what volume fraction would the "infinite" matrix solution be a good one? What do you think would happen if the volume fraction were higher? State all assumptions.
Answer: $\approx$ 5 to 10 vol.% depending on assumptions made.

13.5. (*a*) Is thermal shock more likely to occur as a result of rapid heating or rapid cooling? Explain your answer.

(*b*) A ceramic component with a Young's modulus of 300 GPa and a K_{Ic} of 4 MPa·m$^{1/2}$ is to survive a water quench from 500°C. If the largest flaw in that material is on the order of 10 μm, what is the maximum value of α for this ceramic for it to survive the quench? State all assumptions.
Answer: 5×10^{-6} °C^{-1}.

13.6. (*a*) Derive Eq. (13.11).

(b) Which of the materials listed below would be best suited for an application in which a part experiences sudden and severe thermal fluctuations while in service?

Material	MOR, MPa	k_{th}, W/(m·K)	Modulus, GPa	K_{Ic}, MPa·m$^{1/2}$	α, K^{-1}
1	700	290	200	8	9×10^{-6}
2	1000	50	150	4	4×10^{-6}
3	750	100	150	4	3×10^{-6}

13.7. (*a*) Explain how a glaze with a different thermal expansion can influence the effective strength of a ceramic component. To increase the strength of a component, would you use a glaze with a higher or lower thermal expansion coefficient than the substrate? Explain.

(*b*) Fully dense, 1-cm-thick Al_2O_3 plates are to be glazed with a porcelain glaze ($Y=70$ GPa, $\nu=0.25$) 1 mm thick and with a thermal expansion coefficient of 4×10^{-6} °C. Assuming the "stress-freezing" temperature of the glaze to be 800°C, calculate the stress in the glaze at room temperature.

13.8. Using acoustic emission and thermal contraction data, Ohya et al.[199] measured the functional dependence of the microcracking temperature of aluminum titanate ceramics on grain size as the samples were cooled from 1500°C. The following results were obtained:

Grain size, μm	3	5	9
Microcracking temperature upon cooling, °C	500	720	900

(*a*) Qualitatively explain the trend observed.

(*b*) Are these results consistent with the model presented in Sec. 13.4.1? If so, calculate the value of the constant that appears in Eq. (13.22). You can assume $G_{c,gb}=0.5$ J/m², $Y=250$ GPa and $\Delta\alpha_{max}=15\times10^{-6}$°C.

Answer: ≈337 (°C)$^{-2}$.

(*c*) Based on these results, estimate the grain size needed to obtain a crack-free aluminum titanate body at room temperature. State all necessary assumptions.

Answer: ≈1.47 μm.

13.9. Explain why volume changes as low as 0.5% can cause grain fractures during phase transformations of ceramics. State all assumptions.

13.10. (*a*) If a glass fiber is carefully etched to remove "all" Griffith flaws from its surface, estimate the maximum temperature from which it can be quenched in a bath of ice water without failure. State all assumptions. Information you may find useful: $Y=70$ GPa, $\nu=0.25$, $\gamma=0.3$ J/m² and $\alpha=10\times10^{-6}$°C.

Answer: 5000°C.

(*b*) Repeat part (*a*) assuming 1 μm flaws are present on the surface.

Answer: 82°C.

[199] Y. Ohya, Z. Nakagawa, and K. Hamano, *J. Amer. Cer. Soc.*, 70, C184–C186 (1987).

(c) Repeat part (b) for Pyrex, a borosilicate glass for which $\alpha \approx 3\times10^{-6}$ K. Based on your results, explain why Pyrex is routinely used in labware.

13.11. Qualitatively explain how the following parameters would affect the final value of the residual stresses in a tempered glass pane: (a) glass thickness, (b) thermal conductivity of glass, (c) quench temperature, (d) quench rate.

13.12. Rank the following three solids in terms of their thermal conductivity: MgO, $MgO{\cdot}Al_2O_3$ and window glass. Explain.

13.13. (a) Estimate the heat loss through a 0.5-cm-thick, 1000 cm² window if the inside temperature is 25°C and the outside temperature is 0°C. Information you may find useful: k_{th} conductivity of soda lime is 1.7 W/(m·K).

(b) Repeat part (a) for a porous firebrick that is used to line a furnace running at 1200°C. Typical values of k_{th} for firebricks are 1.3 W/(m·K). State all assumptions.

ADDITIONAL READING

1. W. D. Kingery, H. K. Bowen, and D. R. Uhlmann, *Introduction to Ceramics*, 2nd ed., Wiley, New York, 1976.
2. C. Kittel, *Introduction to Solid State Physics*, 6th ed., Wiley, New York, 1986.
3. W. D. Kingery, Thermal conductivity of ceramic dielectrics, in *Progress in Ceramic Science*, vol. 2, J. E. Burke, Ed., Pergamon Press, New York, 1961.
4. D. P. H. Hasselman and R. A. Heller, Eds., *Thermal Stresses in Severe Environments*, Plenum, New York, 1980.
5. H. W. Chandler, Thermal stresses in ceramics, *Trans. J. Brit. Cer. Soc.*, 80, 191 (1981).
6. Y. S. Touloukian, R. W. Powell, C. Y. Ho, and P. G. Klemens, Eds., *Thermophysical Properties of Matter, vol. 2, Thermal Conductivity—Nonmetallic Solids*, IFI/Plenum Press, New York, 1970.
7. D. W. Richerson, *Modern Ceramic Engineering*, 2nd ed., Marcel Dekker, New York, 1992.
8. C. Barry Carter and M. Grant Norton, *Ceramic Materials*, Springer Verlag, New York, 2013.
9. H. M. Rosenberg, *The Solid State*, Oxford University Press, Oxford, UK, 1988.

OTHER RESOURCES

(1) To appreciate how thermally insulating the space shuttle tiles are see: https://www.youtube.com/watch?v=Pp9Yax8UNoM.

14

LINEAR DIELECTRIC PROPERTIES

It serves to bring out the actual mechanical connexions between the known electro-magnetic phenomena; so I venture to say that any one who understands the provisional and temporary character of this hypothesis will find himself rather helped than hindered by it in his search after the true interpretation of the phenomena.

James Maxwell, *Phil. Mag.*, 21:281 (1861).

14.1 INTRODUCTION

Dielectric materials do not conduct electricity and as such are of critical importance as capacitive elements in electronic applications and as insulators. It could be argued, with some justification, that without the discovery of new compositions with very high charge-storing capabilities, i.e., relative dielectric constants $k' > 10{,}000$, the impressive miniaturization of semiconductor-based devices and circuits would not have been possible. In addition, the traditional use of ceramics as insulators in high-power applications is still a substantial economic activity.

In contrast to electrical conductivity, which involves long-range motion of charge carriers, the dielectric response results from the *short-range* motion of these carriers under the influence of an externally applied electric field, E. Inasmuch as all solids are comprised of positive and negative entities, the application of E to any solid will result in a separation of its charges. This separation of charge is called **polarization**, defined as the *finite displacement of bound charges* of a dielectric in response to an applied electric field. If permanent electric dipoles exist in the dielectric and are mobile, they will orient parallel to the direction of E.

The dielectric properties can vary widely between solids and are a function of temperature, frequency of applied field, humidity, crystal structure, and other external factors. Furthermore, the response can be either *linear* or *nonlinear*. This chapter examines linear dielectric materials from a microscopic point of view. The effects of temperature and frequency on the dielectric response are discussed. Materials for which the dielectric response is nonlinear are discussed in the next chapter.

14.2 BASIC THEORY

Before polarization is discussed, it is imperative to understand how one measures polarization and to gain a qualitative understanding of how readily, or not so readily, polarizable a solid is. Consider two metal parallel plates of area A separated by a distance d in vacuum (Fig. 14.1*a*). Attaching these plates to the simple electric circuit, shown in Fig. 14.1*a*, and closing the circuit will result in a transient surge of current, I that rapidly decays to zero, as shown in Fig. 14.1*b*. Given that

$$Q = \int I dt \tag{14.1}$$

the area under the I versus t curve is the total charge, Q, that has passed through the circuit and is now stored on the capacitor plates.

Repeating the experiment at different voltages V, and plotting Q versus V yields straight lines, as shown in Fig. 14.2. In other words, the well-known relationship

$$Q = CV \tag{14.2}$$

is recovered. The slope of the Q versus V curve is the **capacitance** of the parallel plates in vacuum C_{vac} given by

$$C_{vac} = \frac{\varepsilon_0 A}{d} \tag{14.3}$$

where ε_0 is the permittivity of free space, which is a constant[200] equal to 8.85×10^{-12} F/m. The units of capacitance are farads (F), where 1 F = 1 C/V = 1 C^2/J.

If a dielectric (which can be a gas, solid, or liquid) is introduced between the plates of the capacitor (Fig. 14.1*c*) and the aforementioned experiment is repeated, the current that flows through the external circuit and is now stored on the capacitor plates will increase (Fig. 14.1*d*). Repeating the experiment at different V and plotting Q versus V will again result in a straight line but with a slope that is *larger* than the one obtained in vacuum (Fig. 14.2). In other words, Eq. (14.3) is now modified to read

$$C = \frac{\varepsilon A}{d} \tag{14.4}$$

[200] If cgs electrostatic units are used, then $\varepsilon_0 = 1$ and $\varepsilon = k'$ and a factor $1/4\pi$ appears in all the equations. In SI where $\varepsilon_0 = 8.85 \times 10^{-12}$ F/m, the factor $1/4\pi$ is included in ε_0 and omitted from the equations. Furthermore, in the cgs system, the unit of polarizability is the cubic centimeter. To convert from SI to cgs: α (F·m²) = $4\pi\varepsilon_0 \times 10^{-6} \alpha$(cm³). Mercifully, in this book only SI units are used.

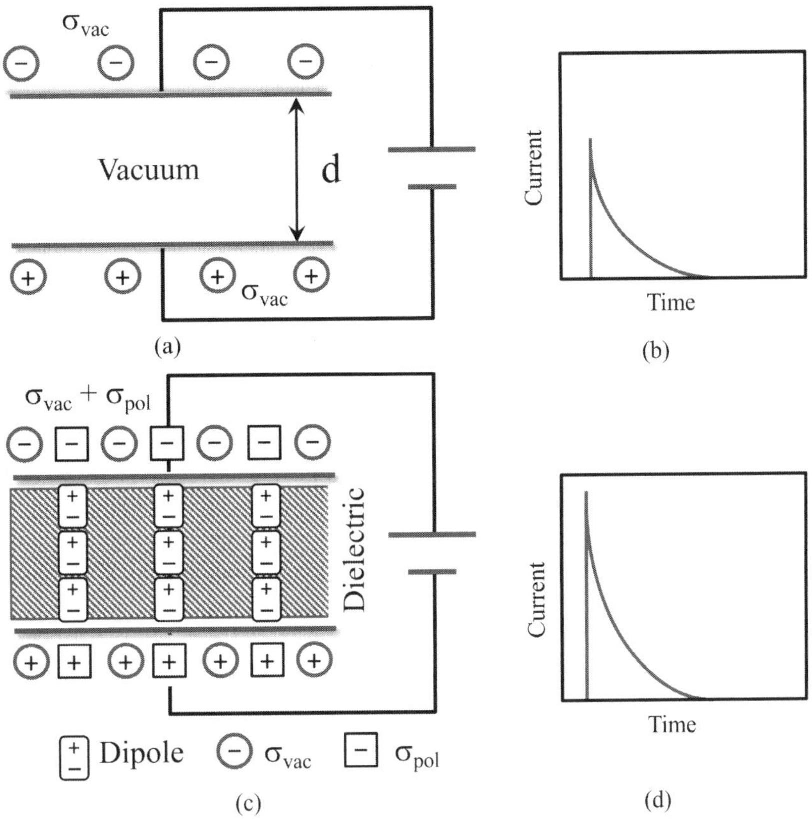

FIGURE 14.1 (*a*) Parallel-plate capacitor of area A and separation d in vacuum attached to a voltage source. (*b*) Closing of the circuit causes a transient surge of current to flow through the circuit. Charge stored on the capacitor is equal to the area under the curve. (*c*) Same as (*a*) except that now a dielectric is placed between the plates. (*d*) Closing of the circuit results in a charge stored on the parallel plates that has to be greater than that stored in (*b*).

where ε is the *dielectric constant* of the material between the plates.

The **relative dielectric constant** of a material k′ is defined as

$$\boxed{k' = \frac{\varepsilon}{\varepsilon_0}} \tag{14.5}$$

Since ε is always greater than ε_0, the minimum value for k′ is 1. By combining Eqs. (14.4) and (14.5), the capacitance of the metal plates separated by the dielectric is

$$C = \frac{k'\varepsilon_0 A}{d} = k'C_{vac} \tag{14.6}$$

Thus k′ *is a dimensionless parameter that compares the charge-storing capacity of a material to that of vacuum.*

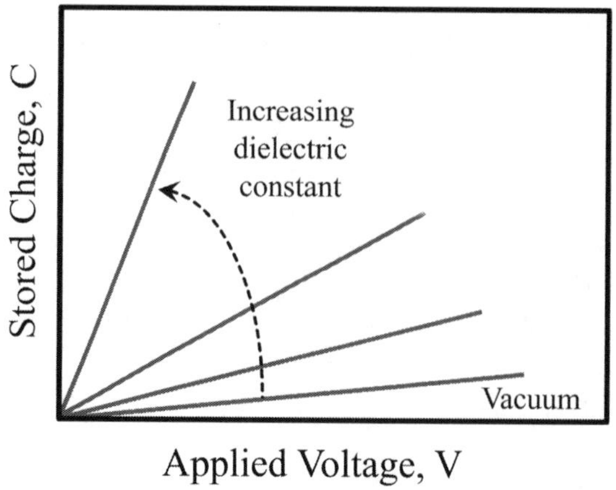

FIGURE 14.2 Functional dependence of Q on applied voltage. Slope of curve is related to the dielectric constant of the material.

The foregoing discussion can be summarized as follows: When a voltage is applied to a parallel-plate capacitor in vacuum, the capacitor will store some charge. In the presence of a dielectric, an additional "something" happens within that dielectric which allows the capacitor to store *more* charge. The purpose of this chapter is to explore the nature of this "something." First, however, a few more concepts need to be clarified.

Polarization charges

By combining Eqs. (14.2) and (14.3), the **surface charge** in vacuum σ_{vac} is

$$\sigma_{vac} = \left[\frac{Q}{A}\right]_{vac} = \frac{\varepsilon_0 V}{d} = \varepsilon_0 E \tag{14.7}$$

where E is the applied electric field. Similarly, by combining Eqs. (14.2) and (14.4), in the presence of a dielectric, the surface charge on the metal plates increases to

$$\left[\frac{Q}{A}\right]_{die} = \frac{\varepsilon_0 k' V}{d} = \sigma_{vac} + \sigma_{pol} \tag{14.8}$$

where σ_{pol} is the excess charge per unit surface area present on the dielectric surface (Fig. 14.1c). Note, σ_{pol} is numerically equal to, and has the same dimensions (C/m^2), as the polarization P of the dielectric, i.e.,

$$\mathbf{P} = \sigma_{pol} \tag{14.9}$$

Electromagnetic theory defines the **dielectric displacement D** as the surface charge on the metal plates, that is, $\mathbf{D} = Q/A|_{die}$. Making use of this definition and combining Eqs. (14.7) to (14.9), one finds that

$$\mathbf{D} = \varepsilon_0 \mathbf{E} + \mathbf{P} \tag{14.10}$$

In other words, the total charge stored on the plates of a parallel-plate capacitor **D** is the sum of the charge that would have been present in vacuum $\varepsilon_0 \mathbf{E}$ and an extra charge that results from the polarization of the dielectric material, **P**. The situation is depicted schematically in Fig. 14.1c. Note that if $\mathbf{P}=0$, **D** is simply given by Eq. (14.7).

Further combining Eqs. (14.7) to (14.10), and noting that **E** is identical in both cases, then

$$\mathbf{P}=(k'-1)\varepsilon_0\mathbf{E}=\chi_{\text{die}}\varepsilon_0\mathbf{E} \tag{14.11}$$

where

$$\chi_{\text{die}}=\frac{\sigma_{\text{pol}}}{\sigma_{\text{vac}}}$$

and χ_{die} is known as the **dielectric susceptibility** of the material. The next task is to relate **P** to what occurs at the atomic scale.

Microscopic approach.

A **dipole moment** μ (Fig. 14.3) is defined as[201]

$$\mu=q\delta$$

where δ is the distance separating the centers of the +ve and –ve charges $q=\pm ze$. As shown in Fig. 14.3, μ is a vector with its positive sense directed from the negative to the positive charges.

If there are N such dipoles per unit volume, it can be shown that **P** is simply

$$\mathbf{P}=N\mu=Nq\delta \tag{14.12}$$

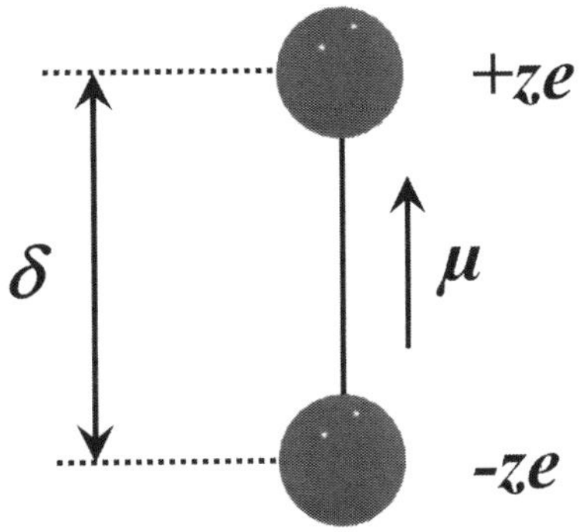

FIGURE 14.3 Definition of an electric dipole moment.

[201] The dipole moment of a charge $q = ze$ relative to a fixed point is defined as the vector $ze\zeta i$, where ζi is the radius vector from the fixed point to the position of the charge. The total dipole moment of a system is the vector sum of all the individual dipoles

$$\mu=\sum z_i e\zeta_i$$

This quantity is independent of the position of the fixed point. In the absence of a field, $\sum z_i e\zeta_0=0$. The application of an electric field results in the displacement of the charges by an amount δ_i from their equilibrium position, that is, $\zeta=\zeta_0+\delta_i$. It follows that $\mu=\sum z_i e\zeta_i=\sum z_i e\delta_i$. Practically, it follows that to calculate the dipole moment of any ion all we need only know is its position relative to its equilibrium position.

Combining Eqs. (14.11) and (14.12), one sees that

$$\boxed{k' - 1 = \frac{P}{\varepsilon_0 E}} \tag{14.13}$$

This result is fundamental to understanding the dielectric response of a solid. It basically says that the greater the separation of the charges of a dipole, δ, for a given **E**, the greater k′. Said otherwise, the more polarizable the medium, the greater its dielectric constant.

One can further define the **polarizability**, α, of an atom or ion as

$$\alpha = \frac{\mathbf{P}}{\mathbf{NE}_{loc}} \tag{14.14}$$

where $\boldsymbol{E}_{loc}$ is the *local electric field* to which the atom is subjected. The SI unit of polarizability is Fm^2 or $C{\cdot}m^2\ V^{-1}$. It is important to note that α is an atomic/ionic property since, as discussed below, it mostly depends on the atomic/ionic radius.

For dilute gases, where the molecules are far apart, $\mathbf{E}_{loc}$ can be assumed to be identical to the externally applied field, **E**, and by combining Eqs. (14.13) and (14.14), it follows that

$$k' - 1 = \frac{N\alpha}{\varepsilon_0} \tag{14.15}$$

However, in a solid, polarization of the surrounding medium can, and will, substantially affect the magnitude of $\mathbf{E}_{loc}$. It can be shown that (see App. 14A), for cubic symmetry, the local field is related to **E** by

$$\mathbf{E}_{loc} = \frac{\mathbf{E}_{loc}}{3}(k' + 2)$$

which when combined with Eqs. (14.13) and (14.14) gives

$$\boxed{k' - 1 = \frac{N\alpha/\varepsilon_0}{1 - N\alpha/(3\varepsilon_0)}} \tag{14.16}$$

This equation can be rearranged to read

$$\boxed{\frac{k' - 1}{k' + 2} = \frac{\alpha N}{3\varepsilon_0}} \tag{14.17}$$

preferred by some. This expression is known as the **Clausius-Mossotti relation.**[202] Note that strictly speaking, Eq. 14.17 is only valid for highly symmetric crystals, such a cubic crystals.

The Clausius-Mossotti relationship is quite useful as it provides a valuable link between the macroscopic k′ and the microscopic α. Based on these expressions a measurement of k′ can, in principle, yield information about the relative displacement of the positive and negative charges making up that solid. It should be

[202] Written in terms of the refractive index $n = \sqrt{k_e}$, this relation is known as the Lorentz–Lorenz relation. It thus should not come as surprise in Chap. 16, when it is shown that the dielectric and optical responses of insulators are intimately related. If one ignores the magnetic component of electromagnetic radiation, the latter is nothing but a time-varying electric field.

emphasized here that this expression is only valid for linear dielectrics and is not applicable to ferroelectrics, discussed in the next chapter. It is also worth noting that whenever the density of atoms—such as in a gas—is low, i.e. when $N\alpha/3\varepsilon_0 << 1$ Eq. (14.16) simplifies to Eq. (14.15), as one would expect.

Up to this point, the discussion was restricted to static electric fields. In most electrical applications, however, the E is far from static—with frequencies that range from 60 Hz for standard ac power to gigahertz and higher for communication networks. It is thus important to introduce a formalism by which one can describe not only the static response of a dielectric which is represented by k′ or α, but also the effect of frequency on both k′ and any losses that occur in the dielectric as a result of the application of a time-varying E. This is typically done by representing the dielectric constant as a complex quantity that depends on frequency, as described in the following section.

14.3 EQUIVALENT CIRCUIT DESCRIPTION OF LINEAR DIELECTRICS

IDEAL DIELECTRIC

The application of a sinusoidal voltage with an angular frequency ω in rad/s[203] given by $\mathbf{V} = \mathbf{V_0} \exp i\omega t$, where $i = \sqrt{-1}$, to an ideal dielectric, i.e., one without losses, will result in a charging current (see Prob. 14.1) given by

$$I_{chg} = \frac{dQ}{dt} = C\frac{dV}{dt} = i\omega CV = \omega CV_0 \exp i\left(\omega t + \frac{\pi}{2}\right)$$

or

$$I_{chg} = -\omega k' C_{vac} V_0 \sin\omega t \qquad (14.18)$$

In other words, the resulting *current* will be $\pi/2$ rad, or 90°, out of phase relative to $\mathbf{V_0}$. It is important to note that this implies that the *oscillating charges* are **in phase** with the applied **V**.[204] This comes about because the current is a sine function, while the applied field is a cosine.

NONIDEAL DIELECTRICS

As noted above, Eq. (14.17) is only valid for an ideal dielectric. In reality, the charges are never totally in phase for two reasons: (1) the dissipation of energy due to the inertia of the moving species, and (2) the long-range hopping of charged species, i.e., ohmic conduction. The total current is thus the vectorial sum of $\mathbf{I_{chg}}$ and a loss current $\mathbf{I_{loss}}$ or

$$\mathbf{I}_{tot} = \mathbf{I}_{chg} + \mathbf{I}_{loss} = i\omega C\mathbf{V} + \{G_L(\omega) + G_{dc}\}\mathbf{V} \qquad (14.19)$$

203 Remember $e^{i\omega} = \cos\omega + i\sin\omega$, where ω is the angular frequency in units of radians per second. To convert to Hz, divide by 2π, since $\omega = 2\pi\nu$, where ν is the frequency in Hz or s^{-1}.

204 When the charges are in phase with the applied field, this automatically implies that the current is $\pi/2$ rad ahead of the applied voltage. This comes about because $I = dQ/dt$. Interestingly enough, the loss current is one in which the charges are oscillating $\pi/2$ out of phase with the applied voltage.

where G is the conductance of the material (see below).

The loss current $\mathbf{I_{loss}}$ is defined as

$$\mathbf{I_{loss}} = \{G_L(\omega) + G_{dc}\}\mathbf{V}$$

and is written in this form to emphasize that G_L is a function of frequency, whereas G_{dc} is not. In the limit of zero frequency, $G_L \Rightarrow 0$ and one recovers Ohm's law or

$$\mathbf{I_{tot}} = \mathbf{I_{loss}} = G_{dc}\mathbf{V}$$

since $G_{dc} = 1/R$, where R is the direct-current (dc) resistance of the material. Note that in this case the current and voltage are in phase, but the *oscillating* ***charges*** are 90° **out-of-phase** with the applied **V**. Here both **V** and $\mathbf{I_{loss}}$ are cosine functions.

The total current in the dielectric subjected to a varying electric field is thus made up of two components that are 90° out of phase with each other and have to be added vectorially, as shown in Fig. 14.4. The total current in a nonideal dielectric will thus lead the applied voltage by an angle of $(90° - \phi)$, where ϕ is known as the **loss angle, loss tangent** or **dissipation factor.**

It is important to note that the dielectric response of a solid can also be succinctly described by expressing the relative dielectric constant as a complex quantity, made up of a real k′ component and an imaginary k″ component, or

$$k^* = k' - ik'' \tag{14.20}$$

Replacing k′ in Eq. (14.6) by k*, making use of Eq. (14.2), and noting that the $\mathbf{I_{chg}} = \mathbf{I_{tot}} - \mathbf{I}(\omega = 0) = dQ/dt$, one obtains

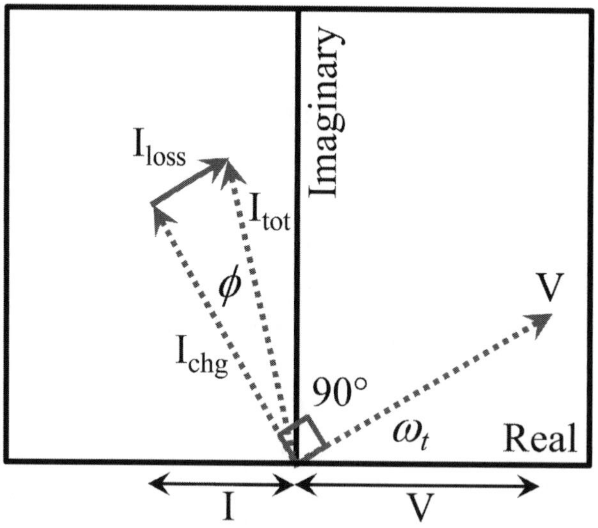

FIGURE 14.4 Vectorial representation of applied voltage, charging, loss and total currents. Note that when $\phi = 0$, $I_{tot} = I_{chg}$, whereas when $\phi = \pi/2$, $I_{tot} = I_{loss}$.

$$\mathbf{I_{tot}} - \mathbf{I}(\omega = 0) = \frac{dQ}{dt} = k^{*}C_{vac}i\omega\mathbf{V} = (k' - ik'')C_{vac}i\omega\mathbf{V} \tag{14.21}$$

which, when rearranged, gives

$$\mathbf{I_{tot}} = i\omega C_{vac}k'\mathbf{V} + (\omega k'' C_{vac} + G_{dc})\mathbf{V} \tag{14.22}$$

Comparing this expression to Eq. (14.19) reveals immediately that

$$G_{ac} = G_{dc} + G_L = G_{dc} + \omega k'' C_{vac}$$

where G_{ac} is the **ac conductance** of the material, with units of Siemens, denoted by S, or Ω^{-1}. The corresponding **ac conductivity** (S/m) is given by

$$\boxed{\sigma_{ac} = \sigma_{dc} + \omega k'' \varepsilon_0} \tag{14.23}$$

To summarize this section to this point: When subjected to time varying electric field, with a frequency ω, the conductivity is comprised of two parts, a dc component that is not a function of ω and one that originates from the oscillating charges. This conclusion is nicely summarized in Fig. 14.4.

Another way to look at the problem is shown in Fig. 14.4, where **V**, $\mathbf{I_{chg}}$ and $\mathbf{I_{loss}}$ are plotted relative to each other. From that figure one can define a tan ϕ as

$$\boxed{\tan\phi = \frac{\mathbf{I_{loss}}}{\mathbf{I_{chg}}} = \frac{G_{dc} + \omega k'' C_{vac}}{\omega k' C_{vac}}} \tag{14.24}$$

Note that for a dielectric for which $G_{dc} \ll \omega k'' C_{vac}$ then $\tan\phi \approx k''/k'$.

POWER DISSIPATION IN A DIELECTRIC

In general, loss currents are a nuisance since they tend to heat up the dielectric and retard electromagnetic signals. The average power dissipated in a dielectric is

$$P_{av} = \frac{1}{T}\int_0^T \mathbf{I_{tot}}\mathbf{V}\,dt$$

where $T = 2\pi/\omega$ is the time period. For an *ideal* dielectric, $\mathbf{I_{tot}} = \mathbf{I_{chg}}$ and making use of Eq. 14.18

$$P_{av} = \frac{1}{T}\int_0^T -\omega k' C_{vac} V_0^2 \sin\omega t \cos\omega t\,dt = 0$$

During one-half of the cycle, the capacitor is being charged and the power source does work on the capacitor; in the second half of the cycle, the capacitor is discharging and does work on the source. Consequently, the average power drawn from the power source is *zero*, which is an important result because it shows that an ideal dielectric is loss-free.

In a non-ideal dielectric, however, $\mathbf{I_{loss}}$ and $\mathbf{V}$ are in phase, and

$$P_{av} = \frac{1}{T}\int_{0}^{T} I_{loss} V\,dt = \frac{1}{T}\int_{0}^{T} (\omega k'' C_{vac} + G_{dc}) V_0^2 \cos\omega t \cos\omega t\,dt$$

$$= \frac{1}{2} G_{ac} V_0^2$$

For dc conditions, $\omega = 0$, and this expression is identical to the well-known expression for the power loss under dc conditions, or I^2R, also known as Joule heating.

The corresponding power loss *per unit volume* is given by

$$\boxed{P_V = \tfrac{1}{2}\sigma_{ac}\mathbf{E_0^2} = \tfrac{1}{2}(\sigma_{dc} + \omega k''\varepsilon_0)\mathbf{E_0^2}} \tag{14.25}$$

where $\mathbf{E_0} = \mathbf{V_0}/d$ is the amplitude of the applied electric field. It follows that the power loss per unit volume (W/m^3) in a dielectric is directly related to σ_{ac} or σ_{dc}, k'' and ω. This apparent sleight of hand in going from **V** to **E** is a natural consequence of moving from a system that is not normalized by a system's geometry to one that is. To appreciate how this is done, divide P_{av} by the volume of the system ($A \times d$).

Although the mathematical representation, at first glance, may not appear to be simple, the physics is more so: When a time-varying electric field is applied to a dielectric, the charges in the material will respond. Some of the bound charges will oscillate in phase with the applied field and result in charge storage and contribute to k'. Another set of charges, both bound and those contributing to the dc conductivity, will oscillate 90° out of phase with the applied voltage and result in energy dissipation in the dielectric.[205] This energy dissipation ends up as heat (the temperature of the dielectric solid will increase). In an ideal dielectric, the loss angle ϕ and energy dissipated are both zero.

The remainder of this chapter is concerned with the various polarization mechanisms operative in ceramics and their temperature and frequency dependencies. Before going there, it is useful to understand how dielectric properties are measured.

EXPERIMENTAL DETAILS: MEASURING DIELECTRIC PROPERTIES

There are several techniques used to measure the dielectric properties of solids. One of the more popular ones is known as *ac impedance spectroscopy,* described below. Another technique compares the response of the dielectric to that of a calibrated variable capacitor. In this method, the capacitance of a parallel-plate capacitor in vacuum is compared with one in the presence of the material for which the dielectric properties are to be measured. Then k' is simply calculated from Eq. (14.6). A typical circuit for

[205] This is a rather simplistic interpretation, but one that is easily visualized. More realistically, the charges will oscillate slightly out of phase, with an angle ϕ out of phase to be exact, with respect to the applied field. It is worth emphasizing once more that k' describes the behavior of the *bound* charges and that σ_{ac} has two contributions to it: the bound charges that are out of phase with the applied field for which the conductance is $k''\omega C_{vac}$ and the "free" charges whose conductance is simply G_{dc}. Whether a charge will jump back when the field reverses sign, and would thus be considered a bound charge, or whether it will continue to drift when the field reverses sign can only be distinguished in response to dc fields.

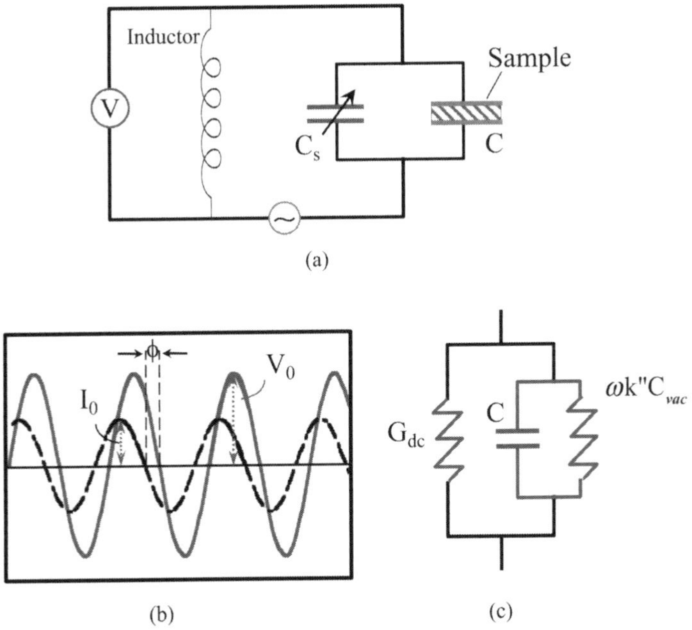

FIGURE 14.5 *(a)* Apparatus for measuring the dielectric constant of a material; L is the inductance of the coil. *(b)* The actual response of a nonideal dielectric to an applied voltage is such that the angle between the current and voltage is not $\pi/2$, but $\pi/2 - \phi$. *(c)* Equivalent circuit used to model the dielectric response of a solid. Here G_{dc} represents the dc response of the material, whereas $\omega k'' C_{vac}$ is the conductance of the bound charges, which vanishes as ω goes to zero.

carrying out such an experiment is shown in Fig. 14.5*a*. Varying the capacitance of the calibrated capacitor to keep the resonance frequency $\omega_0 = \{L(C_s + C)\}^{-1/2}$ constant when vacuum is between the plates, versus when the substance is inserted, allows C_{vac} and C_{solid} to be determined and, in turn, k′.

AC IMPEDANCE SPECTROSCOPY

Here a sinusoidal voltage is applied to the sample, and the magnitude and phase shift of the resulting current are measured by using sophisticated electronics. From the ratio of the magnitude of the resulting current I_0 to the imposed voltage V_0, and the magnitude of the phase difference ϕ between the two, all defined in Fig. 14.5*b*, k′ and k″ can be obtained. It can be shown (see Prob. 14.1*b*) that if one assumes the equivalent circuit shown in Fig. 14.5*c*, k′ and k″ are given by

$$k' = \frac{I_0 d}{V_0 A \omega \varepsilon_0} \sin\left\{\frac{\pi}{2} - \phi(\omega)\right\}$$

and

$$k'' = \frac{\sigma_{ac} - \sigma_{dc}}{\omega \varepsilon_0}$$

where

$$\sigma_{ac} = \frac{I_0 d}{V_0 A}\cos\left\{\frac{\pi}{2} - \phi(\omega)\right\}$$

and d and A are, respectively, the sample's thickness and cross-sectional area. It is important to remember, that both I_0 and ϕ depend on the frequency of the applied field, ω. It is instructive here to look at the dc limit. In that case, the current that passes through the capacitor will be determined by its dc conductivity. As ω increases, more and more of the bound charges will start to oscillate out of phase with the applied voltage and will contribute to σ_{ac}.

In a typical experiment, ω of the applied voltage is varied over a range between a few Hertz and 100 MHz. Measurements in the frequency range between 10^9 and 10^{12} Hz are more complex and beyond the scope of this book. However, in the IR and UV frequencies, the dielectric constant and loss can once again be measured from measurements of the reflectivity of a sample and its refractive index (see Chap. 16).

Given the power and importance of AC impedance spectroscopy, Case Studies 14.1 and 14.2 show how this technique can shed important light on the dielectric response of solids and interfaces.

14.4 POLARIZATION MECHANISMS

Up to this point, the discussion was couched in terms of polarization, or the displacement of charges with respect to each other. In this section, the specifics of charge separation are considered. In solids, especially in ionic ceramics, various charged entities are capable of polarization, such as electrons, protons, cations, anions, and charged defects. The following mechanisms represent the most important polarization mechanisms in ceramics.

1) *Electronic polarization*: This mechanism entails the displacement of the electrons relative to their nucleus (Fig. 14.6*a*).
2) *Ionic polarization:* In this mechanism the displaced ionic charges are *bound elastically* to their equilibrium positions (Fig. 14.9).
3) *Dipolar Polarization*: This mechanism—sometime referred to as orientational polarization—involves ionic displacements that occur between adjacent equivalent, or near-equivalent, lattice sites, with the probability that a site is occupied depending on the strength of the external electric field and its frequency (Fig. 14.10). If the alignment occurs *spontaneously and cooperatively*, **nonlinear polarization** results and the material is termed a *ferroelectric*. Because of the relatively large displacements, relative dielectric constants on the order of 5,000 can be attained in these materials. Nonlinear dielectrics are dealt with separately in Chap. 15. But if the polarization is simply due to the jumping of ions from one adjacent site to another, the polarization response is **linear** with V. These solids are discussed below.
4) *Space charge polarization*: In Chap. 5, the notion of a Debye length was briefly alluded to, and it was argued that whenever two dissimilar phases come into contact with each other, an electrified

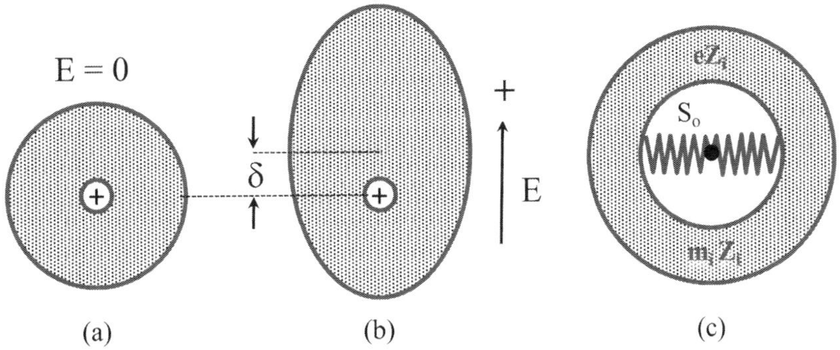

FIGURE 14.6 Electronic polarization of the atomic cloud surrounding a nucleus. (*a*) At equilibrium, i.e., in the absence of an external electric field. (*b*) In the presence of an external electric field. (*c*) Schematic of model assumed in text. S_0 represents the stiffness of the bond between the electrons and their nucleus.

interface will result. This so-called double layer acts as a capacitor with properties and responses different from those of the bulk material. The behavior and interpretation of interfacial phenomena are quite complex and not within the scope of this book; they fall more in the realm of solid-state electrochemistry and will not be discussed further. This comment notwithstanding, Case Study 14.2 briefly discusses interfacial polarization.

The total polarizability is the sum of the contributions from the various mechanisms, or

$$k' = \frac{1}{3\varepsilon_0}\left[N_e\alpha_e + N_{ion}\alpha_{ion} + N_{dip}\alpha_{dip} + N_{space\ chg}\alpha_{space\ chg}\right] \tag{14.26}$$

where N_i represents the number of polarizing species per unit volume. In the remainder of this section, electronic, ionic, and ion jump polarization are discussed in some detail.

14.4.1 ELECTRONIC POLARIZATION

Electronic polarization, shown schematically in Fig. 14.6*a* and *b*, occurs when the electron cloud is displaced relative to the nucleus it is surrounding. It is operative at most frequencies and drops off only at very high frequencies ($\approx 10^{15}$ Hz). Since every atom consist of a nucleus surrounded by electrons, electronic polarization occurs in *all* solids, liquids, and gases. Furthermore, since it does not involve hopping of ions between lattice sites, it is temperature-insensitive.

The simplest classical theory for electronic polarization treats the atom or ion, of atomic number Z_i as an electrical shell of charge Z_ie and mass Z_im_e, attached to an undeformable ion nucleus[206] (Fig. 14.6*c*). If the natural frequency of vibration of the system is ω_e, it follows that the corresponding restoring force is

$$F_{restor} = \mathbf{M_r}\omega_{\mathbf{e}}^2\delta \tag{14.27}$$

[206] Clearly, this is a gross oversimplification. The restoring force, and consequently the resonance frequency of each electron, has to be different. See Eq. (14.40) for a more accurate expression.

where M_r is the **reduced mass** of the oscillating system, defined as

$$M_r = \frac{Z_i m_e m_n}{Z_i m_e + m_n} \tag{14.28}$$

where m_e and m_n are, respectively, the masses of an electron and nucleus. Since in this case $m_e << m_n$, it follows that for electronic polarization $M_r \approx Z_i m_e$.

The application of **E**, as discussed above, will result in the separation of charges and the creation of an electric dipole moment, since now the center of negative charge and the center of positive charge will no longer coincide, as shown schematically in Fig. 14.6*b*.

An oscillator with charge $Z_i e$ (i.e. the entire electron cloud) displaced an amount δ by an oscillatory driving force, $\mathbf{F} = Z_i e\mathbf{E} = Z_i e\mathbf{E_o}\exp(i\omega t)$ with a restoring force given by Eq. 14.27, and a **damping constant** or **friction factor,** ***f***, obeys the equation of motion

$$M_r\left(\frac{d^2\delta}{dt^2} + \mathbf{f}\frac{d\delta}{dt} + \omega_e^2\delta\right) = Z_i e\mathbf{E_o}\exp \mathbf{i}\omega \mathbf{t} \tag{14.29}$$

This equation is nothing but Newton's law with a restoring force and a friction factor **f**. The units of **f** are rad/s. The second term in brackets represents the friction force assumed to be proportional to the velocity of the electron cloud, $d\delta/dt$. If **f** is small, there is little friction, while for large **f**, the frictional forces are large.[207]

It can be shown (see Prob. 14.3) that

$$\delta = \frac{e\mathbf{E_o}}{m_e\sqrt{(\omega_e^2 - \omega^2)^2 + \mathbf{f}^2\omega^2}}\exp \mathbf{i}(\omega \mathbf{t} - \phi) \tag{14.30}$$

and

$$\delta = \frac{e\mathbf{E_o}}{m_e\{(\omega_e^2 - \omega^2) + \mathbf{i}\omega\mathbf{f}\}}\exp \mathbf{i}\omega \mathbf{t} \tag{14.31}$$

are both equally viable solutions to Eq. (14.29), provided ϕ—the phase difference between the forced vibration and the resulting polarization—is given by

$$\tan\phi = \frac{\mathbf{f}\omega}{\omega_e^2 - \omega^2} \tag{14.32}$$

Note that δ is a measure of the displacement of the electron cloud as a whole in the presence of E (Fig. 14.6*b*) relative to its equilibrium position in the absence of a field (Fig. 14.6*a*).

By replacing k_e' in Eq. (14.13) by k_e^*, and substituting for δ, it can be shown that (see Prob. 14.5), if one assumes $\mathbf{E} = \mathbf{E_{loc}}$ the real and imaginary parts of k_e^* are, respectively,

$$k_e'(\omega) = 1 + \frac{Z_i e^2 N(\omega_e^2 - \omega^2)}{\varepsilon_0 m_e\{(\omega_e^2 - \omega^2)^2 + \mathbf{f}^2\omega^2\}} \tag{14.33}$$

[207] The damping constant *f* is related to the anharmonicity of the vibrations — as they become more anharmonic, *f* increases.

and

$$k_e''(\omega) = \frac{Z_i e^2 N \omega f}{\varepsilon_0 m_e \{(\omega_e^2 - \omega^2)^2 + \mathbf{f}^2 \omega^2\}} \tag{14.34}$$

Note Eqs. (14.33) and (14.34) are only valid for dilute gases, since it was assumed that $\mathbf{E} = \mathbf{E}_{\mathbf{loc}}$. To solve the problem more accurately for solids or liquids, $\mathbf{E}_{\mathbf{loc}}$, rather than $\mathbf{E}$, field would have to used in Eq. (14.29). Fortunately, doing this does not change the general forms of the solutions; it only modifies the value of the resonance frequency ω_e (see App. 14A). Making use of 14.17, we can recast the equations for solids to read:

$$\boxed{\frac{k_e'(\omega) - 1}{k_e'(\omega) + 2} = \frac{Z_i e^2 N (\omega_e'^2 - \omega^2)}{3\varepsilon_0 m_e \left\{(\omega_e'^2 - \omega^2)^2 + \mathbf{f}^2 \omega^2\right\}}} \tag{14.33}$$

$$\boxed{k_e''(\omega) = \frac{Z_i e^2 N \omega \mathbf{f}}{\varepsilon_0 m_e \left\{(\omega_e'^2 - \omega^2)^2 + \mathbf{f}^2 \omega^2\right\}}} \tag{14.34}$$

The frequency dependencies of k_e' and k_e'' are plotted in Fig. 14.7 and are characteristic of typical dispersion curves experimentally observed for dielectrics. A note of caution: these relations assume a crystal of high symmetry.

Based on Eqs. (14.33) and (14.34), the frequency response can be divided into three domains:

1. $\omega_e \gg \omega$. Here the charges are oscillating in phase with the applied electric field and contribute to k_e'. Under dc conditions, and assuming Eq. 14.33 is operative, then

$$\frac{k_e' - 1}{k_e' + 2} = \frac{Z_i e^2 N}{3\varepsilon_0 m_e \omega_e'^2} \tag{14.35}$$

   and k_e'' is zero.
2. $\omega_e \approx \omega$. When the frequency of the applied field approaches the natural frequency of vibration of the system, ω_e, the system is said to be at *resonance*. The displacements, were it not for the frictional forces, would go to infinity. It follows that just before resonance, k_e' goes through a maximum. Exactly at resonance the charges are 90° out of phase with the applied field and thus are not contributing to the dielectric constant. At resonance, however, k_e'' and the energy losses are both at their maxima (Fig. 14.7*b*).
3. $\omega_e \ll \omega$. In this region, the electric field is changing direction too fast for the electric charges to respond; no polarization results, and k_e' goes to 1.

To summarize: When a varying electric field is applied to a solid, the charges will start dancing to the tune of that externally applied field—they will oscillate in phase and with the same frequency of vibration as the applied field. The amplitude of the vibrations, however, will vary and will depend on the relative values of ω_e and ω. The charges that are in phase with the applied field will not absorb energy and will contribute to k_e'. Another set of charges will oscillate out of phase with the applied field, will absorb energy, and will contribute to dielectric loss. When $\omega_e \ll \omega$, the charges follow the field nicely, k_e'', and losses are small.

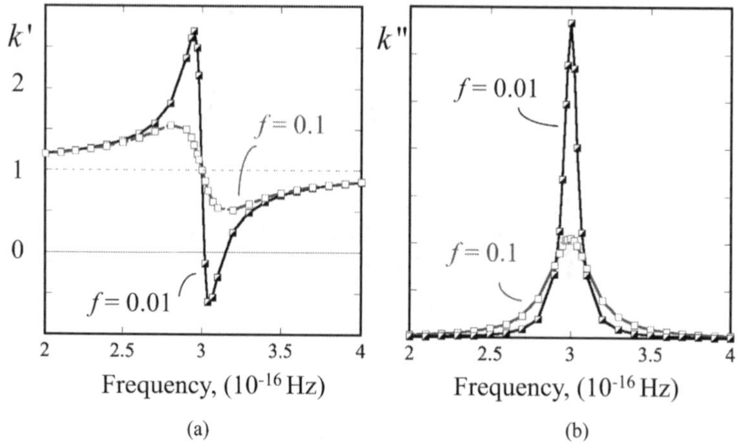

FIGURE 14.7 General frequency dependence of (*a*) k'_e, viz. Eq. (14.33) (note that as f increases, the width of the resonance peak also increases); (*b*) k''_e, viz. Eq. (14.34). Here ω_e was assumed to be 3×10^{16} Hz.

As the frequency of the field becomes comparable to the natural frequency of vibration of the system, the latter goes into resonance; the amplitudes of vibration tend to be very large, and consequently so does k'_e. However, this large increase in the amplitude of vibration results in large losses observed that peak at resonance.[208] At very high applied frequencies, the charges cannot keep up with the applied field, their amplitude of vibration will be uncorrelated, and k'_e will approach 1. In other words, if the frequency of the applied field is so high that the polarization mechanism cannot follow, then no polarization ensues.

MICROSCOPIC FACTORS AFFECTING ELECTRONIC POLARIZATION

Comparing Eqs. (14.17) and (14.35) it follows that

$$\alpha_e = \frac{Z_i e^2}{m_e \omega_e^2} \tag{14.36}$$

It was shown in Sec. 4.4 (Eq. 4.6) that for small displacements, the restoring force can be assumed to be proportional to the displacement, i.e.,

$$F_{restor} = S_0(r - r_0) = S_0 \delta \tag{14.37}$$

where r is the distance between the nucleus and the electron cloud, S_0 is the *stiffness* of that bond and δ is the displacement from its equilibrium position. Here the assumption is made that the electron cloud is attached to its nucleus by a spring of stiffness S_e, as shown in Fig. 14.6c. According to Coulomb's law, the force between Z_i electrons and their nucleus is given by (see Ch. 2)

[208] This description of resonance is applicable to any resonance phenomenon, be it mechanical, electrical, or magnetic. The nature of the resonating species and the driving forces may vary, but the physics and the interpretations do not.

$$F = -\frac{(Z_i e)^2}{4\pi\varepsilon_0 r^2}$$

From which it can be shown that

$$S_e = \left(\frac{dF}{dr}\right)_{r=r_0} = \frac{2(Z_i e)^2}{4\pi\varepsilon_0 r_0^3} \tag{14.38}$$

Combining Eqs. (14.36) to (14.38) together with Eq. (14.27) it can be shown that

$$\alpha_e \approx 2\pi\varepsilon_0 r_{ion}^3 \tag{14.39}$$

According to this result, α_e should scale with the volume of an atom or ion. Simply put, *the larger the atom or ion, the less bound the electrons are to their nucleus and the more amenable they are to polarization.* Table 14.1 lists the values of α_e for a number of common ions.

According to Eq. (14.39) a plot of α_e vs. r_{ion}^3 should yield a straight line with a slope of $2\pi\varepsilon_0$. Figure 14.8 plots α_e vs. r_{ion}^3 of alkali metal cations, halide anions and doubly charged chalcogenides. If our simple model were correct all the points would have fallen on the line labelled Eq. 14.39, i.e. the slope should be $2\pi\varepsilon_0$. It follows that to a first approximation, Eq. 14.39 is indeed borne out by experimental results since we are not too far removed from "theory". However, these results show that in addition to size, two other factors (see also Sec. 4.2.2, where polarizability was first encountered) come into play, namely,

1. *Charge:* The polarizability of an ion is a function of its net charge, as shown in Fig. 14.8. Anions are usually more polarizable than cations, and the effect is greater than a simple volume argument. For example, Cl^-, and S^{2-} are similar in size, yet S^{2-}, with its double negative charge, is almost 3 times more polarizable than Cl^-. This is understandable, since in anions, the outermost electrons are less tightly bound to their nuclei and thus contribute the most to the polarizability.
2. Nucleus shielding: What is not shown in Fig. 14.8, but is generally true, is that *d* electrons do not shield the nucleus as well as *s* or *p* electrons. It follows that the polarizability of atoms or ions with d electrons is less than that of similarly sized atoms with *s* or *p* electrons.

TABLE 14.1 Electronic polarizabilities, α_e, in 10^{40} F·m² and radii of select ions

Alkali metal cations			Halogen anions			Chalcogenide anions		
Ion	α_e	r_0 (pm)	Ion	α_e	r_0 (pm)	Ion	α_e	r_0 (pm)
Na^+	0.22	102	F^-	1.33	133	O^{2-}	3.05	140 pm
K^+	1.00	138	Cl^-	3.33	181	S^{2-}	9.54	184
Rb^+	1.89	152	Br^-	5.00	196	Se^{2-}	12.4	198
Cs^+	2.77	167	I^-	7.77	220	Te^{2-}	17.4	221

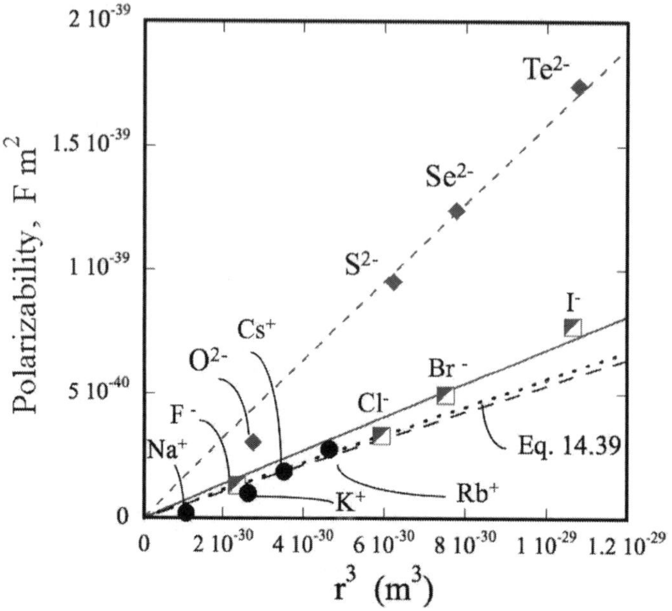

FIGURE 14.8 Relationship between ionic radii and electronic polarizability. (See App. 3A for radii. Polarizabilities were determined empirically from refractive index data.)

Somewhat counterintuitively the natural frequency of the electron cloud, ω_e, is a weak function of polarizability. Why this is the case is best appreciated by noting that in Eq. 14.36, ω_e scales linearly with Z_i and inversely with α_e. And since as Z_i increases so does α_e, the ratio remains more or less constant. This is why the measured values of ω_e for all the noble gases, including radon, vary from a low of 8.9×10^{15} Hz for He to a high of 12.6×10^{15} Hz for Ne. Before moving to the next topic, it is important to point out that ω_e is quite high and typically falls in the 10^{16} Hz range.

Worked Example 14.1

The relative dielectric constant of O_2 gas at 0°C and 1 atm. was measured to be 1.000523. Using this result, predict the dielectric constant of liquid O_2 if its density is 1.19 Mg/m³. Compare your answer with the experimentally determined value of 1.507. State all assumptions and discuss the implications of your results vis-à-vis the assumptions made.

ANSWER

At 0°C, the number of atoms per cubic meter (assuming ideal gas behavior, i.e., $N = N_{AV}P/RT$) is

$$N = \frac{(6.02\times10^{23})(1.013\times10^{5})}{8.31\times273} = 2.69\times10^{25}\,\text{m}^{-3}$$

Substituting this value in Eq. (14.15) and solving for α_e yields

$$\alpha_e = \frac{(1.000523-1)(8.85\times 10^{-12})}{2.69\times 10^{25}} = 1.72\times 10^{-40}\,\mathrm{F\cdot m^2}$$

The number of O_2 molecules in the liquid is:

$$N_{liq} = \frac{1.19\times 10^{6}}{32} 6.02\times 10^{23} = 2.25\times 10^{28}\ \mathrm{m^{-3}}$$

Substituting these values in Eq. 14.17

$$\frac{k'_e - 1}{k'_e + 2} = \frac{(2.25\times 10^{28})(1.72\times 10^{-40})}{3\times 8.85\times 10^{-12}} = 0.145$$

Solving for k'_e yields 1.509, a value is in excellent agreement with the measured value. This implies that i) the assumption that polarizability is *an atomic property* that is not a function of density, viz. N, is an excellent one and, ii) the Clausius–Mossotti relation, Eq. 14.17, is quite valid in this case.

In the discussion so far, and for the sake of simplicity, the electron cloud was treated as a single unit—an obvious oversimplification. In reality, each atom has j ($j = Z_i$) oscillators associated with it, each having an oscillator strength γ_j. The jth oscillator vibrates with its own natural frequency and damping constant $\mathbf{f}_j$. The total electronic polarizability of such an atom or ion is given by the sum of all the oscillators

$$\frac{k_e^* - 1}{k_e^* + 2} = \frac{e^2}{\varepsilon_0 m_e} \sum_j \frac{\gamma_j(\omega_{e,j}^2 - \omega^2)}{\left(\omega_{e,j}^2 - \omega^2\right)^2 + i\omega^2 f_j^2} \tag{14.40}$$

The oscillator strength γ_j is related to—from quantum mechanics—the probability of transition of an electron from one band to the next.

COVALENT INSULATORS

In the foregoing discussion, electronic polarizations were discussed. However, the ideas developed, strictly speaking, only apply to ionic solids, where ions exist. For example, to a very good approximation, α_e of NaCl is the sum of the α_e's of Na^+ and Cl^-. In other words, the α_e of a compound can, to a very good approximation, be taken as the sum of the electronic polarizabilities of the ions making up that compound.

This logic fails, however, when dealing with covalent ceramic insulators such as SiC. In such materials appreciable electron density resides between the atoms. This part of the charge distribution is thus a function of the solid as a whole and not that of the individual atoms. Said otherwise, α_e of SiC is not related to the individual α_e's of Si and C atoms, but is related to the polarizability of the crystal as a whole. It is for this reason that in covalent solids, by far the major contribution to the dielectric constant results from electronic polarization of the electrons in the bonds.

Worked Example 14.2

If the dielectric constant of Si in 11.9 and the polarizability of Ne is 0.45×10^{-40} F m², make the case that the inner electrons in Si do not contribute very much to polarizability. What is the nature of this polarization and would you expect it to be a function of temperature?

ANSWER

Since the electronic structure of Si is [Ne]$3s^2 3p^2$, the easiest way to solve this problem is to estimate the polarizability of Si and compare with that of Ne.

$$N_{Si} = \frac{2.33 \times 10^6}{28.1} 6.02 \times 10^{23} \approx 5 \times 10^{28} m^{-3}$$

Rearranging Eq. 14.35,

$$\alpha_{Si} = \frac{3 \times 8.85 \times 10^{-12}}{5 \times 10^{28}} \frac{11.9 - 1}{11.9 + 2} = 4.16 \times 10^{-40} F \cdot m^2$$

It follows that the inner electrons of Si contribute ≈10% (0.45×10^{-40} F m²) to the total. It is for this reason that typically one can ignore the inner electrons when considering the polarizability of atoms. Since Si is a covalently bonded solids, there are no ions or other diploes that contribute to the dielectric constant. In this case the polarizability is pure electronic in nature and therefore not a function of temperature. It is useful in this context to think of the electrons in Sil to be sloshing about somewhat in response to the applied field.

14.4.2 IONIC POLARIZATION

Electron clouds are not the only entities that can respond to electric fields. Ionic charges in a solid can respond equally well and can in turn contribute to the dielectric constant. *Ionic polarization* is defined as the displacement of positive and negative ions toward the negative and positive electrodes, respectively, as shown schematically in Fig. 14.9. Ionic resonance occurs in the infrared frequency range (10^{12} to 10^{13} Hz) and consequently this phenomenon will be encountered again in Chap. 16.

The equation of motion to solve is similar to Eq. (14.29) except that the:

- ∞ Ions are assumed to be attached to *one another* by a spring having a natural frequency of vibration, ω_{ion}, which, in turn, is directly related to the Coulombic attraction between the ions.
- ∞ Reduced mass of the system is now given by $M_r = m_c m_a/(m_c + m_a)$ where m_c and m_a are the cation and anion masses, respectively.
- ∞ Friction factor, $\mathbf{f}_{ion}$, will also be different in this case and will reflect the energy loss due to the motion of the ions.

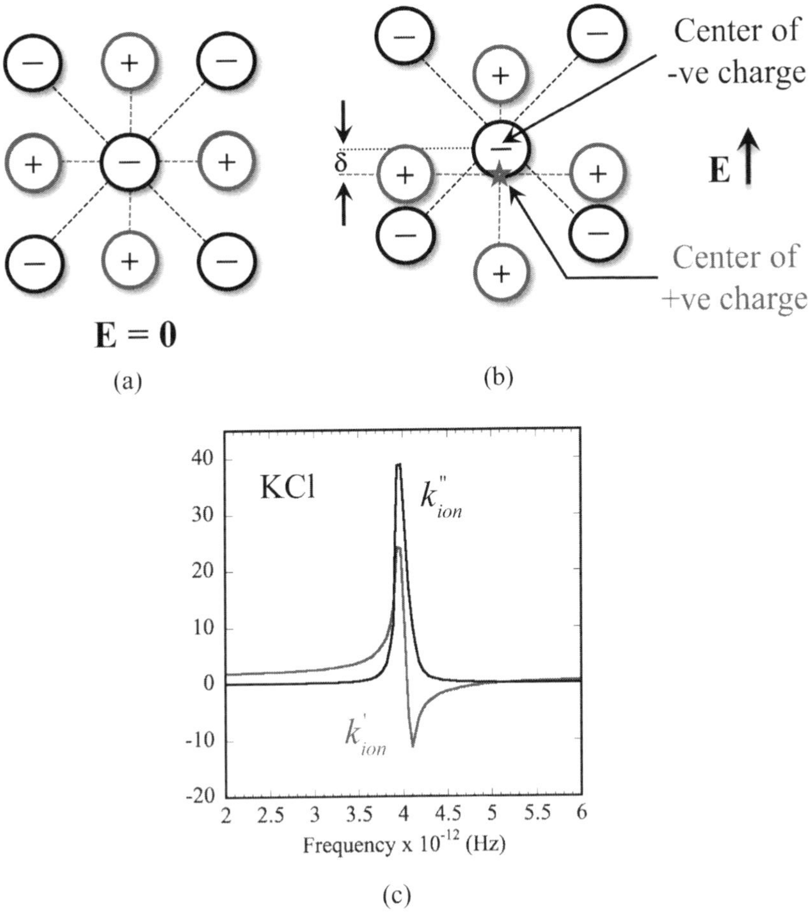

FIGURE 14.9 Ionic polarizability. (*a*) Ion positions at equilibrium. (*b*) Upon the application of a vertical electric field, the center of negative charge is no longer coincident with the center of positive charge, i.e., polarization occurs. (*c*) Real and imaginary parts of the ionic polarization of KCl.

It is thus not surprising that the final result for ionic polarizability is quite similar to that for electronic polarization and is given by

$$\boxed{k'_{ion}(\omega) = 1 + \frac{(ze)^2 N_{ion}\left(\omega_{ion}^2 - \omega^2\right)}{\varepsilon_0 M_r\{(\omega_{ion}^2 - \omega^2)^2 + \mathbf{f}_{ion}^2\omega^2\}}} \quad (14.41)$$

$$\boxed{k''_{ion}(\omega) = \frac{(ze)^2 N_{ion}\omega\mathbf{f}_{ion}}{\varepsilon_0 M_r\{(\omega_{ion}^2 - \omega^2)^2 + \mathbf{f}_{ion}^2\omega^2\}}} \quad (14.42)$$

where N_{ion} is the number of ion pairs per cubic meter. Note the similarity between Eqs. (14.33) and (14.41); the only differences are the appearance of ω_{ion} rather than ω_e, of $\boldsymbol{f}_{ion}$ rather than $\boldsymbol{f}$, and the use of M_r instead of m_e. Moreover, for $\omega \gg \omega_{ion}$, the ions can no longer follow the applied field and drop out, that is, $k'_{ion} \Rightarrow 1$, as expected. Figure 14.9c plots actual results obtained on KCl. In this case, ω_{ion} is in the 10^{12} Hz range.

Under dc conditions, and making use of the Clausius-Mossotti expression, Eq. (14.41) expression reduces to

$$\frac{k'_{ion}-1}{k'_{ion}+2} = \frac{(ze)^2 N_{ion}}{3\varepsilon_0 M_r \omega_{ion}^2} \tag{14.43}$$

In other words, the ionic polarization is inversely proportional to the ionic masses and the square of ω_{ion}, which depends on the strength of the ionic bond. Also, since it does not entail ionic migration, ionic polarization is quite temperature insensitive.

The factors that influence the strength of an ionic bond were discussed in detail in Chap. 4 and will not be repeated here. As discussed in Worked Example 14.2, ceramics in which the bond is predominantly covalent (i.e., the atoms are not charged) exhibit little or no ionic polarization.

Worked Example 14.3

(*a*) In a similar approach used to derive Eq. (14.39), derive an expression for k'_{ion}.

(*b*) Based on the expression derived in part (*a*) calculate k_{ion} of NaCl and MgO, given that the Born exponent for NaCl is ≈8 and that for MgO is ≈7. State all assumptions and compare your answers to experimental results listed in Table 14.2.

ANSWER

(*a*) Replacing ω_{ion}^2 in Eq. (14.43) by S_0/M_r and noting that for the ionic bond (see Prob. 4.2)

$$S_0 \approx \frac{(ze)^2}{4\pi\varepsilon_0 r_0^3}(n-1)$$

where r_0 is the equilibrium interionic spacing and n is the Born exponent, it follows that

$$\alpha_{ion} \approx \frac{4\pi\varepsilon_0 r_0^3}{n-1}$$

which when combined with the Clausius–Mossotti equation yields

$$\frac{k'_{ion}-1}{k'_{ion}+2} \approx \frac{\alpha_{ion} N_{ion}}{3\varepsilon_0} \approx \frac{4\pi N_{ion} r_0^3}{3(n-1)} \tag{14.44}$$

Interestingly, the charges on the ions do *not* appear in the final expression, because both the electric and the restoring forces scale with z_1z_2!

(*b*) Number of ion pairs in NaCl ($\rho = 2.165$ g/cm³) and in MgO ($\rho = 3.6$ g/cm³) are

$$N_{ion}(NaCl) = \frac{2.165 \times 6.02 \times 10^{23} \times 10^{6}}{23 + 35.45} = 2.23 \times 10^{28} \text{ ion pairs/m}^3$$

$$N_{ion}(MgO) = \frac{3.6 \times 6.02 \times 10^{23} \times 10^{6}}{24.31 + 16} = 5.38 \times 10^{28} \text{ ion pairs/m}^3$$

From App. 3A, r_0 (NaCl) = 102 + 181 = 283 pm and r_0 (MgO) = 72 + 140 = 212 pm. Substituting these values in Eq. (14.44) and solving for k'_{ion}, one obtains

$$k'_{ion}(NaCl) = 2.3 \text{ and } k'_{ion}(MgO) = 2.67$$

Experimentally (see Table 14.2), k'_{ion} (NaCl) = 5.89 − 2.41 = 3.48 and k'_{ion} (MgO) = 9.83 − 3 = 6.83. And although the agreement between theory and experiment is not excellent, given the simplicity of the model used (i.e., assuming the ions to be hard spheres, etc.), it is satisfactory. This is especially true when it is appreciated that, as discussed in the following section, the values of the static dielectric constants listed in Table 14.2 can include other contributions in addition to ionic polarization.

14.4.3 DIPOLAR POLARIZATION

In contrast to electronic and ionic polarizations, which occur at high frequencies ($\omega > 10^{10}$ Hz), dipolar polarization occurs at lower frequencies and is thus important because it can greatly affect the capacitive and insulative properties of glasses and ceramics in low-frequency applications. In the remainder of this section, dipolar polarization, is considered, first under a static electric or DC field, followed by the more complicated dynamic case.

STATIC RESPONSE

In this situation, a voltage is applied across our dielectric and held there until equilibrium is achieved. As noted above, **ion jump polarization** is the preferential occupation of equivalent or near-equivalent lattice sites as a result of the applied fields biasing one site over the other.[209] The situation is depicted schematically in Fig. 14.10, where an ion is localized in a deep energy well, but within which two equivalent sites, labeled A and B in Fig. 14.10*b*, exist. The sites are separated from each other by a *jump distance*, λ_s, and an

[209] The discussion in this section is applicable to any polar solid (i.e., one that has a permanent dipole) in which relaxation of the permanent dipoles occurs. In principle this approach could be applicable to piezoelectric and ferroelectric solids at temperatures above their transition temperature as well (see Chap. 15). However, the same response, as shown in Fig. 14.11, can also occur as a result of heavily damped resonance. This is easily seen in Fig. 14.7*a*; as f, which is a measure of the damping, increases, the resultant resonance curves become flatter. Experimentally it is not always easy to distinguish between the two phenomena.

TABLE 14.2 Dielectric properties of some ceramic materials[a]

Compound	k'_{static}	$k'_e = n^2$	$\tan\delta\ (\times 10^4)$	Compound	k'_{static}	$k'_e = n^2$	$\tan\delta\ (\times 10^4)$
Halides							
AgCl	12.3	4.0		LiF	8.9	1.9	2
AgBr	13.1	4.6		LiI	11.0	3.8	
CsBr	6.7	2.4		NaBr	6.4	2.6	
CsCl	7.2	2.6		NaCl	5.9	2.4	2
CsI	5.6	2.6		NaF	5.1	1.7	
KBr	4.9	2.3	2	NaI	7.3	2.9	
KCl	4.8	2.2	10	RbBr	4.8	2.3	
KF	5.5	1.8		RbCl	4.9	2.2	
KI	5.1	2.6		RbF	6.5	2.0	
LiBr	9.0–13.0	3.2		RbI	4.9	2.6	
LiCl	11.9	2.8		TlBr	30.0	5.4	
Binary oxides							
Al_2O_3	9.4	3.1	0.4–2	MnO	18.1		
BaO		3.9		Sc_2O_3		4.0	
BeO	6.8	2.9	2	SiO_2	3.8	2.3	4
CaO	12.0	3.4		SrO	13.0	3.3	
Cr2O3	11.8	6.5		TiO_2 rutile	114.0	6.4–7.4	2–4
Eu_2O_3		4.4		TiO_2 (‖ c)[b]	170.0	8.4	16
Ga_2O_3		3.7		TiO_2 (‖ a)[c]	86.0	6.8	2
Gd_2O_3		4.4		Y_2O_3		3.7	
MgO	9.8	3.0	3	ZnO	9.0	4.0	
Ternary oxides							
$BaTiO_3$	3000.0	5.8	1–200	$MgTiO_3$	16.0		2
$CaTiO_3$	180.0	6.0		$SrTiO_3$	285.0	6.2	
$MgAl_2O_4$	8.2	3.0	5–8				
Glasses							
Pb-silica glass	19.0	57		Soda-lime glass Vycor	7.60	2.3	100
Pyrex	4.0–6.0						8
Others							
AlN	8.80		5–10	α-SiC	9.7	6.7	
C (diamond)	5.7	5.7		Si	11.7	11.7	
				ZnS	8.3	5.1	

[a] The values quoted in the literature are quite variable especially for k'_s, which depends strongly on sample purity and quality.
[b] Parallel to the c-axis in a rutile single crystal.
[c] Parallel to the a-axis.

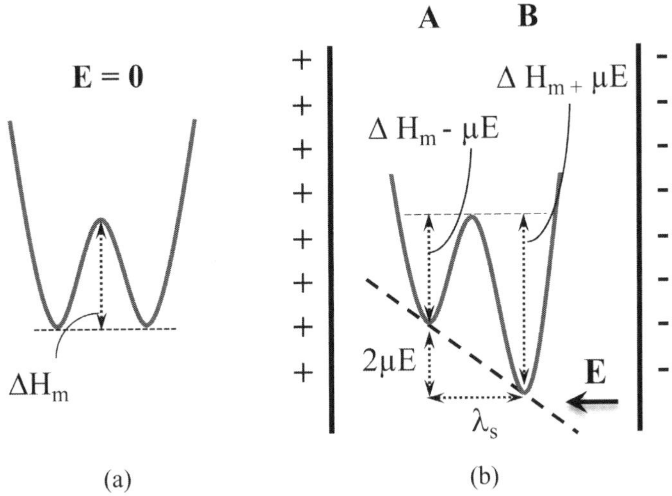

FIGURE 14.10 Dipolar polarization: (*a*) energy versus distance diagram in the absence of applied field, the two sites are equally populated. (*b*) The application of an electric field will bias one site relative to the other.

energy barrier ΔH_m. In the absence of an electric field (Fig. 14.10*a*), each site has an equal probability of being occupied and there is no net polarization. In the presence of a field (Fig. 14.10*b*) the two sites are no longer equivalent—the electric field will bias the B sites, resulting in a net polarization.[210]

As discussed in Chap. 7, the probability of an ion making a jump in the absence of a bias is given by the Boltzmann factor,

$$\Theta = K\exp\left(-\frac{\Delta H_m}{kT}\right) \tag{14.45}$$

The potential energy of a dipole depends on the orientation of the dipole moment μ_{dip} with respect to the applied field E or

$$U = -\mu_{dip} \cdot E = -\mu_{dip} E_{cos}\theta$$

where θ is the angle between μ and E. For $\theta = 0°$ or $180°$, the potential energy is simply $\pm\mu_{dip}E$ depending on whether the moment aligns with, or against, the field. It follows that the energy difference between the two orientations is $2\,\mu_{dip}E$. Assuming N_{dip} bistable dipoles per unit volume, in the presence of a field, the jump probability $A \Rightarrow B$ is

210 The following analogy should help: Think of the crystal as a ship in which the ions are passengers, confined to cabins with two beds each, where the beds are arranged parallel to the ship axis. In the absence of a bias, it is fair to assume the passengers will occupy either bed equally such that the center of gravity of the ship remains at its center. If now the ship tilts to either side, the passengers will tend to favor that side and the center of gravity of the ship will no longer be at its center.

$$\Theta_{A\to B} = K_{exp}\exp\left(-\frac{\Delta H_m - \mu_{dip}E}{kT}\right)$$

For the most part, the applied fields are small enough that $\mu_{dip}E/(kT) \ll 1$ and this equation simplifies to

$$\Theta_{A\to B} = \left(1 - \frac{\mu_{dip}E}{kT}\right)\Theta \quad (14.46)$$

Similarly, the jump probability B → A is given by

$$\Theta_{B\to A} = \left(1 + \frac{\mu_{dip}E}{kT}\right)\Theta \quad (14.47)$$

At steady state,

$$N_A\Theta_{A\to B} = N_B\Theta_{B\to A} \quad (14.48)$$

where N_A and N_B are the number of ions in each well. Combining Eqs. (14.46) to (14.48) and rearranging yields

$$N_B - N_A = (N_B + N_A)\frac{\mu_{dip}E}{kT} = N_{dip}\frac{\mu_{dip}E}{kT} \quad (14.49)$$

The static polarization per unit volume P_s is defined as

$$P_s = (N_B - N_A)\mu_{dip} = N_{dip}\frac{\mu_{dip}^2 E}{kT} \quad (14.50)$$

from which it is obvious that if $N_B = N_A$, there would be no polarization. Combining Eqs. (14.13) and (14.50), and noting that $\mu_{dip} = ze\lambda_s/2$, one obtains

$$k'_{dip} - 1 = \frac{N_{dip}\mu_{dip}^2}{\varepsilon_0 kT} = \frac{N_{dip}(ze)^2\lambda_s^2}{4kT}$$

In deriving this equation, we assumed the diploes to be oriented either with, or against, **E**. In reality we need to average over all angles. The good news is that only introduces a factor of 3 in above equation. Thus

$$\boxed{k'_{dip} - 1 = \frac{N_{dip}\mu_{dip}^2}{3\varepsilon_0 kT} = \frac{N_{dip}(ze)^2\lambda_s^2}{12kT}} \quad (14.51)$$

The following is noteworthy:

- Now k'_{dip} is a function of the total number of dipoles per unit volume, the charge on the ions that are jumping, and the jump distance, λ_s. Neither ΔH_m, nor the frequency of the applied field, ω, play a role because Eq. (14.51) represents the *equilibrium* situation under *static* (i.e., dc) conditions. This expression yields the equilibrium value, but says nothing about how rapidly or slowly equilibrium is reached (see below).

- Increasing the temperature will reduce k'_{dip} as a *result of thermal randomization.* This functionality on temperature is known as **Curie's law** and will be encountered again in Chap. 15. In analogy to paramagnetism (see Chap. 15), any solid for which the susceptibility is proportional to 1/T can be labeled a **paraelectric** solid.

DYNAMIC RESPONSE AND THE DEBYE EQUATIONS

To understand and model the dynamic response of dipolar polarization is quite a complicated affair. Debye, however, rendered the problem tractable by making the following assumptions:

- At high frequencies, that is, $\omega >> 1/\tau$, where τ is the **relaxation time** of the system or the *average residence time of an atom or ion at any given site,* the relative dielectric constant is given by k'_∞, where $k'_\infty = k'_{ion} + k'_e$ (i.e., the sum of the ionic and electronic contributions).
- As $\omega \Rightarrow 0$, the relative dielectric constant is given by k'_S where $k'_S = k'_{dip} + k'_\infty$.
- By assuming that the rate of de-polarization is proportional to the polarization itself, it can be shown (App. 14B) that P decays exponentially as

$$P(t) = P_0 \exp\left(-\frac{t}{\tau}\right) \tag{14.52}$$

From these assumptions it can be shown that:[211]

$$\boxed{k'_{dip} = k'_\infty + \frac{k'_{dip}}{1+\omega^2\tau^2}} \tag{14.53}$$

$$\boxed{k''_{dip} = \frac{\omega\tau}{1+\omega^2\tau^2} k'_{dip}} \tag{14.54}$$

$$\boxed{\tan\phi = \frac{k''_{dip}}{k'_{dip}} = \frac{k'_{dip}\omega\tau}{k'_S + k'_\infty\omega^2\tau^2}} \tag{14.55}$$

These equations are known collectively as the **Debye equations,** and are plotted in Fig. 14.11. At low frequencies, all polarization mechanisms can follow the applied field, and the total dielectric constant is $\approx k'_S$ which includes the dipolar, ionic, and electronic contributions. It can be shown that when $\omega\tau = 1$, k'_{dip} goes through an inflection point and k''_{dip} is at a maximum (see Prob. 14.12). At higher applied frequencies, the dipolar polarization component to k'_{dip} drops out and only the ionic and electronic, or k'_∞, contributions remain.[212]

An implicit assumption made in deriving the Debye equations is that of a *single relaxation time.* In other words, the heights of the barriers shown in Fig. 14.10*a* are assumed to identical for all sites. And while this

[211] See, e.g., L. L. Hench and J. K. West, *Principles of Electronic Ceramics*, Wiley-Interscience, NY, 1990.

[212] If the electric field switches polarity in a time that is much *shorter* than an ion's residence time on a site, the average energies of the two sites become equivalent (i.e., a bias no longer exists—the sites become energetically degenerate).

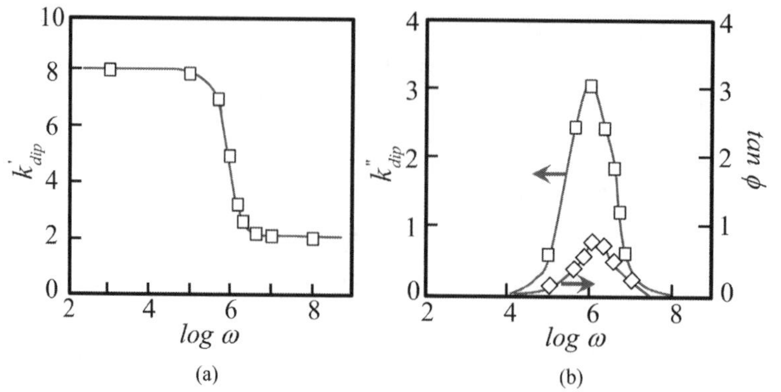

FIGURE 14.11 Frequency dependence of dielectric parameters for dipolar polarization on (*a*) k′, assuming a time constant τ of 10^{-6} s, and (*b*) k″ and tan ϕ. Note that the maximum for k″ is coincident with the inflection point in k′, whereas the maximum for tan δ is shifted to slightly higher frequencies.

may be true for some crystalline solids, it is less likely to be so for an amorphous solid such as a glass, where the random nature of the structure will likely lead to a distribution of relaxation times. In the latter situations, sometimes a distribution of activation barriers is assumed.

TEMPERATURE DEPENDENCE OF DIPOLAR POLARIZATION

As noted above, τ is a measure of the average time an ion spends at any one site. In other words, $\tau \propto 1/\Theta$, which according to Eq. (14.45) renders τ exponentially dependent on temperature, or

$$\tau = \tau_0 \exp\frac{\Delta H_m}{kT} \tag{14.56}$$

and implies that the resonance frequency should also be an exponential function of temperature. As the temperature increases, the atoms vibrate faster and are capable of following the applied field to higher frequencies. This is indeed found to be the case, as shown in Fig. 14.12*a*, where the dielectric loss peaks in a glass are plotted as a function of temperature. As the temperature increases, the maximum in the loss angle shifts to higher frequencies, as expected. Furthermore, when the frequency at which the peaks occur is plotted as a function of reciprocal temperature, the expected Arrhenian relationship is observed (Fig. 14.12*b*).

14.4.4 DIELECTRIC SPECTRUM

From the foregoing discussion, it is clear that the dielectric response is a complex function of frequency, temperature, and type of solid. Under dc conditions, all mechanisms are operative, and the dielectric constant is at its maximum and is given by the sum of all mechanisms. As the frequency increases, various mechanisms will be unable to follow the field and will drop off, as shown in Fig. 14.13. At very high frequencies, none of the mechanisms is capable of following the field, and the relative dielectric constant approaches one.

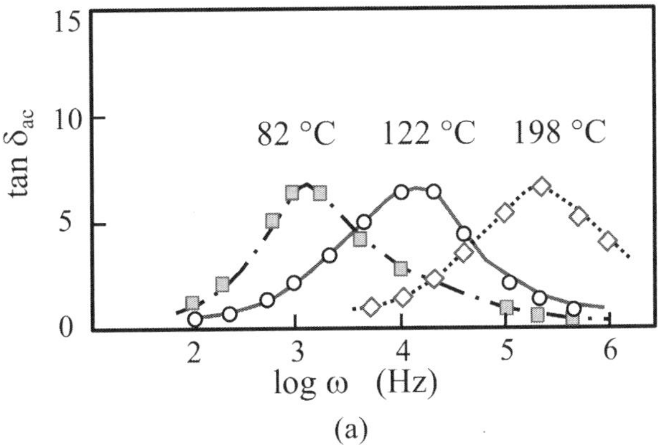

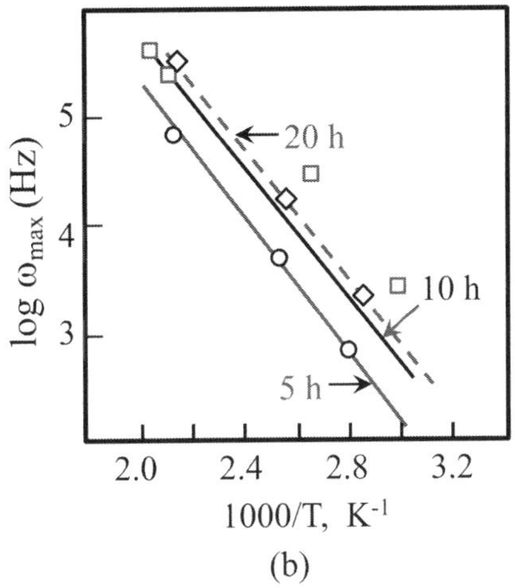

FIGURE 14.12 (*a*) Effect of temperature on dielectric loss peak in a Li_2O–SiO_2 glass. As the temperature increases, the frequency at which the maximum in loss occurs also increases, since the mobilities of the ions increase. (*b*) Temperature dependence of frequency at which maximum in dielectric loss peak occurs.

Temperature will influence only the polarization mechanisms that depend on long-range ionic displacement such as dipolar polarization. Ionic polarization is not strongly affected by temperature since long-range mobility of the ions is not required for it to be operative.[213]

Worked Example 14.4

The static dielectric constant of water is 80 and the high-frequency dielectric constant is 5. At 60°C, k''_{dip} goes through a maximum at 30 GHz, estimate the relaxation time for a water dipole. If in the IR absorbance spectrum of water, a peak is found at 100 cm^{-1}. Do you think the two phenomena are related? State all assumptions.

213 This should not be confused with the effect of temperature on dielectric *loss* (see next section).

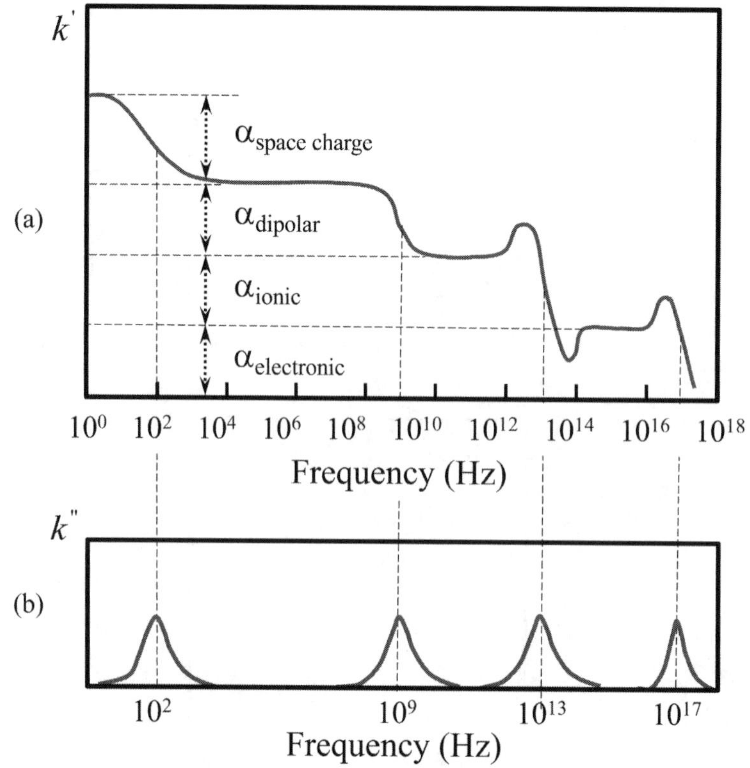

FIGURE 14.13 Variation of (*a*) relative dielectric constant and (*b*) dielectric loss with frequency of applied field.

ANSWER

Since according to Fig. 14.13, 40 GHz is around where dipolar polarization manifests itself, we will assume that this frequency is dipolar in origin. Once that is realized, the simplest way to solve this problem is by appreciating that $\omega_{max}\tau = 1$. In the problem statement, $\omega_{max} = 30$ GHz. To obtain τ, we first we need to convert the frequency in Hz to an angular frequency:

$$\omega_{max} = 30 \times 10^9 \times 2\pi = 1.9 \times 10^{11} \text{ rad/s}$$

and

$$\tau = \frac{1}{1.9 \times 10^{11}} = 5.3 \times 10^{-12} = 5.3 \text{ ps}$$

It follows that the average relaxation time for a water molecule is of the order of 5 ps. This implies that if the frequency of the applied field is say of the order of 10^{10} rad/s, the diploes would have no trouble following the field. However, if on the other hand the frequency is say 10^{12} rad/s, the water molecules will no longer be able to follow the field and will drop out.

To compare these results with the IR absorption result we need to convert the wavenumber, to a frequency. Since in SI units, $k = 10^4\ m^{-1}$, then

$$\omega = 2\pi\nu = 2\pi c(k) = 2\pi \times 3 \times 10^8 (10^4) = 1.9 \times 10^{13}\ rad/s$$

where c is the velocity of light. The difference between the two frequencies is large enough that we can assume them to have different origins, and indeed they do. The first is dipolar in origin, the latter is due to ionic polarization.

Notice that in this problem, like in life, you are given more information than you need!

14.5 DIELECTRIC LOSS

The dielectric loss is a measure of the energy dissipated in the dielectric in unit time when an electric field acts on it. Referring to Eq. (14.25), the power loss per unit volume dissipated in a dielectric is related to k'', the frequency of the applied field, and its dc conductivity.

This power loss represents a waste of energy, typically manifested as attendant heating of the dielectric. If the rate of heat generation is faster than it can be dissipated, the dielectric will heat up, which, as discussed below, could lead to dielectric breakdown and other problems. Furthermore, as the temperature increases, the dielectric constant is liable to change as well, which for finely tuned circuits can create serious problems. Another reason for minimizing k'' is related to the sharpness of the tuning circuit that would result from using a capacitor—lower values of k'' give rise to sharper resonance frequencies (see Fig. 14.7*b*).

From Eq. (14.25) it is apparent that in order to reduce power losses, it is imperative to

- Use solids that are highly insulating, $\sigma_{dc} \Rightarrow 0$. In other words, use pure solids with large band gaps such that the number of free charge carriers—impurity ions, free electrons, or holes—is as low as possible.
- Reduce k''.

Thus, almost by definition, a good dielectric must have low $\sigma_{dc} \Rightarrow 0$ and a low k'' hence the need to understand what contributes to k''. Since temperature usually tends to increase the conductivity of a ceramic exponentially (Chap. 7), its effect on dielectric loss can be substantial. This is demonstrated in Fig. 14.14*a*, where the loss tangent is plotted as a function of temperature for different glasses with different resistivities. In all cases, the increased cationic mobility results in an increase in the dielectric loss tangent. The effect of impurities, inasmuch as they increase the conductivity of a ceramic, can also result in large increases in the dielectric loss. This is shown in Fig. 14.14*b*, for quite pure NaCl (lower curve) and one that contains lattice impurities (top curve).

Needless to say, the frequency at which a dielectric is to be used must be as far removed from a resonance frequency as possible, since near resonance, k'' can increase substantially.

Interestingly, crystal structure can also affect k''. In general, for close-packed ionic solids, the dielectric loss is small, whereas loosely packed structures tend to have higher dielectric losses. This is nicely

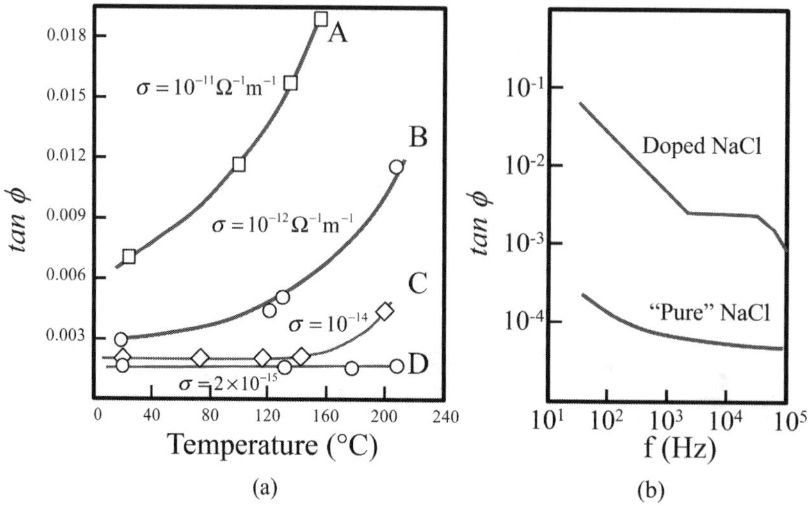

FIGURE 14.14 (*a*) Variation of loss angle with temperature for alkali glasses as a function of their resistivity. The measurements were carried out at 10^6 Hz. (*b*) Effect of impurities and frequency on tan ϕ of NaCl.

demonstrated when values of tan ϕ for α- and γ-aluminas are compared. For α-alumina, tan $\phi < 0.0003$ at 100°C and 10^{-6} s^{-1}, whereas tan ϕ for the less dense γ-modification is greater than 0.1 under the same conditions.

14.6 DIELECTRIC BREAKDOWN

When a dielectric is subjected to an ever-increasing electric field, at some point a short circuit will develop across it. **Dielectric breakdown** is defined as the voltage gradient or electric field sufficient to cause the short circuit. This phenomenon depends on many factors, such as sample thickness, temperature, electrode composition and shape, and porosity.

In ceramics, there are two basic types of breakdown; intrinsic and thermal.

- ∞ *Intrinsic*. In this mechanism, electrons in the conduction band are accelerated to such a point that they start to ionize lattice ions. As more ions are ionized, and the number of free electrons increases, an avalanche effect is created. Clearly, the higher the electric field applied, the faster the electrons will be accelerated and the more likely this breakdown mechanism will be.
- ∞ *Thermal breakdown*. The criterion for thermal breakdown is that the rate of heat generation in the dielectric, as a result of losses, is greater than the rate of heat removal from the sample. Whenever this condition occurs the dielectric will heat up, which in turn will increase its conductivity, which causes further heating, etc. This is termed *thermal breakdown* or *thermal runaway*.

14.7 CAPACITORS AND INSULATORS

Ceramic dielectric materials are typically used in electric circuits as either capacitors or insulators. Capacitors act as electrical buffers, diverting spurious electric signals and storing surges of charge that could damage circuits and disrupt their operation. By blocking dc signals and allowing only ac signals, capacitors can separate also ac and dc signals and couple alternating currents from one part of a circuit to another. They can discriminate between different frequencies, as well as, store charge.

Whether a dielectric solid is to be used as a capacitor or as an insulator will depend on its characteristics. For capacitive functions, high relative dielectric constants are required together with low losses. The perfect dielectric would have a very large k′ and no losses. But if the dielectric is used for its insulative properties, whether in high-power applications or as a substrate for integrated circuits, then it is desirable to have as low a dielectric constant as possible and once again minimal losses. It is worth noting that the need for low-loss insulators has grown significantly recently with the advent of high- frequency telecommunications networks. Since the power losses [Eq. (14.25)] are proportional to ω, the need for lower loss insulators is more crucial than ever.

Table 14.2 lists the values of k'_{static} and k'_e (which, as discussed in Chap. 16, is nothing but the square of the refractive index) together with tan ϕ of a number of ceramics.

In general, dielectrics are grouped into three classes:

Class 1 dielectrics include ceramics with relatively low and medium dielectric constants and dissipation factors of less than 0.003. The low range covers $k'_{static} = 5$ to 15, and the medium k'_{static} range is 15 to 500.

Class II dielectrics are high-permittivity ceramics based on ferroelectrics (see Chap. 15) and have values of k′ between 2000 and 20 000.

Class III dielectrics (not discussed here) contain a conductive phase that effectively reduces the thickness of the dielectric and results in very high capacitances. Their breakdown voltages are quite low, however.

Low-permittivity ceramics are widely used for their insulative properties. The major requirements are good mechanical, thermal and chemical stability; good thermal shock resistance; low-cost raw materials; and low fabrication costs. These include the clay- and talc-based ceramics also known as *electrical porcelains*. A large-volume use of these materials is as insulators to support high tension cables that distribute electric power. Other applications include lead feedthroughs and substrates for some types of circuits, terminal connecting blocks, supports for high-power fuse holders and wire-wound resistors.

Another important low-permittivity, low-loss ceramic is alumina. Alumina has such an excellent combination of good mechanical properties, high thermal conductivity, and ease of metallization that it is widely used today for thick-film circuit substrates and integrated electronic packaging.

Other low-permittivity ceramics that are emerging as likely candidates to replace alumina are BeO and AlN. Broadly speaking, these compounds have properties that are quite comparable to those of alumina, except that their thermal conductivities are roughly five to ten times that of alumina. AlN has a further advantage that its thermal expansion coefficient of 4.5×10^{-6}°C^{-1} is a better match with that of Si (2.6×10^{-6} °C^{-1}) than that of alumina. With these properties, AlN, despite its higher cost, may replace alumina as the size, number and density of chips increase and more heat has to be dissipated.

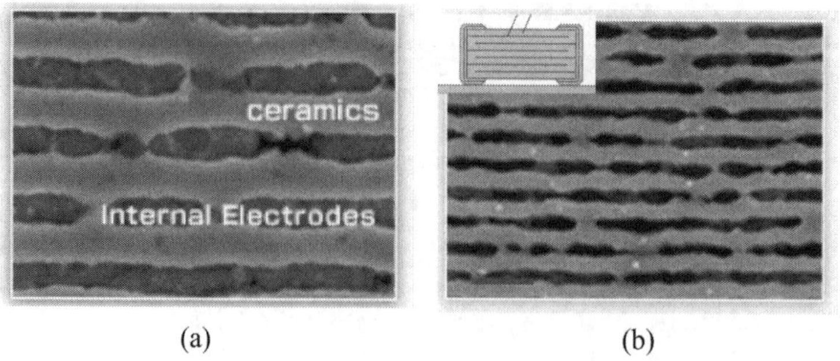

FIGURE 14.15 SEM micrographs of typical multilayered capacitors with ultrathin layers. (*a*) Ceramic layer and metal, Ni, thicknesses are 1 and 0.7 μm, respectively. (*b*) Ceramic layer and metal, thicknesses are 0.5 and 0.4 μm, respectively. Inset is schematic of the overall device. (Courtesy of Murata Corp. https://www.murata.com/products/capacitor/mlcc/strength.)

Medium-permittivity ceramics are widely used as class I dielectrics, but only if they have low dissipation factors. This precludes the use of most ferroelectric compounds that tend to have higher loss tangents. The three principal areas in which these low-loss class I materials are used are high-power transmission capacitors in the megahertz frequency range, stable capacitors for general electronic use and microwave-resonant cavities that operate in the gigahertz range.

Multilayer capacitors

Every *day* literally billions of multilayer capacitors, MLCs—the vast majority of which are $BaTiO_3$-based—are produced. Figure 14.15 shows SEM micrographs of MLCs where the thicknesses of the metal electrodes and dielectric layers are < 1 μm thick. Just about every electronic device incorporates a large number of these MLCs. For example, every cell phone contains over 700 capacitors of all sizes and shapes. The idea of a MLC is to pack the most capacitance in the smallest volume. The worked example explains.

WORKED EXAMPLE 14.5

(*a*) Calculate the capacitance per unit area of a single capacitor with a k′ of 2000 and a distance between the electrodes of 3 mm.

(*b*) If now the distance between the electrodes, *d*, is reduced to 10 μm (see Fig. 14.15) and the electrodes are also 10 μm thick, how many capacitors can be fit in 3 mm? What is their total capacitance and how does it compare to the capacitance calculated in (*a*)?

ANSWER

(*a*) To calculate the capacitance per unit area recast Eq. (14.6) to read

$$\frac{C}{A} = \frac{k'\varepsilon_0}{d} = \frac{2000 \times 8.85 \times 10^{-12}}{0.003} = 5.9\,\mu F/m^2$$

(*b*) In 3 mm one can embed $3\times10^{-3}/20\times10^{-6}=150$ capacitors. The capacitance of each is

$$\frac{C}{A}=\frac{k'\varepsilon_0}{d}=\frac{2000\times8.85\times10^{-12}}{10\times10^{-6}}=1.77\ \text{mF/m}^2$$

Since the capacitors are in parallel, their capacitances add up. It follows that the total capacity of the MLC is $1.77\times10^{-3}\times150=0.265\ \text{F/m}^2$. The MLC can therefore store 45 000 times the charge stored by a single capacitor of the same volume!

CASE STUDY 14.1: ELECTROCHEMICAL IMPEDANCE SPECTROSCOPY

Electrochemical impedance spectroscopy (EIS) sometime referred to as AC impedance was discussed briefly previously. Given the importance and versatility of this technique in elucidating and decoupling bulk and interfacial properties and the quantification of k′ and k″, we revisit it here. In this case study, how this technique can be used to shed important light on what occurs in a solid is illustrated. Before proceeding much further it is imperative to point out that it is impossible to understand/interpret EIS plots without assuming an *equivalent circuit*. The fundamental challenge in interpreting EIS results is thus to conjure an equivalent circuit whose elements represent physical properties of the system being interrogated.

In a typical experiment, the ceramic or glass to be studied is electroded and the impedance and phase shifts are measured. The raw data Z and ϕ can be plotted directly in what is known as a Bode plot (Fig. 14.16*a*). Here Z and ϕ are, respectively, plotted on the y_1 and y_2 axes vs. log ω. Another common plot is to replot the same data in what is known as a Nyquist plot. To plot the latter from the former it is useful to recall that:

$$Z(\omega)=\frac{E}{I}=Z_0\exp(i\phi)=Z_0(\cos\phi+i\sin\phi) \tag{14.57}$$

It follows that to convert the Bode to a Nyquist one first chooses a frequency, find Z_0 at that ω and then find the products $Z_0\cos\phi$ and $Z_0\sin\phi$. These values represent the x and y coordinates of the corresponding point in the Nyquist plot. To illustrate, when ϕ is 0, then $x=Z_0$ and $y=0$ on the Nyquist plot. These values represent R_1 and R_1+R_2 of the equivalent circuit assumed (see inset in Fig. 14.16*b*) and both fall on the x-axis. Now refer to the point on the Bode plot where ϕ is 45° denoted by a thin arrow in Fig. 14.16*a*. To plot this point on the Nyquist plot, the x coordinates would be $Z_0\cos\phi$, and the y-coordinate is $Z_0\sin\phi$. For this point, log $Z_0=3.325$. It is left as an exercise to the reader to show that the corresponding x and y coordinates in the Nyquist plot are ≈1490 Ω. The frequency at the top of the Nyquist semicircle—63 Hz in this case—is quite important since it can be shown that this frequency $=1/R_2C$. Knowing R_2, C can be readily calculated.

Note that the Bode and Nyquist plots plot the *same* data and contain the same information. The lines cutting across Fig. 14.16 link the points on the Bode plot to their location on the Nyquist one. In the Bode plot, the frequency is explicit and small impedances are not swamped by large impedances. In the Nyquist, frequency is implicit and small impedances can be hidden by large impedances.

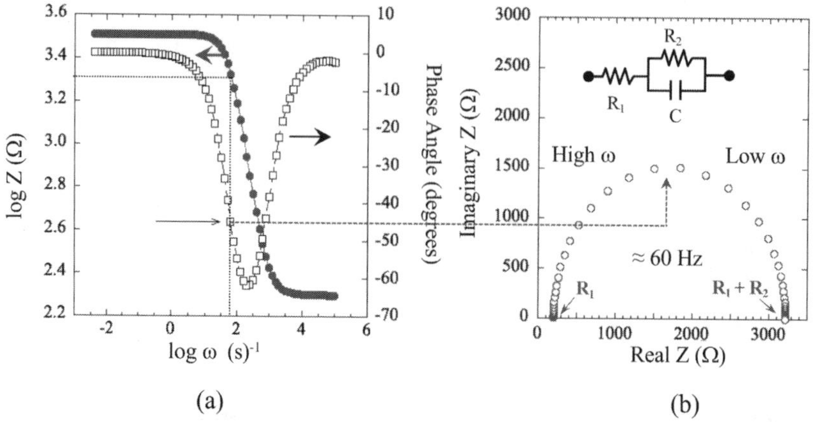

FIGURE 14.16 Typical AC impedance results, (*a*) Bode plot where left y-axis plots the impedance and right y-axis plots the phase angle. (*b*) Nyquist plot of same results. Inset shows equivalent circuit assumed. Note that the point at which the phase angle is 45° in (*a*) corresponds to the maximum in Nyquist plot in (*b*).

To illustrate let us assume that the results shown in Fig. 14.16 were obtained for system in which the interfacial resistance is R_1 and the bulk resistance is R_2 and k′ of the bulk is related to the capacitance C of the system. The equivalent circuit in this case is shown in inset of Fig. 14.16*b*. How R_1 and R_2 relate to the various regions in the diagrams should be obvious at this point. As noted previously, the frequency, ω_{RC}, at the maximum of the semicircle is equal—≈60 Hz—to $1/CR_2$. Since in this case $R_2 \approx 2964\ \Omega$, then $C = 5.6\ \mu F$. Knowing C and the dimensions of the material tested, k′ can be readily calculated as exemplified in previous worked example. Note that if the data points in the Nyquist plot are annotated with the frequencies, then ω_{RC} can be readily determined. In the Bode plot ω_{RC} is the frequency at which $\phi = 45°$ and Z_0 is at a maximum.

In this case we assumed that the interfacial capacitance was negligible and R_1 and R_2 were the resistance of the interface and bulk, respectively. Note that significant work has to be carried out for a given system in order to assign these resistances to the material properties as was done here. This is well illustrated in the following case study.

CASE STUDY 14.2: ELECTRIC POLING AND DEPLETION LAYERS

This case study, while not being of great technological importance, is nevertheless instructive at many levels, since it illustrates (i) the power of EIS to probe what is occurring at the atomic level near electrodes and in the bulk; (ii) the effect of alkali ions on DC conductivity and interfacial capacitances; (iii) what determines k′ of simple glasses, and (iv) migration of ions in glasses under a DC electric field and their effect of conductivity and k′.

A schematic of the experiment performed by McLaren et al. is shown in Fig. 14.17. In this experiment, a glass is placed between two electrodes and subjected to a DC voltage (Fig. 14.17*a*). Simultaneously, EIS is used to probe the changes occurring in the glass as a function of time. Typical results for a Na-silicate

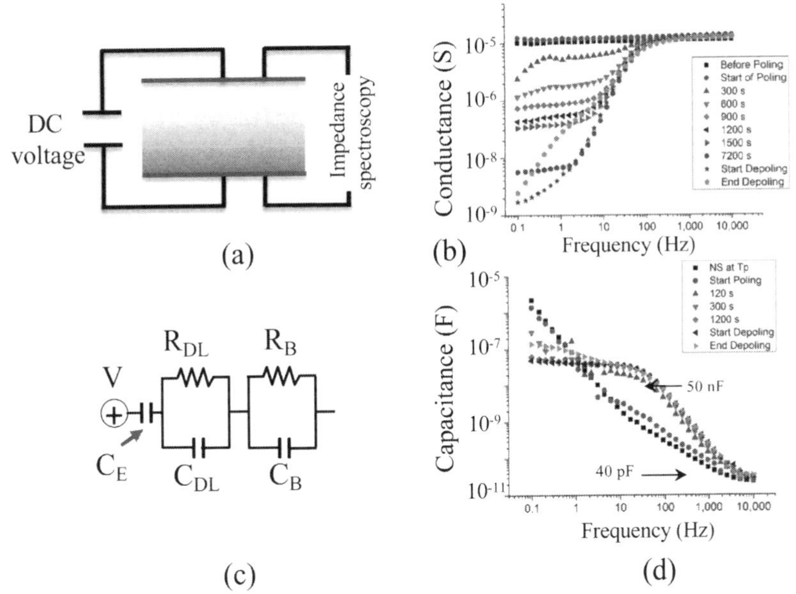

FIGURE 14.17 (*a*) Schematic of setup used to understand the polarization of a sodium silicate glass. (*b*) Bode plot as a function of poling and poling time. (*c*) Equivalent circuit assumed. DL, B and E refer to the depletion layer, bulk and electrode polarization, respectively. (*d*) Dependence of capacitance on frequency, poling and poling time. (Data adapted from C. McLaren et al., *J. Electrochem. Soc.*, 163, H809–H817, 2016.)

($0.33Na_2O{\cdot}0.67SiO_2$) glass tested at a poling temperature, T_p, of 100°C using a poling voltage, V_p, of 25 V, are shown in Fig. 14.17*b*. The T_p was chosen to correspond to the temperature at which the conductivity of the glass is 10^{-6} S/cm. Initially, the alkali ions in the glass are uniformly distributed and the conductivity is $\approx 10^{-6}$ S/cm as shown in Fig. 14.17*b*. With time, the high-frequency conductance does not change, while the low-frequency conductance drops by about 2 orders of magnitude over a time span of 2 h (Fig. 14.17*b*).

Like in all EIS measurements, to make sense of these results an electrical equivalent circuit has to be assumed. The equivalent circuit assumed in this case is shown in Fig. 14.17*c* and assumes the presence of an electrode with capacitance (C_E), a depletion layer (DL), with capacitance (C_{DL}) and resistance (R_{DL}), as well as the corresponding C_B and R_B values of the bulk glass. Initially, the only circuit elements needed to model the impedance spectra (not shown) are C_E, C_B and R_B. During poling, C_{DL} and R_{DL} are added. Once these elements are determined, if the geometry of what is causing them is known, they can be converted to relative dielectric constants and conductivities.

Putting all puzzle pieces together, the conductivity of the DLs was estimated to be of the order of $\approx 2\times10^{-12}$ S/cm. The relative dielectric constants of the bulk glass, k'_B, at RT and 100°C were 14.5 and 35, respectively. At T_p, k'_{DL} of the DLs was of the order of 10. These values were determined from C_B and C_{DL}. The former—given by the high-frequency circuit capacitance (Fig. 14.17*d*)—is ≈40 pF; C_{DL}—given by the low-frequency plateau—is 50 nF. Making use of the facts that k'_B is 8.4 and the glass thickness, d_B, is 0.8 mm, d_{DL}, is calculated from Eq. (14.6), recast as

$$\frac{C_{DL}}{C_B} = \frac{k'_{DL} d_B}{k'_B d_{DL}} \tag{14.58}$$

to be ≈200 nm. The fact that this value is in excellent agreement with the value of ≈200 nm measured by secondary ion mass spectrometry, SIMS—also found to be case for three other glass compositions—nicely confirms that all assumptions made are reasonable.

This slew of properties can now be used to explain what is occurring at the atomic level. Upon application of the electric field, the cations near the anode diffuse away creating a depleted zone. Not surprisingly, at ≈10, k'_{DL} is comparable to that of low alkali silicate glasses. Interestingly, k'_B is only $\approx 3k'_{DL}$. The depleted zone, however, is roughly *6 orders of magnitude* more resistive than the bulk glass. This huge discrepancy between the changes in conductivity and dielectric response is quite instructive because it nicely demonstrates that during polarization only a small fraction of the alkali ions are actually mobile and contribute to the DC conductivity. The rest are trapped and only contribute to k′.

14.8 SUMMARY

1. The application of an electric field **E** across a dielectric material results in polarization **P** or the separation of its positive and negative charges. The relative dielectric constant k′ is a measure of the capacity of a solid to store charge relative to vacuum and is related to the extent to which the charges in a solid polarize. Atomically there are four main polarization mechanisms: electronic, ionic, dipolar and space charge.

   For *linear* dielectrics, it is assumed that **P** scales linearly with **E**, with a proportionality constant related to k′. When a sinusoidal electric field of frequency ω is applied to a dielectric, some of the bound charges move in phase with the applied field and contribute to k′. Another set of bound charges oscillates out of phase with the applied field, result in energy dissipation and contributes to the dielectric loss factor k″. In addition to these bound charges, there will always be a dc component to the total current that contributes to the total conductivity of the sample and registers as a loss current.
2. Electronic polarization involves the displacement of the electrons relative to their nucleus and exhibits a resonance, when the frequency of the applied field is comparable to the natural frequency of vibration of the electronic cloud, ω_e. The latter is determined by several factors, the most important being the volume of the ion involved. Electronic polarization is quite insensitive to temperature and is exhibited by all matter.
3. Ionic polarization involves the displacement of cations relative to anions. Resonance occurs when the frequency of the applied field is close to the natural frequency of vibration of the ions, $\omega_{ion.}$ The latter is determined by the strength of the bond holding the ions together, which in turn is related to, among other things, the net charges on the ions and their equilibrium interatomic distance. Since it does not involve ions jumping from site to site, ionic polarization is also quite insensitive to temperature.
4. For dipolar polarization to occur, two or more adjacent sites separated by an energy barrier must exist. The preferential occupancy of one site relative to the other as a result of the application of an electric field results in solids that can have quite large k′ values. Increasing the temperature increases

the randomness of the system and tends to decrease k''_{dip}. Such solids possess a relaxation time τ that is a measure of the average time an ion spends at any given site. When $\omega\tau = 1$, the loss k''_{dip} is at a maximum and k'_{dip} has an inflection point.

5. Power dissipation in a dielectric depends on both its dc conductivity and k″. In general, a dielectric should be used at temperatures and frequencies that are as far removed as possible from a resonance or relaxation frequency. The composition should also be such as to minimize the dc conductivity.
6. For a capacitor k′ should be maximized; for an insulator it should be minimized. In both cases, however, the losses should be minimized.

APPENDIX 14A: LOCAL ELECTRIC FIELD

To estimate the local field, refer to Fig. 14.18 where a reference atom is surrounded by an imaginary sphere of such an extent that beyond it the material can be treated as a continuum. If the reference atom is removed while the surroundings remain frozen, the total field at point A will stem from three sources:

- E_1, the free charges due to the applied electric field E,
- E_2, the field that arises from the free ends of the dipole chains that line the cavity and
- E_3, the field due to atoms or molecules in the near vicinity of the reference molecule.

In highly symmetric crystals such as cubic crystals, it can be assumed that the additional individual effects of the surrounding atoms mutually cancel, or $E_3 = 0$. By applying Coulomb's law to the surface of the sphere, it can be shown that[214]

$$E_2 = \frac{E}{3}(k' - 1) \tag{14A.1}$$

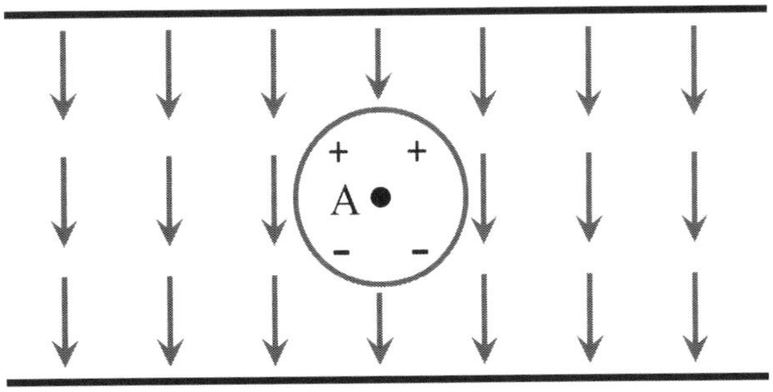

FIGURE 14.18 Model for calculation of internal field.

[214] See, e.g., N. Ashcroft and N. Mermin, *Solid State Physics*, Holt-Saunders, International Ed., 1976, p. 534, or C. Kittel, *Introduction to Solid State Physics*, 6th ed, Wiley, New York, 1988.

The local field is thus given by

$$E_{loc} = E_1 + E_2 = E + \frac{E}{3}(k' - 1) = \frac{E}{3}(k' + 2) \tag{14A.2}$$

When combined with Eq. (14.13), one obtains

$$E_{loc} = E_1 + E_2 = E + \frac{P}{3\varepsilon_0} \tag{14A.3}$$

Substituting the local field instead of the applied field in Eq. (14.29) yields

$$M_r\left\{\frac{d^2\delta}{dt^2} + f\frac{d\delta}{dt} + \left(\omega_e^2 - \frac{Ne^2}{3M_r\varepsilon_0}\right)\delta = Z_i e E_0 \exp i\omega t\right\} \tag{14A.4}$$

It follows that the effect of polarization on the surroundings is to lower the resonance frequency of the individual oscillator from ω_e to

$$\omega_e' = \sqrt{\omega_e^2 - \frac{Ne^2}{3m_e\varepsilon_0}} \tag{14A.5}$$

PROBLEMS

14.1. (*a*) Show that Eq. (14.17) can be written as

$$i\omega CV = \omega CV_0 \exp i\left(\omega t + \frac{\pi}{2}\right)$$

and that consequently $I_{chg} = -\omega k' C_{vac} V_0 \sin \omega t$.

(*b*) The **admittance** of a circuit is defined as

$$Y^* = \frac{I_0}{V_0}\cos\left(\frac{\pi}{2} - \phi\right) - i\frac{I_0}{V_0}\sin\left(\frac{\pi}{2} - \phi\right)$$

where I_0, V_0 and ϕ are defined in Fig. 14.5*b*. Here the first term represents the loss and the second term the charging current. By equating this equation with the charging and loss currents derived in text, show that

$$\sigma_{ac} = \frac{I_0 d}{V_0 A}\cos\left[\frac{\pi}{2} - \phi(\omega)\right]$$

14.2. A parallel-plate capacitor with plates separated by 0.5 cm and with a surface area of 100 cm² is subjected to a potential difference of 1000 V across the plates.

(*a*) Calculate its capacitance.

Answer: $C = 18$ pF.

(*b*) A glass plate with a k′ of 5.6, which just fills the space between the plates, is inserted between them. Calculate the surface charge density on the glass plate.

Answer: 8.14 μC/m².

(*c*) What voltage is required to store a charge of 5×10^{-10} C on a capacitor with plates 20×20 mm² separated by 0.01 mm of (i) vacuum and (ii) $BaTiO_3$? Assume k′ of latter is 2000.

Answer: 1.4×10^{5} V; 47 V

(*d*) If the capacitor in (c) is placed in an electric field of 2000 V/m that causes a polarization of 5×10^{-8} C/m². What is the relative dielectric constant of this material?

Answer: 3.82

(*e*) Repeat Worked Example 14.5 with state-of-the-art numbers, where the overall dimensions are $0.25\times0.125\times0.125$ mm³ and the thicknesses of the $BaTiO_3$ and electrode layers are 0.5 and 0.4 μm, respectively.

14.3. Show that either Eq. (14.30) or Eq. (14.31) is a solution to Eq. (14.29).

14.4. (*a*) When an external field is applied to an NaCl crystal, a 5% expansion of the lattice occurs. Calculate the dipole moment for each Na^+–Cl^- pair. The ionic radii of Na and Cl are 0.116 and 0.167 nm, respectively.

Answer: 2.30×10^{-30} C·m.

(*b*) Calculate the dipole moment of a NaCl molecule in a vapor if the separation between the ions is 2.5 Å. State all assumptions.

14.5. (*a*) Starting with Eq. (14.30) or (14.31), derive Eqs. (14.33) and (14.34).

(*b*) Plot Eqs. (14.33) and (14.34) for various values of **f**.

14.6. (*a*) Discuss possible polarization mechanisms in (*a*) Ar gas, (*b*) LiF, (*c*) water and (*d*) Si.

(*b*) The dielectric constant of a soda-lime glass at very high frequencies (>10¹⁴ Hz) was measured to be 2.3. At low-frequency (≈1 MHz) *k*′ was 6.9. Explain.

(*c*) The static dielectric constants of the following solids are given:

NaCl	5.9	MgO	9.6	SiO_2	3.8	$BaTiO_3$	1600.0	Soda-lime glass	7.0

Give a brief explanation for the different values. Discuss the various contributions to k'_{static}. Would you expect the ranking of these materials to change at a frequency of 10^{14} s⁻¹? Explain.

14.7. The k′ of 1 mole of Ar gas at 0°C and 1 atm pressure was measured to be 1.00056.

(*a*) Calculate the polarizability of Ar.

(*b*) Calculate k′ if the pressure is increased to 2 atm.

(*c*) Estimate the radius of an Ar atom and comment on how your value compares to the Van der Waals radius of ≈0.19 nm.

Answer: 0.218 nm.

14.8. Using the ion positions for the tetragonal $BaTiO_3$ unit cell shown in Fig. 14.19, calculate the electric dipole moment per unit cell and the saturation polarization for $BaTiO_3$. Compare your answer with the observed saturation polarization (that is, $P_s = 0.26$ C/m²). Hint: Use the Ba ions as a reference.

Answer: $P_s = 0.16$ C/m².

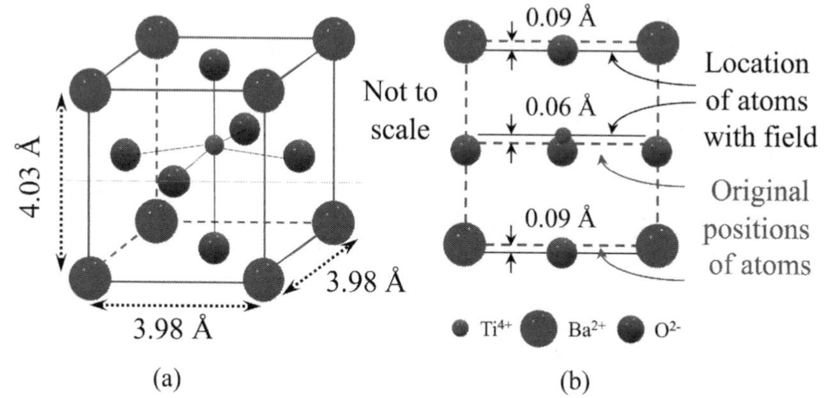

FIGURE 14.19 (*a*) In the tetragonal unit cell of $BaTiO_3$, the Ba ions occupy the unit cell corners, while Ti is near the center of the cell. The oxygens are near the centers of the faces. (*b*) Projection of the (100) face. Because the ions are displaced with respect to the symmetric position, the center of negative charge does not coincide with the center of positive charge, resulting in a net dipole moment per unit cell.

14.9. (*a*) The temperature variation of the static dielectric constant for some gases is shown in Fig. 14.20*a*. Answer the following questions:

(i) Why do the dielectric constants for CCl_4 and CH_4 not vary with T?
(ii) What type of polarization occurs in CCl_4 and CH_4?
(iii) Why is k′ greater for CCl_4 than for CH_4?
(iv) What type of polarization leads to the inverse temperature dependence of k′ shown for CH_3Cl, CH_2Cl_2 and $CHCl_3$?
(v) Why is the temperature variation of k′ greater for CH_3Cl than for $CHCl_3$?

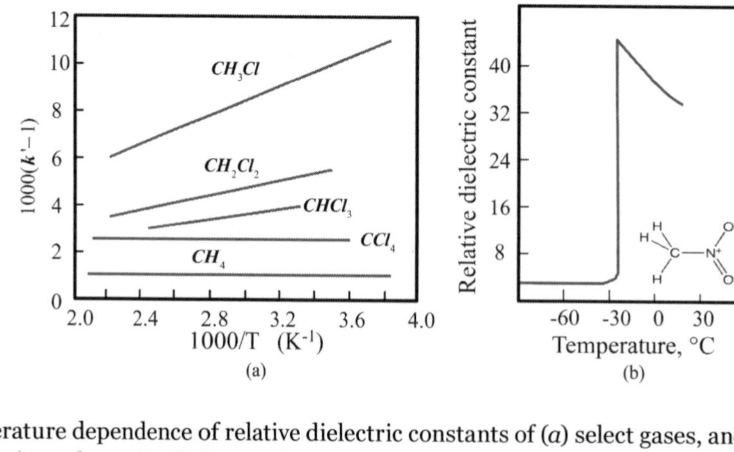

FIGURE 14.20 Temperature dependence of relative dielectric constants of (*a*) select gases, and (*b*) nitromethane measured at 70 kHz. Inset is a schematic of nitromethane.

(*b*) The temperature variation of the static dielectric constant for nitromethane with a melting point of −29°C is shown in Figs. 14.20*b*.

(i) Explain what is happening paying special attention to the polarization mechanism(s) and their temperature dependencies.

(ii) This measurement was carried out at 70 kHz. What do you think will happen if the measuring frequency is increased to say 10 GHz.

14.10. (*a*) Where do you expect to have more ions at equilibrium in Fig. 14.10? Explain in your own words why the unequal distribution of ions in the two sites would give rise to polarization.

(*b*) The static k′ of a ceramic was measured at 200°C and found to be 140. Estimate the jump distance. State all necessary assumptions. Information you may find useful: Molar volume = 10 cm^3/mol; charge on cations is +4.

Answer: 0.036 nm.

14.11. (*a*) The electric dipole moment of water is 6.13×10^{-30} C·m. Calculate the dipole moment of each O-H bond.

Answer: 5.05×10^{-30} C·m.

(*b*) The dielectric constant of water was measured as a function of temperature in such a way that the number of water molecules was kept constant. The results are tabulated below. Do the results behave according to Curie's law, i.e., as 1/T? Explain.

(*c*) From the results estimate the density of the water molecules in the experiment. Does your result make sense?

Temp. (°C)	119	148	171	200
k′	1.004	1.0037	1.0035	1.0032

Answer: $N = 2.62\times10^{25}\ m^{-3}$.

14.12. (*a*) Show that when k''_{dip} is at a maximum, $\tau\omega_{max} = 1$.

(*b*) The dielectric loss for thoria was measured as a function of temperature and frequency, and the results are tabulated below.[215] The static and high-frequency permittivities have been found from other measurements to be

$$k'_s = 19.2 \text{ and } k'_\infty = 16.2$$

Assuming ion jump polarization is responsible for the variation in tan ϕ estimate both τ_0 and ΔH_m [defined in Eq. (14.56)].

ω = 695 Hz		ω = 6950 Hz	
T, K	tan ϕ	T, K	tan ϕ
475	0.029	498	0.010
485	0.042	518	0.025
494	0.063	543	0.055
503	0.081	568	0.086

[215] Data taken from PhD Thesis of J. Wachtman, U. of Maryland, 1962. Quoted in *Lectures on the Electrical Properties of Materials*, 4th ed., Solymar and Walsh, Oxford Science Publications, 1989.

509	0.086	581	0.086
516	0.092	590	0.073
524	0.086	604	0.055
532	0.070	612	0.043
543	0.042	621	0.036
555	0.023	631	0.026

14.13. (*a*) Derive Eq. (14.24).

(*b*) The following data[216] were determined for a technical ceramic as a function of temperature. The ceramic was in the form of a parallel-plate capacitor with thickness 0.5 cm and diameter 2.54 cm.

(i) Plot k′ versus log ω for all three temperatures on same plot.

(ii) Plot k″ versus log ω for all three temperatures on same plot.

(iii) Plot tan ϕ versus log ω for all three temperatures on same plot.

(iv) Calculate the activation energy for the relaxation process and compare it to the activation energy for dc conductivity. How do these activation energies compare? What conclusions can you reach regarding the basic atomic mechanisms responsible for conductivity and those for polarization?

(v) What is the most probable polarization mechanism operative in this ceramic?

	72°C		90°C		112°C	
Frequency, Hz	G_{ac}, $\mu\Omega^{-1}$	C, pF	G_{ac}, $\mu\Omega^{-1}$	C, pF	G_{ac} $\mu\Omega^{-1}$	C, pF
20 K	0.88	4.8	3.8	5.7	6.8	6.4
15 K					6.8	6.8
10 K	0.79	5.4	3.7	6.5	6.7	8.0
5 K	0.75	6.1	3.6	7.9	6.7	12.8
3 K					6.6	21.8
2 K	0.73	7.1	3.6	15.2	6.6	39.4
1.5 K					6.5	62.2
1.2 K					6.5	91.1
1 K	0.71	8.9	3.6	36.1	6.4	118.7
500	0.68	14.5	3.5	104.7	6.1	353.4
340	0.67	22.5				
260	0.60	32.4				
200	0.60	47.1	3.3	455.4		
100	0.57	139.0				

14.14. Clearly stating all assumptions, calculate the power loss for a parallel-plate capacitor (d = 0.02 cm, A = 1 cm²) made of $MgAl_2O_4$ subjected to

(*a*) A dc voltage of 120 V.

[216] Problem adapted from L. L. Hench and J. K. West, *Principles of Electronic Ceramics*, Wiley-Interscience, New York, 1990. The frequencies as reported in Hench and West are incorrect at the low end and have been corrected here.

(*b*) An ac signal of 120 V and a frequency of 60 Hz.

Answer. 1.9×10^{-6} W.

a.(*c*) An ac signal of 120 V and a frequency of 60 MHz.

Answer. 0.05 W.

14.15. (*a*) Explain why microwaves are quite effective at rapidly and efficiently heating water or water-containing substances.

(*b*) In a microwave oven, the food is heated from the inside out, whereas in a regular oven the food is heated from the outside in. Explain.

14.16. (*a*) Why do you think the k' of the glasses in Case Study 14.2 increased from room temperature to 100°C?

(*b*) If the area of the glass slides samples is 10×10 mm, calculate k'_B. Also calculate the conductivity from Fig. 14.17*b*.

ADDITIONAL READING

H. Fröhlich, *Theory of Dielectrics*, 2nd ed., Oxford Science Publications, 1958.

L. L. Hench and J. K. West, *Principles of Electronic Ceramics*, Wiley-Interscience, New York, 1990.

A. J. Moulson and J. H. Herbert, *Electroceramics*, Chapman & Hall, London, 1990.

N. Ashcroft and N. Mermin, *Solid State Physics*, Holt-Saunders International Ed., 1976.

C. Kittel, *Introduction to Solid State Physics*, 6th ed., Wiley, New York, 1988.

J. C. Anderson, *Dielectrics*, Chapman & Hall, London, 1964.

L. Azaroff and J. J. Brophy, *Electronic Processes in Materials*, McGraw-Hill, New York, 1963.

A. Von Hippel, Ed., *Dielectric Materials and Applications*, Wiley, New York, 1954.

R. C. Buchanan, Ed., *Ceramic Materials for Electronics*, Marcel Dekker, New York, 1986.

D. M. Trotter, Capacitors, *Sci. Am.*, July 1988, p. 86.

P. Fulay, *Electronic, Magnetic and Optical Properties of Materials*, CRC Press, Boca Raton, FL, 2010.

15

MAGNETIC AND NONLINEAR DIELECTRIC PROPERTIES

Magnetic Atoms, such as Iron, keep
Unpaired Electrons in their middle shell,
Each one a spinning Magnet that would leap
The Bloch Walls whereat antiparallel
Domains converge. Diffuse Material
Becomes Magnetic when another Field
Aligns domains like Seaweed in a swell.
How nicely microscopic forces yield
In Units growing visible, the World we wield!

John Updike, *The Dance of the Solids**

15.1 INTRODUCTION

The first magnetic material exploited by humankind as a navigational tool was a ceramic—the natural mineral magnetite, Fe_3O_4, also known as *lodestone*. It is thus somewhat ironic that it was as late as the mid-1950s that magnetic ceramics started making significant commercial inroads. Since then magnetic ceramics have acquired an ever-increasing share of world production and have largely surpassed metallic magnets in tonnage. With the advent of the information communication revolution, their use is expected to become even more ubiquitous.

* J. Updike, Midpoint and Other Poems, A. Knopf, Inc., New York, New York, 1969. Reprinted with permission.

In addition to dealing with magnetic ceramics, this chapter also deals with dielectric ceramics, such as ferroelectrics, for which the dielectric response is nonlinear. Ferroelectricity was first discovered in 1921 during the investigation of the anomalous behavior of Rochelle salt. A second ferroelectric material was not found until 1935. The third major ferroelectric material, $BaTiO_3$, was reported in 1944. Ferroelectric ceramics possess some fascinating properties that have been exploited in a number of applications such as high permittivity capacitors, displacement transducers and actuators, infrared detectors, gas igniters, accelerometers, wave filters, color filters, optical switches and the generation of sonic energy, among others.

And while at first glance magnetic and nonlinear dielectric properties may not appear to have much in common, they actually do. The term *ferroelectric*[217] was first used in analogy to ferromagnetism. It is thus instructive at the outset to point out some of the similarities between the two phenomena. In both cases, the properties that result can be traced to the presence of *permanent dipoles, magnetic* in one and *electric* in the other, which respond to externally applied fields. An exchange energy exists between the dipoles that allows them to interact with one another in such a way as to cause their spontaneous alignment, which in turn gives rise to nonlinear responses. The orientation of all the dipoles with the applied field results in saturation of the polarization, and the removal of the field results in a permanent or residual polarization. The concept of domains is valid in both, and they both respond similarly to changes in temperature.

This chapter is structured as follows: Section 15.2 introduces the basic principles and relationships between various magnetic parameters. Section 15.3 deals with magnetism at the atomic level. In Sec. 15.4, the differences and similarities between para-, ferro-, antiferro- and ferrimagnetism are discussed. Magnetic domains and hysteresis curves are dealt with in Sec. 15.5, while Sec. 15.6 deals with magnetic ceramics. The remainder of the chapter deals with the nonlinear dielectric response of ceramics in light of both discussions in Chap. 14 and magnetic properties.

Before delving in the details, it is worth noting that for historical reasons, the units in the field of magnetism can rapidly get very confusing indeed. It is thus crucial that careful attention be paid to what is being discussed.

15.2 BASIC THEORY

A **magnetic field intensity or applied field H** is generated whenever electric charges are in motion. The latter can be simply electrons flowing in a conductor or the orbital motion and spins of electrons around nuclei and/or around themselves. For example, it can be shown that the magnetic field intensity, **H**, at the center of a circular loop of radius r through which a current i is flowing is

$$\mathbf{H} = \frac{i}{2r} \tag{15.1}$$

[217] As will become clear shortly, there is nothing "ferro" about ferroelectricity. It was so named because of the similarities between it and ferromagnetism, which was discovered first.

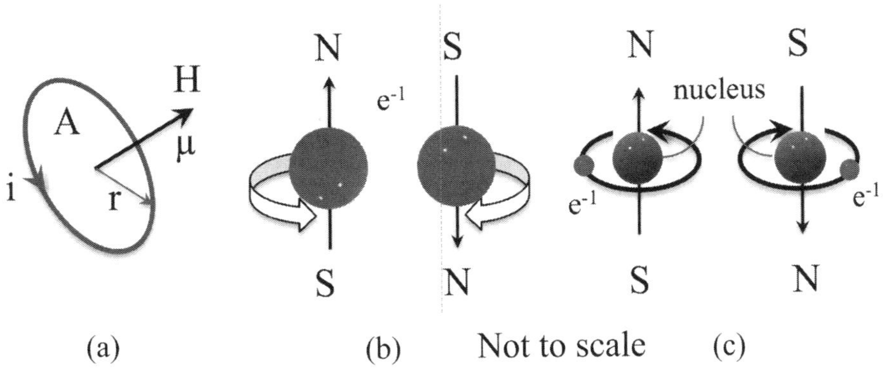

FIGURE 15.1 (*a*) Magnetic field intensity resulting from passage of a current through a loop of radius r or, equivalently, magnetic moment, μ, of current loop of area A. (*b*) Spin of electron around its own axis results in a spin magnetic moment. (*c*) Orbital moment arises from electrons orbiting around the nucleus. In both (*b*) and (*c*) the direction of the moment depends on spin direction. Schematics are not to scale.

H is a vector and in this case points normal to the plane of the loop as shown in Fig. 15.1*a*. The units H are amperes per meter: an H-field of 1 A/m is produced when a current of 1 A flows through loop with a 1 m diameter.

In vacuum, **H** will result in a **magnetic field**[218] **B** given by

$$\mathbf{B} = \mu_0 \mathbf{H} \tag{15.2}$$

where the constant μ_0 is the **permeability of free space** which is $4\pi \times 10^{-7}$ Wb/(A·m). **B** can be expressed in a number of equivalent units, such as V·s/m² = Wb/m² = T (Tesla) 1 = 10⁴ G (Gauss) (see Table 15.1). A magnetic induction of 1 T will generate a force of 1 N on a conductor carrying a current of 1 A perpendicular to the direction of induction.

In the presence of a solid, **B** will be composed of two parts—that which would be observed in the absence of the solid, i.e. in vacuum, plus that due to the solid, or

$$\boxed{\mathbf{B} = \mu_0 (\mathbf{H} + \mathbf{M})} \tag{15.3}$$

where **M** is the **magnetization** of the solid, defined as the **net magnetic moment,** $\boldsymbol{\mu}_{ion}$, per unit volume,

$$\mathbf{M} = \frac{\mu_{ion}}{V} \tag{15.4}$$

[218] Confusingly, the magnetic field B has been refered to by a number of terms: *magnetic induction, magnetic field strength* and *magnetic flux density*. Here B is referred to strictly as the *magnetic field*. Even more confusingly, in many cases, H is also referred to as the magnetic field. This came about historically, because in the cgs system μ_0 was 1 and so B and H were numerically *identical*. To avoid confusion, H is strictly referred to here as the *magnetic field intensity, applied field, magnetizing field or H-field*.

TABLE 15.1 Definitions, dimension of units and symbols used in magnetism[a]

Symbol	Quantity	Value	Units
H	Magnetic field intensity or magnetizing field		A/m[b]
M	Magnetization		A/m
B	Magnetic field		$Wb/m^2 = T = V{\cdot}s/m^2 = 10^4 G$
μ_0	Permeability of free space	$4\pi \times 10^{-7}$	Wb/(A·m) = V·s/(A·m)
μ	Permeability of a solid		Wb/(A·m) = V·s/(A·m)
μ_r	Relative permeability		Dimensionless
χ_{mag}	Relative susceptibility		Dimensionless
μ_{ion}	Net magnetic moment of an atom or ion		$A{\cdot}m^2 = C{\cdot}m^2/s$
μ_s	Spin magnetic moment		$A{\cdot}m^2$
μ_{orb}	Orbital magnetic moment		$A{\cdot}m^2$
μ_B	Bohr magneton	9.274×10^{-24}	$A{\cdot}m^2$

[a] It is unfortunate that both μ_0 and μ_{ion} have the same symbol, but they will be clearly marked at all times to avoid confusion.

[b] 1A/m = 0.0126 Oersted (Oe).

The atomic origins of μ_{ion} are discussed in detail in the next section. The units of μ_{ion} are $A{\cdot}m^2$. A μ_{ion} of 1 $A{\cdot}m^2$ experiences a maximum torque of 1 N·m when oriented perpendicular to a B-field of 1 T.

In paramagnetic and diamagnetic solids (see below), **B** is a linear function of **H** such that

$$\mathbf{B} = \mu \mathbf{H} \tag{15.5}$$

where μ is the permeability of the solid (*not* to be confused with μ_{ion}). For ferro- and ferrimagnets, however, **B** and **H** are no longer linearly related, but as discussed below, μ can vary rapidly with **H**.

The **magnetic susceptibility** is defined as

$$\chi_{mag} = \frac{\mathbf{M}}{\mathbf{H}} \tag{15.6}$$

The **relative permeability** μ_r is given as

$$\mu_r = \frac{\mu}{\mu_0} \tag{15.7}$$

and it compares the permeability of a medium to that of vacuum. This quantity is analogous to the relative dielectric constant k′. It can also be shown that μ_r and χ_{mag} are related by

$$\mu_r = \chi_{mag} + 1 \tag{15.8}$$

Note that **M** and **H** have the same units, but in contradistinction to **H**, which is generated by electric currents or permanent magnets *outside* the material (which is why it is sometimes referred to as the magnetizing field), **M** is generated from the uncompensated spins and angular momenta of electrons *within* the solid. In isotropic media, ***B, H*** and **M** are vectors that point in the same direction, whereas χ and μ are scalars.

EXPERIMENTAL DETAILS: MEASURING MAGNETIC PROPERTIES

Consider the experimental setup[219] shown in Fig. 15.2*a*, which is composed of four elements:

- ∞ A sample for which the magnetic properties are to be measured.
- ∞ A sensitive balance from which the sample is suspended.
- ∞ A permanent bar magnet with its north pole pointing upward.
- ∞ A solenoid of n turns per meter, through which a current i flows so as to produce a magnetic field intensity, **H**, that is in the *same* direction as that of the permanent magnet (i.e., with the north pole pointing up). Here the right-hand rule is used.

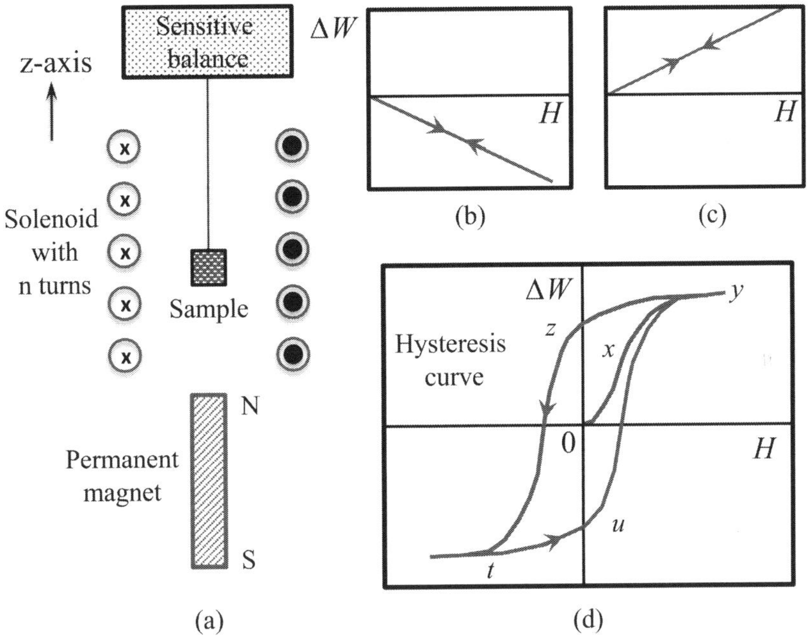

FIGURE 15.2 (*a*) Schematic arrangement by which the magnetic properties of a material can be measured. The solenoid results in a uniform applied magnetic field and hence by itself will not create a force on the sample. It is only the nonuniformity of the magnetic field (that is, $dB/dz \neq 0$) of the permanent magnet that will create the force. (*b*) Typical response of a diamagnet. (*c*) Typical response of a paramagnet. (*d*) Typical response of a ferromagnet. Note that upon removal of the field a remnant magnetization remains.

[219] In reality, a shaped pole piece that creates a uniform magnetic field gradient dB/dz is used instead of the solenoid. The setup shown in Fig. 15.2*a* is simpler and is used to clarify the various concepts.

To measure the magnetic properties, a small cylinder of the solid is suspended from a sensitive balance into the center of the solenoid (Fig. 15.2*a*). According to Ampere's law, passing a current i through the solenoid will create an axial uniform **H** of strength

$$\mathbf{H} = \text{ni} \tag{15.9}$$

This magnetizing field will in turn induce magnetic moments (what that means is discussed later) in the material to align themselves either with or against the applied field. It can be shown that the magnetic force on a material with magnetization **M** and volume V is given by[220]

$$\mathbf{F}_z = \mu_{ion}\frac{d\mathbf{B}}{dz} = V\mathbf{M}\frac{d\mathbf{B}}{dz} \tag{15.10}$$

where d**B**/dz is the gradient in magnetic field along the z-axis due to the *permanent magnet.* Combining Eq. (15.10) with Eq. (15.6), shows that

$$\mathbf{F}_z = V\chi_{mag}\mathbf{H}\frac{d\mathbf{B}}{dz} \tag{15.11}$$

In other words, the force on the sample is directly proportional to its susceptibility and the applied field. Since in this case dB/dz is negative, if the sample is attracted to the permanent magnet, it implies χ_{mag} is positive and vice versa (see Worked Example 15.1).

There are four possible outcomes of such an experiment:

1. The sample is very weakly repelled by the permanent magnet, and the weight of the sample will appear to diminish (Fig. 15.2*b*), implying that χ_{mag} is *negative*. Increasing **H** by increasing the current in the solenoid will linearly increase the repulsive force. Such a material is termed a **diamagnetic** material, and most ceramics fall in this category.
2. The sample is weakly attracted to the permanent magnet, with a force of attraction that is proportional to **H**. The sample will appear to have gained weight (Fig. 15.2*c*), implying a positive χ_{mag}. Such solids are known as **paramagnets.** Repeating the experiment at different temperatures quickly establishes that the force of attraction, or χ_{mag}, decreases with increasing T. When the field intensity is removed, the sample will return to its original weight; in other words, all the changes that occurred in the presence of the field are completely reversible. The same is true for diamagnetic materials.
3. The sample is strongly attracted to the permanent magnet (Fig. 15.2*d*). The shape of the curve obtained, however, will depend on the sample's history. If one starts with a virgin sample, i.e., one that was never exposed to a field before, the weight gain will follow the line Oxy, shown in Fig. 15.2*d*. At low **H**, the sample is initially weakly attracted, but as **H** is further increased, the rate of weight increase will be quite rapid. At high magnetizing fields, the force will appear to saturate and further increases in **H** will have a minimal effect (Fig. 15.2*d*). The plateau is known as the **saturation magnetization**.

[220] If the permanent magnet were not present, no force would be exerted on the sample, since to experience a force, the magnetic field has to have a gradient (i.e., it cannot be uniform).

Furthermore, as **H** is reduced, the sample's response is nonreversible in that it follows the line yz. When **H** = 0, the sample will appear to have permanently gained weight! In other words, a permanent magnet with a **remnant magnetization** $\mathbf{M}_r$ has been created. Upon further cycling, the sample's response will follow the loop yztu. So-called hysteresis loops represent energy losses and are typical of all ferromagnetic materials. The reason for their behavior is described more thoroughly below. Such behavior is termed **ferromagnetic** or **ferrimagnetic**.

Repeating the same experiment at increasing temperatures results in essentially the same behavior except that the response of the material weakens and $\mathbf{M}_r$ diminishs. At a critical temperature T_c, the material will totally lose its ferromagnetism and will behave as a paramagnet instead.

4. An **antiferromagnetic** material behaves similarly to a paramagnetic one, i.e., it is weakly attracted. However, to differentiate between the two, the experiment would have to be carried out as a function of temperature: The susceptibility of an antiferromagnetic material will appear to go through a maximum as the temperature is lowered (see Fig. 15.5*b*); in contrast to that of a paramagnet, which will continually increase with decreasing temperature (Fig. 15.3*c*).

Before this plethora of phenomena can be explained, it is imperative to understand what occurs at the atomic level that gives rise to magnetism, which is the topic of Sec. 15.3.

Worked Example 15.1

A chunk of a magnetic ceramic weighing 10 g is attached to the sensitive balance shown in Fig. 15.2*a* and is suspended in the center of a toroidal solenoid with 10 turns per centimeter. A current of 9 A is passed through the coils. The magnetic field gradient due to the permanent magnet was measured to be 100 G/cm. When the current was turned on, such that H was in the same direction as the permanent magnet, the weight of the sample was found to increase to 10.00005 g. The density of the solid is 5 g/cm³. (*a*) Calculate the susceptibility of this material. (*b*) Calculate the magnetization **M** of the solid. (*c*) What conclusions can be inferred concerning its magnetic properties?

ANSWER

(*a*) The force in the z direction, **H**, dB/dz, and V in SI units are, respectively:

$$F_2 = \Delta W \cdot g = 0.00005 \times 10^{-3} \times 9.8 = -\ 4.9 \times 10^{-7}\ \text{N}$$

$$H = ni = \frac{9 \times 10}{10^{-2}} = 9000\ \text{A/m}$$

$$\frac{dB}{dz} = \frac{100 \times 10^{-4}}{10^{-2}} = 1.0\ \text{T/m}$$

$$V = \frac{10}{5} = 2\ \text{cm}^3 = 2 \times 10^{-6}\ \text{m}^3$$

Substituting these values in Eq. (15.11) and solving for χ_{mag}, one obtains

$$\chi_{mag} = \frac{4.9\times10^{-7}}{2\times10^{-6}\times9000\times1} = 2.7\times10^{-5}$$

(*b*) $\mathbf{M} = \chi_{mag}\mathbf{H} = 2.7\times10^{-5}\times9000 = 0.245$ A/m. Note that in this case, because χ_{mag} is small then with very little loss in accuracy,

$$\mathbf{B} = \mu_o(\mathbf{H}+\mathbf{M}) \approx \mu_o\mathbf{H} \tag{15.12}$$

(*c*) Given that the sample was attracted to the magnet, it must be paramagnetic, ferromagnetic or antiferromagnetic. However, given the small value of χ_{mag}, ferromagnetism can be safely eliminated, and the material must either be paramagnetic or antiferromagnetic. To narrow the possibilities further, the measurement would have to be repeated as a function of temperature.

15.3 MICROSCOPIC THEORY

For a solid to interact with a magnetizing field, **H**, it must possess a net magnetic moment, which, as discussed momentarily, is related to the angular momentum of the electrons, as a result of either their revolution around themselves (Fig. 15.1b) and/or their nuclei (Fig. 15.1c). The latter gives rise to an **orbital angular moment $\boldsymbol{\mu}_{orb}$**, whereas the former is the **spin angular moment $\boldsymbol{\mu}_s$**.The sum of these two contributions is the **total angular moment** of an atom or ion, μ_{ion}.

If when an atom is placed in a magnetizing H-field, it experiences a torque then the atom possesses a magnetic moment. Note that in the remainder of this chapter, the vector notation will sometimes be dropped.

15.3.1 ORBITAL MAGNETIC MOMENT

From elementary magnetism, a current i going around in a loop of area A′ (πr^2) will produce an orbital magnetic moment μ_{orb} given by

$$\mu_{orb} = iA' = i\pi r^2 \tag{15.13}$$

that points normal to the plane of the loop (Fig. 15.1*a*). A single electron, with a charge e, rotating with an angular frequency ω_o gives rise to a current

$$i = \frac{e\omega_o}{2\pi} \tag{15.14}$$

Assuming the electron moves in a circle of radius r, then combining Eqs. (15.13) and (15.14), one obtains

$$\mu_{orb} = \frac{e\omega_0 r^2}{2} \tag{15.15}$$

But since $m_e \omega_0 r^2$ is nothing but the **orbital angular momentum,** Π_0, of the electron, it follows that

$$\mu_{orb} = \frac{e\Pi_0}{2m_e} \tag{15.16}$$

where m_e is the rest mass of the electron. This relationship clearly indicates that it is the angular momentum that gives rise to magnetic moments.

Equation (15.16) can be slightly recast in units of $h/(2\pi)$ to read

$$\mu_{orb} = \frac{eh}{4\pi m_e}\frac{2\pi\Pi_0}{h} = \frac{eh}{4\pi m_e} l \tag{15.17}$$

where the integer l ($=2\pi\Pi_0/h$) is the orbital angular momentum quantum number (see Chap. 2). Note that this result is consistent with quantum theory predictions that the angular momentum has to be an integral multiple of $h/(2\pi)$. Said otherwise, the angular momentum is quantized.

The ratio $eh/4\pi m_e$ occurs quite frequently in magnetism and has a numerical value of 9.27×10^{-24} A·m². This value is known as the **Bohr magneton**, μ_B, and, as discussed below, it is the value of the orbital angular momentum of a single electron spinning around the Bohr atom. In terms of μ_B, Eq. (15.17) can be succinctly recast as:

$$\mu_{orb} = \mu_B l \tag{15.18}$$

In deriving Eq. (15.18), it was assumed that the angular momentum was an integral multiple of l; quantum mechanically, however, it can be shown[221] (not very easily, one should add) that the relationship is more complicated. The more accurate expression given here without proof is

$$\boxed{\mu_{orb} = \mu_B\sqrt{l(l+1)}} \tag{15.19}$$

15.3.2 SPIN MAGNETIC MOMENT

This moment arises from the spin of the electrons around themselves (Fig. 15.1b. Quantitatively, the **spin magnetic moment** μ_s is given by

$$\mu_s = \frac{e\Pi_s}{m_e} \tag{15.20}$$

where Π_s is the **spin angular momentum.** Given that Π_s has to be an integer multiple of $h/2\pi$, it can be shown that

[221] See, e.g., R. P. Feynman, R. B. Leighton, and M. Sands, *The Feynman Lectures on Physics*, vol. 2, Chap. 34, Addison-Wesley, Reading, Massachusetts, 1964.

$$\mu_s = 2\mu_B s \tag{15.21}$$

where s is the *spin quantum number*. Since $s = \pm 1/2$, it follows that the μ_s of one electron is one Bohr magneton. It is important to note, however, that Eq. (15.21) is not quite accurate. Quantum mechanically, it can be shown that the correct relationship—and the one that should be used—is

$$\boxed{\mu_{ion} = 2\mu_B \sqrt{s(s+1)}} \tag{15.22}$$

15.3.3 TOTAL MAGNETIC MOMENT OF A POLYELECTRONIC ATOM OR ION

The total angular moment of an ion, with *one* unpaired electron, $\mu_{ion,1}$, is simply

$$\mu_{ion,1} = \mu_s + \mu_{orb} = \frac{e\Pi_s}{m_e} + \frac{e\Pi_o}{2m_e} \tag{15.23}$$

The subscript 1 was added to emphasize this expression is only valid for *one* electron. Combining terms and introducing a factor *g*, known as the **Lande splitting factor,** one arrives at

$$\mu_{ion,1} = \mu_s + \mu_{orb} = g\left(\frac{e}{2m_e}\right)\Pi_{tot} \tag{15.24}$$

where Π_{tot} is the total angular momentum of an atom or ion. If only the spin is contributing to $\mu_{ion,1}$, then $g = 2$. Conversely, if only the orbital momentum is contributing to the total, then $g = 1$. Thus in general, g lies between 1 and 2 depending on the relative contribution of μ_s and μ_{orb} to $\mu_{ion,1}$.

Equations (15.23) and (15.24) are only valid for ions that possess one electron. If an atom, or ion, has more than one electron the situation is more complicated. For one, it can be shown (App. 15A) that the orbital magnetic momenta of the electrons add vectorially such that

$$\mathbf{L} = \sum \mathbf{m}_l \tag{15.25}$$

where m_l is the orbital magnetic quantum number (see Chap. 2). Similarly, the spins add such that the total spin angular momentum of the ion is given by

$$\mathbf{S} = \sum \mathbf{s} \tag{15.26}$$

where $s = \pm 1/2$.

The total angular momentum **J** of the atom is then simply the *vector* sum of the two noninteracting momenta **L** and **S** such that

$$\mathbf{J} = \mathbf{L} + \mathbf{S}$$

This is known as **Russell–Saunders coupling**, and will not be discussed any further because, for the most part, the orbital angular momentum of the transition-metal cations of the 3d series that are responsible for most of the magnetic properties exhibited by ceramic materials is *totally quenched,* that is, $\mathbf{L}=0$. Needless to add, this greatly simplifies the problem at hand because now $\mathbf{J}=\mathbf{S}$ and μ_{ion} is simply given by

$$\boxed{\mu_{ion}=2\mu_B\sqrt{\mathbf{S}(\mathbf{S}+1)}} \tag{15.27}$$

Thus the total magnetic moment of an ion—provided that the angular momentum is quenched—is related to the sum of the individual contributions of unpaired electrons. To predict the moment, one simply needs to know how many electrons are unpaired, or which quantum states are occupied in a given atom or ion. The following examples should help clarify this important point.

Worked Example 15.2

Calculate the spin and total magnetic moment of an isolated Mn^{2+} cation, assuming that the orbital angular momentum is quenched, that is, $\mathbf{L}=0$.

ANSWER

Mn^{2+} has five d-electrons (see Table 15.2) that occupy the following orbitals:

m_l	2	1	0	−1	−2	2	1	0	−1	−2
m_s	1/2	1/2	1/2	1/2	1/2	−1/2	−1/2	−1/2	−1/2	−1/2
s	↓	↓	↓	↓	↓					

It follows that $\mathbf{S}=\Sigma\mathbf{m}_s=5\times1/2=2.5\mu_B$. Since the angular momentum is quenched, $\mathbf{J}=\mathbf{S}$, and according to Eq. (15.27), the total magnetic moment for the ion is

$$\mu_{ion}=2\mu_B\sqrt{\mathbf{S}(\mathbf{S}+1)}=5.92\mu_B$$

which is in excellent agreement with the measured value of $5.9\mu_B$ (see Table 5.2).

Worked Example 15.3

Show that the angular momentum of an atom or ion with a closed-shell configuration is zero.

Answer

A good example is Cu^{+}. It has 10 d-electrons arranged as follows:

m_l	2	1	0	−1	−2	2	1	0	−1	−2
m_s	1/2	1/2	1/2	1/2	1/2	−1/2	−1/2	−1/2	−1/2	−1/2
s	↓	↓	↓	↓	↓	↑	↑	↑	↑	↑

Thus $\mathbf{L}=\sum\mathbf{m}_l=0$ and $\mathbf{S}=\sum\mathbf{m}_s=0$, and consequently, $J=0$.

TABLE 15.2 Magnetic moments of *isolated* cations of *3d* transition series. All moments are given in multiples of μ_B. When not isolated the numbers are different

Cations	Electronic configuration	Calculated quantum moments $2\sqrt{S(S+1)}$	Classical moments	Measured moments
Sc^{3+}, Ti^{4+}	$3d^0$ [ ][ ][ ][ ][ ]	0.00	0	0.0
V^{4+}, Ti^{3+}	$3d^1$ [↑][ ][ ][ ][ ]	1.73	1	1.8
V^{3+}	$3d^2$ [↑][↑][ ][ ][ ]	2.83	2	2.8
V^{2+}, Cr^{3+}	$3d^3$ [↑][↑][↑][ ][ ]	3.87	3	3.8
Mn^{3+}, Cr^{2+}	$3d^4$ [↑][↑][↑][↑][ ]	4.90	4	4.9
Mn^{2+}, Fe^{3+}	$3d^5$ [↑][↑][↑][↑][↑]	5.92	5	5.9
Fe^{2+}	$3d^6$ [↓↑][↑][↑][↑][↑]	4.90	4	5.4
Co^{2+}	$3d^7$ [↓↑][↓↑][↑][↑][↑]	3.87	3	4.8
Ni^{2+}	$3d^8$ [↓↑][↓↑][↓↑][↑][↑]	2.83	2	3.2
Cu^{2+}	$3d^9$ [↓↑][↓↑][↓↑][↓↑][↑]	1.73	1	1.9
Cu^+, Zn^{2+}	$3d^{10}$ [↓↑][↓↑][↓↑][↓↑][↓↑]	0.00	0	0.0

Two important conclusions can be drawn from these worked examples:

1. When an electronic shell is completely filled, all electrons are paired, their magnetic moments cancel and, consequently, their net magnetic moment vanishes. Hence in dealing with magnetism, only partially filled orbitals need to be considered. Said otherwise, the *existence of unpaired electrons is a necessary condition for magnetism to exist.*
2. The fact that the calculated magnetic moment assuming *only* spin orbital momentum for the isolated cations of the *3d* transition series compares favorably with the experimentally determined values (see Table 15.2) implies that the orbital angular momentum for these ions is indeed quenched.

15.4 PARA-, FERRO-, ANTIFERRO-, AND FERRIMAGNETISM

As noted previously, the vast majority of ceramics are diamagnetic with negative susceptibilities of the order of $\approx 10^{-6}$. The effect is thus small and of little practical significance, and will not be discussed further. Instead, in the following sections, the emphasis will be on by far the more useful, and technologically important, classes of magnetic ceramics, namely, the ferro- and ferrimagnetic ones. Before considering these, however, it is important to understand the behavior of another class of materials, namely, paramagnetic

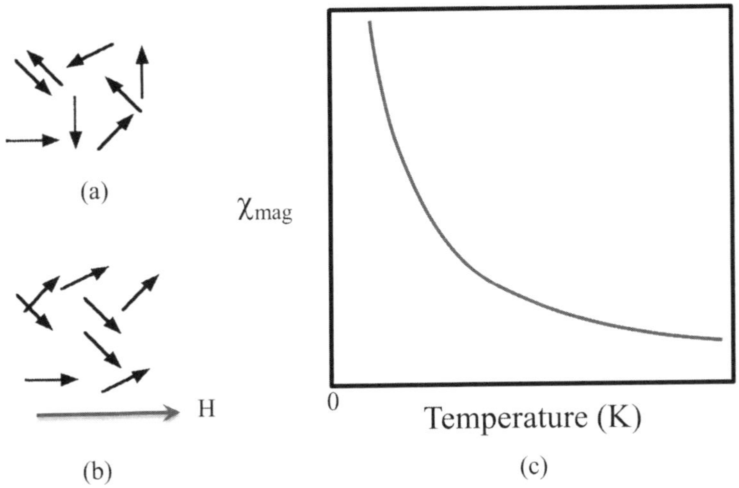

FIGURE 15.3 (*a*) In the absence of an applied magnetic field intensity, the magnetic moments of the individual atoms are pointing in random directions, resulting is a net magnetic moment of zero for the solid. (*b*) The application of a magnetizing field tends to align the moments in the direction of the field, resulting in a net moment. (*c*) Variation of χ_{mag} with temperature for a paramagnetic solid.

materials, not because of their practical importance but because they represent an excellent model system for understanding the other, more complex ones.

15.4.1 PARAMAGNETISM

Paramagnetic solids are those in which the atoms have a permanent magnetic dipole (i.e., unpaired electrons). In the absence of a magnetizing field, the magnetic moments of the electrons are randomly distributed, and the net magnetic moment per unit volume is zero (Fig. 15.3*a*). A magnetizing H-field tends to orient these moments in the direction of the field such that a net magnetic moment in the same sense as H develops (Fig. 15.3*b*). The susceptibilities are thus positive but small, usually in the range of 10^{-3} to 10^{-6}. This tendency for order is naturally counteracted, as always, by thermal motion. It follows that the susceptibility decreases with increasing T, and the relationship between the two is given by

$$\boxed{\chi_{mag} = \frac{C}{T}} \tag{15.28}$$

where C is a constant known as the **Curie constant**. This 1/T dependence, shown schematically in Fig. 15.3*c*, is known as **Curie's law**; the remainder of this section is devoted to understanding the origin of this dependence.

The situation is almost identical to that worked out in Chap. 14 for dipolar polarization, and the problem can be tackled by following a derivation nearly identical to the derivation of Eq. (14.51). To simplify the problem, it is assumed here that the ions in the solid possess a total magnetic moment μ_{ion} given by Eq. (15.27) (i.e., it is assumed that the orbital angular momentum is quenched). Furthermore, it is assumed that these moments can align themselves either parallel or antiparallel to the magnetic B-field. The magnetic energy is thus $\pm\mu_{ion}B$, with the plus sign corresponding to the case where the electrons are aligned against the field and the minus sign when they are aligned with the field. For $\mu_{ion}B/(kT) << 1$, it can be shown that (Prob. 15.2*a*) the net magnetization, **M**, is given by[222]

$$\mathbf{M} = (N_1 - N_2)\mu_{ion} = \frac{N\mu_{ion}^2\mathbf{B}}{kT} \tag{15.29}$$

where N is total number of magnetic atoms or ions per unit volume, that is, $N = N_1 + N_2$. Here N_1 and N_2 represent, respectively, the number of electrons per unit volume aligned with, and against, the **B**-field, respectively. In other words, the net magnetization is proportional to the *net* number of electrons aligned with the field, that is, $N_1 - N_2$. For paramagnetic solids, one can further assume that $\mathbf{B} \approx \mu_o\mathbf{H}$ [Eq. (15.12)] which, when combined with Eqs. (15.29) and (15.6), results in

$$\chi_{mag} = \frac{N\mu_o\mu_{ion}^2}{kT} = \frac{C}{T} \tag{15.30}$$

which is the sought-after result, since it predicts that χ_{mag} should vary as 1/T. It is important, however, to point out that Eq. (15.30) is slightly incorrect because the electrons were assumed to be aligned either with, or opposite to, the **B**-field. In reality, the electron momenta may have any direction in between, and an angular dependence should be accounted for in deriving Eq. (15.30). This problem was tackled by Langevin, with the final result being

$$\boxed{\chi_{mag} = \frac{N\mu_o\mu_{ion}^2}{3kT} = \frac{C}{T}} \tag{15.31}$$

which, but for a factor of 3 in the denominator—which comes about from averaging the moments over all angles—is identical to Eq. (15.30). When analyzing paramagnetic results, it is this equation that should be used.

15.4.2 FERROMAGNETISM

In a certain class of magnetic materials, namely, ferromagnets, the temperature dependence of χ_{mag} does no obey Curie's law, but rather the modified version

$$\boxed{\chi_{mag} = \frac{C}{T - T_c}} \tag{15.32}$$

known as the **Curie–Weiss law,** a plot of which is shown in Fig. 15.4*a*. Above a critical temperature T_C, known as the **Curie temperature,** the material behaves paramagnetically; below T_C *spontaneous* magnetization, M, sets in. Furthermore, the extent of M is a function of temperature and reaches a maximum at absolute zero, as shown in Fig. 15.4*b*.

[222] The assumption that $\mu_{ion}B/(kT) \ll 1$ is for the most part an excellent one. See Prob. 15.3.

Qualitatively, this is explained as follows: At high temperatures, thermal disorder rules and the solid is paramagnetic. As the temperature is lowered, however, a **magnetic interaction energy** comes into play that tends to align the magnetic moments parallel to one another and produce a macroscopic magnetic moment with maximum ordering occurring at 0 K. The temperature at which this ordering appears is called the *Curie temperature,* with high Curie temperatures corresponding to strong interactions and vice versa.[223]

Given that ferromagnetism exists up to a finite temperature above absolute zero and then disappears, one is forced to postulate that in these materials:

- ∞ Some of the spins on the atoms must be unpaired.
- ∞ There is some *interaction* between neighboring electronic spins that tends to align them and keep them aligned even in the *absence* of an external field.
- ∞ This ordering energy, at a sufficiently high temperature, is no longer capable of counteracting the thermal disordering effect, at which point the material loses its ability to spontaneously magnetize and becomes paramagnetic.

Before proceeding much further, it should be emphasized that the nature of this interaction energy, also known as the **exchange energy**, is nonmagnetic and originates from quantum mechanical electrostatic interactions between neighboring atoms. Suffice it to say here that ferromagnetism is caused by a strong internal or local magnetic field aligning the magnetic moments on individual ions.

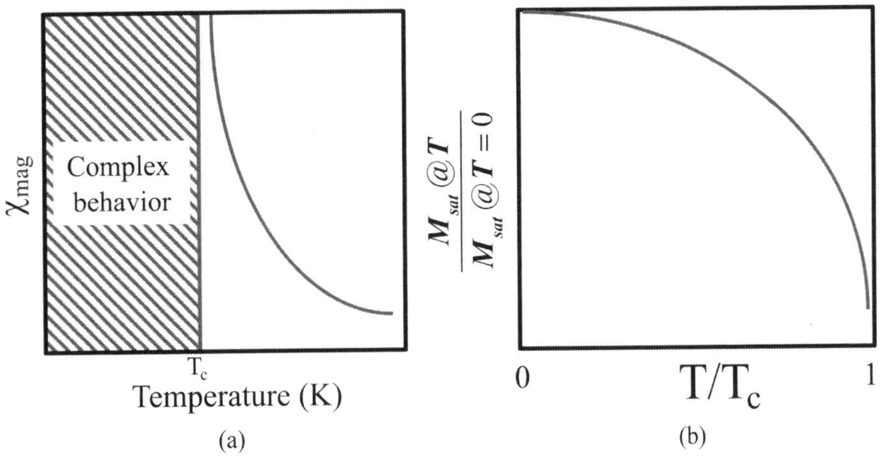

FIGURE 15.4 (*a*) Temperature dependence of χ_{mag} for a ferromagnetic solid that undergoes a transition at the Curie temperature T_C. (*b*) Spontaneous magnetization (H = 0) of a ferromagnetic crystal as a function of temperature.

223 The situation is quite analogous to melting. The stronger the bond between the atoms the higher the melting points. The Curie temperature can be considered to be the temperature at which the magnetic ordering "melts."

To understand the temperature dependence of ferromagnetic materials, one needs to find an expression for the local field B_{loc} that an electron inside a ferromagnetic material placed in a magnetizing field H experiences. This is a nontrivial problem and only the end result, namely,

$$B_{loc} = \mu_o(H + \lambda M) \tag{15.33}$$

is given here. λ is the known as the **mean field constant** or **coupling coefficient** and is a measure of the strength of the interaction between neighboring moments—as noted above, the larger λ, the stronger the interaction.

Replacing B in Eq. (15.29) by B_{loc} and making use of Eq. (15.33) one obtains

$$M = N\mu_{ion}^2 \frac{B_{loc}}{kT} = N\mu_{ion} \frac{\mu_{ion}\mu_o H + \lambda_{loc}\mu_{ion}\mu_o M}{kT} \tag{15.34}$$

By further noting that $M_{sat} = N\mu_{ion}$ (see below) and defining

$$T_c = \frac{\lambda\mu_{ion}\mu_o M_{sat}}{k} \tag{15.35}$$

Eq. (15.34) can be recast as

$$\frac{M}{M_{sat}} = \frac{\mu_{ion}\mu_o H}{kT} + \frac{MT_c}{M_{sat}T} \tag{15.36}$$

which can be further simplified to

$$\frac{M}{M_{sat}} = \frac{\mu_{ion}\mu_o H}{k(T - T_c)} \tag{15.37}$$

Lastly, since by definition $M/H = \chi_{mag}$, one obtains the final sought-after temperature dependence of χ_{mag}, namely, the Curie–Weiss law:

$$\boxed{\chi_{mag} = \frac{\mu_{ion}\mu_o M_{sat}}{3k(T - T_c)} = \frac{C}{(T - T_c)}} \tag{15.38}$$

The factor of 3 in the denominator comes from averaging over all angles. Note that if one neglects the interaction between neighbors, i.e., if λ, and consequently T_C, are assumed to be zero, then Eq. (15.28) is recovered, as one would expect. It follows then that it is the *interaction of neighboring electrons that gives rise to T_C* and ferromagnetism.

Also, note that Eq. (15.38) only applies above T_C. Below T_C the material behaves quite differently in that it will spontaneously (i.e., even when H = 0) magnetize. The behavior below T_C is shown schematically in Fig. 15.4*b*, where M/M_{sat} is plotted versus T/T_C. As T approaches absolute zero, M approaches M_{sat}.

The underlying physics of the discussion so far can be summarized as follows: When the atomic thermal motions are small enough, the coupling between the tiny atomic magnets, viz. unpaired electrons, causes them to all line up parallel to one another, even in the absence of an externally applied field, which gives rise to a permanently magnetized material.

To recap: By invoking that neighboring unpaired electrons interact in such a way as to keep their spins pointing all in the same direction and in the same direction as the applied magnetizing H-field, it is possible to, at least qualitatively, explain the general response of ferromagnetic materials to temperature and magnetizing fields.

15.4.3 ANTIFERROMAGNETISM AND FERRIMAGNETISM

In some materials, the coupling coefficient is *negative,* which implies that the magnetic moments on adjacent ions are antiparallel, as shown schematically in Fig. 15.5*a*. If these moments are equal, they cancel and the net moment is zero—such solids are known as **antiferromagnets**. According to Eq. (15.35), a negative λ gives rise to a negative T_C, and the resulting susceptibility versus temperature curve is shown in Fig. 15.5*b*, where a maximum in susceptibility is observed at a temperature T_N, known as the **Néel temperature**. Above T_N, the material is paramagnetic, the Curie–Weiss law holds again, except that it is modified to read

$$\chi_{mag} = \frac{C}{T + T_N} \tag{15.39}$$

which takes into account the fact that T_C is negative.

Antiferromagnetism has been established in a number of compounds, primarily the fluorides and oxides of Mn, Fe and Co such as MnF_2, MnO, FeF_2, CoF_2, NiO, CoO, FeO, MnS, MnSe and Cr_2O_3. The type of ordering that occurs depends on the crystal structure of the compound in question. For instance, it is now well established that the spins in NiO and MnO are arranged as shown in Fig. 15.5*c*, where the spins in a single (111) plane are parallel, but adjacent (111) planes are antiparallel.

A variation of antiferromagnetism is seen in the situation depicted in Fig. 15.5*d*, where the coupling is negative, but the *adjacent moments are unequal,* which implies that they do not cancel, and a net moment equal to the difference between the two submoments results. Such materials are termed **ferrimagnets** and, for reasons discussed later, include most if not all magnetic ceramics. Interestingly enough, the temperature dependence of their properties is the same as that for ferromagnets.

15.4.4 CRYSTAL FIELD THEORY

Before moving on to describe magnetic domains and their role in magnetism, it is important to outline an important idea that has been quite powerful in accounting for the colors and magnetic properties of many transition-metal, TM, containing ceramics. For the most part, the d-electrons of isolated TM cations behave quite differently when they are in placed a crystal. In the crystal, and depending on symmetry, the d-orbitals of the TM interact directionally with the ligands surrounding it in such a fashion that its energy levels are split. A **ligand** is a negatively charged, nonspherical environment surrounding the centrally located TM ion and is partially covalently bonded with it (open red circles in Fig. 15.6*b*).

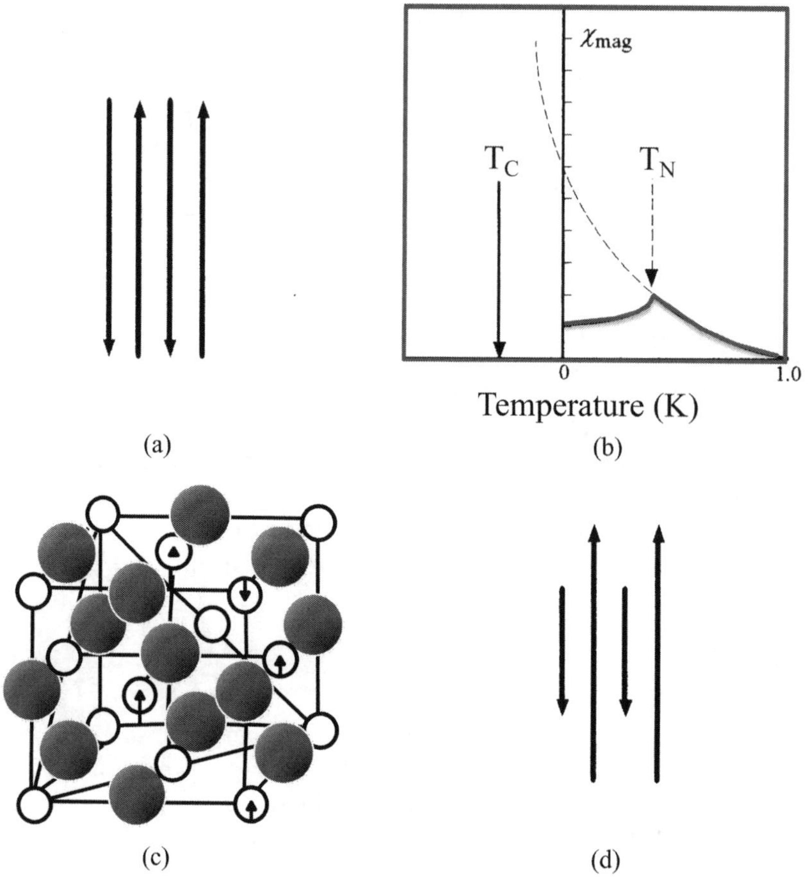

FIGURE 15.5 (*a*) In antiferromagnetic materials, nearest-neighbor moments are aligned antiparallel to one another, and the net moment is zero. (*b*) Temperature dependence of the susceptibility of an antiferromagnetic material. Maximum in χ_{mag} occurs at the Néel temperature T_N. (*c*) Antiparallel magnetic spins in MnO. (*d*) Unequal magnetic moments on adjacent sites give rise to a net magnetic moment. This is termed *ferrimagnetism*.

A TM ion will have five d-orbitals which when placed in a spherical environment are degenerate in energy and in which the electrons can be found with equal probability (Fig. 15.6*a*). However, if that ion is placed in a field where the ligands surround it octahedrally, symmetry dictates that the d_{z^2} and $d_{x^2-y^2}$ (Fig. 15.6*a*) orbitals will be repelled more strongly than either of the d_{xy}, d_{xz} or d_{zx} orbitals (Fig. 15.6*b*). This in turn will result in a splitting of the orbital energies such that the repulsion of the surrounding ligands increases the energy of the $\mathbf{e_g}$ orbitals relative to their $\mathbf{t_{2g}}$ counterparts. This splitting in energy, Δ_o, is called **crystal field splitting.** For historical reasons, when the d_{xy}, d_{xz} or d_{zx} orbitals are in *octahedral* coordination they are sometime referred to collectively as $\mathbf{t_{2g}}$; the d_{z^2} and $d_{x^2-y^2}$, on the other hand, are referred to as $\mathbf{e_g}$.

If the surrounding ligands are tetrahedrally coordinated (see Fig. 15.6*c and d*), the opposite occurs—the d_{xy}, d_{xz} and d_{zx} orbitals are now the ones that are repelled more strongly by the crystal, or ligand, field and

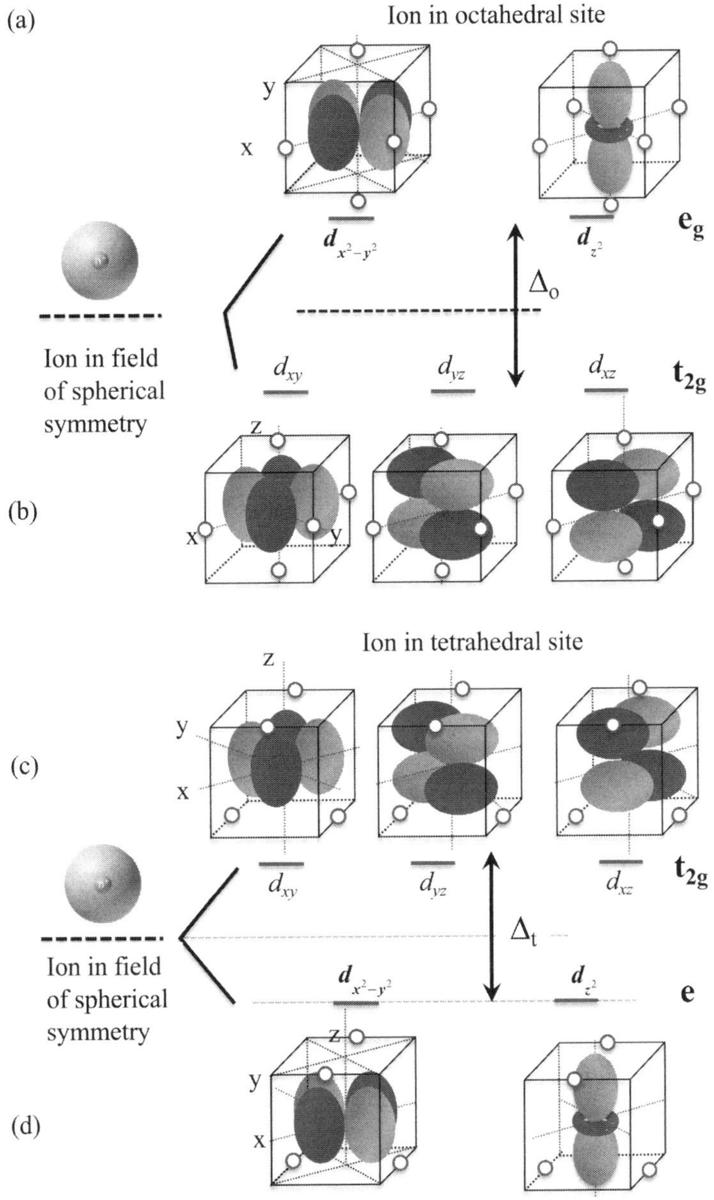

FIGURE 15.6 Interaction of octahedral and tetrahedral ligands with d-orbitals. (*a*) If the symmetry of the ligands is spherical all d-orbitals have the same energy. (*b*) In an octahedral field with six ligands, the d_{z^2} and $d_{x^2-y^2}$ orbitals are higher in energy (top) and the d_{xy}, d_{xz} and d_{zx} orbitals are lower (bottom). The difference in energy is Δ_O. (*c*) In a tetrahedral field with four ligands, the d_{z^2} and $d_{x^2-y^2}$ orbitals are lower in energy (bottom) and the d_{xy}, d_{xz} and d_{zx} orbitals are higher (top). The difference in energy is Δ_t.

consequently have the *higher* energies (Fig. 15.6c). The other two orbitals have the *lower* energy levels (Fig. 15.6*d*.) Here again there is a crystal field energy splitting. This one is labeled Δ_t to emphasize the tetrahedral symmetry. Again, for historical reasons, when the d_{xy}, d_{xz} or d_{zx} orbitals are in *tetrahedral* coordination they are sometimes referred as $\mathbf{t_2}$ and the d_{z^2} and $d_{x^2-y^2}$ orbitals are referred to simply as **e**.

The magnitudes of Δ_O and Δ_t depend on (i) the extent of overlap between the oxygen p-orbitals and the TM d-orbitals, (ii) the charges on the TM cation, and (iii) whether the metal is 3d, 4d or 5d. In all cases, $\Delta_O > \Delta_T$.

Knowing that these orbitals exist, the next step is to determine their occupancy. Hund developed a few rules governing under what conditions electrons will, or will not, pair up. And while the rules can be somewhat complicated to follow, the end result is that electrons will only pair up under great duress. In this context, larger values of Δ_o would constitute such a situation. Said otherwise, electrons, for the most part, prefer to be unpaired.

In metals, adjacent moments interact directly with each other to yield FM or AF order. In magnetic oxides, on the other hand, the moments of the TM ions interact ***indirectly*** via the oxygen ions between them. Two examples are shown in Figs. 15.9*a* and *b*. This interaction is called **superexchange** and is discussed below.

15.5 MAGNETIC DOMAINS AND HYSTERESIS CURVES

15.5.1 MAGNETIC DOMAINS

In the foregoing discussion of ferromagnetism, it was concluded that the material behaves parmagnetically above some temperature T_C, with M being proportional to H (or B), but that below a T_C spontaneous magnetization occurs. However, in the experiment described at the beginning of this chapter, it was explicitly stated that a virgin slab of magnetic material had a net magnetization of zero. At face value, these two statements appear to contradict each other. The way out of this apparent dilemma is to appreciate that spontaneous magnetization occurs only within small regions ($\approx 10^{-5}$ m) within a solid. These are called **magnetic domains,** defined as regions where all the spins are pointing in the same direction. As discussed in greater detail below, these domains form in order to reduce the overall energy of the system and are separated from one another by **domain** or **Bloch walls,** which are high-energy areas,[224] defined as a transition layer that separate adjacent regions magnetized in different directions (Fig. 15.8*d*). The presence of domain walls and their mobility, both reversibly and irreversibly, are directly responsible for the B–H hysteresis loops discussed below.

The reason magnetic domains form is best understood by referring to Fig. 15.7*a*–*c*. The single domain configuration (Fig. 15.7*a*) is a high-energy configuration because the magnetic field has to exit the crystal and close back on itself. By forming domains that close on themselves, as shown in Fig. 15.7*b* and *c*, the net macroscopic field is zero and the system has a lower energy. However, this reduction in energy is partially

[224] The situation is not unlike grain boundaries in a polycrystalline material, with the important distinction that whereas a polycrystalline solid will always attempt to eliminate these areas of excess energy, in a magnetic material an equilibrium is established.

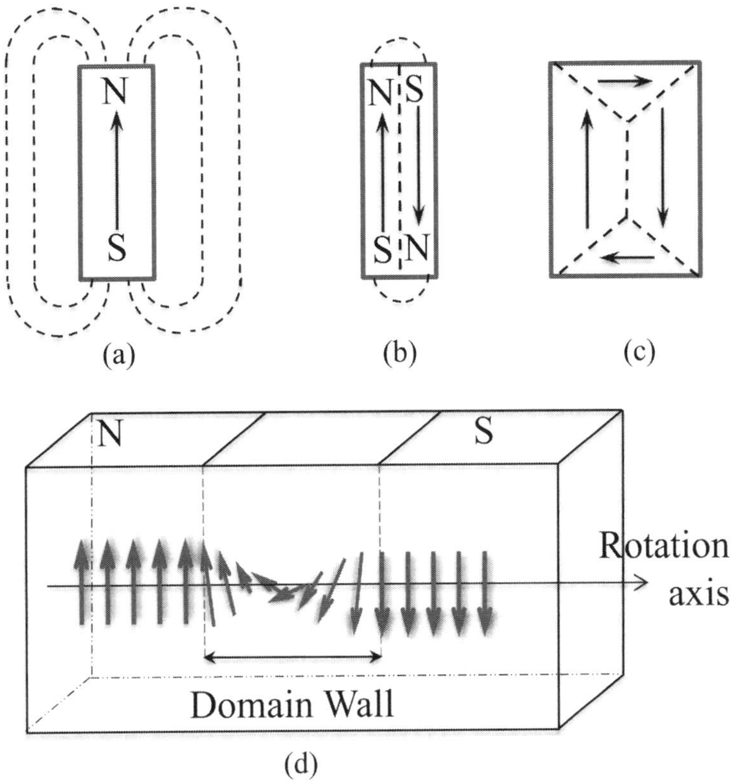

FIGURE 15.7 Schematic showing how formation of domains lowers the energy of the system. (*a*) No domains. (*b*) Two domains separated by a 180° wall. (*c*) The 90° domains are called *closure domains* because they result in the flux lines being completely enclosed within the solid. Closure domains are much more common in cubic crystals than in hexagonal ones because of the isotropy of the former. (*d*) Alignment of individual magnetic dipoles within a 180° wall.

offset by the creation of domain walls. For instance, the structure of a 180° domain wall is shown schematically in Fig. 15.7*d*. Some of the energy is also offset by the **anisotropy energy**, which is connected with the energy difference that arises when the crystal is magnetized in different directions. As noted below, the energy to magnetize a solid is a function of crystallographic direction—there are "easy" and "difficult" directions.

15.5.2 HYSTERESIS LOOPS

The relationship between the existence of domains and hysteresis is shown in Fig. 15.8. The dependence of M on H for a virgin sample is shown in Fig. 15.8*a*. What occurs at the microscopic level is depicted in Fig. 15.8*b*–*e*. Initially the virgin sample is not magnetized, because the moments of the various domains cancel (Fig. 15.8*b*). The change in M very near the origin represents magnetization by reversible Bloch wall

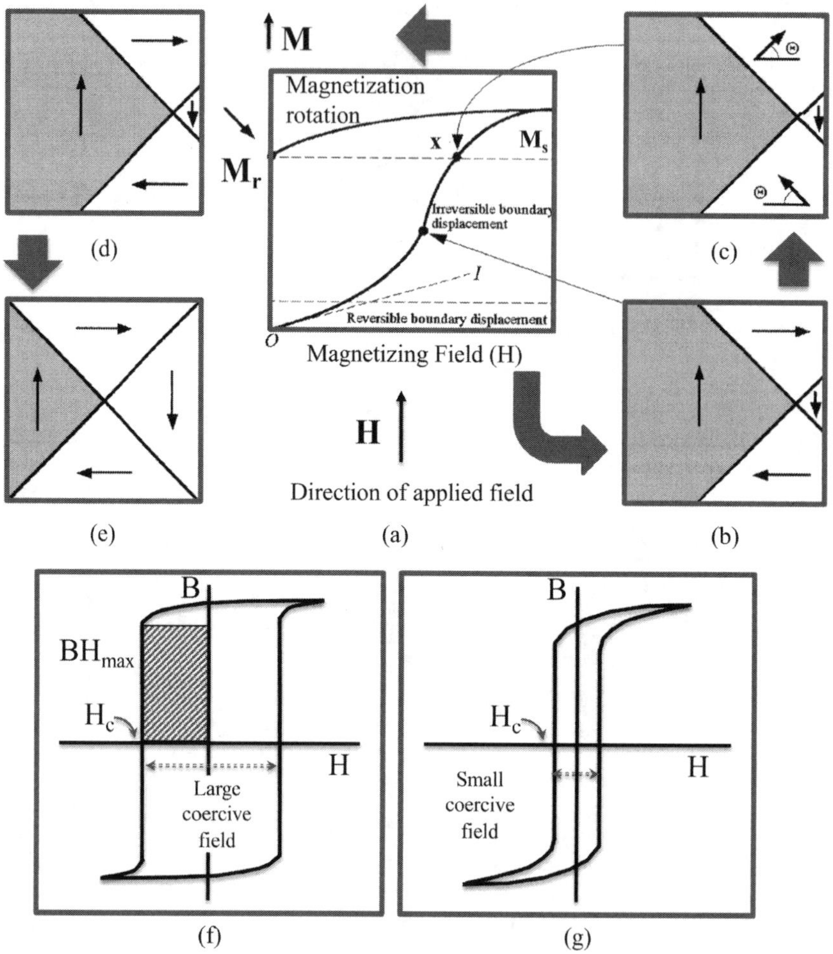

FIGURE 15.8 Relationship between domains and hysteresis. (*a*) Typical hysteresis loop for a ferromagnet. (*b*) For a virgin sample, H = 0 and M = 0 due to closure domains. (*c*) With increasing H, the shaded domain which was favorably oriented to H grows by the irreversible movement of domain walls up to point X. (*d*) Beyond point X, magnetization occurs only by the rotation of the moments. (*e*) Upon removal of the field, the irreversibility of the domain wall movement results in a remnant magnetization, i.e., the solid is now a permanent magnet. Hysteresis loops for (*f*) hard and (*g*) soft magnets. Note high H_c values for hard magnets.

displacements,[225] and the tangent OI to this initial permeability is called the **initial relative permeability** μ_i. As H is increased, the domains in which the moments are favorably oriented with the applied

[225] At low fields, the movement of the domain walls is very much like an elastic band (or a pinned dislocation line) that stretches reversibly. If the field is removed at that point, the "unloading" curve is coincident with the loading curve and the process is completely reversible.

field grow at the expense of those that are not. (Compare shaded areas in Fig. 15.8*b* and *c*.) This occurs by the *irreversible* movement of the domain walls, up to a certain point (point X in Fig. 15.8*a*), beyond which wall movement, more or less, ceases. Further magnetization occurs by the rotation of the moments within the domains that are not aligned with the field, as shown in Fig. 15.8*d*. At very high H, all the domains will have rotated, and magnetization is said to have *saturated* at M_s. This **saturation magnetization** is simply the product of the magnetic moment on each ion μ_{ion} and the total number of *magnetic* atoms N per unit volume, or

$$M_{sat} = N\,\mu_{ion} \tag{15.40}$$

Upon the removal of H, M does not follow the original curve, but instead follows the line $M_s - M_r$ and intersects the y-axis at the point labeled M_r, which is known as the **remnant magnetization**. Note that upon removal of the magnetizing field, the size of the domains and their orientation do not change. The difference between M_s and M_r simply reflects the recovery of the rotational component of the domains.

In order to completely rid the material of M_r, the polarity of M has to be reversed. The value of H at which M goes to zero is called the **coercive magnetic field**, H_c (Fig. 15.8*f*).

Based on the shape of their hysteresis loops, magnetic materials have been classified as either soft or hard. The B–H hysteresis loops for each are compared in Fig. 15.8*f* and *g*. Broadly speaking, **soft magnetic** materials (Fig. 15.8*g*) have coercive fields below about 1 kA/m, whereas **hard magnetic** materials (Fig. 15.8*f*) have H_c above about 10 kA/m. In addition to this classification, the shape of the hysteresis curve is also used to estimate the magnetic energy stored per unit volume in a permanent magnet. This value is given by the product $(BH)_{max}$ shown as the hatched area in Fig. 15.8*f*.

This difference in behavior between the two types of materials is related, once again, to the presence of domains and the ease, or difficulty, with which they can be induced to migrate and/or demagnetize. In the discussion up to this point, **M** was treated as if it were a unique function of **H**, but the actual situation is more complicated; **M** depends on the relative orientation of the various crystallographic planes to the direction of **H**. In other words, it exhibits **orientation anisotropy**. Also, **M** depends on the shape of the crystal being magnetized; i.e., it exhibits **shape anisotropy**. This shape factor is quite important; for example, it is much easier to magnetize a thin, long needle if its long axis is aligned parallel to **H** than if it is perpendicular to it.

Orientational anisotropy is related to **magnetostriction.** When a material is magnetized, it changes shape slightly, which in turn introduces elastic strains in the material. And since elastic properties are tensors, it follows that the penalty for magnetizing crystals in different crystallographic directions is not equivalent. Some crystallographic directions are easier to magnetize than others. One measure of this effect is the **magnetostriction constant** λ_m, defined as the strain induced by a saturating magnetic field. λ_m is positive if the field causes an increase in dimension in the field direction. Table 15.3 lists some λ_m values for a number of polycrystalline ferrites. The values listed represent the average of single-crystal values.

The details of these intriguing phenomena are beyond the scope of this book, but the interested reader is referred to Additional Reading for references. What is important here is to appreciate how these phenomena can, and have, been exploited to increase magnetic energy density by an order of magnitude roughly for every decade since the turn of the last century!

TABLE 15.3 Saturation magnetostriction constants of some polycrystalline ferrites

Composition	λ_m
Fe_3O_4	$+40\times10^{-6}$
$MnFe_2O_4$	-5×10^{-6}
$CoFe_2O_4$	-110×10^{-6}
$NiFe_2O_4$	-26×10^{-6}
$MgFe_2O_4$	-6×10^{-6}

15.6 MAGNETIC CERAMICS AND THEIR APPLICATIONS

As noted above, soft magnetic materials are characterized by large saturation magnetizations at low H values and low coercive fields (Fig. 15.8f) and are typically used in applications where rapid reversal of the magnetization is required, such as electromagnets, transformer cores and relays. The major advantage of soft magnetic ceramics, compared to their metal counterparts, is the fact that they are electrical insulators. This property is fundamental in keeping eddy current losses low and is one of the main reasons why the major applications of magnetic ceramics has been in areas where such losses have to be minimized.

Hard magnets, however, are characterized by high saturation magnetization, as well as high coercive forces, i.e., they are not easily demagnetized (Fig. 15.8e). Hard magnetic solids are thus used for example, to make permanent magnets and recording media among other applications.

Magnetic ceramics are further classified according to their crystal structures into spinels, garnets and hexagonal ferrites. Typical compositions and some of their magnetic properties are listed in Table 15.4.

15.6.1 SPINELS OR CUBIC FERRITES

Spinels were first encountered in Chap. 3 (Fig. 3.10), and their structure was described as having an oxygen ion sublattice arranged in a cubic close-packed arrangement with the cations occupying various combinations of the octahedral and tetrahedral sites. The cubic unit cell is large, comprising 8 formula units and containing 32 octahedral, or O, sites and 64 tetrahedral or T-sites (see Fig. 3.10). In normal spinels, for which the general chemical formula is $A^{2+}B^{3+}O_4$ (or equivalently $AO{\cdot}B_2O_3$), the divalent cations A are located on T-sites and the trivalent cations B on the O-sites. In inverse spinels, the A cations and one-half of the B cations occupy the O-sites, with the remaining B cations occupying the T-sites.[226]

As noted in Chap. 6, the spinel structure is quite amenable to large substitutional possibilities that, in turn, has led to considerable technological exploitation of ferrites. The simplest magnetic oxide, magnetite,[227] or Fe_3O_4, is a naturally occurring ferrite that has been used for hundreds of years as a lodestone for navigational purposes. There are quite a number of other possible compositions with the general formula $MeO{\cdot}Fe_2O_3$, some of which are listed in Table 15.4. The Me ion represents divalent ions such as Mn^{2+}, Co^{2+}, Ni^{2+} and Cu^{2+}

[226] It is the crystal field energy or ligand field splitting (Fig. 15.6) that stabilizes the inverse spinel.

[227] Its structural relationship to spinel becomes apparent when its formula is rewritten as $FeO \cdot Fe_2O_3$.

TABLE 15.4 Magnetic properties of a number of magnetic ceramics. Magnetic moments are given in μ_B per formula unit at 0 K

Material	Curie T (K)	B_{sat} (T) @ RT	Calculated moments[a]			Experimental
			T site	O site	Net	
Fe[b]	1043	2.14			2.14	2.22
Spinel ferrites [AO B_2O_3]						
$Zn^{2+}[Fe^{3+}Fe^{3+}]O_4$			0	5−5	0	Antiferro.
$Fe^{3+}[Cu^{2+}Fe^{3+}]O_4$	728	0.20	−5[c]	1+5	1	1.30
$Fe^{3+}[Ni^{2+}Fe^{3+}]O_4$	858	0.34	−5[c]	2+5	2	2.40
$Fe^{3+}[Co^{2+}Fe^{3+}]O_4$	1020	0.50	−5[c]	3+5	3	3.70–3.90
$Fe^{3+}[Fe^{2+}Fe^{3+}]O_4$	858	0.60	−5[c]	4+5	4	4.10
$Fe^{3+}[Mn^{2+}Fe^{3+}]O_4$	573	0.51	−5[c]	5+5	5	4.60–5.0
$Fe^{3+}[Li_{0.5}Fe_{1.5}]O_4$	943		−5[c]	5+2.5	2.5	2.60
$Mg_{0.1}Fe_{0.9}[Mg_{0.9}Fe_{1.1}]O_4$	713	0.14	0–4.5	0+5.5	1	1.10
Hexagonal ferrites						
$BaO{\cdot}6Fe_2O_3$	723	0.48				1.10
$SrO{\cdot}6Fe_2O_3$	723	0.48				1.10
$Y_2O_3{\cdot}5Fe_2O_3$	560	0.16				5.00
$BaO{\cdot}9Fe_2O_3$	718	0.65				
Garnets						
YIG$\{Y_3\}[Fe_2]Fe_3O_{12}$	560	0.16			5	4.96
$\{Gd_3\}[Fe_2]Fe_3O_{12}$	560				16	15.20
Binary oxides						
EuO	69					6.80
CrO_2	386	0.49				2.00

[a] For the sake of simplicity, the moments were calculated by using the classical expression [Eq. (15.21)] rather than the more accurate quantum mechanical result given in Eq. (15.27).
[b] Fe is included for comparison purposes.
[c] The minus sign denotes antiferromagnetic coupling.

or a combination of ions with an average valence of + 2. In general, the divalent ions prefer the O-sites, and thus most ferrites form inverse spinels. However, Zn and Cd ions prefer the T-sites, forming normal spinels.

In spinels the interaction between the A and B sublattices is *almost* always antiferromagnetic[228] (i.e., they have opposite spins). The spin ordering is of the **superexchange** type because it occurs via the agency intermediary nonmagnetic ions (e.g. O^{2-}). The ordering follows some basic rules named after the names of the scientists who developed, viz. Goodenough–Kanamori–Anderson or GKA rules. There are several rules we consider two.

a) When the lobes are d_{z^2} orbitals in the octahedral case, in the 180° position in which these lobes point directly toward a ligand and each other (Fig. 15.9*a*), particularly large AFM superexchanges are obtained.
b) If the d-orbitals are arranged as shown in Fig. 15.9*b*, the interaction is FM.

[228] There are exceptions, however. (See Worked Example 15.4c.)

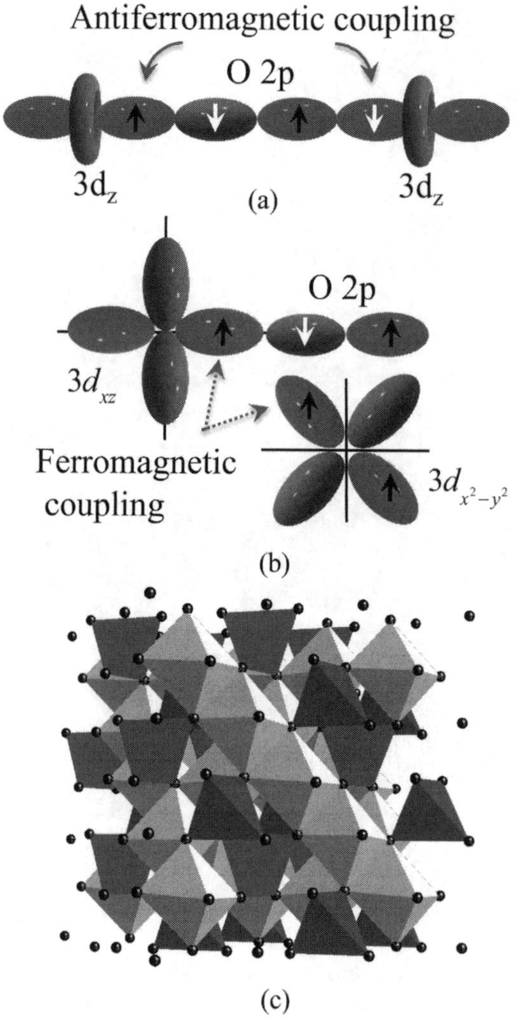

FIGURE 15.9 Superexchange configurations of d-electrons with O 2p, (*a*) d_{z^2}, (*b*) $d_{x^2-y^2}$ and t_{2g}.

Figure 15.9c shows a schematic of the spinel unit cell with special emphasis on the nature of the bonds between the atoms in the tetrahedral (red) and octahedral (gray) sites. It is left as an exercise to the reader to convince themselves that the arrangements of the these cations viz-a-viz the oxygen atoms will indeed lead to AFM coupling between the A and B sublattices.

Worked Example 15.4

(*a*) Calculate the net magnetic moment[229] of the inverse spinel Fe_3O_4. Also calculate M_{sat} and B_{sat}. Information you may find useful. lattice parameter of unit cell, comprised of 32 oxygen ions, is 837 pm.

(*b*) The addition of nonmagnetic ZnO to a spinel ferrite, such as Ni ferrite, leads to an *increase* in the saturation magnetization. Explain.

(*c*) Repeat part (*a*) for the normal spinel $ZnO{\cdot}Fe_2O_3$. Note that in this spinel the octahedral, or O-sites, are antiferromagnetically coupled.

ANSWER

(*a*) Fe_3O_4 can be written as $FeO{\cdot}Fe_2O_3$ or $Fe^{3+}[Fe^{2+}Fe^{3+}]O_4$. Because it is an inverse spinel, one-half the Fe^{3+} cations occupy the T-sites; the other half the O-sites. These cations interact *antiferromagnetically*, which implies that their net moment is zero. The Fe^{2+} cations occupy the remaining O-sites and their net magnetic moment is (see Table 15.2) 4.9 μ_B. The calculated net moment is thus 4.9 μ_B, which is in reasonably good agreement with the measured value of 4.1 (Table 15.4). Note that this agreement implies that, for the most part, the orbital angular momenta of the ions are indeed quenched.

Since each unit cell contains eight Fe^{2+} ions (see Fig. 15.9), the saturation magnetization is given by Eq. (15.40), or

$$M_S = \frac{8 \times 4.9 \times 9.274 \times 10^{-24}}{\left(8.37 \times 10^{-10}\right)^3} = 6.2 \times 10^5 \text{ A/m}$$

It follows that the saturation magnetic field is given by

$$B_{sat} = \mu_0 M_s = 4\pi \times 10^{-7} \times 6.2 \times 10^5 = 0.78 \text{ T}$$

which compares favorably with the measured value of 0.6 T (see Table 15.4).

Interestingly, and for reasons that are not entirely clear, even better agreement between the measured and theoretical values is obtained if the classical expression for μ_{ion}, that is, $\mu_{ion} = 2\mu_B S$, is used instead of the more exact expression, viz. Eq. (15.27).

(*b*) According to Table 5.4, the saturation magnetization of $NiO{\cdot}Fe_2O_3$ is $2\mu_B$. The substitution of Ni by Zn, which prefers the T-sites, results in an occupancy given by

$$\left(Fe^{3+}_{1-d}Zn^{2+}_{d}\right)\left(Fe^{3+}_{1+d}Ni^{2+}_{1-d}\right)O_4$$

[229] In a typical spinel, solid solutions, quenching and redox equilibria can very rapidly complicate the simple analysis presented here. Additional Reading contains a number of references that the interested reader can consult.

which results in diminishing the number of magnetic moments on the T-sites (first set of parentheses) and increasing the number on O-sites, resulting in a higher magnetization. Furthermore, as the occupancy of the A sites by magnetic ions decreases, the antiparallel coupling between the A and B sites is reduced, which lowers the Curie temperature.

(c) Being a normal spinel implies that the Zn^{2+} ions occupy the tetrahedral A sites and the Fe^{3+} ions occupy the O-sites. The Zn ions are diamagnetic and do not contribute to the magnetic moment (Table 15.2). Given that the Fe^{3+} cations on the O-sites couple antiferromagnetically, their moments cancel and the net magnetization is zero, as observed.

In short, O-T coupling is good for magnetic properties but O-O is bad.

In commercial and polycrystalline ferrites processing variables and resultant microstructures have important consequences on measured properties. Only a few will be mentioned here. For example, the addition of a few percent of cobalt to Ni ferrite can increase its resistivity by several orders of magnitude by ensuring that the iron is maintained in the Fe^{3+} state. Similarly, it is important to sinter MnZn ferrites under reducing atmospheres to ensure that the manganese is maintained in the Mn^{2+} state but not too reducing so as to convert the Fe^{3+} to Fe^{2+}.

Changes in the microstructure in the form of additional inclusions such as second-phase particles, or pores, introduce pinning sites that impede domain wall motion and thus lead to increased coercivity and hysteresis loss. Conversely, for high-permeability ceramics, very mobile domain wall motion is required. It is now well established that one of the most significant microstructural factors that influence domain motion, and hence the shape of the hysteresis curves in magnetic ceramics, are grain boundaries. For example, increasing the average grain size of some Mn–Zn ferrites from 10 to 30 μm increases μ_i from 10^4 to 2.5×10^4.

Applications of spinel ferrites can be divided into three groups: (i) low-frequency, high-permeability applications, (ii) high-frequency, low-loss applications and (iii) microwave applications. It is important to note that the properties of magnetic materials are as much a function of frequency as dielectric materials. The ideas of resonance and loss and their frequency dependencies, which were discussed in detail in Chap. 14, also apply to magnetic materials. The coupling in this case is between the applied magnetic field frequency and the response of the magnetic moment vectors. And while these topics are beyond the scope of this book, it is important to be cognizant of them when choosing magnetic materials for various applications.[230]

15.6.2 GARNETS

The general formula of magnetic garnets,[231] is $P_3Q_2R_3O_{12}$ or $3RE_2O_3 \cdot 5Fe_2O_3$, where RE is typically yttrium but can also be other rare earth, RE, ions. The basic crystal structure is cubic with an octahedron, a

230 For more information see, e.g., A. J. Moulson and J. M. Herbert, *Electroceramics*, Chapman & Hall, London, 1990.

231 Magnetic garnets are isostructural with the semiprecious garnet mineral $Ca_3Al_2(SiO_4)_3$. In natural garnets, which are nonmagnetic, the R ions are always Si^{4+}, the divalent cations such as Ca or Mn are the P cations, whereas the trivalent cations such as Al^{3+} or Fe^{3+} are the Q cations.

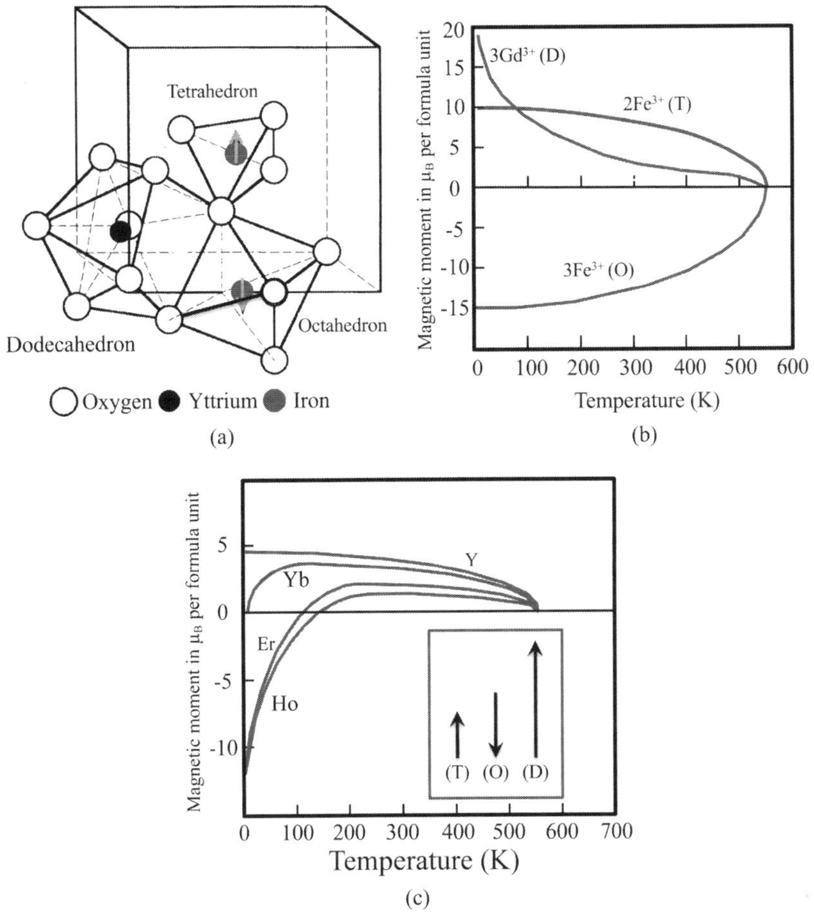

FIGURE 15.10 (*a*) Structural units and positions of cations in a, c and d sites in magnetic garnets. The oxygen ions are shared by four polyhedra—an octahedron, a tetrahedron, and two (one not shown) dodecahedra. (*b*) Magnetization of the sublattices in Gd-iron garnet or GdIG as a function of temperature. Because of weak coupling between the Gd and the Fe, its magnetization drops more rapidly with temperature. Note compensation point at ≈290 K. (*c*) Resulting magnetization versus temperature curves for some Fe garnets.

tetrahedron and two dodecahedra (a distorted or skewed cube) as building blocks arranged as shown in Fig. 15.10*a* (here only one dodecahedron is shown for clarity's sake). The Q and R cations occupy the O- and T-sites, respectively; the dodecahedra are occupied by the P or RE cations. Each oxygen lies at the vertex that is common to four polyhedra, i.e., one tetrahedron, one octahedron and two dodecahedra.

The most investigated, and probably the most important magnetic garnet, is the yttrium–iron garnet, $\{Y_3\}[Fe_2]Fe_3O_{12}$ or $3Y_2O_3{\cdot}5Fe_2O_3$, commonly referred to as YIG. In YIG, the Y^{3+} cations occupy the dodecahedral sites and because of their closed-shell configuration are diamagnetic. The Fe^{3+}

cations—shown in red in Fig. 15.10*a*—are distributed on the O- and T-sites, and the net magnetization is due to the difference between their respective moments. Given that there are 3 Fe^{3+} ions on the O-sites for every 2 Fe^{3+} ions on the T-sites, the net magnetic moment per formula unit (at T = 0 K) is $3 \times 5.92 - 2 \times 5.92 = 5.9\ \mu_B$, which is in reasonable agreement with the measured value of 4.96 μ_B. Here once again that if the classical μ_B value, 5 μ_B, is used, even better agreement between theory and experiment is obtained.

The situation becomes more complicated when magnetic RE ions are substituted for the Y, as shown in Fig. 15.10*b*. For these so-called RE garnets, the RE^{3+} ions are trivalent ions that occupy the dodecahedral sites. The magnetization of these ions is opposite to the *net* magnetization of the ferric ions on the O- **plus** T-sites (see inset in Fig. 15.10*c*). At low T's, the net moment of the RE ions can dominate the moment of the Fe^{3+} ions (Fig. 15.10*b*). But because of coupling between the dodecahedral sites with the T- and O-sites, the RE lattice loses its magnetization rapidly with increasing temperature (Fig. 15.10*b*). The total moment can thus pass through zero, switch polarity and increase once again as the Fe^{3+} moment starts to dominate. Typical M vs. T curves for various iron garnets are shown in Fig. 15.10*c*. The point at which the magnetization goes to zero is known as a **compensation point**. One consequence of having this compensation point is that the magnetization is quite stable with temperature, an important consideration in microwave devices. It is this property that renders Fe garnets unique and quite useful.

15.6.3 HEXAGONAL FERRITES

Here the material is not ferromagnetic in the sense that all adjacent spins are parallel; rather, all the spins in one layer are parallel and lie in the plane of the layer.[232] In the adjacent layer, all the spins are once again parallel within the layer, but pointing in a different direction from the first layer, etc. Commercially the most important hexagonal ferrite is $BaO{\cdot}6Fe_2O_3$, which is isostructural with a mineral known as magnetoplumbite and is the reason hexagonal ferrites are sometimes called **magnetoferrites**.

The crystallographic structure of hard ferrites is such that the magnetically preferred (i.e., easy) orientation is the c-axis, i.e., perpendicular to the basal plane. Consequently, hexagonal ferrites have been further classified as being either isotropic or anisotropic depending on whether the grains are arranged randomly or aligned. The latter is attained by compacting the powder in a magnetic field. The effect of aligning the grains on the *B–H* loop is shown in Fig. 15.11, from which it is obvious that the energy product is significantly improved by the particle orienting during the fabrication process. Other important microstructural factors include particle size and shape and the volume fraction of the ferrite phase. Typical values for the latter are 0.9 for sintered materials and 0.6 for plastic-bonded materials.

One of the major attributes of hexagonal ferrites is their very high crystal anisotropy constants and, consequently, they are used to fabricate hard magnets with high coercive fields. Typical hexagonal ferrites have $(BH)_{max}$ values in the range of 8 to 27 kJ/m^3, quite a bit lower than those of good metallic permanent magnets ($\approx 80\ kJ/m^3$). Despite this disadvantage, their low conductivity (10^{-18} S/m), together with high coercive forces (0.2 to 0.4 T), low density, ease of manufacturing, availability of raw materials and especially low

[232] For more details, see L. L Hench and J. K. West, *Principles of Electronic Ceramics*, Wiley, New York, 1990, p. 321.

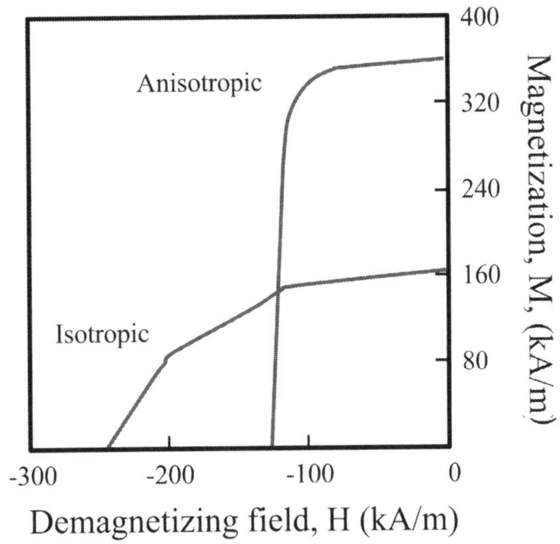

FIGURE 15.11 Demagnetization curves for oriented (top) and isotropic or random (bottom) hexagonal ferrites. (Adapted from F. Esper, in *High Tech Ceramics*, G. Kostorz, ed., Academic Press, London, 1989.)

cost per unit of available magnetic energy renders them one of the most important permanent magnetic materials available. They are mainly used in applications where large demagnetizing fields are present such as flat loudspeakers and compact dc motors. They are also used, for example, to produce "plastic" magnets, in which the magnetic particles are embedded in a polymer matrix.

15.7 PIEZO- AND FERROELECTRIC CERAMICS

The solids discussed in the remainder of this chapter have one thing in common: They exhibit various polar effects, such as piezoelectricity, pyroelectricity and ferroelectricity.

Piezoelectric crystals are those that become electrically polarized or undergo a change in polarization when subjected to a stress. The application of a compressive stress results in the flow of charge in one direction in the measuring circuit and in the opposite direction for tensile stresses. Conversely, the application of an electric field, E, stretches or compresses the crystal depending on the orientation of E relative to the polarization in the crystal.

Pyroelectric crystals are ones that are spontaneously polarizable (see below) and in which a change in temperature produces a change in that spontaneous polarization. A limited number of pyroelectric crystals have the additional property that the direction of spontaneous polarization can be reversed by application of an electric field, in which case they are known as ferroelectrics. Thus, a **ferroelectric** is a spontaneously polarized material with reversible polarization. Before proceeding much further it is important to appreciate that not all crystal classes can exhibit a polar effect.

15.7.1 CRYSTALLOGRAPHIC CONSIDERATIONS

Of the 32 crystal classes or point groups (see Chap. 1), 11 are centrosymmetric and thus nonpolar (e.g., Fig. 15.12*a* and *b*). Of the remaining 21 noncentrosymmetric, 20 have one or more polar axes and are piezoelectric, and of these 10 are polar.[233] Crystals in the latter group are called **polar** crystals because they are spontaneously polarizable, with the magnitude of the polarization, P, dependent on temperature. In the polar state, the center of positive charge does **not** coincide with the center of negative charge (Fig. 15.12*c*); i.e., the crystal possesses a **permanent electric dipole**. Each of these 10 classes is pyroelectric with a limited number of them being ferroelectric. It thus follows that all ferroelectric crystals are pyroelectric and all pyroelectric crystals are piezoelectric, but not vice versa. To appreciate the difference between a piezoelectric and a ferroelectric crystal it is instructive to compare Fig. 15.12*c* and *d*. An unstressed piezoelectric (Fig. 15.12*c*) crystal only develops a dipole when stressed (Fig. 15.12*d*) as a consequence of its symmetry. A ferroelectric crystal, on the other hand, possesses a permanent dipole moment even in the unstressed state (Fig. 15.12*e*). The application of a stress only changes the value of P (Fig. 15.12*f*). For example, quartz is piezoelectric but not ferroelectric, whereas $BaTiO_3$ is both. In the remainder of this chapter, ferro- and piezoelectricity are described in some detail.

15.7.2 FERROELECTRIC CERAMICS

Given the definition of ferroelectricity as the *spontaneous* and reversible polarization of a solid, it is not surprising that ferroelectricity and ferromagnetism have a lot in common (Table 15.5). Ferroelectricity usually disappears above a certain critical temperature T_C, above that temperature the crystal is said to be in a paraelectric (in analogy with paramagnetism) state and obeys a Curie–Weiss law. Below T_C, spontaneous polarization occurs in domains. A typical plot of polarization versus electric field will exhibit a hysteresis loop (see Fig. 15.14).

STRUCTURAL ORIGIN OF THE FERROELECTRIC STATE

Commercially, the most important ferroelectric materials are the titania-based ceramics with the perovskite structure such as $BaTiO_3$ and $PbTiO_3$. In contradistinction to magnetism, ferroelectric materials go through a phase transition from a centrosymmetric nonpolar lattice to a noncentrosymmetric polar lattice at T_C. Typically, these perovskites are cubic at elevated temperatures and become tetragonal as the temperature is lowered. The crystallographic changes that occur in $BaTiO_3$ as a function of temperature and the resulting polarization are shown in Fig. 15.13*a*. In the cubic structure, the TiO_6 octahedron has a center of symmetry, and the six Ti-O dipole moments cancel in antiparallel pairs. Below T_C, the position of the Ti ions moves off center, which in turn results in a permanent dipole for the unit cell (see Worked Example 15.5). The resulting changes in the dielectric constant are shown in Fig. 15.13*b*, the most salient feature of which is the sharp increase in k' around the same temperature that the phase transition from cubic to tetragonal phase occurs.

[233] One of the noncentrosymmetric point groups (cubic 432) has symmetry elements, which prevent polar characteristics.

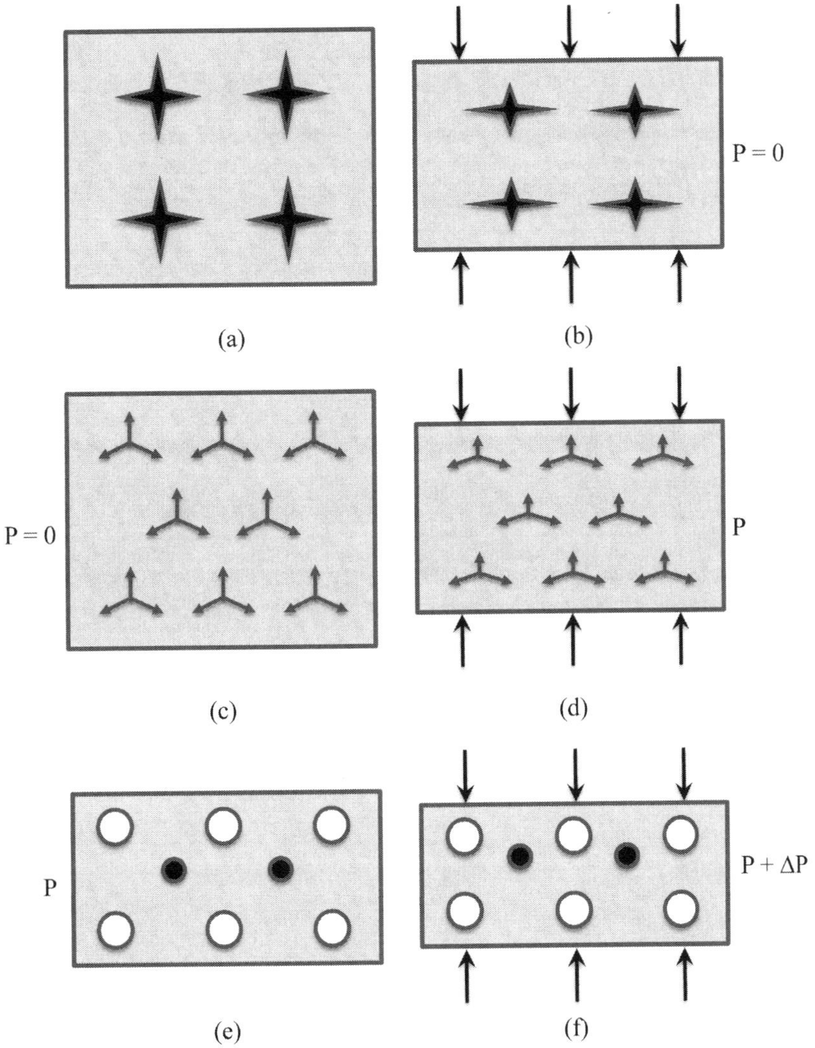

FIGURE 15.12 (*a*) Unstressed centrosymmetric crystal. The arrows represent dipole moments. (*b*) Applying a stress to such a crystal *cannot* result in polarization. (*c*) Unstressed noncentrosymmetric crystal, i.e., piezoelectric. Note that this structure is *not* ferroelectric because it does not possess a permanent dipole. (*d*) Stressed crystal develops a polarization as shown. (*e*) Unstressed *polar* crystal, i.e., a ferroelectric, possesses a permanent dipole even in the unstressed state. (*f*) Stressed ferroelectric crystal. The applied stress changes the polarization by ΔP.

In order to understand the origin of the paraelectric to ferroelectric transition and the accompanying structural phase transitions it is important to understand how the local field is affected by the polarization of the lattice. Equation (14A.3), which relates the local field E_{loc} to the applied field E, and the polarization P can be generalized to read

TABLE 15.5 Comparison of dielectric and magnetic parameters

	Magnetic	Dielectric
General		
Applied field	H (A/m)	E (V/m)
Material response	*M* (A/m)	P (C/m²)
Field equations	$B=\mu_0(H+M)$	$D=\varepsilon_0 E+P$
	$B_{loc}=\mu_0(H+\lambda M)$	$E_{loc}=E+\dfrac{\beta P}{\varepsilon_0}$
Susceptibility	$\mu_r-1=\chi_{mag}=\dfrac{M}{H}$	$k'-1=\chi_{die}=\dfrac{P}{\varepsilon_0 E}$
Energy of moment	$U=-\mu_{ion}\bullet B$	$U=-\mu_{dip}\bullet E$
	Paramagnetism	***Dipolar polarization***
	$M=\dfrac{N\mu_{ion}^2 B}{3kT}$	$P=\dfrac{N\mu_{dip}^2 E}{3kT}$
Curie constant (K)	$C=\dfrac{\mu_0\mu_{ion}^2 N}{3k}$ Eq. (15.31)	$C=\dfrac{N_{dip}\mu_{dip}^2}{3k\varepsilon_0}$ Eq. (14.51)
	Ferromagnetic	***Ferroelectric***
Curie–Weiss law ($T>T_c$)	$\chi_{mag}=\dfrac{C}{T-T_C}$	$\chi_{die}=\dfrac{C}{T-T_C}$
Saturation	$M_s=N\mu_{ion}$	$P_s=N_{dip}\mu_{dip}$

$$\mathbf{E}=\mathbf{E}_0+\beta\frac{\mathbf{P}}{\varepsilon_0} \tag{15.41}$$

where β is a measure of the enhancement of the local electric field.[234] By postulating that the polarizability α varies inversely with temperature, i.e., $N\alpha=C'/T$, and combining Eqs. (15.41), (14.13) and (14.14), one can show that:

$$k'-1=\frac{N\alpha}{\varepsilon_0-\beta N\alpha}=\frac{(T_C/\beta)}{T-T_C} \tag{15.42}$$

where $T_c=\beta C'/\varepsilon_0$. Comparing this expression to the Curie–Weiss law [Eq. (15.32)], it follows that

$$T_c=\beta C' \tag{15.43}$$

[234] In App. 14A, β was found to be 1/3, a value only valid, however, if the material is a linear dielectric with cubic symmetry.

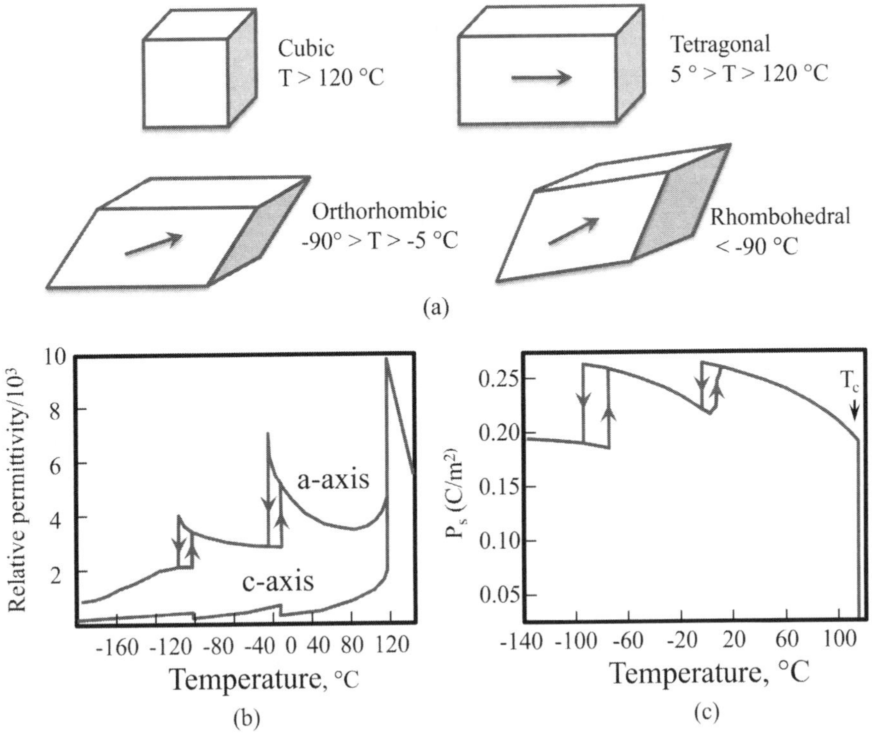

FIGURE 15.13 (*a*) Crystallographic changes in $BaTiO_3$ as a function of temperature. (*b*) Temperature dependence of relative dielectric constant of $BaTiO_3$ single crystal, for the c and a axis. (*c*) Temperature dependence of the saturation polarization for polycrystalline $BaTiO_3$. Note similarity between this figure and Fig. 15.4*b*.

where C′ is the Curie constant for ferroelectric ceramics. A perusal of Table 15.6 shows that for titanate- and niobate-based ceramics, the Curie constant, C′, is of the order of 10^5 K.[235]

Equation (15.42) is important because it predicts that in the absence of a phase transition, the crystal would fly apart as T approached T_C, or equivalently when $\varepsilon_0 = \beta N\alpha$. As discussed in Chap. 2, for every bond there is an attractive component and a repulsive component to the total bond energy. If, as the temperature is lowered, the repulsive component becomes weaker or softer, it follows that the anharmonicity of the bond will increase, which, as seen in Chap. 4, increases the magnitude of the displacements of the ions, which in turn increases the dielectric constant as observed. The anharmonicity of the bond cannot increase indefinitely, however, and at some critical temperature the energy well for the Ti^{4+} ions in the center of the unit cell bifurcates into two sites, as shown in Fig. 14.10*b* As the ions populate one site or the other,

[235] It is instructive once again to compare ferromagnetic and ferroelectric solids. In both cases, an exchange energy or interaction energy is responsible for the spontaneous polarization. For ferromagnetic solids typical values of the exchange coupling coefficient λ are of the order of 350 (see Prob. 15.5). For ferroelectric solids, on the other hand, the interaction, as measured by β, is of the order of 2×10^{-3} (see Prob. 15.13). It follows that the interaction factor is about 5 orders of magnitude larger in ferromagnets than in ferroelectrics.

TABLE 15.6 Summary of dielectric data for a number of ferroelectric ceramics

Material	T_C (°C)	Curie constant (K)	k′ at T_C	P_{sat} (C/m²)
Rochelle salt	24	2.2×10^2	5000	0.25 (RT)
$BaTiO_3$	120	1.7×10^5	1600	0.26 (RT) 0.18@ T_C
$SrTiO_3$ (paraelectric down to ≈ 1 K)		7.8×10^5		
$PbTiO_3$	490	1.1×10^5		0.50 (RT)
$PbZrO_3$ (antiferroelectric)	230		3500	
$LiNbO_3$	1210			0.71–3.00 (RT)
$NaNbO_3$	−200			0.12 @ T_C
$KNbO_3$	434	2.4×10^5	4200	0.26 @ T_C
$LiTaO_3$	630			0.50 @ T_C
$PbTa_2O_5$	260			0.10 @ T_C
$PbGeO_{11}$	178			0.05 @ T_C
$SrTeO_3$	485			0.40@ T_C

the interaction between them ensures that all other ions occupy the same site, giving rise to spontaneous polarization.

Worked Example 15.5

Using the ion positions for the tetragonal $BaTiO_3$ unit cell shown in Fig. 14.16, calculate the electric dipole moment per unit cell and the saturation polarization for $BaTiO_3$. Compare your answer with the observed saturation polarization (that is, $P_s = 0.26$ C/m²).

ANSWER

The first step is to calculate the moment of each ion in the unit cell. Taking the corner Ba ions as reference ions one obtains:

Ion	Q, C	d, m	$\mu = Qd$ (C m)
Ba^{2+} (reference)	+2e	0	0.0
Ti^{4+}	+4e	$+0.006\times10^{-9}$	3.84×10^{-30}
$2O^{2-}$	−4e	-0.006×10^{-9}	3.84×10^{-30}
O^{2-}	−2e	-0.009×10^{-9}	2.88×10^{-30}
		Sum	10.56×10^{-30}

Thus the dipole moment per unit cell is 10.56×10^{-30} C·m.

Since $P_s = \mu/V$, where V is the volume of the unit cell, it follows that the saturation polarization is given by:

$$P_s = 10.56\times10^{-30}/(0.403\times0.398\times0.398)\times10^{-27} = 0.16\,\text{C/m}^2$$

This value is less than the observed value of 0.26 C/m² because it does not take the contribution of electronic polarization of the ions into account.

HYSTERESIS

In addition to resulting in large k′ at T_C, spontaneous polarization will result in hysteresis loops, as shown in Fig. 15.14. At low applied fields, the polarization is reversible and almost linear with the applied field. At higher field strengths, the polarization increases considerably due to switching of the ferroelectric domains.

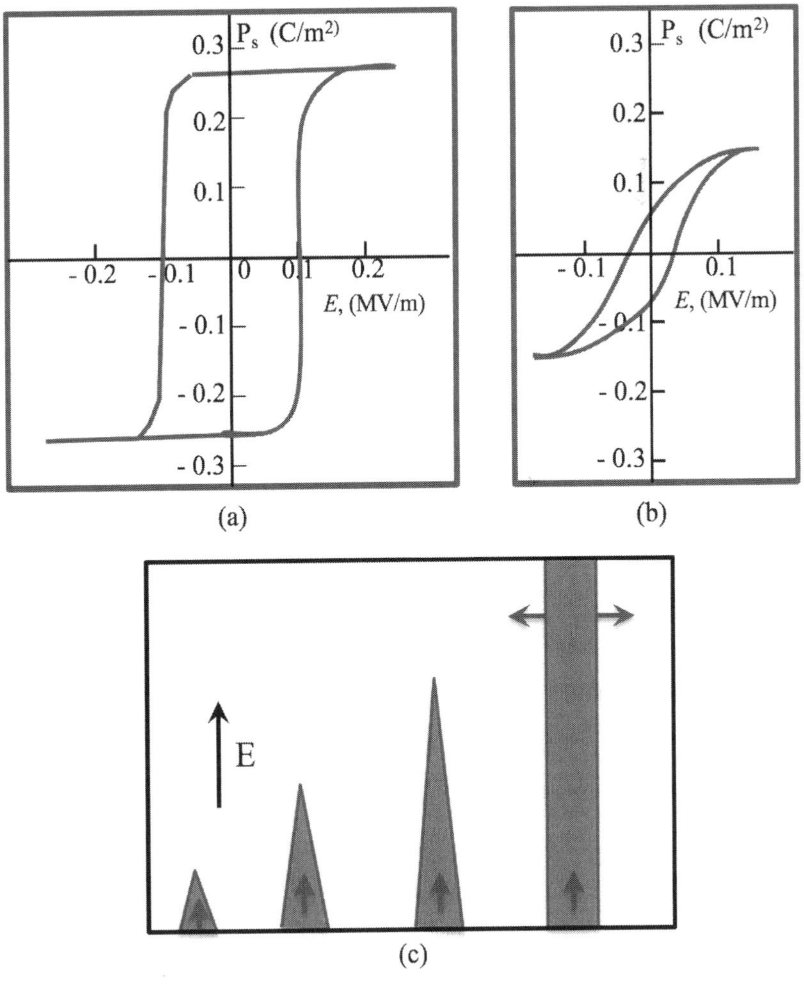

FIGURE 15.14 (*a*) Ferroelectric hysteresis loop for a single-crystal sample. (*b*) Polycrystalline sample. (*c*) Growth of ferroelectric domains favorably oriented to the applied E field.

Further increases in E continue to increase the polarization as a result of further distortions of the TiO_6 octahedra.[236]

Upon removal of the applied E field, P does not go to zero but remains at a finite value called the **remnant polarization P_r**. As in the ferromagnetic case, this remnant is due to the fact that the oriented domains do not return to their random state upon removal of the applied field.[237] In order to do that, the field has to be reversed to a **coercive field E_c**.

The origin of the hysteresis loops is the same as in the ferromagnetic case; viz. upon the application of an electric field, domains whose dipoles are favorably oriented with respect to E grow at the expense of those that are not (Fig. 15.14*c*). At saturation, most of the dipoles are oriented with E. More specifically, the Ti^{4+} ions move to the locations along the c-axis favored by E (see Fig. 14.10*a*).

FERROELECTRIC DOMAINS

As defined above, a domain is a microscopic region in a crystal in which the polarization is homogeneous. However, in contrast to domain walls in ferromagnetic materials that can be relatively thick (Fig. 15.7*d*), the ferroelectric domain walls are exceedingly thin (Fig. 15.15). Consequently, the wall energy is highly localized and the walls do not move easily.

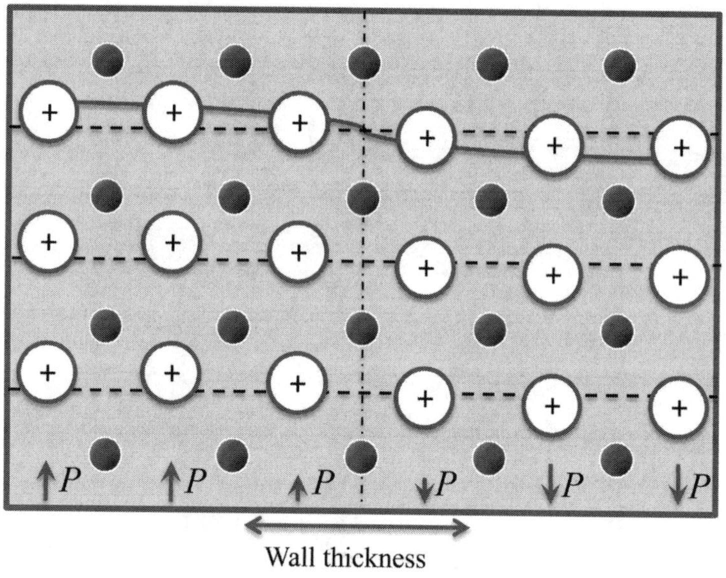

FIGURE 15.15 Ferroelectric domain wall thickness.

[236] In contrast to ferromagnetic materials, ferroelectrics exhibit true saturation.

[237] It seems natural to assume that polar crystals would be a source of electric fields around them just as magnets are a source of magnetic fields. In practice, however, the net dipole moment is not detectable in ferroelectrics because the surface charges are usually rapidly neutralized by ambient charged particles.

Practically, it is important to reduce the sharp dependence of k′ on temperature. In other words, it is important to broaden the permittivity versus temperature peaks as much as possible. One significant advantage of ceramic ferroelectrics is the ease with which their properties can be modified by adjusting composition and/or microstructure. For example, the substitution of Ti by other cations results in a shift in T_C, as shown in Fig. 15.16. Replacing Ti^{4+} by Sr^{2+} ions reduces T_C, while the substitution of Pb^{2+} increases it. This is very beneficial because it allows for the tailoring of the peak permittivity in the temperature range for which the ferroelectric device is to be used. Furthermore, certain additions, e.g., $CaZrO_3$, to $BaTiO_3$ can result in regions of variable composition that contribute a range of Curie temperatures so that the high permittivity is spread over a wider temperature range.

Sintering conditions can also have an important effect on the permittivity. The replacement of various aliovalent cations such as Nb^{5+} in $BaTiO_3$ has also been shown to inhibit grain growth, which, as seen in Fig. 15.17, has the effect of increasing the permittivity below T_C. Finally, lower-valency substitutions such as Mn^{3+} on Ti^{4+} sites act as acceptors and enable high-resistivity dielectrics to be sintered in low-partial-pressure atmospheres.

Examples of a number of ceramic ferroelectric crystals and some of their properties are listed in Table 15.6.

As discussed toward the end of Chap. 14, every *day* literally billions of multilayer capacitors—the vast majority of which are $BaTiO_3$ based—are produced. The choice of $BaTiO_3$ for this application is obvious given its large dielectric constant, compositional versatility, good frequency response and the fact that it is relatively inexpensive.

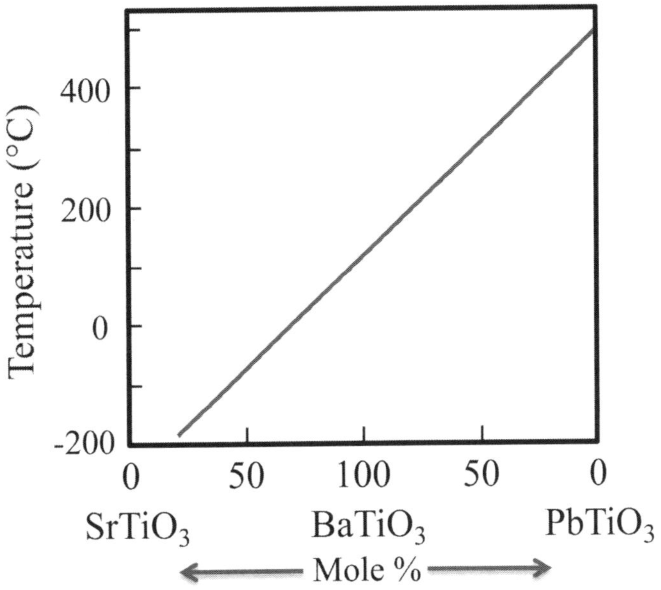

FIGURE 15.16 Effect of cationic substitutions in $BaTiO_3$ on T_C.

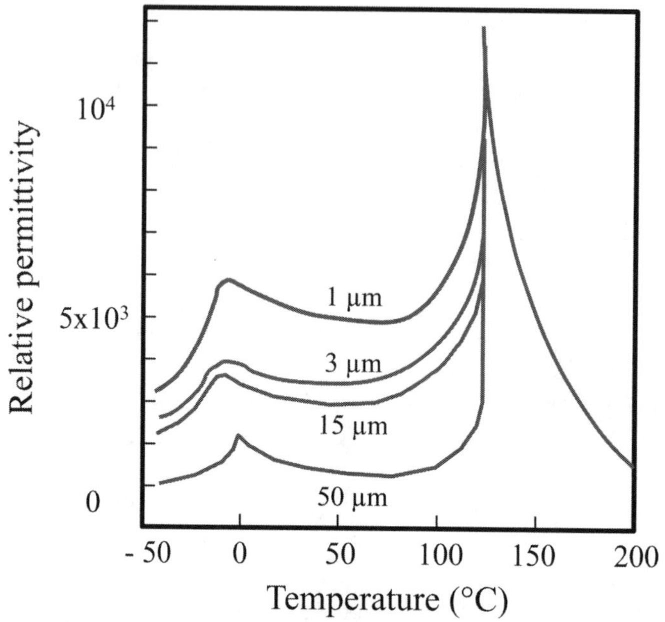

FIGURE 15.17 Effect of grain size on permittivity of $BaTiO_3$.

EXPERIMENTAL DETAILS: MEASURING FERROELECTRIC PROPERTIES

The signature of a ferroelectric material is the hysteresis loop. This loop can be measured in a variety of ways, one of which is by making use of the electric circuit shown schematically in Fig. 15.18. A circuit voltage across the ferroelectric crystal is applied to the horizontal plates of an oscilloscope. The vertical plates are attached to a linear capacitor in series with the ferroelectric crystal. Since the voltage generated across the linear capacitor is proportional to the polarization of the ferroelectric, the oscilloscope will display a hysteresis loop.

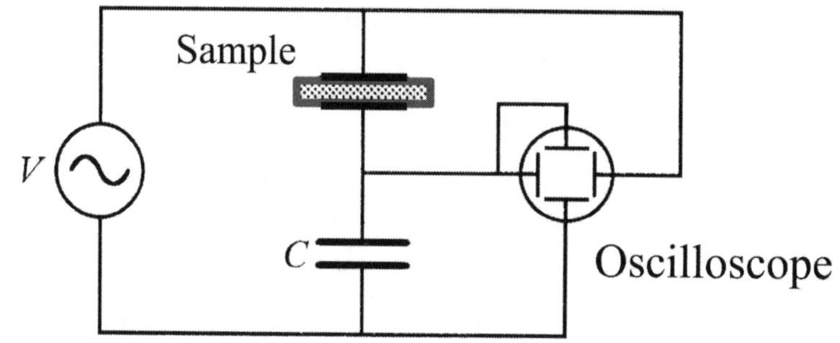

FIGURE 15.18 Circuit used to measure ferroelectric hysteresis.

15.7.3 ANTIFERROELECTRIC CERAMICS

In some perovskite ceramics, the instability that occurs at the Curie temperature is not ferroelectric but rather antiferroelectric. In antiferroelectric crystals, the neighboring lines of ions are displaced in opposite senses, which creates two alternating dipole sublattices of equivalent but opposite polarization. Consequently, the net polarization is zero. Examples of antiferroelectric crystals include WO_3, $NaNbO_3$, $PbZrO_3$ and $PbHfO_3$.

In general, the difference in energies between the ferroelectric and antiferroelectric states is quite small (a few joules per mole); consequently, phase transitions between the two states occur readily and can be brought about by slight variations in composition or the application of strong electric fields.

15.7.4 PIEZOELECTRIC CERAMICS

Piezoelectric materials are solids that are capable of converting mechanical energy to electrical energy and vice versa. This is shown schematically in Fig. 15.19*a*, where the application of a stress changes P. When an external force is applied to produce a compressive or tensile strain, a change is generated in the dipole moment, and a voltage is developed across the ceramic (Fig. 15.19*a*). The opposite is also true; application of an electric field results in dimensional changes (Fig. 15.19*b*).

The main uses of piezoceramics are in the generation of charge at high voltages, detection of mechanical vibrations, control of frequency, and generation of acoustic and ultrasonic vibrations. Most, if not all, commercial piezoelectric materials are based on ferroelectric crystals. The first commercially developed piezoelectric material was $BaTiO_3$. One of the most widely exploited piezoelectric materials today, however, is based on the $Pb(Ti,Zr)O_3$ or a PZT solid solution system.

To produce a useful material, a permanent dipole has to be frozen in the piezoelectric. This is usually done by applying an electric field as the specimen is cooled through the Curie temperature. This process is known as **poling** and results in the alignment of the dipoles, and an electrostatic permanent dipole results.

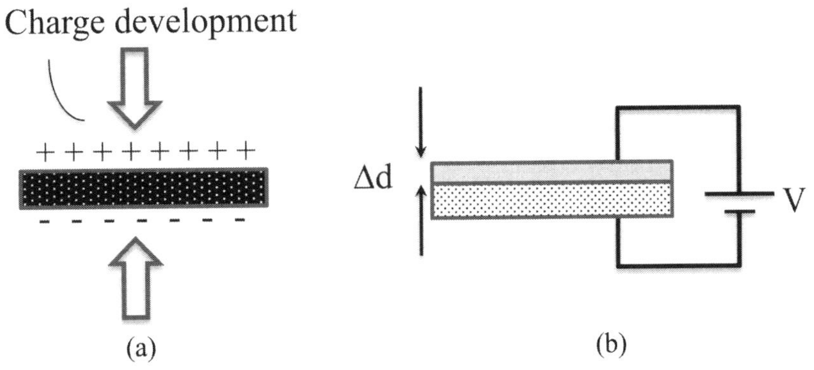

FIGURE 15.19 (*a*) In the direct piezoelectric effect the polarization charges are created by stress. (*b*) In the inverse effect, a strain is produced as a result of the applied voltage.

CASE STUDY 15.1: COBALITES, MANGANITES AND COLOSSAL MAGNETORESISTANCE

The lanthanum cobalites and manganites have interesting and potentially useful transport properties that can be influenced by external magnetic fields, H. However, to understand their transport properties requires that we first understand the electron configurations of the Mn and Co cations. Consider each separately:

Cobalites: Of special interest here in the ternary $LaCoO_3$—which crystalizes in the spinel structure. At relatively low temperatures, the 6 d-electrons of Co^{3+} cations are in a low spin (LS) configuration (Fig. 15.20*a*). As the temperature is increased, some of the electrons are promoted and the intermediate (Fig. 15.20*b*) and high (Fig. 15.20*c*) spin states become more probable. Consequently, transport changes from semiconductor at lower T to a metallic at higher T.

Manganites: Interest in $LaMnO_3$ was triggered when it was shown that when this material is doped it exhibited **colossal magnetoresistance (CMR)**. Magnetoresistance (MR) is a measure of the changes in resistance as a function of H and is defined as:

$$MR = \frac{R(H) - R(o)}{R(o)} \qquad (15.44)$$

R(H) is the resistance of the material in the presence of an external H-field and R(o) is the resistance when $H = o$. And while many materials' resistivities will change slightly in the presence of a magnetic field, CMR refers to mostly Mn-based perovkites where the resistance changes by orders of magnitude. CMR is not to be confused with giant MR, observed in thin metallic multilayers.

The prototypical CMR compound is derived from the parent perovskite, $LaMnO_3$, that is hole doped by an alkaline earth cation—such as Ca, Mg or Sr—that substitutes for the La^{3+} cation. A typical reaction is:

$$CaO + \frac{1}{2}O_2 + MnO \xrightarrow[LaMnO_3]{} Ca'_{La} + 3O_o^x + Mn_{Mn}^x + h^{\bullet} \qquad (15.45)$$

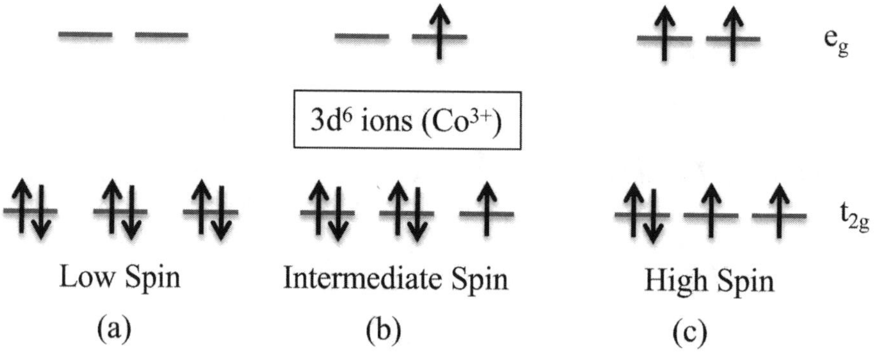

FIGURE 15.20 Spin states in $3d^6$ electrons of Co^{3+} cations in cobalites: (*a*) low spin, (*b*) intermediate spin and (*c*) high spin. Note the HS configuration is a good conductor.

The solid's final composition is thus $Ca_xLa_{(1-x)}MnO_3$. Focusing on the oxidation states of the Mn cations, the final chemistry can be written as:

$$(La_{La}^{3+})_{1-x}(Ca_{La}^{2+})_x(Mn_{Mn}^{3+})_{1-x}(Mn_{Mn}^{4+})_x(O_o^{2-})_3 \tag{15.46}$$

Note the extent of doping, x, is a direct measure of the hole concentration. Said otherwise, the Mn^{4+}/Mn^{3+} ratio for the most part is *fixed by the dopant*, which is why the electrical conductivity is not a function of P_{O2} in the high P_{O2} regime (see Fig. 7.21c).

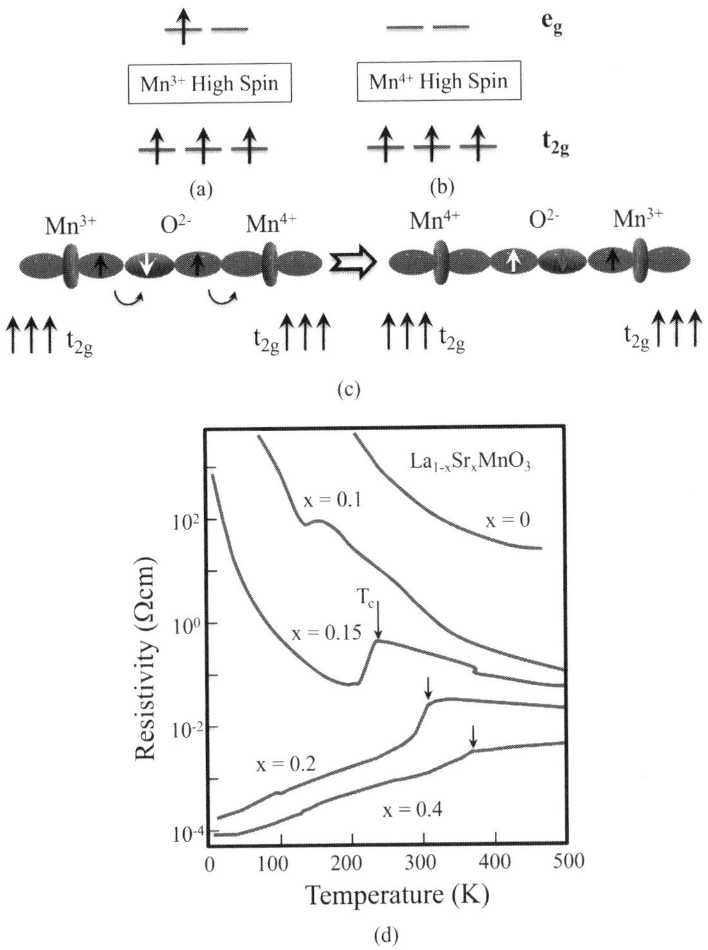

FIGURE 15.21 High spin states of (*a*) Mn^{3+} and (*b*) Mn^{4+} cations, (*c*) Double exchange between two adjacent Mn^{3+} and Mn^{4+} cations mediated by p-orbital of the O shared between them. (*d*) Effect of x on the nature of the conductivity in $La_{1-x}Sr_xMnO_3$. (Adapted from Urushibara et al. *Phys. Rev. B* 51, 1995.)

The key to CMR is the hole doping, which converts Mn^{3+} to Mn^{4+}. When the hole concentration, or x, is 20–40% holes/Mn ion, the material displays a transition from a high-temperature paramagnetic insulator to a low-temperature ferromagnetic metal (Fig. 15.21*e*).

As noted previously, the magnetic ground, or lowest, energy state of TM oxides is almost always determined by indirect magnetic "superexchange" interactions via the O^{2-} ions between them (see Fig. 15.7). In hole-doped manganates, a new mechanism referred to as **double exchange** is believed to play an important role. To shed more light on the subject, first refer to Figs. 15.20*a* and *b*, in which the HS configuration of the Mn^{3+} and Mn^{4+} are sketched, respectively. In the former case, one electron is in a e_g orbital. This electron can hop via the O-2p_z orbital to the 3d_{z2} orbital of Mn^{4+} (Fig. 15.20*c*). The latter will only occur if the core spins of the respective Mn ions are aligned parallel to each other as shown in Fig. 15.20*c*. If they are not, however, charge transfer is hampered. It thus follows that there is a strong correlation between the FM ground state and the electrical conductivity by hopping between the Mn^{3+} and Mn^{4+} cations (see Chap. 7). How an H-field can also influence the resistivity is also not difficult to appreciate. Note that the end member $LaMnO_3$, in which all the Mn ions are in the +3 state, exhibits pure semiconducting behavior. At this stage, the exact mechanisms responsible for CMR are not fully understood, but the aforementioned interaction between Mn^{3+} and Mn^{4+} cations are believed to play an important role.

15.8 SUMMARY

1. The presence of uncompensated or unpaired electron spins and their revolution around themselves (spin magnetic moment) and around their nuclei (orbital magnetic moment) endow the atoms or ions with a net magnetic moment. The net magnetic moment of an ion is the sum of the individual contributions from all unpaired electrons.
2. These magnetic moments can
   (i) not interact with one another, in which case the solid is a paramagnet and obeys Curie's law where the susceptibility is inversely proportional to temperature. Thermal randomization at higher temperature reduces the susceptibility.
   (ii) interact in such a way that adjacent moments tend to align themselves in the same direction as the applied field intensity, in which case the solid is a ferromagnet and will spontaneously magnetize below a certain critical temperature T_C. The solid will also obey the Curie–Weiss law (above T_C, it will behave paramagnetically). To lower the energy of the system the magnetization will not occur uniformly, but will occur in domains. It is the movement of these domains, which are separated by domain walls, that is responsible for the hysteresis loops typical of ferromagnetic materials.
   (iii) interact in such a way that the adjacent moments align themselves in opposite directions. If the adjacent moments are exactly equal they cancel each other out, and the solid is an *antiferromagnet*. However, if the adjacent moments are unequal they will *not* cancel, and the solid will possess a net magnetic moment. Such materials are known as *ferrimagnets* and constitute all known magnetic ceramics. Phenomenologically ferrimagnets are indistinguishable from ferromagnets.

3. Magnetic ceramics are ferrimagnetic and are classified according to their crystal structure into spinels, hexagonal ferrites and garnets.
4. The interaction between, and the alignment of, adjacent dipoles in a solid give rise to ferroelectricity.
5. The interaction between adjacent moments (magnetic as well as dipolar) causes the solid to exhibit a critical temperature below which spontaneous magnetization or polarization sets in, where all the moments are aligned parallel to one another in small microscopic domains. The rotation and growth of these domains in externally applied fields give rise to hysteresis loops and remnant magnetization or polarization, whichever the case may be.

APPENDIX 15A: ORBITAL MAGNETIC QUANTUM NUMBER

In Eq. (15.25), **L** is the summation of m_l, rather than l. The reason is shown schematically in Fig. 15.22, for the case where $l = 3$; in the presence of a magnetizing field H, the electron's orbital angular momentum is quantized along the direction of H. The physical significance of m_l and why it is referred to as the orbital magnetic quantum number should now be more transparent—it is the projection of the orbital angular momentum l along the direction of the applied magnetic field intensity.

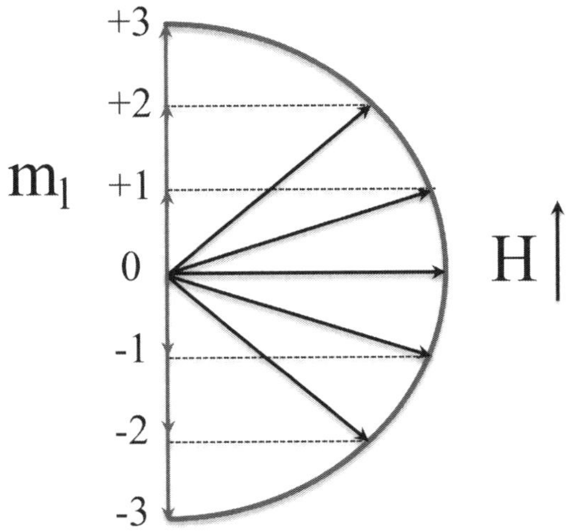

FIGURE 15.22 Total number of quantized allowed projections of the orbital angular momentum for $l = 3$ in the direction of the applied magnetic field is determined by m_l. Note that for $l = 3$ there are seven allowed projections: 3, 2, 1, 0, −1, −2 and −3.

PROBLEMS

15.1. In the Bohr model for the hydrogen atom,[238] the electron travels in a circular orbit of radius 5.3×10^{-11} m. Calculate the current and magnetic moment associated with this electron. How does the moment calculated compare to μ_B? *Hint*: Calculate the potential energy of the electron and equate it to the vibration energy.

Answer: $i \approx 1.05$ mA, $\mu_{orb} \approx 9.25 \times 10^{-24}$ A·m².

15.2. (*a*) Derive Eq. (15.29).

(*b*) A solid with electron spins is placed in a magnetizing field H of 1.6×10^6 A/m. If the number of spins parallel to the field was three times as large as the number of antiparallel spins, what was the temperature of the system? State all assumptions.

Answer: T ≈ 2.54 K.

(*c*) At what temperature would you expect all spins be aligned to the field? At what temperature would the number of spins upward exactly equal the number of spins downward?

15.3. (*a*) Show that if the assumption $\mu_{ion}B \ll kT$ is not made, then the susceptibility will be given by

$$M = (N_1 - N_2)\mu_{ion} = N\mu_{ion} \tanh\left(\frac{\mu_{ion}B}{kT}\right)$$

Plot this function as a function of $\mu_{ion}B / kT$. What conclusions can you reach about the behavior of the solid at high fields or very low temperatures?

(*b*) For a solid placed in a field of 2 T, calculate the error in using this equation as opposed to Eq. (15.29) at 300 K. You may assume $\mu_{ion} = \mu_B$.

Answer: ≈ 0.001%.

(*c*) Repeat part (*b*) at 10 K.

Answer: ≈ 0.6%.

15.4. A beam of electrons enters a uniform magnetic field of 1.2 T. What is the energy difference between the electrons whose spins are parallel to and those whose spins are antiparallel to the field? State all assumptions.

Answer: 1.4×10^{-4} eV.

15.5. (*a*) For a ferromagnetic, FM, solid, derive the following expression:

$$T_C = \frac{\lambda \mu_{ion}^2 \mu_0 N}{k} = 3\lambda C$$

(*b*) For FM iron, C ≈ 1 and T_C is 1043 K. Calculate the values of μ_{ion} and λ.

Answer: $\mu_{ion} = 2.13\mu_B$; $\lambda \approx 350$.

---

[238] It is important to note that, this problem notwithstanding, s electrons do not have an orbital moment. Quantum mechanics predicts that their angular momentum is zero, since l = 0.

15.6. The susceptibility of a Gd^{3+}-containing salt was measured as a function of temperature. The results are shown below. Are these results consistent with Curie's law? If yes, then calculate the Curie constant (graphically) and the effective Bohr magnetons per ion.

T, K	100	142	200	300
χ, cm³/mol	6.9×10^{-5}	5×10^{-5}	3.5×10^{-5}	2.4×10^{-5}

Information you may find useful: Molecular weight of salt = 851 g/mol and its density is 3 g/cm³. Hint: Convert χ to SI units first. To do that you may have to consult the internet.

15.7. $ZnO{\cdot}Fe_2O_3$ is antiferromagnetic. If it is known that this compound is a regular spinel, suggest a model that would explain the antiferromagnetism.

15.8. Sketch the magnetization versus temperature curve for Fe_3O_4.

15.9. Each ion in an iron crystal contributes on average 2.22 μ_B. In Fe_3O_4, however, each Fe ion contributes an average of 4.08 μ_B. How can you rationalize this result?

15.10. Consider $Fe(Ni_xFe_{2-x})O_4$. What value of x would result in a net moment per formula unit of exactly $2\mu_B$. *Hint*: Ni^{2+} occupies the octahedral sites.
Answer: x = 1.

15.11. (*a*) Calculate the spin-only magnetic moments of Ni^{2+}, Zn^{2+} and Fe^{3+}.
(*b*) Nickel ferrite ($NiO{\cdot}Fe_2O_3$) and zinc ferrite ($ZnO{\cdot}Fe_2O_3$) have inverse and normal spinel structures, respectively. The two compounds form mixed ferrites. Assuming that the coupling between the ions is the same as in magnetite and that the orbital momenta are quenched, calculate the magnetic moment per formula unit for $(Zn_{0.25}Ni_{0.75}O){\cdot}Fe_2O_3$.

15.12. Figure 15.23 presents magnetization curves for a magnetic ceramic fired at two different temperatures, shown on the plot.

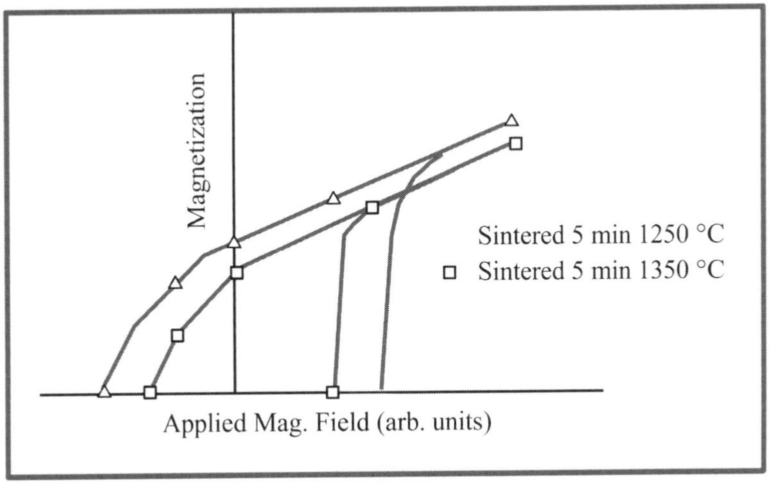

FIGURE 15.23 Upper half of the hysteresis loop for a barium ferrite as a function of processing time and temperature.

(*a*) Explain the general reasons for the shape of these curves. In other words, what processes determine the shape of the curves at low, intermediate and large values of the applied magnetic field?
(*b*) What changes in the material might explain the change in magnetic behavior with increased firing temperatures?

15.13. Using the data given in Table 15.6, estimate the value of the interaction factor β in $BaTiO_3$.
Answer: 2.3×10^{-3}.

15.14. (*a*) Calculate the dc capacitance of a $BaTiO_3$ capacitor 1 mm thick and 1 cm^2 in area that is operating near T_C.
Answer: 1.4 μF.
(*b*) Calculate the total dipole moment for a 2 mm thick disk of $BaTiO_3$ with an area of 1 cm^2.
(*c*) Calculate the unit cell geometry of a unit cell of $BaTiO_3$ that is subjected to an electrical field applied in such a way as to increase the polarization of the unit cell to 0.18 C/m^2.

15.15. Estimate the density of Ar at which its polarization would go to infinity. State all assumptions. The atomic weight of Ar is 39.94 g/mol.
Answer: 9.57 g/cm^3.

ADDITIONAL READING

D. Jiles, *Introduction to Magnetism and Magnetic Materials*, Chapman & Hall, London, 1991.

L. L. Hench and J. K. West, *Principles of Electronic Ceramics*, Wiley, New York, 1990.

B. Jaffe, W. R. Cook, and H. Jaffe, *Piezoelectric Ceramics*, Academic Press, New York, 1971.

E. Fatuzzo and W. J. Merz, *Ferroelectricity*, North-Holland, Amsterdam, 1967.

H. Frohlich, *Theory of Dielectrics*, 2nd ed., Oxford Science Publications, 1958.

M. E. Lines and A. M. Glass, *Ferroelectrics and Related Materials*, Oxford Science Publications, 1977.

N. Ashcroft and N. Mermin, *Solid State Physics*, Holt-Saunders Int. Ed., 1976.

C. Kittel, *Introduction to Solid State Physics*, 6th ed., Wiley, New York, 1988.

K. J. Standley, *Oxide Magnetic Materials*, 2nd ed., Oxford Science Publications, 1992.

H. Morrish, *The Physical Principles of Magnetism*, Wiley, New York, 1965.

J. Moulson and J. M. Herbert, *Electroceramics*, Chapman & Hall, London, 1990.

D. W. Richerson, *Modern Ceramic Engineering*, 2nd ed., Marcel Dekker, New York, 1992.

J. M. Herbert, *Ceramic Dielectrics and Capacitors*, Gordon & Breach, London, 1985.

J. M. Herbert, *Ferroelectric Transducers and Sensors*, Gordon & Breach, London, 1982

J. C. Burfoot and G. W. Taylor, *Polar Dielectrics and Their Applications*, Macmillan, London, 1979.

R. C. Buchanan, Ed., *Ceramic Materials for Electronics*, Marcel Dekker, New York, 1986.

L. M. Levinson, Ed., *Electronic Ceramics*, Marcel Dekker, New York, 1988.

J. C. Burfoot, *Ferroelectrics, An Introduction to the Physical Principles, Nostrand*, London, 1967.

P. Fulay, *Electronic, Magnetic and Optical Properties of Materials*, CRC Press, Boca Raton, FL, 2010.

16

OPTICAL PROPERTIES

White sunlight Newton saw, is not so pure;
A Spectrum bared the Rainbow to his view.
Each Element absorbs its signature:
Go add a negative Electron to
Potassium Chloride; it turns deep blue,
As Chromium incarnadines Sapphire.
Wavelengths, absorbed are reemitted through
Fluorescence, Phosphorescence, and the higher
Intensities that deadly Laser Beams require.

John Updike, *The Dance of the Solids**

16.1 INTRODUCTION

Since the dawn of civilization, gems and glasses have been prized for their transparency, brilliance and colors. The allure was mainly aesthetic, fueled by the rarity of some of these gems. With the advent of optical communications and computing, the optical properties of glasses and ceramics have become crucially important, an importance that cannot be overemphasized. For example, in its 150th anniversary issue devoted to the key technologies for the twenty-first century, *Scientific American*[239] devoted an article to all-optical networks. Today commercial fiber-optic networks are based on the ability of very thin, cylindrical conduits of glass to transmit information at hundreds of gigabits of information per second.[240] Today

* J. Updike, *Midpoint and Other Poems*, A. Knopf, Inc., New York, 1969. Reprinted with permission.
[239] V. Chan, *Scientific American*, September 1995, p. 72.
[240] A gigabit is 1 billion bits; a terabit is 1 trillion bits.

one fiber can transmit 13 terra bits/s! (a terra is 10^{12}). It follows that two fibers can transport 25 terabits of information, an amount sufficient to *simultaneously* carry all the telephone calls in the United States on Mother's Day (one of the busiest days of the year).

Alternating currents, infrared radiation, microwaves, visible light, X-rays, ultraviolet light, etc. all produce oscillating electromagnetic fields differing only in their frequencies. And although sometimes they are thought of as being distinct, they do constitute a continuum known as the *electromagnetic* (EM) spectrum that spans 24 orders of magnitude of frequencies, ν, or wavelengths, λ, (see Fig. 16.1). Within that spectrum, the visible range occupies a tiny window from 0.4 to 0.7 μm or 1.65 to 3.0 eV.[241]

All EM radiation will interact with solids in some fashion or other. Understanding the nature of this interaction has been and remains invaluable in deciphering and unlocking the mysteries of matter. For instance, it is arguable, and with good justification, that one of the most important techniques to study the solid state has been X-ray diffraction. Other spectroscopic techniques are as varied as radiation sources and what is being monitored, i.e., reflected, refracted, absorbed rays, etc.

In what follows, the various interactions between EM radiation and ceramics will be discussed. However, the phenomena described in this chapter relate mostly to the "optical" region of the EM spectrum, which includes wavelengths, λ, from 50 nm to 100 μm (25 to 0.1 eV). In other words, the discussion will be limited to the part of the spectrum shown at the top of Fig. 16.1. Furthermore, only insulating ceramics will be dealt with here—the cases where the concentration of free electrons is large will not be considered.

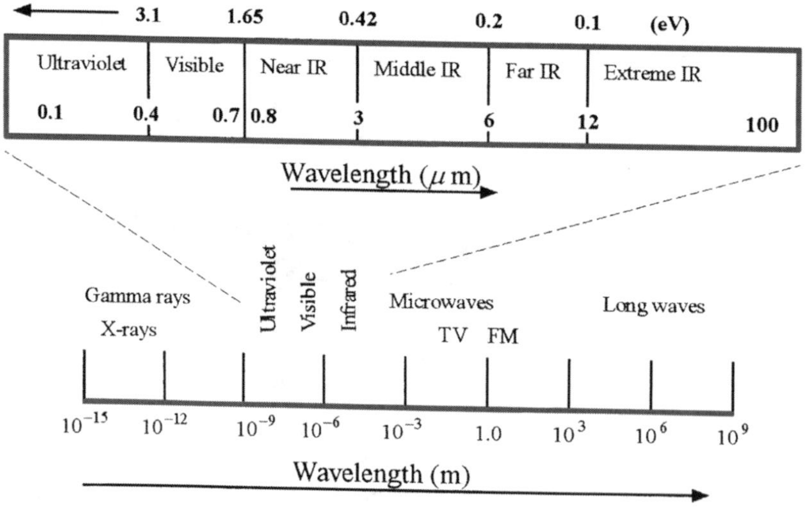

FIGURE 16.1 Electromagnetic spectrum. The visible spectrum constitutes a small window between 0.4 and 0.7 μm or 1.7 and 3.1 eV.

[241] In vacuum, the wavelength λ and frequency v, in Hz, are related by $v = c/\lambda$, where c is the velocity of light. The energy is given by $E = hv$, where h is Planck's constant. Also recall $\omega = 2\pi v$.

16.2 BASIC PRINCIPLES

When a beam of light or EM radiation impinges on a solid (Fig. 16.2), that radiation can be:

- ∞ *Transmitted* through the sample
- ∞ *Absorbed* by the sample
- ∞ *Scattered* at various angles

The scattered waves can be coherent or incoherent (see App. 16A for more details). When the scattered waves constructively interfere with one another, the scattering is termed *coherent*. Light scattered in the opposite direction of the incident beam leads to reflection. Light scattered in the same direction as the incident beam and recombining with it gives rise to refraction. The recombination of the scattered beams

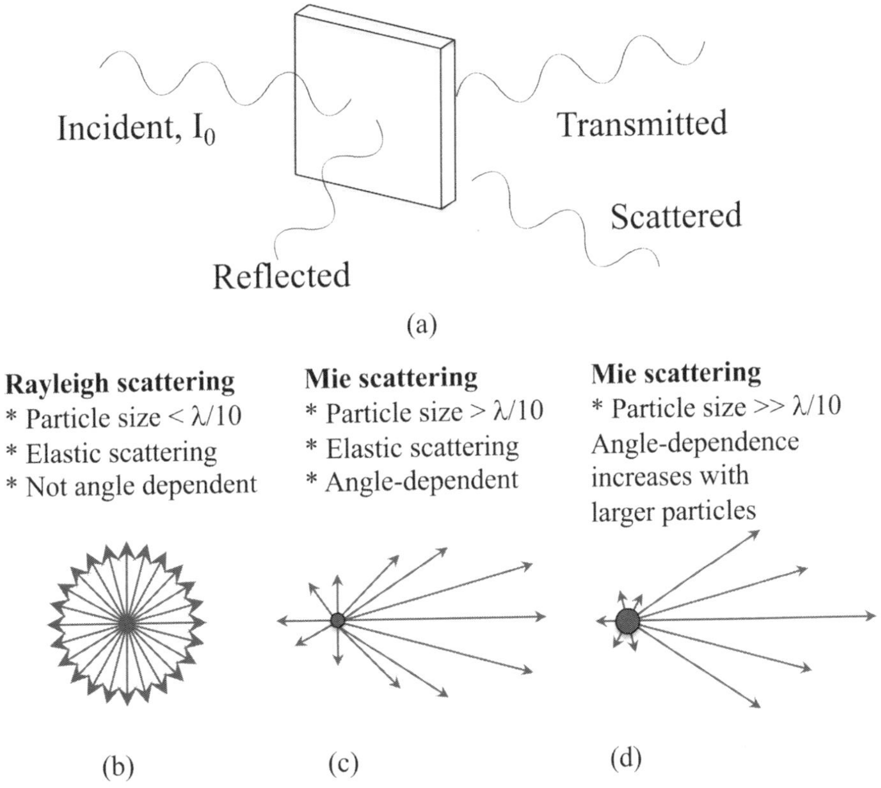

FIGURE 16.2 Various interactions between radiation and solids. *(a)* monochromatic ray with frequency ν and intensity I_0 can be transmitted (with the same frequency but a reduced intensity), scattered, absorbed or reflected. *(b)* Schematic of Rayleigh, and (c) and (d) Mie scattering for two different sized particles. As the particle size r_s increases relative to λ, the scattering becomes less isotropic.

can also give rise to diffraction, where a diffracted beam's intensity depends on the relative positions of the atoms and is thus used to determine the position of atoms in a solid (e.g., X-ray diffraction, see Chap. 3). Incoherent interference, on the other hand, gives rise to other forms of scattering. Chief among them are:

- **Rayleigh scattering** is the elastic scattering of light by particles with dimensions of the order of r_s such that $r_s << \lambda$. As shown in Fig. 16.2*b*, the scattering is more or less isotropic. It is this type of scattering that renders the sky blue during the day and red at sunset. It is also the main cause of signal loss in optical fibers. This scattering is discussed in Sec. 16.4.
- **Mie scattering** is the elastic scattering of light by spherical particles of any radius, but generally reserved to particles that are larger than those responsible for Rayleigh scattering, or $r_s > \lambda/10$. As r_s increases, the scattering is more asymmetric (compare Fig. 16.2c and d). When $r_s << \lambda$, Mie scattering is indistinguishable from Rayleigh scattering. Under those conditions, the shape of the particle is immaterial. This scattering is also discussed in Sec. 16.4. It is this scattering that gives clouds and fog their white color.
- **Tyndall scattering** is similar to Mie scattering, but the scattering particles need not be spherical.
- **Brillouin scattering** is the inelastic scattering of light by acoustic phonons in a solid. This scattering can be used to measure sound velocities in solids and can, in principle, be used to calculate the elastic constants of single crystals.
- **Raman scattering** is the inelastic scattering of light by optical phonons. Because the scattering is inelastic, the scattered phonons have an energy different – usually lower – than the incident photons. This difference in energy corresponds to the energy required to excite a molecule to a higher vibrational mode. The Raman effect forms the basis for Raman spectroscopy which is used by scientists to gain information about solids.

For a total incident flux of photons I_0, energy conservation requires that

$$I_0 = I_T + I_R + I_A$$

where I_T, I_R and I_A represent the transmitted, reflected and absorbed intensities, respectively. The intensity I is the energy flux per unit area and has units of $J/(m^2 \cdot s)$. Dividing both sides of this equation by I_0 yields

$$1 = T + R + A \tag{16.1}$$

where T, R and A represent, respectively, the fraction of light transmitted, reflected and absorbed.

In the following sections and throughout this chapter, the relationship between the makeup of a solid and its optical properties is discussed. The optical properties of greatest interest here are the refractive index n, which for low-loss materials determines the reflectivity and transmissivity, and the various processes responsible for absorption and/or scattering.

16.2.1 REFRACTION

A common example of refraction is the apparent bending of light rays as they pass from one medium to another, e.g., a rod immersed in a fluid will appear bent. The extent of this effect is characterized by a fundamental property of all materials, namely, their refractive index n. When light encounters a boundary between two materials with different refractive indices, for reasons that are touched upon subsequently, its velocity and direction will change abruptly, a phenomenon called *refraction*. The physics behind what gives rise to n is intimately related to the *electronic polarizability* of the atoms or ions in a solid. To understand the physical origin of n, it is useful to make the following two simplifying assumptions: First, the frequency of the applied field, ω, is much greater than ω_{ion} but smaller than ω_o, the natural frequency of vibration of the ***electronic cloud***. Second, $k_e'' = 0$; in other words, the electronic charges are all oscillating in phase with the applied field (see Chap. 14). As these charged particles oscillate, they in turn reradiate an EM waves of the same frequency, creating their own electric field, which interacts with, and slows, down the incident field.[242]

As noted, the major effect of the interaction of the incident and reradiated waves is to make the velocity of the transmitted light appear to have traveled through the solid (v_{sol}) more slowly than through vacuum (v_{vac}), which leads to perhaps the simplest definition of n, namely,

$$n = \frac{v_{vac}}{v_{sol}} \tag{16.2}$$

Refer to Fig. 16.3. Another equivalent definition is

$$n = \frac{\sin i}{\sin r} \tag{16.3}$$

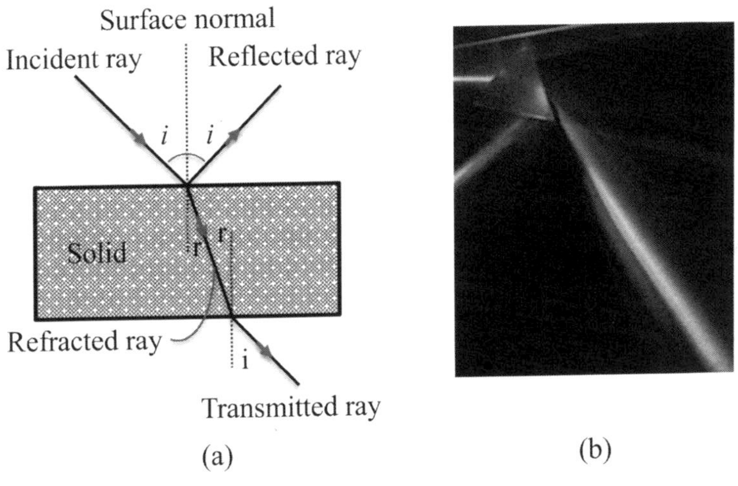

FIGURE 16.3 (*a*) Light refraction and reflection. Each interface reflects and refracts a portion of the incident beam. (*b*) Light refraction by a prism.

[242] The exact details of how this occurs are beyond the scope of this book but are excellently described in R. P. Feynman, R. B. Leighton, and M. Sands, *The Feynman Lectures on Physics*, vol. 1, Chap. 31, Addison-Wesley, Reading, MA, 1963.

TABLE 16.1 Refractive indices of select ceramic materials

Material	n	Material	n
Halides and Sulfides			
CaF_2	1.430	NaF	1.330
BaF_2	1.480	NaI	1.770
KBr	1.560	PbF_2	1.780
KCl	1.510	PbS	3.910
LiF	1.390	TlBr	2.370
NaCl	1.550	ZnS	2.200
Oxides			
Al_2O_3 (Sapphire)	1.760	PbO	2.610
$3Al_2O_3 \cdot 2SiO_2$	1.640	TiO2	2.710
$BaTiO_3$	2.400	SrO	1.810
BaO	1.980	$SrTiO_3$	2.490
BeO	1.720	Y_2O_3	1.920
$CaCO_3$	1.658, 1.486	ZnO	2.000
$MgAl_2O_4$	1.720	ZrSiO4	1.950
MgO	1.740	ZrO_2[a]	2.190
Covalent Ceramics			
C (diamond)	2.424	α-SiO_2 (quartz)	1.544, 1.553
α-SiC	2.680		
Glasses			
Pb-silicate glasses	2.500	Soda-lime silicate glass	1.510
Fused quartz	1.458	$Na_2O-CaO-SiO_2^{\dagger}$	1.458
Pyrex®[b]	1.470	Vycor®[c]	1.458

[a] Dense optical flint.
[b] Borosilicate

Typical values of n are listed in Table 16.1, from which it is obvious that for most ceramics n lies between 1.2 and 2.6. Note that all the values are greater than 1, indicating that for all materials, $v_{solid} < v_{vac}$.

In the more general case, where k_e'' cannot be neglected, n has to be complex, i.e.,[243]

$$\hat{n} = n + i\kappa \tag{16.4}$$

where κ is called the **extinction coefficient** or **absorption index** and is a measure of the absorbing capability of a material. Kappa (κ) should not be confused with the k_e' or k_e'', although they are related (see below).

As discussed in Chap. 14, at frequencies greater than about $10^{15}\ s^{-1}$, only electrons can follow the field and all other polarization mechanisms including ionic polarization drop out. In this situation, it can be

[243] It is instructive to note the similarity between this equation and Eq. (14.20).

shown (not too easily) that the following relationships between the electronic polarizability parameters k'_e and k''_e, on one hand, and n and κ, on the other, hold:

$$k'_e = n^2 - \kappa^2 = 1 + \frac{e^2 N}{\varepsilon_0 m_e} \frac{\omega_e^2 - \omega^2}{(\omega_e^2 - \omega^2)^2 + f^2\omega^2} \tag{16.5}$$

and

$$k''_e = 2n\kappa = \frac{e^2 N}{\varepsilon_0 m_e} \frac{\omega f}{(\omega_e^2 - \omega^2)^2 + f^2\omega^2} \tag{16.6}$$

Furthermore, from these two equations it can be shown that (see Prob. 16.1)

$$n = \frac{1}{\sqrt{2}}\sqrt{(k_e'^2 + k_e''^{\,2})^{1/2} + k'_e} \tag{16.7}$$

and

$$\kappa = \frac{1}{\sqrt{2}}\sqrt{(k_e'^2 + k_e''^{\,2})^{1/2} - k'_e} \tag{16.8}$$

Equations (16.5) to (16.8) are important for the following reasons:

1. They clearly demonstrate the one-to-one correspondence between electronic polarization and n. Typically, for ceramics k''_e is on the order of 0.01 to 0.0001 (see Table 14.1); consequently, without much loss in accuracy, the k''_e term can be neglected with respect to k'_e in Eq. (16.7), in which case

$$\boxed{n = \sqrt{k'_e}} \tag{16.9}$$

2. Since k'_e is a function of ω, it follows that n also has to be a function of ω (see Worked Example 16.1). This change in n with ω, or wavelength, is called **dispersion**. Typical dispersion curves for a number of ceramics are shown in Fig. 16.4, where it is clear that n increases as the ω of light increases.
3. Even at the high end of the range of k''_e (i.e., $\approx$ 0.01) and assuming the lowest value of k'_e possible, that is, 1, the κ calculated from Eq. (16.8) is on the order of 0.005 and thus can be ignored for most applications. Note that this conclusion is valid only as long as the system is far from resonance.

Worked Example 16.1

Based on Eq. (16.5), explain the phenomenon of light refraction by a prism shown in Fig. 16.3*b*. What does that say about the nature of white light?

ANSWER

According to Eq. (16.5), as long as $\omega < \omega_0, k'_e$ increases with increasing ω of the incident light. Consequently, according to Eq. (16.9), n should also increase with increasing ω. In other words,

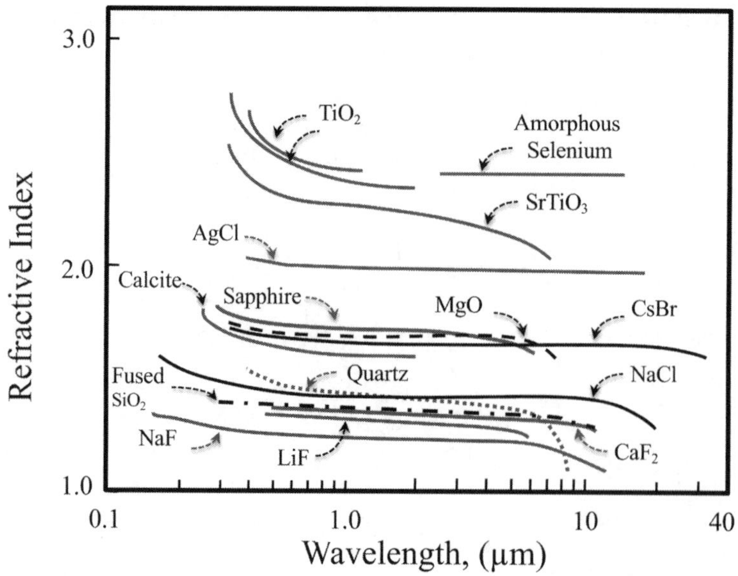

FIGURE 16.4 Change in refractive index with frequency for select glasses and crystals.

higher-frequency light (e.g., violet, blue) should be refracted, or deflected, by a larger angle than lower-frequency light (e.g., red), as observed in Fig. 16.3*b*. This simple experiment makes it clear that "white" light is composed of a spectrum of frequencies and that n is indeed a function of ω.

16.2.2 REFLECTIVITY

Not all light that is incident on a surface is refracted; as shown in Fig. 16.3*a*, a portion of it can be reflected. It can be shown, again not too easily, that the reflectivity of a perfectly smooth solid surface at *normal* incidence is given by

$$R = \frac{(n-1)^2 + \kappa^2}{(n+1)^2 + \kappa^2} \tag{16.10}$$

which is known as **Fresnel's** formula.

Again, given that for most ceramics and glasses $\kappa \ll 1$, it follows that R is simply related to n by:

$$R = \frac{(n-1)^2}{(n+1)^2}$$

and

$$T = R - 1 = \frac{(n-1)^2}{(n+1)^2} - 1$$

For example, lead-silicate glasses, also known as crystal glasses,[244] with refractive indices of about 2.6, reflects about 20% of the incident light (which explains why crystal glass sparkles). By contrast, a typical soda-lime silicate glass, with an n of ≈1.5, only reflects about 4%.

Interestingly enough, near resonance, n will increase dramatically, and so, according to Eq. (16.10), will reflectivity. This occurs because the various secondary waves from the surface atoms cooperate to produce a reflected wave front traveling at an angle equal to the angle of incidence. *Selective* reflection is thus a phenomenon of resonance and occurs strongly near those wavelengths corresponding to natural frequencies of bound charges in the substance, i.e., near resonance. The substance will *not* transmit light of these wavelengths; instead, it reflects strongly. True absorption (see below), where the light is converted to heat (i.e., processes associated with k_e'' or κ), also occurs at these frequencies to a greater or lesser extent because of the large amplitudes of the vibrating charges involved. If true absorption were entirely absent, however, the reflecting power would be 100% at the λ in question.

16.2.3 ABSORBANCE AND TRANSMITTANCE

The transmittance T through a transparent medium is proportional to the amount of light that is neither reflected nor absorbed. For low-loss (low-absorbing) materials, the absorption A in Eq. (16.1) can be neglected and $T \approx 1 - R$. In other words, the fraction of light not reflected is transmitted.

In general, however, as light passes through a medium in a given direction, say x, it is attenuated, or lost, by one of two mechanisms: Either it is absorbed, i.e., the light is transformed to heat, or it is scattered, i.e., a portion of the beam is deflected and is scattered from the x-direction.[245]

Intrinsic absorption.

In Chap. 14 the power dissipation per unit volume in a dielectric was shown to be [Eq. (14.25)]

$$P_V = \frac{1}{2}\sigma_{ac}E_0^2 = \frac{1}{2}(\sigma_{dc} + \omega k_e''\varepsilon_0)E_0^2 \tag{16.11}$$

where E_0 is the applied electric field. Energy conservation dictates that in the absence of any other energy-dissipating mechanisms, this loss will result in a decrease in the intensity I of the light, that is, $P_V = -dI/dx$, passing through a material of thickness dx. Furthermore, it can be shown[246] that the intensity of light in a medium of refractive index n is given by

$$I = \frac{n\varepsilon_0 c E_0^2}{2} \tag{16.12}$$

[244] This is an unfortunate nomenclature. There is nothing crystalline about this glass, or any other glass for that matter.

[245] A good example of scattering is the way that rays of sunlight from a window are made visible by very fine dust particles suspended in air.

[246] See, e.g., R. P. Feynman, R. B. Leighton, and M. Sands, *The Feynman Lectures on Physics*, vol. 1, pp. 31–110, Addison-Wesley, Reading, MA, 1963.

where c is the velocity of light in vacuum. Ignoring σ_{dc} in Eq. (16.11), which for most insulators and optical materials is an excellent assumption (see Prob. 16.1), noting that $k_e'' = 2n\kappa$, and combining Eqs. (16.11) and (16.12), one obtains

$$\frac{dI}{dx} = -\frac{2I\omega\kappa}{c} = -\alpha_a I \tag{16.13}$$

Integrating from the initial intensity I_0 to the final, or transmitted, intensity I_T gives

$$\boxed{\frac{I_T}{I_0} = \exp(-\alpha_a x)} \tag{16.14}$$

where x is the optical path length and α_a is the **absorption constant**, given by $2\omega\kappa/c$. Here α_a is measured in m^{-1} and is clearly a function of ω.

Note that α_a is proportional to k_e'', which in turn reflects the fact that the oscillating charges *not* in phase with the applied EM field are the ones responsible for the absorption. For an ideal dielectric, k_e'', κ and α_a all vanish, and no energy is absorbed (see Worked Example 16.2). Finally, note that when ω of the incident radiation approaches the resonance frequency of either the bonding electrons or the ions, then strong absorptions occur, absorptions that, as discussed subsequently, are ultimately responsible for delineating the frequency range over which a material is transparent.

Absorbance by impurity ions. As discussed in greater detail subsequently, impurity ions in a material can *selectively* absorb light at specific wavelengths. Such a chemical species is called a **chromophore.** Attenuation is proportional to the path traveled dx and the concentration of absorbing centers c_i, as described by the **Beer–Lambert law**:

$$-\frac{dI}{dx} = \varepsilon_{BL} c_i I \tag{16.15}$$

where ε_{BL} is a constant that depends on the impurities and the medium in which they reside. ε_{BL} is referred to sometimes as the **linear absorption coefficient** and sometimes as the **extinction coefficient.** Once again, integrating this expression yields

$$\boxed{\frac{I_T}{I_0} = \exp(-\varepsilon_{BL} c_i x)} \tag{16.16}$$

This is an important result because it predicts that the reading of a radiation detector (which measures the rate of flow of energy per unit area and unit time) will decrease exponentially with the thickness of the medium *and* the concentration of absorbing centers.

An implicit assumption made in deriving Eqs. (16.14) and (16.16) was that scattering could be neglected. In general, however, the loss coefficient must account for all losses, hence

$$\alpha_{tot} = \alpha_a + \varepsilon_{BL} c_i + \alpha_s \tag{16.17}$$

where α_s is the absorption coefficient due to scattering (Sec. 16.4). In the most general case

$$\boxed{I_T = I_0 \exp(-\alpha_{tot} x)} \tag{16.18}$$

In many cases some of these mechanisms are negligible with respect to the others, but it is important to realize their existence and the fact that more than one mechanism may be operating.[247]

EXPERIMENTAL DETAILS: MEASURING OPTICAL PROPERTIES

Clearly two important optical properties are n and κ. There are several techniques to measure them Here we describe probably one of the simplest which entails measuring the transmission and the reflectivity of a thin slab of material, which is usually carried out in a device known as a *spectrophotometer*.[248]

Figure 16.5 schematically illustrates the four major components of such a device: a source of radiation, a monochromator, the sample and a number of detectors. In a typical experiment, both T and R are measured, preferably simultaneously. Because of multiple reflections at the various planes of the crystal, T is not given by Eq. (16.14), but rather by[249]

$$T = \frac{I_{out}}{I_0} = \frac{(1-R)^2 e^{-\alpha_a x}}{1-R^2 e^{-2\alpha_a x}} \tag{16.19}$$

where R is given by Eq. (16.10) and x is the thickness of the sample. Note this expression is only valid for normal incidence and is identical to Eq. (16.14) when $R=0$.

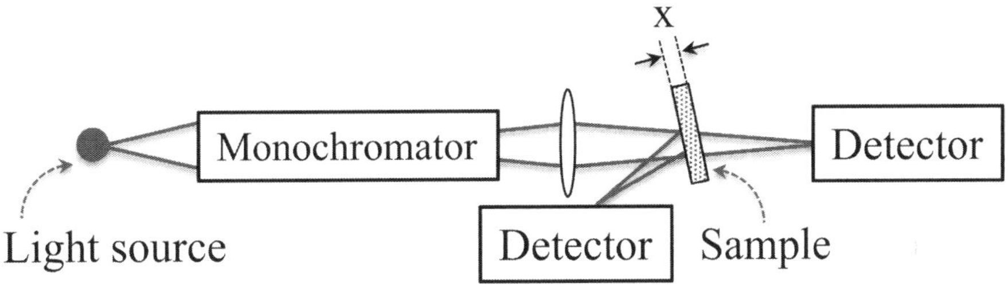

FIGURE 16.5 Schematic of apparatus used for measuring the optical constants of a solid.

247 It is not possible to distinguish between absorption and scattering losses from a simple measurement of the attenuation; both phenomena cause attenuation. One method to differentiate between the two, however, is to measure the light intensity at all angles. If the measurements show that all the light taken away from the original beam reappears as scattered light, the conclusion is that scattering—not absorption—is responsible for the attenuation. If the energy is absorbed, it will not disappear, but reappears at a different frequency, i.e., as heat.

248 Implicit in this discussion is that the material is fully dense and pore-free, with a grain size that is either much smaller than, or much greater, than the wavelength of the incident radiation. If that were not the case, as discussed in greater detail subsequently, scattering would have to be taken into account. Furthermore, it is assumed that the material is pure enough that the $\varepsilon_{BL} c_i$ term can be neglected.

249 J. C. Slater, *Electromagnetic Theory,* McGraw-Hill, New York, 1941.

In principle, the observed reflectivity R_{obs} can be measured by providing a second detector position, as shown in Fig. 16.5. The observed reflectivity is related to R by

$$R_{obs} \approx \left(1 + \frac{I_T}{I_0} e^{-\alpha_a x}\right) \tag{16.20}$$

Thus, by measuring both the reflectivity and the transmittivity of a sample, n and κ can be calculated from Eqs. (16.10), (16.19) and (16.20) (see Worked Example 16.2).

An alternate approach is to measure the transmission of two different samples of different thicknesses with identical reflectivities.[250]

Worked Example 16.2

If 80% of a Na lamp light, λ = 0.59 μm, incident on a 1 mm glass panel is transmitted and 4% is reflected, determine the n and κ for this glass.

ANSWER

Given that R = 0.04, applying Eq. (16.10) and ignoring κ (see below) yields n = 1.5. To calculate α_a use is made of Eq. (16.19). However, given that the loss is small, the second term in the denominator can be neglected, and Eq. (16.19) simplifies to

$$T = (1 - R)^2 e^{-\alpha_a x}$$

$$0.8 = (1 - 0.04)^2 e^{-\alpha_a (0.001)}$$

Solving for α_a yields 141 m^{-1}.

If $\lambda = 0.59$ μm, then $\nu = c/\lambda = 5.1 \times 10^{14}$ s. Furthermore, given that $\alpha_a = 2\omega\kappa/c$, it follows that

$$\kappa = \frac{\alpha_a c}{4\pi\nu} = \frac{141(3\times10^8)}{(4\times3.14)(5.1\times10^{14})} = 6.6\times10^{-6}$$

It follows that the error in ignoring κ in Eq. (16.10) is fully justified.

Worked Example 16.3

Carefully describe the changes that occur to an EM ray when it impinges on a solid for which (*a*) n = 2 and $\alpha_{tot} = \alpha_a = 0$, and (*b*) n = 2 and $\alpha_{tot} = \alpha_a = 0.4$.

[250] Interestingly enough (and it is left as an exercise to the reader to show that) strongly absorbing samples are also quite reflective, and vice versa.

ANSWER

(a) When a wave propagating in vacuum impinges normally on a solid for which n = 2 and $\alpha_a = 0$, the outcome is shown schematically in Fig. 16.6*a*. Since n = 2, the velocity of the wave is halved, which implies that λ also is halved in the solid. Note that because $\alpha_a = 0$, the intensity of the transmitted light and its *frequency* remain constant throughout.

(b) When, α_a is nonzero, the intensity of the transmitted wave is reduced as a result of absorption by the solid (Fig. 16.6*b*). For simplicity in both these cases, reflection was neglected.

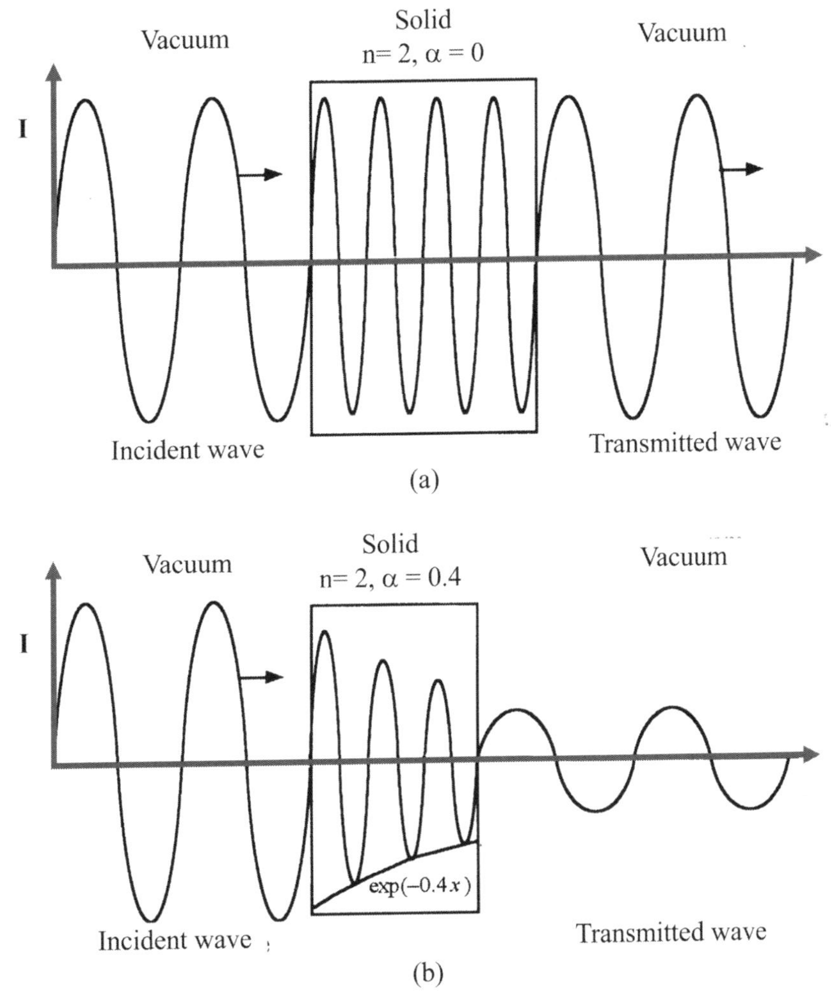

FIGURE 16.6 Schematic of changes that occur to an EM wave transmitted through a solid with (*a*) n = 2 and $\alpha_a = 0$. Note the halving in wavelength. Since $\alpha_a = 0$, there is no loss in intensity, or energy, of the wave as it passes through the solid. (*b*) Here n = 2 and $\alpha_a = 0.4$. The sample absorbs a portion of the energy of the incident wave, and the intensity of the transmitted wave is thereby reduced. Since n = 2, the wavelength in the solid is again half that in vacuum.

16.3 ABSORPTION AND TRANSMISSION

In the previous section, the relationships between transmittance and absorbance were described, with few details given. In this section, some of the specifics are elucidated. Scattering is dealt with separately in Sec. 16.4.

The complexity of the situation is depicted in Fig. 16.7, where the reflectance[251] of KBr at 10 K is plotted as a function of incident photon energy over a wide range from the IR to the UV. The salient features are an IR absorption edge at 0.03 eV, an absorption peak at about 2 eV, and a number of absorption peaks in the UV part of the spectrum around 7 eV.

From the previous discussion it is clear that the requirements for a material to be transparent are the absence of strong absorption and/or scattering in the visible range. The range over which a solid is transparent is called the **transmission range** and is bounded on the high-frequency (low-λ) side by UV absorption phenomena and on the low-frequency side by IR absorption. The spectral transmission ranges of a number of ceramics are compared in Figs. 16.8 and 16.9, from which it is obvious that most ceramics are indeed transparent over a wide range of frequencies. For example, window glass transmits light from 1×10^{15} to $\approx 7.5\times10^{13}$ s^{-1}, which is why, not surprisingly, it is used for windows. It is interesting to note that typical semiconductor materials, such as Si and GaAs, are only transparent in the IR range.

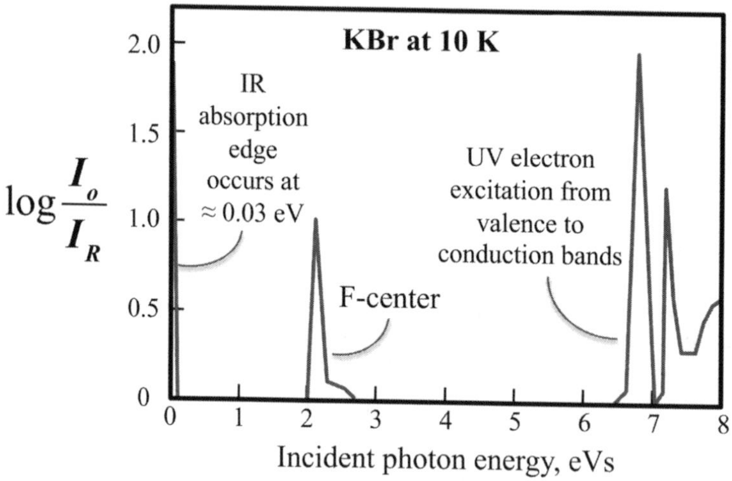

FIGURE 16.7 Spectral reflectance of KBr over a wide energy range of incident radiation. Note that at resonance, the solid becomes very reflective, whereas away from resonance, most of the light is transmitted. (Note that the information embedded in this graph, between the IR absorption edge and the UV spectrum, is the same information that appears in Fig. 14.13*b* over the same range.)

[251] As noted earlier, near resonance, crystals become considerably more reflective.

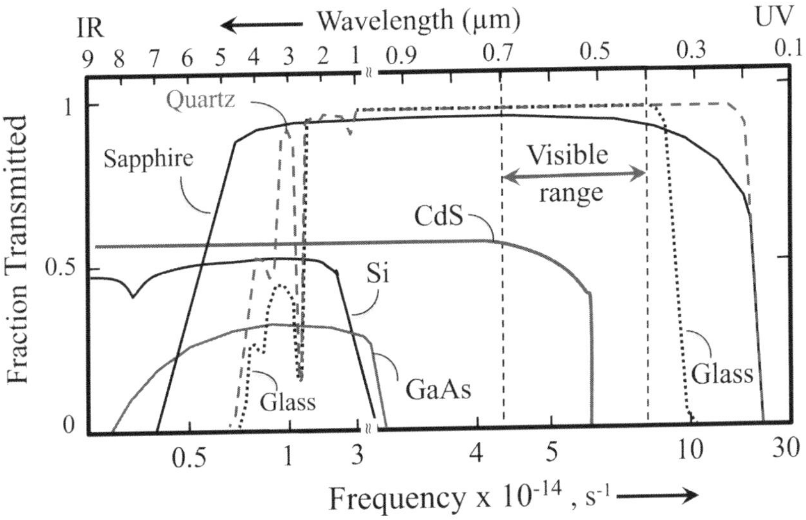

FIGURE 16.8 Spectral transmission of a number of select materials. Note change of scale in λ between 0.9 and 1 μm.

16.3.1 UV RANGE

ELECTRONIC RESONANCE

This was discussed in Sec. 16.2. The factors that affect the frequency at which resonance occurs were discussed in Chap. 14 and will not be repeated here, except to point out that in glasses the formation of non-bridging oxygens NBO, tends to decrease the frequencies (increase λ) at which resonance will occur. This is clear from Fig. 16.8—quartz is transparent to higher frequencies than window glass that contains NBO. It was also noted in Chap. 14 (see Fig. 14.8) that S^{2-}, Te^{2-} and Se^{2-} are some of the more polarizable ions. It is thus not surprising that ceramics containing these ions tend to be opaque in the visible spectrum and have absorption edges that are shifted into the IR range; CdS (Fig. 16.8) is a good example.

PHOTOELECTRIC EFFECT

As discussed in Chap. 2, insulating crystalline materials exhibit an energy gap E_g between their valence and conduction bands. When the incident photon with energies is greater than E_g, that is,

$$h\nu > E_g \tag{16.21}$$

it will be absorbed by promoting an electron from the valence band to the conduction band. This is known as the **photoelectric effect**, and in addition to increasing the conductivity of the solid, it results in the absorption of the incident wave. It is important to note that if the light has an energy less than E_g, absorption will not occur. Hence a well-established technique for measuring a material's band gap is to measure the conduction of the sample as a function of the ν of incident light—the ν at which the onset of photoconductivity is related to E_g through Eq. (16.21).

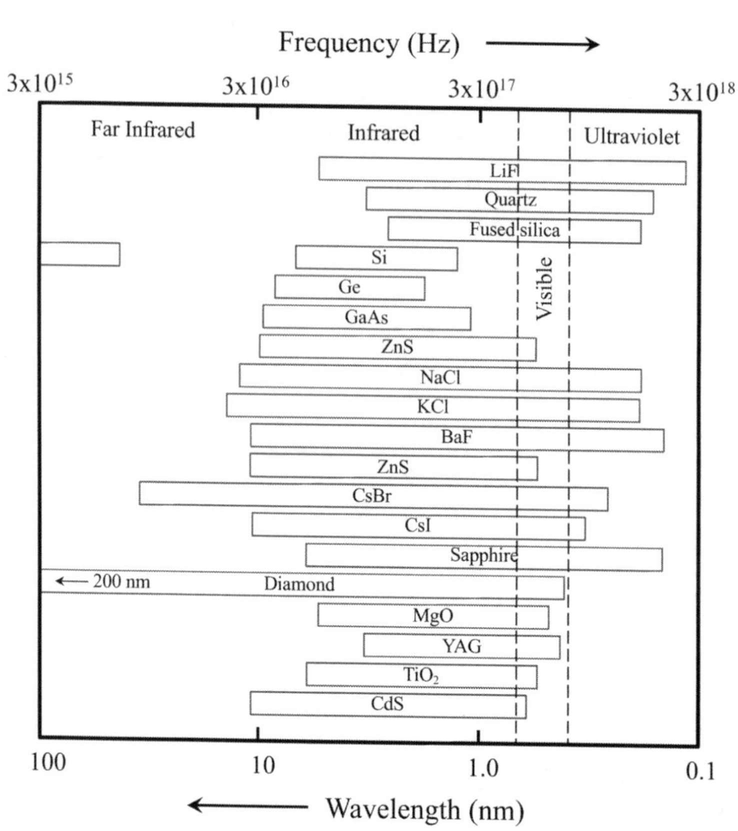

FIGURE 16.9 Useful transmission ranges for select materials. T > 10% for a 2 mm thick sample.

16.3.2 VISIBLE RANGE

For appreciable absorption to occur in the visible range, electronic transitions must occur. The nature of these transitions can result from various sources, as described now.

TRANSITION-METAL CATIONS

It is well known that the colors of crystals and minerals are a strong function of the type of dopant or impurity atoms, especially transition-metal cations, present. For example, rubies are red and some sapphires are blue, yet both are essentially Al_2O_3. It is only by doping Al_2O_3 with parts per million of Cr ions that the magnificent red color develops. Similarly, sapphires develop their blue color as a result of Ni doping. Since rubies are red and pure alumina is transparent, it follows that the Cr ions must absorb blue light and transmit red light, that is registered by the eyes.

To explain this phenomenon the **ligand field theory** has been proposed, which successfully accounts for the coloring and magnetic properties of many transition-metal-containing ceramics. This theory was discussed in detail in Sec. 15.4.4 and will not be repeated here. Between them Figs. 15.6 and 16.10

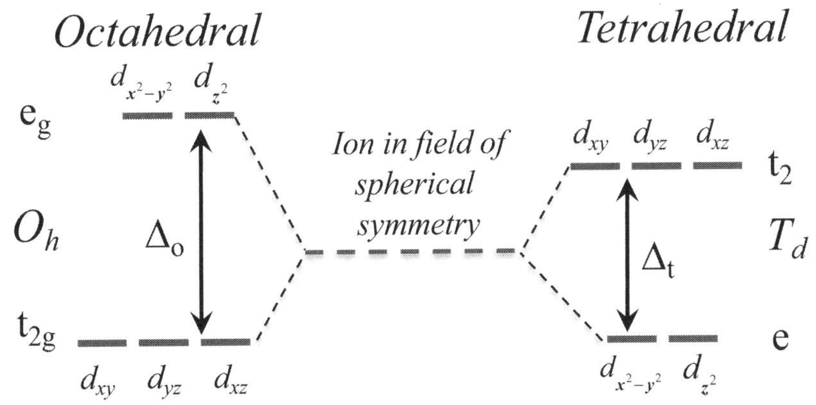

FIGURE 16.10 Crystal field splitting of d orbitals. If the ion is in a spherical cavity, all the orbitals have the same energy (center). If the ligands are in an octahedral arrangement, the energy of the orbitals split—by an energy of Δ_O—as shown on left. In this case the lower orbitals are labeled t_{2g} and the higher ones e_g. If the ligands are in a tetrahedral arrangement, the energy of the orbitals split—by an energy of Δ_t— as shown on right. In this case, the lower ones are labeled e and the higher ones are labeled t_2.

summarize the basic idea which is that ligands—depending on symmetry—split the energies of the d orbitals. It is this energy split that gives rise to the various colors observed. If the energy of the incoming photon is close to the energy difference between the d orbitals it is absorbed, and the electrons will be promoted from the lower to the higher level. The magnitude of the energy split, and consequently the resulting color, depends on the strength of the interaction between the transition ion and the host crystal, as well as on the coordination number of the central ions. This can be clearly seen in Table 16.2, in which various absorption bands for transition-metal ions in soda-lime silicate glasses are summarized.

Note, the probability of the transition decreases as the energy of the incident light differs significantly from the energy split between the d orbitals. In other words, the maximum probability for transition occurs when the energy of the incident light is identical to the energy split between the levels. Furthermore, in most cases objects remain colored as they are observed, because the electrons that are excited rapidly lose

TABLE 16.2 Absorption maxima and colors of transition-metal ions in soda-lime silicate glasses

Ion	Number of d electrons	Absorption maximum, μm	Coordination number with oxygen	Color
Cr^{3+}	1	0.66	6	Green
V^{3+}	2	0.64	6	Green
Fe^{2+}	6	1.10	4 or 6?	Blue
Mn^{3+}	4	0.50	6	Purple
Mn^{2+}	5	0.43	4 or 6?	Brown
Ni^{2+}	8	1.33	6	Purple
Cu^{2+}	9	0.79	6	Blue

their energy to their immediate surroundings as heat (i.e., at a different frequency) and hence the number of ions available for excitation remains approximately constant with time even though absorption occurs.

ABSORPTION BY COLOR CENTERS PRODUCED BY RADIATION OR REDUCTION

It is often observed that ceramics, especially oxides, will turn black when they are heavily reduced, or exposed to strong radiation for extended periods. In either case, the formation of **color centers** is responsible for the observed phenomena. A color center is an impurity or a defect onto which an electron or a hole is locally bound. For example, how the reduction of an oxide can result in the formation of both $V_O^{\bullet}$ and $V_O^{\times}$ defects was discussed in some detail in Chap. 6 (see Fig. 6.4*a* and *b*). In the context of this chapter, both are considered color centers. If E_d for these defects is the energy needed to liberate the electron into the conduction band (see Fig. 7.12*c*), it follows that light of that frequency will be absorbed. Note here that *all* incident wavelengths with energies equal to or greater than E_d, and not just the ones that are $\approx E_d$, will be absorbed because electrons can be promoted into any level in the conduction band, because of the latter's finite width. How defects can result in color is discussed in more detail in Case Study 16.1.

ABSORPTION BY MICROSCOPIC SECOND PHASES

Small metallic particles dispersed in glasses scatter light and can create striking colors. This phenomenon is essentially a scattering effect and is discussed in greater detail in the next section.

Worked Example 16.4

Given the band structure shown in Fig. 16.11*a*, sketch the optical absorption spectrum as a function of incident photon energy. Also discuss the expected photoconductivity. Arrows in Fig. 16.11*a* denote allowable transitions. All others are not allowed.

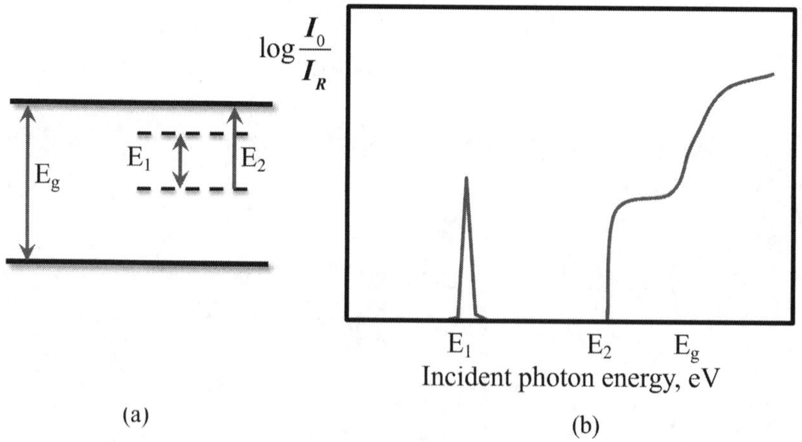

FIGURE 16.11 (*a*) Energy-level diagram. (*b*) Corresponding optical response.

ANSWER

The absorption spectrum is shown in Fig. 16.11*a* is characterized by the following: below E_1 there is no absorption. The first absorption peak centered on E_1 corresponds to the excitation of an electron from the ground state to the excited state of the imperfection. This transition does not affect the photoconductivity, however, because the electron is still localized. The next absorption centered on E_2 is due to the excitation of an electron from the ground state of the imperfection to the *conduction* band. This will give rise to a current. Finally, the last absorption centered on E_g is due to the intrinsic transitions across the bandgap.

16.3.3 IR RANGE

IONIC POLARIZATION

This phenomenon was dealt with in detail in Sec. 14.4.2. As ω of the incident light approaches the natural frequency of vibration of the ions in a solid ω_{ion}, resonance occurs and energy is transferred from the incident light to the solid. In other words, the incident wave is absorbed. The ω at which this occurs, ω_{ion}, is called the **IR absorption edge**. As discussed in Sec. 14.4.2, it depends on the strength of the ionic bond as reflected by the natural frequency of vibration of the ions ω_{ion}, their charges and their mass. These factors and their effect on the IR absorption edge are clearly demonstrated in Fig. 16.12, where the IR absorption edge is plotted as a function of ω for a number of ceramic crystals.

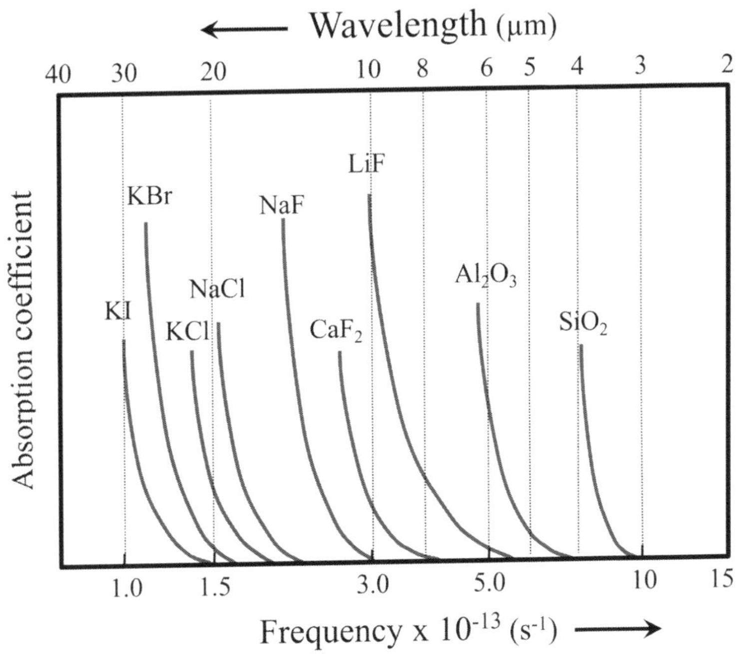

FIGURE 16.12 Infrared absorption edges of select ceramic crystals. Note correlation between IR edge and melting points of the solids.

16.4 SCATTERING AND OPACITY

Given that most ceramics have band gaps in excess of 1 eV (see Table 2.6), and based on the foregoing discussion, the inevitable conclusion is that most ceramics are intrinsically transparent. However, everyday experience indicates that with the notable exception of glasses, most ceramics are *not* transparent but opaque. As noted previously and elaborated on in this section, the reason for this state of affairs is not related to any absorption mechanisms *per se*, but is due to scattering of incident light by pores and/or grain boundaries present within the ceramic. Note that dense, single crystals of most ceramics are indeed transparent, with gems being excellent examples.

Systems that are optically heterogeneous scatter light. Optically heterogeneous solids are those in which there are density variations, such a pores. Scattering is probably most easily described as resulting from reflections from internal surfaces. Figure 16.13 schematically illustrates how a light beam is scattered by an isolated spherical void. Note that the emerging rays are no longer parallel.

By neglecting multiple and intrinsic scattering and absorption due to impurities, Eq. (16.18) simplifies to

$$\frac{I_T}{I_0} = \exp(-\alpha_s x) \tag{16.22}$$

where α_s was defined earlier as the scattering coefficient, sometimes referred to as the **turbidity** or **extinction coefficient**.

Assuming there are N_s scatterers per unit volume, each with a radius r_s, it follows that the intensity scattered per unit volume in any given direction is simply proportional to N_s times the intensity scattered by one particle. In other words

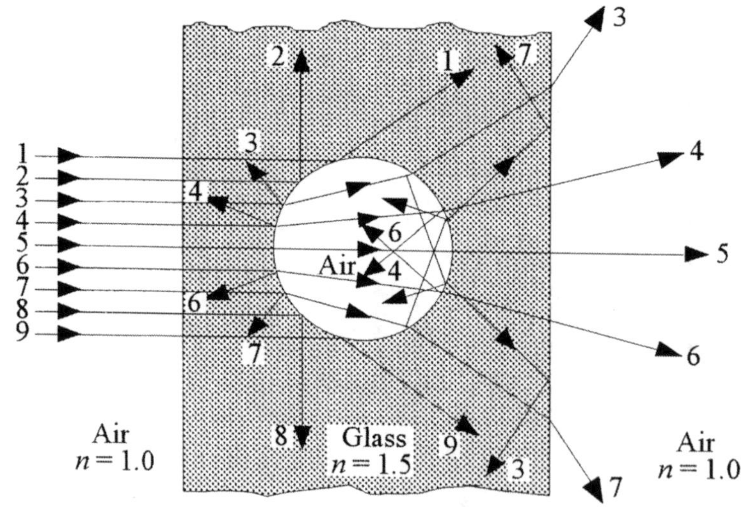

FIGURE 16.13 Light scattering by a spherical pore in an otherwise homogeneous medium.

$$\alpha_s = Q_s N_s \pi r_s^2 \tag{16.23}$$

where Q_s is a dimensionless constant that depends on the angle between the incident and scattered light, as well as the relative size of r_s to λ of the incident light. At this point it is useful to consider two limiting cases.

First are the particles whose dimensions are small with respect to the wavelength of the light, that is, $r_s << \lambda$. In this case, the scattering in the forward direction is equal to the scattering in the backward direction (see Fig. 16.2*b*), and it can be shown that[252]

$$Q_s = (\text{const})\left(\frac{r_s}{\lambda}\right)^4 \left(n_{\text{matrix}}^2 - n_{\text{scatter}}^2\right)^2 \tag{16.24}$$

where the n_i represent the refractive indices of the matrix and scattering particles. This type of scattering is known as **Rayleigh scattering** and pertains to single scattering by independent spheres of identical size. In other words, this is under experimental conditions in which the particles are so far from one another that each is subjected to a parallel beam of light and has sufficient room to form its own scattering pattern, undisturbed by the presence of other particles (see App. 16B for more details).

Second are particles that are very large compared to the wavelength of light, or $r_s >> \lambda$. This is referred to as Mei scattering (Fig. 16.2*c* and *d*). Here it can be shown that the total energy scattered is simply twice the amount it can intercept, or

$$Q_s = 2 \tag{16.25}$$

In other words, the total light scattered by a particle of radius r_s is simply twice the cross-sectional area of that particle.[253]

Finally, note that if the volume fraction of the scattering phase is f_p, then

$$f_p = \frac{4}{3}\pi r_s^3 N_s \tag{16.26}$$

Based on the preceding discussion, the following points are noteworthy:

1. Scattering of small particles scales as $1/\lambda^4$ (Eq.16.24). Consequently, blue light is scattered much more strongly than red light. This phenomenon is responsible for blue skies and red sunsets. At sunset, the sun is observed directly, and it appears red because the blue light has been selectively scattered away from the direct beams. During the day, the molecules and dust particles in the atmosphere scatter the blue light through various angles, rendering the sky blue.

---

[252] See, e.g., H. C. van de Hulst, *Light Scattering by Small Particles*, Dover, New York, 1981.

[253] That a particle of area A removes *twice* the energy it can intercept is known as the *extinction paradox*. After all, common experience tells us that the shadow of an object is usually equal to the object—not twice as large! The paradox is removed when the assumptions made to derive Eq. (16.25) are taken into account, namely, that (1) all scattered light including that at small angles is removed and (2) the observation is made at a very great distance, i.e., far beyond the zone where a shadow can be distinguished.

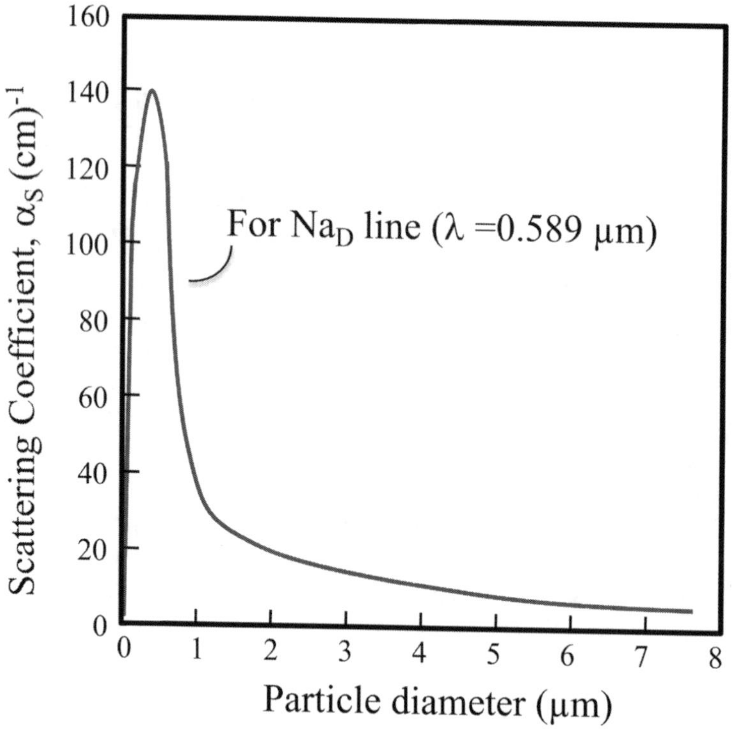

FIGURE 16.14 Effect of particle size on the scattering coefficient of a fixed volume of particles. The light used was monochromatic, with a wavelength of 0.589 μm.

2. Scattering by small particles occurs only to the extent that there is a difference between the refractive indices of the matrix and of the scatterers. In ceramics, pores, with $n = 1$, are very potent scatterers. It is also for the same reason that TiO_2 with the relatively high n of 2.5 (Fig. 16.4) is added to latex to create white paint.
3. Scattering is a strong function of r_s. By assuming a fixed volume of particles and combining Eqs. (16.23), (16.24) and (16.26), it is not difficult to show that for very small particles, α_s is proportional to r_s^3. Conversely, by combining Eqs. (16.23), (16.25) and (16.26), the result that α_s scales as $1/r_s$ is readily obtainable. The effect of r_s on α_s is illustrated in Fig. 16.14. Maximum scattering occurs when $r_s \approx \lambda$.

EXPERIMENTAL DETAILS: MEASURING LIGHT SCATTERING

A typical arrangement for the study of light scattering is shown in Fig. 16.15. The detector is mounted so that it can measure the angular dependence θ of the intensity of the scattered light from the direction of the incident beam. The scattering coefficient α_s is determined by the integration of the scattered intensity at all angles to the incident beam. Note that such an arrangement is needed to differentiate between scattering and absorption.

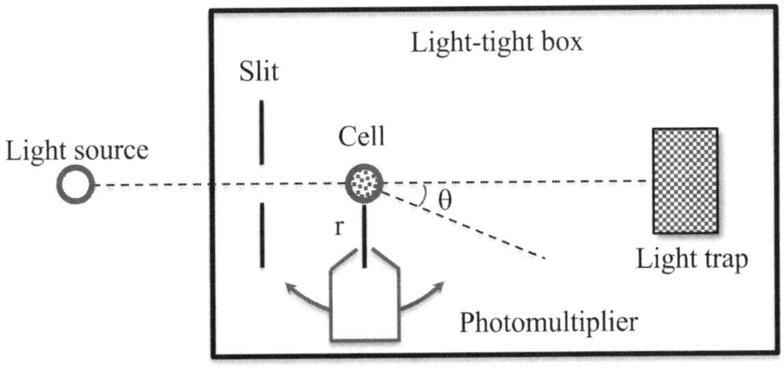

FIGURE 16.15 Basic construction of light-scattering apparatus.

Worked Example 16.5

(*a*) Why is the sky blue? Why is the sun red at sunset?
(*b*) Why are clouds white?
(*c*) Why is visibility diminished in a fog? If the water droplets in the fog are all the same size, what renders one fog light and the other heavy?

ANSWER

(*a*) The diameters of the molecules in our atmosphere, N_2 and O_2, are much smaller than the wavelengths of light coming from the sun in the visible spectrum. It follows that, $r_s << \lambda$ and thus Eq. 16.24 is applicable. Now since λ_{blue} (0.4 μm) $< \lambda_{red}$ (0.7 μm) the blue light will be scattered more. Quantitatively,

$$\frac{Q_{blue}}{Q_{red}} = \left(\frac{\lambda_{red}}{\lambda_{blue}}\right)^4 = 9.4$$

Blue light is thus scattered roughly 10 times more than red light. Another way to look at the problem is to think of red light going in more straight lines – relative to blue. This is why when you look directly at the sun it is yellow; the blue is scattered in all directions. This multiscattering process results in blue skies. At sunset, the sunlight traverses more atmosphere which causes it loses intensity (you can now stare at it all you want), and now even more blue is scattered and the deep red color remains.

(*b*) In the case of clouds the size of the scattering particles – droplets of water/ice – is larger than the λ's of the light coming from the sun in the visible spectrum and Eq. 1.6.25, applies, $Q = 2$. The scattering is *no* longer a function of λ. So all λ's are scattered equally and the scattering is weak which is why we can see across the street when it is raining.

(c) Fog is a cloud that has landed. In this case, the water particles are close to 0.5 μm in diameter and thus the scattering is maximized (see Fig. 16.14). Since the scattered rays are no longer coherent the light appears white. If you are looking directly at a light source, its light will be scattered and re-scattered in many directions and appear dim and unfocused. If, on the other hand, you are driving down a road with your headlights on, the water molecules simply back-scatter, or reflect, the light back into your eyes, which is unhelpful. If r_s of all the water droplets are the same, then the only difference between a light and a heavy fog is N_s, the number of scattering centers per unit volume.

CASE STUDY 16.1: COLORS FROM DEFECTS

In attempts to understand the photographic process (before digital cameras), the formation of color centers in the alkali halides, especially silver, was studied extensively and in great detail. At least half a dozen color centers have been identified in these materials, of which the most widely studied and best understood is probably the ***F-center***, defined as an electron trapped at an anion vacancy (Fig. 16.16). The name comes from the German word for color: *Farbe*.

When alkali halides are exposed to X-ray or γ-rays (that produce a multitude of defects) they develop various colors. The same occurs when they are heated in vapors of their own metals. To understand this intriguing phenomenon, it is important to note that after exposure to their metal vapors the densities of the crystals decrease as the concentration of F-centers increases. We thus need to postulate a defect reaction in which the density decreases. The simplest is:

$$Na(vap.) = V_{Cl}^{x} + Na_{Na}^{x} \tag{16.27}$$

This reaction is written in this form to emphasize that the electron, introduced into the crystal by the Na atoms, is *localized* at the anion vacancy site as shown schematically in Fig. 16.16.

The F-center can be modeled by assuming the electron is trapped in a box of side d, which scales with the lattice parameter, a, of the alkali halide. The F-center transition is believed to be between the ground and first excited state of this "particle in a box." If a particle with mass, m, is trapped in a cube of dimensions, d, its energy levels will be given by:

$$E_n = \frac{h^2 n^2}{8md^2} \tag{16.28}$$

and

$$\Delta E = E_{n+1} - E_n = \frac{h^2(2n+1)}{8md^2} \tag{16.29}$$

It follows that if one assumes that d is proportional to a, and n = 1, then a color center will absorb a photon when $E_{cc} = \Delta E$, or

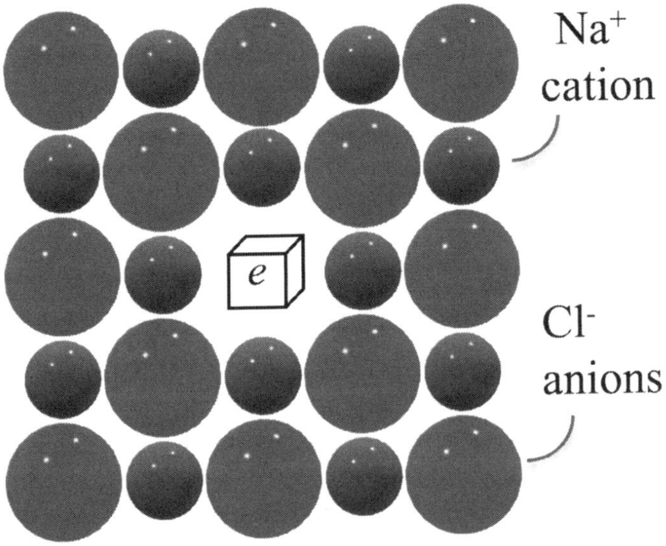

FIGURE 16.16 Schematic of a F-center in NaCl.

$$E_{cc} \propto \frac{3h^2}{8ma^2} \tag{16.30}$$

Table 16.3 lists some experimental E_{cc} results for a number of alkali halide salts and a few oxides, all with the rock salt structure. Plotting E_{cc} in eV vs. a, in Å it can been shown that (see Problem 16.15)

$$E_{cc} = \frac{17.7}{a^{1.84}} \tag{16.31}$$

This relationship was first obtained by Mollwo and later modified by Ivey, which is why it is sometimes referred to as the **Mollwo–Ivey** relation. This is a very elegant example of how confining a small particle, such as an electron, in this case to a vacant lattice site, invariably results in the quantization of its energies as predicated by Schrödinger's equation.

TABLE 16.3 Optical absorption peak positions, in eV, of F-centers for select halides and oxides

Crystal	E_{cc}	Crystal	E_{cc}	Crystal	E_{cc}	Crystal	E_{cc}
LiF	5.08	NaCl	2.75	KBr	2.06	RbI	1.70
LiCl	3.25	NaBr	2.34	KI	1.87	BaO	2.0
LiBr	2.77	NaI	2.06	RbF	2.41	MgO	4.9
LiI	3.18	KF	2.87	RbCl	2.04	CaO	3.7
NaF	3.7	KCl	2.29	RbBr	1.85	SrO	3.0

CASE STUDY 16.2: FIBER OPTICS AND OPTICAL COMMUNICATION

A fiber-optic waveguide is a thin device composed of a high-refractive-index substance that is completely surrounded by a lower-refractive-index one. The situation is depicted in Fig. 16.17*a*, where according to **Snell's law**,

$$n \sin\phi = n' \sin\phi' \tag{16.32}$$

If the angle of incidence is greater than a critical angle ϕ_c total internal reflection will occur rather than refraction, as shown in Fig. 16.18*b*. This angle is given by Snell's law when $\varnothing' = 90°$, or

$$\sin\phi_c = \frac{n'}{n} \tag{16.33}$$

Thus, in an optical waveguide, some of the light that is launched into the high-index core is carried along that region by reflecting off the interface with the low-index cladding, as shown in Fig. 16.17*c*.

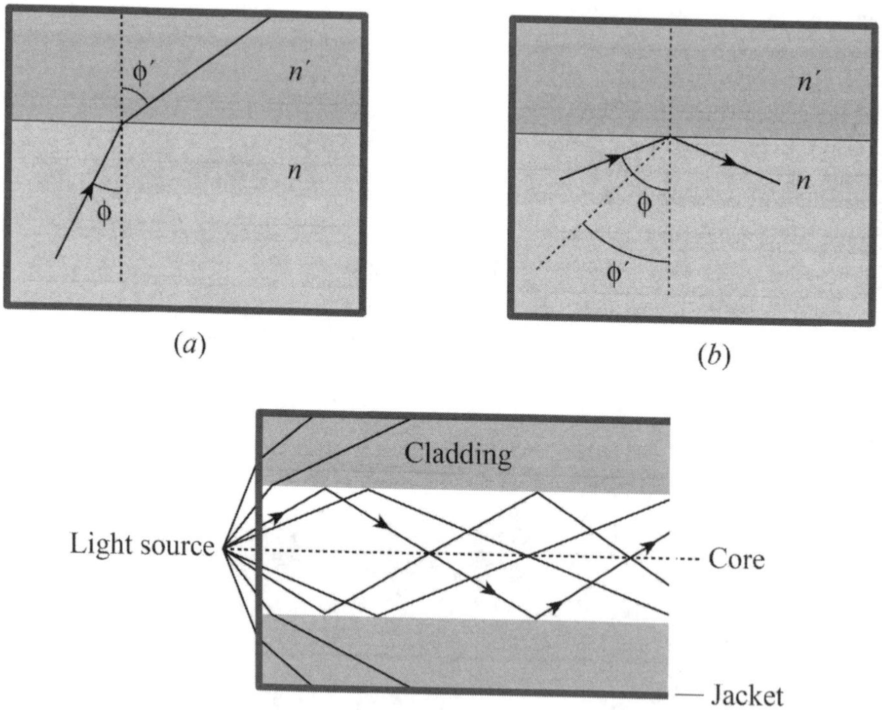

FIGURE 16.17 (*a*) Snell's law of refraction. (*b*) Total internal reflection. (*c*) Light rays from point source enter at many angles. Rays that impinge at an angle that is less than the critical angle are guided down the optical waveguide by total internal reflection.

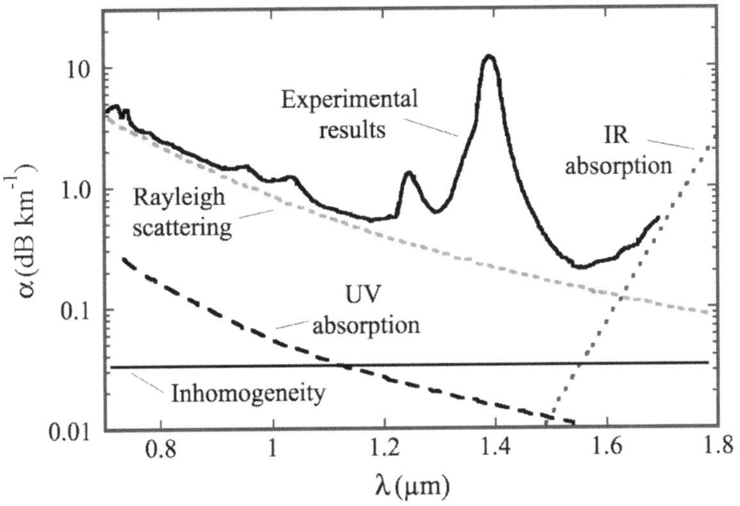

FIGURE 16.18 Sources of optical losses in fused silica.

The process of optical telecommunications consists of four parts:

1. The electric signal is coded digitally and converted to an optical signal.
2. The optical signal consisting of high-frequency laser pulses is sent along the waveguide, which is typically an ultrapure silica fiber with a core and a cladding. The core carries the light, and the cladding guides the light through the core.
3. As the light travels along the fibers, it broadens and weakens; hence, the signals have to be reamplified periodically.
4. The signal is received and decoded by converting the light pulses back to electric signals in a form that a phone or computer can interpret.

What is of interest in this section is the transmission medium and what limits the length over which the light pulses that represent the digital 0s or 1s can be transmitted without distortion or attenuation.

For short distances of roughly 1 km or less, polymer waveguides can be used. For longer distances, however, the losses are unacceptable and inorganic glasses have to be used. To date, the material of choice is extremely pure silica glass fibers. Ideally an optical fiber should be loss-free, in which case the signal would not be attenuated. The attenuation is usually expressed in decibels (dB) as

$$dB = 10\log\left(\frac{\text{power output}}{\text{power output}}\right) = 10\log\frac{I_T}{I_0} \tag{16.34}$$

where I_T/I_0 is the ratio of the intensity at the detector to that at the source.[254]

[254] Note 1 dB/km = 0.23 km^{-1} (see Prob. 16.23).

A number of phenomena contribute to the scattering, absorption, and overall deterioration of an optical signal as it travels down a waveguide. These are discussed in some detail in the following sections. Typical absorption or loss of data for a typical silica optical fiber are shown in Fig. 16.18, where the following points are noteworthy:

1. Above 5 μm, absorptions of the Si–O–Si bond network, i.e., ionic polarizations, become important.
2. Trace amounts of impurities, particularly transition-metal oxides, can have a profound effect on absorption due to electronic transitions alluded to earlier. Furthermore, the presence of Si–OH in the glass can cause significant absorption due to overtones of the O–H bond vibrations. It is important to control these impurities since they lie in the useful transmission window. For example, it has been estimated that one part per billion of these impurities could lead to 1 dB/km loss in silicate glasses (see Prob. 16.25). The effect of these impurities is why the absorption of silica fibers peaks around 1.4 µm in Fig. 16.18.
3. Electronic transitions of the glass become important at wavelengths shorter than 0.5 μm.
4. Density and composition fluctuations are inherent in glasses and lead to scattering. For example, Rayleigh scattering in fused silica amounts to about 0.7 dB/km at 1 μm. Since scattering scales as $1/\lambda^4$, it becomes most important at shorter wavelengths.

In addition to these mechanisms, scattering due to defects such as pores, inclusions or dust particles introduced in the fiber during processing have to be eliminated. Another source of scattering is irregularities in fiber diameter; this is especially important if the diameter fluctuations are regularly and closely spaced (<1 mm apart).

Another limitation to the information-carrying capacity of a fiber waveguide is how close the light pulses can be transmitted without overlapping. This is usually determined by pulse broadening. One reason for this is called *differential delay*, which results from light traveling different paths through the fiber. To avoid the problem, waveguides are often constructed with a graded index, with the composition at the center being silica-doped with germania and the amount of germania decreasing radially outward. Since germania has a higher n than silica, n will decrease with increasing distance from the center of the fiber. Given that light travels faster in low-index media, the light waves that travel off center travel at a faster rate than those transmitted down the center of the fiber, which tends to minimize the undesirable broadening.

The transmission capacity, defined as the highest bit rate times the maximum transmission length, has increased by roughly an order of magnitude every 4 years since 1975. By 1978, 1 billion bits (1 Gbit) could be transmitted each second through a system 10 km long. The transmission capacity was thus 10 gigabit-kilometers per second. During the next 3 years, improved technology increased capacity to 100 Gbit·km/s. This was done by reducing the size of the core to create "single-mode" fibers, which forced the light to travel at nearly uniform velocity, which greatly reduced dispersion. The second advance was in developing transmitters and receivers that could handle light at 1.3 μm, a wavelength in which silica is more transparent (Fig. 16.18). In 1982, the third generation began to appear as researchers developed processing techniques that increased the purity of the silica fibers in the 1.2 to 1.6 μm range. This improvement raised the transmission capacity to hundreds of gigabites.

The development of erbium-doped silica glasses in the late 1980s ushered in a new generation of light wave communication systems with transmission capacities on the order of thousands of gigabit-kilometers per second. The Er ions embedded in the glass amplify the signal as they absorb infrared radiation produced by a laser diode chip at a wavelength of 1.48 or 0.98 μm. The light is absorbed by the Er atoms by pumping them to a higher energy level. When a weakened signal enters the Er-doped fiber, the excited Er atoms transfer their energy to the weakened optical signal, which in turn is regenerated. This was a major breakthrough because it eliminated the need for signal regenerators or repeaters. The repeaters convert the light to an electric current, amplify the current and transform it back to light. The Er-doped fibers do not interrupt the path of the light as it propagates through the fiber, and much longer cables are now possible.

16.6 SUMMARY

When electromagnetic radiation proceeds from one medium to another, some of it can be reflected, absorbed, scattered, and/or transmitted.

1. Electronic polarization results in the retardation of EM radiation, which is directly responsible for refraction. The index of refraction n quantifies the degree of bending or retardation. n is directly related to electronic polarization, which in turn is determined by the polarizability of the atoms or ions in the solid. The more polarizable the ions or atoms, the larger n.
2. The reflectivity of a surface depends on n as well. Insulators with high indices of refraction are more reflective than ones with low n.
3. The processes by which light is absorbed by solids are several and include:
   - ∞ The photoelectric effect where electrons absorb the incident light and are promoted into the conduction band. For this process to occur the energy of the incident light has to be greater than the materials' band gap. For many ceramics this energy is typically in the UV range.
   - ∞ The split in energy of transition-metal ion d and f orbitals, as a result of their interaction with their local environment, gives rise to selective absorption in the visible range. It is this absorption that is responsible for the striking colors exhibited by some glasses and gems.
   - ∞ Reduction and radiation can give rise to color centers defined as an impurity or defect onto which an electron or hole is locally bound. The localization of the electron and its promotion to higher energy levels give rise to absorption.
4. In the IR range, absorption is usually associated with ionic polarization in which the ionic lattice as a whole absorbs the radiation and starts vibrating in resonance with the applied field. The most important factor determining the frequency at which IR absorption occurs is the strength of the ionic bond and the masses of the ions involved. Stronger bonds and lighter atoms/ions result in higher resonance frequencies and vice vera.
5. In addition to absorption, light can be scattered in different directions. Scattering is distinguishable from absorption in that the energy of the incident light is not absorbed by the sample but simply scattered in various directions. Scattering is a complex function of the density of scatterers, their relative

size with respect to the wavelength of the incident light and the relative values of the refractive indices of the scatterers and the medium in which they reside. In general scattering is at a maximum when the size of the scatterers is of the order of the wavelength of the incident light.

6. Optical communication depends on the ability of very thin silica fibers to transmit light signals over large distances with little attenuation. The glass fiber is designed such that its outside surface has a lower refractive index than its center, which results in the total internal reflection of the optical signal within the fiber. In other words, the light signal is confined within the fiber with little loss and essentially acts as an optical waveguide.

APPENDIX 16A: COHERENCE

For refraction, reflection and diffraction to be observed it is important that the incoming light beams be coherent. Typically, light from common sources such as the sun or incandescent lamp filaments is incoherent because the emitting atoms in such sources act independently rather than cooperatively. Coherent and incoherent light are treated differently. For completely coherent light, the amplitudes of the waves are added vectorially, and the resultant amplitude is squared to obtain a quantity proportional to the resultant intensity. For completely incoherent light beams, first the amplitudes of the light are squared to obtain a quantity proportional to the intensities and then the intensities are added to obtain the resultant intensity. This procedure is consistent with the fact that for completely independent light sources, the intensity at every point is greater than the intensity due to either of the light sources acting alone.

APPENDIX 16B: ASSUMPTIONS MADE IN DERIVING EQ. (16.24)

Four assumptions are made in deriving Eq. (16.24):

1. The scattered light has the same frequency as the incident light, which in turn is monochromatic, i.e., confined to one frequency.
2. The scatterers are assumed to be independent; i.e., there are no cooperative effects between scatterers, hence there is no systematic relation between the phases of the scattered beams. To ensure independent scattering, it is estimated that the distance between scatterers should be about three times the radius of the particles. This assumption allows for the intensities scattered by the various particles to be simply added without regard to phase. In other words, intensities rather than amplitudes are added, as noted above.
3. Multiple scattering is neglected. In other words, it is assumed that each particle is exposed to the light of the original beam. Scattering where a particle is exposed to light scattered by other particles is termed *multiple scattering* and is neglected. To ensure that this condition is met, the sample has to be thin or dilute. This implies that if there are N_s scattering centers, the intensity of the scattered beam is simply N_s times that removed by a simple particle.
4. The scattering centers are isotropic and the same size.

PROBLEMS

16.1. Typical values for k_e'' for ceramics range from 0.01 to 0.0001. Estimate the value of σ_{dc} below which it can be safely neglected when one is dealing with optical properties. State all assumptions.
Answer: ≈ 0.01 S/m.

16.2. (*a*) Refer to Table 16.1. Identify the materials with the highest and lowest values of n. Explain the differences in terms of what you know about the polarizabilities of the constituent ions.
(*b*) Do you expect LiF or PbS to have a higher n? Explain.
Answer: PbS.
(*c*) Do you expect MgO or BaO to have a higher n? Explain.
Answer: BaO.

16.3. If a highly reflective surface is required, should one use a material with a high or low n? Explain.

16.4. One way to tell whether a glass plate is made of pure silica or soda-lime silica glass is to view it on edge. The silica plate is clear, whereas the window glass is green. Explain.

16.5. It was noted in Experimental Details that it is possible to measure α_a by measuring the transmission of two different samples of different thicknesses that have identical reflectivities. Describe an experimental setup you would use to carry out such measurements, what would you measure and how would you extract α from the results you obtain. Why is it important that the two samples have the same reflectivity?

16.6. For an ion for which $\varepsilon_{BL} = 10\ m^{-1}\%^{-1}$, answer the following questions:
(*a*) If the concentration of the light-absorbing ions in a solution is tripled, how does the transmission change if the thickness is 1 cm?
Answer: $I_1 = 1.22\ I_2$.
(*b*) How must the thickness of the sample be altered to keep the transmission invariant through the two solutions?
Answer: $x_1 = 3\ x_2$.

16.7. A 40-cm glass rod has an absorption coefficient α_a of 0.429 m^{-1}. If 50% of the light entering one end of the rod is transmitted, determine
(*a*) The scattering coefficient α_s
Answer: 1.304 m^{-1}.
(*b*) The total coefficient α_{tot}
Answer: 1.733 m^{-1}.

16.8. (*a*) Experimentally in IR absorption, two absorption bands are measured at 3000 and 750 cm^{-1}. One is suspected to be due to a C–H stretching vibration, while the other is suspected to be due to a C–Cl stretching vibration. Assign each absorption band to its appropriate bond. Explain your answer.
(*b*) Repeat part (*a*) for C–O and C=O; the absorption bands measured were at 1000 and 1700 cm^{-1}. Which band corresponds to which bond? Explain your answer.

16.9. (*a*) Weaker bonds and heavier ions are preferable for extended IR transmission. Is this statement true or false? Explain, using examples from Fig. 16.9.
(*b*) Which of the following three materials will transmit IR radiation to the longest wavelength: MgO, SrO or BaO? Explain.

16.10. Which of the following materials do you anticipate to be transparent to visible light? Explain.

Material	Diamond	ZnS	CdS	PbTe
Band gap, eV	5.4	3.54	2.42	0.25

16.11. (*a*) What material would you use for a prism for infrared investigations?
(*b*) Which material would you use in making lenses for an ultraviolet spectrograph?

16.12. The transmitted light through a 5 mm sample of CdS, which has a band gap of 2.4 eV is observed. Under these conditions, what is
(*a*) The color of the sample?
Answer: Reddish orange.
(*b*) Cu can dissolve in CdS as an impurity and has an energy level that is normally, in the dark, electron-occupied and lies 1.0 eV above the valence band of CdS. What color changes, if any, would you expect as the Cu concentration increases from 1 to 1000 ppm?
(*c*) The E_g of CdS changes with temperature according to: $E_g = 2.56 - 5.2 \times 10^{-4}$ T. What color changes do you expect in the transmitted light as CdS is heated from 0 to 1000 K?

16.13. Crystals of NaCl show strong absorption of EM radiation at a wavelength of about 0.6 μm. Assume this is due to the vibration of individual atoms.
(*a*) Calculate the frequency of vibrations.
Answer: 5×10^{12} Hz.
(*b*) Calculate the potential energy of a sodium ion as a function of distance r from its equilibrium position, assuming the vibration to be simple harmonic.
Answer: 1.89×10^{-4} r^2 J, where r is in Å.

16.14. Calculate the ratio of molecules in a typical excited rotational, vibrational and electronic energy level to that in the lowest energy state at 25 and 1000°C, taking the levels to be 30, 1000 and 40,000 cm^{-1}, respectively, above the lowest energy state.

16.15. The experimental values for absorption energies, in eV, of *F*-centers in various alkali halides and a few oxides are listed in Table 16.3. Do you think alkali halide results support the Mollwo–Ivey relation, Eq. (16.31)? Explain.

16.16. Using sketches, explain why s and p orbitals are unaffected (i.e., do not split) by ligands in octahedral fields.

16.17. Rayleigh scattering is a strong function of particle size, r. Plot the functional dependence of the scattering coefficient as a function of r for a given wavelength of light and volume fraction of scattering particles. Said otherwise plot a figure similar to Fig. 16.14

16.18. Typically, TiO_2 particles are added to latex, a polymeric base with a refractive index of 1.5, to make white paint.
(*a*) Discuss why TiO_2 is a good candidate for this application.
(*b*) On the market you find three TiO_2 particle sizes with narrow distributions and average particle sizes of 0.2, 2.0 or 20 μm. Which would you use to make white paint, and why?

16.19. A 40-cm glass rod absorbs 15% of the light entering at one end. When it is subjected to intense radiation, tiny particles are produced in it that give rise to Rayleigh scattering. After radiation the rod transmits 55% of the light. Calculate the

(*a*) Absorption coefficient α_a

Answer: 0.406 m^{-1}.

(*b*) Scattering coefficient α_s

Answer: 1.09 m^{-1}.

16.20. The surface of a glass plate is rough on the scale of the incident light wavelength. Use a sketch to show what happens when the beam of light strikes the surface at a glancing incidence. Show what happens when the surface is wet with a liquid of equal refractive index.

16.21. Why do car headlights appear brighter when the road is wet?

16.22. (*a*) Explain why optical waveguides often have a refractive index gradient.

(*b*) Show how n can be calculated given knowledge of the critical angle.

16.23. Show that 1 dB/km = 0.23 km^{-1}

16.24. (*a*) In an optical communications network, the ratio of the light intensity at the source to that at the detector is 10^{-6}. What is the loss in decibels in this system?

Answer: −60 dB.

(*b*) The attenuation of ordinary soda-lime silicate glass is ≈−3000 dB/km. What fraction of the light signal will be lost in 1 m?

Answer: 50%.

16.25. A certain glass containing 500 ppm of Cr^{3+} ions absorbs 10% of the incident light in 10 cm. Assume the Cr^{3+} ions are responsible for the absorbance.

(*a*) What is the absorbance loss in dB/km of the original glass?

Answer: 4576 dB/km.

(*b*) What must the concentration of Cr^{3+} be so that the absorbance is 10% in 100 m?

Answer: 0.5 ppm.

(*c*) Calculate the loss (dB/km) for the 100 m sample.

Answer: 4.6 dB/km.

16.26. (*a*) What is the critical angle for total internal reflection for an optical fiber with a core refractive index of 1.52 and a cladding of 1.46?

Answer: 74°.

(*b*) Repeat part (*a*) for a system for which the core refractive index is 1.46 and that for the cladding is 1.46.

16.27. (*a*) A ceramic body containing 0.25 vol% spherical pores transmits 50% of the incident light and scatters 50% in 1 mm thickness. Estimate the average diameter of the pores. State all necessary assumptions.

Answer: 10.8 μm.

(*b*) Calculate the fraction of light transmitted if the average pore diameter is 1 μm. What does this result imply about the requirements for obtaining transparent polycrystalline ceramics?

Answer: 0.05%.

16.28. (*a*) Assuming the constant in Eq. (16.24) is 30, calculate the fraction of light transmitted through 5 cm of a solution with a concentration of 10^{25} m^{-3} of scatterers for which the diameter is 1.2 nm. You can assume the incident light is monochromatic with $\lambda = 0.6$ μm. You can further assume that the relative dielectric constant of the solution at this wavelength is 2.25, while that of the particles is near 1.

Answer: 99.97%.

(*b*) Repeat part (*a*) for particles with a diameter of 6 nm.

Answer: ≈15%.

16.29. (*a*) Explain why low-fat milk is more translucent than regular milk?

(*b*) Why do you think some fog headlights are yellow? Explain.

16.30. ZnS has a band gap of 3.64 eV. When doped with Cu^{2+}, it emits radiation at 670 nm. When zinc vacancies are produced by the incorporation of Cl^- ions, the radiation is centered on 440 nm.

(*a*) Write the incorporation reaction that results in the formation of the zinc vacancies.

(*b*) Using a sketch, locate the impurity levels in the band gap in relation to the valence band.

16.31. (*a*) You are asked to compare the values of elastic moduli, thermal conductivities and thermal expansion coefficients in the temperature range of 50 to 800°C, of optical-quality polycrystalline MgF_2 from different sources ranging in impurity levels from 300 to 5000 ppm impurity content. Do you expect the results to be almost the same or markedly different for these properties and samples? Explain.

(*b*) List other physical properties that you expect to be much more variable between the those listed in (*a*). Explain.

ADDITIONAL READING

1. J. N. Hodgson, *Optical Absorption and Dispersion in Solids*, Chapman & Hall, London, 1976.
2. F. Wooten, *Optical Properties of Solids*, Academic Press, New York, 1972.
3. L. L. Hench and J. K. West, *Principles of Electronic Ceramics*, Wiley-Interscience, New York, 1990.
4. A. J. Moulson and J. H. Herbert, *Electroceramics*, Chapman & Hall, London, 1990.
5. R. P. Feynman, R. B. Leighton, and M. Sands, *The Feynman Lectures on Physics*, vols. 1 and 3, Addison-Wesley, Reading, MA, 1963.
6. A. Javan, The optical properties of materials, *Sci. Am.*, 1967.
7. W. D. Kingery, H. K. Bowen, and D. R. Uhlmann, *Introduction to Ceramics*, 2nd ed., Wiley, New York, 1976.
8. J. S. Cook, Communications by optical fibers, *Sci. Am.*, 229, 28–35, 1973.
9. M. E. Lines, The search for very low loss fiber-optics materials, *Science*, 226, 663, 1984.
10. E. Dusurvire, Lightwave communications: the fifth generation, *Sci. Am.*, 266 (1), 114, 1993.
11. B. E. A. Saleh and M. C. Teich, *Fundamentals of Photonics*, Wiley, New York, 1991.
12. F. A. Jenkins and H. E. White, *Fundamentals of Optics*, 4th ed., McGraw-Hill, New York, 1976.
13. H. C. van de Hulst, *Light Scattering by Small Particles*, Dover, New York, 1981.
14. B. A. Leyland, *Introduction to Laser Physics*, Wiley, New York, 1966.
15. P. Fulay, *Electronic, Magnetic and Optical Properties of Materials*, CRC Press, 2010.

INDEX

D

J

K

L

N

O

P

Q

R

T

U

V

W

X

Y

Z